AF815128

Springer Geophysics

The Springer Geophysics series seeks to publish a broad portfolio of scientific books, aiming at researchers, students, and everyone interested in geophysics. The series includes peer-reviewed monographs, edited volumes, textbooks, and conference proceedings. It covers the entire research area including, but not limited to, geodesy, planetology, geodynamics, geomagnetism, paleomagnetism, seismology, and tectonophysics.

More information about this series at http://www.springer.com/series/10173

Amir Khan · Frédéric Deschamps
Editors

The Earth's Heterogeneous Mantle

A Geophysical, Geodynamical, and Geochemical Perspective

Springer

Editors
Amir Khan
Institute of Geophysics
ETH Zurich
Zurich
Switzerland

Frédéric Deschamps
Institute of Earth Sciences
Academia Sinica
Taipei
Taiwan

Springer Geophysics
ISBN 978-3-319-15626-2 ISBN 978-3-319-15627-9 (eBook)
DOI 10.1007/978-3-319-15627-9

Library of Congress Control Number: 2015934945

Springer Cham Heidelberg New York Dordrecht London
© Springer International Publishing Switzerland 2015
This work is subject to copyright. All rights are reserved by the Publisher, whether the whole or part of the material is concerned, specifically the rights of translation, reprinting, reuse of illustrations, recitation, broadcasting, reproduction on microfilms or in any other physical way, and transmission or information storage and retrieval, electronic adaptation, computer software, or by similar or dissimilar methodology now known or hereafter developed.
The use of general descriptive names, registered names, trademarks, service marks, etc. in this publication does not imply, even in the absence of a specific statement, that such names are exempt from the relevant protective laws and regulations and therefore free for general use.
The publisher, the authors and the editors are safe to assume that the advice and information in this book are believed to be true and accurate at the date of publication. Neither the publisher nor the authors or the editors give a warranty, express or implied, with respect to the material contained herein or for any errors or omissions that may have been made.

Cover design: Name of original series cover designer (key designer): deblik, Berlin.

Printed on acid-free paper

Springer International Publishing AG Switzerland is part of Springer Science+Business Media (www.springer.com)

Preface

The Earth's mantle is not homogeneous. This has become more than evident from the analysis of seismic waves, which are strongly affected by changes in temperature, mineralogical phase, and chemical composition as they travel through Earth's mantle. The message from geochemistry is similarly complex. For instance, analysis of geochemical isotopes showed that the plumes from which Ocean Island Basalts are thought to originate are sampling at least two distinct chemical reservoirs. Recovering detailed maps of mantle heterogeneities at the various scales of interest to Earth scientists, for the purpose of unraveling their nature and origin, raises several challenges that Earth scientists attempt to solve by combining data, observations, theoretical and numerical models, and experimental results from several fields including geophysics, geochemistry, geodynamics, and mineral physics. This monograph aims to discuss recent developments that have contributed to improved understanding of the physico–chemical structure and dynamics of Earth's mantle through a series of topical reviews and original research contributions with emphasis on interdisciplinary studies. *In fine*, our ability to impose constraints on accretion, differentiation, and early evolution of our planet hinges crucially on our ability to elucidate its internal structure.

Over the past decade, considerable observational and experimental evidence from seismology, geochemistry, mineral physics, and geodynamical studies have accumulated attesting to the presence of radial and lateral heterogeneities that permeate Earth's mantle. These heterogeneities cover length scales from microscopic (chemistry) over patches of lithospheric slabs that scatter seismic waves in the lower mantle (~10–100 km) to cold subducted lithosphere, transition-zone topography, and large low shear-wave velocity provinces in the deep mantle (~1000 km) and are evidence of the dynamic evolution that Earth's mantle has undergone and continues to undergo. Variations in composition, temperature, and mineralogy are all believed to play a role in explaining the observed heterogeneity, but their relative contributions remain elusive. However, recent advances in creating comprehensive mineral physics databases combined with improved geophysical imaging techniques, greater accuracy of geochemical analyses, and more complex geodynamic models, has greatly improved our ability to make quantitative inferences on the thermo–chemical state of Earth's mantle, which ultimately holds the key to understanding the evolution of our planet.

Of all methods available to Earth scientists seismology has proved the most important because it affords the highest resolution. Accordingly, seismology has been the main source of information on mantle structure. For decades radial seismic profiles have formed the basis for inferring structure and constitution of the Earth. The variations in seismic wave-speed of the upper mantle and transition-zone have been recognized as arising from phase transformations undergone by the silicate minerals that constitute the mantle as pressures and temperatures increase. However, the role and extent of any compositional layering between upper and lower mantle, is yet to be fully resolved. The question of whether phase transitions in the olivine system alone are sufficient, or indeed, whether a chemical change is needed to explain the observed discontinuous increase in, e.g., seismic wave-speeds, remains to be understood.

Seismic tomography has provided information on both lateral and radial structure of Earth and has done much to advance our understanding of its dynamics. The large-scale global velocity structure is relatively well-resolved, as is apparent from the current consensus among studies that employ different data and modeling techniques, and correlates well with surface tectonics. This is demonstrated in the opening chapter of this monograph by Schaeffer and Lebedev, who present a global surface-wave tomographic model of the upper mantle and compare it to existing models. Increased resolution in seismic imaging on regional scales has been made possible through analysis of data from high-density seismic networks. An example of such a regional model is presented and discussed in Chap. 2 by Rawlinson and coauthors who address the seismic wave-speed structure of the upper mantle beneath Australia. Heterogeneities in the upper mantle associated with discontinuities that arise as a result of phase changes are identified and discussed by Schmerr in Chap. 3.

In spite of advances in tomographic techniques and a tremendous increase in the volume of seismic data that has become available, accurate mapping of mantle structure is still mired by a number limiting factors including trade-off between seismic heterogeneities and topography of interfaces, separation of isotropic and anisotropic anomalies, finite-frequency effects, and details of wave propagation. In Chap. 4, Bodin and coauthors consider the trade-off that occurs between radial anisotropy and small-scale radial heterogeneities, and show how this problem might possibly be solved by adding high-frequency data.

In the lowermost mantle, tomographic models published since the 1990s show that the dominant structures are two large low shear-wave velocity provinces (LLSVPs), whose detailed nature is still debated. In the past decade, waveform modeling and travel-time data from specific seismic phases provided interesting details on LLSVPs structure. Besides LLSVPs, small-scale heterogeneities are also present in the deep mantle and may be detected using scattered seismic waves as shown by Rost and coauthors (Chap. 12).

Another challenge related to imaging the lower mantle is detection of the mineral post-perovskite (pPv), a high-pressure phase of perovskite (pv) discovered experimentally in 2004. Laboratory experiments and ab-initio calculations indicate that the stability field of pPv is likely to oscillate around the depth of the core-mantle-boundary because of local variations in temperature and composition. Therefore, pPv may only be present on local or regional scales, but is unlikely to make up a global layer at the bottom of the lower mantle. Seismology, however, has yet to identify such pPv-dominated regions unambiguously, although D″ figures as a prominent candidate. In Chap. 13, Cobden and coauthors review the different seismological data and techniques used to detect the pv-pPv phase transition, its properties, and discuss the seismic observations that point to its presence.

Seismic data and models are capable of imaging mantle heterogeneities, but can only provide clues about the nature of these heterogeneities. Interpreting seismic data and models in terms of variations in mantle composition, thermal state, phase transitions, water content, or a combination thereof, as sources of seismic heterogeneities, requires additional information. This missing information is supplied in the form of mineral physics data, which encompasses a wide range of data determined both experimentally or from first-principles (ab initio) numerical computations, including phase equilibria and stability regions of mantle minerals, sensitivities of thermodynamical, thermo-elastic properties, and transport properties of mantle rocks to changes in pressure, temperature, chemical composition, and water concentration. Data for most of the upper and lower mantle minerals are now available and gathered in thermodynamic databases that, when combined with equation-of-state modeling, allow us to construct elastic and transport properties (e.g., seismic wave-speeds, density, and electrical conductivity) in the mantle at pressure and temperature conditions spanning most of the mantle. In turn, these "synthetic" profiles can be compared to field-derived estimated and thereby provide a means of testing various hypotheses for the structure, constitution, and chemical make-up of Earth's mantle.

As an illustration of this approach, Kawai and Tsuchyia (Chap. 8) use recent data determined from *ab initio* calculations to estimate the seismic signature of recycled granitic material in the mid-mantle (400–1200 km depth), and subsequently compare it to the seismic signature of other mantle materials as a means of assessing the possible origin of heterogeneities in the mid-mantle. In a related approach Khan and coauthors (Chap. 5) employ a free-energy minimization method to interpreting seismic surface-wave dispersion data for the thermochemical structure of the mantle beneath the Australian continent. The advantage of free-energy minimization methods is that profiles of physical properties (e.g., seismic wave-speeds and density) can be computed self-consistently as a function of temperature, pressure, and composition. This allows for joint interpretation of diverse geophysical, petrological, and mineral physics data sets as considered in the contribution of Khan and coauthors.

Interpretation of seismic observations in terms of mantle chemistry and thermal state is nonunique and notoriously difficult to separate. An illustration of this concerns the nature of LLSVPs observed in the deep mantle. Two hypotheses are discussed in the present volume. The purely (or mostly) thermal origin scenario, which possibly involves the presence of the mineral pPv, is reviewed by Davies and coauthors in Chap. 14, while a different explanation for the LLSVPs is offered by Deschamps and coauthors (Chap. 15), who favour a thermochemical origin.

An alternative means of addressing mantle heterogeneity is to investigate transport properties such as electrical conductivity that, in principle, are more sensitive to parameters such as composition and temperature than is elasticity. The importance of electrical conductivity arises because of the strong dependence on temperature, major element composition, water content, partial melt, and oxygen fugacity. This is illustrated in Chap. 6, where Katsura and Yoshino consider variations in electric conductivity observed in the oceanic upper mantle to show that the high conductivity observed beneath mid-oceanic ridges could be caused by partial melting. The contribution by Khan and coauthors in Chap. 5 provides another example of using electrical conductivity to recover the thermo-chemical structure beneath Australia.

Geochemistry, like geophysics, has provided a wealth of data that bear on mantle heterogeneities, albeit at different scales. Important insights on the chemical and dynamical processes that rocks have undergone during their formation, and the physical conditions under which rocks have formed are obtained from analysis of rocks originating in the uppermost mantle. A good example of this approach is given in Chap. 7 by Ma and coauthors, where a suite of spongy clinopyroxene and melt-pockets from the Al Ghab volcanic field is investigated. Their analysis shows that compositional heterogeneities occured at micro-scale, and suggests that the spongy clinopyroxene and melt pockets were formed as a result of decompressional melting most probably associated with the recent development of the pull-apart basin in this region. Several hints point to distinct reservoirs coexisting in the deeper mantle, including recycled oceanic crust and a source of undegassed material. In Chap. 11, Caro presents updated results for the Sm–Nd, Lu–Hf, and Rb–Sr systems suggesting that Earth accreted from non-chondritic material depleted in incompatible elements, and that pristine material may be preserved in the deep mantle up until now. Kaminski and Javoy also challenge the chondritic model of the Earth in Chap. 10 by proposing that Earth's mantle may have formed from enstatite chondrites. Combined with a two-stage formation scenario involving a giant impact, their compositional model result in a large-scale chemically heterogeneous deep mantle.

Finally, geodynamics is a key ingredient to understand the formation and evolution of mantle heterogeneities. Earth's mantle is continuously being stirred by convection, which is the dominant mechanism for heat and mass transport throughout the mantle. However, the detailed mode of mantle convection is still a matter of debate. In this context, it is unclear whether stirring induced by the flow is able to efficiently mix large-scale chemical heterogeneities with ambient mantle. Several parameters may influence mantle flow and its ability to mix (or non-mixing), including viscosity and density contrasts between heterogeneous regions and ambient mantle. One obvious source of chemical heterogeneity is the continuous recycling of oceanic crust through slab subduction. In Chap. 9, Ishikawa and coauthors investigate the influence of the water content on slab subduction. Another possible source of heterogeneity, which may have survived until now, is early partial differentiation of the mantle. In Chap. 15, Deschamps and coauthors investigate the stability of an initial basal layer of dense material and identify important parameters controlling this stability. If, by contrast, convection is efficient enough to mix chemical heterogeneities, the mantle may appear isochemical at medium-to-large scales. As a result, mantle dynamics may be described by purely thermal models of convection, of which a few examples are discussed in Chap. 14 by Davies and coauthors.

This monograph is divided into two parts: Part I focuses on the upper mantle and transition zone, while Part II is dedicated to the lower mantle. As will become clear to the reader, many contributions collected in this monograph are based on multidisciplinary approaches, blending results from several fields. Since data, numerical modeling techniques, and analyses typically differ depending on the region addressed and on the scale of the heterogeneities investigated, it was deemed more appropriate to group contributions according to the part of the mantle being addressed rather than topically. In closing, we would like to acknowledge all of the people who contributed to the development and production of this volume. Above all, we would like to thank all the contributing authors for their participation. Without their efforts this volume would clearly not have been possible. We are also indebted to the numerous reviewers for their crucial assessment of the various contributions making up this volume. Finally, we would like to thank the editorial staff at Springer, in particular Naomi Portnoy and Elodie Tronche, who contributed to making this volume possible.

Amir Khan
Frédéric Deschamps

Contents

Contributors

Thomas Bodin Berkeley Seimological Laboratory, UC Berkeley, Berkeley, CA, USA; Laboratoire de Géologie de Lyon: Terre, Planètes et Environnement, CNRS, Université de Lyon 1, Villeurbanne, France

Yann Capdeville Laboratoire de Planétologie et Géodynamique de Nantes, CNRS, Université de Nantes, Nantes, France

Guillaume Caro CRPG-CNRS, Université de Lorraine, Vandoeuvre les Nancy, France

Gavin H.-N. Chan Department of Earth Sciences, Oxford University, Oxford, UK; SRK Consulting, Wanchai, Hong Kong

Queenie H.-S. Chan Department of Earth Sciences, The University of Hong Kong, Pokfulam, Hong Kong; ARES, NASA Johnson Space Center, Houston, USA

Laura Cobden Institut für Geophysik, Westfälische Wilhelms-Universität, Münster, Germany; Department of Earth Sciences, Utrecht University, Utrecht, The Netherlands

J.A.D. Connolly Institute of Geochemistry and Petrology, ETH Zürich, Zürich, Switzerland

D.R. Davies Research School of Earth Sciences, The Australian National University, Canberra, ACT, Australia

Frédéric Deschamps Institute of Earth Sciences, Academia Sinica, Nangang, Taipei, Taiwan

Paul S. Earle United States Geological Survey, DFC, Denver, CO, USA

Daniel A. Frost Institute of Geophysics and Tectonics, School of Earth and Environment, University of Leeds, Leeds, UK

R.A. Glen Geological Survey of New South Wales, NSW Department of Trade and Investment, Maitland, NSW, Australia

S. Goes Department of Earth Sciences and Engineering, Imperial College London, London, UK

Hiroki Ichikawa Geodynamics Research Center, Ehime University, Matsuyama, Ehime, Japan; Earth-Life Science Institute, Tokyo Institute of Technology, Meguro, Tokyo, Japan

Yoshiyuki Iizuka Institute of Earth Sciences, Academia Sinica, Taipei, Taiwan

Marc Javoy Institut de Physique du Globe de Paris, Sorbonne Paris Cité, CNRS, Université Paris Diderot, Paris, France

Masanori Kameyama Geodynamics Research Center, Ehime University, Matsuyama, Ehime, Japan

Edouard Kaminski Institut de Physique du Globe de Paris, Sorbonne Paris Cité, CNRS, Université Paris Diderot, Paris, France

Tomoo Katsura Bayerisches Geoinstitut, University of Bayreuth, Bayreuth, Germany

Kenji Kawai Earth-Life Science Institute, Tokyo Institute of Technology, Meguro, Tokyo, Japan; Department of Earth and Planetary Sciences, Graduate School of Arts and Sciences, Tokyo Institute of Technology, Meguro, Tokyo, Japan; Department of Earth Science, University of California, Santa Barbara, CA, USA

B.L.N. Kennett Research School of Earth Sciences, The Australian National University, Canberra, ACT, Australia

A. Khan Institute of Geophysics, ETH Zürich, Zürich, Switzerland

S. Koch Institute of Geophysics, ETH Zürich, Zürich, Switzerland

H.C.P. Lau Department of Earth Sciences and Engineering, Imperial College London, London, UK; Department of Earth and Planetary Sciences, Harvard University, Cambridge, MA, USA

S. Lebedev Geophysics Section, School of Cosmic Physics, Dublin Institute for Advanced Studies, Dublin, Ireland

Yang Li Institute of Geophysics, Swiss Federal Institute of Technology Zurich, Zurich, Switzerland

George S.-K. Ma Institute of Earth Sciences, Academia Sinica, Taipei, Taiwan; Dragon Mining Consulting, Hong Kong, Hong Kong; Department of Earth Sciences, The University of Hong Kong, Pokfulam, Hong Kong

John Malpas Department of Earth Sciences, The University of Hong Kong, Pokfulam, Hong Kong

Jean-Paul Montagner Institut de Physique du Globe de Paris, Paris, France

N. Rawlinson School of Geosciences, University of Aberdeen, Aberdeen, Scotland

Barbara Romanowicz Berkeley Seimological Laboratory, UC Berkeley, Berkeley, CA, USA; Institut de Physique du Globe de Paris, Paris, France

Sebastian Rost Institute of Geophysics and Tectonics, School of Earth and Environment, University of Leeds, Leeds, UK

M. Salmon Research School of Earth Sciences, The Australian National University, Canberra, ACT, Australia

A.J. Schaeffer Geophysics Section, School of Cosmic Physics, Dublin Institute for Advanced Studies, Dublin, Ireland

Nicholas Schmerr Department of Geology, University of Maryland, College Park, MD, USA

Neil D. Selby AWE Blacknest, Reading, UK

T.J. Shankland Geophysics Group, Los Alamos National Laboratory, Los Alamos, USA

Peter M. Shearer Institute of Geophysics and Planetary Physics, Scripps Institution of Oceanography, University of California, San Diego, USA

P.J. Tackley Institute of Geophysics, Swiss Federal Institute of Technology Zurich, Zurich, Switzerland

Christine Thomas Institut für Geophysik, Westfälische Wilhelms-Universität, Münster, Germany

Jeannot Trampert Department of Earth Sciences, Utrecht University, Utrecht, The Netherlands

Taku Tsuchiya Geodynamics Research Center, Ehime University, Matsuyama, Ehime, Japan

Abdulsalam A. Turkmani General Establishment of Geology and Mineral Resources, Ministry of Petroleum and Mineral Resources, Damascus, Syria

Tadashi Usuki Institute of Earth Sciences, Academia Sinica, Taipei, Taiwan

Kuo-Lung Wang Institute of Earth Sciences, Academia Sinica, Taipei, Taiwan

Costas Xenophontos Department of Earth Sciences, The University of Hong Kong, Pokfulam, Hong Kong

Shinji Yamamoto Department of Earth Science and Astronomy, Graduate School of Arts and Sciences, University of Tokyo, Meguro, Tokyo, Japan

Takashi Yoshino Institute for Study of the Earth's Interior, Okayama University, Misasa, Japan

A. Zunino Niels Bohr Institute, University of Copenhagen, Copenhagen, Denmark

Part I
Upper Mantle and Transition Zone

Chapter 1
Global Heterogeneity of the Lithosphere and Underlying Mantle: A Seismological Appraisal Based on Multimode Surface-Wave Dispersion Analysis, Shear-Velocity Tomography, and Tectonic Regionalization

A.J. Schaeffer and S. Lebedev

Abstract Heterogeneity of the composition and physical state of the rocks within the Earth is reflected in variations in seismic wave speeds at depth. This seismic heterogeneity can be observed in a number of different ways, each yielding a complementary perspective on the Earth's bulk properties, structure, and dynamics. A surface-wave dispersion diagram, constructed from millions of fundamental-mode and higher mode dispersion measurements around the world, shows variability around global averages for all modes and all frequencies that are included in it, with the largest variations seen for the fundamental-mode phase and group velocities at short periods (less than 30 and 40 s, respectively) that sample the highly heterogeneous crust and uppermost mantle. Seismic tomography turns large sets of measurements into models of three-dimensional wave speed variations at depth. Global shear-wave speed models have been in agreement since 1990s regarding heterogeneity in the upper mantle at thousands-of-kilometres scales. The rapid recent increase in global data sampling facilitated an increase in the tomographic resolution, and a number of today's models show close agreement in the upper 200 km of the mantle at much shorter, hundreds-of-kilometres scale lengths. Greater disagreements between different models remain in the mantle transition zone. Our new model SL2013sv, constrained by an unprecedentedly large new data set of multimode waveform fits, demonstrates increased resolution compared

A.J. Schaeffer (✉) · S. Lebedev
Geophysics Section, School of Cosmic Physics, Dublin Institute for Advanced Studies, Dublin, Ireland
e-mail: aschaeff@cp.dias.ie

© Springer International Publishing Switzerland 2015

A. Khan and F. Deschamps (eds.), *The Earth's Heterogeneous Mantle*,
Springer Geophysics, DOI 10.1007/978-3-319-15627-9_1

to other existing models for a variety of features; it captures regional-scale heterogeneity globally, within both the upper mantle and the crust. A global stack of shear-velocity profiles extracted from SL2013sv shows a monotonic decrease in the amplitude of wave speed variations with depth, mirrored by a decrease in RMS variations in SL2013sv and other current models, from largest in the top 150–200 km to much smaller below 250 km. Regionalization of SL2013sv by means of cluster analysis, with no *a priori* information, provides an accurate tectonic regionalization of the entire Earth. The three oceanic and three continental types that naturally come out of the clustering differ by the age of the deep lithosphere. The results give a new perspective on the "depth of tectonics"—the depths down to which shear speed profiles (and, by inference, geotherms) beneath oceanic and continental regions of different ages are different. Old oceanic plates are underlain by higher shear-wave speeds compared to young- and intermediate-age oceans down to 200 km depth. At 200–250 km, all type-average mantle profiles converge, except for the Archean craton profile that shows distinctly higher velocities down to 250–280 km depths.

Keywords Lithosphere dynamics · Mantle heterogeneity · Waveform inversion · Seismic tomography

1.1 Introduction

The speeds of seismic waves travelling through the Earth vary across a broad range of scales, from the crystals that make up rocks, the tectonic blocks that make up the continents, up to the scale of entire tectonic plates. This wave-speed heterogeneity is a consequence of heterogeneity in the composition and physical state of the rocks at depth. Mapping the variations of seismic wave speeds within the Earth—with seismic tomography or other methods—thus provides snapshots of the three-dimensional structure of the Earth's interior, most of it inaccessible for direct sampling, and yields important clues on the dynamic processes within the planet.

One-dimensional (1D) reference seismic models (PREM: Dziewonski and Anderson 1981; AK135: Kennett et al. 1995) capture the key changes of seismic velocities with depth. These radial changes reflect the compositional stratification within the Earth, phase transformations in minerals to their higher pressure polymorphs with increasing depth, and the gradual increases of elastic moduli with pressure.

Seismic-velocity profiles beneath different points at the surface are different from the reference profiles and from one another. This lateral heterogeneity is manifested clearly in seismic observations. Seismic tomography uses these observations to constrain three-dimensional (3D) models of seismic-velocity variations at regional or global scales, offering insight into the basic mechanisms of plate tectonics and patterns of mantle convection.

Tomographic studies using surface and body waves in the late 1970s and early 1980s (Aki et al. 1977; Dziewónski et al. 1977; Kovach 1979; Woodhouse and Dziewónski 1984) demonstrated lateral variations in seismic velocities within the lithosphere and asthenosphere globally. Since about two decades ago, the long-wavelength (several thousands of kilometres) features of the Earth's lithospheric and asthenospheric mantle at depths down to ~300 km have been well established and resolved consistently by various global models (Becker and Boschi 2002). Short- and intermediate-wavelength heterogeneities (Kennett 1987; Wu and Flatte 1990; Nolet et al. 1994) have been more difficult to resolve uniformly at a global scale, although many of the prominent features of upper-mantle heterogeneity—including subducting slabs, spreading centres, or hot spots—have been imaged in numerous regional- and larger-scale studies.

At the global scale, variations in seismic velocity indicate and distinguish the main different types of lithosphere: oceanic versus continental. Within these two types, further variability is clearly seen, including the evolution of the oceanic plates from younger to older ocean basins (the latter with colder and thicker lithosphere, with higher seismic wave speeds within it) and differences between stable Precambrian cratons (high seismic velocities; particularly thick and cold lithosphere) and younger, Phanerozoic units. At regional scales, seismic imaging reveals deep expressions of active tectonic process, as well as preserved signatures of ancient ones. Beneath oceans, strong heterogeneity is observed in the vicinity of mid-ocean ridges (MOR), where new basaltic crustis generated through the extraction of melt from the mantle beneath the spreading centre and the newly created oceanic lithosphere cools rapidly as it moves further away from the ridge. At the opposite end of the plate cycle, strong heterogeneity can be observed beneath convergent plate boundaries, in subduction zones (e.g. around the Pacific Rim) or zones of continental collision (e.g. Tibet, Zagros).

The heterogeneity of seismic velocity in the Earth's upper mantle can be observed, first of all, directly in seismic measurements, for example, in variations of travel times of body waves or of phase and group velocities of surface waves as a function of the station and event locations. Secondly, heterogeneity can be examined in seismic tomographymodels constrained by the measurements. Bulk properties of these models, such as the RMS velocity anomaly as a function of depth, provide insight into the overall character of the heterogeneity. Visual and quantitative comparison of different tomographic models, computed by different groups using different methods, can provide further insight into the heterogeneous nature of the Earth's upper mantle. For example, the lithospheric mantle of several recent global tomographic models (Lebedev and van der Hilst 2008; Lekić and Romanowicz 2011a; Debayle and Ricard 2012; Schaeffer and Lebedev 2013) show agreement at significantly shorter length scales, compared to models of only a few years ago. Finally, regionalization analysis of seismic tomography models provides an objective subdivision of the Earth into regions with different properties within a given depth range. In the lithosphere–asthenosphere depth range, regionalization groups together geologic domains with similar deep structure, which—most often but with important exceptions—reflects similar tectonic evolution.

In this paper, we present views of the heterogeneity within the Earth's crust and upper mantle from different perspectives, from measurements of phase and group velocities of multimode surface waves (particularly sensitive to the crust and upper mantle) to tomographic models and inferences from them. We utilize the very large phase- and group-velocity data set created in the course of the construction of our global tomographic model SL2013sv (Schaeffer Lebedev 2013), as well as the model itself. SL2013sv offers increased resolution globally, approaching that of regional-scale studies. The vertical component data set used to compute it consists of almost three quarters of a million waveform fits, from which more than half a million of the most mutually consistent seismograms were selected to constrain the model. Multimode phase velocities (the fundamental–mode and higher mode vertical component surface waves), measured as a by-product of the waveform fitting, span the broad period range from ~10 s to more than 400 s. Their distributions with period and their variability with source and station locations provide an empirical sampling of the bulk dispersion properties of the Earth's upper mantle. We also explore the heterogeneity of the Earth's upper mantle through examination of the salient features in the model SL2013sv, compare this view of the upper mantle with that from other recent tomographic models, and discuss inferences from global tomography regarding the structure and dynamics of the crust and upper mantle.

1.2 Heterogeneity in Measurements of Phase and Group Velocities of Multimode Surface Waves

Surface waves are highly sensitive to the structure of the crust and upper mantle. Due to the frequency dependence of their depth sensitivity—and, consequently, frequency dependence of their propagation speed (dispersion)—surface waves have two different measurable velocities: phase velocity and group velocity. Phase velocity ($C(\omega)$) is the speed at which the phase at a particular frequency travels, whereas the group velocity ($U(\omega)$) is the speed at which the energy of the wave packet propagates. With decreasing frequency (increasing period), surface waves sample greater depths within the subsurface: short periods (high frequencies) are sensitive to crustal structure, whereas longer periods are sensitive to depths near the base of the upper mantle (see Fig. 1.1). The fundamental modeof Rayleigh or Love surface waves is typically the most energetic arrival on a vertical- or horizontal-component seismogram, respectively. Higher surface-wave modes interfere constructively to produce the arrivals of body waves (*P* and *S* waves) and their multiple surface reflections (e.g. SS, SSS, SSSS)—a manifestation of the ray–mode duality (Dahlen and Tromp 1998). Each of the modes has a different frequency-dependent depth sensitivity; together, they provide rich information regarding the structure within the Earth.

Variations in surface-wave dispersion measured along different source-station paths provide a direct indication of structural heterogeneity and lateral variations

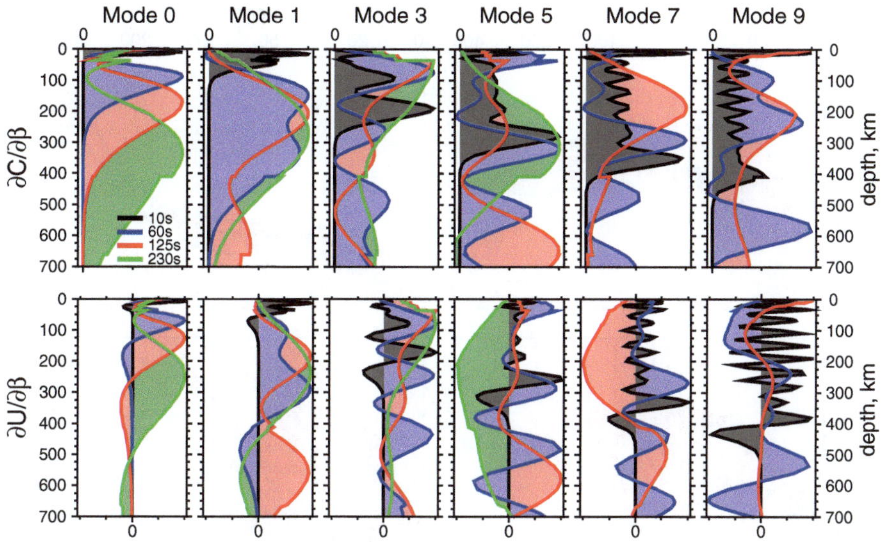

Fig. 1.1 Depth sensitivity kernels of the Rayleigh-wave phase and group velocities, computed for AK135. The derivatives are with respect to shear velocity, $\partial C(\omega)/\partial\beta$ and $\partial U(\omega)/\partial\beta$, at four periods 10 s (*dark grey*), 60 s (*blue*), 125 s (*red*), and 230 s (*green*), for the fundamental mode and the first, third, fifth, seventh, and ninth overtones. The amplitude of each kernel is scaled independently to its maximum within the depth interval. Note that for modes seven and nine at 230 s, there is no sensitivity within this depth range

in the elastic properties of the crust and mantle. As the depth sensitivity of the modes is known, the measurements at different periods can be viewed as expressions of heterogeneity at different depths.

In order to examine bulk distributions of phase and group velocities globally, we use the data set generated using the Automated Multimode Inversion (AMI; Lebedev et al. 2005) of surface and *S* waveforms in the course of construction of the global tomographic model of Schaeffer and Lebedev (2013) (Sect. 1.3).

As a by-product of successfully fitting a seismogram, AMI measures phase velocities for the modes which contribute significantly to the waveform fit within the time–frequency windows that were selected. The group velocity ($U(\omega)$) can then be computed from the phase velocity ($C(\omega)$) using:

$$U(\omega) = \frac{C(\omega)}{1 - \left(\frac{\omega}{C(\omega)}\right)\left(\frac{\mathrm{d}C}{\mathrm{d}\omega}\right)}, \tag{1.1}$$

where $\omega = 2\pi/T$ is the angular frequency and T is the period.

The data set used to constrain the model SL2013sv consists of 712,504 waveform fits computed for vertical component seismograms recorded up to March 2010. Here, we augmented it with an additional 239,462 fits computed for seismograms recorded until June 2012. In total, the data set is formed by more than

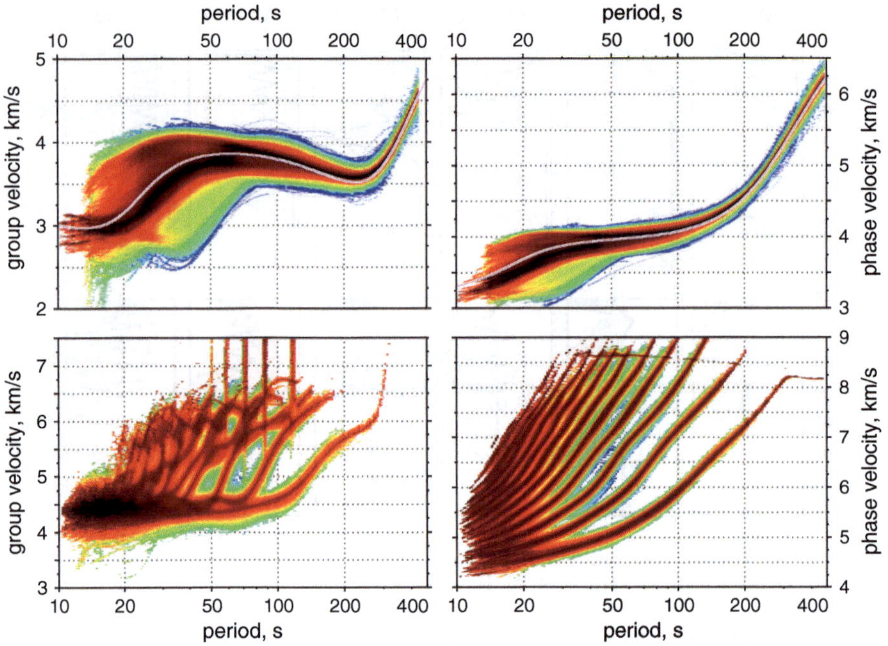

Fig. 1.2 Empirical dispersion diagrams of Rayleigh waves on the Earth. Fundamental-mode (*top row*) and higher mode (*bottom row*) Rayleigh-wave group-velocity (*left column*) and phase-velocity (*right column*) *curves* were measured by AMI after (successful) waveform fitting. Increased sampling density of phase or group velocity is indicated by warming colours from *blues* through *reds* to *black*; the sampling density is normalized for each period. Over-plotted in the fundamental-mode panel are the dispersion *curves* computed for AK135

951,000 vertical component waveform fits (951,000 seismograms). It yields more than 951 thousand fundamental-mode and ~480 thousand higher mode, Rayleigh-wave, phase-velocity—and therefore group-velocity—curves, across a broad period range of 10–450 s. The criteria required to generate a waveform fit result in a fundamental-mode dispersion curve for every successfully fit seismogram; the total frequency band of each dispersion curve varies on a case-by-case basis.

Although these dispersion curves are not used in the generation of the tomographic model, they can be used in other imaging studies, such as array-based teleseismic interferometery utilizing the fundamental mode (e.g. Deschamps et al. 2008; Darbyshire and Lebedev 2009; Zhang et al. 2009; Endrun et al. 2011; Adam and Lebedev 2012, to name a few) and, potentially, higher modes.

In Fig. 1.2, we plot the fundamental-mode and higher mode phase- and group-velocity curves, as in Nolet (2008), from our data set of ~512 thousand multimode Rayleigh-waveform fits used to constrain SL2013sv. The individual dispersion curves were binned, such that the blue colours represent the lowest density sampling (white means not sampled), whereas warmer colours indicate increasing density; black represents the highest density velocity at each period.

The group-velocity curves were computed from the phase velocity ones, using Eq. 1.1. In the fundamental-mode windows, the superimposed lines indicate the dispersion curves computed for AK135.

The inclusion of higher modes is critical for maintaining resolving power at the base of the upper mantle and in the transition zone. This sensitivity is given by *S* and multiple *S* waves that bottom throughout this depth range and represents a superposition of multiple higher modes. Our complete data set contains many tens of thousands to a hundred thousand dispersion curves for each of higher modes one through eight, thousands for modes up to ten, and relatively minor contributions from those modes beyond (<0.05 % from modes 11–20). The impact of the inclusion of these higher modes (i.e. *S* and multiple *S* waves) on the tomographic resolution in the deep upper mantle and transition zone is illustrated in Figs. 4 and 5 of Lebedev et al. (2005), who performed the waveform fitting first for the fundamental mode only, second for the fundamental mode and *S* waves, and finally for the fundamental mode and *S* and multiple *S* waves. As more *S* wave information is added, the eigenvalues drop off less rapidly, indicating greater structure in the resulting path-average model, as can also be seen in the shear-velocity models in their Fig. 4.

The global variability of the fundamental-mode and higher mode dispersion curves offers a perspective on the bulk heterogeneity within the Earth's upper mantle. At a given period, the total spread in velocity indicates an empirical range of Earth properties, whereas the relative distribution in colour across the velocity band indicates which velocities are most common—subject to the sampling filters imposed by the distribution of the stations and events and by the averaging over the Frésnel zones between the sources and stations.

For the fundamental-mode group velocities (top left, Fig. 1.2), the strongest variability is at periods less than 45 s, with a transitional band between 45 and 80 s, and reduced variability at periods beyond 80–100 s. The total spread in vesensitivityvolume average plocity decreases by a factor of 2 at long periods compared to short periods. This results from the increased sensitivity of the long periods to greater depths, at which heterogeneity is weaker than at shallow, crustal depths. A similar pattern can be observed for the fundamental-mode phase velocity (top right panel), however, with the transition shifted to slightly shorter periods due to the differing depth sensitivities of group and phase velocities (e.g. Lebedev et al. 2013).

At the shorter periods sampling the crust and uppermost mantle, the relatively strong heterogeneity is readily apparent. In continental regions, this shorter period band is most sensitive to the crustal structure; we observe that the densest sampled velocities at these periods less than 50–60 s (35–45 s for phase velocity) are slower than AK135. In regions where the Moho is at greater depths (beneath orogenic belts, for example), the low velocities extend to longer periods, which is seen as the yellow-green colours well below AK135 at 30–75 s and 30–50 s periods for group and phase velocities, respectively. In oceanic regions, the Moho is much shallower and the fundamental mode at 15–30 s period samples the uppermost mantle rather than the crust; it is thus faster than in AK135.

At periods shorter than 50 s, the dispersion-curve representation of heterogeneity is affected by a sampling bias, with the most sampled velocities similar to or slower than AK135. This is not what would be expected when looking at the relative surface areas of oceans versus continents. In fact, one would expect greater sampling of high velocities (relative to AK135) in this period band, representative of the fast uppermost mantle beneath oceans, as there is a larger proportion of oceans than continents. However, this is opposite to what we observe, with the densest sampling at velocities slower than AK135. This discrepancy is explained simply by the distribution of stations and events, dominated by paths shorter than 4000–5000 km, which results in preferential sampling of continental and backarc regions.

1.3 Global Heterogeneity of the Upper Mantle and the Crust from Multimode Surface-Wave Tomography

We now explore the heterogeneity of the Earth's upper mantle using the tomographic model SL2013sv (Schaeffer and Lebedev 2013). This model is parameterized on a global triangular grid (Wang and Dahlen 1995) with an average spacing of ~280 km (minimum 250 km, maximum 290 km) and is constrained by the ~520 thousand vertical component seismograms selected from more than 710 thousand successful waveform fits, as the most mutually consistent data.

The model was computed using the Automated Multimode Inversion (AMI; Lebedev et al. 2005) that performs automated, accurate processing of large numbers of vertical- and horizontal-component broadband seismograms. The result of each successful waveform inversion is a set of linear equations with uncorrelated uncertainties (Nolet 1990, 2008) that describe 1D perturbations in elastic structure within a finite-width sensitivity volume between the source and receiver relative to a three-dimensional (3D) reference model (Lebedev and van der Hilst 2008; Schaeffer and Lebedev 2013). Synthetic seismograms are computed using the JWKB approximation by summing over modes (m), with the phase velocity of mode m given by $C_m^0(\omega) + \overline{\delta C_m(\omega)}$, where $C_m^0(\omega)$ is the average initial phase velocity within the sensitivity volume and $\overline{\delta C_m(\omega)}$ is the average phase velocity perturbations. These are expressed as a function of the sensitivity-volume average perturbations in P and S velocity ($\overline{\delta\alpha(r)}$ and $\overline{\delta\beta(r)}$, respectively):

$$\overline{\delta C_m(\omega)} = \int_0^R \frac{\partial C_m^0(\omega)}{\partial\beta(r)} \overline{\delta\beta(r)} \mathrm{d}r + \int_0^R \frac{\partial C_m^0(\omega)}{\partial\alpha(r)} \overline{\delta\alpha(r)} \mathrm{d}r, \tag{1.2}$$

where R is the radius of the earth and $\partial C_m^0(\omega)/[\partial\beta(r), \partial\alpha(r)]$ are the Fréchet derivatives. The sensitivity-volume-average perturbations in seismic velocity ($\overline{\delta\beta(r)}$, $\overline{\delta\alpha(r)}$) are thus related to the synthetic waveform $s(\omega)$ through the average phase velocity perturbation $\overline{\delta C_m(\omega)}$. The $\overline{\delta\beta(r)}$ (and equivalently the $\overline{\delta\alpha(r)}$) are parameterized with M model parameters γ_i on a set of 1D triangular basis functions $h_i(r)$ which span from the crust to the lower mantle (7–1600 km), $\overline{\delta\beta(r)} = \gamma_i h_i(r)$. AMI then produces a set of linear equations with uncorrelated uncertainties:

$$\overline{\delta\beta(r)} = \eta_j g_j(r), \tag{1.3}$$

where each $g_j(r)$ is a linear combination of the original basis functions (see Nolet 1990, 2008; Lebedev et al. 2005; Lebedev and van der Hilst 2008; Schaeffer and Lebedev 2013, for further details).

A large system is constructed using the independent linear equations generated by AMI (the η_j from Eq. 1.3 for each path), with the waveform structural information mapped into the model parameters using the same approximate sensitivity kernels as used in the waveform inversion by AMI, and is then solved for 3D perturbations in isotropic S and P velocity with respect to the 3D reference model and 2Ψ azimuthal anisotropy of S velocity (see Schaeffer and Lebedev 2013 for details on the data set and model construction).

In Figs. 1.3 and 1.4, we plot 12 horizontal slices through the model, at depths of 36, 56, 80, 110, 150, and 200 km (Fig. 1.3), and 260, 330, 410, 485, 585, and 660 km (Fig. 1.4). The perturbations (as indicated beneath each slice) are with respect to a 3D reference model. At depths greater than the Moho (Moho depth is position dependent), perturbations are in per cent, relative to the global 1D mantle reference model (a modified AK135, Lebedev and van der Hilst 2008; Schaeffer and Lebedev 2013). At depths shallower than the Moho (some regions in the slices at 36 and 56 km depth), model perturbations are indicated in $\mathrm{m\,s^{-1}}$ and are relative to the 3D crustal model (modified CRUST2); in these regions, only the perturbation in $\mathrm{m\,s^{-1}}$ is correct. Although this makes interpretation of the strength of velocity perturbations in crustal regions more complex, the relative variations are still readily interpreted in terms of heterogeneity and structure.

This new model provides improvements, compared to other existing models, in the resolution of fine-scale regional features through much of the upper mantle and transition zone. At the longer wavelengths, lithospheric-depth structures broadly agree with observations from numerous past models utilizing varying methodologies, parameterizations, and data sets (both type and size) (e.g. Debayle et al. 2005; Zhou et al. 2006; Houser et al. 2008; Kustowski et al. 2008; Lebedev and van der Hilst 2008; Nettles and Dziewónski 2008; Panning et al. 2010; Ferreira et al. 2010; Lekić and Romanowicz 2011a; Ritsema et al. 2011; Debayle and Ricard 2012). At greater depths in the sub-lithospheric mantle and transition zone however, larger variations between models are evident even at the longer wavelengths (thousands of kilometres, e.g. Ritsema et al. 2011).

In the following sections, we examine some of the prominent features in SL2013sv that exhibit deep expressions of tectonic structures and processes, including spreading ridges and subduction zones. The strongest shear-speed anomalies are observed beneath mid-ocean ridges (MORs) and rift systems (negative), backarcs and active orogens (negative), stable continental cratons (positive), and within subducted lithosphere (positive).

1.3.1 Spreading Ridges

As observed in the shallow mantle in the horizontal sections (Fig. 1.3), spreading ridges globally are resolved as narrow anomalies directly beneath the ridge axis.

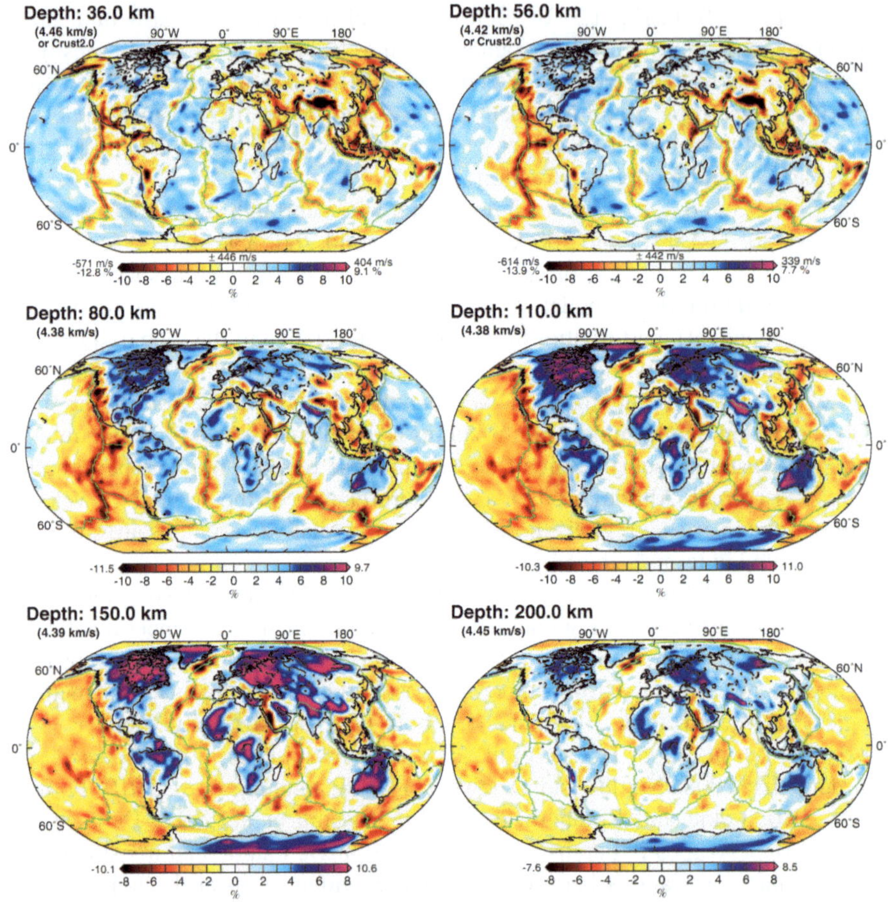

Fig. 1.3 Horizontal slices through SL2013sv at six depths in the crust and uppermost mantle: 36, 56, 80, 110, 150, and 200 km. The reference velocity is indicated in brackets beneath each depth, and perturbations are plotted with respect to this value, for locations beneath the Moho. The minimum and maximum perturbations are indicated at the edges of the saturated-scale bar. For the depths 36 and 56 km, some locations are still in the crust, and therefore, perturbations are instead in m s^{-1} with respect to our 3D crustal reference model (modified CRUST2), as indicated above the colour scales. Plate boundaries are indicated by *green lines*

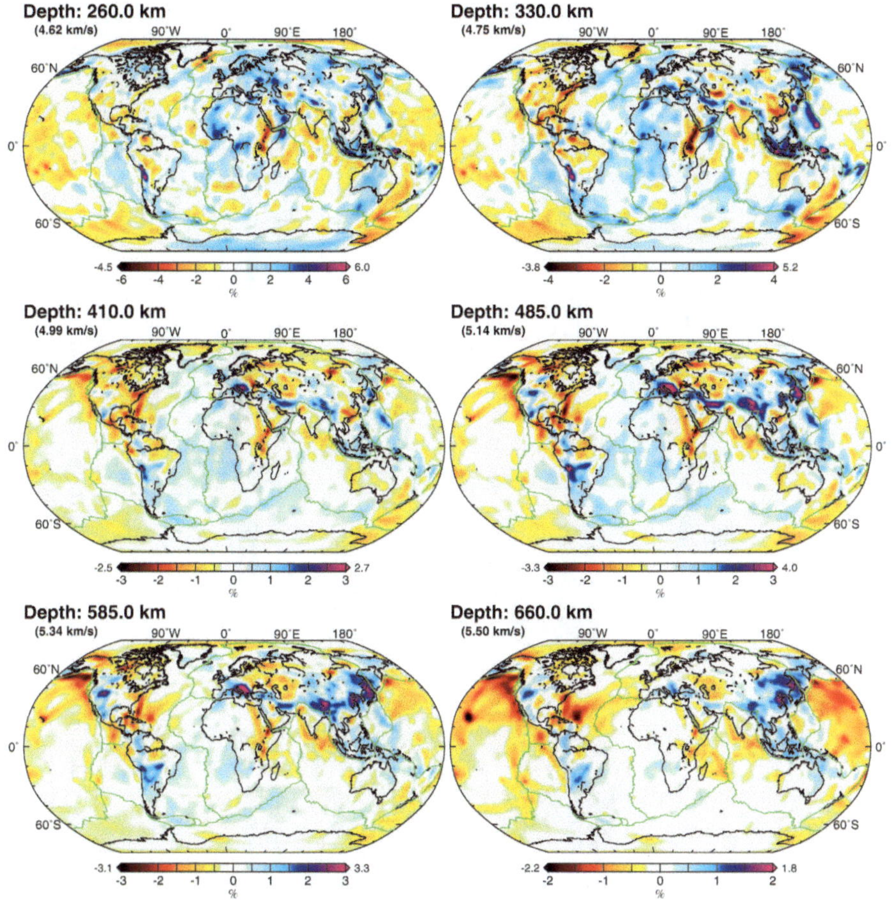

Fig. 1.4 Horizontal slices in the sub-lithospheric mantle and transition zone: 260, 330, 410, 485, 585, and 660 km. Reference velocity and perturbations are as in Fig. 1.3

With increasing depth, from 36 to 80 km, the anomalies increase in amplitude and also broaden in extent from the spreading centre. The very strong negative anomalies are indicative of the presence of partial melt. Continuing to greater depths, from 110 to 150 km, the anomalies broaden and also decrease in amplitude. By 150 km depth, the central low-velocity zone no longer stands out from the surrounding oceanic asthenosphere. At depths greater than 120–130 km, partial melting may still take place, but with a lower degree, no longer visible in the vertically polarized shear velocity. This is in agreement with some past studies (e.g. Zhang and Tanimoto 1992; Forsyth et al. 1998), but does not confirm the inferences in other studies (e.g. Su et al. 1992), suggesting that MOR processes extend into much deeper upper mantle.

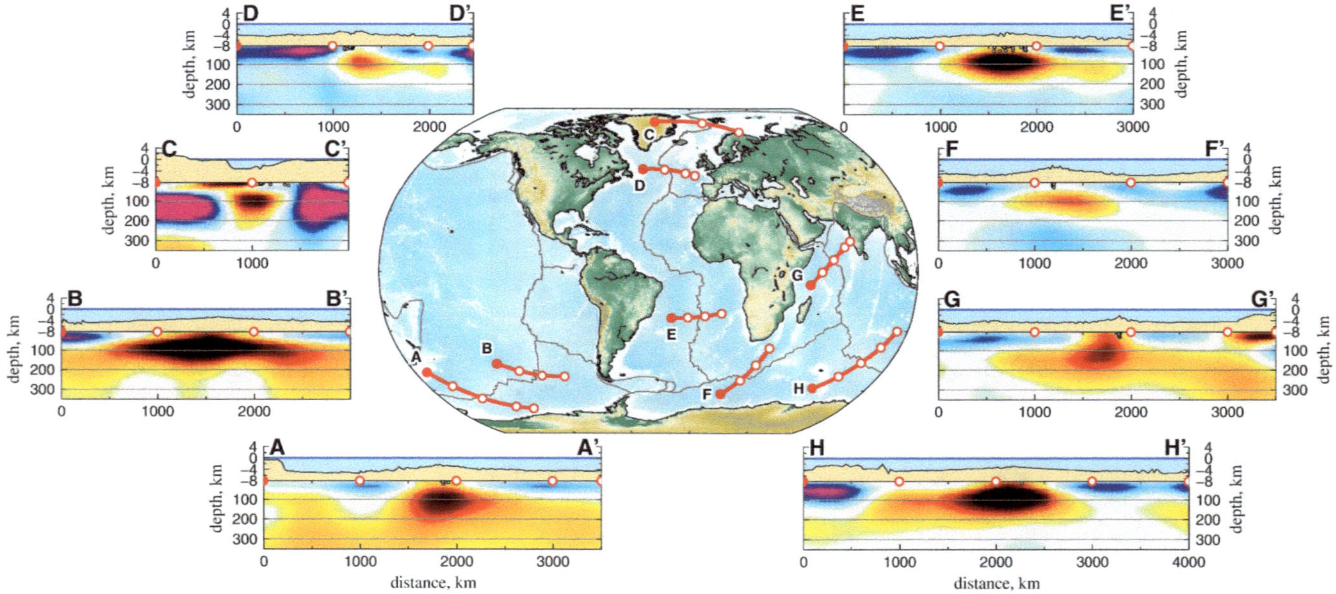

Fig. 1.5 Vertical cross sections through "normal" spreading ridge segments. Above each slice through the model SL2013sv, topography is plotted with tick marks denoting elevations (or bathymetry) spaced at 4 km. Horizontal distance between *red circles* (both on map and each cross section) is 1000 km; the *closed red circle* marks one end of each section to indicate orientation. The (saturated) colour scale is the same in each cross section, with ± 220 m s^{-1} limits

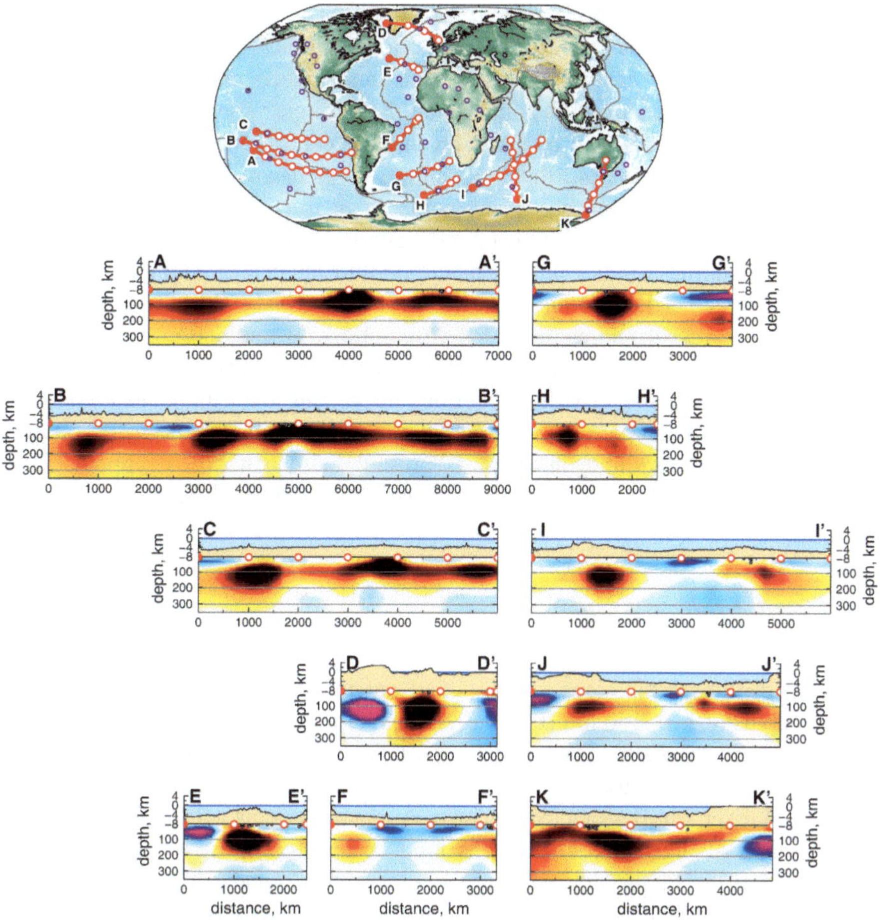

Fig. 1.6 Vertical cross sections through spreading ridges with nearby hot spots. Above each slice through the model SL2013sv, topography is plotted with tick marks denoting elevations (or bathymetry) spaced at 4 km. Horizontal distance between *red circles* (both on map and each cross section) is 1000 km; the *closed red circle* marks one end of each section to indicate orientation. The (saturated) colour scale is the same in each cross section, with ± 220 m s^{-1} limits. Hot spots from the catalog of Steinberger (2000) are indicated by *purple circles* on the map

In Figs. 1.5 and 1.6, we plot vertical cross sections through spreading ridges around the globe. Each section spans from 7 to 350 km depth, with grid lines every 100 km. Topography/bathymetry is indicated above each section (extracted from ETOPO2), with tick marks every 4 km. Open red circles on both the map and each cross section are every 1000 km, and the closed red circle marks matching ends of the profiles to clarify the orientation. Seismicity within 40 km perpendicularly of the cross section is extracted from the EHB catalog (Engdahl et al. 1998) and plotted with white circles. Figure 1.5 shows "normal" cross sections, which demonstrate the expected triangular region of decompression melting beneath the ridge axis. The cross sections in Fig. 1.6 highlight sections of ridges with non-triangular melting regions, as well as interactions with nearby hot spots.

In comparing the "normal" cross sections and indeed many of the asymmetric interacting cross sections, it is clear that the strongest spreading anomalies are confined to depths, ~150 km or less, and that the central anomaly clearly widens as a function of depth. Note that the saturations here are relatively higher than those in the horizontal cross sections and are in $m\,s^{-1}$ rather than percentage. Beneath most of the ridges, sharp lateral boundaries between the ridge anomaly and the surrounding asthenospheric mantle delimit the central anomaly associated with high-degree partial melting.

Beneath the rapidly spreading EPR (Figs. 1.5A, B and 1.6A, B, and C), the velocity anomalies are the strongest and spread the widest, compared to other ridges around the globe. For those sections interacting with the nearby hot spots, several connected low-velocity anomalies are present. In profile 1.5B crossing the southern EPR, a single anomaly is present, with a width of approximately 2000 km, significantly wider than other "normal" spreading ridges.

1.3.2 Subduction Zones

Subducting lithospheric slabs clearly stand out in the horizontal sections, beginning at depths of 110–150 km (Fig. 1.3), beneath the Aleutians, Kuriles, Japan, Izu-Bonin, Mariana, Sumatra, Tonga-Kermadec-Hikurangi, and portions of the Andean. Additionally, smaller anomalies can be observed associated with subduction at the Cascadia, Lesser Antilles, and Scotia arcs. At greater depths and into the transition zone, a band of high velocities is observed extending almost continuously from Tibet through the Pamirs and Hindu Kush to Anatolia, and westwards to Central Europe. This material likely represents fragments of ocean basins, continental lithosphere, and continental margins, subducted in the final stages of the closure of the Tethys Ocean (Hafkenscheid et al. 2006; Schaeffer and Lebedev 2013).

In Fig. 1.7, we plot nine vertical cross sections through different subduction zones around the globe, including (A) Tonga and New Hebrides, (B) Sumatra–Java, (C) Ryuku and the Marianas, (D) Ryuku and Izu-Bonin, (E) Japan, (F) the Kuriles, (G) the Aleutians, (H) Central America and the Lesser Antilles, and (I) the Chilean segment of the Andean subduction zone.

In each cross section, a high-velocity anomaly can be seen dipping well into the upper mantle, correspondingwith the overplotted seismicity (white circles). In addition to the high velocities associated with the subduction of fast oceanic lithosphere, low velocities are clearly observed in the backarc regions, reflecting the high water content in the mantle there. The strength and depth distribution of these anomalies varies between different subduction zones.

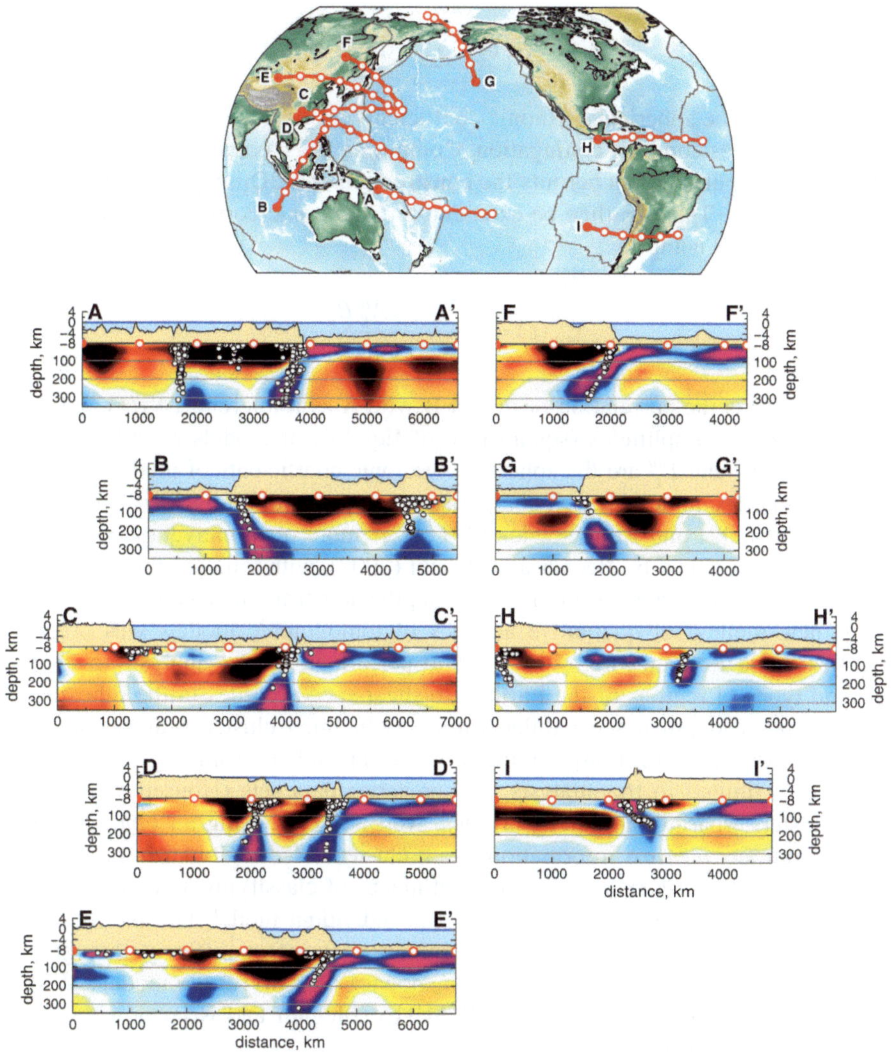

Fig. 1.7 Vertical slices across subduction zones. Plotting set-up and conventions are as in Figs. 1.5 and 1.6

1.4 Tectonic Regionalization of the Upper Mantle

Over the last several decades, the number of available global tomography models has increased steadily; however, there has not been a comparable proliferation in tools to compare them (Lekić and Romanowicz 2011b). A common methodology was to compare harmonic spectra and examine at their correlation. The spectrum of model *A* is computed as follows:

$$S_A^2(\ell) = \sum_{m=-\ell}^{\ell} A_\ell^m A_\ell^{m*}, \tag{1.4}$$

where A_ℓ^m are the spherical harmonic coefficients for degree ℓ and azimuthal order m, and $*$ denotes complex conjugation. Computing the spectral power at a range of depths and angular orders permits the construction of a 2D image depicting where the energy lies. This can then be extended to a harmonic correlation between two different models using

$$C_A^B(\ell) = \frac{\sum_{m=-\ell}^{\ell} A_\ell^m B_\ell^{m*}}{S_A(\ell) S_B(\ell)}, \tag{1.5}$$

where $C_A^B(\ell)$ is the correlation of model B with A as a reference, $S_A(\ell)$ and $S_B(\ell)$ are the spectral amplitudes (square root of Eq. 1.4) of models A and B, respectively, and A_ℓ^m and B_ℓ^m are the spherical harmonic coefficients of models A and B, respectively. Examples of such comparisons are presented in Debayle and Ricard (2012) and Lekić and Romanowicz (2011a). A more quantitative analysis of this form was undertaken by Becker and Boschi (2002), with tomographic and geodynamic models correlated as a function of depth and harmonic degree.

A new method of qualitative analysis of tomographic models was proposed by Lekić and Romanowicz (2011b), who performed cluster analysis to identify several types of regions of the Earth with shear-speed profiles in the lithosphere–asthenosphere depth range similar within each. Such cluster analysis has been used in various applications for many years; in global seismology, its use was previously limited to the analysis of time-series data sets (Houser et al. 2008). The clusters, sharing similar V_S profiles, were selected using an objective criteria without any *a priori* information. Such cluster-based regionalization of tomographic models provides not only a novel means of classifying and characterizing the upper mantle structure identified by an individual model, but also is readily extended to the comparison of the prominent features of different tomographic models, as carried out by Lekić and Romanowicz (2011b) for the models SEMum (Lekić and Romanowicz 2011a), SAW24B16 (Mégnin and Romanowicz 2000), and S362ANI (Kustowski et al. 2008).

We perform regionalization of our model SL2013sv following the procedure of Lekić and Romanowicz (2011b). We resample the tomographic model into 1D shear-speed profiles, sampled every 10 km with depth between 30 and 350 km, and on a finer lateral grid than the original model, to avoid spatial aliasing. The triangular grid provides knots with approximately equal areas, ensuring that the polar regions are not over-represented as would occur in an un-weighted global cartesian grid. The clustering analysis is carried out using the MATLAB implementation of the *k-means* algorithm. Given a set of N observations $\mathbf{x}_p$, $(\mathbf{x}_1, \mathbf{x}_2, \ldots, \mathbf{x}_N)$, we wish to sort them into k sets $[S_1, S_2, \ldots, S_k]$ with corresponding centroids $[\mathbf{m}_1, \mathbf{m}_2, \ldots, \mathbf{m}_k]$. The basic algorithm operates in two steps. First, the Euclidean distance between an $\mathbf{x}_p$ and each $\mathbf{m}_i$ is computed, with $\mathbf{x}_p$ assigned to only one S_i with the minimum sum of squares:

$$S_i^{(t)} = \left[\mathbf{x}_p : \left\| \mathbf{x}_p - \mathbf{m}_i^{(t)} \right\|^2 \leq \left\| \mathbf{x}_p - \mathbf{m}_j^{(t)} \right\|^2 \quad \forall\ 1 \leq j \leq k \right], \tag{1.6}$$

In the first iteration ($t = 0$), the initial centroids $\mathbf{m}_i^{(0)}$ are selected using a random subset of observations. In the second step, new centroids $\mathbf{m}_i^{(t+1)}$ are computed based on the distribution of observations $\mathbf{x}_p$ in each S_i:

$$\mathbf{m}_i^{(t+1)} = \frac{1}{|S_i^{(t)}|} \sum_{\mathbf{x}_j \in S_i^{(t)}} \mathbf{x}_j \tag{1.7}$$

This two step process is repeated until the assignments of $\mathbf{x}_p$ into S_i no longer change.

In our case, the N shear-speed profiles (the $\mathbf{x}_p$, with $p = 1, \ldots, N$) are partitioned into k arbitrary sets such that each intra-cluster variance is minimized. We stress that all the cluster types and their geographic distribution are not predetermined, but result directly from the minimization. In particular, oceans and continents are not separated from each other *a priori* but fall into different clusters naturally. (We select $k = 6$ and use 5 replicates in part to ensure consistent comparison with the results of Lekić and Romanowicz (2011b).

The results of the clustering of SL2013sv are shown in Fig. 1.8. The six clusters are readily distinguished: young oceans, ridges, and backarcs (which we label O1, identified with red); intermediate-aged oceans (O2, orange); oldest oceanic (O3, brown); continental regions presently deforming or having undergone deformation during the Phanerozoic (C1, light blue grey); stable continental regions of Precambrian fold belts and cratons modified in the Proterozoic (C2, blue); and stable Archean cratons (C3, green). The areas of the clusters are given in Table 1.1, along with their per cent coverage of the Earth's surface, for both SL2013sv and SEMum.

The three "ocean" clusters make up ~67 % of the Earth's surface area. Locations that belong to the slowest cluster (O1, red) are predominantly beneath MORs and backarc basins. Along most MOR segments, the width of the cluster is not constant, but depends on the spreading rate and is confined to within ~2.0 Ma of the ridge axis. This cluster also includes several continental locations, including south-western North America, the northern portion of the East African Rift, the region including the Red Sea, the Dead Sea Transform and the East Anatolian Fault regions, the Pannonian Basin, and localized points in close proximity to known continental intra-plate volcanism. On its own, this cluster takes up almost 20 % of the Earth's surface, with much of it over the East Pacific Rise and surroundings.

The cluster O3 (brown) includes the oldest oceanic regions. It is characterized by very high velocities in the uppermost mantle and makes up ~21 % of the Earth's surface area.

The intermediate-age oceanic cluster (O2, orange) has the largest single surface area of all clusters, encompassing more than 26 % of the Earth's surface.

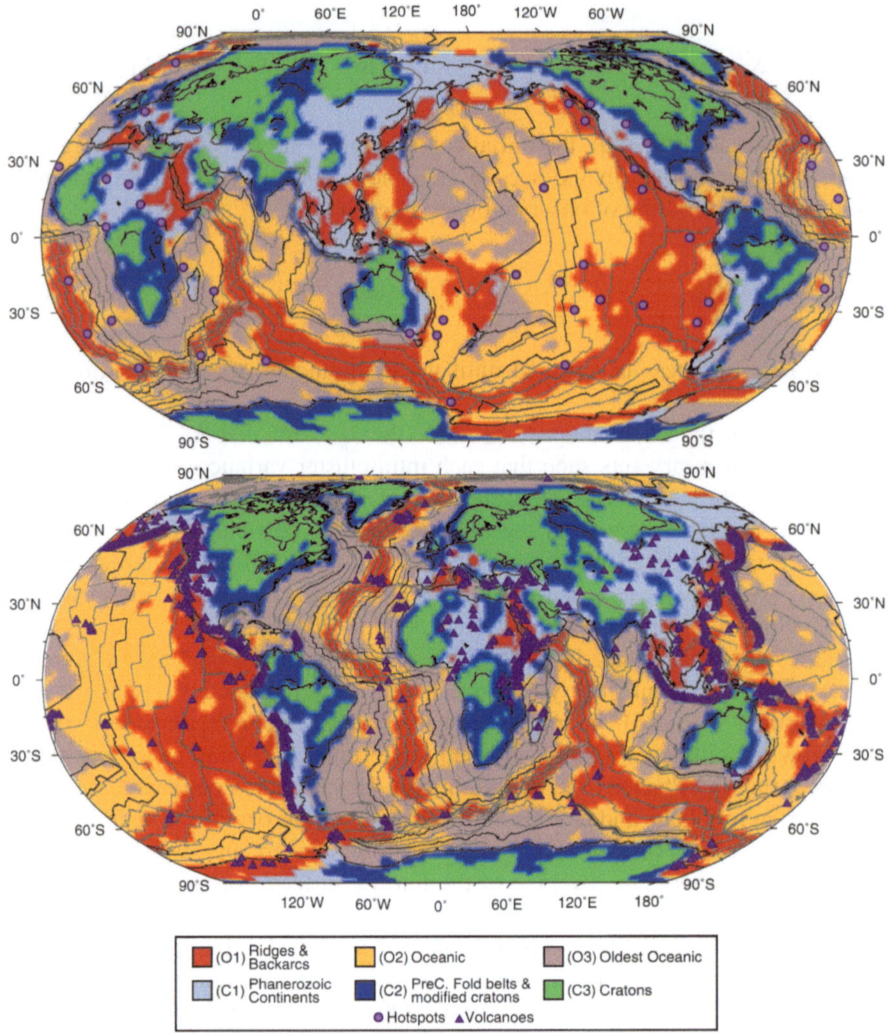

Fig. 1.8 Tectonic regionalization of the Earth. The regionalization was computed from the model SL2013sv using the *k-means* clustering (Lekić et al. 2010). The classification and colour of each region are given in the legend. The two maps present the same data, with the difference that the top one is centred at 150°E and the bottom one is centred at 5°E. Known hot spots (Steinberger 2000) are plotted as *purple circles* (*top*) and volcanoes from the Smithsonian Global Volcanism database (Siebert and Simkin 2002) as *purple triangles* (*bottom*). Ocean-age contours (Muller et al. 1997) are indicated every 20 Ma (*thin light grey lines*) and 60 Ma (*thin darker grey lines*)

Of particular, interest is the signature of lithospheric rejuvenation. This is evident along the Hawaiian and Emperor Islands hot spot track. Stretching to the north-west from Hawaii, a band of this cluster bisects through that of older oceanic lithosphere (O3) parallel with the string of islands. The variations in plate age do not

Table 1.1 Summarizing the areal extent of the 6 different cluster types identified by the *k-means* algorithm applied to SL2013sv and SEMum

Region name	Cluster ID	Colour	SL2013sv SA		SEMum SA	
			km^2	%	km^2	%
Ridges and backarcs	O1	Red	9.71×10^7	19.09	1.10×10^8	21.55
Intermediate oceans	O2	Orange	1.36×10^8	26.82	8.80×10^7	17.28
Oldest oceanic	O3	Brown	1.09×10^8	21.41	1.31×10^8	25.83
Phanerozoic continents	C1	Light grey	5.60×10^7	11.00	6.71×10^7	13.18
PreCambrian fold belts and modified cratons	C2	Blue	5.36×10^7	10.53	7.18×10^7	14.10
Cratons	C3	Green	5.77×10^7	11.33	4.20×10^7	8.25

explain this pattern; the lithospheres to the west and east of this band are older and younger, respectively. Therefore, this band of oceanic lithosphere classified as "younger" than the surrounding lithosphere represents the lasting imprint due to the passage over the hot spot, which has rejuvenated the lithosphere, leaving a velocity structure more characteristic of a younger oceanic lithosphere.

A similar feature can be seen off the western margin of Africa. The Cameroon Line volcanic belt bisects the African continent, separating the west African craton from the Congo craton. The present locus of the volcanism (and possibly a hot spot, according to the Steinberger (2000) catalog) is indicated on the bottom (and the top) maps in Fig. 1.8, respectively. Extending south-westwards into the Atlantic Ocean towards the MOR is a thin band of oceanic lithosphere clustered as type O2, which cuts through type O3, old eastern Atlantic lithosphere. Further south, a similar band of O2 lithosphere may be associated with several hot spots in the south Atlantic (Tristan and Verna).

Continents are characterized by the other three cluster types. The cluster with the highest shear speeds includes the cratons (C3, green). It makes up 11 % of the Earth's surface area and includes all the major and minor Archean cratons making up the continental cores. In addition to the large cratons within each of the continents, we also map smaller cratonic blocks, such as the Sichuan block within the Yangtze craton. In the African continent, each of the three main cratons is clearly identified. Great improvement compared to previous tomographic models can be seen in South America, where the three main cratons are mapped clearly: the large Amazonia in the north east; the south-eastern Río de la Plata and eastern central São Francisco, together.

Somewhat lower shear speeds characterize the cluster that includes Precambrian fold belts and modified cratons (C2, blue). This cluster takes up the smallest proportion of the Earth's surface, at just over 10 % of surface area. Tectonic units that fall within it often occupy the spaces between Archean cratons, for example, in South America, southern Africa, and in the West Siberian platform between the east European and Siberian cratons.

The last cluster includes predominantly Phanerozoic portions of the continents (C1, light grey) and makes up just over 11 % of the Earth's surface. It is characterized by seismic-velocity profiles slower than those of the stable continental regions. It is observed stretching along the entire western margin of North America, from the Aleutians and Alaska southwards through the Canadian and US Cordillera into Mexico and Central America. A similar band of C1 can be seen in the Cordillera of South America and extending southwards through the Antarctic Peninsula. It is also dominant across much of East Asia and extends through the Zagros Mountains and Anatolia to the Alps and Iberia. Finally, it is prevalent in much of north-central Africa. There and elsewhere, the portions of continents that have thin, warm lithosphere and comprise the cluster C1 include almost all the locations of intra-plate continental volcanism (purple triangles, Fig. 1.8), as shown previously for Europe by Legendre et al. (2012).

In Fig. 1.9, we compare different tectonic regionalizations. Two of them were computed from surface-wave tomographic models, with no *a priori* information. The other three, GTR1 (Jordan 1981), RUM (Gudmundsson and Sambridge 1998), and 3SMAC (Nataf and Ricard 1996), were constructed using compilations of *a priori* geologic and geophysical information. The SEMum clusters are exactly those from Lekić and Romanowicz (2011b), obtained directly from the authors.

The regionalizations based solely on tomography (left panel) have strong similarities at large scales. The biggest difference between SL2013sv and SEMum is the relative partitioning of the oceanic clusters. In SEMum, the O2 cluster is reduced in area by >9 %, whereas O1 and O3 increase by 2.5 and 4.5 %, respectively. Lekić and Romanowicz (2011b) suggest that the presence of plate-motion-aligned "finger-like" structures in the O2 cluster is associated with small-scale convection cells known as Richter Rolls, previously inferred in the Pacific from gravity measurements. Similar fingers are not observed in the clustering of SL2013sv; however, as pointed out above, a section of the O2 cluster following both the Hawaiian and Emperor Islands hot spot tracks and that associated with the Cameroon Line volcanic belt bisects two regions of O3, suggesting a lasting imprint in the oceanic lithosphere due to rejuvenation by an impinging hot spot. The ridge cluster (O1) is very similar between SL2013sv and SEMum, with those in the former being slightly narrower at many ridge segments. Additionally, much of the tectonically active western USA is classified as O1 (young ocean) in SEMum, whereas in SL2013sv, most of this region is instead classified as C1 (tectonic continent), with O1 seen only in close proximity to the rifting and spreading within the Gulf of California.

In the right hand panels of Fig. 1.9, we also include the regionalizations based on *a priori* information: GTR1, RUM, and 3SMAC. Note that for 3SMAC, we only plot the continental regions. We briefly highlight several key differences between the regionalizations based on *a priori* information and on tomography. First, in the *a priori* models, Australia is made up of cratons in the west and a

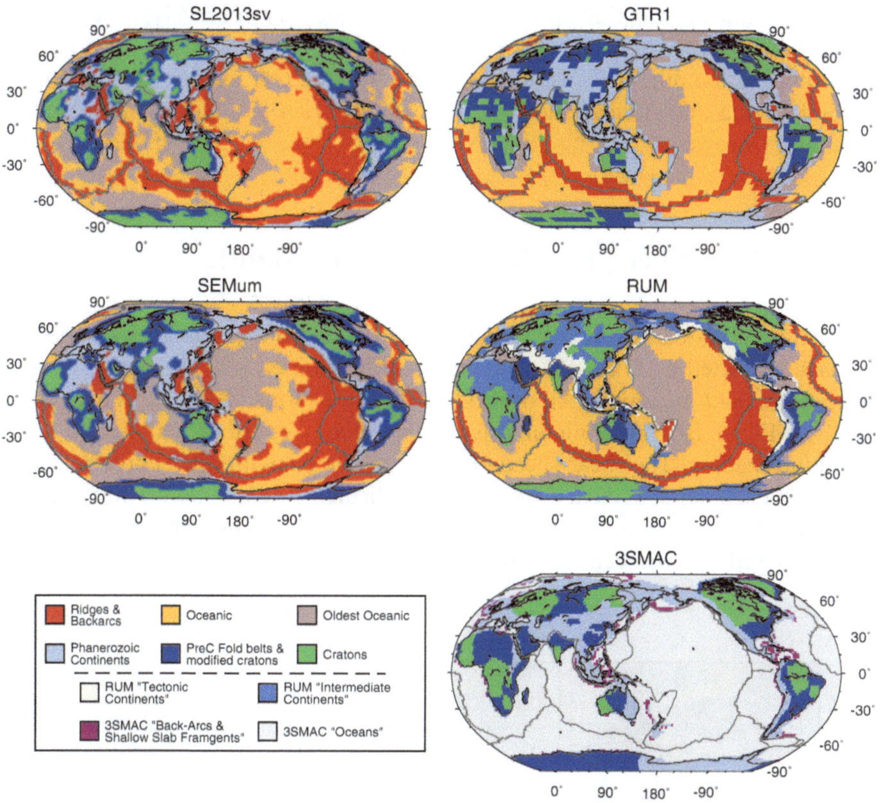

Fig. 1.9 Comparison of tectonic regionalization for the multimode surface wave tomography models SL2013sv (Schaeffer and Lebedev 2013) and SEMum (Lekić and Romanowicz 2011a) with the a priori models 3SMAC (Nataf and Ricard 1996), RUM (Gudmundsson and Sambridge 1998), and GTR1 (Jordan 1981). The results for SEMum were obtained directly from the authors. The same colours for the six *k-means* clusters (from Fig. 1.8) are shown in the legend. The additional colours for 3SMAC and RUM (which have 8 and 9 cluster types, respectively) are also indicated. Note that for the RUM model, our current classification of "Phanerozoic Continents (C1)" is equivalent to the RUM "Young Continent", "Precambrian fold belts and modified cratons (C2)" are equivalent to RUM "Old Continent", and "Cratons (C3)" match the RUM "Ancient Continents". (The tectonic regionalization plotted for 3SMAC is as in the 3SMAC model distribution, which sets all the oceans to the same type, although regionalization of oceans based on their age is implied.)

platform type in the centre; in tomography-based regionalizations, the fast and deep cratonic root is imaged to occupy most of central Australia, extending almost to the coast in the west, into the Arafura and Timor Seas to the north, and much further eastward across the continent. Second, the tomography-based regionalizations identify relatively larger proportions of ancient oceanic (O3) clusters.

Outside of the Pacific Ocean, *a priori* models identify relatively little O3 (old ocean) lithosphere. The tomography models, in contrast, consistently identify such lithosphere in the other oceans, including the Atlantic (off the coasts of North and South America and Africa) and Indian (eastern Africa and west of Sumatra and Australia) Oceans.

1.5 1D Shear-Speed Profiles of the Upper Mantle

To look at the global distribution of shear velocities from yet another perspective, we have extracted a 1D absolute velocity profile from each of the model knots of SL2013sv (7842 locations). All of these profiles are plotted together in each of the panels of Fig. 1.10. The minimum/maximum velocity envelope is indicated by the light-blue shading. The shear-speed profiles are binned, with the relative density of profiles at each depth and each speed value indicated by colour, with blues indicating minimal and warmer colours through to black indicating increasing sampling. Superimposed are the centroids of the clusters given by the empirical tectonic

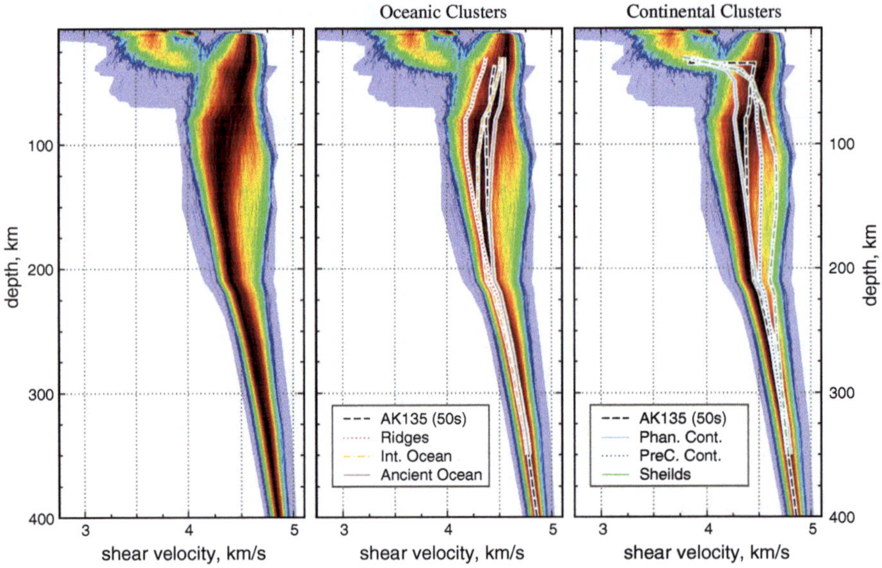

Fig. 1.10 Global variability in 1D shear-wave speed profiles in the crust and upper mantle. The 1D profiles beneath the 7842 model knots of the model SL2013sv, distributed on the sphere with approximately equal spacing, are binned. The relative density of profiles at each depth and each shear-speed value is shown by colour, with *blues* indicating minimal sampling and warmer colours through to *black* indicating increasing sampling. The minimum/maximum velocity envelope is indicated by the *light-blue shading*. The *centre panel* has the cluster centroids associated with oceanic lithosphere plotted, whereas the *right panel* indicates those for clusters associated with continental regions. These centroid profiles were generated for the depth range 30–350 km

regionalization: the oceanic ones (O1–O3) in the centre panel and the continental ones (C1–C3) in the right panel. In the centre and right panels, the mantle reference model (AK135, recomputed at 50 s reference period) is also plotted, with the dashed black line.

The strongest heterogeneity in the oceans (Fig. 1.10, middle panel) is confined to the upper 200 km of the mantle. At 30–50 km depths, the O2 and O3 clusters are quite similar and correspond with the most densely sampled velocities (darkest colours). With increasing depth, these two clusters diverge, with the oldest oceanic type remaining the fastest down to 200 km depth. In regions associated with ridges and backarcs (O1), the lowest velocity anomaly occurs in the depth range 80–115 km and lies at the slower (left) side of the most densely sampled region (reds and blacks). Wave speeds within this cluster converge with those of the middle-aged oceans cluster (O2) at a depth of 150 km. The cooling of oceanic upper mantle with the age of the plate is fast at shallow depths (wave speeds beneath the middle-aged oceans nearly reach old-oceans values in the shallow mantle) and eventually affects the top 200 km of the mantle. Beneath 200 km depth, all three profiles are nearly indistinguishable.

The continental profiles (C1–C3) are dominated by two distinct patterns. As expected, stable continental interiors (C2 and C3) show mantle velocities much faster than the reference, typical of cratonic and stable-continent environments. In contrast, the C1 cluster associated with Phanerozoic portions of continents (recent or active continental deformation) is characterized by an uppermost mantle slower than the reference, down to depths of 150 km. The C3 cluster demonstrates the typically high-velocity lithospheric root characteristic of cratonic regions; the high velocities persist to a depth of ~250 km, before converging with the reference velocity. In comparison, the velocity of the Precambrian fold belts and modified cratons (C2) approaches the reference at a shallower depth of ~200 km. Although the C3 cluster remains faster than C2 to approximately 250–275 km depth, we consider the base of the continental lithosphere to occur closer to 200 km depth, beneath which the relative velocity perturbation with respect to surrounding continental regions is substantially reduced.

We can also examine shear-wave velocity distribution in the shallow mantle (<30 km depth), which is not included in the clustering analysis (the clusters become dominated by the strong variations in crustal structure instead of the lithospheric mantle). In the depth range 10–30 km, the sampling has a bi-modal distribution, with a broad slow region representative of continental crust, and a narrower fast region indicative of the uppermost oceanic lithosphere. Since the knots of the model grid represent roughly equal areas, the entire globe is sampled uniformly, contrary to the observations made for the fundamental-mode group-velocity distribution, where sampling at the shortest periods was influenced by the distributions of sources and stations.

In Fig. 1.11, we plot histograms of shear velocity at ten different depths in the crust, upper mantle, and transition zone; these represent cross sections through the binned profiles of Fig. 1.10. It is clear that the strongest shear-wave speed heterogeneity is largely at depths less than ~300–350 km. (For this comparison, each

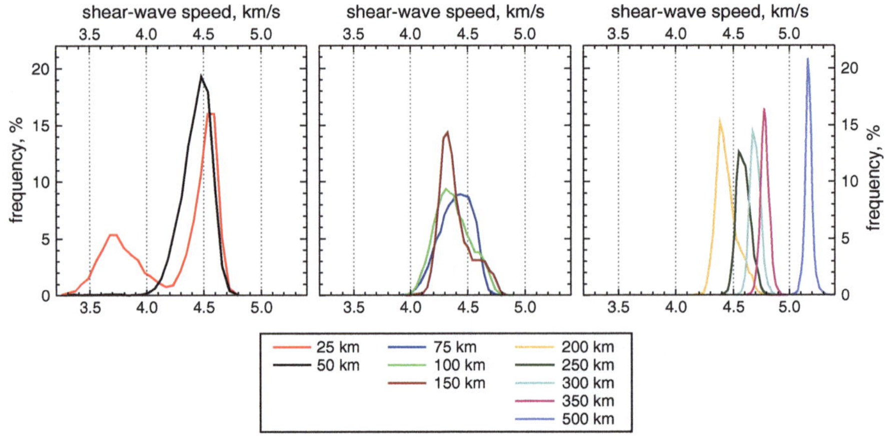

Fig. 1.11 Histograms of global shear-wave speed distributions at different depths within the Earth's crust and upper mantle. Each histogram was computed over all 7482 horizontal shell knots in the model SL2013sv and with the same number of bins (so that the bin width decreases with depth)

histogram was computed with the same number of bins, so that the bin width decreases with depth.)

The 25 km histogram shows a bi-modal distribution of velocity, already mentioned above, with a broad band of lower velocities centred at ~3.7 km s^{-1} and a second much narrower region at ~4.6 km s^{-1}. These represent the continental crust and oceanic mantle, respectively. The centre of each histogram (highest rate of occurrence) shifts to lower velocity from 25 to 100 km depth, before beginning to move to higher velocities at 150 km depth. At 150 and 200 km depth, the tails of the distributions extend to velocities up to ~350–500 m s^{-1} higher than that at the distribution maximum, due to the high-velocity anomalies within the deep lithospheric roots of cratons; at 250 km, this high-velocity tail is nearly gone. With increasing depths (300–500 km), the central velocity increases and the distributions look much more Gaussian, with standard deviation decreasing.

1.6 Comparison of Recent Global Models

Much of our current understanding of the large-scale heterogeneity in deep Earth structure has come from the analysis of global tomographic models. Each model is constructed using different data sets, data types, and methodologies. Body-wave travel-time tomography, for example, is very effective for high-resolution imaging of subduction zones, due to the high density of crossing rays given by the stations and number of events in these regions (e.g. Bijwaard et al. 1998; Karason and van der Hilst 2001; Grand 2002; Montelli et al. 2004; Amaru 2006; Li et al.

2008; Simmons et al. 2012). However, global body-wave tomography models tend to lack resolution in the lithospheric mantle in regions where seismicity and station coverage are sparse or absent.

Surface-wave models are constrained by the waves that propagate along the Earth's surface through the crust and upper mantle and can recover the shear-velocity structure in the lithosphere–asthenosphere depth range globally, with relatively uniform resolution. The addition of multimode *S* waves (higher mode surface waves make up *S* and multiple *S* waves) augments the fundamental-mode sensitivity to constrain structure at depths down to the base of the transition zone (see Fig. 1.1). Important insights into the structure of deep lithospheric roots of continental cratons and the structure and evolution of oceanic lithosphere have been obtained from models constrained with surface waves (e.g. Montagner and Tanimoto 1991; Zhang and Tanimoto 1993; Boschi and Ekström 2002; Shapiro and Ritzwoller 2002; Zhou et al. 2006; Nettles and Dziewónski 2008; Ekström 2011) and surface waves combined with body waves (or fundamental and higher modes, e.g. Woodhouse and Dziewónski 1984; Li and Romanowicz 1996; Mégnin and Romanowicz 2000; Debayle et al. 2005; Panning and Romanowicz 2006; Lebedev and van der Hilst 2008; Ferreira et al. 2010; Panning et al. 2010; Lekić and Romanowicz 2011a; Debayle and Ricard 2012; Schaeffer and Lebedev 2013). A number of surface-wave models have also been computed with teleseismic travel times and normal mode splitting functions incorporated into the data sets (e.g. Su et al. 1994; Masters et al. 1996; Ritsema et al. 1999; Masters et al. 2000; Gu et al. 2001; Ritsema et al. 2004; Houser et al. 2008; Kustowski et al. 2008; Ritsema et al. 2011).

In addition to the data sensitivities and basic methodologies, properties of each tomographic model also depend on its regularization (damping, smoothing), chosen by its creators. A comparison of a range of different models reveals the features that are resolved and confirmed by most. In addition, this helps establish the comparative advantages of different models. In this section, we compare SL2013sv with eight other global tomography models: DR2012 (Debayle and Ricard 2012), SEMum (Lekić and Romanowicz 2011a), S40RTS (Ritsema et al. 2011), SAW642ANb Panning et al. (2010), LH2008 (Lebedev and van der Hilst 2008), S362ANI (Kustowski et al. 2008), S20RTS (Ritsema et al. 1999, 2004), and CUB (Shapiro and Ritzwoller 2002). Each of these models is parameterized differently and is constrained by a different data set, with different sensitivity to upper mantle structure.

SEMum is a Voigt-average shear speed and radial anisotropy model, $V_S^2 = \frac{1}{3}\left(2V_{SV}^2 + V_{SH}^2\right)$ (Babuska and Cara 1991) and $\xi = V_{SH}^2/V_{SV}^2$, respectively, constrained by long-period waveforms (multimode Rayleigh and Love waves and long-period body waves), and group-velocity dispersion maps; SAW642ANb is derived from a similar data set. The CUB model (CU_SDT1.0) is an isotropic shear-velocity and radial anisotropy model of the crust and uppermost mantle derived from fundamental-mode Rayleigh and Love group and phase measurements. LH2008 is a predecessor of the SL2013sv model generated using the same methodology but constrained by fewer seismograms and computed with lower target resolution. S362ANI is a Voigt-average whole-mantle isotropic shear-velocity

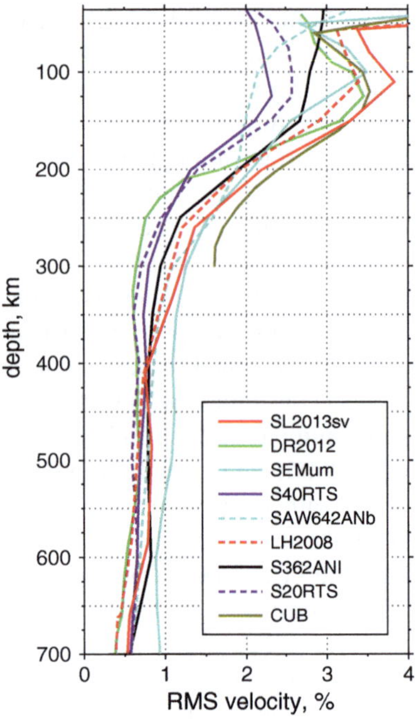

Fig. 1.12 Root mean square (RMS) of shear-wave speed variations in the upper mantle, according to selected models of the last decade: SL2013sv (Schaeffer and Lebedev 2013), DR2012 (Debayle and Ricard 2012), SEMum (Lekić and Romanowicz 2011a), S40RTS (Ritsema et al. 2011), SAW642ANb (Panning et al. 2010), LH2008 (Lebedev and van der Hilst 2008), S362ANI (Kustowski et al. 2008), S20RTS (Ritsema et al. 1999, 2004), and CUB (Shapiro and Ritzwoller 2002)

model generated from surface-wave dispersion measurements, mantle and body-wave waveforms, and body-wave traveltimes. S40RTS (and S20RTS) are isotropic shear-velocity models of the mantle constrained by three data sets: minor and major arc Rayleigh wave dispersion, teleseismic body-wave traveltimes, and spheroidal-mode splitting functions; S40RTS is a newer model constructed with more data. Finally, DR2012 is an upper-mantle S_V-wave model constrained by multi-mode Rayleigh waveforms.

In Figs. 1.12, 1.13, 1.14, 1.15, 1.16, 1.17, 1.18, and 1.19, we perform a consistent comparison of the models. We re-sample each to the same 0.5° latitude–longitude grid and extract or compute only the V_{SV} component for comparison. At each depth for each model, the mean value (computed as a weighted average using the cell areas) is removed. Where required, the grid values are converted to per cent variation from this mean. We first compare the root-mean-square (RMS) velocity perturbations of each model as a function of depth and then compare map views at different depths in the crust and upper mantle.

1.6.1 RMS of Shear-Wave Speed Perturbations

The RMS of wave speed perturbations as a function of depth provides a simple statistical measure of the magnitude of anomalies in each model. For each model, a weighted RMS is computed as follows:

$$\langle x_i \rangle = \sqrt{\frac{\sum_{j=1}^{N} \left(x_{ij} - \overline{x_i}\right)^2 w_j}{\sum w_j}}, \tag{1.8}$$

where $\langle x_i \rangle$ is the RMS at the ith depth, x_{ij} is the model perturbation at the jth grid cell (at depth i), w_j is the area of each cell (the same at each depth), N is the number of horizontal cells (the same at each depth), and $\overline{x_i} = \sum \left(x_{ij} w_j\right) / \sum w_j$ is the mean value at the given depth. As the mean has already been removed ($\overline{x_i} = 0$), the RMS reduces to:

$$\langle x_i \rangle = \sqrt{\frac{\sum \left(x_{ij}\right)^2 w_j}{\sum w_j}}. \tag{1.9}$$

In Fig. 1.12, the RMS calculated using Eq. 1.9 for each of the nine models is plotted as a function of depth over the upper mantle and transition zone depth range. For all models, the RMS reaches a mantle maximum at a depth of less than 150 km, most commonly between 100 and 130 km. Between 150 and 300 km depths, the RMS drops off rapidly, to ~1 % at 300 km. Through the rest of the upper mantle and transition zone, this gradually decreases to an average value of ~0.6 %. At the shallowest depths (i.e. ~50 km and less), SL2013sv, LH2008, and CUB have the highest RMS, which reflects their recovery of the global crustal structure (although not obvious, SEMum's RMS also increases at similar shallow depths). The other models use different methods of crustal corrections (aimed at isolating and removing the effect of the crust and focussing on the mantle only) and do not resolve crustal structure.

In the upper 200 km, SL2013sv shows close to the highest RMS, indicative of greater structural heterogeneity in it compared to other models; SEMum, CUB, DR2012, and LH2008 attain RMS values close to that of SL2013sv. At 125 km, these five models have an average RMS of 3.4 %, whereas S40RTS, S20RTS, and SAW642ANb have an average of 2.3 %. Based on this difference, these two groups can be separated into two classes of models: the latter is a "lower amplitude" class, whereas the former represents a "higher amplitude" class. The model S362ANI falls between these two classes, with an RMS of 2.75 %.

From the 100–150 km depths to just above the transition zone, the RMS of the models drops by a factor of 2–3, with the bulk of this decrease taking place between 150 and 300 km depths. The CUB model RMS becomes higher than that of all other models in the mid-upper-mantle depth range. However, we do not plot it to depths greater than 300 km, as its uncertainty estimates identify robust

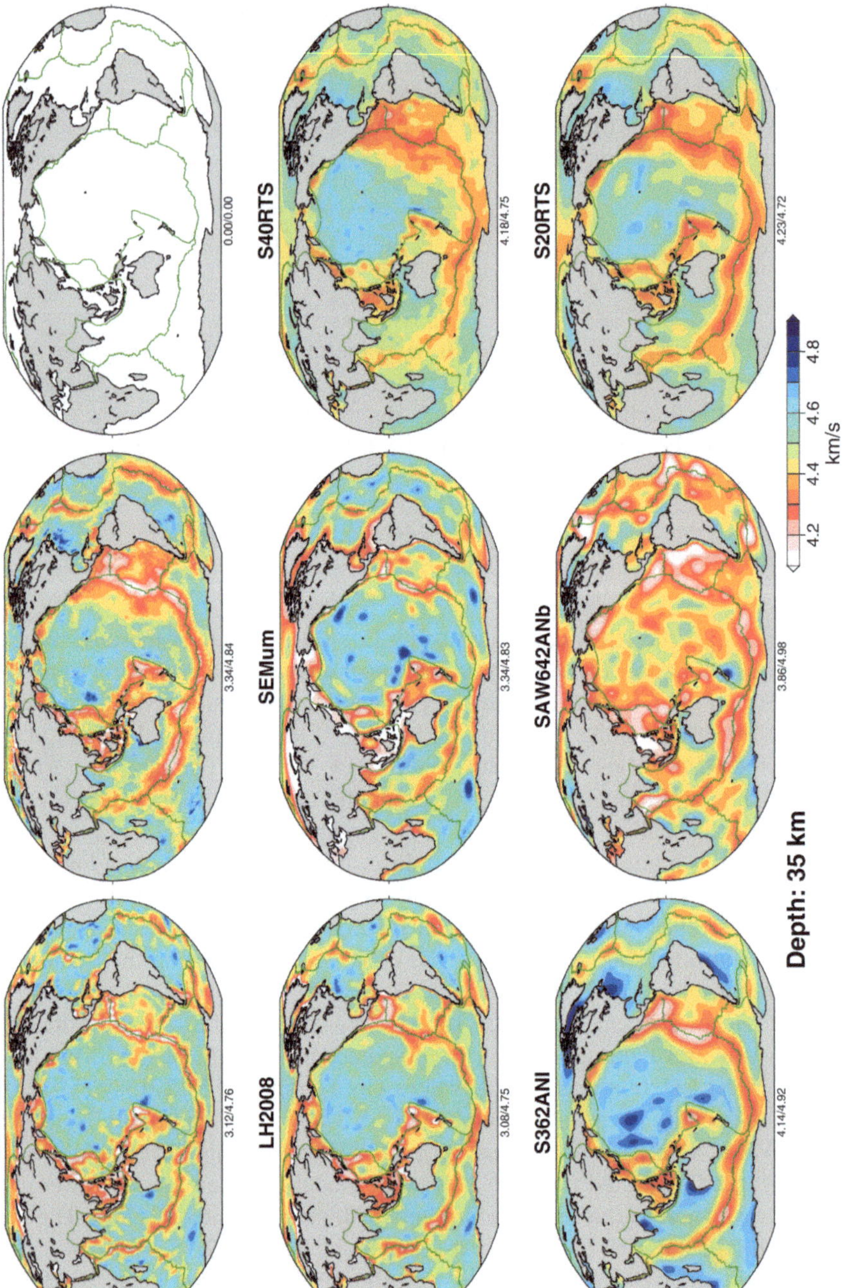
DR2012
0.00/0.00
CUB
3.34/4.84
SL2013sv
3.12/4.76
S40RTS
4.18/4.75
SEMum
3.34/4.83
LH2008
3.08/4.75
S20RTS
4.23/4.72
SAW642ANb
3.86/4.98
S362ANI
4.14/4.92
4.2
4.4
4.6
4.8
km/s
Depth: 35 km

◀ **Fig. 1.13** Comparison of SL2013sv and eight other global tomographic models at a depth of 35 km. The models are as follows: DR2012 (Debayle and Ricard 2012), SEMum (Lekić and Romanowicz 2011a), S40RTS (Ritsema et al. 2011), SAW642ANb (Panning et al. 2010), LH2008 (Lebedev and van der Hilst 2008), S362ANI (Kustowski et al. 2008), S20RTS (Ritsema et al. 1999, 2004), and CUB (Shapiro and Ritzwoller 2002). The absolute velocity (V_{SV}) of each model is plotted, with the colour scale saturating for velocities less than 4.1 km s^{-1} to highlight structure in the mantle instead of the crust. Continental regions were *greyed out* to focus the comparison on the oceanic mantle, as the continental crust was not meant to be resolved accurately in some of these models (a 36 km cross section through SL2013sv, which does resolve anomalies in the continental crust, relative to a 3D reference model, is plotted in Fig. 1.3). Each models' global minimum and maximum absolute velocity (over both continents and oceans) is indicated beneath the maps, and the same linear colour scale is used in each

features persisting only to a depth of ~250 km (Shapiro and Ritzwoller 2002). The RMS of the model DR2012 decreases most rapidly in this depth range and has the smallest RMS by 225 km depth. SL2013sv, LH2008, SAW642ANb, and SEMum maintain an overall higher RMS to a depth of 400 km.

In the transition zone, seven of the eight remaining models (excluding SEMum) have very similar RMS values, with variations of around 15 % from the average RMS of 0.6 %. SL2013sv has an RMS in the middle of this range through most of the transition zone. SEMum stands out with a significantly higher RMS (on average ~1 %) between 400 and 700 km; the high amplitudes that give rise to this increased RMS are clearly apparent in the SEMum model (Fig. 1.19 at 500 km depth).

1.6.2 Map View Comparisons

In Figs. 1.13, 1.14, 1.15, 1.16, 1.17, 1.18, and 1.19, we compare the shear-wave speed structure in the Earth's upper mantle as resolved by the nine recent global tomographic models constrained using data sets that included surface-wave measurements. The depths plotted are 35 km in Fig. 1.13, 50 km in Fig. 1.14, 100 km in Fig. 1.15, 150 km in Fig. 1.16, 250 km in Fig. 1.17, 350 km in Fig. 1.18, and 500 km in Fig. 1.19. Each model is re-sampled to the same 0.5° grid, colour scales are indicated (absolute or relative) at the lower right of each figure, and the total range for each model is indicated in small print beneath each map. For depths 50 km and greater, the mean absolute velocity of each model is removed, with perturbations plotted from this value in per cent. The location of the main plate boundaries is indicated by a thin green line. A model is plotted with no colours (DR2012 at 35 km and CUB at 350 and 500 km) when that depth is not included in the model files.

In a few of the models, the strong crustal heterogeneity in the Earth was accounted for through the use of "crustal corrections", which remove the effects of a predetermined crustal model from the measurements (see Boschi and Ekström 2002; Gu et al. 2003; Chevrot and Zhao 2007; Kustowski et al. 2007; Marone

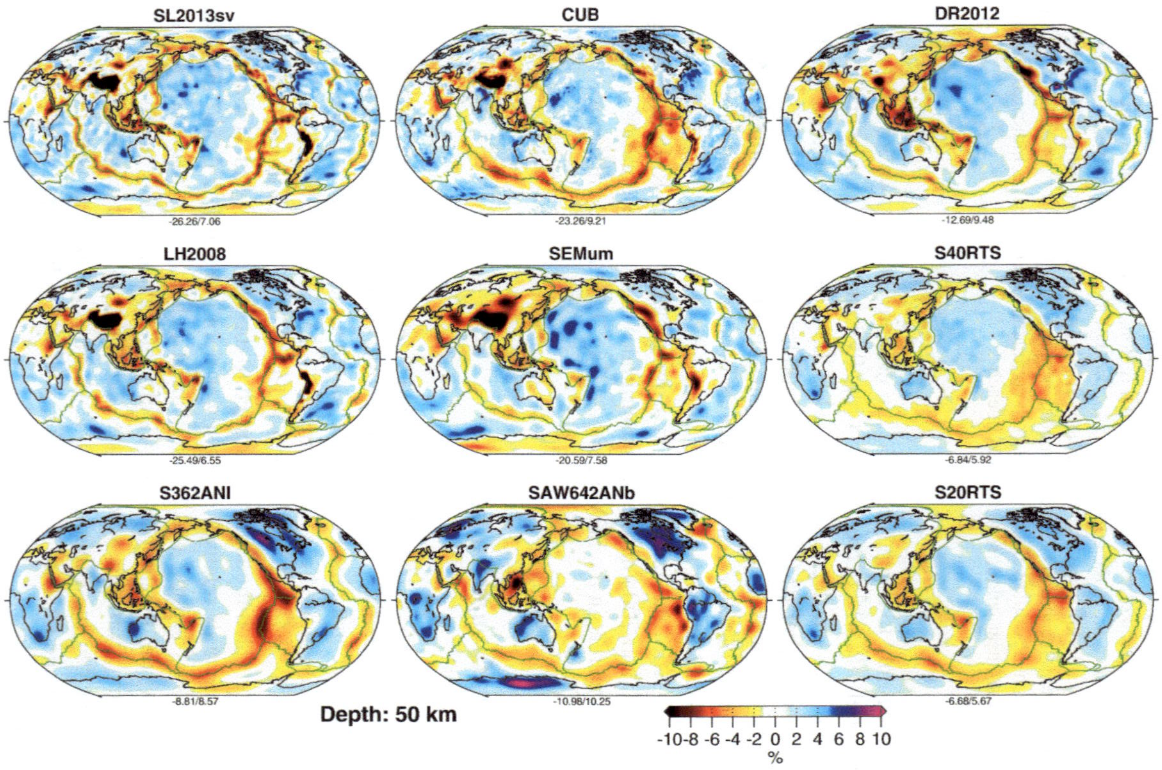

Fig. 1.14 Comparison of SL2013sv and eight other global tomographic models at a depth of 50 km. Models are the same as those in Fig. 1.1. Perturbations are plotted in per cent with respect to the mean absolute velocity of each model. The minimum and maximum perturbations are indicated beneath the maps; the same linear colour scale is used for each map)

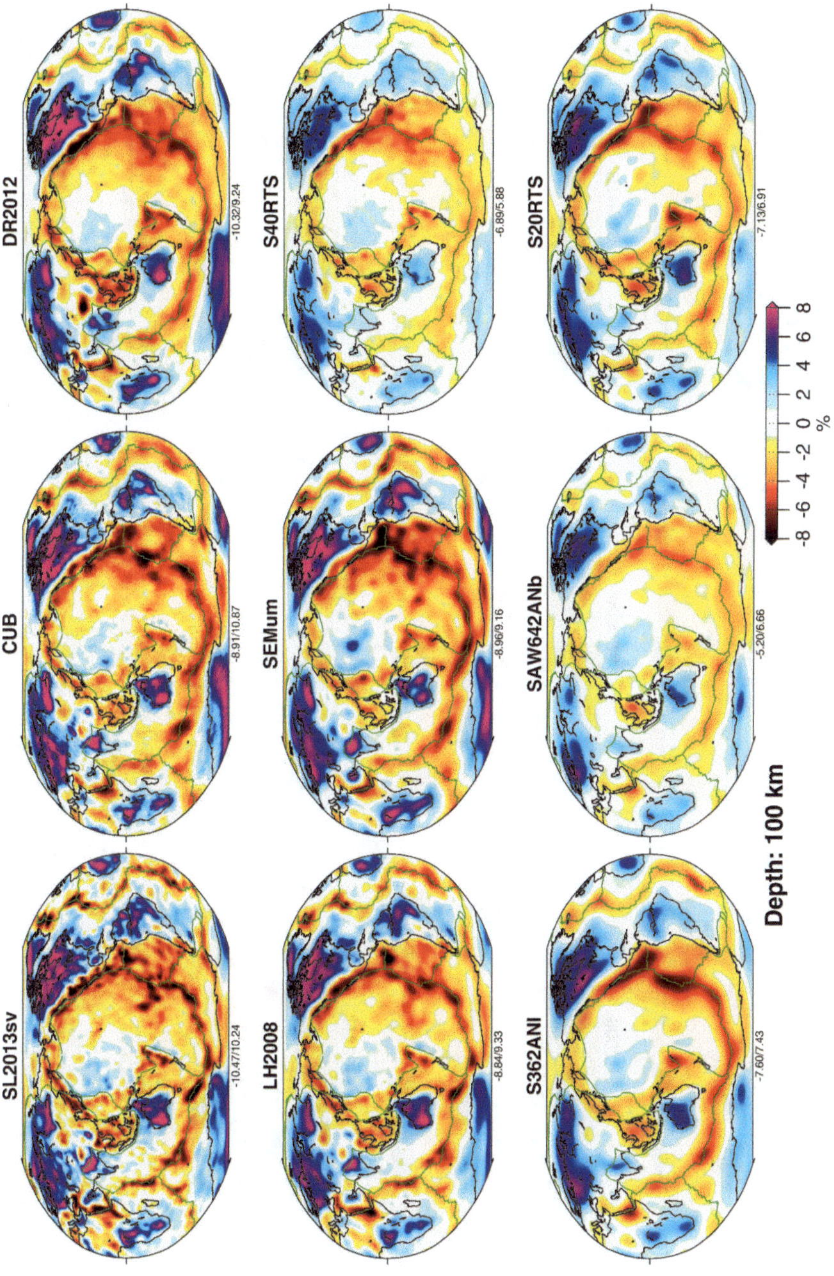

Fig. 1.15 Comparison of SL2013sv and eight other global tomographic models at a depth of 100 km depth. The models and plotting conventions are as in Fig. 1.14

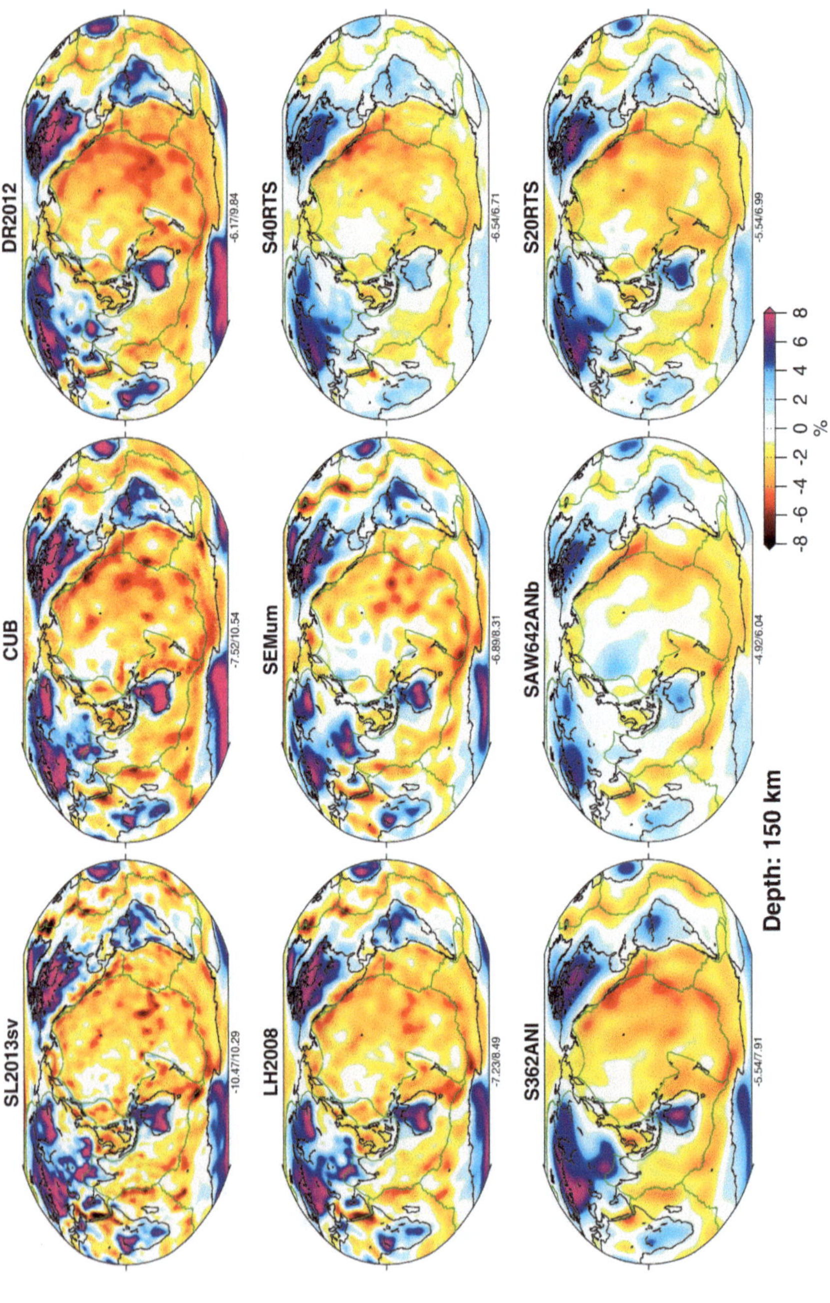

Fig. 1.16 Comparison of SL2013sv and eight other global tomographic models at a depth of 150 km depth. The models and plotting conventions are as in Fig. 1.14

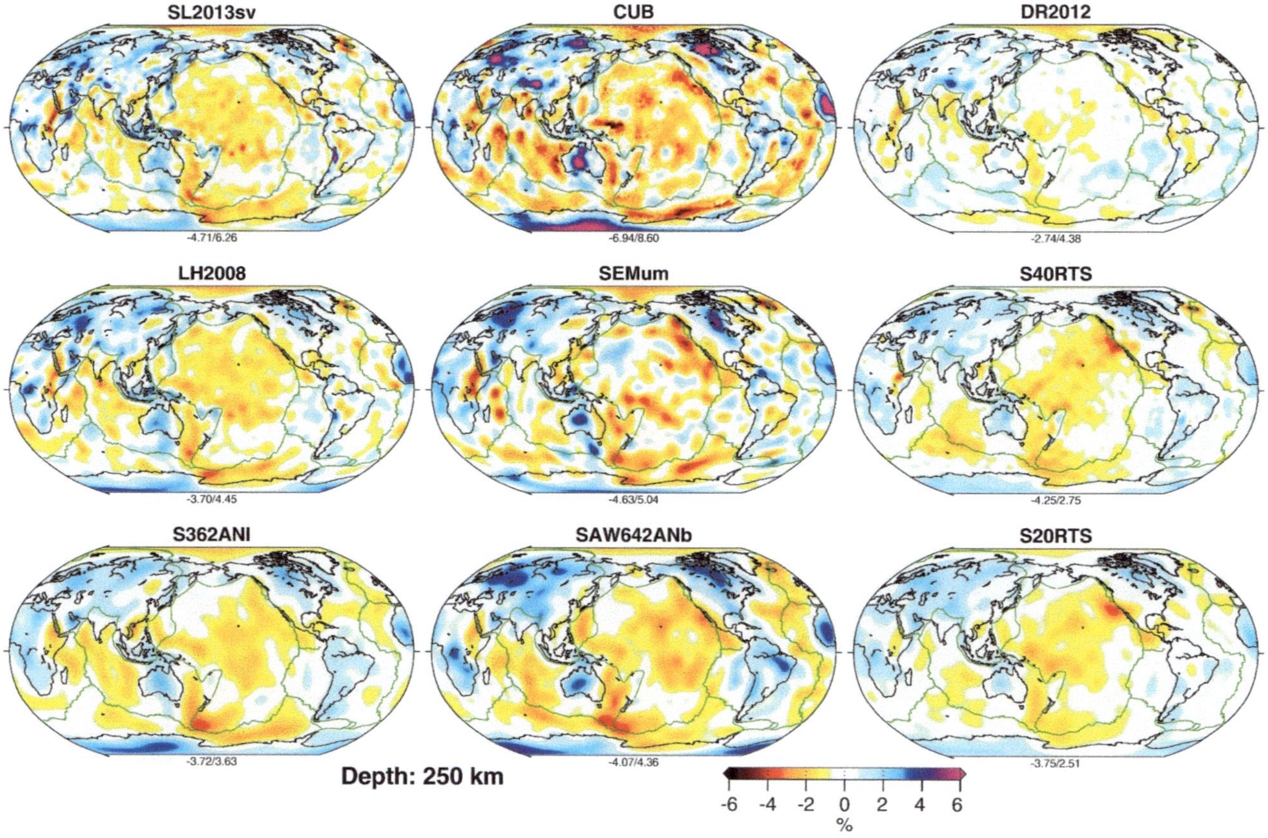

Fig. 1.17 Comparison of SL2013sv and eight other global tomographic models at a depth of 250 km depth. The models and plotting conventions are as in Fig. 1.14

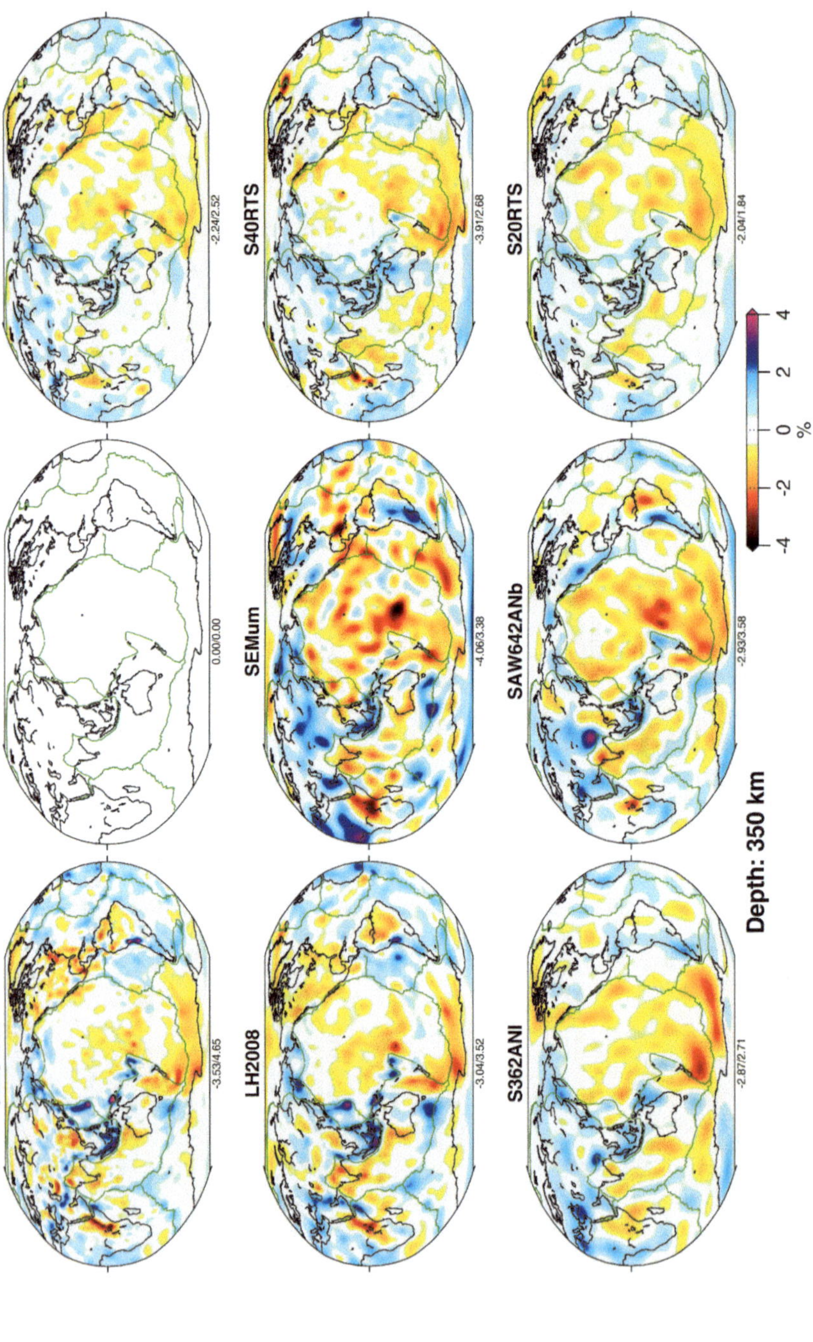

Fig. 1.18 Comparison of SL2013sv and eight other global tomographic models at a depth of 350 km depth. The models and plotting conventions are as in Fig. 1.14

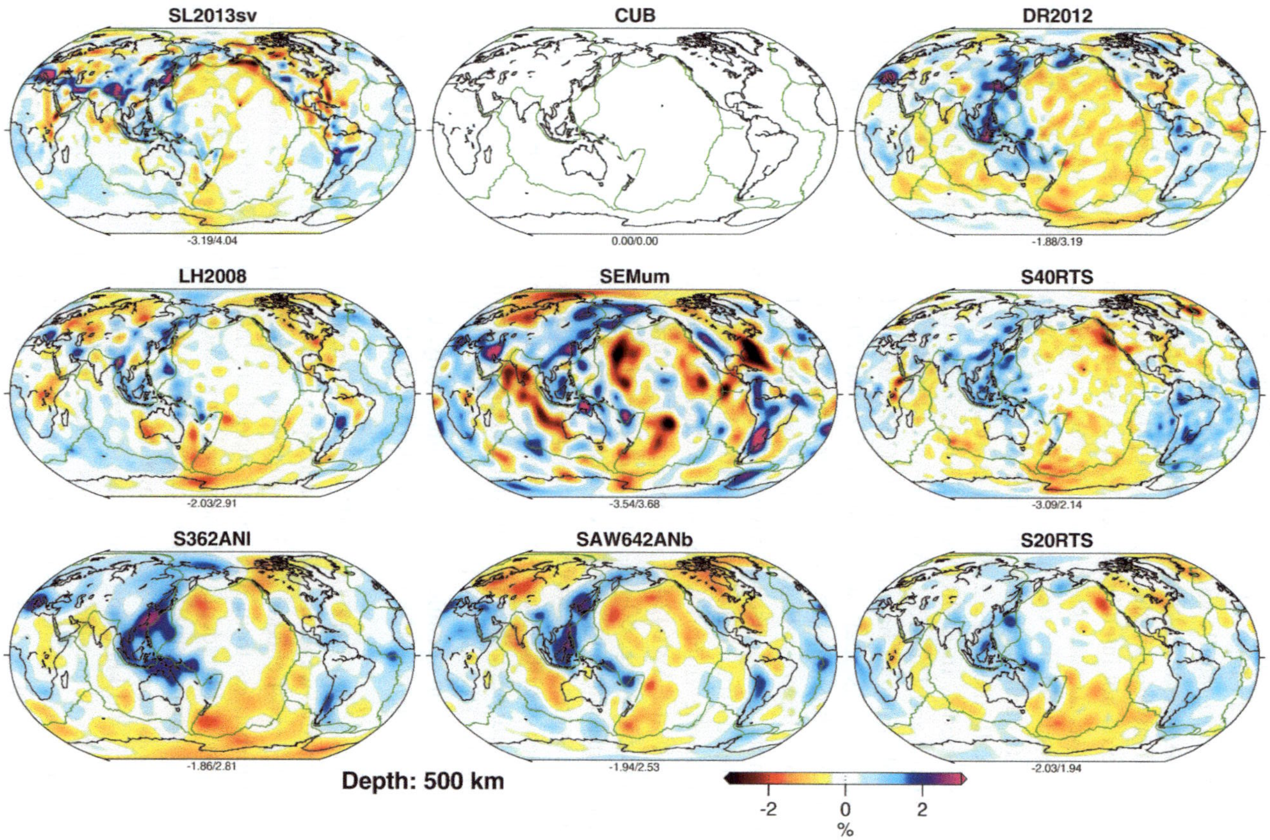

Fig. 1.19 Comparison of SL2013sv and eight other global tomographic models at a depth of 500 km depth. The models and plotting conventions are as in Fig. 1.14

and Romanowicz 2007; Bozdăg and Trampert 2008; Lekić et al. 2010, for further information). As a result, it is not straightforward to interpret what the structure at crustal depths within these models represent. We plot a 35 km map view for 8 of the 9 models in Fig. 1.13 (shallowest knot of DR2012 is at 40 km) but focus on the uppermost mantle and, thus, grey-out the continents, with their Moho normally near to or below 35 km depth (we note that in many cases, the authors of the different models would not interpret the structures at these shallow depths). Although we do not intend to describe in detail the relative merits of the different approaches, we do summarize those of each model for insight into the comparisons. For details on the treatment of the crust in each model, we refer the interested reader to each of the model papers.

In S362ANI, the CRUST2 (Bassin et al. 2000) reference model is assumed as an *a priori* model. In S40RTS and S20RTS, crustal corrections were performed using CRUST2 and CRUST5.1 (Mooney et al. 1998), respectively. DR2012 uses a fixed crustal structure based on 3SMAC and solves for perturbations in regions where the Moho is shallower than 40, 50, and 70 km depth. The authors only show tomographic maps for depths larger than 100 km in order to avoid possible contamination by crustal structure; here, we include DR2012 at 50 km for completeness in the comparison. SAW642ANb utilizes a linear correction methodology (developed from Lekić et al. 2010) to account for crustal structure, which they demonstrate improves variance reduction compared to nonlinear corrections as used in SAW642AN (Panning and Romanowicz 2006). SEMum use a 3D, smoothly varying, 60-km-thick effective elastic medium as a crustal model, which is constrained by group-velocity dispersion data, such that the effective crust produces the same waveforms as a thinner layered crust. The CUB model uses a number of different inputs to constrain the crustal reference model, including various sediment thickness models, crustal thickness models (CRUST5.1 and maps from the Russian Institute of Physics of the Earth), as well as thickness estimates based on seismic profiles. Additionally, CUB includes crustal parameters in the inversion, which allows perturbations from the reference model to be determined.

In SL2013sv and LH2008, a 3D crustal reference model based on CRUST2 is used in both the waveform fitting and tomographic inversion (for more detail, see Schaeffer and Lebedev 2013; Lebedev and van der Hilst 2008; Lebedev et al. 2005). Perturbations from this crustal model are then *solved for* directly. In continental regions, three depths of the model grid knots (at 7, 20, and 36 km depth) often fall within the crust; in the oceans, this is normally limited to one knot at 7 km depth. Although CRUST2 is known to be imperfect at many locations, inclusion of the crustal knots into the inversion enable perturbations relative to the 3D reference to be determined in regions where the data require them. As a result, in addition to preventing artefacts at depth in the mantle due to unaccounted for or assumed crustal structure, a global crustal model is generated (beginning at a depth of 7 km). SL2013sv clearly recovers strong perturbations from the reference model that are required by the data; a prominent example of this is the Tibetan

Plateau, where boundaries of a strong low-velocity anomaly in the model follow closely the boundaries of the plateau at the surface (this is illustrated in Fig. 1.3).

To facilitate easier comparison of the shallowest structure in the different models (Fig. 1.13), we plot the absolute velocity variations in each of the models. However, to reduce the effect of the strongly varying crustal structure (and correction terms), the colour scale saturates to white for velocities less than 4.1 km s^{-1}, which emphasizes the mantle structure within each model (essentially the oceanic regions), and the continents have been coloured grey.

The degree of heterogeneity as observed in each of the nine models varies substantially at depths of 35 and 50 km depth (Figs. 1.13 and 1.14). The dominant features common between all models are the low-velocity anomalies associated with the spreading ridges, back arcs, and regions of continental deformation (at 50 km), as well as relatively high velocities in the uppermost oceanic lithosphere. SL2013sv recovers these features at a finer resolution. For instance, at both 35 and 50 km depth, the central low-velocity anomalies associated with the MORs are very tightly confined to the region around the ridge axis, for most of the ridge segments globally. Although this signature of MORs is observed in all models to varying degrees, the highest amplitude anomalies are not as continuous and extend much greater distances away from the spreading centre. Similarly, the very low-velocity anomaly associated with the partially molten Tibetan crust is observed in SEMum, CUB, SL2013sv, DR2012, and LH2008. However, the low-velocity anomaly in SL2013sv has a closer correlation with the regional tectonic boundaries. The northern boundary corresponds with the surface expression of the Altyn Tagh Fault; to the west, low velocities are also present in the Hindu Kush and Pamir Mountains; in the east, the low-velocity anomaly extends south-east around the eastern Himalayan syntaxis.

At a depth of 100 km (Fig. 1.15), the signature of the spreading centres is still one of the dominant features in all the models, although there remain variations in the width and resolution of the central anomaly underlying the ridge centre. At 150 km depth (Fig. 1.16), the strongest anomaly of the ridge is gone in all the models. Instead, most of the oceans now show a relatively uniform, low-velocity asthenosphere. Smaller-scale heterogeneities in the Pacific, Atlantic, and Indian Oceans remain in SEMum, CUB, SL2013sv, DR2012, and LH2008. A strong low-velocity anomaly remains beneath Iceland in SL2013sv, as well as in LH2008, CUB, and SEMum. In SL2013sv, the strong, well-defined anomaly here is seen down to 250–270 km depth (Figs. 1.4 and 1.17).

The 100–150 km map views are dominated by the strong positive anomalies associated with the stable lithospheric roots of cratons, which can be seen in each of the models beneath each of the continents. At shorter wavelengths (<500–1000 km), there are significant differences between the different models. Those identified as "low-amplitude" models in this depth range (S40RTS, S20RTS, SAW642ANb, and S362ANI) resolve the cratons, particularly in the northern hemisphere. However, little short wavelength structure either within the cratons or along their boundaries is resolved. The "higher amplitude" class of models captures stronger heterogeneity, both within and along the craton boundaries.

SL2013sv captures the greatest detail of these strong structural boundaries: the cratons in South America (which are also clear in the tectonic regionalization), the cratonic blocks in southern Africa, the clear signature of the Indian lithosphere underthrusting beneath the Himalaya and western Tibet, and the very narrow high-velocity subducting oceanic lithosphere along the trenches around much of the western Pacific. In addition, we highlight finer scale structural heterogeneity resolved along the western boundary of the North American craton, which improves upon past continental-scale models (Frederiksen et al. 2001; van der Lee and Frederiksen 2005; Nettles and Dziewónski 2008; Bedle and van der Lee 2009) and correlates well with the results of regional-scale high-resolution imaging studies using the USArray (e.g. Sigloch 2011; Tian et al. 2011; Shen et al. 2013; Obrebski et al. 2011; Burdick et al. 2012).

At depths near the base of the continental lithospheric mantle, the strength of heterogeneity is beginning to decrease, as is indicated by the RMS of the shear-wave velocity distribution (Fig. 1.12). At 250 km depth in SL2013sv, LH2008, S40RTS, S20RTS, S362ANI, and DR2012, most anomalies associated with the deep cratonic lithospheric roots have disappeared; what remains is only slightly elevated velocities beneath the continents and larger slightly negative velocity anomalies beneath the oceans. Alternatively, SEMum, CUB, and SAW642ANb show strong fast anomalies persisting to greater depths beneath most continents. These disappear by 350 km in SAW642ANb and, mostly, in SEMum (although an anomaly remains beneath the western African craton). In this depth range, subducting oceanic lithosphere is seen clearly, particularly around the margin of the western Pacific. Broad high-velocity anomalies can be seen in this region in all the models, although they are more focused in SL2013sv and LH2008. In SL2013sv, the highest velocity anomalies correlate well with the superimposed plate boundaries, especially along the Izu-Bonin and Marianas, and south along Sumatra.

The low velocities associated with the East African Rift (and nearby Afar and Red Sea) are consistent between most of the different models. At depths down to 150 km, SL2013sv, LH2008, DR2012, CUB, and SEMum all display a relatively strong, linear negative anomaly underlying the Red Sea; S40RTS, S362ANI, SAW642ANb, and S20RTS also show a negative anomaly, although it is somewhat more circular to oblate. At greater depths down 250–350 km, the low-velocity anomaly stretches from afar southwards along the East African Rift. It is most clear in SL2013sv and SEMum, although still present in the other models, with much lower amplitude.

In the transition zone, the length-scale of heterogeneity imaged increases. At 500 km depth (Fig. 1.19), high-velocity anomalies are imaged in the western Pacific, related to the subduction of the Pacific oceanic lithosphere. There are large differences in the amplitude of the anomalies observed in each of the models and the location of slab anomalies, even at the longer wavelengths (i.e. $\geq$3000 km). The inclusion of many *S* and multiple *S* body waves in SL2013sv has enabled improved resolution of high-velocity slabs in the upper mantle, from lithospheric depths through to the transition zone (Schaeffer and Lebedev 2013). We can see the effect of this increase by comparing LH2008 and SL2013sv, which are generated using the same methodology; however, SL2013sv uses an order of magnitude more seismograms and applies less smoothing at transition zone depths.

1.7 Conclusions

Variations in seismic wave velocities provide insight into heterogeneity in the composition and physical state of the rocks within the Earth. The seismic-velocity heterogeneity can be observed in a number of different ways, each offering a complementary perspective on the bulk properties, structure, and dynamics of the Earth.

Our large new data set of global measurements of phase and group velocities permits construction of the dispersion diagram of surface waves on the Earth, spanning a broad period range from ~10 to >400 s. In the Rayleigh-wave diagram presented in this paper, 12 higher modes (overtones) are clearly mapped, in addition to the fundamental mode. The greatest variability is observed at shorter periods for the fundamental mode ($T < 30$ s for phase velocities; $T < 40$ s for group velocities), reflecting the higher heterogeneity in the crust–uppermost mantle depth range compared to that in deeper upper mantle.

Global tomographic models constrained with data sets including surface waves have shown agreement at long wavelengths (thousands of kilometres) since 1990s (e.g. Becker and Boschi 2002). Today, a number of recent models show close agreement in the upper 200 km of the mantle at much shorter scale lengths (several hundreds of kilometres). The agreement includes the large amplitudes of shear-speed variations (over ± 10 %) in the lithosphere–asthenosphere depth range. In the deep upper mantle and transition zone, greater disagreements between different available models still remain.

The model SL2013sv (Schaeffer and Lebedev 2013), constrained by an unprecedentedly large data set of multimode waveform fits, displays increased resolution compared to other existing models for many upper-mantle and crustal features across the globe. For many cratons, the boundaries of their high-velocity lithospheres at mantle depths are seen closely following their boundaries at the surface. Important exceptions include low velocities beneath cratons whose lithosphere was modified (e.g. NE China, Wyoming) and very high velocities in cratonic mantle–lithosphere recently underthrust beneath thick orogenic crust (Tibet, Zagros) (e.g. Agius and Lebedev, manuscript in revision, 2013). The model also resolves in detail the structure of the mid-ocean ridges, with the partial-melting induced low-velocity anomalies being very narrow in the uppermost mantle beneath the ridge axis and broadening with depth down to 100–120 km. Sharp shear-speed contrasts at crustal and mantle–lithosphere depths closely match tectonic boundaries at the surface in regions undergoing active deformation, such as Tibet and surroundings or western North America. The model thus captures regional-scale heterogeneity globally, within both the upper mantle and the crust.

Regionalization of SL2013sv by means of clustering (Lekić and Romanowicz 2011b) offers another perspective on upper-mantle heterogeneity. Sorting all locations around the globe into six clusters according to the shear-wave speed profiles at 30–350 km depths beneath them, with no *a priori* information, we obtain an accurate tectonic regionalization of the entire Earth. Oceanic locations fall into three clusters distinguished by the lithospheres maturity: youngest oceans, ridges and backarcs (hot or wet mantle with partial melt likely at many places; very low V_S), ancient oceanic basins (cold, high-velocity lithospheres), and intermediate-aged oceans, including portions of old oceanic lithosphere that were "rejuvenated" when passing over hot spots. The continents also split into three clusters: cratons (cold, deep, high-velocity lithospheric roots), intermediate continents, including Precambrian fold belts or cratons modified in the Proterozoic, and Phanerozoic continents (the latter containing almost all recent continental volcanism).

A global stack of 1D shear-speed profiles through the crust and upper mantle shows a monotonic decrease in the amplitude of wave speed variations with depth. This is mirrored by a decrease in RMS variations—from large in the top 150–200 km to much smaller below 250 km—seen in all the tomographic models compared.

The "depth of tectonics"—depths down to which differences between shear-wave speed profiles under oceanic and continental lithospheres of different ages persist (Jordan and Paulson 2013)—can be estimated from the type-average profiles given by the cluster analysis of the global shear-wave speed heterogeneity (Fig. 1.8). Young oceans display the lowest velocities among oceanic regions in the uppermost mantle. The intermediate-age-ocean average profile does not differ substantially from the young ocean one at depths of 150 km and greater but is similar to the old ocean profile in the upper 50 km of the mantle. The oldest oceans show higher shear-wave speeds than young and intermediate ones down to ~200 km depth. Phanerozoic and intermediate-age continents' profiles converge with each other and with the oceanic profiles at 200–250 km depth. The oldest continents (cratons) stand out to the greatest depth, with shear-wave speeds beneath them higher than beneath all other regions, on average, down to 250–280 km depths.

Acknowledgments Insightful comments by Li Zhao and Frédéric Deschamps have helped us to improve the manuscript. We thank the creators of the tomographic models compared in this study for making them available. Waveform data used for the construction of the model SL2013sv were obtained from the facilities of IRIS, ORFEUS, GFZ, and CNSN. We are grateful to the operators of the many networks used in this study. All figures were generated using Generic Mapping Tools (GMT; Wessel and Smith 1995). This work was supported by Science Foundation Ireland (Grant 09/RFP/GEO2550), with additional support by Science Foundation Ireland and the Marie-Curie Action COFUND (Grant Number 11/SIRG/E2174). Our tomographic model, SL2013sv, can be downloaded from http://www.dias.ie/~aschaeff/SL2013sv.html.

References

Adam JM-C, Lebedev S (2012) Azimuthal anisotropy beneath southern Africa from very broadband surface-wave dispersion measurements. Geophys J Int 191(1):155–174

Aki K, Christoffersen A, Husebye ES (1977) Determination of the three-dimensional seismic structure of the lithosphere. J Geophys Res 82:(277–296)

Amaru ML (2006) Global travel time tomography with 3-D reference models. PhD thesis, Universiteit Utrech

Babuska V, Cara M (1991) Seismic anisotropy in the earth. Kluwer Academic Press, Boston

Bassin C, Laske G, Masters G (2000) The current limits of resolution for surface wave tomography in North America. EOS 81:F897

Becker TW, Boschi L (2002) A comparison of tomographic and geodynamic mantle models. Geochem Geophys Geosys 3

Bedle H, van der Lee S (2009) S velocity variations beneath North America. J Geophys Res 114(B7)

Bijwaard H, Spakman W, Engdahl ER (1998) Closing the gap between regional and global travel time tomography. J Geophys Res 103(B12):30055–30078

Boschi L, Ekström G (2002) New images of the Earths upper mantle from measurements of surface wave phase velocity anomalies. J Geophys Res 107(B4)

Bozdag E, Trampert J (2008) On crustal corrections in surface wave tomography. Geophys J Int 172:1066–1082

Burdick S, van der Hilst RD, Vernon FL, Martynov V, Cox T, Eakins J, Karasu G, Tylell J, Astiz L, Pavlis GL (2012) Model update March 2011: upper mantle heterogeneity beneath North America from traveltime tomography with global and USArray Transportable Array data. Seismol Res Lett 83(1):23–28

Chevrot S, Zhao L (2007) Multiscale finite-frequency Rayleigh wave tomography of the Kaapvaal craton. Geophys J Int 169:201–215

Dahlen FA, Tromp J (1998) Theoretical global seismology. Princeton University Press, Princeton

Darbyshire FA, Lebedev S (2009) Rayleigh wave phase-velocity heterogeneity and multilayered azimuthal anisotropy of the Superior Craton, Ontario. Geophys J Int 176:215–234

Debayle E, Ricard Y (2012) A global shear velocity model of the upper mantle from fundamental and higher Rayleigh mode measurements. J Geophys Res 117(B10):1–24

Debayle E, Kennett BLN, Priestley K (2005) Global azimuthal seismic anisotropy and the unique plate-motion deformation of Australia. Nature 433(7025):509–512

Deschamps F, Lebedev S, Meier T, Trampert J (2008) Stratified seismic anisotropy reveals past and present deformation beneath the East-central United States. Earth Planet Sci Lett 274(3–4):489–498

Dziewonski AM, Anderson DL (1981) Preliminary reference Earth model. Phys Earth Planet 25:297–356

Dziewónski AM, Hager B, O'Connell RJ (1977) Large-scale heterogeneities in the lower mantle. J Geophys Res 82:239–255

Ekström G (2011) A global model of Love and Rayleigh surface wave dispersion and anisotropy, 25–250 s. Geophys J Int 187:1668–1686

Endrun B, Lebedev S, Meier T, Tirel C, Friederich W (2011) Complex layered deformation within the Aegean crust and mantle revealed by seismic anisotropy. Nat Geosci 4(3):203–207

Engdahl ER, van der Hilst RD, Buland R (1998) Global teleseismic earthquake relocation with improved travel times and procedures for depth determination. B Seismo Soc Am 88(3):722–743

Ferreira AMG, Woodhouse JH, Visser K, Trampert J (2010) On the robustness of global radially anisotropic surface wave tomography. J Geophys Res 115(B4):1–16

Forsyth DW, Scheirer D, Webb S, The MELT Seismic Team (1998) Imaging the deep seismic structure beneath a mid-ocean ridge: the MELT experiment. Science 280(5367):1215–8

Frederiksen AW, Bostock MG, Cassidy JF (2001) S-wave velocity structure of the Canadian upper mantle. Phys Earth Planet 124:175–191

Grand SP (2002) Mantle shear-wave tomography and the fate of subducted slabs. Philos T R Soc Lond 360:2475–2491

Gu YJ, Dziewónski AM, Su W, Ekström G (2001) Models of the mantle shear velocity and discontinuities in the pattern of lateral heterogeneities. J Geophys Res 106(B6):11169–11199

Gu YJ, Dziewónski AM, Ekström G (2003) Simultaneous inversion for mantle shear velocity and topography of transition zone discontinuities. Geophys J Int 154:559–583

Gudmundsson O, Sambridge M (1998) A regionalized upper mantle (RUM) seismic model. J Geophys Res 103:7121–7136

Hafkenscheid E, Wortel MJR, Spakman W (2006) Subduction history of the Tethyan region derived from seismic tomography and tectonic reconstructions. J Geophys Res 111(B8):1–26

Houser C, Masters G, Shearer PM, Laske G (2008) Shear and compressional velocity models of the mantle from cluster analysis of long-period waveforms. Geophys J Int 174:195–212

Jordan TH (1981) Global tectonic regionalization for seismological data analysis. B Seismo Soc Am 71(4):1131–1141

Jordan TH, Paulson EM (2013) Convergence depths of tectonic regions from an ensemble of global tomographic models. J Geophys Res 118 (B8):4196–4225

Karason H, van der Hilst RD (2001) Improving global tomography models of P-wavespeed I: incorporation of differential times for refracted and diffracted core phases (PKP, Pdiff). J Geophys Res 106:6569–6587

Kennett BLN (1987) Observational and theoretical constraints on crustal and upper-mantle heterogeneity. Phys Earth Planet 47:319–332

Kennett BLN, Engdahl ER, Buland R (1995) Constraints on seismic velocities in the Earth from traveltimes. Geophys J Int 122(1):108–124

Kovach RL (1979) Seismic surface waves and crustal and upper mantle structure. Rev Geophys 1(16):1–13

Kustowski B, Dziewónski AM, Ekström G (2007) Nonlinear crustal corrections for normal-mode seismograms. B Seismo Soc Am 97(5):1756–1762

Kustowski B, Ekström G, Dziewónski AM (2008) Anisotropic shear-wave velocity structure of the Earth's mantle: a global model. J Geophys Res 113(B6):1–23

Lebedev S, van der Hilst RD (2008) Global upper-mantle tomography with the automated multimode inversion of surface and S-wave forms. Geophys J Int 173:505–518

Lebedev S, Nolet G, Meier T, van der Hilst RD (2005) Automated multimode inversion of surface and S waveforms. Geophys J Int 162:951–964

Lebedev S, Adam J, Meier T (2013) Mapping the Moho with seismic surface waves: a review, resolution analysis, and recommended inversion strategies. Tectonophysics (Moho special edition)

Legendre CP, Meier T, Lebedev S, Friederich W, Viereck-Götte L (2012) A shear wave velocity model of the European upper mantle from automated inversion of seismic shear and surface waveforms. Geophys J Int 191:282–304

Lekić V, Romanowicz B (2011a) Inferring upper-mantle structure by full waveform tomography with the spectral element method. Geophys J Int 185(2):799–831

Lekić V, Romanowicz B (2011b) Tectonic regionalization without a priori information: a cluster analysis of upper mantle tomography. Earth Planet Sci Lett 308(1–2):151–160

Lekić V, Panning M, Romanowicz B (2010) A simple method for improving crustal corrections in waveform tomography. Geophys J Int 182:265–278

Li X-D, Romanowicz B (1996) Global mantle shear velocity model developed using nonlinear asymptotic coupling theory. J Geophys Res 101(B10):22245–22272

Li C, van der Hilst RD, Engdahl ER, Burdick S (2008) A new global model for P wave speed variations in Earth's mantle. Geochem Geophys Geosys 9(5):Q05018

Marone F, Romanowicz B (2007) Non-linear crustal corrections in high-resolution regional waveform seismic tomography. Geophys J Int 170(1):460–467

Masters G, Johnson S, Laske G, Bolton H (1996) A shear-velocity model of the mantle. Philos T R Soc Lond 354(1711):1385–1411

Masters G, Laske G, Bolton H, Dziewónski AM (2000) The relative behavior of shear velocity, bulk sound speed, and compressional velocity in the mantle: implications for chemical and thermal structure, Earth's deep interior: mineral physics and tomography from the atomic to the global scale, pp 63–87

Mégnin C, Romanowicz B (2000) The three-dimensional shear velocity structure of the mantle from the inversion of body, surface and higher-mode waveforms. Geophys J Int 143(3):709–728

Montagner J-P, Tanimoto T (1991) Global upper mantle tomography of seismic velocities and anisotropies. J Geophys Res 96(B12):20337–20351

Montelli R, Nolet G, Masters G, Dahlen FA, Hung S-H (2004) Global P and PP traveltime tomography: rays versus waves. Geophys J Int 158(2):637–654

Mooney WD, Laske G, Masters G (1998) Crust 5.1: a global crustal model at 5° × 5°. J Geophys Res 103(B1):727—747

Muller RD, Roest WR, Royer J-Y, Gahagan LM, Sclater JG (1997) Digital isochrons of the world's ocean floor. J Geophys Res 102(B2):3211–3214

Nataf H, Ricard Y (1996) 3SMAC: an a priori tomographic model of the upper mantle based on geophysical modeling. Phys Earth Planet 9201(95)

Nettles M, Dziewónski AM (2008) Radially anisotropic shear velocity structure of the upper mantle globally and beneath North America. J Geophys Res 113(B2):1–27

Nolet G (1990) Partitioned waveform inversion and two-dimensional structure under the network of autonomously recording seismographs. J Geophys Res 95(B6):8499–8512

Nolet G (2008) A breviary of seismic tomography. Cambridge University Press, Cambridge

Nolet G, Grand SP, Kennett BLN (1994) Seismic heterogeneity in the upper mantle. J Geophys Res 99(B12):23753–23766

Obrebski M, Allen RM, Pollitz F, Hung S-H (2011) Lithosphere-asthenosphere interaction beneath the western United States from the joint inversion of body-wave traveltimes and surface-wave phase velocities. Geophys J Int 185(2):1003–1021

Panning MP, Romanowicz B (2006) A three-dimensional radially anisotropic model of shear velocity in the whole mantle. Geophys J Int 167(1):361–379

Panning MP, Lekić V, Romanowicz Ba (2010) Importance of crustal corrections in the development of a new global model of radial anisotropy. J Geophys Res 115(B12):B12325

Ritsema J, van Heijst HJ, Woodhouse JH (1999) Complex shear wave velocity structure imaged beneath Africa and Iceland. Science 286:1925–1928

Ritsema J, van Heijst HJ, Woodhouse JH (2004) Global transition zone tomography. J Geophys Res 109(B2)

Ritsema J, Deuss A, van Heijst HJ, Woodhouse JH (2011) S40RTS: a degree-40 shear-velocity model for the mantle from new Rayleigh wave dispersion, teleseismic traveltime and normal-mode splitting function measurements. Geophys J Int 184(3):1223–1236

Schaeffer AJ, Lebedev S (2013) Global shear speed structure of the upper mantle and transition zone. Geophys J Int 194(1):417–449

Shapiro NM, Ritzwoller MH (2002) Monte-Carlo inversion for a global shear-velocity model of the crust and upper mantle. Geophys J Int 151(1):88–105

Shen W, Ritzwoller MH, Schulte-Pelkum V (2013) A 3-D model of the crust and uppermost mantle beneath the central and western US by joint inversion of receiver functions and surface wave dispersion. J Geophys Res 118:1–15

Siebert L, Simkin T (2002) Volcanoes of the World: an Illustrated catalog of holocene volcanoes and their eruptions

Sigloch K (2011) Mantle provinces under North America from multifrequency P wave tomography. Geochem Geophys Geosys 12(2):1–27

Simmons NA, Myers SC, Johannesson G, Matzel E (2012) LLNL-G3Dv3: Global P wave tomography model for improved regional and teleseismic travel time prediction. J Geophys Res 117(B10):1–28

Steinberger B (2000) Plumes in a convecting mantle: models and observations for individual hotspots. J Geophys Res 105(B5):11127–11152
Su W-J, Woodward R, Dziewónski AM (1992) Deep origin of mid-ocean-ridge seismic velocity anomalies. Nature 360:149–152
Su W-J, Woodward R, Dziewónski AM (1994) Degree 12 model of shear velocity heterogeneity in the mantle. J Geophys Res 99(B4):6945–6980
Tian Y, Zhou Y, Sigloch K, Nolet G, Laske G (2011) Structure of North American mantle constrained by simultaneous inversion of multiple-frequency SH, SS, and Love waves. J Geophys Res 116(B2):1–18
van der Lee S, Frederiksen AW (2005) Surface wave tomography applied to the North American upper mantle. In: Levander A, Nolet G (eds) Seismic earth: array analysis of broadband seismograms, vol 157, pp 67–80. AGU Geophysical Monograph Series, Washington, DC
Wang Z, Dahlen FA (1995) Spherical-spline parameterization of three-dimensional Earth models. Geophys Res Lett 22(22):3099–3102
Wessel P, Smith W (1995) New version of the generic mapping tools released. EOS 76:329
Woodhouse JH, Dziewónski AM (1984) Mapping the upper mantle: three-dimensional modeling of earth structure by inversion of seismic waveforms. J Geophys Res 89(B7):5953–5986
Wu RS, Flatte SM (1990) Transmission fluctuations across an array and heterogeneities in the crust and upper mantle. Pure Appl Geophys 132:175–196
Zhang Y, Tanimoto T (1992) Ridges, hotspots and their interaction as observed in seismic velocity maps. Nature 355:45–49
Zhang Y, Tanimoto T (1993) High-resolution global upper mantle structure and plate tectonics. J Geophys Res 98(B6):9793–9823
Zhang X, Paulssen H, Lebedev S, Meier T (2009) 3D shear velocity structure beneath the Gulf of California from Rayleigh wave dispersion. Earth Planet Sci Lett 279(3–4):255–262
Zhou Y, Nolet G, Dahlen Fa, Laske G (2006) Global upper-mantle structure from finite-frequency surface-wave tomography. J Geophys Res 111(B4):1–24

Chapter 2
Origin of Lateral Heterogeneities in the Upper Mantle Beneath South-east Australia from Seismic Tomography

N. Rawlinson, B.L.N. Kennett, M. Salmon and R.A. Glen

Abstract We use teleseismic body wave tomography to reveal anomalous P wave velocity variations in the upper mantle beneath south-east Australia. Data are sourced from the WOMBAT transportable seismic array, the largest of its kind in the southern hemisphere, which enables horizontal resolution of approximately 50 km to be achieved over a large region that includes Victoria, New South Wales and southern South Australia. In order to account for long-wavelength structure that is lost due to the use of multiple teleseismic datasets from adjacent arrays with non-overlapping recording periods, the AuSREM mantle model is included as prior information in the inversion. Furthermore, AuSREM crust and Moho structure is explicitly included in the initial model in order to account for the presence of shallow heterogeneity which is poorly constrained by the teleseismic dataset. The P wave velocity model obtained from the joint inversion of WOMBAT teleseismic data represents a vast new resource on the seismic structure of the upper mantle beneath south-east Australia. One of the most striking features of the model is the presence of a north-dipping low-velocity anomaly beneath the Newer Volcanics province, a Quaternary intraplate basaltic province in western Victoria. The anomaly appears to terminate at approximately 200 km depth and has a structure that is more suggestive of a source confined to the upper mantle rather

N. Rawlinson (✉)
School of Geosciences, University of Aberdeen, Aberdeen AB24 3UE, Scotland
e-mail: nrawlinson@abdn.ac.uk

B.L.N. Kennett · M. Salmon
Research School of Earth Sciences, The Australian National University,
Canberra, ACT 0200, Australia
e-mail: brian.kennett@anu.edu.au

M. Salmon
e-mail: michelle.salmon@anu.edu.au

R.A. Glen
Geological Survey of New South Wales, NSW Department of Trade and Investment,
Maitland, NSW 2310, Australia
e-mail: dick.glen@industry.nsw.gov.au

© Springer International Publishing Switzerland 2015
A. Khan and F. Deschamps (eds.), *The Earth's Heterogeneous Mantle*,
Springer Geophysics, DOI 10.1007/978-3-319-15627-9_2

than a deeply rooted mantle plume. Other features that can be observed include a high-velocity anomaly beneath the Curnamona province and a high-velocity salient beneath the New England Orogen. Of particular interest is an extensive high-velocity anomaly beneath the Lachlan Orogen, which coincides almost exactly with the surface expression of the Hay–Booligal Zone in the south, and extends northwards beneath the Macquarie Arc. The higher velocities beneath the Hay–Booligal Zone are consistent with the idea that it may be floored by a fragment of Proterozoic continental lithosphere that was once part of the east Gondwana margin, while the higher velocities beneath the Macquarie Arc may be related to its origin as an intra-oceanic arc.

Keywords Australia · Continental lithosphere · Seismic tomography · Accretionary orogen · Gondwana

2.1 Introduction

The seismic structure of the upper mantle beneath Australia has been progressively revealed over the last half century using a variety of passive and active source techniques. Some of the earliest measurements of mantle velocities come from recordings of nuclear tests undertaken by the British Atomic Weapons Research Establishment in South Australia (Doyle 1957; Bolt et al. 1958). As a result of the foresight of Sir Edward Bullard and Prof. John Jaeger, seven seismometers were placed along a transect that followed the Trans-Australian railway westwards away from the test site at Maralinga out to an angular distance of nearly 11° towards Perth. This allowed refraction and wide-angle reflection phases generated by the nuclear explosions to be recorded and analysed. Doyle (1957) used simple refraction analysis that assumed lateral homogeneity to estimate Pg (6.12 km/s), Sg (3.56 km/s), Pn (8.23 km/s), Sn (4.75 km/s) and Moho depth (35–40 km) beneath cratonic central and Western Australia.

In the mid-1960s, active sources were again used to interrogate crust and upper mantle structure when obsolescent depth charges were exploded on the sea floor beneath Bass Strait as part of BUMP (Bass Strait Upper Mantle Project). Using recording stations in Tasmania and mainland Australia, Underwood (1969) and Johnson (1973) exploited the move-out characteristics of refraction and wide-angle reflection phases to show that the Moho varied in depth from 37 km beneath the Snowy Mountains in New South Wales, to 25 km beneath Bass Strait and to 30–35 km beneath Tasmania, with an average Pn velocity of 7.8 km/s. Despite the use of relatively low fidelity equipment (by modern standards), analogue records and simple methods of analysis that relied on several major assumptions (e.g. constant velocity crust), these early measurements are at a very broad scale consistent with more recent results; in particular that the P wave velocity of the upper mantle beneath south-east Australia is markedly lower than that of cratonic central and Western Australia (Kennett and Salmon 2012).

Following the early work described above, active source seismology began to be used more widely, although largely with a view to constraining crustal structure. However, measurements of mantle velocities were still made if the sources were large enough (e.g. Finlayson et al. 1974). Over time the use of explosive sources was gradually replaced with vibroseis and marine airguns. Although still capable of penetration into the upper mantle, most studies confined analysis of the mantle to measurement of Pn and in some cases vertical P wave velocity gradient (Finlayson et al. 1980, 1984, 1998; Rawlinson et al. 2001). As such, it is probably fair to say that most of our current knowledge of upper mantle seismic structure beneath south-east Australia has come from passive seismic imaging. Early studies tended to use surface wave recordings and measure basic properties such as path average group and phase velocities over a number of frequency bands (e.g. De Jersey 1946; Bolt and Niazi 1964). More complete dispersion analysis and inversion for 1-D shear-wave models soon followed (Thomas 1969). The advent of digital seismographs and more powerful computing facilities allowed increasingly sophisticated models of the upper mantle beneath Australia to be constructed (Lambeck and Penny 1984; Bowman and Kennett 1990, 1993; Kennett and Bowman 1990), although it was not until the continent-wide SKIPPY project (van der Hilst et al. 1994) that a coherent picture of the Australian mantle began to emerge.

SKIPPY was a transportable array project that can be regarded as the progenitor of both USArray and WOMBAT. It used a moveable array of broadband seismometers to gradually achieve continent-wide coverage of Australia with a station spacing of approximately 400 km, sufficient to record the large number of surface wave paths from regional events required for 3-D surface wave tomography. The SKIPPY dataset, plus data collected by subsequent more targeted arrays placed in regions of particular geological interest, has formed the basis for many surface wave studies of the Australian region over the last two decades (Zielhuis and van der Hilst 1996; Debayle 1999; Simons et al. 1999, 2002; Debayle and Kennett 2000; Yoshizawa and Kennett 2004; Fishwick et al. 2005, 2008; Fishwick and Rawlinson 2012). These results have proven instrumental in gaining a first-order understanding of the heterogeneous mantle structure beneath the Australian continent, with one of the principal outcomes being the ability to discriminate between older Precambrian shield regions of Australia and younger Phanerozoic terranes (Kennett et al. 2004), much of which are masked by younger sediments. Continent-wide body wave studies of the Australian mantle have also been carried out using the same dataset and include velocity tomography (Kennett 2003; Kennett et al. 2004), attenuation tomography (Kennett and Abdullah 2011), receiver functions (Clitheroe et al. 2000; Reading and Kennett 2003) and shear wave splitting (Heintz and Kennett 2005).

The goal of this paper is to illuminate the upper mantle beneath south-east Australia using teleseismic data from the WOMBAT transportable array and attempt to explain the origins of the lateral heterogeneitiesthat are observed. To date, WOMBAT has involved 15 contiguous array deployments over the last 15 years and has achieved cumulative coverage of most of Tasmania, Victoria,

New South Wales and southern South Australia with approximately 650 sites and interstation spacings varying from 15 km in Tasmania to 50 km on mainland Australia. Compared to the continent-wide broadband experiments discussed above, WOMBAT passive seismic data have the potential to resolve much finer-scale variations in mantle structure, as has been demonstrated over the last decade by a number of teleseismic body wave studies (Graeber et al. 2002; Rawlinson et al. 2006a, b, 2010b, 2011; Rawlinson and Urvoy 2006; Clifford et al. 2008; Rawlinson and Kennett 2008; Rawlinson and Fishwick 2012; Fishwick and Rawlinson 2012). The current study differs from those that precede it by using a much larger dataset that includes recent arrays from northern New South Wales and southern South Australia, and explicitly accounting for near surface unresolved crust and Moho structure via inclusion of the recently released AuSREM (Australian Seismological Reference Earth Model) model.

2.1.1 Tectonic Setting

The WOMBAT array overlies the southern half of the Tasmanides, a complex series of Palaeozoic—early Mesozoic orogens that formed along the eastern margin of Gondwana during the retreat of the Pacific plate and encroaches onto the eastern edge of cratonic central and Western Australia (Figs. 2.1 and 2.3). The current state of knowledge on how this region of the Australian plate evolved is well described by a number of recent papers (e.g. Foster et al. 2009; Glen et al. 2009, 2012; Cayley 2011; Gibson et al. 2011; Glen 2013), so the brief overview given here will only focus on aspects that are particularly relevant to this study.

One view of the Tasmanides is that it progressively formed along a monotonically eastward rolling Pacific plate, which is consistent with its largely south-west–north-east younging direction (Foster and Gray 2000; Spaggiari et al. 2003, 2004; Foster et al. 2009). However, others have argued for alternative scenarios which involve significant continental transform faults (Glen et al. 1992; VandenBerg 1999) and the presence of sizable fragments of Precambrian continental crust (Cayley et al. 2002, Cayley 2011; Glen et al. 2012), relics of the break-up of the supercontinent Rodinia. In a recent paper, Glen (2013) argues that the palaeo-Pacific plate was segmented, with major transform faults propagating into the Tasmanides and leading to the formation of supra-subduction zone systems, which are bounded to the north and south by these faults. The differential rollback implied by this model explains why Palaeozoic orogenesis in the south (eastern South Australia, Victoria and New South Wales) occurred over a plate margin-perpendicular distance of over 1500 km, compared to only a few hundred km in the north. The transform faults which segment the palaeo-Pacific plate may manifest as E–W trending accommodation zones, one candidate being the curvilinear east–west trending structures at the southern margin of the Thomson Orogen.

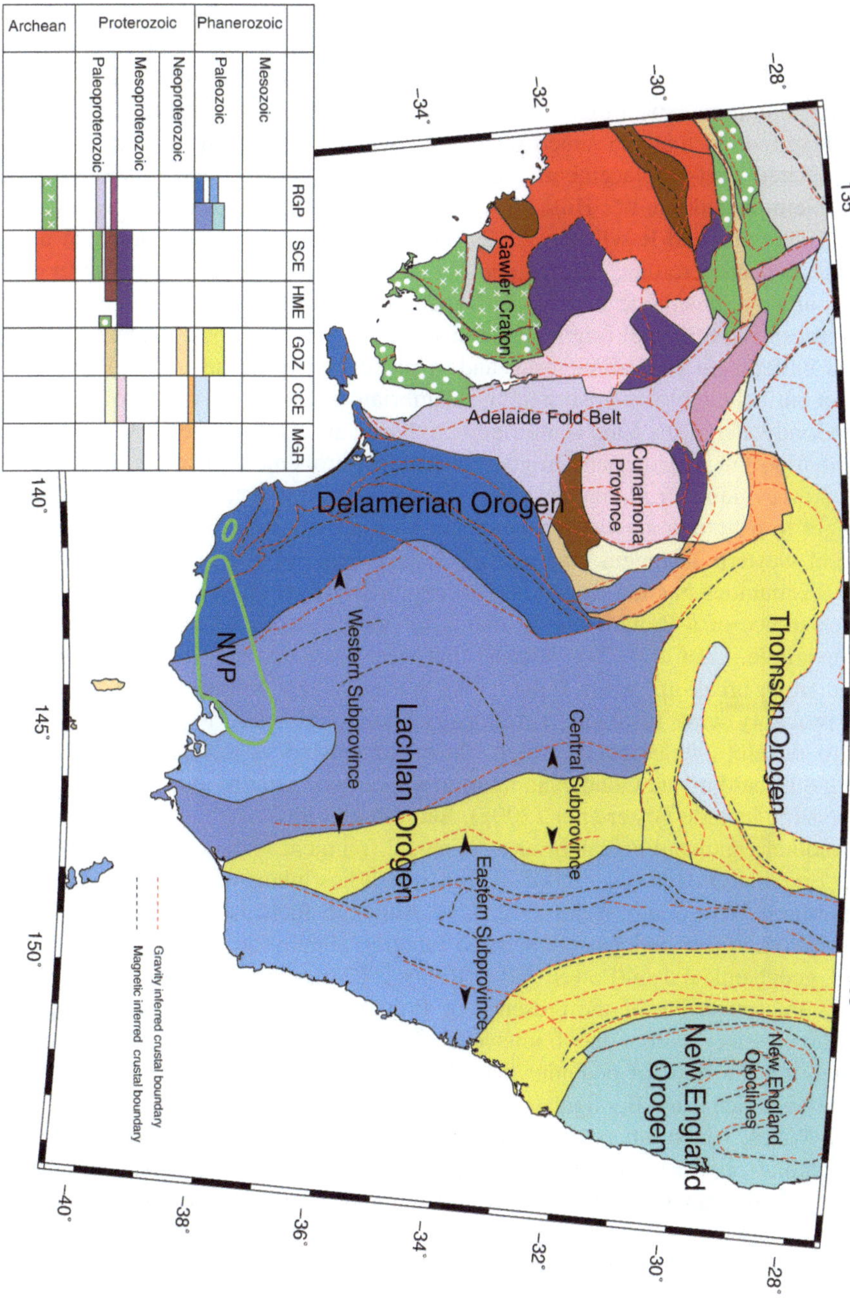

Fig. 2.1 The crustal elements of south-east Australia, as inferred from potential field data (based on data contained in Shaw et al. 1996). *RGP* relict geophysical pattern; *SCE* standard crustal element; *HME* highly magnetic element; *GOZ* geophysically overprinted zone; *CCE* covered crustal element; *MGR* subelement with muted geophysical response (see Shaw et al. 1996 for more details). *NVP* Newer Volcanics province. *Thick green line* denotes approximate surface expression of NVP

Precambriancontinental fragments within the Tasmanides have proven difficult to identify and delineate, partly due to the extensive Mesozoic and Cenozoic sedimentary cover sequences that mask large regions of the Tasmanides. However, if present, one may expect them to have a distinctive seismic signature in the upper mantle provided that the subcrustal component of the lithosphere has not been delaminated post-emplacement. Previous seismic studies of the Australian plate (e.g. Zielhuis and van der Hilst 1996; Debayle and Kennett 2000; Fishwick et al. 2005; Fishwick and Rawlinson 2012) show that older, colder and depleted upper mantle beneath cratonic central and Western Australia is distinctly faster than the younger lithospheric mantle of Phanerozoic provenance. One candidate for a Precambrian continental fragment is the so-called Selwyn Block, which is inferred to lie underneath central Victoria, and is postulated to be an exotic Proterozoic microcontinental block—overlain by Cambrian greenstones—that collided with east Gondwana in the Late Cambrian (Cayley et al. 2002; Cayley 2011), possibly terminating the Delamerian Orogeny, the oldest orogenic belt in the Tasmanides. Based on evidence from magnetic and seismic reflection data, the Selwyn Block is thought to extend southwards beneath Bass Strait, where it forms the Proterozoic core of western Tasmania. The relative lack of Precambrian outcrop in the mainland Tasmanides has led researchers to explore various allochthonous models for the existence of a Proterozoic Tasmania, of which the Selwyn Block is but one. For example, Li et al. (1997) suggest that west Tasmania may be a by-product of Rhodianian break-up, while Calvert and Walter (2000) propose that King Island, and probably west Tasmania, rifted away shortly after 600 Ma. Other researchers to develop scenarios in which Tasmania existed as a separate continental block outboard of the Gondwana margin before becoming re-attached in the early Palaeozoic include Foster et al. (2005), Berry et al. (2008).

Another candidate Precambrian continental fragment is the core of the southern New England Orogen (Glen 2005, 2013), which lies inland of the eastern seaboard of northern New South Wales (Fig. 2.1). Although there is no outcrop, evidence drawn from active and passive seismic imaging (Finlayson 1993; Fishwick et al. 2008, geochemical data (Powell and O'Reilly 2007) and geochronology (Shaw et al. 2011) are suggestive of the presence of Proterozoic lower crust and lithospheric mantle, possibly within the framework of an emplaced microcontinental block. Largely on the basis of potential field data, Hallet et al. (2005) infer the presence of the so-called Hay–Booligal Block beneath the Murray Basin in southern New South Wales, which others have inferred to be a possible rifted fragment from the Curnamona province to the west (Glen 2013). Although somewhat speculative, the oroclinal wrapping of Ordovician to Cambrian strata around this region distinguishes it from other parts of the Lachlan Orogen. Another possible Precambrian continental fragment was recently identified by Glen et al. (2013) beneath the southern margin of the Thomson Orogen, which is heavily masked by sediments. On the basis of deep seismic reflection data, supported by potential field data and inferences from zircon measurements, they infer that a sliver of continent became stranded during the Neoproterozoic rifting of Rodinia and now sits juxtaposed against Palaeozoic oceanic crust to the north (Thomson Orogen) and to the south (Lachlan Orogen).

Despite many attempts, the boundary between the Australian craton and the Tasmanides—often referred to as the Tasman Line—has proven difficult to identify (Hill 1951; Scheibner 1974; Wellman 1976; Shaw et al. 1996; Scheibner and Veevers 2000; Li 2001), partly due to the presence of younger cover sequences which mask the Palaeozoic and Precambrian basement, and partly due to differences in definition (Direen and Crawford 2003). For instance, is the onset of the Tasmanides defined by the earliest Palaeozoic basin systems, the western-most limit of deformation associated with Palaeozoic fold belts, or the onset of Palaeozoic oceanic substrate? Direen and Crawford (2003) discuss many attempts at defining a Tasman Line, and come to the conclusion that the various geophysical signatures present cannot be explained by the presence of a simple boundary and that attempts to define a Tasman Line should be abandoned. Passive seismic data have played a role in helping to understand the transition from cratonic Australia to the Tasmanides; for example, Kennett et al. (2004) carry out a synthesis of surface wave and body wave tomography and find a strong wavespeed contrast in the mantle beneath eastern and central Australia, which can be interpreted as a change from Precambrian cratonic to Palaeozoic lithosphere. However, they also found that the transition from one region to the other varied in complexity across strike and showed that the various interpretations of the Tasman Line coincide with different aspects of the mantle structures that are present.

The region under consideration in this study (see Figs. 2.1 and 2.3) crosses from the Tasmanides in the east into cratonic central Australia. In Fig. 2.1, the Delamerian Orogen represents the western most orogen of the Tasmanides and incorporates the Adelaide fold belt, Curnamona province and Kanmantoo fold belt. The Adelaide fold belt was formed by inversion of the Neoproterozoic–Cambrian Adelaide Rift Complex (Glen 2005; Foden et al. 1999, 2006), while the Curnamona province was thought to have been originally connected to the Gawler Craton prior to the Neoproterozoic (Belousova et al. 2006; Hand et al. 2008; Wade et al. 2012). Although originally regarded as cratonic (Thomson 1970), the Curnamona province underwent Middle-to-Late Cambrian deformation and metamorphism associated with the Delamerian Orogeny (e.g. Conor and Preiss 2008). The Kanmantoo fold belt developed from the inversion of the deep water Kanmantoo Trough (Glen 2005). The basement to the Delamerian Orogen appears to comprise reworked Precambrian continental rocks (Direen and Crawford 2003), although a transition to Palaeozoic oceanic crust beneath the south-east region of the orogen has been inferred (Glen 2013). Previous teleseismic studies of the region (e.g. Graeber et al. 2002; Rawlinson and Fishwick 2012) indicate that the lithospheric mantle beneath the Delamerian Orogen is characterized by higher velocities than that beneath the Lachlan Orogen, which is thought to be largely floored by Oceanic crust (Foster et al. 2009; Glen 2013).

Although the orogenic cycles that gave rise to the Tasmanides were largely complete by 227 Ma (Glen 2005), several subsequent tectonic events have significantly affected the nature of the mantle beneath south-east Australia. The main change was the termination of subduction off the eastern margin of Australia at about 100 Ma (Glen 2013) followed by the break-up of Australia and Antarctica, and the opening of the Tasman Sea around 80–90 Ma (Gaina et al. 1998), which resulted in significant lithospheric thinning towards the passive margins and the formation of the Bass Basin as a result of failed intracratonic rifting (Gunn et al. 1997). The Southern Highlands, which are located in eastern New South Wales and Victoria and have a maximum elevation of over 2 km, may have formed as a result of the rifting process, although the exact timing and uplift mechanism remain a subject of debate. For example, if the eastern Australian passive margin formed as a result of asymmetric rifting, uplift beneath the upper plate margin is expected due to delamination of the mantle lithosphere (Lister et al. 1986, 1991; Lister and Etheridge 1989). However, others have argued that the Southern Highlands are the remnants of an older Palaeozoic Orogen (Lambeck and Stephenson 1986; van der Beek et al. 1999).

The widespread coverage of Cenozoic volcanism along the eastern seaboard of the Australian continent (Johnson 1989) has undoubtedly had an impact on the composition and thermal structure of the upper mantle in this region. The cause of this volcanism is often attributed to the so-called East Australia Plume System (Wellman 1983; Sutherland 1983), the origin of which lies offshore, but has been linked to 65 Ma rifting of the Coral Sea (Sutherland 1983). It was most active after 35 Ma and gets progressively younger further south in a manner that is consistent with the northward movement of the Australian continent over stationary hot spots. However, a consensus on the mechanism of formation has yet to be reached, with alternatives including mantle upwelling associated with slab detachment and temporal variations in crustal stresses coupled with underlying mantle convection (see Sutherland et al. 2012, for an overview).

The most recent example of Cenozoic volcanism in Australia occurs in western Victoria and south-east South Australia, where the Newer Volcanics province (NVP) contains evidence of eruptive activity that is less than 5 ka (Johnson 1989). Compared to the Cenozoic volcanic chains that populate the eastern seaboard of Australia, the NVP is unique in that it has not migrated over time. Moreover, it appears that the NVP is the latest phase of an eruption cycle that has operated intermittently since the early Eocene when fast northern motion of the Australian continent commenced (Demidjuk et al. 2007). Coupled with modest surface topographic response (~100 m) and a relatively low eruption volume (~20,000 km^3) researchers have begun to suspect that the source of the NVP does not fit the mould of a traditional mantle plume model, but instead may be a phenomenon localized to the upper mantle (Demidjuk et al. 2007; Rawlinson and Fishwick 2012).

2.1.2 AuSREM Model

The AuSREM represents a synthesis of structural information on the Australian lithosphere that has been derived from various active and passive source seismic imaging experiments over the last few decades (Kennett and Salmon 2012; Kennett et al. 2013, Salmon et al. 2012). AuSREM defines seismic structure on a

0.5° grid in both latitude and longitude, with depth variations determined by major boundaries such as the Moho and sediment-basement interface, and information obtained from smooth tomographic models. P and S wavespeed as well as density is defined in the crust, mantle lithosphere and sublithospheric mantle to a depth of 350 km. Laterally, the crustal model spans 110°E to 160°E, and 10°S to 45°S and is built from observations that have a five-layer representation which include sediments, basement, upper crust, middle crust and lower crust. However, the AuSREM crust is not explicitly layered and is defined by a regular grid of points with a depth discretization of 5 km. In the mantle, the grid is defined at 25-km intervals in depth and spans a geographic region from 105°E to 180°E, and 0°S to 50°S. Beyond this region and at depths greater than 350 km, AuSREM is linked to S40RTS (Ritsema et al. 2011), which is constructed from a large global dataset of surface, normal mode and body wave observations.

Seismic data exploited by AuSREM come from a wide variety of Australian-centric sources and include over 12,000 km of deep crustal reflection profiles, a number of major refraction experiments that used explosive sources between the 1960s and 1980s and a large number of passive portable seismic deployments that have sampled various parts of the continent since the early 1990s. The active source experiments have largely constrained the crustal architecture of the continent and have allowed major boundaries such as the Moho to be delineated. A degree of control over crustal properties is also possible, particularly from the refraction data, although in general P wave velocity is much better constrained than S wave velocity. A wide variety of techniques have been applied to the passive source data, which has provided valuable constraints on both crustal and mantle structure. One of the main sources of information comes from distant earthquake sources that surround the Australian plate, particularly from the north and east. The frequency of these earthquakes has allowed detailed images of mantle P wave and S wave velocity to be built up from sequences of portable deployments using both surface wave and body wave tomography. In the upper mantle, the primary source of information comes from surface wave tomography, although this is supplemented by body wave traveltime tomography which provides important constraints on the relationship between P and S wavespeeds. Receiver function studies, which provide important information on crustal and upper mantle discontinuities, have also been used. Ambient noise tomography, which exploits relatively high-frequency surface waves generated by oceanic and atmospheric disturbances, provides valuable constraints on crustal shear wave velocity variations,and is an important complement to the active source studies. A more detailed description of the data sources and how they are combined to form the AuSREM model can be found in Kennett and Salmon (2012).

Figure 2.2 illustrates the lateral heterogeneity that is inherent to the AuSREM model at crustal and mantle depths within the confines of the study area. The geometry of the Moho is also shown. Within the mantle, the first-order variation in P wavespeed is a gradual decrease from the NW to the SE, which reflects a transition from Precambrian cratonic Australia, to the younger fold belts of the Tasmanides, and then the thinned and possibly hotter and more enriched lithosphere beneath the eastern seaboard. Although the horizontal sampling of

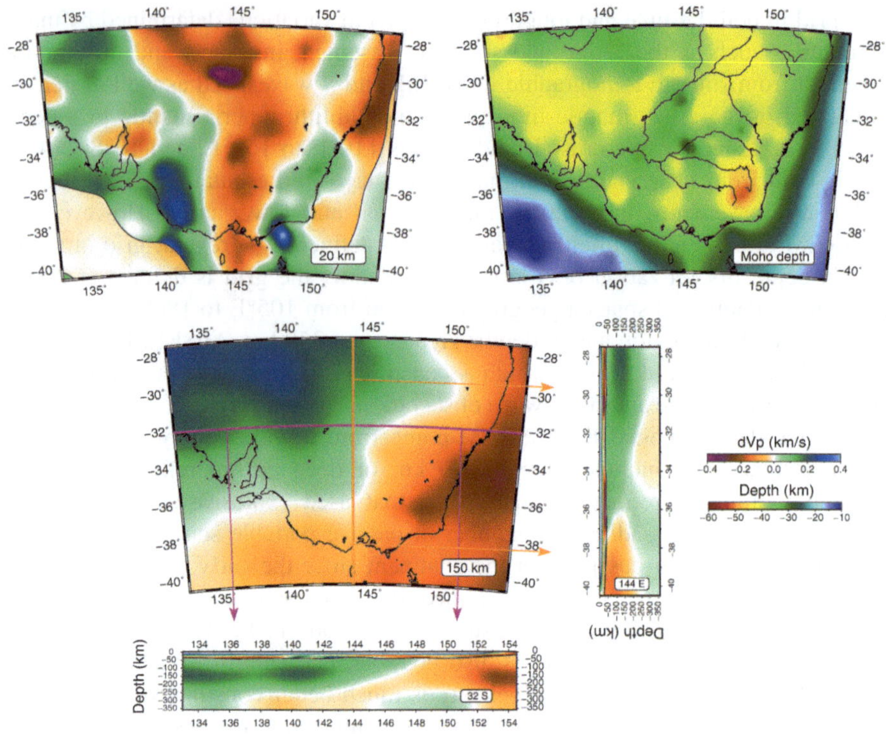

Fig. 2.2 P wave velocity variations (relative to laterally averaged 1-D model) and Moho geometry of the AuSREM model in the study region. Note that the depth scale refers only to the Moho depth variations displayed in plan view in the *top right*; all other horizontal and vertical slices display P wave velocity perturbations (dVp). The velocity scale saturates at ±0.4 km/s in order to properly visualize mantle variations; consequently, the full detail of crustal variations, particularly in cross-sectional view, is not represented in its entirety. Lateral discontinuities in velocity in the 20-km-depth slice are due to the undulating Moho, which becomes shallower than 20 km in several regions

AuSREM in this region is 0.5° in latitude and longitude, the maximum resolution in the mantle is of the order of 200–250 km, a limit inherited from the body and surface wave datasets used to constrain seismic structure at these depths. The Moho varies in depth from about 10 to 55 km, although it is almost entirely below 25 km depth inboard of the coastline. Nonetheless, variations in Moho depth beneath WOMBAT are sufficient to make sizable contributions to measured relative arrival time residuals, which should be accounted for in the inversion for 3-D velocity structure. The Moho is deepest beneath the Southern Highlands, which lie inboard of the east coast, and reflects the presence of a crustal root beneath up to 2 km of elevation. The shallow regions of the Moho lie offshore and are associated with the continental margin and the transition to oceanic lithosphere. The predominant strike of long-wavelength structures at 20 km depth (Fig. 2.2, top left) is N–S, which is largely consistent with the crustal elements plot shown in Fig. 2.1.

2.2 Data and Method

Data for this study are taken from the mainland stations of the WOMBAT transportable seismic array (Fig. 2.3), which has been in continuous operation in south-east Australia since 1998. To date, over 650 sites have been occupied during the course of 15 deployments, with individual arrays varying in size between 20 and 72 stations. Recording durations have also varied, with earlier arrays (e.g. LF98) operational for 4–5 months, while more recent deployments (e.g. EAL3) have run for more than 12 months. Prior to 2006, all seismometers used by WOMBAT were vertical component L4C sensors with a natural frequency of 1 Hz. Subsequent deployments have exclusively used 3-component Lennartz LE3Dlites which also have a natural frequency of 1 Hz. The use of short-period sensors has meant that the longer period component of teleseismic body waves and the bulk of surface wave energy from regional earthquakes, are not detected. However, teleseismic P phases such as direct P, PcP, ScP, PKiKP, PP, etc., are relatively ubiquitous in the long-term records, which makes them ideal for constraining relative variations in P wave velocity in the lithosphere beneath WOMBAT.

Figure 2.4 plots the locations of all 2985 teleseismic sources used in this study. Although earthquakes completely encircle WOMBAT, the azimuthal distribution varies quite considerably, with large concentrations lying to the north and east of Australia. In order to reduce the influence of uneven source distribution, we bin the data using the approach described in Rawlinson et al. (2011), which reduces the effective number of sources to 1115. The adaptive stacking method of Rawlinson and Kennett (2004) is used to measure relative arrival time residuals on a source by source basis. For a particular phase, the ak135 global reference model

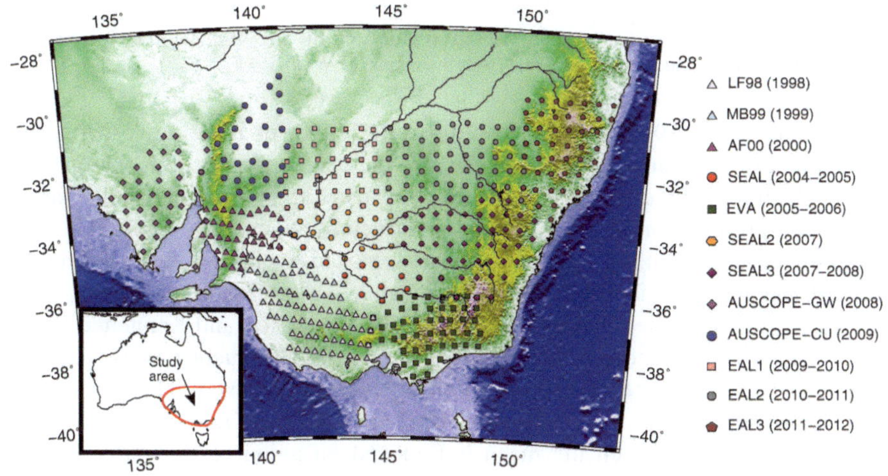

Fig. 2.3 Coverage of the WOMBAT transportable seismic array experiment in south-east mainland Australia

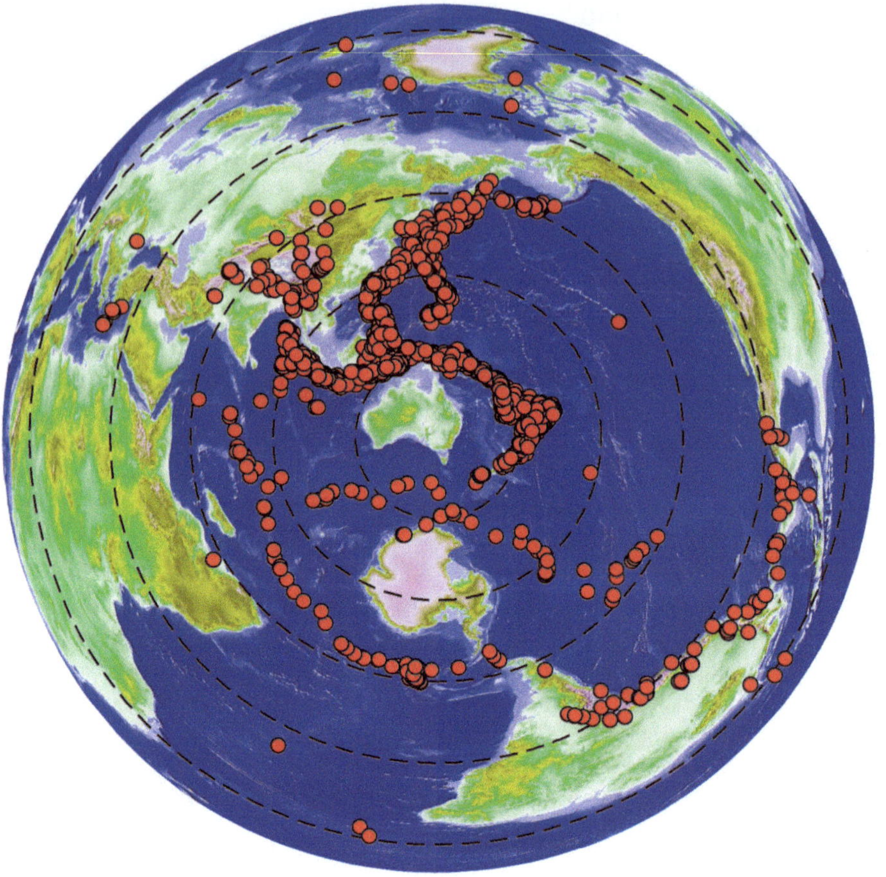

Fig. 2.4 Distribution of seismic sources used to constrain the 3-D P wave model. Concentric *dashed circles* are plotted at 30° intervals from the centre of WOMBAT

of Kennett et al. (1995) is used to remove the effect of trace move-out, which results in a preliminary alignment of the arriving wave trains at each station in the array. The remaining differences represent the arrival time residuals, which largely reflect the presence of lateral variations in structure beneath the array. Following preliminary alignment, all traces are linearly stacked, which produces a reference trace. Each trace is then optimally aligned with the reference trace using an L_p measure of misfit, where in our case $p = 3$ (see Rawlinson and Kennett 2004, for more details). The traces are then re-stacked and the alignment process repeated. This can be applied iteratively until convergence is achieved. The perturbation applied to each trace in order to achieve final alignment is a measure of the arrival time residual, from which the mean is removed on a source by source basis to help minimize temporal and spatial errors in source location, and long-wavelength variations in velocity in the deep mantle. In general, adaptive stacking is effective for teleseismic data, although problems can arise if the waveform is not coherent across the array or lateral velocity variations are large enough to cause cycle skipping. Following binning of the data, the total number of arrival time residuals is 41,380.

The FMTOMO package (de Kool et al. 2006; Rawlinson and Urvoy 2006; Rawlinson et al. 2010b) is used to invert the measured traveltime residuals for variations in P wavespeed beneath the mainland component of the WOMBAT array. FMTOMO uses an eikonal solver known as the fast marching method (Sethian 1996; Rawlinson and Sambridge 2004a, b) to solve the forward problem of source–receiver traveltime prediction, and a subspace inversion method (Kennett et al. 1988) to adjust model parameters in order to satisfy the data. The subsurface can be characterized by both smooth variations in wavespeed and discontinuities, which allows interfaces such as the Moho to be explicitly included. Phases can be composed of any number of reflection and refraction branches, and sources may be local or teleseismic. This level of flexibility means that FMTOMO can be used with reflection, wide-angle, local earthquake and/or teleseismic data. Model unknowns that can be constrained include velocity, interface depth and source location. Although FMTOMO could be used for quasi-global applications (it is written in spherical coordinates but does not allow for periodicity or the presence of the poles), its strength compared to conventional ray tracing lies in its ability to robustly compute two point traveltimes in the presence of sizable lateral heterogeneity. At the global scale, lateral heterogeneities are not significant enough to strongly perturb ray paths, which means that iterative two-point ray tracing schemes such as pseudo-bending (Koketsu and Sekine 1998) are much more efficient. However, within the upper mantle and particularly the crust, departures from lateral homogeneity can be very large, which makes eikonal solvers such as fast marching much more attractive.

We define our 3-D model structure beneath WOMBAT with a crustal layer and a mantle layer, separated by an undulating Moho. P wave variations in the crust and Moho depth variations are defined by AuSREM (Kennett and Salmon 2012) and are not permitted to vary during the inversion, since the shallow regions of the model are poorly constrained by the data. Although it would be possible to simultaneously invert for crust and upper mantle structure, we are unlikely to achieve any advantage as there is no crossing ray path coverage in the crust owing to the ~50-km station separation and use of teleseismic phases. Of particular concern would be trying to manage the trade-off between perturbations in crustal velocity and uppermost mantle velocity across the Moho discontinuity. The AuSREM mantle model (Kennett et al. 2013) defines initial P wave velocities in the mantle, which are subsequently adjusted during the inversion process. The complete model is defined by a total of 459,000 velocity nodes and 7650 interface nodes. During the iterative nonlinear inversion process, the data misfit component of the objective function is based on the difference between the predictions through the current model and the ak135 reference model (with receivers at zero elevation); thus, contributions to the measured residuals from both AuSREM and variations in receiver elevations are accounted for. The appropriate level of damping and smoothing regularization is determined by examining trade-off curves between data fit, model smoothness and model perturbation (see Rawlinson et al. 2006a, b, for details);

ideally, the optimum model is one that is as smooth and as close to the initial model as possible but still satisfies the data. The results of the synthetic testing shown in the next section help to illustrate the effect of the imposed regularization.

The use of FMTOMO together with AuSREM allows us to address several weaknesses in the traditional teleseismic tomography method. First, by explicitly including crustal and Moho structure, contributions to measured arrival time residuals from shallow unresolved structure are accounted for. The significant lateral heterogeneity in crustal velocity and thickness beneath WOMBAT (Fig. 2.2) has the potential to impart traveltime perturbations of a similar magnitude to those generated by lateral perturbations in the upper mantle. As such, a crustal correction is mandatory for the reliable recovery of deep structure. Traditionally, this has been done using station correction terms that are treated as unknowns in the inversion (Frederiksen et al. 1998; Graeber et al. 2002; Rawlinson et al. 2006b), but the potential trade-off with mantle velocity perturbations can be difficult to quantify. Due to this difficulty, the use of a priori crustal models is becoming more popular (Waldhauser et al. 2002; Martin et al. 2005, Lei and Zhao 2007, Rawlinson et al. 2010b). Another weakness of teleseismic tomography is its reliance on relative arrival time residuals. Although absolute arrival time residuals could be used, contributions from source time uncertainty and structure outside the model region would negate any potential advantage. When using data from multiple arrays that do not record simultaneously, the removal of the mean on a source by source basis means that residuals from one array are not comparable to those from another unless the laterally averaged velocity structure beneath each array is identical. If this is not the case, then long-wavelength structure with a scale length similar to or larger than the aperture of the array will be lost. This phenomenon is explained in more detail in Evans and Achauer (1993), Rawlinson et al. (2011, 2013) and can be remedied in seismic tomography by using a starting model that contains the long-wavelength information that would otherwise be lost. Hence, by including the AuSREM mantle model in the initial model used by FMTOMO, we hope to end up with a final model that captures both the short- and long-wavelength features of the upper mantle. The principle assumption we are working with here is that the AuSREM model is correct; errors in crustal velocity will result in the introduction of artefacts into the recovered mantle structure, while errors in mantle velocity will translate as errors in the absolute velocities of the recovered model and, potentially, the larger scale patterns of lateral heterogeneity that extent between adjacent arrays.

2.3 Results

2.3.1 Synthetic Tests

Synthetic reconstruction tests are the most commonly used approach for assessing solution robustness for large tomographic inverse problems (Rawlinson et al. 2010a). Rather than directly quantifying solution uncertainty, this approach essentially measures how variations in data coverage influence the recovery of structure (Rawlinson et al. 2010a). To do so, a synthetic dataset is constructed by solving the forward problem in the presence of a known model with preset velocity variations using the same sources, receivers and phase types as the field dataset. The inversion scheme is applied to this synthetic dataset, and a comparison of the recovered model and the true model provides insight into which regions of the model are well constrained by the data. A checkerboard pattern of alternating high and low velocities is often favoured as the input test model, as it makes visual comparison a relatively simple task. The limitations of synthetic reconstruction tests (Lévêque et al. 1993; Nolet 2008; Rawlinson et al. 2010a) are well known, such as the results varying according to the input structure used. Common alternatives include the calculation of formal estimates of posterior covariance and resolution from linear theory (Tarantola 1987), which provide quantitative measures of model uncertainty. However, these too suffer from several limitations such as decreasing validity with increasing nonlinearity of the inverse problem; inversion of a potentially very large matrix, although approximation methods have been developed to overcome this problem (Zhang and McMechan 1995; Zhang and Turber 2007); implicit regularization imposed by the chosen parameterization not being accounted for; a priori model covariance and data uncertainty usually poorly known. This coupled with the use of variable damping and smoothing often make the absolute value of posterior uncertainty rather meaningless. Rawlinson et al. (2010a, b) perform numerical experiments to show that the usefulness of synthetic reconstructions and posterior covariance and resolution estimates appear similar. In recent years, a number of other methods for assessing solution reliability of large tomographic systems have emerged, including dynamic objective functions (Rawlinson et al. 2008), generalized ray density tensors (Fichtner and Trampert 2011) and indirect methods for computing the resolution operator (MacCarthy et al. 2011; Trampert et al. 2013).

Here, we carry out two synthetic tests to help asses the reliability of P wavespeed patterns recovered in the mantle (Fig. 2.5). The first is a checkerboard reconstruction test in which the input checkerboard pattern is restricted to the mantle, where it is superimposed on the AuSREM P wave model (see Fig. 2.5a). Crustal P wave velocity structure and Moho geometry are defined by AuSREM and have not been perturbed. The peak amplitudes of the checkerboard anomalies are ±0.4 km/s, which are similar to the peak amplitudes of velocity anomalies recovered from the WOMBAT dataset, as will be shown in the next section. The synthetic dataset that is generated from the model shown in Fig. 2.5a is contaminated with Gaussian noise with a standard deviation of 59 ms in order to simulate the noise content of the observational dataset, as estimated by the adaptive stacking procedure. Figure 2.5b shows horizontal and vertical slices through the recovered or output model that can be compared directly with Fig. 2.5a. Overall, the recovery of the checkerboard pattern is good within the horizontal bounds of the receiver array. Smearing is present towards the edge of the model where crossing ray path coverage is less dense. There is also more smearing in the N–S direction as a consequence of the source distribution (Fig. 2.4), which results in many more

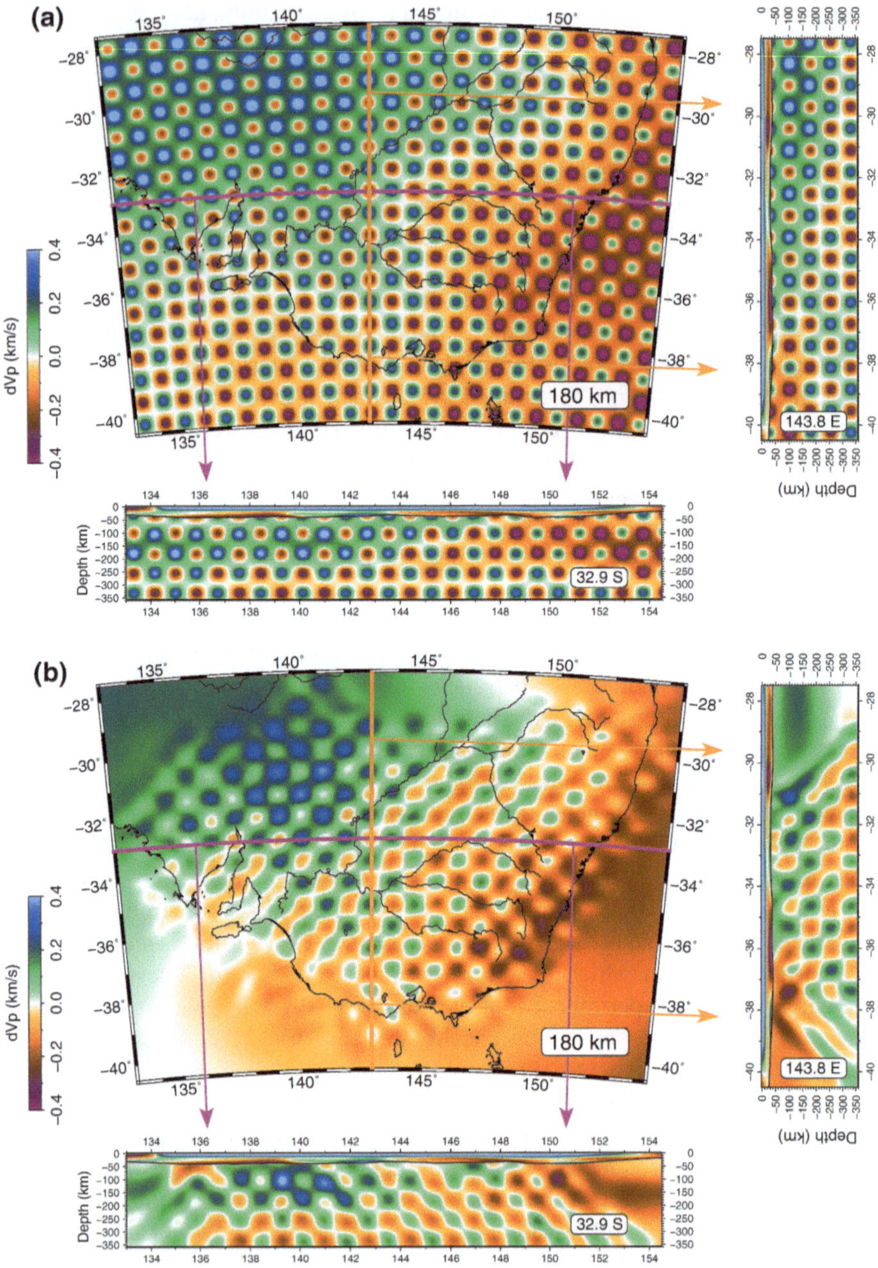

Fig. 2.5 Results of the synthetic checkerboard reconstruction test. **a** Input model; **b** recovered model. Velocity perturbation is relative to a laterally averaged 1-D model

paths impinging on the model from the north compared to the south. Moreover, the amplitudes of anomalies also tend to be underestimated, a feature typical of inversions that are regularized using smoothing and damping (Rawlinson et al. 2010a). However, the results do indicate that overall, the path coverage is sufficient to constrain features with a scale length upwards of 50 km within the horizontal bounds of the receiver array, although some caution is needed when interpreting the geometry of anomalies in the N–S direction. It also appears that structural recovery is uniformly good between the Moho and the base of the model at approximately 350 km depth.

In addition to the checkerboard reconstruction test, we also carry out a synthetic test in which the input anomalies are random patterns with a Gaussian distribution and standard deviation of 0.3 km/s. Compared to a checkerboard test,

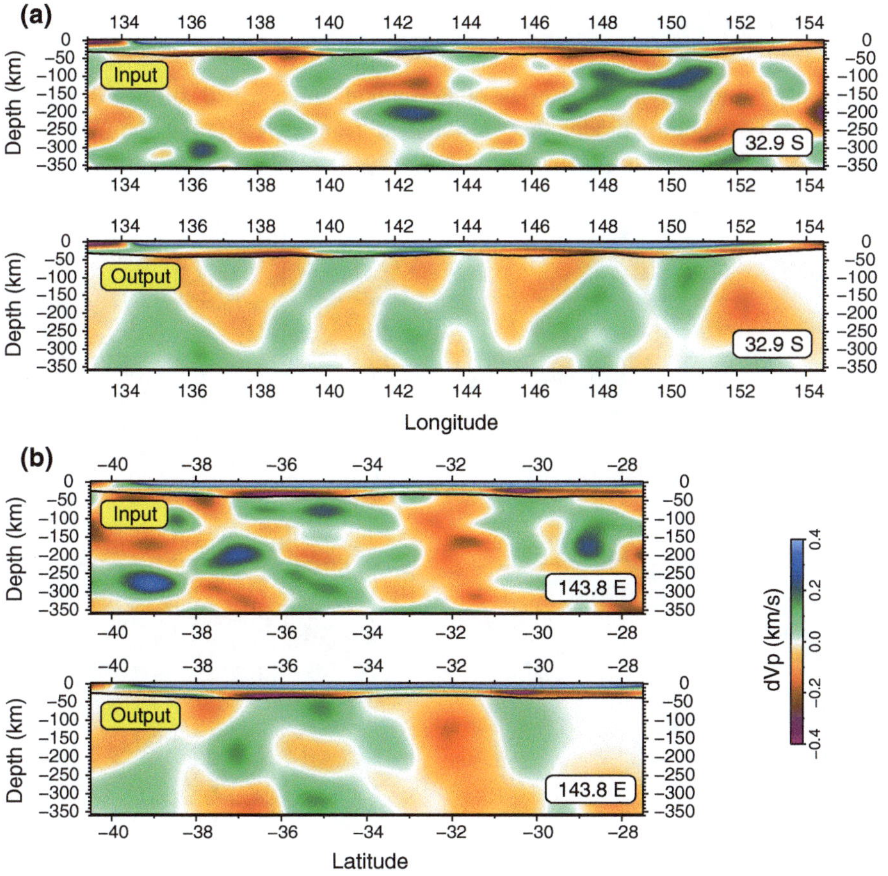

Fig. 2.6 Results of the synthetic reconstruction test using random velocity patterns with a Gaussian distribution. Vertical sections are taken at the same locations as shown in Fig. 2.5 checkerboard test. **a** E–W slice; **b** N–S slice. Velocity perturbation is relative to a laterally averaged 1-D model

this has the advantage that there is no regularity to the velocity variations and a range of scale lengths are present, which helps assess the extent of smearing. Figure 2.6a, c show cross sections through the input model, while Fig. 2.6b, d shows the corresponding output. As with the checkerboard model, Gaussian noise with a standard deviation of 59 ms has been added to the synthetic dataset to simulate the picking uncertainty associated with observational dataset. In comparing the input and output, the presence of vertical smearing of structure is the most obvious artefact of the recovery. This effect is typical in teleseismic tomography due to the steep inclination angles of impinging ray paths and imposes a limit on the extent to which vertical continuity of structure can be inferred.

2.3.2 P Wave Velocity Structure Beneath South-east Australia

Figures 2.7 and 2.8 show a series of horizontal and vertical slices through the final P wave velocity model that is obtained from inversion of the WOMBAT teleseismic dataset. The horizontal slices (Fig. 2.7) show that both short- and

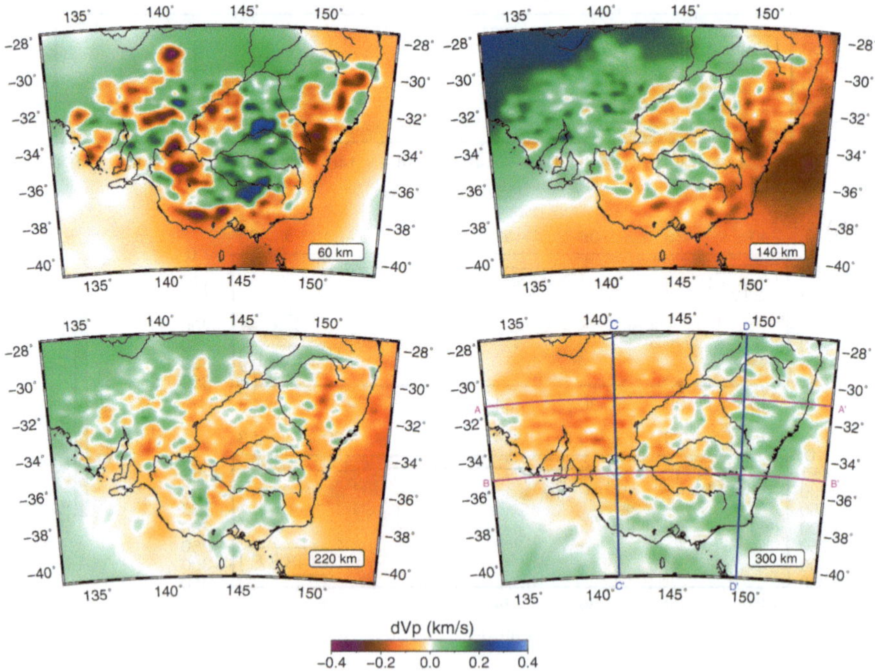

Fig. 2.7 Horizontal slices through the final WOMBAT teleseismic model. Velocity perturbation is relative to a laterally averaged 1-D model. Thick magenta and *blue lines* superimposed on the 300-km-depth slice (*bottom right*) denote locations of vertical slices shown in Fig. 2.8

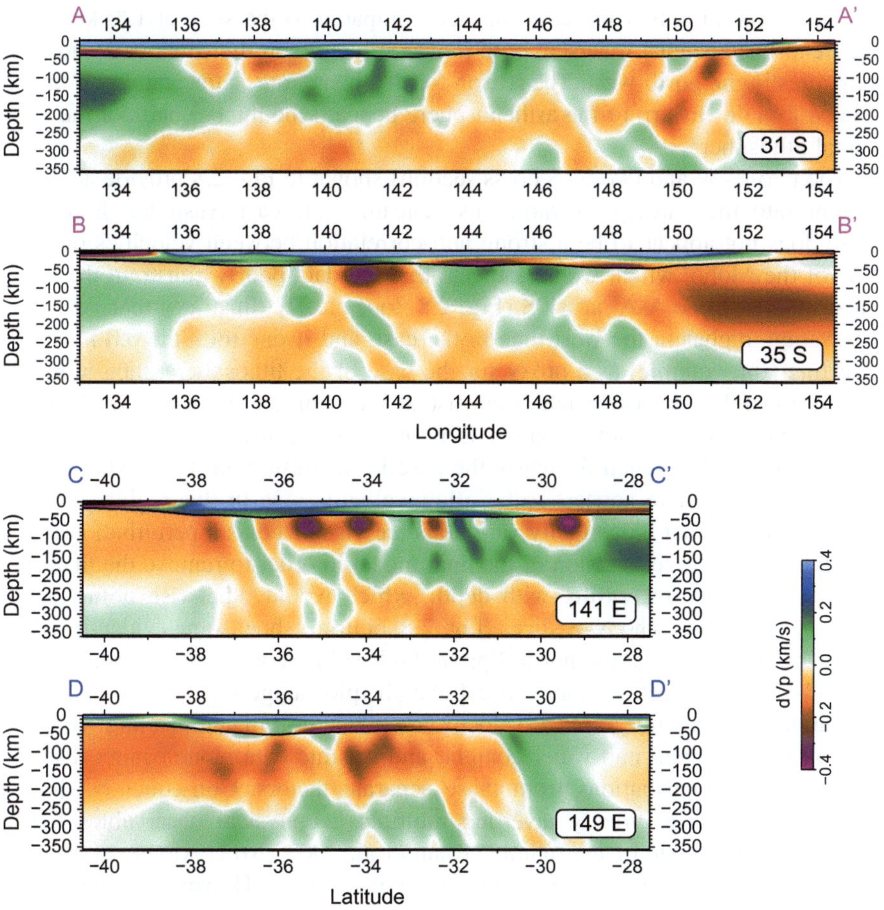

Fig. 2.8 Vertical slices through the final WOMBAT teleseismic model. Velocity perturbation is relative to a laterally averaged 1-D model. Refer to Fig. 2.7 for slice locations in plan view

long-wavelength patterns of lateral heterogeneity change markedly with depth. In the uppermost mantle at 60 km depth, the amplitude of anomalies is larger compared to the slices at greater depth; this may in part be due to less crossing path coverage, but it does make sense to have a more heterogeneous mantle in the immediate proximity of a very heterogeneous crust, given that the Moho does not necessarily represent a discontinuity in processes (e.g. deformation, intrusions, chemical alteration etc.). At 140 km depth, the general trend is a decrease in wavespeed to the south-east, which probably represents a decrease in the thickness of the continental lithosphere, which eventually transitions to oceanic outboard of the coastline. At 220 km depth, there is still evidence of higher wavespeeds in the north-west that are associated with Australia's cratonic interior, but by 300 km depth, this is no longer the case. In fact, compared to the slice at 120 km depth, the trend is somewhat reversed, with higher velocities in the south-east and lower velocities in the north-west. The reason for this behaviour is unclear, but it may be related to a downwarp of the asthenosphere beneath the thicker parts of the continental lithosphere.

The two E–W and two N–S cross sections shown in Fig. 2.8 provide additional insight into the wavespeed variations beneath south-east Australia. In general, there does not appear to be a strong anti-correlation between velocities on either side of the Moho, which would indicate that the data poorly constrain the uppermost mantle. Slice A-A′ at 31°S (Fig. 2.8, top) clearly illustrates that velocity is on average higher in the west than in the east, and if one took the 0.0 km/s contour line (for example) as a proxy for the base of the lithosphere, then it appears to thin towards the east which is consistent with surface wave results (Fishwick et al. 2008). Where path coverage is poor or non-existent, the model defaults back to AuSREM, which is clearly the case in the eastern sector of slice B-B′ at 39°S. Although there appears to be a low-velocity zone in this region, it should be remembered that the velocity variations are visualized as a perturbation from a 1-D laterally averaged version of the final model in order to remove the first-order effect of velocity increasing with depth, which would otherwise obfuscate the velocity anomalies. The N–S slices (Fig. 2.8, bottom) show some evidence of N–S smearing; for example, the north-dipping high-velocity anomaly at about 37°S in slice C–C′ may in fact be more vertical than the plot suggests.

Variations in mantle P wave velocity are a function of temperature, composition, grain size, melt and water content, and seismic data alone cannot discriminate between contributions from these separate factors. Consequently, there is growing interest in directly inverting multiple geophysical observables for the thermochemical state of the mantle (Khan et al. 2007, 2011; Afonso et al. 2008, 2013a, b; Simmons et al. 2009; Cammarano et al. 2011). However, in the absence of such methods and data, it is still possible to make reasonable inferences regarding the relationship between seismic velocity and physical and chemical properties of the mantle. Temperature appears to be a dominant factor, with a 1 % perturbation in velocity caused by as little as a 100 °C change in temperature (Cammarano et al. 2003). Nevertheless, the influence of composition has been argued to be equally as great in some circumstances, with realistic changes in composition producing anything between 1 and 5 % velocity variation (Griffin et al. 1998). Within the Australian continent, Faul and Jackson (2005) argue that lateral velocity variations can be explained by reasonable changes in temperature: thus, older cratonic Australia is simply colder than the younger warmer Tasmanides, which are characterized by lower velocities. More recently, Dalton and Faul (2010) found that while lateral variations in seismic velocity in the upper mantle are largely controlled by temperature variations in dry melt-free olivine, temperature alone cannot explain the high velocities that are sometimes observed above 200 km depth, particularly in cratonic regions. One possible explanation for this is the presence of chemically depleted lithosphere (higher Mg#), which will have the effect of increasing velocity.

2.4 Discussion and Conclusions

Compared to previous teleseismic studies which exploit WOMBAT data to image the lithosphere beneath south-east Australia (Graeber et al. 2002; Rawlinson et al. 2006b, 2011, 2013; Clifford et al. 2008; Rawlinson and Kennett 2008; Fishwick and Rawlinson 2012), the work presented here uses a geographically more extensive dataset and explicitly accounts for crustal velocity variations, Moho structure and long-wavelength anomalies in the mantle by adopting AuSREM as the starting model. Although the tomography results contained in Rawlinson et al. (2013) use a similar dataset, a less sophisticated imaging technique is used which relies on station terms to account for near-surface structure. Consequently, we regard the tomography results presented in the current study to be the most reliable produced so far. That said, the broad-scale variations in wavespeed in regions which overlap with previous studies are in general agreement; this can be regarded as a positive sign when sequential studies are performed in the same region using increasingly large datasets and/or different tomographic imaging schemes (Fishwick and Rawlinson 2012).

Figure 2.9 shows a 120-km-depth slice with crustal element and gravity and magnetic lineament information from Fig. 2.1 superimposed, as well as several regions of interest marked as A, B, C, D and E. Alongside, this horizontal slice is an E–W cross section taken at 37.5°S, an E–W cross section taken at 32.5°S, and a N–S cross section taken at 144.0°E. Perhaps the most interesting feature of the 3-D model is anomaly A in south-western Victoria which underlies the Quaternary NVP. The NVP is an extensive (~15,000 km^2) intraplate basaltic province which is characterized by extensive lava flows, scoria cones and small shield volcanos of tholeiitic to alkalic composition (Vogel and Keays 1997; Price et al. 1997). It represents the largest and youngest expression of Late Tertiary to Quaternary basaltic volcanism in Australia, with an estimated eruption volume of 20,000 km^3 and ages as recent as 5 Ka (Johnson 1989). Arguably the most outstanding question surrounding the NVP is the nature of the mantle source that gave rise to its existence. There has been no recent extension, nor is there strong evidence for a mantle plume source (Demidjuk et al. 2007). In the latter case, the elongated shape of the province in the E–W direction appears at odds with a plume source lying beneath a rapidly N–NE migrating Australian plate; in fact, over the approximately 4.5 M year duration of the eruptive sequences which characterize the NVP, the Australian plate may have moved over 300 km northwards, assuming current plate motion. Furthermore, Demidjuk et al. (2007) point out that this region of the Australian plate is in a state of compression and that the subdued topographic uplift (order of 100 m) associated with the NVP appears more consistent with a shallow mantle source rather than one which originates deep in the mantle. Figure 2.9 clearly shows the presence of a low-velocity zone beneath the surface outcrop of the NVP, both in plan view and in cross-sectional view. We attribute these lower velocities to the presence of increased temperature and melt in the uppermost mantle that gave rise to the intraplate basaltic province at the surface.

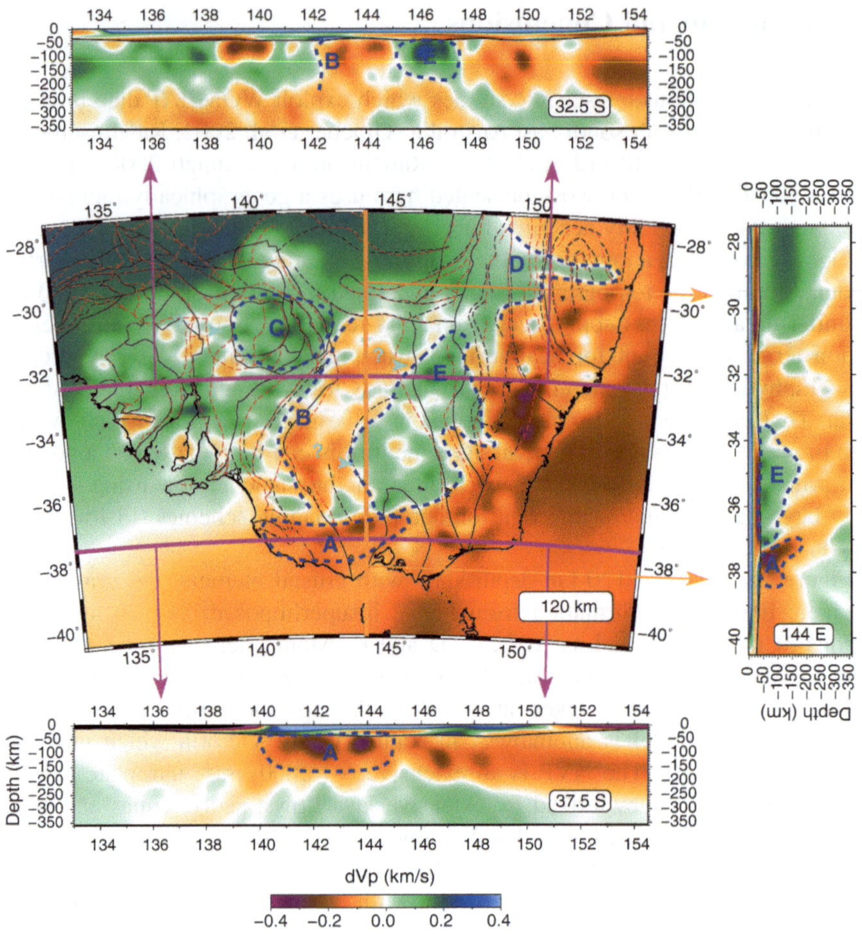

Fig. 2.9 Horizontal and vertical slices through WOMBAT teleseismic model with several features of interest highlighted in *blue*. See text for discussions of anomaly *A*, *B*, *C*, *D* and *E*

Both cross sections conclusively show that the low-velocity zone—and by inference the mantle source of the NVP—does not extend below approximately 200 km depth, which appears to rule out the possibility of a plume source. The N–S cross section (Fig. 2.9, right) shows that the low-velocity zone beneath the NVP appears to dip slightly to the north, something which can also be seen in the 60- and 140-km-depth sections of Fig. 2.7. However, as the checkerboard tests show (Fig. 2.5), the irregularity of the ray path geometry may contribute to this effect.

One alternative to a mantle plume source for the NVP that has been suggested previously is that of edge-driven convection (Demidjuk et al. 2007), in which the topography of the base of the lithosphere combines with its rapid northward migration to drive a localized convection process. More specifically, Demidjuk et al. (2007) argue that, based on surface wave tomography results, a step in lithospheric thickness occurs some 300–400 km north of the NVP with a trend of ~80°E. This inference is based on a N–S contrast from higher to lower velocities at 125 km depth. They use numerical modelling (assuming a 100 km change in lithospheric thickness across the step) to show that a convective cell with an E–W axis can be driven by the existence of such a step in the presence of a northward moving plate. Our results (Fig. 2.9) clearly reveal a N–S change from higher to lower velocities at around 30°–31°S, which could be interpreted as a change from thicker to thinner lithosphere, although a change in lithospheric composition and temperature may also induce such a velocity contrast. Perhaps more significantly, sizable velocity anomalies exist to the south of this "step", which suggest that a simple convection model involving a single marked change in lithospheric thickness probably does not tell the whole story. Further work is evidently needed to combine these new seismic results with those from geochemistry and geodynamic modelling in order to refine our understanding of melting processes in the upper mantle.

In addition to the low-velocity zone beneath the NVP, low velocities also encroach several 100-km inboard of the east coast. Unlike the NVP, these low velocities are likely a consequence of lithospheric thinning associated with the adjacent passive margin, which has brought the sublithospheric mantle closer to the surface. However, it is also possible that extensive Cenozoic volcanism in the region (Johnson 1989), which is a manifestation of processes in the mantle, may have effected the compositional and thermal structure of the lithosphere. Given that much of this volcanism occurred between 10 and 30 Ma, it is unclear whether remnant temperature anomalies or melt may contribute to the lower velocities, as is the case for the much younger NVP.

The extent of Precambrian continental basement beneath the southern Tasmanides has been the ongoing focus of much attention over the last few decades (Rutland 1976; Chappell et al. 1988; VandenBerg 1999; Foster and Gray 2000; Willman et al. 2002; Spaggiari et al. 2004; Glen 2005; Glen et al. 2009; Glen 2013), not only with regard to how far east parts of cratonic Australia—possibly deformed during the formation of the Tasmanides—may underpin the Tasmanides, but the extent to which Precambrian continental fragments may have become entrained within the Palaeozoic orogenic process. Feature B in Fig. 2.9, which highlights a W–E contrast from higher to lower velocities, has a very similar trend to the crustal element boundary in Fig. 2.1, and we interpret this change to mark the approximate location of the eastern boundary of the Delamerian Orogen in the upper mantle. It has been argued that the origin of the substrate in this south-east region of the Delamerian Orogen is dominantly Palaeozoic oceanic (Glen 2005, 2013; Foden et al. 2006), yet at upper mantle depths it appears to have a different seismic structure to the adjoining Lachlan Orogen, which is widely regarded as Palaeozoic oceanic in origin (Foster and Gray 2000, Spaggiari et al. 2003; Foster et al. 2009; Glen 2013). This difference could be attributed to thicker lithosphere beneath the Delamerian Orogen compared to the Lachlan Orogen, and/or involve some compositional change. Handler and Bennet (2001) use Re–Os isotopic data from spinel peridotite xenoliths to show that the eastern edge of the Precambrian interior (at upper mantle depths) of Australia appears to extend as far east as the Delamerian–Lachlan boundary at the surface. Our seismic tomography results appear to be consistent with this finding, although it may be that remnant Proterozoic continental mantle lithosphere—perhaps a product of passive margin formation—underlies or is intermixed with Palaeozoic oceanic crust or lithosphere, respectively. If this was the case, it would help reconcile what is observed at the surface with what is inferred at depth.

Another distinctive feature of the Delamerian Orogen is the Palaeoproterozoic–Mesoproterozoic Curnamona province in the north, which recent evidence suggests formed part of a coherent and contiguous crustal system with the Gawler Craton by the late Palaeoproterozoic to early Mesoproterozoic (Hand et al. 2008). In Fig. 2.8, there is a clear zone of elevated velocity beneath the Curnamona province (denoted as anomaly C), which is characteristic of ancient depleted lithosphere (Dalton and Faul 2010). Although one could argue that translating this anomaly some 250–300 km westward would slot it back into the re-entrant in the Gawler Craton, it is a little difficult to substantiate such a claim on the basis of seismic structure alone.

North of about 30°S, and eastwards as far as 147°E, the lithosphere at 120 km depth (Fig. 2.9) is characterized by the same high velocities that are found throughout much of cratonic central and Western Australia. This strongly implies that the Thomson Orogen is underlain by Precambrian continental crust, a result that appears consistent with the recent results of Glen et al. (2013), who argue that the lower crust beneath the southern Thomson Orogen comprises continental crust that is older than 580 Ma. Anomaly D in Fig. 2.9, which lies to the east of the high-velocity substrate of the southern Thomson Orogen, indicates that a high-velocity salient protrudes beneath the New England Orogen and appears to terminate beneath the eastern edge of the New England oroclines. Whether this "finger" is actually so narrow in the N–S direction is difficult to answer due to the lack of resolution in the northernmost part of the model, but given the much lower velocities to the south, there is evidently a very strong N–S change in lithospheric character that takes place beneath the New England oroclines. The New England oroclines were formed between 310–230 Ma as a result of deformation applied to a pre-Permian convergent margin assemblage (Cawood et al. 2011; Rosenbaum 2012). The exact mechanism of formation is still unclear, with several recent tectonic models invoking different convergent margin behaviour to explain the existence of a northern and southern oroclinal fold pair. For example, Cawood et al. (2011) propose a model in which buckling about a vertical axis is accomplished by having the southern part of the convergent margin elements moving northwards as a result of oblique sinistral strike-slip motion between the Pacific and Gondwana plates, with the northern part pinned relative to cratonic Gondwana. Other models appeal to irregularities along strike of the convergent margin, such as amplified buckle folds (Glen and Roberts 2012) or differential subduction rollback (Rosenbaum 2012). It is unclear whether the spatial correlation of the New England oroclines and the high-velocity salient is coincidental or provides insight into the localization of the oroclines. Indeed, if the higher velocities do point to possibly thicker and stronger lithosphere, then this may have the opposite effect of resisting such severe deformation. Of greater certainty is that anomaly D in Fig. 2.9 represents the presence of Precambrian lithosphere of continental origin beneath the New England Orogen. This has been inferred to exist on the basis of Re–OS analysis of peridotitic xenoliths (Powell and O'Reilly 2007), seismic reflection data (Finlayson 1993) and Geochronology (Hensel et al. 1985).

Anomaly E in Fig. 2.9 spans the western, central and eastern subprovinces of the Lachlan Orogen and in some regions is characterized by velocities as high as those observed beneath the Curnamona province. The similarity in the geometry of the western margin of anomaly E and the east Delamerian margin indicated by transition zone B in Fig. 2.9 raises the possibility that the two regions were once joined, in which case anomaly E may represent lithosphere with a similar provenance to the eastern Delamerian Orogen. In the study by Hallet et al. (2005), the authors use potential field data to identify a new zone within the Lachlan Orogen which they call the Hay–Booligal Zone, and others have inferred to be a fragment of crust of continental origin (e.g. Musgrave and Rawlinson 2010). The southern half of anomaly E corresponds almost exactly with the location of the Hay–Booligal Zone at the surface; for instance, the western edge of the anomaly almost coincides with the curved magnetic lineament (black dashed line) which marks the near-surface boundary of the zone. Given the strong correlation in shape between the western edge of this anomaly and the geometry of anomaly B, it would appear more likely that the Hay–Booligal Zone is a rifted remnant of the east Gondwana margin (Glen 2013) than a more exotic microcontinent. The northern part of anomaly E lies beneath parts of the Ordovician Macquarie Arc, which various lines of evidence, including a lack of continent-derived detritus and the primitive nature of Pb isotopes, point to it forming as an intra-oceanic arc (Glen et al. 2011). This would rule out a Precambrian continental origin for the northern part of anomaly E; instead, the higher velocities that are observed may be the result of local thickening of oceanic lithosphere and magmatic intrusion.

While it appears that several high-velocity upper mantle bodies beneath the Tasmanides correlate quite well with the presence of crustal blocks of Precambrian continental origin, there is by no means a perfect match, as was demonstrated by the northern part of anomaly E which lies beneath the Macquarie Arc. Another example is the Selwyn Block, which has been inferred to lie beneath central Victoria (Cayley et al. 2002; Cayley 2011); in Fig. 2.9, there is no evidence of such a feature in the upper mantle. This may be due in part to the overprinting effects of the Newer Volcanics province, or the fact that near the southern limit of the WOMBAT array, resolution decreases as a consequence of reduced path coverage. Other possibilities include that the Precambrian mantle lithosphere is no longer present (e.g. due to delamination), or that the composition of the Selwyn Block lithospheric mantle is such that it is not characterized by anomalously high velocities in comparison with the surrounding Lachlan Orogen.

Acknowledgments This work was supported by Australian Research Council Discovery Grant DP120103673.

References

Afonso JC, Fernàndez N, Ranalli G, Griffin WL, Connolly JAD (2008) Integrated geophysical-petrological modelling of the lithospheric-sublithospheric upper mantle: methodology and applications. Geochem Geophys Geosyst 9. doi:10.1029/2007GC001834

Afonso JC, Fullea J, Griffin WL, Yang Y, Jones AG, Connolly JAD, O'Reilly SY (2013a) 3D multi-observable probabilistic inversion for the compositional and thermal structure of the lithosphere and upper mantle. I: a priori petrological information and geophysical observables. J Geophys Res (in press)

Afonso JC, Fullea J, Yang Y, Connolly JAD, Jones AG (2013b) 3D multi-observable probabilistic inversion for the compositional and thermal structure of the lithosphere and upper mantle. II: General methodology and resolution analysis. J Geophys Res (in press)

Belousova E, Preiss W, Schwarz M, Griffin W (2006) Tectonic affinities of the Houghton Inlier, South Australia: U-Pb and Hf-isotope data from zircons in modern stream sediments. Aust J Earth Sci 53:971–989

Berry RF, Steele DA, Meffre S (2008) Proterozoic metamorphism in Tasmania: implications for tectonic reconstructions. Precambr Res 166:387–396

Bolt BA, Niazi M (1964) Dispersion of Rayleigh waves across Australia. Geophys J Royal Astr Soc 9:21–35

Bolt BA, Doyle HA, Sutton DJ (1958) Seismic observations from the 1956 atomic explosions in Australia. Geophys J Royal Astr Soc 1:135–145

Bowman R, Kennett BLN (1990) An investigation of the upper mantle beneath northwestern australia using a hybrid seismograph array. Geophys J Int 101:411–424

Bowman R, Kennett BLN (1993) The velocity structure of the Australian shield from seismic traveltimes. Bull Seism Soc Am 83:25–37

Calvert CR, Walter MR (2000) The Late Neoproterozoic Grassy Group of King Island, Tasmania: correlation and palaeogeographic significance. Precambr Res 100:299–312

Cammarano F, Goes S, Vacher P, Giardini D (2003) Inferring upper-mantle temperatures from seismic velocities. Phys Earth Planet Inter 138:197–222

Cammarano F, Tackley P, Boschi L (2011) Seismic, petrological and geodynamical constraints on thermal and compositional structure of the upper mantle: global thermo-chemical models. Geophys J Int 187:1301–1318

Cawood PA, Pisarevsky SA, Leitch EC (2011) Unraveling the New England orocline, East Gondwana accretionary margin. Tectonics 30. doi:10.1029/2011TC002864

Cayley R (2011) Exotic crustal block accretion to the Eastern Gondwana margin in the Late Cambrian-Tasmania, the Selwyn Block, and implications for the Cambrian-Silurian evolution of the Ross, Delamerian and Lachlan orogens. Gondwana Res 19:628–649

Cayley R, Taylor DH, VandenBerg AHM, Moore DH (2002) Proterozoic—early Palaeozoic rocks and the Tyennan Orogeny in central Victoria: the Selwyn Block and its tectonic implications. Aust J Earth Sci 49:225–254

Chappell BW, White AJR, Hine R (1988) Granite provinces and basement terranes in the Lachlan Fold Belt, Southeastern Australia. Aust J Earth Sci 35:505–521

Clifford P, Greenhalgh S, Houseman G, Graeber F (2008) 3-d seismic tomography of the Adelaide fold belt. Geophys J Int 172:167–186

Clitheroe G, Gudmundsson O, Kennett BLN (2000) The crustal thickness of Australia. J Geophys Res 105:13697–13713

Conor CHH, Preiss WV (2008) Understanding the 1720-1640 Ma Palaeoproterozoic Willyama Supergroup, Curnamona province, Southeastern Australia: Implications for tectonics, basin evolution and ore genesis. Precambr Res 166:297–317

Dalton CA, Faul UH (2010) The oceanic and cratonic upper mantle: Clues from joint interpretation of global velocity and attenuation models. Lithos 120:160–172

De Jersey NI (1946) Seismological evidence bearing on crustal thickness in the southwest Pacific. University of Queensland, papers 3: no 2

Debayle E (1999) *SV*-wave azimuthal anisotropy in the Australian upper mantle: preliminary results from automated Rayleigh waveform inversion. Geophys J Int 137:747–754

Debayle E, Kennett BLN (2000) The Australian continental upper mantle: Structure and deformation inferred from surface waves. J Geophys Res 105:25423–25450

de Kool M, Rawlinson N, Sambridge M (2006) A practical grid based method for tracking multiple refraction and reflection phases in 3D heterogeneous media. Geophys J Int 167:253–270

Demidjuk Z, Turner S, Sandiford M, Rhiannon G, Foden J, Etheridge M (2007) U-series isotope and geodynamic constraints on mantle melting processes beneath the newer volcanic province in south australia. Earth Planet Sci Lett 261:517–533

Direen NG, Crawford AJ (2003) The Tasman Line: where is it, what is it, and is it Australia's Rodinian breakup boundary? Aust J Earth Sci 50:491–502

Doyle HA (1957) Seismic recordings of atomic explosions in australia. Nature 180:132–134

Evans JR, Achauer U (1993) Teleseismic tomography using the ach method: theory and application to continental scale studies. In: Iyer HM, Hirahara K (eds) Seismic tomography: theory and practice. Chapman & Hall, London, pp 319–360

Faul UH, Jackson I (2005) The seismological signature of temperature and grain size variations in the upper mantle. Earth Planet Sci Lett 234:119–134

Fichtner A, Trampert J (2011) Resolution analysis in full waveform inversion. Geophys J Int 187:1604–1624

Finlayson DM (1993) Crustal architecture across Phanerozoic Australia along the Eromanga-Brisbane geoscience transect: evolution and analogues. Tectonophysics 219:191–211

Finlayson DM, Cull JP, Drummond BJ (1974) Upper mantle structure from the trans-Australia seismic survey (TASS) and other seismic refraction data. J Geol Soc Aust 21:447–458

Finlayson DM, Collins CDN, Denham D (1980) Crustal structure under the Lachlan Fold Belt, Southeastern Australia. Phys Earth Planet Inter 21:321–342

Finlayson DM, Collins CDN, Lock J (1984) P-wave velocity features of the lithosphere under the Eromanga Basin, Eastern Australia, including a prominent MID-crustal (Conrad?) discontinuity. Tectonophysics 101:267–291

Finlayson DM, Collins CDN, Lukaszyk I, Chudyk EC (1998) A transect across Australia's southern margin in the Otway Basin region: crustal architecture and the nature of rifting from wide-angle seismic profiling. Tectonics 288:177–189

Fishwick S, Rawlinson N (2012) 3-D structure of the Australian lithosphere from evolving seismic datasets. Aust J Earth Sci 59:809–826

Fishwick S, Kennett BLN, Reading AM (2005) Contrasts in lithospheric structure within the Australian craton—insights from surface wave tomography. Earth Planet Sci Lett 231:163–176

Fishwick S, Heintz M, Kennett BLN, Reading AM, Yoshizawa K (2008) Steps in lithospheric thickness within Eastern Australia, evidence from surface wave tomography. Tectonics 27. doi:10.1029/2007TC002116

Foden J, Sandiford M, Dougherty-Page J, Williams I (1999) Geochemistry and geochronology of the Rathjen Gneiss: implications for the early tectonic evolution of the Delamerian Orogen. Aust J Earth Sci 46:377–389

Foden J, Elburg MA, Dougherty-Page J, Burtt A (2006) The timing and duration of the Delamerian Orogeny: correlation with the Ross Orogen and implications for Gondwana assembly. J Geol 114:189–210

Foster DA, Gray DR (2000) Evolution and structure of the Lachlan Fold Belt (Orogen) of Eastern Australia. Ann Rev Earth Planet Sci 28:47–80

Foster DA, Gray DR, Spaggiari CV (2005) Timing of subduction and exhumation along the Cambrian East Gondwana margin and formation of Paleozoic back arc basins. Geol Soc Am Bull 117:105–116

Foster DA, Gray DR, Spaggiari C, Kamenov G, Bierlein FP (2009) Palaeozoic Lachlan orogen, Australia; accretion and construction of continental crust in a marginal ocean setting: isotopic evidence from Cambrian metavolcanic rocks. Geol Soc Lond Spec Publ 318:329–349

Frederiksen AW, Bostock MG, VanDecar JC, Cassidy JF (1998) Seismic structure of the upper mantle beneath the northern Canadian Cordillera from teleseismic traveltime inversion. Tectonophysics 294:43–55

Gaina C, Müller D, Royer JY, Stock J, Hardebeck J, Symonds P (1998) The tectonic history of the Tasman Sea: A puzzle with 13 pieces. J Geophys Res 103:12413–12433

Gibson GM, Morse MP, Ireland TR, Nayak GK (2011) Arc-continent collision and orogenesis in western Tasmanides: Insights from reactivated basement structures and formation of an ocean-continent transform boundary off western Tasmania. Gondwana Res 19:608–627

Glen RA (2005) The Tasmanides of Eastern Australia. In: Vaughan APM, Leat PT, Pankhurst RJ (eds) Terrane processes at the margins of Gondwana. Geological Society, London, pp 23–96

Glen RA (2013) Refining accretionary orogen models for the Tasmanides of eastern Australia. Aust J Earth Sci (in press)

Glen RA, Roberts J (2012) Formation of Oroclines in the New England Orogen, Eastern Australia. J Virtual Explorer 43. doi:10.3809/jvirtex.2012.00305

Glen RA, Scheibner E, Vandenberg AHM (1992) Paleozoic intraplate escape tectonics in Gondwanaland and major strike-slip duplication in the Lachlan Orogen of South-eastern Australia. Geology 20:795–798

Glen RA, Percival IG, Quinn CD (2009) Ordovician continental margin terranes in the Lachlan Orogen, Australia: implications for tectonics in an accretionary orogen along the east Gondwana margin. Tectonics 28:doi:10.1029/2009TC002446

Glen RA, Saeed CDA, Quinn Griffin WL (2011) U-Pb and Hf isotope data from zircons in the Macquarie Arc, Lachlan Orogen: implications for arc evolution and Ordovician palaeogeography along part of the east Gondwana margin. Gondwana Res 19:670–685

Glen RA, Quinn CD, Cooke DR (2012) The Macquarie Arc, Lachlan orogen, New South Wales; its evolution, tectonic setting and mineral deposits. Episodes 35:177–186

Glen RA, Korsch RJ, Hegarty R, Costello RD, Saeed A, Poudjom Djomani RD, Costello RD, Griffin W (2013) Geodynamic significance of the boundary between the Thomson Orogen and the Lachlan Orogen, northwestern New South Wales and the implications for Tasmanide tectonics. Austr J Earth Sci (submitted)

Graeber FM, Houseman GA, Greenhalgh SA (2002) Regional teleseismic tomography of the western Lachlan Orogen and the Newer Volcanic province, southeast Australia. Geophys J Int 149:249–266

Griffin WL, O'Reilly SY, Ryan CG, Gaul O, Ionov DA (1998) Secular variation in the composition of subcontinental lithospheric mantle: geophysical and geodynamic implications. In: Braun J, Dooley J, Goleby B, van der Hilst R, Kllotwijk C (eds) Structure and evolution of the Australian continent, vol 26. American Geophysical Union Geodynamic Series, pp 1–26

Gunn PJ, Mackey TE, Yeates AN, Richardson RG, Seymour DB, McClenaghan MP, Calver CR, Roach MJ (1997) The basement elements of Tasmania. Explor Geophys 28:225–231

Hallet M, Vassallo J, Glen R, Webster S (2005) Murray–Riverina region: an interpretation of bedrock Palaeozoic geology based on geophysical data. Q Notes Geol Surv NSW 118:1–16

Hand M, Reid A, Szpunar M, Direen n, Wade B, Payne j, Barovich K (2008) Crustal architecture during the early Mesoproterozoic Hiltaba-related mineralisation event: are the Gawler Range Volcanics a foreland basin fill? MESA J 51:19–24

Handler MR, Bennet VC (2001) Constraining continental structure by integrating Os isotopic ages of lithospheric mantle with geophysical and crustal data: an example from Southeastern Australia. Tectonics 20:177–188

Heintz M, Kennett BLN (2005) Continental scale shear-wave splitting analysis: investigation of seismic anisotropy underneath the Australian continent. Earth Planet Sci Lett 236:106–119

Hensel HD, Mcculloch MT, Chappell BW (1985) The New England Batholith: constraints on its derivation from Nd and Sr isotopic studies of granitoids and country rocks. Geochim Cosmochim Acta 49:369–384

Hill D (1951) Geology. In: Mack G (ed) Handbook of Queensland. Australian Association for the Advancement of Science, Brisbane, pp 13–24

Johnson BD (1973) A time term analysis of the data obtained during the Bass Strait upper mantle project (operation BUMP). Bull Aust Soc Explor Geophys 4:15–20
Johnson RW (1989) Intraplate volcanism in Eastern Australia and New Zealand. Cambridge University Press, New York
Kennett BLN (2003) Seismic structure in the mantle beneath Australia. In: Hillis RR, Müller RD (eds) The evolution and dynamics of the Australian PLATE, pp 7–23
Kennett BLN, Abdullah A (2011) Seismic wave attenuation beneath the Australasian region. Aust J Earth Sci 58:285–295
Kennett BLN, Bowman JR (1990) The structure and heterogeneity of the upper mantle. Phys Earth Planet Inter 59:134–144
Kennett BLN, Salmon M (2012) AuSREM: Australian seismological reference model. Aust J Earth Sci 59:1091–1103
Kennett BLN, Sambridge MS, Williamson PR (1988) Subspace methods for large scale inverse problems involving multiple parameter classes. Geophys J 94:237–247
Kennett BLN, Engdahl ER, Buland R (1995) Constraints on seismic velocities in the earth from traveltimes. Geophys J Int 122:108–124
Kennett BLN, Fishwick S, Reading AM, Rawlinson N (2004) Contrasts in mantle structure beneath Australia: relation to Tasman Lines? Aust J Earth Sci 51:563–569
Kennett BLN, Fichtner A, Fishwick S, Yoshizawa K (2013) Australian seismological reference model (AuSREM): mantle component. Geophys J Int 192:871–887
Khan A, Connolly JAD, Maclennan J, Mosegaard K (2007) Joint inversion of seismic and gravity data for lunar composition and thermal state. Geophys J Int 168:243–258
Khan A, Boschi L, Connolly JAD (2011) Mapping the Earths thermochemical and anisotropic structure using global surface wave data. J Geophys Res 116. doi:10.1029/2010JB007828
Koketsu K, Sekine S (1998) Pseudo-bending method for three-dimensional seismic ray tracing in a spherical earth with discontinuities. Geophys J Int 132:339–346
Lambeck K, Penny C (1984) Teleseismic traveltime anomalies and crustal structure in central Australia. Phys Earth Planet Inter 34:46–56
Lambeck K, Stephenson R (1986) The post-Palaeozoic uplift history of Southeastern Australia. Aust J Earth Sci 33:253–270
Lei J, Zhao D (2007) Teleseismic P-wave tomography and the upper mantle structure of the central Tien Shan orogenic belt. Phys Earth Planet Inter 162:165–185
Lévêque JJ, Rivera L, Wittlinger G (1993) On the use of the checker-board test to assess the resolution of tomographic inversions. Geophys J Int 115:313–318
Li ZX (2001) An outline of the palaeogeographic evolution of the Australasian region since the beginning of the Neoproterozoic. Earth Sci Rev 53:237–277
Li ZX, Baillie PA, Powell CM (1997) Relationship between northwestern tasmania and East Gondwanaland in the late cambrian/early ordovician: paleomagnetic evidence. Tectonics 16:161–171
Lister GS, Etheridge MA (1989) Detachment models for uplift and volcanism in the Eastern Highlands, and their application to the origin of passive margin mountains. In: Johnson RW (ed) Intraplate Volcanism in Eastern Australia and New Zealand. Cambridge University Press, New York, pp 297–312
Lister GS, Ethridge MA, Symonds PA (1986) Detachment faulting and the evolution of passive continental margins. Geology 14:246–250
Lister GS, Ethridge MA, Symonds PA (1991) Detachment models for the formation of passive continental margins. Tectonics 10:1038–1064
MacCarthy J, Brochers B, Aster R (2011) Efficient stochastic estimation of the model resolution matrix diagonal and generalized cross-validation for large geophysical inverse problems. J Geophys Res 116. doi:10.1029/2011JB008234
Martin M, Ritter JRR (2005) High-resolution teleseismic body-wave tomography beneath SE Romania—I. Implications for the three-dimensional versus one-dimensional crustal

correction strategies with a new crustal velocity model. Geophys J Int 162:448–460 (the CALIXTO working group)

Musgrave R, Rawlinson N (2010) Linking the upper crust to the upper mantle: comparison of teleseismic tomography with long-wavelength features of the gravity and magnetic fields of Southeastern Australia. Explor Geophys 41:155–162

Nolet G (2008) A breviary of seismic tomography: imaging the interior of the earth and sun. Cambridge University Press, Cambridge

Powell W, O'Reilly S (2007) Metasomatism and sulfide mobility in lithospheric mantle beneath Eastern Australia: implications for mantle ReOs chronology. Lithos 94:132–147

Price RC, Gray CM, Frey FA (1997) Strontium isotopic and trace element heterogeneity in the plains basalts of the newer Volcanic province, Victoria, Australia. Geochim Cosmochim Acta 61:171–192

Rawlinson N, Fishwick S (2012) Seismic structure of the Southeast Australian lithosphere from surface and body wave tomography. Tectonophysics 572:111–122

Rawlinson N, Houseman GA, Collins CDN, Drummond BJ (2001) New evidence of Tasmania's tectonic history from a novel seismic experiment. Geophys Res Lett 28:3337–3340

Rawlinson N, Kennett BLN (2004) Rapid estimation of relative and absolute delay times across a network by adaptive stacking. Geophys J Int 157:332–340

Rawlinson N, Kennett BLN (2008) Teleseismic tomography of the upper mantle beneath the southern Lachan Orogen, Australia. Phys Earth Planet Inter 167:84–97

Rawlinson N, Kennett BLN, Heintz M (2006a) Insights into the structure of the upper mantle beneath the Murray Basin from 3D teleseismic tomography. Austr J Earth Sci 53:595–604

Rawlinson N, Reading AM, Kennett BLN (2006b) Lithospheric structure of Tasmania from a novel form of teleseismic tomography. J Geophys Res 111. doi:10.1029/2005JB003803

Rawlinson N, Kennett B, Vanacore E, Glen R, Fishwick S (2011) The structure of the upper mantle beneath the Delamerian and Lachlan orogens from simultaneous inversion of multiple teleseismic datasets. Gondwana Res 19:788–799

Rawlinson N, Pozgay S, Fishwick S (2010a) Seismic tomography: a window into deep Earth. Phys Earth Planet Inter 178:101–135

Rawlinson N, Salmon N, Kennett BLN (2013) Transportable seismic array tomography in Southeast Australia: illuminating the transition from Proterozoic to Phanerozoic lithosphere. Lithos (submitted)

Rawlinson N, Sambridge M (2004a) Multiple reflection and transmission phases in complex layered media using a multistage fast marching method. Geophysics 69:1338–1350

Rawlinson N, Sambridge M (2004b) Wavefront evolution in strongly heterogeneous layered media using the fast marching method. Geophys J Int 156:631–647

Rawlinson N, Sambridge M, Saygin E (2008) A dynamic objective function technique for generating multiple solution models in seismic tomography. Geophys J Int 174:295–308

Rawlinson N, Tkalčić H, Reading AM (2010b) Structure of the Tasmanian lithosphere from 3-D seismic tomography. Aust J Earth Sci 57:381–394

Rawlinson N, Urvoy M (2006) Simultaneous inversion of active and passive source datasets for 3-D seismic structure with application to Tasmania. Geophys Res Lett 33:doi:10.1029/2006GL028105

Reading AM, Kennett BLN (2003) Lithospheric structure of the Pilbara Craton, Capricorn orogen and Northern Yilgarn craton, Western Australia, from teleseismic receiver functions. Aust J Earth Sci 50:439–445

Ritsema J, Deuss A, van Heijst HJ, Woodhouse JH (2011) S40RTS: a degree-40 shear velocity model for the mantle from new Rayleigh wave dispersion, teleseismic traveltime and normal-mode splitting function measurements. Geophys J Int 184:1223–1236

Rosenbaum G (2012) Oroclines of the southern New England Orogen, eastern Australia. Episodes 35:187–194

Rutland RWR (1976) Orogenic evolution of australia. Earth Sci Rev 12:161–196

Salmon K, Kennett BLN, Saygin E (2012) Australian seismological reference model (AuSREM): crustal component. Geophys J Int 192:190–206

Scheibner E (1974) Fossil fracture zones (transform faults), segmentation and correlation problems in the Tasman Fold Belt system. In: Demead AK, Tweedale GW, Wilson AF (eds) Tasman geosyncline: a symposium in honour of Professor Dorothy Hill, Geological Society of Australia, Brisbane, pp 65–98

Scheibner E, Veevers JJ (2000) Tasman Fold Belt System. In: Veevers JJ (ed) Billion-year earth history of Australia and neighbours in Gondwanaland. GEMOC Press, Macquarie Univ, Sydney, pp 154–234

Sethian JA (1996) A fast marching level set method for monotonically advancing fronts. Proc Nat Acad Sci 93:1591–1595

Shaw RD, Wellman P, Gunn P, Whitaker AJ, Tarlowski C, Morse M (1996) Australian Crustal Elements based on the distribution of geophysical domains (1:5 000 000 scale map; version 2.4, ArcGIS dataset). Geoscience Australia, Canberra

Shaw SW, Flood RH, Pearson NJ (2011) The New England Batholith of eastern Australia: evidence of silicic magma mixing from ^{176}Hf/^{177}Hf ratios. Lithos 126:115–126

Simmons NA, Forte AM, Grand SP (2009) Joint seismic, geodynamic and mineral physical constraints on three-dimensional mantle heterogeneity: implications for the relative importance of thermal versus compositional heterogeneity. Geophys J Int 177:1284–1304

Simons F, Zielhuis A, van der Hilst RD (1999) The deep structure of the Australian continent from surface wave tomography. Lithos 48:17–43

Simons FJ, van der Hilst RD, Montagner JP, Zielhuis A (2002) Multimode Rayleigh wave inversion for heterogeneity and azimuthal anisotropy of the Australian upper mantle. Geophys J Int 151:738–754

Spaggiari CV, Gray DR, Foster DA, McKnight S (2003) Evolution of the boundary between the western and central Lachlan Orogen: implications for Tasmanide tectonics. Aust J Earth Sci 50:725–749

Spaggiari CV, Gray DR, Foster DA (2004) Lachlan Orogen subduction-accretion systematics revisited. Aust J Earth Sci 51:549–553

Sutherland FL (1983) Timing, trace and origin of basaltic migration in eastern Australia. Nature 305:123–126

Sutherland FL, Graham IT, Meffre S, Zwingmann H, Pogson RE (2012) Passive-margin prolonged volcanism, East Australian Plate: outbursts, progressions, plate controls and suggested causes. Aust J Earth Sci 59:983–1005

Tarantola A (1987) Inverse problem theory. Elsevier, Amsterdam

Thomas L (1969) Rayleigh wave dispersion in Australia. Bull Seism Soc Am 59:167–182

Thomson BP (1970) A review of the Precambrian and lower Palaeozoic tectonics of South Australia. Transactions of the Royal Society of South Australia 94:193–221

Trampert J, Fichtner A, Ritsema J (2013) Resolution tests revisited: the power of random numbers. Geophys J Int 192:676–680

Underwood R (1969) A seismic refraction study of the crust and upper mantle in the vicinity of Bass Strait. Aust J Phys 22:573–587

van der Beek PA, Braun J, Lambeck K (1999) Post-palaeozoic uplift history of Southeastern Australia revisited: results from a process-based model of landscape evolution. Aust J Earth Sci 46:157–172

VandenBerg AHM (1999) Timing of orogenic events in the Lachlan orogen. Aust J Earth Sci 46:691–701

van der Hilst R, Kennett B, Christie D, Grant J (1994) Project SKIPPY explores the lithosphere and mantle beneath Australia. EOS, Trans Amer Geophys Union 75:177–181

Vogel DC, Keays RR (1997) The petrogenesis and platinum-group element geochemistry of the newer Volcanic province, Victoria, Australia. Chemical Geology 136:181–204

Wade C, Reid A, Wingate M, Jagodzinski Barovich K (2012) Geochemistry and geochronology of the c. 1585 ma Benagerie Volcanic Suite, southern Australia: relationship to the Gawler Range Volcanics and implications for the petrogenesis of a Mesoproterozoic silicic large igneous province. Precambr Res 206:17–35

Waldhauser F, Lippitsch R, Kissling E, Ansorge J (2002) High-resolution teleseismic tomography of upper-mantle structure using an *a priori* three-dimensional crustal model. Geophys J Int 150:403–414

Wellman P (1976) Gravity trends and the growth of Australia: a tentative correlation. J Geol Soc Aust 23:11–14

Wellman P (1983) Hotspot volcanism in Australia and New Zealand: Cainozoic and mid-Mesozoic. Tectonophysics 96:225–243

Willman CE, VandenBerg AHM, Morand VJ (2002) Evolution of the southeastern Lachlan Fold Belt in Victoria. Aust J Earth Sci 49:271–289

Yoshizawa K, Kennett BLN (2004) Multimode surface wave tomography for the Australian region using a three-stage approach incorporating finite frequency effects. J Geophys Res 109. doi:10.1029/2002JB002254

Zhang H, Thurber CH (2007) Estimating the model resolution matrix for large seismic tomography problems based on Lanczos bidiagonalization with partial reorthogonalization. Geophys J Int 170:337–345

Zhang J, McMechan GA (1995) Estimation of resolution and covariance for large matrix inversions. Geophys J Int 121:409–426

Zielhuis A, van der Hilst RD (1996) Upper-mantle shear velocity beneath eastern Australia from inversion of waveforms from SKIPPY portable arrays. Geophys J Int 127:1–16

Chapter 3
Imaging Mantle Heterogeneity with Upper Mantle Seismic Discontinuities

Nicholas Schmerr

Abstract We use underside reflections of S wave seismic energy arriving as precursors to the seismic phase SS to image the depth and impedance contrast present across mantle discontinuities in the depth range of 230–380 km beneath the Pacific basin. A number of past studies have identified seismic discontinuities at these depths, known as the X-discontinuities, ascribing the interfaces to a variety of mineral physical mechanisms, including the coesite to stishovite phase transition, the formation of hydrous Phase A, and/or the reorganization of orthopyroxene into a C2/c monoclinic structure. Thus, the presence of the X-discontinuity (abbreviated here as the X) may be indicative of the nature of mantle heterogeneity. This study finds discontinuities associated with the X Pacific-wide, with SS precursory reflections present beneath the subduction, hot spots, and ridges. Where detected, the X is at an average depth of 293 $\pm$ 65 km and the precursor amplitudes indicate a mean shear impedance contrast of 2.3 $\pm$ 1.6 %. We model mantle heterogeneity by comparing the depth and the impedance contrasts at the X with predictions for seismic structure from a mineral physics model in which the mantle is considered to be a mechanically mixed bulk assemblage of subducted basalt and harzburgite. In this model, the average mantle composition of the Pacific is fit by a mixture of ~20 % basalt and 80 % harzburgite, roughly consistent with the bulk chemistry of mantle peridotite (18 % basalt, 82 % harzburgite). In some regions beneath the hot spots and subduction, there is evidence for a bulk chemistry enriched in basalt (basalt fraction ~30–35 %), lending evidence to the hypothesis that the mantle is laterally heterogeneous and that dynamics are stirring an enriched component, perhaps from the deep or shallow Earth, into the upper mantle.

Keywords SS precursors · X-discontinuity · Coesite · Stishovite · Mantle heterogeneity · Basalt enrichment

---

N. Schmerr (✉)
Department of Geology, University of Maryland, College Park, MD 20742-4211, USA
e-mail: nschmerr@umd.edu

© Springer International Publishing Switzerland 2015

A. Khan and F. Deschamps (eds.), *The Earth's Heterogeneous Mantle*,
Springer Geophysics, DOI 10.1007/978-3-319-15627-9_3

3.1 Introduction

The mantle of the Earth is characterized by discontinuous vertical changes over depth of shear and compressional-wave speeds and density, called seismic discontinuities. The largest and most well-known seismic discontinuities occur at major changes in material composition, for example, the boundary between the atmosphere and Earth's surface (the free surface), the Mohorovičić (Moho) discontinuity at the base of the mafic crust and top of the ultramafic mantle, or the interface present between the iron of the core and the silicate of the mantle at the core–mantle boundary (CMB). In addition to these major compositional interfaces, there exist a number of weaker seismic discontinuities throughout the silicate mantle that arise from a variety of mechanisms related to changes in mineral structure, the introduction of new mineral phases, the presence or production of melts, or more subtle changes in mantle composition. Many of these weaker boundaries are situated in the upper mantle, the region of the deep Earth extending from the base of the crust (the Moho) to approximately 670 km depth. For simplicity, it is common to refer to each discontinuity by a reference depth, though their actual depths may vary within the Earth. In this chapter, we focus on how upper mantle discontinuities can be used to seismologically interrogate the nature of both thermal and chemical mantle heterogeneity (for reviews, see Bina and Helffrich 1994; Helffrich 2000; Shearer 2000).

Three primary properties describe the nature of a seismic discontinuity, namely depth, sharpness, and impedance contrast (Fig. 3.1). Discontinuity depth is defined simply as the radial position of the boundary from the center of the Earth (or from the surface) and can vary regionally. Some seismic discontinuities are intermittent, meaning that they occur only beneath a few geographic regions, while other boundaries are laterally continuous and global interfaces. The sharpness of a discontinuity is defined as the width of the depth interval of the vertical gradient over which the transition in seismic properties occurs. In general, a sharp seismic discontinuity has a gradient thickness on the order of 1–10 km. The exact definition of sharpness is dependent upon the relative wavelength of seismic energy used to interrogate the boundary and depth interval of the gradient; high-frequency seismic waves (>1 Hz) are sensitive to thinner velocity and density gradients, while lower frequency waves (<0.1 Hz) will be sensitive to thicker changes in velocity and density. Gradients do not have to be linear and may take different forms depending upon the underlying physical mechanism producing the discontinuity. The final parameter that describes a seismic discontinuity is the magnitude of the change in shear and compressional velocities, as well as density across the interface. This change can be positive (increase with depth) or negative (decrease with depth), and is often described using a percentile change in the boundary properties [e.g., seismic velocity where v_1 is the velocity in the top layer and v_2 is the velocity in the underlying layer]:

$$\delta v(\%) = \frac{1}{2} \frac{(v_2 - v_1)}{(v_2 + v_1)} * 100\ \% \tag{3.1}$$

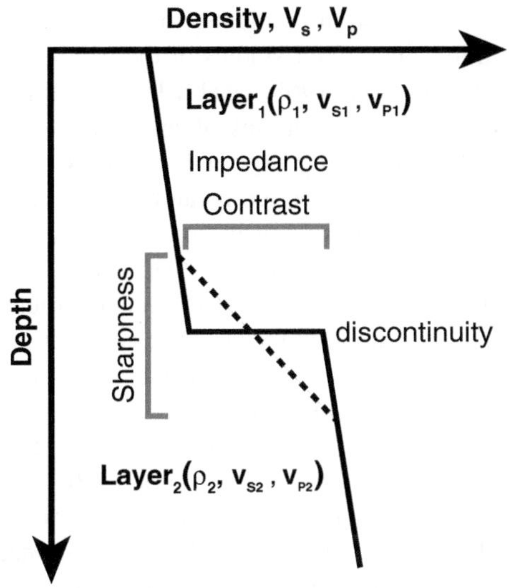

Fig. 3.1 Generalized structure of a seismic discontinuity. The depth of a discontinuity is defined as the radial position of the change in material properties from the center of the Earth. At this boundary, there exists a few percent increase (or decrease) in the density and seismic velocities (Eq. 3.1), resulting in an impedance contrast (Eq. 3.2). For this example, the impedance contrast is positive. The sharpness of the seismic gradient (not necessarily linear) is defined as the depth interval over which the impedance contrast increases (or decreases) and deviates from the background velocity

The transition in these properties can also be described in terms of impedance contrast, where impedance (I) is the product of velocity (v) and density (ρ) across the two layers, shear impedance contrast is the product of density and shear velocity, whereas compressional impedance contrast is the product of density and compressional velocity:

$$I(\%) = \frac{1}{2}\frac{(\rho_2 v_2 - \rho_1 v_1)}{(\rho_2 v_2 + \rho_1 v_1)} * 100\ \% \tag{3.2}$$

A seismic wave can reflect, transmit, or diffract along a seismic interface with the partitioning of energy dependent upon the incidence angle, relative velocity, and density changes at the discontinuity (see Chap. 5.2 in Aki and Richards 2002). The depth, sharpness, and impedance contrast are directly relatable to the underlying mechanisms that give rise to the boundaries.

Early estimates of seismic profiles for the Earth by Jeffreys (1939) revealed abnormally strong velocity and density gradients extending over the depths from 400 to 1000 km, producing triplications of seismic waves around 20° in epicentral distance. This observation was recognized early on by Birch (1952) to be inconsistent with a purely homogeneous and adiabatic interior, and it was postulated that this region represented either a compositional change in the mantle or solid-to-solid phase transformation of mantle minerals to denser and more compact, high-pressure equivalents. Later studies of the triplication of refracted S and P waves by Niazi and Anderson (1965) established that this region was bounded by two first-order seismic discontinuities at approximately 350–450 and 650–700 km depth. The early pioneering experimental work of Ringwood (1962) on analog germinates at low pressure and temperature led to the proposal that materials such as ferromagnesium silicates, including olivine $[Fe,Mg]_2SiO_4$, pyroxenes $[Ca,Na][Mg,Fe,Al][Si,Al]_2O_6$, and garnets $[Fe^{2+},Mg,Ca]_3[Al,Fe^{3+},Si]_2(SiO_4)_3$ transform into the more compact and dense spinel structures at mantle pressures and temperatures. Two mineralogical phase changes in the olivine (ol) system were discovered in pioneering work with laboratory apparatuses capable of achieving mantle temperatures and pressures: a modified spinel form, wadsleyite (wd) at 13.6 GPa, and a true spinel form, ringwoodite (rw) at 19 GPa (e.g., Akimoto and Fujisawa 1966; Ringwood and Major 1966) (Fig. 3.2). An additional higher-pressure phase transformation at 23 GPa of rw to perovskite (pv) [now named bridgmanite (br) as recently approved by the International Mineralogical Association] plus ferropericlase or magnesiowüstite (fp) was revealed by the advent of laser-heated diamond anvil cells (Liu 1976). An excellent review of work on these boundaries can be found in the following references: Bina and Helffrich (1994), Stixrude (1997), Shearer (2000).

Additional deep discontinuities have been proposed in a variety of studies including the 220 km discontinuity (e.g., Lehmann 1961; Dziewonski and Anderson 1981), the X-discontinuities at 250–330 km depth, discussed here (e.g., Revenaugh and Jordan 1991), as well as a 520 km discontinuity (e.g., Shearer 1990). The 220 is attributed to several different mechanisms; the tectosphere hypothesis of Jordan (1975) suggested the 220 is a chemically distinct boundary layer beneath the cratons, while Karato (1992) attributes the 220 to a switch from anisotropy to isotropy in the mantle, resulting from a change in the deformation mechanism from dislocation creep to diffusion creep. Seismically, the 220 is observed regionally, with the most robust observations occurring beneath the continents (e.g., Gu et al. 2001; Deuss and Woodhouse 2002) and a few detections beneath the oceans (Rost and Weber 2001). For the 520 km discontinuity, Ringwood (1975) predicted a discontinuity near this depth arising from the mineralogic phase transformation of wadsleyite to ringwoodite, though a global seismic study by Deuss and Woodhouse (2001) has suggested multiple 520 km discontinuities that may also be attributed to phase changes in the garnet system. The 520 has been argued to be a weak global discontinuity (e.g., Shearer 1991; Flanagan and Shearer 1998b), though in some seismic studies it is only detected beneath the continents (e.g., Gu et al. 1998).

The primary focus of this chapter, the X-discontinuities, or more simply the X, is a set of intermittently observed interfaces with seismic detections at 250–330 km depth beneath the continents and oceanic regions, spanning a wide variety of seismic probes, including PP precursors (Wajeman 1988), ScS reverberations (Revenaugh and Jordan 1991; Williams and Revenaugh 2005; Courtier et al. 2007; Bagley and Revenaugh 2008), receiver functions (Eagar et al. 2010; Shen et al. 2014), and SS

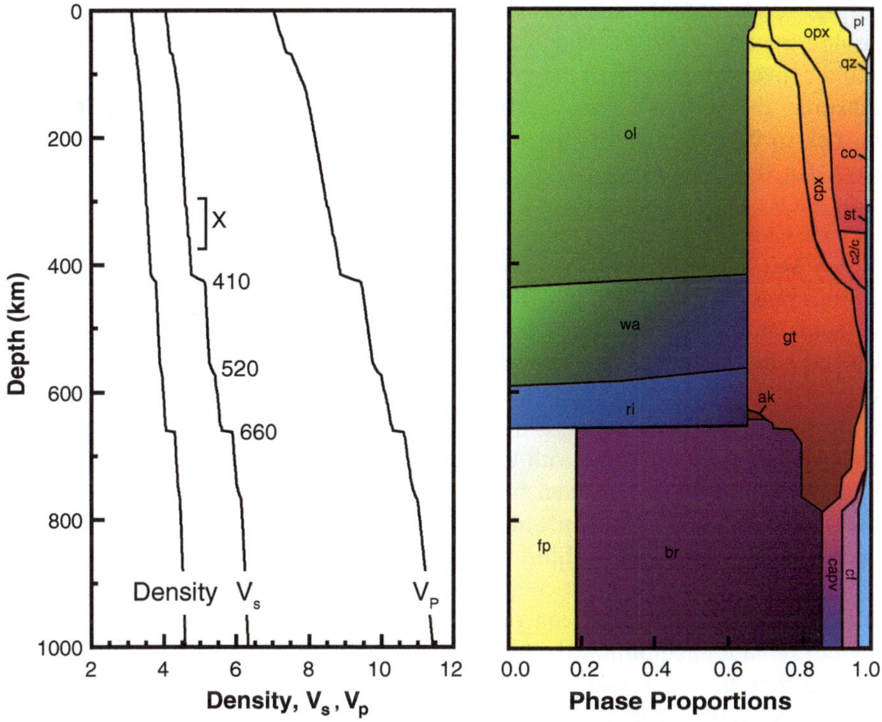

Fig. 3.2 Seismic discontinuities related to phase changes of minerals within a mechanical mixture of basalt and harzburgite. The left panel shows the change in elastic parameters with depth for a theoretical mantle composition with a 20 % basalt fraction and potential temperature of 1800 K (Xu et al. 2008). The right panel shows the phase proportions of minerals at depth for the seismic model. Phase labels are as follows: olivine (ol), wadsleyite (wa), ringwoodite (ri), ferropericlase (fp), Mg-perovskite or bridgmanite (br), majorite garnet (gt), akimotoite (ak), clinopyroxene (cpx), orthopyroxene (opx), plagioclase (pl), quartz (qz), coesite (co), stishovite (st), high-pressure clinopyroxene (c2/c), Ca-perovskite (capv), Ca-ferrite (cf)

precursors (Zhang and Lay 1993; Deuss and Woodhouse 2002; Schmerr et al. 2013). These seismic studies determined that the X possesses a relatively low positive shear impedance contrast (3–5 %) but is seismically sharp, with the gradient constrained to be <5 km across the discontinuities (Revenaugh and Jordan 1991; Zheng et al. 2007; Bagley and Revenaugh 2008). The X is primarily observed beneath the regions of subduction (Zhang and Lay 1993), continents (Wajeman 1988), and a few hot spots (Courtier et al. 2007; Bagley et al. 2009).

Several different mineral physical and petrologic mechanisms have been proposed to explain the X-discontinuities near 300 km depth (for a review, see Williams and Revenaugh 2005). Mechanisms ascribed to the X include the following: the formation of hydrous phase A in subduction zones (Akaogi and Akimoto 1980; Revenaugh and Jordan 1991), a crystallographic transition in pyroxene from orthorhombic to monoclinic C2/c structure (Woodland 1998), and the solid-to-solid phase transformation of coesite to stishovite in subducted basaltic materials (Williams and Revenaugh 2005). At high pressure, this basalt would convert to eclogite in the mantle. The extra SiO_2 producing the co–st phase transition results from the conversion of feldspar (plagioclase) within basaltic rock to both aluminous diopside and garnet via the following reaction pathways: $CaAl_2Si_2O_8$ (plagioclase) + n(Ca,Mg)Si_2O_6 (diopside) → $nCaMgSi_2O_6 \cdot CaAl_2SiO_6$ (aluminous diopside) + SiO_2 and 2(Mg,Fe)SiO_3 (orthopyroxene) + $CaAl_2Si_2O_8$ (plagioclase) → Ca(Mg, Fe)$_2Al_2Si_3O_{12}$ (garnet) + SiO_2 (Green and Ringwood 1967; Williams and Revenaugh 2005). The formation of hydrous phase A is feasible in regions of subduction, although relatively low mantle temperatures are required to stabilize this phase (Revenaugh and Jordan 1991). Similarly, the formation of stishovite requires the presence of unequilibrated basalt (eclogite) in the mantle (Liu et al. 1996). The crystallographic reorganization of pyroxene produces a 0.7 % positive shear impedance contrast (Woodland 1998), which is in poor agreement with the higher contrast detected in previous seismic observations and may not even be the stable phase in the mantle (Zhang et al. 2012). Regardless, thermodynamical modeling of whole mantle compositions for mechanically mixed and equilibrium assemblages of the mantle calculated by Xu et al. (2008) indicates that the coesite to stishovite transformation is present where eclogitized grains of mid-ocean ridge basalt (MORB) and harzburgite (depleted mantle) remain unequilibrated in the mantle due to incomplete mixing processes. For this reason, we investigate the hypothesis that the presence of the X may be used an indicator of lateral variations in mantle heterogeneity.

3.2 Relationship of Discontinuities to Mantle Heterogeneity

The pressures at which phase changes occur in the Earth correspond to the depth at which discontinuities are observed (Fig. 3.2); a set of well-known examples include the ol to wd transition near 410 km depth, and the rw to br + fp dissociation near 660 km depth (Ringwood 1975). The nature of a phase transformation is sensitive to the thermal and chemical state of the mantle, and thus, discontinuities can be used as an indicator of mantle heterogeneity (e.g., Stixrude 1997), (Fig. 3.3). An additional consequence is that regional variation in mantle composition and temperature will produce topography on a phase-change-related discontinuity. The Clapeyron slope of a phase change is defined as the response of the phase change to a perturbation in the surrounding pressure and temperature; a positive Clapeyron slope results in a phase change moving to greater pressure (depth) in response to an increase in temperature, while a negative Clapeyron slope results in the phase change migrating to lower pressure (depth) for an increase in temperature (and vice versa). This is classically recognized in the phase changes of the olivine system, where the ol to wd phase transition is exothermic (Katsura and Ito 1989), giving

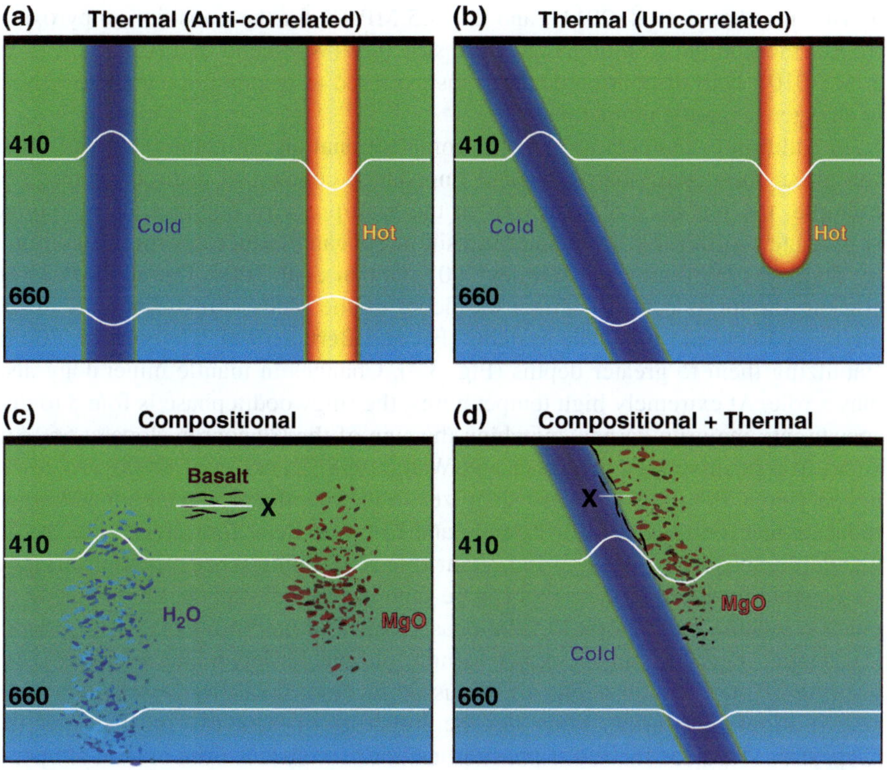

Fig. 3.3 The response of seismic discontinuity depth to changes in mantle temperature and composition arising from mantle dynamics. **a** Continuous thermal anomalies such as plumes and slabs. **b** Dipping and non-continuous anomalies in the mantle producing topography on discontinuities with uncorrelated behavior. **c** Scenarios in which the composition of the mantle varies laterally; hydration (Smyth and Jacobsen 2006), the enrichment of Mg (Fei and Bertka 1999), and a component of unequilibrated basalt (Xu et al. 2008). **d** Combination of thermal and chemical effects; in this case subducting lithosphere entraining depleted and enriched material from the upper mantle into the deep Earth

it a positive Clapeyron slope (dP/dT), while the rw to br + fp phase transition is endothermic (Ito and Takahashi 1989), possessing a negative Clapeyron slope (Fig. 3.2). In the presence of a continuous thermal anomaly across the mantle transition zone (Fig. 3.3), the 410 and 660 km discontinuities will move in opposite directions: toward each other in hot regions, and away from each other in colder areas. This has led to the proposition of using the 410 and 660 km discontinuities, as well as other discontinuities, as mantle thermometers (e.g., Jeanloz and Thompson 1983; Bina and Wood 1987; Revenaugh and Jordan 1989; Stixrude 1997; Helffrich 2000; Cammarano et al. 2003; Lebedev et al. 2003).

In the case of the X, the proposed mechanisms all have positive Clapeyron slopes, 3.2 ± 0.1 MPa/K (1100–1600 K) for the coesite to stishovite phase transition (Akaogi et al. 2011), and 2.9–4.5 MPa/K for the transition in pyroxene from orthorhombic to monoclinic C2/c structure (Woodland and Angel 1997; Yu et al. 2010); both discontinuities are thus expected to respond in a similar fashion to changes in mantle temperature.

In addition to temperature effects, compositional and mineralogical variations can affect the depth, sharpness, and impedance contrast at a discontinuity. For example, the presence of hydrogen in the high-pressure polymorphs of olivine is predicted to thicken the mantle transition zoneand reduce seismic discontinuity sharpness (Williams and Hemley 2001; Ohtani et al. 2004; Karato 2006). Iron content variations can also affect the depth of the phase transformations in the olivine, garnet, and pyroxene systems (Fei and Bertka 1999; Thybo et al. 2003), stabilizing them to greater depths (Fig. 3.3). Changes in mantle mineralogy also play a role. At extremely high temperatures, the ringwooditephase is found to dissociate into majorite garnet, switching the sign of the Clapeyron slope at 660 km depth to a positive sign (Weidner and Wang 1998; Deuss et al. 2006). At lower temperatures, the ilmenite phase stabilizes in the mantle, producing an additional reflector near 660 km depth (Stixrude and Lithgow-Bertelloni 2007). For the X, mantle composition plays an important role in defining what mechanism(s) are responsible for the interface; this can be seen in Fig. 3.2, the coesite to stishovite phase transition andthe opx to C2/c phase transition are both capable of producing seismic interfaces in the upper mantle, ultimately depending upon the composition of the rocks at depth, as discussed in more detail by Deschamps et al. (2015). Figure 3.4 further illustrates the properties of these two mechanisms; the Clapeyron slope for the co–st phase transition is constant over mantle temperatures, while the Clapeyron slope of the opx to C2/c phase transition has a temperature dependence, steepening at higher mantle potential temperatures (Xu et al. 2008). In addition, the predicted shear impedance contrast for opx to C2/c transition is significantly weaker than the co–st phase transition. For high fractions of harzburgite, the opx to C2/c transition produces a stronger impedance contrast, while at high fractions of basalt, the co–st is the stronger impedance contrast. An interesting consequence of this behavior is the possibility for the presence of multiple discontinuities near the depth of the X. Ultimately, the relative depth and impedance contrast of the X may serve as a probe sensitive to mantle composition and temperature.

Seismic discontinuities can play a role in mantle dynamics, ultimately affecting how compositional and thermal heterogeneities are distributed through the mantle. At present, our understanding of how the mechanisms underlying the X would affect mantle flow is poorly understood. Geodynamical considerations predict that the Clapeyron slopes of discontinuities help to organize the nature of convection, by either assisting or impeding rising, and hot or cold and sinking anomalies. This is exemplified by the exothermic reaction at 410 km depth, which allows materials to pass unimpeded, with the possible exception of rapidly sinking, cold, and dry subducted lithosphere (e.g., Bina and Wood 1987), while the endothermic phase change at 660 km depth serves as an imperfect barrier to mantle flow (Christensen and Yuen 1985; Brunet and Yuen 2000). However, both the Clapeyron slope at the

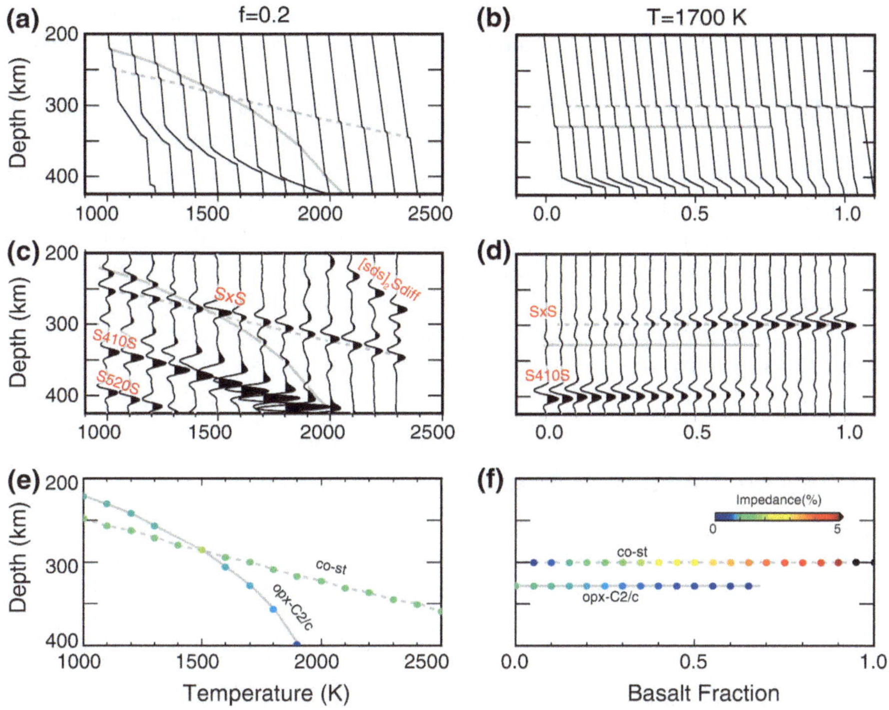

Fig. 3.4 The effect of thermal and compositional heterogeneity on the shear impedance contrast of the X-discontinuities for select models in a mechanically mixed mantle assemblage (Xu et al. 2008). The gray lines indicate the predicted behavior of the co–st (dashed) and opx–C2/c phase transitions (solid). **a** Shear-wave velocity models as a function of mantle potential temperature, showing thermal effects on the depth of the X-discontinuities for a fixed basalt/harzburgite fraction. **b** Shear-wave velocity models as a function of mantle composition, showing compositional effects on the impedance contrast of the X mechanisms; depth is constant across a fixed temperature for the two boundaries. In A and B, shear velocities are fixed to the model velocity at 200 km depth, and offset every 100 K in part A and 0.1 basalt fraction in part B. **c** and **d** Reflectivity synthetic waveforms calculated for the velocity and density models in parts A and B. The S410S and S520S precursors, precursors from the X-discontinuities (SxS), and an non-precursory seismic phase, [s410s, s660 s]Sdiff, are labeled in red. **e** and **f** Shear impedance values calculated from the velocity models in parts A and B. The detection threshold for the SS precursors of the X is ~1 % shear impedance contrast

410 and especially the 660 are predicted to be small, weakening their roles (Bina and Helffrich 1994) and it is unlikely the X will have large effects. The role of the 660 in convection is still non-negligible, as viscosity modeling predicts that the mantle becomes 30–50 times more viscous below 660 km depth (Forte and Woodward 1997) which could potentially change the scale length of convection across this boundary. Geodynamical models incorporating a large viscosity change at the 660 suggests materials can "pool" or stagnate at the 660 before continuing onward to greater or shallower depths in the mantle (Farnetani and Samuel 2005).

This generates a mechanism for isolating subducted materials into the upper mantle, at least temporarily, or perhaps sequestering volatiles or compositional heterogeneities into the mantle transition zone (Bercovici and Karato 2003). Dynamical studies by Richard et al. (2006) posed that material stalled at the base of the transition zone can be remobilized upward into the mantle. Likewise, deeper heterogeneities from the lower mantle can be entrained into the upper mantle (Farnetani and Samuel 2005). Geodynamical modeling of a hot, compositionally heterogeneous mantle plume containing an enriched, denser component finds that this material would tend to pool at 300–410 km depth, before rising to feed a shallower plume conduit (Ballmer et al. 2013). Such models could be influenced by the mechanisms underlying the X, with the implication that detections of the X may be sensitive to upper mantle heterogeneity and the nature of mantle convection processes.

3.3 Seismic Imaging of Mantle Discontinuities

Seismic mapping of mantle discontinuities reveals a wealth of information about the thermal and chemical properties of the mantle, as well as the nature and mode of mantle convection. A number of seismic techniques have been developed for imaging discontinuity structure both at the regional and global scales. The two primary methods are (1) seismic refraction studies of triplicated waves (e.g., Grand and Helmberger 1984; Ryberg et al. 1998; Song et al. 2004) and (2) secondary phases either converted or reflected at a discontinuity, such as, but not limited to: precursors to PKPPKP from underside reflections from discontinuities e.g., Rost et al. (2015), P-to-SV conversions at the discontinuity beneath the receiver, underside reflections occurring as precursors to SS and PP, and top- and bottom-side reflections of ScS reverberations off the discontinuities (for an extensive review see Shearer 2000). Most measurements of global discontinuity topography utilize long-period seismic datasets (>25 s or 0.04 Hz), limiting their resolution to structures on the order of several thousand kilometers (Flanagan and Shearer 1998a; Gu et al. 1998; Gu and Dziewonski 2002). Higher resolution techniques, such as the receiver function method (Ammon 1991), provide much more detailed pictures of discontinuity structure, but are generally limited to regional scales by a lack of sampling away from seismic stations. Regardless, these regional studies reveal evidence for a variety of thermal and chemical anomalies, including high-temperature thermal anomalies beneath the hot spot volcanism (Li et al. 2000), the possible presence of hydrogen in the mantle transition zone (e.g., Song et al. 2004), and a multitude of other chemical anomalies (Zheng et al. 2007). High-resolution studies indicate that the long wavelength global models smooth small-scale regional variations. Furthermore, there is also evidence that other phase changes outside the olivine system can produce seismic discontinuities and that multiple boundaries are possible (Deuss and Woodhouse 2002). Much work remains to elucidate the smaller scale structure of discontinuities around the globe.

To investigate the structure of the X, we use secondary seismic phases arriving as precursors to the seismic phase SS that form as underside reflections of seismic energy from the upper mantle discontinuities (Fig. 3.5). The precursors have previously been used to produce global maps of 410 and 660 km discontinuity topography (e.g., Shearer and Masters 1992; Shearer 1993; Gossler and Kind 1996; Flanagan and Shearer 1998b, 1999; Gu et al. 1998; Gu and Dziewonski 2002; Chambers et al. 2005). SS precursors have also been used to map regional discontinuity topography in study areas spanning several thousand kilometers (e.g., Lee and Grand 1996; Niu et al. 2000, 2002; Deuss and Woodhouse 2002). The SS precursory phase is denoted SdS, where *d* is the depth of the underside reflection, located beneath the SS surface bounce point (halfway between source and receiver). SS precursor studies have established the 410- and 660-km discontinuities as global features, though using relatively long-period energy (periods > 25 s), providing maps with lateral resolution on the order of several thousands of kilometers (e.g., Flanagan and Shearer 1998b; Gu et al. 1998). The precursors have also been used to study the X-discontinuity (Deuss and Woodhouse 2002; Deuss 2009; Schmerr et al. 2013), although the X precursory arrival is much lower amplitude than the precursors from the 410- and 660-km discontinuities.

The precursors are low-amplitude phases, on the order of 1–10 % of the reference SS phase, and require stacking of many hundreds of seismograms to bring the precursory arrivals out of the background noise. Seismic stacking enhances low-amplitude coherent signals relative to the background noise level by using the constructive interference and deconstructive interference. Coherent signals, such as the SS precursors, constructively interfere, enhancing the amplitude of each arrival, while uncorrelated noise will deconstructively interfere, canceling the noise signal out. The relative travel times and amplitudes of the stacked precursors are used to investigate the depth, sharpness, and impedance contrast present at mantle discontinuities. A thorough description of stacking techniques and array methodologies used for obtaining these parameters here can be found in Schmerr and Garnero (2006) and Schmerr and Thomas (2011).

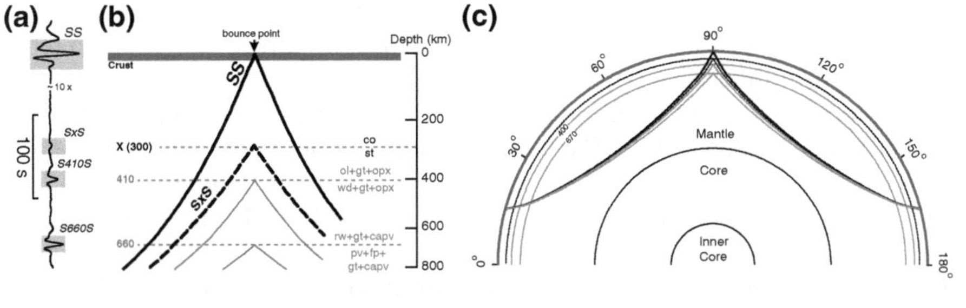

Fig. 3.5 The geometry of SS and its precursors that sample the upper mantle discontinuities. **a** Synthetic seismogram waveforms of SS and the precursors (Fuchs and Müller 1971). Arrivals from the free surface (SS) and upper mantle discontinuities are highlighted in gray. The waveform is amplified by a factor of 10 before the SS arrival to better illustrate the lower amplitude precursor waveforms. **b** Raypaths of SS and the precursors in the upper mantle. Mineral phases are labeled to the right and follow the convention given in Fig. 3.2c global raypath for a source/event pair at an epicentral distance of 150° showing the paths of SS (*black*), SxS at 300 km depth (*dashed line*) and S410S and S660S (*gray*)

Owing to the wide distribution of sources and seismometers over the globe combined with a fairly broad distance range in which SS can be observed, global and regional bounce point coverage for these phases is relatively dense compared to other seismic phases used to study the discontinuities; thus SS precursory phases are ideal tools for investigating discontinuity structure at a variety of scales.

3.4 Detections of the X-Discontinuity

The Pacific basin is an ideal region to investigate the relationship between thermal and chemical heterogeneity at the X-discontinuity. It is host to a number of presumably deeply seated hot spots (e.g., Hawaii, Louisville, Samoa, and the South Pacific Superswell), an extensive length of mid-ocean spreading ridge (the East Pacific Rise), and is circumscribed by multiple subduction zones. The SS precursors have dense sampling across this entire region, making it possible to look with a high degree of confidence at lateral variations in discontinuity structure.

To observe the X, we use the broadband displacement seismogram dataset of (Schmerr et al. 2010) to search for SS precursors arriving from the depth range of 230–350 km that sample beneath the Pacific basin. This dataset was collected from earthquakes with a moment magnitude >5.8 to ensure sufficient energy for SS arrivals, source depths $\leq$75 km to avoid confusion of sS410S with X precursory reflections, and source–receiver geometries falling in the epicentral distance range of 80–165°. The final dataset consisted of over 130,000 SS waveforms recorded at broadband seismic stations that densely sample the mantle beneath the Pacific Ocean. All seismograms were Butterworth low-pass-filtered with a corner at 10 s to exclude the high-frequency noise band present from 5–8 s. Low-pass-filtered data were required to have a signal-to-noise ratio of >2.5 (the "noise" window is defined by measuring the enveloped maximal amplitude in a 100-s window preceding the theoretical P wave arrival, and "signal" as the enveloped maximal amplitude in a $\pm$50 s window around the theoretical SS arrival), resulting in retention of 17,000 of the highest quality data.

We then use stacking of seismograms to bring coherent arrivals out of the background noise (Shearer 1990). There exist a variety of approaches for studying discontinuity structure using the SS precursors (Deuss and Woodhouse 2002; An et al. 2007; Houser et al. 2008; Lawrence and Shearer 2008), including sensitivity tests and synthetic modeling of precursor behavior in response to 3-D heterogeneities (Chaljub and Tarantola 1997; Zhao and Chevrot 2003; Bai and Ritsema 2013). Here, we implement the slowness stacking and distance exclusion procedures of Schmerr and Garnero (2006), Schmerr and Thomas (2011), which is similar to the methodology of Flanagan and Shearer (1998b), except that epicentral distances with interfering seismic waves are excluded from the stack. Data falling in the distance range of 105–125° and 140–165° are included, as they are separated from any interfering phases, such as ScS660ScS or ScS410ScS and s410ss660sSdiff that could masquerade as SxS (Fig. 3.4). As a further check, stacking is done with data and reflectivity synthetic seismograms (Fuchs and Müller 1971), where synthetics were made for each source and receiver combination, to examine predictions for the waveforms of the stacked results. This approach evaluates stack uncertainty using a bootstrap-resampling algorithm (Efron and Tibshirani 1986) that determines the 95 % confidence level (2σ) in travel time (depth) and amplitude (impedance) of every stack. To investigate regional discontinuity structure, as well as approximate the averaging effects of the SS Fresnel zone, the dataset is organized into geographic bins that stack all data falling within a 1000 km radius of nodes spaced every 500 km across the Pacific.

The stacks of SS precursors are converted to depth using theoretical reflectors introduced into the PREM background velocity model. These depths are corrected for the travel time perturbations introduced by crustal structure, topography/bathymetry using CRUST2.0 (Bassin et al. 2000), and for the differential path of the SS wave in the upper mantle by raytracing through the S40RTS shear-wave tomography model (Ritsema et al. 2011). The impedance contrast is estimated at the observed discontinuity by measuring the relative amplitude of the stacked SS to the precursory stack and corrected for geometric spreading, attenuation, stack defocusing, and other systematic effects using a synthetic seismograms generated for PREM and stacked alongside the dataset. The resulting measurements are translated into discontinuity depth and impedance contrast as well as the associated uncertainties provided by the bootstrap resampling; prior modeling efforts of 10–15 s period waves by Schmerr et al. (2013) indicated that the sharpness of the discontinuity must have a thickness <20 km to resolve the precursor from discontinuities near 300 km depth. The locations and elastic contrasts of precursor observations beneath the Pacific of the X-discontinuity can be found in Fig. 3.6 and are compared to earlier results (Deuss and Woodhouse 2002; Schmerr et al. 2013).

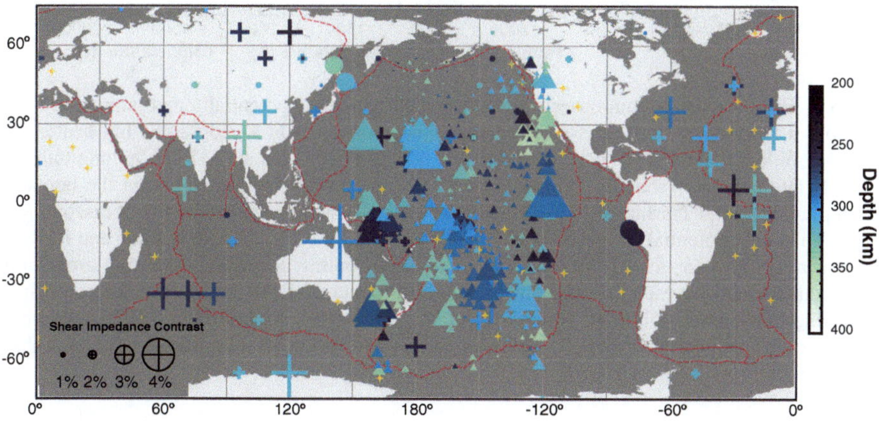

Fig. 3.6 Detections of the X-discontinuity by the SS precursors. The symbols are scaled to the size of the shear impedance contrast at the interface; (Deuss and Woodhouse 2002) (*crosses*), (Schmerr et al. 2013) (*circles*) and this work (*triangles*). The locations of hot spots (*gold stars*) (Courtillot et al. 2003), plate boundaries (*red lines*) (Bird 2003), and continents indicate tectonic and volcanic features. The results here are focused on X observations beneath the Pacific and surrounding plates

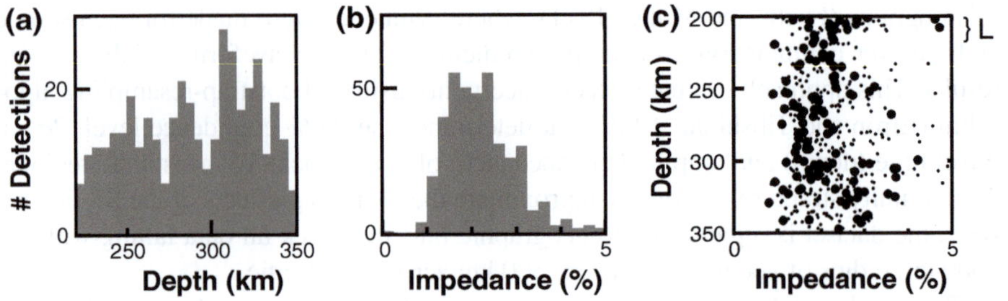

Fig. 3.7 Statistical information on X-discontinuity detections beneath the Pacific basin. Histograms of **a** X-discontinuity depth and **b** shear impedance contrast. **c** Correlation between discontinuity depth and impedance contrast. In part **c**, the *large black circles* indicate detections with a depth uncertainty ≤5 km, while *smaller circles* indicate uncertainties >5 km. The dip in detections at ~230 km depth is used to define the separation of the Lehmann (L) discontinuity from detections the X-discontinuity

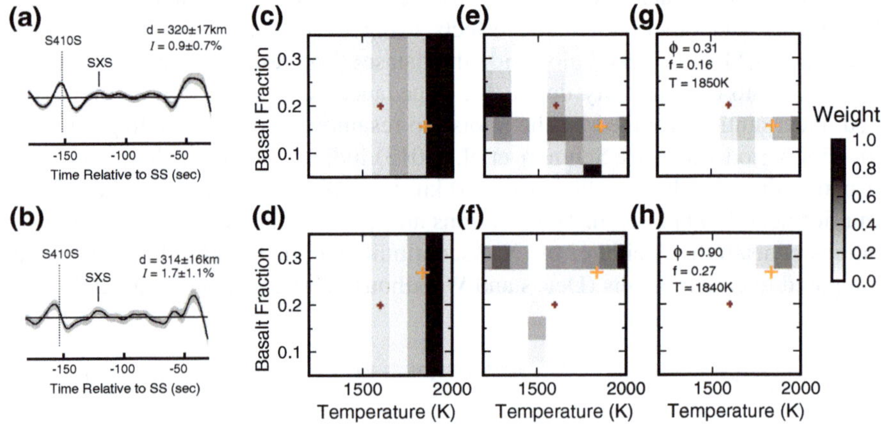

Fig. 3.8 Example fits of data observations to models of mantle potential temperature and basalt fraction. **a** Data stacks for the X-discontinuity precursor (SxS) in a geographic bin centered at -179.11° W longitude and 35.99° N latitude, and **b** for a bin located at -137.08° W longitude and 3.75° N latitude. Gray shading indicates the 95 % (2σ) confidence interval from the bootstrap resample (Efron and Tibshirani 1986). **c** and **d** Inverse chi-square weights for fits of discontinuity depths to mantle potential temperature and basalt fraction for each example stack. **e** and **f** Inverse chi-square weights for fits of discontinuity impedance contrast to mantle potential temperature and basalt fraction for each example stack. **g** and **h** Product of depth and impedance contrast fit, indicating overlapping solutions. In each panel, the reference a priori structure is indicated by a *red cross*, and the weighted best fit solutions for basalt fraction (f), temperature (T), and phase coexistence (ϕ) by the *orange cross*

The SS precursor analysis reveals discontinuities falling throughout the upper mantle beneath the Pacific (Fig. 3.7). The 3–5 % drop in shear impedance contrast that is observed near 230 km depth (Fig. 3.7c) is used to define detections of the Lehmann discontinuity, labeled as "L" in Fig. 3.7c. The L only appears sub-regionally in our study area and is primarily present beneath the mid-ocean ridges and subduction zones, and sporadically elsewhere. Boundaries beneath this cutoff depth are assigned to the X-discontinuity; the X is detected in 76 % of the geographic bins for the Pacific (two example stacks are shown in Fig. 3.8a, b). The average depth of the X falls at 293 ± 65 km (Fig. 3.7a), with a mean shear impedance contrast of 2.3 ± 1.6 % (Fig. 3.7b). Approximately 20 % (n = 77) of the X detections take the form of multiple geographically co-located boundaries that are separated in depth by an average of 77.5 ± 14.5 km. The average depth for the shallower boundary is 251 ± 15 km, while the deeper interface is at 327 ± 14 km depth (2 peaks in Fig. 3.7a). These detections of multiple X are found near the East Pacific Rise, the Big Island of Hawaii, and the Aleutian, Cascadia, and Tonga subduction zones (Fig. 3.9).

This range of depths and shear impedance contrasts falls within the predicted ranges for the coesite to stishovite and opx to C2/c pyroxene as shown in Fig. 3.4. To unravel the complexity in discontinuity depth and shear impedance contrast, the results from our dataset are compared to the mineral physics model of Xu et al. (2008), and lateral variations in discontinuity depth and shear impedance (Fig. 3.9a, b) are assigned to potential temperature and basalt fraction of the

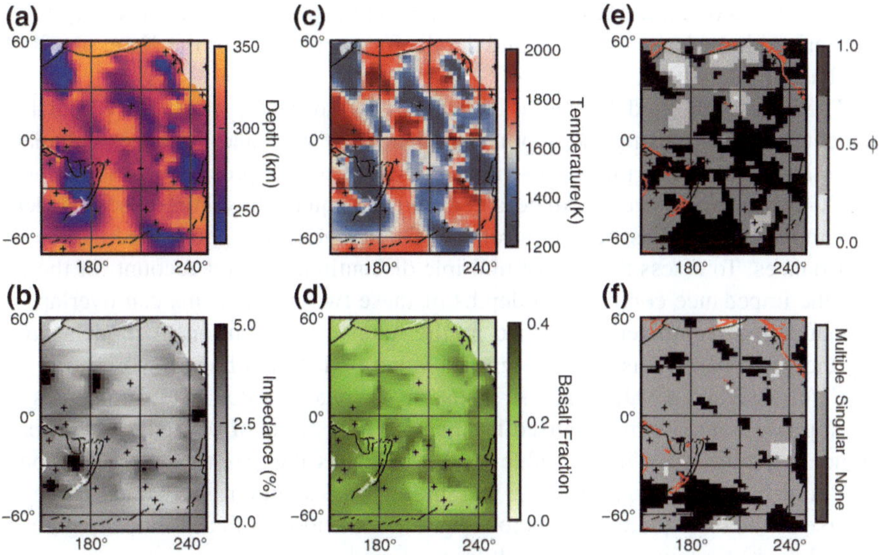

Fig. 3.9 Lateral variations in Pacific mantle temperature and composition as recovered from the mechanical mixture assemblage model of (Xu et al. 2008) (see text). **a** X-discontinuity depth, and **b** impedance contrast across the Pacific study region. **c** Average weighted model solutions for temperatures. **d** Average weighted model solutions for the basalt fraction of the mantle. All maps were smoothed using a minimum curvature algorithm with a tension factor of 0.25 (Wessel and Smith 1998). **e** Average weighted model solutions for ϕ value. **f** The locales of multiple X detections in the Pacific. In parts E and F, the continental coastlines in outlined in *red*. In all panels, hot spots (*black crosses*) (Courtillot et al. 2003) and plate boundaries (Bird 2003) indicate the location of surface tectonic and volcanic features

mantle (Fig. 3.9c, d), respectively. In this modeling approach, the mantle's composition is treated as a mixture of basalt (oceanic crust) and harzburgite (depleted oceanic lithospheric mantle) forming a mechanical mixture or equilibrium assemblage of thermodynamically stable minerals.

3.5 Modeling of Heterogeneity

To model mantle heterogeneity with our results, we use the mechanical mixture assemblage, allowing for both the coesite to stishovite and opx to C2/c pyroxene to coexist and investigate the relationship between our observables (discontinuity depth and impedance contrast) and mantle temperature and composition. We require the mechanical mixture assemblage over the equilibrium model as quartz would react with olivine to form orthopyroxene, eliminating the possibility of having coesite to stishovite phase transition that is necessary to explain the large impedance contrasts observed in the data (Stixrude and Lithgow-Bertelloni 2005). To compare the observations from data to the models, we generate synthetic seismograms using the reflectivity method (Fuchs and Müller 1971). The synthetic travel times and amplitudes are then stacked and used to construct a set of precursor-derived impedance contrasts and depths for the models for the Xu et al. (2008) database (Fig. 3.4b, c).

The best fitting models (and associated errors) to the Xu et al. (2008) database are determined by using the depth, impedance contrast, and variances calculated from the stacks of SS precursors (Fig. 3.8). In every SS precursor observation of the X-discontinuity, we calculate a reduced chi-square fit of the observed depth and impedance (and associated errors) to the synthetically determined depths and impedances. To assess the role of multiple discontinuities, and account for the fact that the impedance contrast and depths of these two mechanisms can overlap, we introduce the parameter ϕ, defined as the relative contribution of models favoring the cost phase transition (assigned a value of 1.0) and those favoring the opx to C2/c (assigned a value of 0.0). Thus, an average ϕ value of 1.0 indicates the co–st phase transition is preferred by the data fit, a value of 0.5 indicates no preference (equally fit), while a value of 0.0 indicates the opx to C2/c transition is preferred by the data. An average weighted model solution is obtained using the inverse of the chi-square value is to weight the likelihood of each model parameter (Fig. 3.8c–h) for temperature, basalt fraction, and ϕ.

In the average weighted model solution (red crosses in Fig. 3.8c–h), we impose the following a priori constraints: (1) The mantle potential temperature must fall within ± 400 K of a value of 1600 K. This reference temperature is consistent with the average discontinuity depth (293 km) and range of mantle potential temperature estimated for the Pacific (Putirka 2005). Owing to the positive amplitude of the Clapeyron slopes, it is implicit that shallower discontinuities will map to models with lower temperatures, while deeper discontinuities will map to higher temperatures (Fig. 3.9a, c). (2) We assume a reference bulk composition of 20 % basalt and 80 % harzburgite, similar to the basalt/harzburgite fraction calculated for the bulk chemistry of mantle peridotite (18 % basalt, 82 % harzburgite) (Ringwood 1975; Hirschmann and Stolper 1996; Xu et al. 2008), and allow basalt fraction to vary from 5 to 35 %. This spread of values is broadly consistent with bulk mantle compositions ranging from depleted mantle harzburgite to that of mantle eclogites (Ritsema et al. 2009; Ballmer et al. 2013; Khan et al. 2015). The final solution(s) for each data stack are obtained by finding the weighted average temperature and basalt fraction(s) for the models that satisfy both of these a priori limits; two example solutions are shown in Fig. 3.8. Note that in most cases, multiple solutions are found for basalt fraction, temperature, and ϕ (Fig. 3.9); this weighted average corresponds to where the various parameters overlap (orange crosses; Fig. 3.8g, h). Across the entire dataset, there are consistent bimodal distributions of preferred parameters. Temperature solutions occur most frequently at 1250 and 1750 K ($\pm$124 K; min = 16 K, max = 239 K), basalt fraction has two modes at 0.18 and 0.25 ($\pm$7 %; min 3 %, max 11 %), and ϕ has two modes 0.6 and 0.8 (Fig. 3.10).

It is important to recognize that there are numerous inherent uncertainties associated with this approach, and the solutions shown here are not unique, but rather represent the field of probable models for the underlying assumptions in Xu et al. (2008) that best match the observables. As can be seen in the examples in Fig. 3.8, and the maps in Fig. 3.9a, c, discontinuity depth largely defines the solution for temperature, while the solution for basalt fraction is closely tied to the measured impedance contrast and hence SS precursor amplitude ratio (Fig. 3.8e, f and Fig. 3.9b, d). Increasing the uncertainties associated with these values in the stacks opens up a much wider variety of models that can satisfy the constraints provided by the Xu et al. (2008) database. We discuss the patterns of these models and the associated uncertainties in more detail in the following section.

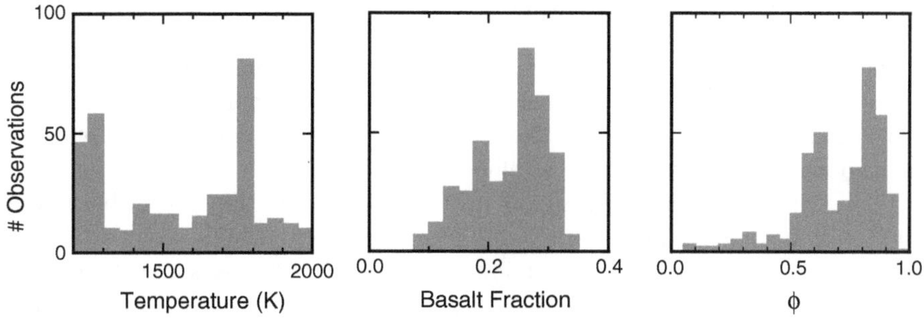

Fig. 3.10 Histograms of the average weighted model solutions for temperature, basalt fraction, and ϕ in the Pacific. Temperatures are bin in 50 K increments, basalt fractions in 0.025 increments, and ϕ in 0.05 increments

3.6 Patterns of Heterogeneity

The purpose of this section is to discuss the patterns of thermal and chemical heterogeneity shown in Fig. 3.9, and to examine the uncertainties that are present in the seismic observations and mineral physical modeling that may bias our interpretation of the X.

A major source of uncertainty in our interpretations derives from the corrections applied to the data prior to conversion to depth and impedance contrasts. The travel time corrections for upper mantle heterogeneity used to correct precursor arrival times are almost certainly underestimated, as many body wave tomography models are overly smooth and do not fully recover the amplitude of velocity anomalies in the mantle (Zhao and Chevrot 2003; Ritsema et al. 2011; Bai and Ritsema 2013; Schaeffer and Lebedev 2015; Rawlinson et al. 2015). This bias in upper mantlevelocity heterogeneity will translate into excess topography on the mantle discontinuity and ultimately an overestimation of the variation in mantle potential temperature (Gu et al. 2003; Deuss 2009; Houser and Williams 2010). For example, the addition of ± 1 s of travel time in the correction for upper mantle structure and crustal thickness and topography introduces ± 3 km of uncertainty in depth, resulting in an additional uncertainty of ± 20–50 K in the temperature estimates. The actual uncertainty in travel time corrections can be as high as 3–4 s, translating to potential temperature uncertainty of ~200 K. The accuracy of depth mapping to temperature is also highly dependent upon the choice for the Clapeyron slope of each phase transition (Stixrude 1997). If the Clapeyron slope used to map depth perturbation to temperature is too low and/or assigned to the wrong phase transition at the interface, this can produce a temperature mismap. In addition, the efficacy of the correction factor applied to precursor amplitude ratios (and conversion to shear impedance) to remove systematic effects such as differential attenuation between SS and the precursor, source radiation pattern, geometric spreading, interfering phases, and stack defocusing is reliant upon the accuracy of the 1-D model used to construct the synthetics for computation of the correction factor. SS precursor amplitude ratios are also affected by 3-D structures including attenuation, discontinuity topography, and scattering effects that are not accounted for in the 1-D synthetics (Chaljub and Tarantola 1997; Zhao and Chevrot 2003; Bai and Ritsema 2013). An increase in the precursor amplitude ratio of 25 % translates to a similar increase in the recovered impedance contrast, and results in the mapping of basalt fraction to a value ~7–8 % higher.

Even with these uncertainties, in many regions there appears to be a weak correlation of our modeled thermal and chemical heterogeneity with surface tectonics and volcanism. This correlation is not unexpected, as subduction zones, ridges, and hot spots are the surface expression of deeper processes that can introduce thermal and chemical heterogeneity into the mantle. Resolution of structure related to these surface features is typically averaged or smoothed by the long-period SS waves sampling the mantle over scale lengths of thousands of kilometers. Synthetic tests show that the SS precursor is more sensitive to long wavelength structure on the mantle discontinuities and would not resolve many of the smaller scale features associated with plume conduits or subducting crust (e.g., Chaljub and Tarantola 1997). In addition, the large Fresnel zone of SS leads to smearing of structure into neighboring regions. Thus, more emphasis should be placed on interpreting the long wavelength variations in structure (>1000 km) over numerous small-scale variations seen in Fig. 3.9.

Beneath the subduction of the southwestern Pacific, the X is detected at shallower depths; this is consistent with the presence of lowered mantle potential temperature in these regions. The modeling also indicates these regions favor a bulk composition that is basalt-enriched (Fig. 3.9e). The presence of the cost phase transition in the vicinity of subduction would result from the introduction of sediments, fluids, metasomatized mantle, and basaltic melts that are emplaced in the mantle wedge and entrained downward during subduction (e.g., Revenaugh and Jordan 1991; Williams and Revenaugh 2005; Zheng et al. 2007; Akaogi et al. 2011). A less coherent pattern of heterogeneity lies beneath the Aleutian and Cascadian subduction zones, with a range of lowered and elevated temperatures and bulk basalt fractions associated with these subduction zones. These areas fall at the margins of our study region, with a geographic bias toward the sub-slab structure and less sampling of the mantle wedges, thus their patterns of heterogeneity are not fully resolved here.

With the hot spots and ridges, the pattern is complex and many locales indicate some thermal perturbation (± 100 K) from the average mantle potential temperature, while others indicate colder temperatures associated with upwelling, and others registering hotter mantle (Fig. 3.10). Indeed, it is intriguing that many of the hot spots appear to fall at colder potential temperatures or along the margins of hotter anomalies. For example, at Hawaii, there is a trend perpendicular to the hot spot track in both the modeled basalt fraction and temperature. To the northeast, there is a lower basalt fraction, lower potential temperatures, and $\phi < 0.5$, indicating the preference of the opx to C2/c pyroxene mechanism in the modeling, while to the southwest, there is a higher basalt fraction, elevated temperatures, and $\phi > 0.5$, indicating the cost phase transition is preferred in the models. This bimodal trend in bulk mantle composition across the Hawaiian hot spot is not unexpected; the hot spot has classically been characterized by two geochemical trends, the Loa and Kea (e.g., Sobolev et al. 2005) that are argued to delineate separate sampling of an enriched and depleted chemical reservoir at depth (Weis et al. 2011). Similar trends have been found at other hot spots through the Pacific (Huang et al. 2011); many hot spots in Fig. 3.9 are associated with plate motion perpendicular trends in bulk mantle composition. However, such models do not explain the large variation in mantle potential temperature.

There are several possibilities that may explain the inconsistency in potential temperature and hot spot location; the geometry of plume conduits may be tilted or non-vertical structures and the underlying upwelling may be deflected into the mantle flow pattern (e.g., Sleep 1990; van Keken and Gable 1995; Nataf 2000; Zhong and Watts 2002; Montelli et al. 2004; Zhao 2004), leading to decorrelation of surface features and deeper structures (An et al. 2007; Houser and Williams

2010; Schmerr et al. 2010). However, it is unlikely the SS precursors would be sensitive to narrow upwellings with small diameters <500 km (Schmerr et al. 2013) that are expected at a plume conduit, or that temperature variation from a plume upwelling could produce such broad features in the upper mantle (Wolfe et al. 2009). Alternatively, this complexity of the X-discontinuity beneath the hot spots may not be the result from a thermal anomaly interacting with mantle phase transitions. For example, the onset of carbonated silicate melting in the underlying mantle is expected to occur at 250–300 km depth (Dasgupta et al. 2013). The depth and impedance contrast of the X for such the onset of carbonated silicate melting would be sensitive to variations in mantle volatile content and melt fraction. Another possible hypothesis is that the X demarcates the base of hot, channelized, finger-like flow in the low-velocity zone (French et al. 2013). The base of such regions would potentially give rise to anisotropic discontinuities oriented parallel to mantle flow; this is a hypothesis difficult to test adding azimuthal coverage across the Pacific (e.g., Rychert et al. 2012).

All these possibilities reveal a primary limitation in our modeling with the Xu et al. (2008) database. The model does not incorporate the effects of small-scale partial melts, volatiles, and/or mantle fabrics, though efforts are underway to add such parameters (e.g., Hier-Majumder and Abbott 2010). For example, dynamical simulations of entrainment of a fertile, presumably subducted basalt component from the deepest mantle, show that compositional zonation can be produced in an upwelling plume, similar to what is observed near Hawaii (Ballmer et al. 2013). Likewise, dynamical simulations of small fractions of melt concentrated by flow in the mantle are still needed to understand the seismic signature of melt and volatile transport in the mantle wedge beneath the subduction zones and mid-ocean ridges (Hirschmann 2010; Dasgupta et al. 2013). Regardless if deeply sourced or entrained into the mantle from the surface, a localized enrichment of basalt (eclogite) or volatiles in the upper mantle (Hirschmann and Stolper 1996; Sobolev et al. 2005) may provide an essential link for combining the geophysical and geochemical observables in the mantle. Thus, the regional results presented here suggest the X-discontinuities provide an important window on the mantle mixing processes, and future work is needed to characterize the global presence, depth, and impedance contrast of these mantle interfaces.

3.7 Conclusions

Mantle discontinuities provide valuable information on the thermal and chemical state of Earth's interior. In particular, the X-discontinuities, if related to phase changes in coesite/stishovite and pyroxene, are capable of providing information on the temperature and chemical state of the upper mantle. This investigation of the X-discontinuities used underside reflected shear-waves, arriving as precursors to the seismic phase SS, to reveal that X-discontinuities are a consistent feature of the upper mantle beneath the Pacific Ocean. These boundaries are present beneath the multiple tectonic and volcanic features such as subduction zones, hot spots, and ridges with an average depth of 293 ± 65 km and mean shear impedance contrast of 2.3 ± 1.6 %. By comparing our results to a mineral physics model of mantle composition and temperature (Xu et al. 2008), we find that the depth and shear impedance contrast of the X-discontinuities are consistent with predictions for the coesite to stishovite phase transition and the transition of orthopyroxene to C2/c pyroxene, though in some regions other mechanisms, such as volatiles and/or melt, may warrant further investigation. The presence of the coesite to stishovite phase transition in the mantle is indicative a bulk composition with 20–30 % basalt mechanically mixed into the upper mantle. The presence of excess basalt implies that the mantle is compositionally heterogeneous, i.e., the metasomatism of the mantle at subduction zones and sinking of slabs are entrained/transported back to the surface, either through hot spots returning slabs into the upper mantle or whole mantle convection. Alternatively, the X-discontinuities may be indicative of carbonated silicate melting or the presence of volatiles in the upper mantle. Thus, the X-discontinuities are a powerful probe for accessing mantle chemical heterogeneity, and further work is needed to globally map the topography, sharpness, and impedance contrast of these mantle interfaces.

Acknowledgments This work was supported by NSF-EAGER 1247608. The author wishes to thank the editor Amir Khan, as well as Sebastian Rost and 2 anonymous reviewers for providing extensive constructive feedback that greatly improved the quality of the manuscript.

References

Akaogi M, Akimoto SI (1980) High-pressure stability of a dense hydrous magnesian silicate Mg23Si8O42H6 and some geophysical implications. J Geophys Res 85(NB12):6944–6948

Akaogi M, Oohata M, Kojitani H, Kawaji H (2011) Thermodynamic properties of stishovite by low-temperature heat capacity measurements and the coesite-stishovite transition boundary. Am Mineral 96(8–9):1325–1330

Aki K, Richards P (2002) Quantitative seismology, 2nd edn. University Science Books, Sausalito, 700 pp

Akimoto S, Fujisawa H (1966) Olivine-spinel transition in system Mg_2SiO_4-Fe_2SiO_4 at 800 °C. Earth Planet Sci Lett 1(4):237–240

Ammon CJ (1991) The isolation of receiver effects from teleseismic p-wave-forms. Bull Seismol Soc Am 81(6):2504–2510

An Y, Gu YJ, Sacchi MD (2007) Imaging mantle discontinuities using least squares Radon transform. J Geophys Res Solid Earth 112(B10)

Bagley B, Revenaugh J (2008) Upper mantle seismic shear discontinuities of the pacific. J Geophys Res Solid Earth 113(B12)

Bagley B, Courtier AM, Revenaugh J (2009) Melting in the deep upper mantle oceanward of the Honshu slab. Phys Earth Planet Inter 175(3–4):137–144

Bai L, Ritsema J (2013) The effect of large-scale shear-velocity heterogeneity on SS precursor amplitudes. Geophys Res Lett 40(23):6054–6058

Ballmer MD, Ito G, Wolfe CJ, Solomon SC (2013) Double layering of a thermochemical plume in the upper mantle beneath Hawaii. Earth Planet Sci Lett 376:155–164

Bassin C, Laske G, Masters G (2000) The current limits of resolution for surface wave tomography in North America. Eos Trans 81:F897

Bercovici D, Karato S (2003) Whole-mantle convection and the transition-zone water filter. Nature 425(6953):39–44

Bina CR, Wood BJ (1987) Olivine-spinel transitions—experimental and thermodynamic constraints and implications for the nature of the 400-Km seismic discontinuity. J Geophys Res Solid Earth Planets 92(B6):4853–4866

Bina CR, Helffrich G (1994) Phase-transition Clapeyron slopes and transition zone seismic discontinuity topography. J Geophys Res Solid Earth 99(B8):15853–15860

Birch F (1952) Elasticity and constitution of the Earth's interior. J Geophys Res 37(2):227–286

Bird P (2003) An updated digital model of plate boundaries. Geochem Geophys Geosystems 4. doi: 10.1029/2001GC000252

Brunet D, Yuen DA (2000) Mantle plumes pinched in the transition zone. Earth Planet Sci Lett 178(1–2):13–27

Cammarano F, Goes S, Vacher P, Giardini D (2003) Inferring upper-mantle temperatures from seismic velocities. Phys Earth Planet Inter 138(3–4):197–222

Chaljub E, Tarantola A (1997) Sensitivity of SS precursors to topography on the upper-mantle 660-km discontinuity. Geophys Res Lett 24(21):2613–2616

Chambers K, Deuss A, Woodhouse JH (2005) Reflectivity of the 410-km discontinuity from PP and SS precursors. J Geophys Res Solid Earth 110(B2):B02301. doi:10.1029/02004JB003345

Christensen UR, Yuen DA (1985) Layered convection induced by phase-transitions. J Geophys Res Solid Earth Planets 90(NB12):291–300

Courtier AM, Bagley B, Revenaugh J (2007) Whole mantle discontinuity structure beneath Hawaii. Geophys Res Lett 34(17)

Courtillot V, Davaille A, Besse J, Stock J (2003) Three distinct types of hotspots in the Earth's mantle. Earth Planet Sci Lett 205(3–4):295–308

Dasgupta R, Mallik A, Tsuno K, Withers AC, Hirth G, Hirschmann MM (2013) Carbon-dioxide-rich silicate melt in the Earth's upper mantle. Nature 493(7431):211–222

Deschamps F, Li Y, Tackley PJ (2015) Large-scale thermo-chemical structure of the deep mantle: observations and models, this volume

Deuss A (2009) Global observations of mantle discontinuities using SS and PP precursors. Surv Geophys 30(4–5):301–326

Deuss A, Woodhouse J (2001) Seismic observations of splitting of the mid-transition zone discontinuity in Earth's mantle. Science 294(5541):354–357

Deuss A, Woodhouse JH (2002) A systematic search for mantle discontinuities using SS-precursors. Geophys Res Lett 29(8)

Deuss A, Redfern SAT, Chambers K, Woodhouse JH (2006) The nature of the 660-kilometer discontinuity in Earth's mantle from global seismic observations of PP precursors. Science 311(5758):198–201

Dziewonski AM, Anderson DL (1981) Preliminary reference Earth model. Phys Earth Planet Inter 25(4):297–356

Eagar KC, Fouch MJ, James DE (2010) Receiver function imaging of upper mantle complexity beneath the Pacific Northwest. U.S. Earth Planet Sci Lett 297(1–2):141–153

Efron B, Tibshirani R (1986) Bootstrap methods for standard errors, confidence intervals, and other measures of statistical accuracy. Stat Sci 1(1):54–75

Farnetani CG, Samuel H (2005) Beyond the thermal plume paradigm. Geophys Res Lett 32(7)

Fei Y, Bertka C (1999) Phase transitions in the Earth's mantle and mantle mineralogy. In: Fei Y, Bertka C, Mysen B (eds) Mantle petrology: field observations and high pressure experimentation. The Geochemical Society, Houston, pp 189–207

Flanagan MP, Shearer PM (1998a) Topography on the 410-km seismic velocity discontinuity near subduction zones from stacking of sS, sP, and pP precursors. J Geophys Res Solid Earth 103(B9):21165–21182

Flanagan MP, Shearer PM (1998b) Global mapping of topography on transition zone velocity discontinuities by stacking SS precursors. J Geophys Res Solid Earth 103(B2):2673–2692

Flanagan MP, Shearer PM (1999) A map of topography on the 410-km discontinuity from PP precursors. Geophys Res Lett 26(5):549–552

Forte AM, Woodward RL (1997) Seismic-geodynamic constraints on three-dimensional structure, vertical flow, and heat transfer in the mantle. J Geophys Res Solid Earth 102(B8):17981–17994

French S, Lekic V, Romanowicz B (2013) Waveform tomography reveals channeled flow at the base of the oceanic asthenosphere. Science 342(6155):227–230

Fuchs K, Müller G (1971) Computation of synthetic seismograms with the reflectivity method and comparison with observations. Geophys J Roy Astron Soc 23:417–433

Gossler J, Kind R (1996) Seismic evidence for very deep roots of continents. Earth Planet Sci Lett 138(1–4):1–13

Grand SP, Helmberger DV (1984) Upper mantle shear structure of North-America. Geophys J Roy Astron Soc 76(2):399–438

Green DH, Ringwood AE (1967) An experimental investigation of gabbro to eclogite transformation and its petrological applications. Geochim Cosmochim Acta 31(5):767–833

Gu Y, Dziewonski AM, Agee CB (1998) Global de-correlation of the topography of transition zone discontinuities. Earth Planet Sci Lett 157(1–2):57–67

Gu YJ, Dziewonski AM (2002) Global variability of transition zone thickness. J Geophys Res Solid Earth 107(B7):2135. doi:10.1029/2001JB000489

Gu YJ, Dziewonski AM, Ekstrom G (2001) Preferential detection of the Lehmann discontinuity beneath continents. Geophys Res Lett 28(24):4655–4658

Gu YJ, Dziewonski AM, Ekstrom G (2003) Simultaneous inversion for mantle shear velocity and topography of transition zone discontinuities. Geophys J Int 154(2):559–583

Helffrich G (2000) Topography of the transition zone seismic discontinuities. Rev Geophys 38(1):141–158

Hier-Majumder S, Abbott ME (2010) Influence of dihedral angle on the seismic velocities in partially molten rocks. Earth Planet Sci Lett 299(1–2):23–32

Hirschmann MM (2010) Partial melt in the oceanic low velocity zone. Phys Earth Planet Inter 179(1–2):60–71

Hirschmann MM, Stolper EM (1996) A possible role for garnet pyroxenite in the origin of the "garnet signature" in MORB. Contrib Miner Petrol 124(2):185–208

Houser C, Williams Q (2010) Reconciling pacific 410 and 660 km discontinuity topography, transition zone shear velocity patterns, and mantle phase transitions. Earth Planet Sci Lett 296(3–4):255–266

Houser C, Masters G, Flanagan MP, Shearer PM (2008) Determination and analysis of long-wavelength transition zone structure using SS precursors. Geophys J Int 174:178–194

Huang S, Hall PS, Jackson MG (2011) Geochemical zoning of volcanic chains associated with Pacific hotspots. Nat Geosci 4(12):874–878

Ito E, Takahashi E (1989) Postspinel transformations in the system Mg_2SiO_4-Fe_2SiO_4 and some geophysical implications. J Geophys Res Solid Earth Planets 94(B8):10637–10646

Jeanloz R, Thompson AB (1983) Phase-transitions and mantle discontinuities. Rev Geophys 21(1):51–74

Jeffreys H (1939) The times of P, S and SKS, and the Velocities of P and S. Mon Notations R Astron Soc Geophys Suppl 4:498–533

Jordan TH (1975) Continental tectosphere. Rev Geophys 13(3):1–12

Karato S (1992) On the Lehmann discontinuity. Geophys Res Lett 19(22):2255–2258

Karato S (2006) Remote sensing of hydrogen in Earth's mantle. Water Nominally Anhydrous Miner 62:343–375

Katsura T, Ito E (1989) The system Mg_2SiO_4-Fe_2SiO_4 at high-pressures and temperatures—precise determination of stabilities of olivine, modified spinel, and spinel. J Geophys Res Solid Earth Planets 94(B11):15663–15670

Khan A, Koch S, Shankland TJ, Zunino A, Connolly JAD (2015) Relationships between seismic wave-speed, density, and electrical conductivity beneath australia from seismology, mineralogy, and laboratory-based conductivity profiles, this volume

Lawrence JF, Shearer PM (2008) Imaging mantle transition zone thickness with SdS-SS finite-frequency sensitivity kernels. Geophys J Int 174:143–158
Lebedev S, Chevrot S, van der Hilst RD (2003) Correlation between the shear-speed structure and thickness of the mantle transition zone. Phys Earth Planet Inter 136(1–2):25–40
Lee DK, Grand SP (1996) Depth of the upper mantle discontinuities beneath the East Pacific Rise. Geophys Res Lett 23(23):3369–3372
Lehmann I (1961) S and the structure of the upper mantle. Geophys J R Astron Soc 4:124–138
Li X, Kind R, Priestley K, Sobolev SV, Tilmann F, Yuan X, Weber M (2000) Mapping the Hawaiian plume conduit with converted seismic waves. Nature 405(6789):938–941
Liu L (1976) The high-pressure phases of MgSiO3. Earth Planet Sci Lett 31(2):200–208
Liu J, Topor L, Zhang J, Navrotsky A, Liebermann RC (1996) Calorimetric study of the coesite stishovite transformation and calculation of the phase boundary. Phys Chem Miner 23(1):11–16
Montelli R, Nolet G, Dahlen FA, Masters G, Engdahl ER, Hung SH (2004) Finite-frequency tomography reveals a variety of plumes in the mantle. Science 303(5656):338–343
Nataf HC (2000) Seismic imaging of mantle plumes. Annu Rev Earth Planet Sci 28:391–417
Niazi M, Anderson DL (1965) Upper mantle structure of western North America from apparent velocities of P waves. J Geophys Res 70(18):4633–4640
Niu FL, Inoue H, Suetsugu D, Kanjo K (2000) Seismic evidence for a thinner mantle transition zone beneath the South Pacific Superswell. Geophys Res Lett 27(13):1981–1984
Niu FL, Solomon SC, Silver PG, Suetsugu D, Inoue H (2002) Mantle transition-zone structure beneath the South Pacific Superswell and evidence for a mantle plume underlying the society hotspot. Earth Planet Sci Lett 198(3–4):371–380
Ohtani E, Litasov K, Hosoya T, Kubo T, Kondo T (2004) Water transport into the deep mantle and formation of a hydrous transition zone. Phys Earth Planet Inter 143–44:255–269
Putirka KD (2005) Mantle potential temperatures at Hawaii, Iceland, and the mid-ocean ridge system, as inferred from olivine phenocrysts: Evidence for thermally driven mantle plumes. Geochem Geophys Geosyst 6
Rawlinson N, Kennett BLN, Salmon M, Glen RA (2015) Origin of lateral heterogeneities in the upper mantle beneath south-east australia from seismic tomography, this volume
Revenaugh J, Jordan TH (1989) A study of mantle layering beneath the Western Pacific. J Geophys Res Solid Earth Planets 94(B5):5787–5813
Revenaugh J, Jordan TH (1991) Mantle layering from Scs reverberations 3. The upper mantle. J Geophys Res Solid Earth 96(B12):19781–19810
Richard G, Bercovici D, Karato SI (2006) Slab dehydration in the Earth's mantle transition zone. Earth Planet Sci Lett 251(1–2):156–167
Ringwood AE (1962) A model for the upper mantle. J Geophys Res 67(2):857–867
Ringwood AE (1975) Composition and petrology of the Earth's interior. McGraw-Hill, New York, pp 618
Ringwood AE, Major A (1966) Synthesis of Mg2SiO4-Fe2SiO4 spinel solid solutions. Earth Planet Sci Lett 1(4):241–245
Ritsema J, Xu WB, Stixrude L, Lithgow-Bertelloni C (2009) Estimates of the transition zone temperature in a mechanically mixed upper mantle. Earth Planet Sci Lett 277(1–2):244–252
Ritsema J, Deuss A, van Heijst HJ, Woodhouse JH (2011) S40RTS: a degree-40 shear-velocity model for the mantle from new Rayleigh wave dispersion, teleseismic traveltime and normal-mode splitting function measurements. Geophys J Int 184(3):1223–1236
Rost S, Weber M (2001) A reflector at 200 km depth beneath the northwest Pacific. Geophys J Int 147(1):12–28
Rost S, Earle PS, Shearer PM, Frost DA, Selby ND (2015) Seismic detections of small-scale heterogeneities in the deep earth, this volume
Ryberg T, Wenzel F, Egorkin AV, Solodilov L (1998) Properties of the mantle transition zone in northern Eurasia. J Geophys Res Solid Earth 103(B1):811–822

Rychert CA, Schmerr N, Harmon N (2012) The Pacific lithosphere-asthenosphere boundary: Seismic imaging and anisotropic constraints from SS waveforms. Geochem Geophys Geosyst 13
Schaeffer AJ, Lebedev S (2015) Global heterogeneity of the lithosphere and underlying mantle: a seismological appraisal based on multimode surface-wave dispersion analysis, shear-velocity tomography, and tectonic regionalization, this volume
Schmerr N, Garnero E (2006) Investigation of upper mantle discontinuity structure beneath the central Pacific using SS precursors, J Geophys Res Solid Earth 111(B8). 10.1029/2005JB004197
Schmerr N, Thomas C (2011) Subducted lithosphere beneath the Kuriles from migration of PP precursors. Earth Planet Sci Lett 311(1–2):101–111
Schmerr N, Garnero E, McNamara AK (2010) Deep mantle plumes and convective upwelling beneath the Pacific Ocean. Earth Planet Sci Lett 294:143–151
Schmerr NC, Kelly BM, Thorne MS (2013) Broadband array observations of the 300 km seismic discontinuity. Geophys Res Lett 40(5):841–846
Shearer PM (1990) Seismic imaging of upper-mantle structure with new evidence for a 520-km discontinuity. Nature 344(6262):121–126
Shearer PM (1991) Imaging global body wave phases by stacking long-period seismograms. J Geophys Res Solid Earth 96(B12):20353–20320, 20364
Shearer PM (1993) Global mapping of upper-mantle reflectors from long-period SS precursors. Geophys J Int 115(3):878–904
Shearer PM (2000) Upper mantle discontinuities. In: Karato S, Forte AM, Liebermann RC, Masters G, Stixrude L (eds) Earth's deep interior: mineral physics and tomography from the atomic to the global scale. AGU, Washington D.C, pp 115–131
Shearer PM, Masters TG (1992) Global mapping of topography on the 660-km discontinuity. Nature 355(6363):791–796
Shen X, Yuan X, Li X (2014) A ubiquitous low velocity layer at the base of the mantle transition zone. Geophys Res Lett n/a-n/a
Sleep NH (1990) Hotspots and mantle plumes—some phenomenology. J Geophys Res Solid Earth Planets 95(B5):6715–6736
Smyth JR, Jacobsen SD (2006) Nominally anhydrous minerals and Earth's deep water cycle. In: Jacobsen SD, Van der Lee S (eds) Earth's deep water cycle. American Geophysical Union, Washington, DC, pp 1–11
Sobolev AV, Hofmann AW, Sobolev SV, Nikogosian IK (2005) An olivine-free mantle source of Hawaiian shield basalts. Nature 434(7033):590–597
Song TRA, Helmberger DV, Grand SP (2004) Low-velocity zone atop the 410-km seismic discontinuity in the northwestern United States. Nature 427(6974):530–533
Stixrude L (1997) Structure and sharpness of phase transitions and mantle discontinuities. J Geophys Res Solid Earth 102(B7):14835–14852
Stixrude L, Lithgow-Bertelloni C (2005) Thermodynamics of mantle minerals—I. Phys Prop Geophys J Int 162(2):610–632
Stixrude L, Lithgow-Bertelloni C (2007) Influence of phase transformations on lateral heterogeneity and dynamics in Earth's mantle. Earth Planet Sci Lett 263(1–2):45–55
Thybo H, Nielsen L, Perchuc E (2003) Seismic scattering at the top of the mantle transition zone. Earth Planet Sci Lett 216(3):259–269
van Keken PE, Gable CW (1995) The interaction of a plume with a rheological boundary—a comparison between 2-dimensional and 3-dimensional models. J Geophys Res Solid Earth 100(B10):20291–20302
Wajeman N (1988) Detection of underside P-reflections at mantle discontinuities by stacking broad-band DATA. Geophys Res Lett 15(7):669–672
Weidner DJ, Wang YB (1998) Chemical- and Clapeyron-induced buoyancy at the 660 km discontinuity. J Geophys Res Solid Earth 103(B4):7431–7441

Weis D, Garcia MO, Rhodes JM, Jellinek M, Scoates JS (2011) Role of the deep mantle in generating the compositional asymmetry of the Hawaiian mantle plume. Nat Geosci 4(12):831

Wessel P, Smith WHF (1998) New, improved version of generic mapping tools released. EOS Trans AGU 79(47):579

Williams Q, Hemley RJ (2001) Hydrogen in the deep earth. Annu Rev Earth Planet Sci 29:365–418

Williams Q, Revenaugh J (2005) Ancient subduction, mantle eclogite, and the 300 km seismic discontinuity. Geology 33(1):1–4

Wolfe CJ, Solomon SC, Laske G, Collins JA, Detrick RS, Orcutt JA, Bercovici D, Hauri EH (2009) Mantle shear-wave velocity structure beneath the Hawaiian hot spot. Science 326(5958):1388–1390

Woodland AB (1998) The orthorhombic to high-P monoclinic phase transition in Mg-Fe pyroxenes: can it produce a seismic discontinuity? Geophys Res Lett 25(8):1241–1244

Woodland AB, Angel RJ (1997) Reversal of the orthoferrosilite-high-P clinoferrosilite transition, a phase diagram for FeSiO3 and implications for the mineralogy of the Earth's upper mantle. Eur J Mineral 9(2):245–254

Xu WB, Lithgow-Bertelloni C, Stixrude L, Ritsema J (2008) The effect of bulk composition and temperature on mantle seismic structure. Earth Planet Sci Lett 275(1–2):70–79

Yu YG, Wentzcovitch RM, Angel RJ (2010) First principles study of thermodynamics and phase transition in low-pressure (P2(1)/c) and high-pressure (C2/c) clinoenstatite MgSiO3. J Geophys Res Solid Earth 115

Zhang JS, Dera P, Bass JD (2012) A new high-pressure phase transition in natural Fe-bearing orthoenstatite. Am Mineral 97(7):1070–1074

Zhang Z, Lay T (1993) Investigation of upper mantle discontinuities near northwestern Pacific subduction zones using precursors to SSH. J Geophys Res Solid Earth 98(B3):4389–4405

Zhao DP (2004) Global tomographic images of mantle plumes and subducting slabs: insight into deep Earth dynamics. Phys Earth Planet Inter 146(1–2):3–34

Zhao L, Chevrot S (2003) SS-wave sensitivity to upper mantle structure: Implications for the mapping of transition zone discontinuity topographies. Geophys Res Lett 30(11):1590. doi: 10.1029/2003GL017223

Zheng YC, Lay T, Flanagan MP, Williams Q (2007) Pervasive seismic wave reflectivity and metasomatism of the Tonga mantle wedge. Science 316(5826):855–859

Zhong SJ, Watts AB (2002) Constraints on the dynamics of mantle plumes from uplift of the Hawaiian Islands. Earth Planet Sci Lett 203(1):105–116

Chapter 4
Interpreting Radial Anisotropy in Global and Regional Tomographic Models

Thomas Bodin, Yann Capdeville, Barbara Romanowicz and Jean-Paul Montagner

Abstract We review the present status of global and regional mantle tomography and discuss how resolution has improved in the last decade with the advent of full waveform tomography and exact numerical methods for wave-field calculation. A remaining problem with full waveform tomography is computational cost. This leads seismologists to only interpret the long periods in seismic waveforms and hence only constrain long-wavelength structure. In this way, tomographic images do not represent the true Earth, but rather a smooth effective, apparent, or equivalent model that provides a similar long-wavelength data fit. In this paper, we focus on the problem of apparent radial anisotropy due to unmapped small-scale radial heterogeneities (e.g., layering). Here, we propose a fully probabilistic approach to sample the ensemble of layered models equivalent to a given smooth tomographic profile. We objectively quantify the trade-off between isotropic heterogeneity and strength of anisotropy. The non-uniqueness of the problem can be addressed by adding high-frequency data such as receiver functions, sensitive to first-order discontinuities. We show that this method enables us to distinguish between intrinsic and artificial anisotropy in 1D models extracted from tomographic results.

Keywords Seismic tomography · Inverse theory · Bayesian inference · Monte Carlo methods · Seismic anisotropy

T. Bodin (✉) · B. Romanowicz
Berkeley Seimological Laboratory, UC Berkeley,
215 McCone Hall, Berkeley, CA 94720-4760, USA
e-mail: thomas.bodin@berkeley.edu

T. Bodin
Laboratoire de Géologie de Lyon: Terre, Planètes et Environnement, CNRS,
Université de Lyon 1, Ecole Normale Supérieure de Lyon, Villeurbanne, France

Y. Capdeville
Laboratoire de Planétologie et Géodynamique de Nantes, CNRS,
Université de Nantes, Nantes, France

B. Romanowicz · J.-P. Montagner
Institut de Physique du Globe de Paris, 4 Place Jussieu, 75252 Paris Cedex 05 Paris, France

© Springer International Publishing Switzerland 2015

A. Khan and F. Deschamps (eds.), *The Earth's Heterogeneous Mantle*,
Springer Geophysics, DOI 10.1007/978-3-319-15627-9_4

4.1 Introduction

For more than thirty years, seismologists have imaged the earth's interior using seismic waves generated by earthquakes and traveling through different structures of the planet. A remaining challenge in seismology is to interpret the recovered Earth models in terms of physical properties (e.g., temperature, density, mineral composition) that are needed for understanding mantle dynamics and plate tectonics. For example, a region of slow wave speed can be either interpreted as anomalously warm, or rich in water, or iron.

Although seismic waves are sensitive to a large number of viscoelastic parameters as well as density, the mantle models constructed from seismic tomography are only parameterized with a few physical parameters, for example, average isotropic shear-wave velocity and radial anisotropy (e.g., French et al. 2013). This is because given the available information observed at the surface, there is not enough resolution to entirely describe the local elastic tensor. In addition to the limited number of resolvable elastic (and anelastic) parameters, there is also the question of spatial resolution, namely the smallest spatial scale at which heterogeneities can be imaged.

The number of independent elastic parameters that can be constrained is intrinsically associated with the level of spatial resolution. For example, it is well known that a stack of horizontal isotropic layers will be equivalent, at large scales, to a homogeneous anisotropic medium (Backus 1962). As we increase the scale at which we "see" the medium (the minimum period in the observed waveforms), we lose the ability to distinguish different layers, as well as the ability to distinguish between isotropy and anisotropy. The anisotropy observed at large scales may be artificial and simply the effect of unmapped fine layering. In other words, whether a material is heterogeneous (and described by a number of spatial parameters) or anisotropic (described by different elastic parameters) is a matter of the scale at which we analyze its properties (Maupin and Park 2014).

Therefore, there is a trade-off between spatial roughness and anisotropy when inverting long-period seismic data. By introducing anisotropy as a free parameter in an inversion, tomographers are able to fit seismic data with smoother models and fewer spatial parameters (Montagner and Jobert 1988; Trampert and Woodhouse 2003).

In this manuscript, we will first describe the issues that limit resolution in seismic imaging at regional and global scales (uneven data sampling, limited frequency band, data noise, etc), with a focus on the significance of observed seismic anisotropy and on the problem of distinguishing its different possible causes. Following Wang et al. (2013) and Fichtner et al. (2013a), here, we make the distinction between intrinsic anisotropy and extrinsic (i.e., artificial) anisotropy induced by structure. In the last section, we propose a method to separate these two effects in a simplified 1D case with vertical transverse isotropy (i.e., ignoring azimuthal anisotropy).

4.2 The Resolving Power of Regional and Global Seismic Tomography

It can be proven that if one had an unlimited number of sources, receivers, and an unlimited frequency band, one would be able to entirely describe an elastic medium from the displacement of elastic waves propagating through it and observed at its surface. That is, the function linking an elastic medium subjected to excitation by a source to the displacement measured at its boundaries is bijective. For detailed mathematical proofs, see Nachman (1988), Nakamura and Uhlmann (1994), and Bonnet and Constantinescu (2005).

However, in seismology, there are a number of elements that limit the resolving power of seismic observations, i.e., the ability to image structure. Firstly, the seismic records are limited both in time and frequency, and the number of sources and receivers is limited. Furthermore, there are a number of observational and theoretical errors that propagate into the recovered images. Finally, the earth is not entirely elastic, and seismic energy is dissipated along the path.

In this section, we give a brief description of these limiting factors which directly condition the level of resolution. Note that here, the phrase "level of resolution" or "resolving power" will be used in a broad sense and defined as the quantity of information that can be extracted from the data (the maximum number of independent elastic parameters or the minimum distance across which heterogeneities can be mapped). Here, we do not consider the resolution as it is mathematically defined in linear inverse theory and represented by a resolution matrix (e.g., Backus and Gilbert 1968; Aki et al. 1977), which, for example, does not depend on data noise or theoretical errors.

4.2.1 Different Seismic Observables

There are a multitude of ways of extracting interpretable information from seismograms. Due to practical, theoretical, and computational considerations, imaging techniques often only involve a small part of the seismic record. Different parts of the signal can be used, such as direct, reflected, and converted body waves, surface waves, or ambient noise. Different components of the signal can be exploited such as travel times, amplitudes, shear-wave splitting measurements, waveform spectra, full waveforms, or the entire wave-field (for comprehensive reviews, see Rawlinson and Sambridge 2003; Romanowicz 2003; Liu and Gu 2012).

Each observable has its own resolution capabilities. For example, analysis of converted body waves, now widely called the "receiver function", is used as a tool to identify horizontal discontinuities in seismic velocities (small-scale radial heterogeneities), but fails at determining long-wavelength anomalies. Conversely, surface wave measurements are sensitive to 3D absolute S wave velocities, but cannot constrain sharp gradients, and are poor at locating interfaces. Surface wave-based imaging usually involves only the relatively low-frequency component of seismograms and is particularly effective in mapping the large-scale pattern of upper mantle structure. We will show how the seismic discontinuities that can be constrained with converted and reflected body waves are sometimes seen by surface waves as anisotropic structure.

Hence, the gaps between existing models can be described in terms of seismic wavelengths. The difficulty of assembling different databases with different sensitivities that sample the earth at different scales and the differences in the theory relating earth structure to seismic data of different nature have resulted in most models being based only on a limited portion of potentially available observations.

4.2.2 An Uneven Sampling of the Earth

One of the most important causes of poor resolution in seismic tomography is limited sampling of the volume of interest. In global seismic mantle tomography, there is no control on the distribution of the earthquake sources, which mostly occur at plate boundaries. Moreover, most receivers are located on continents, which cover only about one-third of the surface of the planet. This results in an uneven distribution of sources and receivers, especially in the southern hemisphere.

Traditional tomography relies primarily on the information contained in the travel times of seismic phases that are well separated on the seismic record: first arriving P and S body waves on the one hand and fundamental mode surface waves on the other. For the latter, which are dispersive, the measured quantity is the phase or group velocity as a function of period, in a period range accessible for teleseismic observations, typically ~30 to ~250 s.

The theoretical framework is typically that of infinite frequency ray theory for body waves, or its equivalent for surface waves, the "path average approximation" (PAVA) [see reviews by Romanowicz (2002), Romanowicz et al. (2008)]. Below, we briefly discuss how body and surface waves sample the earth differently, and then discuss how waveform tomography allows us to compensate for the non-uniform distribution of sources and receivers by exploiting more fully the information contained in each seismogram.

4.2.2.1 Body Wave Tomography

Because of the lack of stations in the middle of the oceans, body wave tomography based on first arrival travel times achieves best resolution in regions where the density of both sources and stations is high, typically in subduction zone regions around the Pacific Ocean and in the Mediterranean region (e.g., Bijwaard et al. 1998; Kárason and Van Der Hilst 2000; Fukao et al. 2001; Rawlinson et al. 2015). Much progress has been made in the last few years, owing to improvements in both quality and quantity of

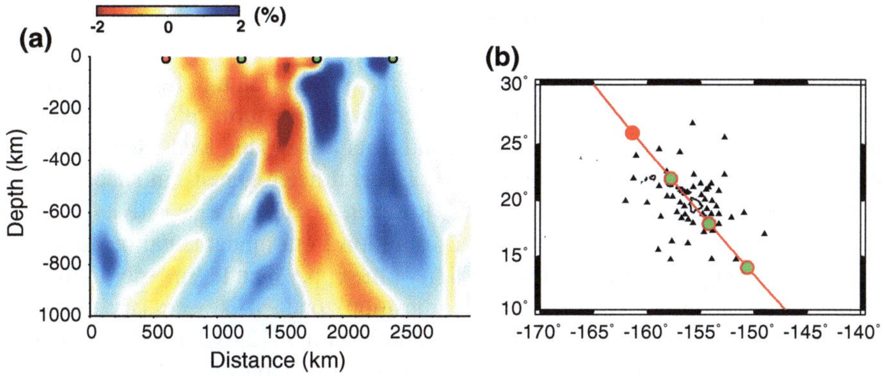

Fig. 4.1 Example of teleseismic body wave tomography under Hawaii with poor vertical resolution, i.e., vertical smearing. **a** Vertical cross sections (parallel to the Pacific plate motion) through the HW13 model (Cheng et al. 2015). **b** Locations and orientation of the cross section, along with the distribution of stations. This is a typical example of limited resolution due to poor data sampling. Modified from Cheng et al. (2015)

seismic data. Some technical improvements have also been made, such as the introduction of finite frequency kernels that take into account the sensitivity of the body wave to a broader region around the infinitesimal ray path (e.g., Dahlen et al. 2000). These improvements have led to increasingly high-resolution images in the last ten years indicating different behaviors of slabs in the transition zone, with some ponding on the 660 km discontinuity, and/or around 1000 km depth, while others appear to penetrate deep into the lower mantle (e.g., Li et al. 2008; Fukao and Obayashi 2013).

In other parts of the world, where only teleseismic data can be used, resolution in body wave travel time tomography depends strongly on the density of stations. In the oceans and in poorly covered continental regions, there is very poor vertical and horizontal resolution in the upper mantle, even when considering finite frequency effects, because of smearing effects due to the lack of crossing paths. In Fig. 4.1, we show an example of regional body wave tomography under Hawaii, where only teleseismic events originating at subducting zones around the Pacific are used (Cheng et al. 2015). Seismic rays arrive almost vertically under the array of stations, which results in poor vertical resolution as velocity anomalies are "smeared" along seismic rays. In this context, interpretation of the vertical plume-like low-velocity anomalies must be done with caution, and extra constraints from surface waves are needed (Cheng et al. 2015).

On the other hand, in some continental regions, such as in North America, owing to the recent dense USArray deployment, improved resolution is progressively achieved (e.g., Burdick et al. 2008; Obrebski et al. 2011; Sigloch and Mihalynuk 2013). Nevertheless, at the global scale, resolution from body wave tomography remains uneven, even when surface or core reflected teleseismic phases are added. Also, these tomographic models generally provide high-resolution information on *P* velocity, since S wave travel times are more difficult to measure accurately.

4.2.2.2 Surface Wave Tomography

Because their energy is concentrated near the surface along the source station great circle path, fundamental mode surface waves, in turn, allow the sampling of the upper mantle under oceans and continents alike. This leads to robust resolution of the long-wavelength component of lateral heterogeneity in shear velocity in the upper mantle at the global scale. However, because the sensitivity to structure decreases exponentially with depth, resolution from fundamental mode surface wave tomography is best in the first 300 km of the upper mantle. In order to improve resolution at larger depths, i.e., into the transition zone , it is necessary to include surface wave overtone data (e.g., Debayle and Ricard 2012; Schaefer and Lebedev 2015). These have similar group velocities, and hence, sophisticated approaches are required to separate and measure dispersion on individual overtone branches [see review by Romanowicz (2002)]. This presents a challenge for achieving comparable coverage to fundamental mode surface waves at the global scale.

This is why the recent global whole-mantle shear-velocity models that provide the best resolution in the transition zone (Kustowski et al. 2008; Ritsema et al. 2011) are based on a combination of different types of data which provide complementary sampling of the mantle: (1) fundamental mode surface waves and overtones, which provide resolution across the upper mantle; (2) for the lower mantle, body wave travel times, which generally include, in addition to first arriving S waves, surface reflected SS and core reflected ScS waves, sometimes complemented by core-propagating SKS travel time data. Some models, based on secondary travel time observables, also consider another type of data, normal mode "splitting functions", which provide constraints on the longest wavelength structure throughout the mantle (e.g., Ritsema et al. 2011).

4.2.2.3 Global Waveform Tomography Based on Asymptotic Methods

Since body and surface waves sample the earth differently , a powerful way to improve the sampling of the mantle is to combine them by exploiting the information contained in the entire seismogram (i.e., seismic waveforms). This idea was first introduced in global tomography by Woodhouse and Dziewonski (1984), where observed and synthetic seismograms were directly compared in the time domain. Introducing long-period seismic waveform tomography allowed these authors to include information from overtones in a simple way and thus to improve resolution in the transition zone. Synthetic seismograms were computed in a 3D earth using normal mode summation and the "path average" approximation (PAVA). A similar type approach has also been developed (Nolet 1990) and applied to upper mantle tomography at the continental (Van der Lee and Nolet 1997) and global scales (Lebedev and Van Der Hilst 2008; Schaeffer and Lebedev 2013).

In standard body wave tomography, the ensemble of body wave phases available through travel time measurements is largely limited. For example, the study of Kustowski et al. (2008) mentioned above was limited to measurements of SS,

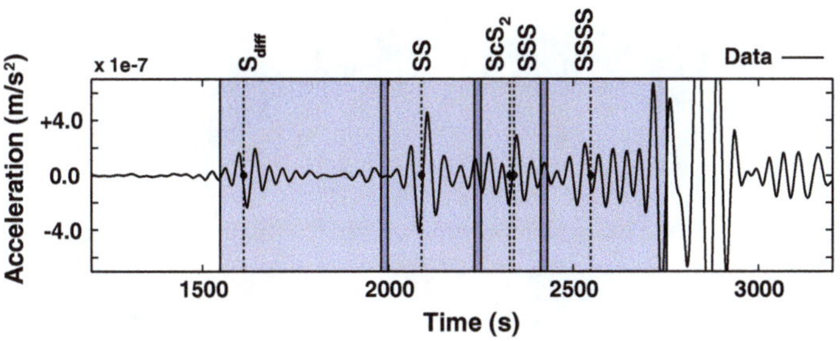

Fig. 4.2 Example of wave packet selection procedure for time-domain waveforms, as used in Mégnin and Romanowicz (2000) and following models from the Berkeley group, that are based on time-domain waveform inversion. *Shaded areas* indicate wave packets picked. Note in the third wave packet, the combination of two body wave phases (SSS, ScS2) that are not separable for travel time computation, but that sample very different parts of the mantle (courtesy of Scott French)

ScS, and SKS phases. A clear advantage of waveform tomography is that one can include body phases that cannot be separated in the time domain such as, for example, ScS_2 and SSS, as well as diffracted waves, whose propagation cannot be well described by ray theory (see Fig. 4.2).

However, when using body waveforms, the path average approximation (PAVA) may not be valid anymore. Indeed, the drawback of the PAVA is that it assumes that sensitivity of the waveforms is limited to the average 1D structure between the epicenter and the receiver, which is clearly inappropriate for body waves, whose sensitivity is concentrated along the ray path (Romanowicz 1987). In order to take into account the concentration of sensitivity along the ray path of body waves, across-branch coupling needs to be included (e.g., Li and Tanimoto 1993). Li and Romanowicz (1995) developed NACT (nonlinear asymptotic coupling theory), which introduced an additional term to PAVA that accounted for coupling across normal mode dispersion branches, bringing out the ray character of body waveforms [see Romanowicz et al. (2008) for details and a comparison of mode-based methods for modeling seismic waveforms]. This approach has been applied to the development of waveform-based global long-wavelength shear-velocity models since the mid-1990s (e.g., Li and Romanowicz 1996; Mégnin and Romanowicz 2000; Panning and Romanowicz 2006; Panning et al. 2010).

Comparing models obtained by different groups using different datasets and methodologies is one way to evaluate the robustness of the retrieved structure. The advantage of using full waveform tomography is that, by including a variety of phases that illuminate the mantle in different ways, the sampling is improved in ways that cannot be attained using only travel times of well isolated phases, largely because the distribution of earthquake sources and receivers is limited resulting in many redundant paths even as new data are added. Thus, at the very least, the same resolution can be achieved using considerably fewer source station paths. This is illustrated in Fig. 4.3 which shows a comparison of three recent global shear-velocity tomographic models at a depth of 100 km. Models (a and c)

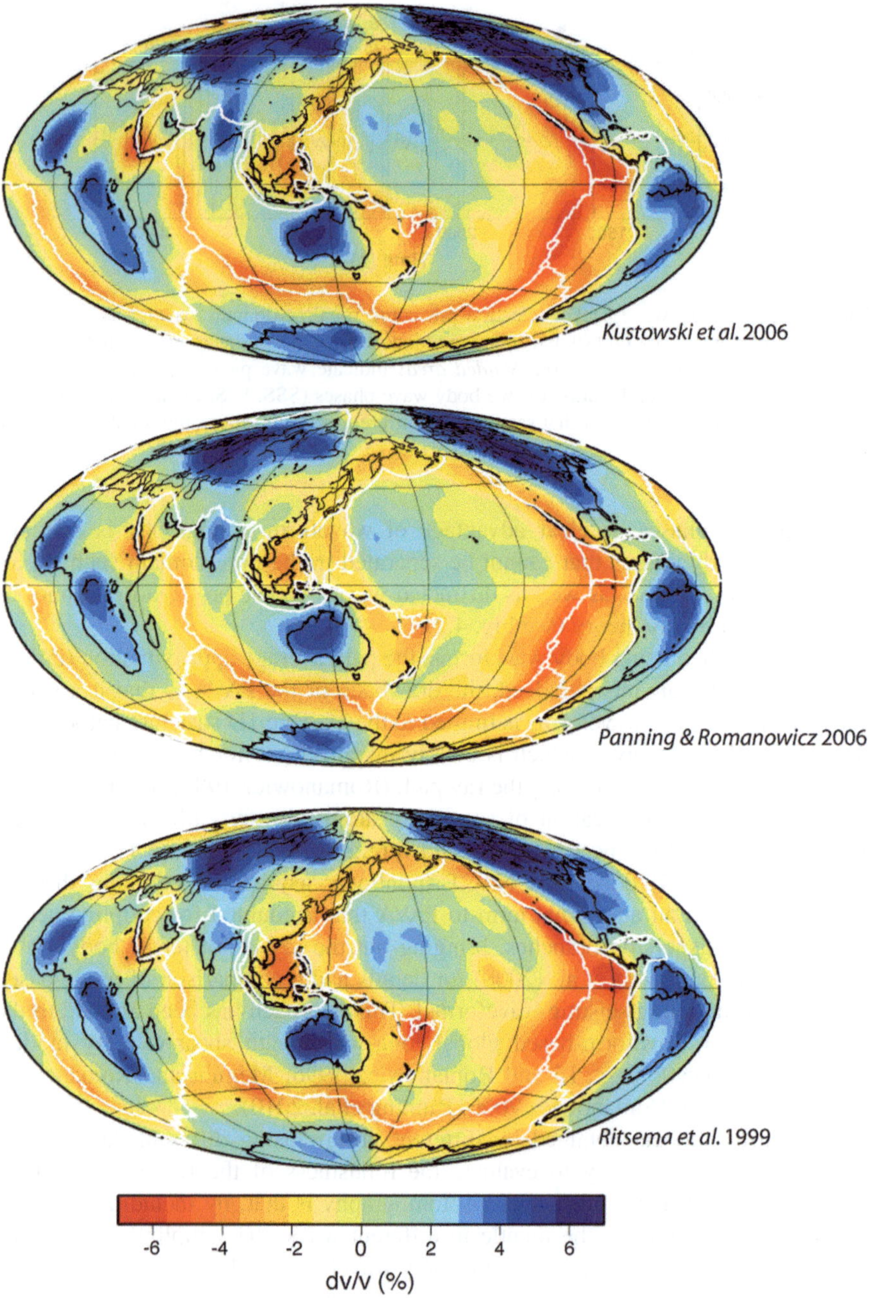
Kustowski et al. 2006
Panning & Romanowicz 2006
Ritsema et al. 1999
-6
-4
-2
0
2
4
6
dv/v (%)

Fig. 4.3 Comparison of maps of isotropic versus at a depth of 100 km from three whole-mantle tomographic models: **a** S362ANI (Kustowski et al. 2008); **b** SAW642AN (Panning and Romanowicz 2006); and **c** S20RTS (Ritsema et al. 1999). Model **a** was constructed using a combination on body wave travel times, surface wave dispersion, and long-period waveforms, albeit with the PAVA approximation; Model **c** was constructed using a combination of surface wave dispersion and body wave travel times. Both models used over 200,000 data. Model **b** was constructed using time-domain waveforms exclusively and the NACT theoretical framework, obtaining an equivalent resolution to the two other models, albeit with an order of magnitude fewer data (20,000 waveform packets)

were obtained using a conventional approach: Ritsema et al. (1999) used over 2M fundamental mode and overtone measurements combined with over 20,000 body wave travel time measurements to construct model S20RTS (a), while Kustowski et al. (2008) used several million dispersion measurements and about 150,000 body wave travel time measurements to construct model 362ANI. In contrast, Panning and Romanowicz (2006) used "only" 20,000 long-period time-domain seismograms (i.e., waveforms) and NACT to construct model SAW36ANI and were able to resolve the long-wavelength structure in the upper mantle just as well. With the ability to include increasingly shorter periods, i.e., constraints from phases that sample the mantle in yet other ways, as well as improving the accuracy with which the interactions of the wave-field with heterogeneity are computed, this opens the way to increased resolution in the future, as will be discussed in the next section. For now, beyond details of the datasets and theories used, Figs. 4.3 and 4.4 indicate that the level of agreement between global shear-velocity models is presently excellent up to at least degree 12 in a spherical harmonics expansion of the model, both in the upper and the lowermost mantle (e.g., Lekic et al. 2012).

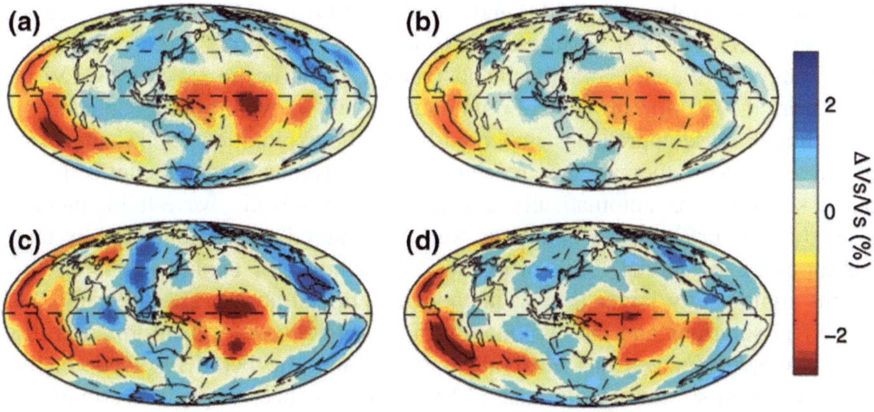

Fig. 4.4 Comparison of 4 recent shear-wave tomographic models at a depth of 2800 km. **a** Kustowski et al. (2008). **b** Ritsema et al. (2011). **c** Mégnin and Romanowicz (2000). **d** Houser et al. (2008). Model **c** was developed using only time-domain waveforms (about 20,000), while all other models are based on a combination of secondary observables (travel times of body waves and surface waves), except model **a** which includes long-period waveforms, albeit in the surface wave (PAVA) approximation. After Lekic et al. (2012)

4.2.2.4 Global Waveform Inversion Based on Direct Numerical Solvers

In the previous section, we have described how, in principle, full waveform tomography provides access to more of the information contained in seismograms than a collection of travel times of a limited number of seismic phases. As mentioned above, normal mode summation has provided a successful theoretical approach for computation of waveforms and led to several generations of whole-mantle shear-velocity models in the last 20 years. However, asymptotic normal mode perturbation theory (Li and Romanowicz 1995) is only valid for earth models for which the wavelength of the structure is large compared to that of the seismic waves considered (i.e., smooth models) and heterogeneity is weak (nominally, lateral variations of up to $\sim$10 %). Yet, in the earth's boundary layers, i.e., in the upper mantle and in the D'' region, there is ample evidence for the presence of stronger heterogeneity, whereas throughout the mantle, heterogeneity at many different scales may be present. First-order mode perturbation theory is not appropriate in this case, and more accurate numerical methods must be used. The challenge then is how to compute the synthetic seismograms in a 3D earth model without the weak heterogeneity approximation.

Finite difference methods are the traditional approach used for numerical calculation of seismograms (Kelly et al. 1976; Virieux 1986). In the 1990s, pseudo-spectral methods have also become a popular alternative and have been applied to regional (Carcione 1994) and global (Tessmer et al. 1992) problems. However, both finite difference and pseudo-spectral schemes perform poorly at representing surface waves. This issue can be addressed with the spectral element method (SEM) where the wave equation is solved on a mesh that is adapted to the free surface and to the main internal discontinuities of the model. The SEM was first introduced by Priolo et al. (1994) and Seriani and Priolo (1994) for wave-field calculation in 2D and later perfected by Komatitsch and Vilotte (1998), Komatitsch and Tromp (1999), and Komatitsch and Tromp (2002) for the 3D case. See Virieux and Operto (2009) for a review of numerical solvers in exploration geophysics.

Although these approaches started earlier in the exploration community than in global seismology, they are now reaching similar advance levels. Numerically computed seismograms automatically contain the full seismic wave-field, including all body and surface wave phases as well as scattered waves generated by lateral variations of the model Earth properties. The amount of exploitable information is thus significantly larger than in methods mentioned above. The accuracy of the numerical solutions and the exploitation of complete waveform information result in tomographic images that are both more realistic and better resolved (Fichtner et al. 2010). In seismology, the use of SEM has now been applied to tomographic inversions for crustal structure at the local scale (e.g., Tape et al. 2010) and upper mantle structure at regional scales (e.g., Fichtner et al. 2009, 2010; Rickers et al. 2013; Zhu et al. 2012; Zhu and Tromp 2013).

The forward numerical computation is generally combined with an "adjoint" formulation for the numerical computation of the kernels for inversion (Tromp et al. 2005; Fichtner et al. 2006) or, alternatively, with a "scattering integral formalism" (e.g., Chen et al. 2007). In this context, Fichtner and Trampert (2011) showed how a local quadratic approximation of the misfit functional can be used for resolution analysis.

Here, we note that the inverse step is currently approached differently by different investigators. Following the nomenclature of the geophysical exploration community, the term FWI (full waveform inversion) is often used synonymously to "adjoint inversion," which relies, at each iteration, on the numerical computation of the gradient followed by a conjugate-gradient step. An alternative method, which has been used so far in SEM-based global waveform inversions, is to compute an approximate Hessian using mode-coupling theory in the current 3D model, followed by a Gauss–Newton (GN) inversion scheme. While it might be argued that the partial derivatives computed in this manner are more "approximate," the GN scheme is much faster converging (less than 10 iterations typically, compared to 30–40 or more) and can now take advantage of efficient methods for the assembly and inversion (ScalaPak) of large full matrices.

At the global scale, because the wave-field needs to be computed for a long time interval, in order to include all seismic phases of interest, the use of the SEM is particularly challenging computationally (Capdeville et al. 2005). Furthermore, computational time increases as the fourth power of frequency and limits the frequency range of waveforms to relatively long periods (typically longer than 40 or 50 s). The first global shear-velocity models developed using SEM (Lekić and Romanowicz 2011; French et al. 2013) are limited to the upper mantle due to the use of relatively long periods (longer than 60 s). In these models, the numerical computation of the forward step is restricted to the mantle and coupled with 1D mode computation in the core (CSEM, Capdeville et al. 2003). For the inverse step, kernels are computed using a mode-based approximation.

These modeling efforts have demonstrated the power of the SEM to sharpen tomographic images at the local, regional, and global scales and have led to the discovery of features previously not detected, such as the presence of low-velocity channels in the oceanic asthenosphere (e.g., French et al. 2013; Colli et al. 2013; Rickers et al. 2013). This is shown in Fig. 4.5 where model SEMum2 (French et al. 2013) is compared to other global shear-velocity models. SEMum2 more accurately recover both the depth and strength of the low-velocity minimum under ridges. It also shows stronger velocity minima in the low-velocity zone, a more continuous signature of fast velocities in subduction zones and stronger, clearly defined, low-velocity conduits under the Pacific Superswell, while confirming the robust long-wavelength structure imaged in previous studies, such as the progressive weakening and deepening of the oceanic low-velocity zone with overlying plate age. Of course, a more objective way to compare tomographic methods would be to conduct a blind test using numerically generated data, but this is beyond the scope of this study.

Because the frequency range of global inversions remains limited and because features smaller than the shortest wavelength cannot be mapped, this approach is not able, however, to resolve sharp discontinuities. The resulting tomographic images can therefore be seen as a smooth representation of the true Earth.

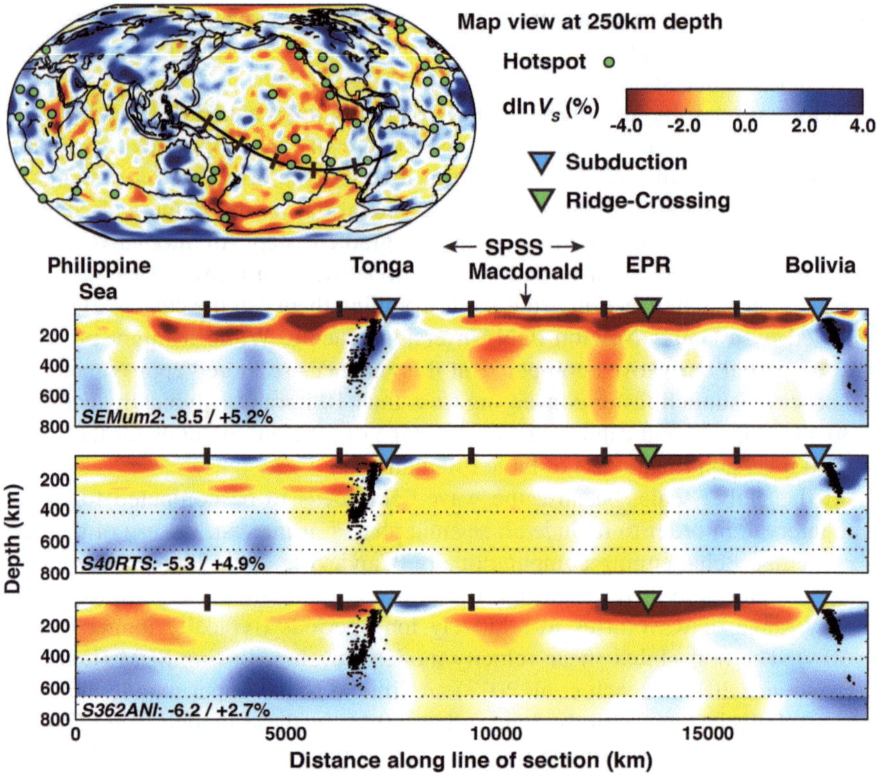

Fig. 4.5 Upper mantle depth cross sections across the Pacific superswell, comparing two recent global models obtained using classical approaches based on a combination of travel times, dispersion measurements, and approximate wave propagation theories (S362ANI, Kustowski et al. 2008; S40RTS, Ritsema et al. 2011) and a recent model constructed using waveforms and wavefield computations using SEM (SEMum2, French et al. 2013). While all three models agree in their long-wavelength structure in the transition zone, model SEMum2 shows more sharply delineated structures, both in subduction zones (highlighted by seismicity) and in the central Pacific, where the large low-velocity region is now resolved into two separate vertically oriented features. Model SEMum2 also exhibits stronger low-velocity minima in the uppermost mantle low-velocity zone. After French et al. (2013), courtesy of Scott French

However, they are not a simple spatial average of the true model, but rather an effective, apparent, or equivalent model that provides a similar long-wave data fit (Capdeville et al. 2010a, b). Hence, the geological interpretation of global tomographic models is limited, mainly due to two reasons:

1. The constructed images are smooth and do not contain discontinuities that are crucial to understand the structure and evolution of the earth.
2. The relations that link the true Earth to the effective (and unrealistic) earth that is seen by long-period waves are strongly nonlinear, and their inverse is highly non-unique. As a result, it is difficult to quantitatively interpret the level of imaged

anisotropy in tomographic models, as it may be the effect of either "real" local anisotropy or unmapped velocity gradients, or a combination of both.

4.3 Seismic Anisotropy

4.3.1 Observation of Anisotropy

It is well known that anisotropic structure is needed to predict a number of seismic observations such as:

1. Shear-wave splitting (or birefringence), the most unambiguous observation of anisotropy, particularly for SKS waves (Vinnik et al. 1989).
2. The Rayleigh–Love wave discrepancy. At global as well as at regional scale, the lithosphere appears faster to Love waves than to Rayleigh waves. It is impossible to simultaneously explain Rayleigh and Love wave dispersion by a simple isotropic model (Anderson 1961).
3. Azimuthal variation of the velocity of body waves. For example, Hess (1964) showed that the azimuthal dependence of Pn-velocities below oceans can be explained by anisotropy.

The goal here is not to provide a review of seismic anisotropy, but to address the issue of separating intrinsic and extrinsic anisotropy in apparent (observed) anisotropy. There are excellent review papers and books that have been written on anisotropy. For example, the theory of seismic wave propagation in anisotropic media has been described in Crampin (1981), Babuška and Cara (1991), and Chapman (2004). See also Maupin and Park (2014) for a review of observations of seismic anisotropy. Montagner (2014) gives a review of anisotropic tomography at the global scale. Montagner (1994); Montagner and Guillot (2002) give a review of geodynamic implications of observed anisotropy. Finally, a review of the significance of seismic anisotropy in exploration geophysics has been published by Helbig and Thomsen (2005).

Seismic waves are sensitive to the full elastic tensor (21 parameters), density, and attenuation. As seen above, it is not possible to resolve all 21 components of the anisotropic tensor at every location. Therefore, seismologists rely on simplified (yet reasonable) assumptions on the type of anisotropy expected in the earth's upper mantle, namely hexagonal symmetry. This type of anisotropy (commonly called transverse isotropy) is defined by the 5 Love parameters A, C, F, L, and N (love 1927), and two angles describing the tilt of the axis of symmetry (Montagner and Nataf 1988). In this manuscript, we will limit ourselves to the case of radial anisotropy, which corresponds to transverse isotropy with a vertical axis of symmetry and no azimuthal dependence.

It can be shown (Anderson 1961; Babuška and Cara 1991) that for such a vertically transversely isotropic (VTI) medium, long-period waveforms are primarily sensitive to the two parameters:

$$V_{SH} = \sqrt{\frac{N}{\rho}} \tag{4.1}$$

$$V_{SV} = \sqrt{\frac{L}{\rho}} \tag{4.2}$$

where ρ is density, V_{SV} is the velocity of vertically traveling S waves or horizontally traveling S waves with vertical polarization, and V_{SH} is the velocity of horizontally traveling S waves with horizontal polarization. The influence of other parameters A (related to V_{VH}), C (related to V_{PV}), and F can be large (Anderson and Dziewonski 1982) and is usually taken into account with petrological constraints (Montagner and Anderson 1989). That is, once V_{SH} and V_{SV} are constrained from long-period seismic waves, the rest of the elastic tensor and density is retrieved with empirical scaling laws (e.g., Montagner and Anderson 1989). Globally, SH waves propagate faster than SV waves in the upper mantle. The velocity difference is of about 4 % on average in the preliminary reference Earth model (PREM) of PREM Dziewonski and Anderson (1981) in the uppermost 220 km of the mantle.

Although early global radially anisotropic models were developed in terms of V_{SH} and V_{SV}, more recent models are parameterized in terms of an approximate Voigt average isotropic shear velocity (Montagner 2014) and radial anisotropy as expressed by the ξ parameter (e.g., Gung et al. 2003; Panning and Romanowicz 2006):

$$V_S = \frac{2V_{SV} + V_{SH}}{3} = \sqrt{\frac{2L + N}{3\rho}} \tag{4.3}$$

$$\xi = \frac{V_{SH}^2}{V_{SV}^2} = \frac{N}{L} \tag{4.4}$$

4.3.2 Anisotropy of Minerals: Intrinsic Anisotropy

Anisotropy can be produced by multiple physical processes at different spatial scales. It exists from the microscale (crystal scale) to the macroscale, where it can be observed by seismic waves that have wavelengths up to hundreds of kilometers. We name intrinsic anisotropy, the elastic anisotropy still present whatever the scale of investigation, down to the crystal scale. Most minerals in the earth's upper mantle are anisotropic. Olivine, the most abundant mineral in the upper mantle, displays a P wave anisotropy larger than 20 %. Other important constituents such as orthopyroxene or clinopyroxene are anisotropic as well (>10 %). Under finite strain accumulation, plastic deformation of these minerals can result in a preferential orientation of their crystalline lattices. This process is usually referred to as LPO (lattice-preferred orientation) or CPO (crystalline-preferred orientation). This phenomenon is often considered as the origin of the observed large-scale seismic anisotropy in the upper mantle. With increasing the depth, most of minerals undergo a series of phase transformations. There is some tendency (though not systematic) that with increasing pressure, the crystallographic structure evolves toward a more closely packed, more isotropic structure, such as cubic structure. For example, olivine transforms into β-spinel and then γ-spinel in the upper transition zone (410–660 km of depth) and into perovskite and magnesiowustite in the lower mantle and possibly into post-perovskite in the lowermost mantle. Perovskite, post-perovskite $(Mg,Fe)SiO_3$, and the pure end-member of magnesiowustite MgO are still anisotropic. That could explain the observed anisotropy in some parts of the lower mantle and D''-layer.

Mantle rocks are assemblages of different minerals which are more or less anisotropic. The resulting amount of anisotropy is largely dependent on the composition of the aggregates. The relative orientations of crystallographic axes in the different minerals must not counteract in destroying the intrinsic anisotropy of each mineral. For example, the anisotropy of peridotites, mainly composed of olivine and orthopyroxene, is affected by the relative orientation of their crystallographic axes, but the resulting anisotropy is still larger than 10 %.

In order to observe anisotropy due to LPO at very large scale, several conditions must be fulfilled. The crystals must be able to re-orient in the presence of strain and the deformation due to mantle convection must be coherent over large scales to preserve long-wavelength anisotropy. These processes are well known for the upper mantle, and in oceanic plates, and anisotropy remains almost uniform on horizontal length scales in excess of 1000 km. The mechanisms of alignment are not so well known in the transition zone and in the lower mantle. In addition, a significant water content such as proposed by Bercovici and Karato (2003) in the transition zone can change the rheology of minerals, would make the deformation of the minerals easier, and change their preferential orientation. A complete discussion of these different mechanisms at different scales can be found in Mainprice (2007).

At slightly larger scale (but smaller than the seismic wavelength), a coherent distribution of fluid inclusions or cracks (Crampin and Booth 1985) can give rise to apparent anisotropy due to shape-preferred orientation (SPO). This kind of anisotropy related to stress field can be considered as the lower lilt of extrinsic anisotropy.

Anisotropic properties of rocks are closely related to their geological history and present configuration and reveal essential information about the earth's structure and dynamics (Crampin 1981). This justifies the great interest of geophysicists in all seismic phenomena which can be interpreted in the framework of anisotropy. However, the observation of large-scale anisotropy is also due to other effects such as unmapped velocity gradients.

4.3.3 Apparent Anisotropy Due to Small-Scale Inhomogeneities

It has been known for a long time in seismology and exploration geophysics that small-scale inhomogeneities can map into apparent anisotropy (Postma 1955; Backus 1962). The problem is very well described in the abstract by Levshin and Ratnikova (1984): "In homogeneities in a real material may produce a seismic wave-field pattern qualitatively indistinguishable from one caused by anisotropy. However, the quantitative description of such a medium as an apparently anisotropic elastic solid may lead to geophysically invalid conclusions."

The scattering effect of small-scale heterogeneities on seismograms has been extensively studied in seismology (e.g., Aki 1982; Richards and Menke 1983; Park and Odom 1999; Ricard et al. 2014). As an example, Kennett and Nolet (1990) and Kennett (1995) demonstrated the validity of the great circle approximation when modeling long-period waveforms. However, despite all these studies, poor attention has been given to the theoretical relations between small-scale heterogeneities and equivalent anisotropy. By definition, an anisotropic material has physical properties which depend on direction, whereas a heterogeneous material has properties which depend on location. But the distinction between heterogeneity and anisotropy is a matter of the scale at which we analyze the medium of interest. Alternating layers of stiff and soft material will be seen at large scales as a homogeneous anisotropic material. At the origin of any anisotropy, there is a form for heterogeneity. In this way, the most basic form of anisotropy, related to the regular pattern made by atoms in crystals, can also be seen as some form of heterogeneity at the atomic scale (Maupin and Park 2014).

Although poorly studied theoretically, this phenomenon has been recognized in a number of studies. Maupin (2001) used a multiple-scattering scheme to model surface waves in 3D isotropic structures. She found that the apparent Love–Rayleigh discrepancy ($V_{SH} - V_{SV}$) varies linearly with the variance of isotropic S wave velocity anomalies. In the case of surface wave phase-velocity measurements done at small arrays, Bodin and Maupin (2008) showed that heterogeneities located close to an array can introduce significant biases which can be mistaken for anisotropy. For the lowest mantle, Komatitsch et al. (2010) numerically showed that isotropic velocity structure in D$''$ can explain the observed splitting of S_{diff}, traditionally interpreted as LPO intrinsic anisotropy due to mantle flow.

In the context of joint inversion of Love and Rayleigh waveforms, a number of studies acknowledged that the strong mapped anisotropy is difficult to reconcile with mineralogical models. This discrepancy may be explained in part by horizontal layering, or by the presence of strong lateral heterogeneities along the paths, which are simpler to explain by radial anisotropy (Montagner and Jobert 1988; Friederich and Huang 1996; Ekström and Dziewonski 1998; Debayle and Kennett 2000; Raykova and Nikolova 2003; Endrun et al. 2008; Bensen et al. 2009; Kawakatsu et al. 2009).

Bozdağ and Trampert (2008) showed that the major effect of incorrect crustal corrections in surface wave tomography is on mantle radial anisotropy. This is because the lateral variation of Moho depth trade-offs with radial anisotropy [see also Montagner and Jobert (1988), Muyzert et al. (1999), Lebedev et al. (2009), Lekić et al. (2010), and Ferreira et al. (2010)].

Therefore, it is clear that both vertical and lateral isotropic heterogeneities can contribute to the observed radial anisotropy. The problem of separating intrinsic and apparent anisotropy is too complex in full generality. We can, however, examine a simple and illustrative problem. Following the recent work of Wang et al. (2013) and Fichtner et al. (2013a), we will place ourselves in the 1D radially symmetric case (VTI medium) and assume that apparent radial anisotropy is only due to vertical gradients, i.e., layering. Indeed, apart from the crust, the D' layer and around subducting slabs, to first order the earth is radially symmetric, with sharp horizontal seismic discontinuities separating different "layers" PREM (Dziewonski and Anderson 1981). In such a layered earth, vertical velocity gradients are much stronger than lateral ones and will significantly contribute to apparent anisotropy.

4.4 The Elastic Homogenization

We have seen that the limited resolution of long-wavelength seismic tomography only allows us to probe a smooth representation of the earth. However, this smooth equivalent Earth is not a simple spatial average of the true Earth, but the result of highly non-linear "upscaling" relations. In solid mechanics, these "upscaling" relations that link properties of a rapidly varying elastic medium to properties of the effective medium as seen by long waves have been the subject of extensive research (e.g., Hashin and Shtrikman 1963; Auriault and Sanchez-Palencia 1977; Bensoussan et al. 1978; Sanchez-Palencia 1980; Auriault et al. 1985; Murat and Tartar 1985; Sheng 1990; Allaire 1992, and many others).

In global seismology, upscaling schemes, also called elastic homogenization, have been recently developped for different kinds of settings (Capdeville and Marigo 2007; Capdeville et al. 2010a, b; Guillot et al. 2010; Capdeville et al. 2015). This class of algorithms enables to compute the effective properties of complex media, thus reducing the meshing complexity for the wave equation solver and hence the cost of computations. Elastic homogenization has been used to model complex crustal structures in full waveform inversions (Fichtner and Igel 2008; Lekić et al. 2010) and to combine results from different scales (Fichtner et al. 2013b).

4.4.1 The Backus Homogenization

Following the pioneering work by Thomson (1950), Postma (1955), and Anderson (1961), it was shown by Backus (1962) that a vertically transversely isotropic (VTI) medium is a "long-wave equivalent" to a smoothly varying medium of same nature (i.e., transversely isotropic). For parameters concerning shear-wave velocities, the smooth equivalent medium is simply described by the arithmetic and harmonic spatial average of elastic parameters N and L:

$$\tilde{N} = \langle N \rangle \tag{4.5}$$

$$\tilde{L} = \langle 1/L \rangle^{-1} \tag{4.6}$$

where $\langle \cdot \rangle$ refers to a spatial average with length scale given by the shortest wavelength defining our "long-wave." In the rest of the manuscript, the symbol ~ will be used to describe long-wave equivalent parameters. Note that these two relations are analogous to computing the equivalent spring constant (or equivalent resistance) when multiple springs (or resistors) are mounted either in series or parallel. In simple words, a horizontally traveling wave V_{SH} will see a set of fine horizontal layers "in parallel" (5), whereas a vertically traveling wave V_{SV} will see them "in series" (6). The apparent density $\tilde{\rho}$ is also given by the arithmetic mean of the local density:

$$\tilde{\rho} = \langle \rho \rangle \tag{4.7}$$

In the case of a locally isotropic medium ($N = L$), i.e., with no intrinsic anisotropy, the homogeneous anisotropy is simply given by the ratio of arithmetic to harmonic mean:

$$\tilde{\xi} = \frac{\tilde{N}}{\tilde{L}} = \langle N \rangle \langle 1/N \rangle \tag{4.8}$$

It can be easily shown that the arithmetic mean is always greater than the harmonic mean, which results in having artificial anisotropy in (8) always greater than unity in the case of an underlying isotropic model. In the case where the underlying layered model contains anisotropy ($N \neq L$), the observed anisotropy is given by

$$\tilde{\xi} = \frac{\tilde{N}}{\tilde{L}} = \langle N \rangle \langle 1/L \rangle \tag{4.9}$$

Here, it is clear that when inverting waveforms with a minimum period of ~40 s (i.e., with minimum wavelength is 160 km) that sample a medium with velocity gradients occurring at much smaller scales, the observed apparent anisotropy $\tilde{\xi}$ is going to be different from the intrinsic anisotropy $\xi = N/L$. Therefore, as shown by Wang et al. (2013) and Fichtner et al. (2013a), interpreting the observed effective $\tilde{\xi}$ in terms of ξ may lead to misinterpretations.

4.4.2 The Residual Homogenization

In this study, the goal was to interpret smooth tomographic models in terms of their layered and hence more realistic equivalent. However, tomographic models are not completely smooth and they are instead constructed as smooth anomalies around a discontinuous reference model. This is because the function linking the unknown model to the observed waveforms is linearized around a local point in the model space. This reference model often contains global discontinuities such as the Moho, or transition zone discontinuities at 410 and 660 km, which are fixed in the inversion and preserved in the model construction.

In the previous section, we have summarized an absolute homogenization for which no small scale is left in the effective medium. To account for the presence of a reference model, Capdeville et al. (2013), Capdeville and Cance (2015) recently described a modified homogenization, carried out with respect to a reference model, which we refer to as the residual homogenization. It allows us to homogenize only some interfaces of a discontinuous medium, while keeping the others intact.

Let us define the reference earth model by its density and elastic properties: (ρ_{ref}, A_{ref}, C_{ref}, F_{ref}, L_{ref}, and N_{ref}). Capdeville et al. (2013) showed that an equivalent model to the layered (A, C, F, L, and N) medium can be constructed with simple algebraic relations. For elastic parameters related to shear-wave velocities, we have:

$$\tilde{N} = N_{\text{ref}} + \langle N - N_{\text{ref}} \rangle \tag{4.10}$$

$$\frac{1}{\tilde{L}} = \frac{1}{L_{\text{ref}}} + \left\langle \frac{1}{L} - \frac{1}{L_{\text{ref}}} \right\rangle \tag{4.11}$$

Note that no particular assumption on the reference model is made, which can contain any wavelengths, and can be discontinuous. Furthermore, there is no linearity assumption, and this results holds for large differences between the reference and the layered model.

We show in Fig. 4.6 an example of residual homogenization. The layered VTI Model is shown in red with layers either isotropic ($\xi = 1$) or anisotropic ($\xi \neq 1$). A smooth equivalent model (for long waves of minimum wavelength of 100 km) that preserves the small scales of the reference model is shown in blue. The homogenization is done on the difference between the layered model in red and the reference model in thick light blue. After homogenization, we lose information about both the number and locations of discontinuities which are not in the reference model, as well as the location and level of intrinsic anisotropy.

It can be verified numerically that waveforms computed in the residual effective model and in the true layered model are identical when filtered with minimum period of 25 s (which corresponds to a minimum wavelength of 100 km). Figure 4.7 shows an example of seismogram computed by normal mode summation (Gilbert and Dziewonski 1975) in the residual effective model and compared with the solution computed in the true layered model. In both cases, the reference and homogenized traces show an excellent agreement.

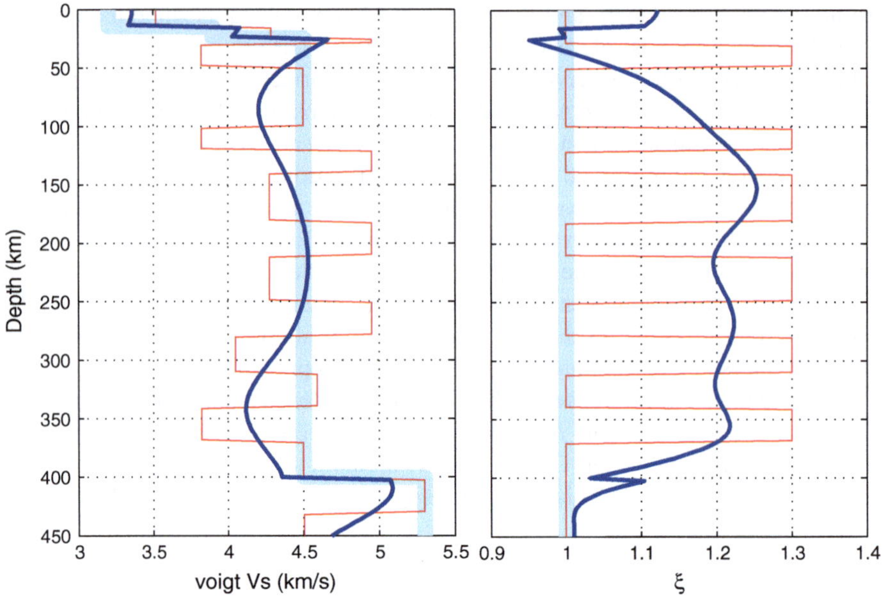

Fig. 4.6 Example of residual homogenization. *Left* Voigt average shear-wave velocity. *Right* Radial anisotropy. The layered model in *red* is homogenized around a reference model in light *blue*. The homogenized model is plotted in *blue*

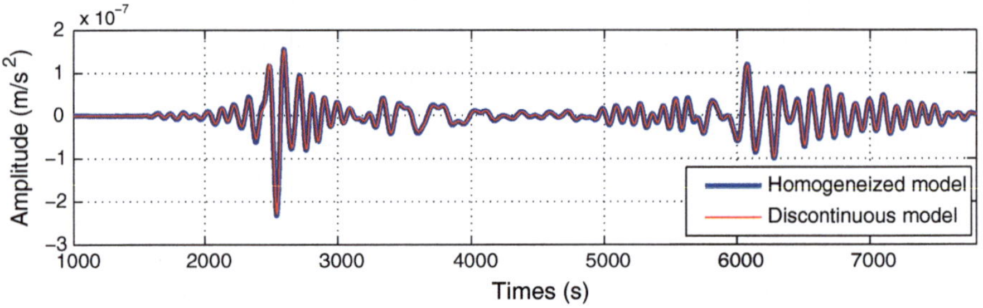

Fig. 4.7 Waveforms computed for the layered and homogenized models in Fig. 4.6. This is the radial component for an event with epicentral distance 82° and depth 150 km. The computation was done by normal mode summation (Gilbert and Dziewonski 1975)

4.4.3 An Approximation of "the Tomographic Operator"

Global full waveform tomography is always carried out with frequency band limited data. Intuitively, it makes sense to assume that such inversions can retrieve, at best, what is "seen" by the wave-field, i.e., an homogenized equivalent, and not the real medium.

Although it is difficult to mathematically prove this conjecture in general, Capdeville et al. (2013) numerically showed with synthetic examples that this is indeed the case for VTI media. That is, the inverted medium coincides with the residual homogenized version of the target model. Given a radially symmetric Earth and given enough stations and earthquakes, an inversion of full waveforms carried out around a reference model will therefore produce the residual homogeneous model defined above.

In this way, for any given layered model, one is able to predict with simple nonlinear algebraic smoothing operations what an inversion will find, without actually running the inversion. Therefore, we can view the residual homogenization as a first-order approximation of the "tomographic operator."

In practice, several practical issues complicate the situation: The real inversions are damped, producing unknown uncertainties in the recovered model, which can potentially bias our results. Furthermore, as seen above, ray coverage is not perfect and tomographic schemes may actually recover less than the effective medium.

4.5 Downscaling Smooth Models: The Inverse Homogenization

As we have seen, a tomographic inversion of long-period waves can only retrieve at best a homogenized model (and less in the case of an incomplete data coverage). Homogenization can lead to non-trivial and misleading effects that can make the interpretation difficult. We propose to treat the interpretation of tomographic images in terms of geological structures (discontinuities in our layered case) as a separate inverse problem, allowing to include a priori information and higher frequency data.

We call this inverse problem the inverse homogenization: For a given smooth 1D profile extracted from a tomographic model, what are the possible fine-scale (i.e., layered) models that are equivalent to this smooth 1D profile? Since the upscaling relations are based on nonlinear smoothing operators, it is not trivial to invert them to derive the true Earth from tomographic images, i.e., from its residual equivalent. In this section, we show that, although there is an infinite number of layered models that are equivalent to the smooth model in blue (Fig. 4.6), these models share common features, and Bayesian statistics can be used to constrain this ensemble of possible models. Furthermore, higher frequency data sensitive to discontinuities in radially symmetric models, such as receiver functions, can be used to constrain the location of horizontal discontinuities and reduce the space of possible earths.

4.5.1 Major Assumptions

Given the simple machinery presented in previous sections, there are obvious limitations to the proposed procedure. Let us here acknowledge a few of them.

1. We will assume here that long-period waves are only sensitive to the elastic parameters N and L (i.e., V_{SH} and V_{SV}). However, in a VTI medium, long-period seismograms, and hence the observed radial anisotropy, are also sensitive to the 3 other Love parameters (i.e., A, C, and F). Fichtner et al. (2013a) recently showed that P wave anisotropy is also important to distinguish between intrinsic and extrinsic anisotropy. Here, P wave anisotropy will be ignored.
2. Here, we restrict ourselves to transverse isotropy with a vertical axis of symmetry. Although this simple parameterization in terms of radial anisotropy is widely used in global seismology, it clearly represents an over-simplification, adopted for convenience in calculation. This is because the separation of intrinsic and apparent anisotropy can be studied analytically. The Earth is certainly not transversely isotropic, and there are indisputable proofs of azimuthal anisotropy. Azimuthal anisotropy might map into radial anisotropy in global models. These effects could be analyzed using the 3D version of non-periodic homogenization (Capdeville et al. 2010a, b).
3. We assume that 1D vertical profiles extracted from 3D tomographic models are the true Earth that has been homogenized with Backus relations. However, the smoothing operator applied to the true Earth during an inversion, namely the "tomographic operator," is determined by an ensemble of factors such as poor data sampling, the regularization and parameterization imposed, the level of data noise, the approximations made on the forward theory, and limited frequency band. It is very difficult to estimate how these averaging processes are applied to the true Earth during a tomographic inversion. What we assume here is that all these effects are negligible compared to the last one (limited frequency band), for which the smoothing operator is simply given by elastic homogenization. This only holds if data sampling is perfect, if no strong regularization has been artificially applied, and if the forward theory is perfect. Therefore, it is going to be most true in the case of full waveform inversion, and full waveform tomographic models are the most adequate for such a procedure. However, it is clear that other types of observations could be used as any tomographic method unavoidably produces apparent anisotropic long-wavelength equivalents. For example, our proposed procedure could be used to describe the ensemble of discontinuous models that fit a set of dispersion curves as in Khan et al. (2011).

4.5.2 Bayesian Inference

Using the notation commonly employed in geophysical inversion, the problem consists in finding a rapidly varying model m, such that its homogenized equivalent profile $g(m)$ is "close" to a given observed smooth model d. Here, the forward function g is the residual homogenization procedure in (4.10) and (4.11). Since the long-period waveforms are sensitive to smooth variations of the

"Backus parameters" (Capdeville et al. 2013), the observed tomographic profile is parameterized as $d = \left[\tilde{N}, 1/\tilde{L}\right]$.

This takes the form of a highly nonlinear inverse problem, and a standard linearized inversion approach based on derivatives is not adequate since the solution would strongly depend on the initial guess. Furthermore, the problem is clearly under-determined and the solution non-unique, and hence, it does not make sense to look for a single best fitting model that will minimize a misfit measure $\|d - g(m)\|$. For example, one can expect strong correlations and trade-offs between unknown parameters as homogeneous anisotropy can be either explained by discontinuities or intrinsic anisotropy. An alternative approach is to embrace the non-uniqueness directly and employ an inference process based on parameter space sampling. Instead of seeking a best model within an optimization framework, one seeks an ensemble of solutions and derives properties of that ensemble for inspection. Here, we use a Bayesian approach and tackle the problem probabilistically (Box and Tiao 1973; Sivia 1996; Tarantola 2005). We sample a posterior probability distribution $p(m|d)$, which describe the probability of having a discontinuous model m given an observed tomographic homogeneous profile d.

An important issue is the degree of freedom in the layered model. Since the inverse homogenization is a downscaling procedure, the layered model may be more complex (i.e., described with more parameters) than its smooth equivalent. As discussed above, the smooth model may be equivalent to either isotropic models with a large number of spatial parameters (layers), or anisotropic models described with more than one parameter per layer. This raises the question of the parameterization of m. How many layers should we impose on m? Should the existence (or not) of anisotropy be a free parameter? If yes, how many isotropic and anisotropic layers?

We propose to rely on Occam's razor, or the principle of parsimony, which states that simple models with the least number of parameters should be preferred (Domingos 1999). The razor states that one should favor simpler models until simplicity can be traded for greater explanatory power. Although we acknowledge that the definition of "simplicity" is rather subjective, in our problem, we will be giving higher probability to layered models described with fewer parameters.

We impose on m to be described with constant velocity layers separated by infinite gradients. As shown in Fig. 4.8 , we use a transdimensional parameterization, where the number of layers, as well as the number of parameters per layer, is free variables, i.e., unknown parameters (Sambridge et al. 2013). In this way, the number of layers will be unknown in the inversion, as well as the number of parameters in each layer: 1 for isotropic layers (V_S) and 2 for anisotropic layers (V_S and ξ). The goal here is not to describe the algorithm and its implementation in detail, but instead to give the reader a general description of the procedure, and show how it can be used to distinguish between intrinsic and extrinsic anisotropy. For a details on the algorithm, we refer the reader to Bodin et al. (2012b).

Bayes' theorem (1763) is used to combine prior information on the model with the observed data to give the posterior probability density function:

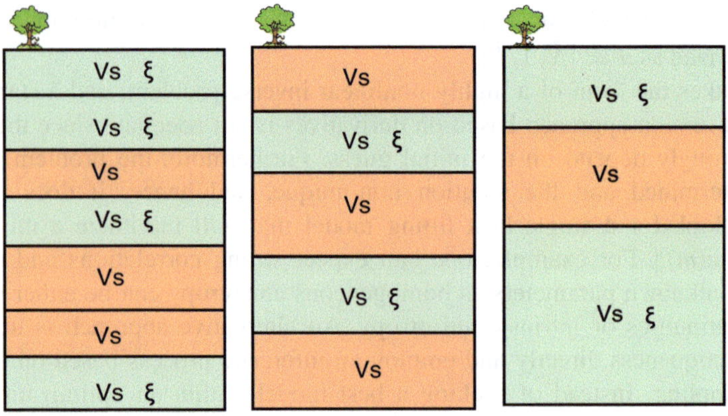

Fig. 4.8 Adaptive parameterization used for the inverse homogenization. The number of layers as well as the number of parameter in each layer (one for isotropic layers and two for anisotropic layers) is unknown in the inversion. This is illustrated here with three different models with different parameterizations. The parameterization is itself an unknown to be inverted for during the inversion scheme. Of course, data can always be better fitted as one includes more parameters in the model, but within a Bayesian formulation, preference will be given to simple models that explain observations with the least number of model parameters

$$\text{posterior} \propto \text{likelihood} \times \text{prior} \tag{4.12}$$

$$p(m|d) \propto p(d)|m)p(m) \tag{4.13}$$

$p(m)$ is the a priori probability density of m, that is, what we (think we) know about the model m before considering d. Here, we use poorly informative uniform prior distributions, and let model parameters vary over a large range of possible values.

The likelihood function $p(d|m)$ quantifies how equivalent a given discontinuous model is to our observed smooth profile d. The form of this probability density function is given by what we think about uncertainties on d. In our case, the form of the error statistics for a tomographic profile must be assumed to formulate $p(d|m)$. A problem with tomographic images is that they are obtained with linearized and regularised inversions, which biases uncertainty estimates. Therefore, we adopt a common and conservative choice (supported by the central limit theorem) and assume Gaussian-distributed errors. Since the data vector d is smooth, its associated errors must be correlated, and the fit to observations, $\Phi(m)$, is no longer defined as a simple 'least-square' measure but is the Mahalanobis distance between observed, d, and estimated, $g(m)$, smooth profiles:

$$\Phi(m) = (g(m) - d)^T C_e^{-1}(g(m) - d) \tag{4.14}$$

where C_e represents the covariance matrix of errors in d. In contrast to the Euclidean distance, this measure takes in account the correlation between data (equality being obtained where C_e is diagonal). Note that there is no user-defined regularization terms in (4.14) such as damping or smoothing constraints. This misfit function only depends on the observed data.

The general expression for the likelihood probability distribution is hence:

$$p(d|m) = \frac{1}{\sqrt{(2\pi)^n |C_e|}} \times \exp\left\{\frac{-\Phi(m)}{2}\right\}. \tag{4.15}$$

This is combined with the prior distribution to construct the posterior probability density function, which is thus defined in a space of variable dimension (transdimensional).

4.5.3 *Sampling a Transdimensional Probability Density Function*

Since the problem is transdimensional and nonlinear, there is no analytical formulation for the posterior probability density function, and instead we approximate it with a parameter search sampling algorithm (Monte Carlo). That is, we evaluate the posterior at a large number of locations in the model space. We use the reversible jump Markov chain Monte Carlo (rj-McMC) algorithm (Geyer and Møller 1994; Green 1995, 2003), which is a generalization of the well-known Metropolis–Hastings algorithm (Metropolis et al. 1953; Hastings 1970) to variable dimension models. The solution is represented by an ensemble of 1D models with variable number of layers and thicknesses, which are statistically distributed according to the posterior distribution. For a review of transdimensional Markov chains, see Sisson (2005). For examples of applications in the Earth sciences, see Malinverno (2002), Dettmer et al. (2010), Bodin et al. (2012a), Ray and Key (2012), Iaffaldano et al. (2012, 2013), Young et al. (2013), Tkalčić et al. (2013), Pilia et al. (2015) and Choblet et al. (2014).

In order to illustrate the power of the proposed Bayesian scheme, we applied it to the synthetic homogenized profile shown in Fig. 4.6, polluted with some Gaussian random correlated (i.e., smooth) noise. The solution is a large ensemble of models parameterized as in Fig. 4.8, for which the statistical distribution approximates the posterior probability distribution. As will be shown below, there are a number of ways to look at this ensemble of models. Here, in Fig. 4.9, we simply plot the 2D marginal distribution on the number of layers and number of anisotropic layers. This allows us to quantify the trade-off between anisotropy and heterogeneity. The distribution is clearly bimodal, meaning that the smooth equivalent profile can either be explained by many isotropic layers or a few anisotropic ones. From this, it is clear that we have not been able to distinguish between real and artificial anisotropy. However, we are able (given a layered parameterization) to quantify probabilistically the non-uniqueness of the problem.

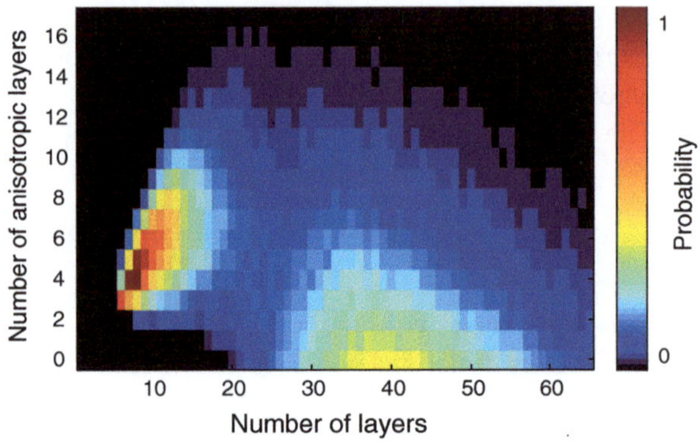

Fig. 4.9 Posterior probability distribution for the number of layers and number of anisotropic layers. This 2D marginal distribution allows us to quantify the trade-off between heterogeneity and anisotropy. Indeed, the smooth model in Fig. 4.6 can be either be explained with a large number of isotropic layers or a few anisotropic layers

This trade-off may be "broken" by adding independent constraints from other disciplines such as geology, mineral physics, or geodynamics. Here, we will show how higher frequency seismic data can bring information on the number and locations of discontinuities and hence enable us to investigate the nature of radial anisotropy in tomographic models.

4.6 Incorporating Discontinuities with Body Waves—Application to the North American Craton

A smooth equivalent profile brings little information about location of discontinuities, and extra information from higher frequency data is needed. Here, we show in a real case how adding independent constraints from converted P to S phases can help locating interfaces. Again, here, we place ourselves in the simplest case and assume horizontal layering when modeling converted phases. We acknowledge that dipping interfaces, or a tilted axis of anisotropy would produce apparent azimuthal anisotropy. Accounting for these effects is the subject of current work. We construct a 1D probabilistic seismic profile under northwest Canada, by combining in a joint Bayesian inversion a full waveform tomographic profile (SEMum2, French et al. 2013) with receiver functions. The goal here is to incorporate horizontal lithospheric discontinuities into a smooth image of the upper mantle and thus investigate the structure and history of the North American craton.

Archean cratons form the core of many of Earth's continents. By virtue of their longevity, they offer important clues about plate tectonic processes during early geological times. A question of particular interest is the mechanisms involved in cratonic assembly. The Slave province is one of the oldest Archaen cratons on Earth. Seismology has provided detailed information about the crust and upper mantle structure from different studies, such as reflection profiling (e.g., Cook et al. 1999), receiver function analysis (e.g., Bostock 1998), surface wave tomography (e.g., Van Der Lee and Frederiksen 2005), or regional full waveform (Yuan and Romanowicz 2010).

Recent studies (Yuan et al. 2006; Abt et al. 2010) have detected a structural boundary under the Slave craton at depths too shallow to be consistent with the lithosphere–asthenosphere boundary. Yuan and Romanowicz (2010) showed that this mid-lithospoheric discontinuity (MLD) may coincide with a change in the direction of azimuthal anisotropy and thus revealed the presence of two distinct lithospheric layers throughout the craton: a top layer chemically depleted above a thermal conductive root. On the other hand, Chen et al. (2009)showed that this seismic discontinuity as seen by receiver functions overlapped with a positive conductivity anomaly and interpreted it as the top of an archean subducted slab.

This type of fine structure within the lithosphere is not resolved in global tomographic models such as SEMum2 and hence may be mapped into radial anisotropy. Here, we will explore whether lithospheric layering as seen by scattered body waves (receiver functions) is compatible with the radial anisotropy imaged from global tomography.

4.6.1 Long-Period Information: A Smooth Tomographic Profile

We used the global model recently constructed by the Berkeley group: SEMum2 (Lekić and Romanowicz 2011; French et al. 2013). This model is the first global model where the synthetic waveforms are accurately computed in a 3D Earth with the spectral element method. Sensitivity kernels are calculated approximately using nonlinear asymptotic coupling theory (NACT: Li and Romanowicz (1995)). The database employed consists of long-period ($60 < T < 400$ s) three-component waveforms of 203 well-distributed global earthquakes ($6.0 < M_w < 6.9$), as well as global group-velocity dispersion maps at $25 < T < 150$ s.

Compared to other global shear-velocity models, the amplitudes of velocity anomalies are stronger in SEMum2, with stronger velocity minima in the low-velocity zone (asthenosphere) and a more continuous signature of fast velocities in subduction zones.

Here, we extract a 1D profile (Fig. 4.10) under station YKW3, located in the southern Slave craton, northwest Canada. As seen in Fig. 4.10, the crustal structure in SEMum2 is replaced with a single, smooth equivalent anisotropic layer, valid for modeling long-period waves. Note also that the high amplitude of radial anisotropy below the crust may be due to unmapped layering at these depths.

4.6.2 Short-Period Information: Teleseismic Converted Phases

In order to bring short-wavelength information to the tomographic profile, we analyzed waveforms for first P arrivals on teleseismic earthquake records at the

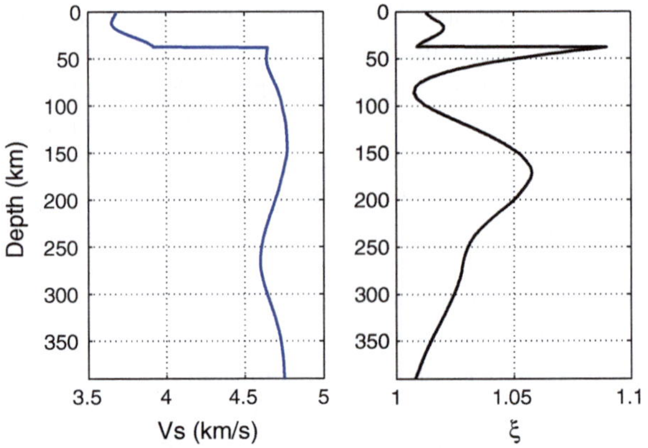

Fig. 4.10 Tomographic profile under station YKW3 for model SEMum2

broadband station YKW3 of the Yellowknife seismic array. The station was installed in late 1989 and has collected a large amount of data. Receiver function analysis consists of deconvolving the vertical from the horizontal component of seismograms (Vinnik 1977; Burdick and Langston 1977; Langston 1979). In this way, the influence of source and distant path effects are eliminated, and hence, one can enhance conversions from P to S generated at boundaries beneath the recording site. This is a widely used technique in seismology, with tens of papers published each year (e.g., Ford et al. 2010; Hopper et al. 2014). For a recent and comprehensive review, see Bostock (2014).

Algorithms for inversion of receiver functions are usually based on optimization procedures, where a misfit function is minimized. Traditionally, this misfit function is constructed by comparing the observed receiver function with a receiver function predicted for some Earth model m:

$$\Phi(m) = \left\| \frac{H(t)}{V(t)} - \frac{h(t,m)}{v(t,m)} \right\|^2 \tag{4.16}$$

where $V(t)$ is the vertical and $H(t)$ the horizontal (radial) component of the observed seismogram, and where $v(t,m)$ and $h(t,m)$ are predicted structure response functions for the unknown Earth model m. The fraction refers to a deconvolution (or spectral division).

A well-known problem is that the deconvolution is an unstable numerical procedure that needs to be damped, which results in a difficulty to correctly account for uncertainties. Therefore, for Bayesian analysis, we choose an alternative misfit function based on a simple cross-product that avoids deconvolution (Bodin et al. 2014):

$$\Phi(m) = \|H(t) * v(t,m) - V(t) * h(t,m)\|^2 \tag{4.17}$$

This misfit function is equivalent to the distance between the observed and predicted receiver functions in (4.16). Since discrete convolution in time is

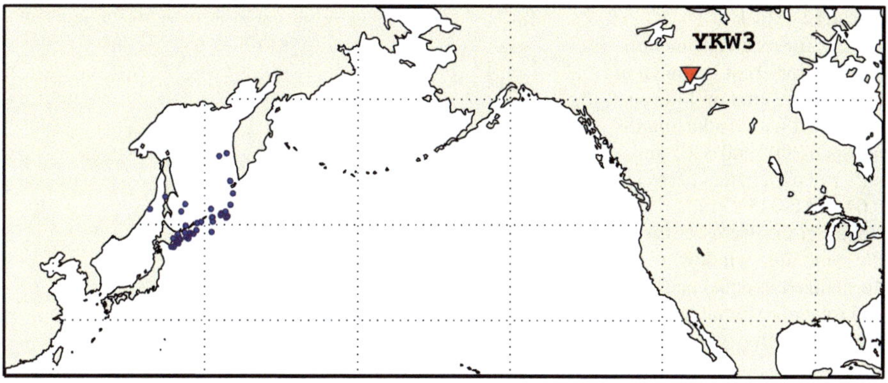

Fig. 4.11 Station YKW3 with the set of events used for receiver function analysis

a simple summation and since seismograms can be seen as corrupted by random errors, each sample of the signal obtained after discrete convolution is then a sum of random variables, whose statistics are straightforward to calculate with algebra of random variables. This is not the case with deconvolution schemes.

Assuming that $V(t)$ and $H(t)$ contain independent and normally distributed random errors with standard deviation σ, a likelihood probability function can be constructed:

$$p(d_{\mathrm{RF}}|m) = \frac{1}{\sqrt{(2\pi\sigma^2)^n}} \times \exp\left(\frac{-\Phi(m)}{2\sigma^2}\right) \tag{4.18}$$

The observed vertical $V(t)$ and horizontal $H(t)$ waveforms needed for inversion were obtained by simply stacking a number of events measured for a narrow range of backazimuths and epicentral distances (see Figs. 4.11 and 4.12). Influence of the receiver structure is common to all records and is enhanced by summation (Shearer 1991; Kind et al. 2012). We refer to Bodin et al. (2014) for details of the procedure.

This likelihood function thus defined for receiver functions $p(d_{\mathrm{RF}}|m)$ can be combined with the likelihood function defined above for the Inverse homogenization problem $p(d_{\mathrm{tomo}}|m)$ for joint inversion of short- and long-wavelength information. Since the observations given by the tomographic model d_{tomo} are independent of the receiver function observations d_{RF}, the complete posterior probability function is then defined as follows:

$$p(m|d_{\mathrm{tomo}}, d_{\mathrm{RF}}) \propto p(d_{\mathrm{tomo}}|m) \times p(d_{\mathrm{RF}}|m) \times p(m) \tag{4.19}$$

and can be sampled with the reversible jump algorithm described above.

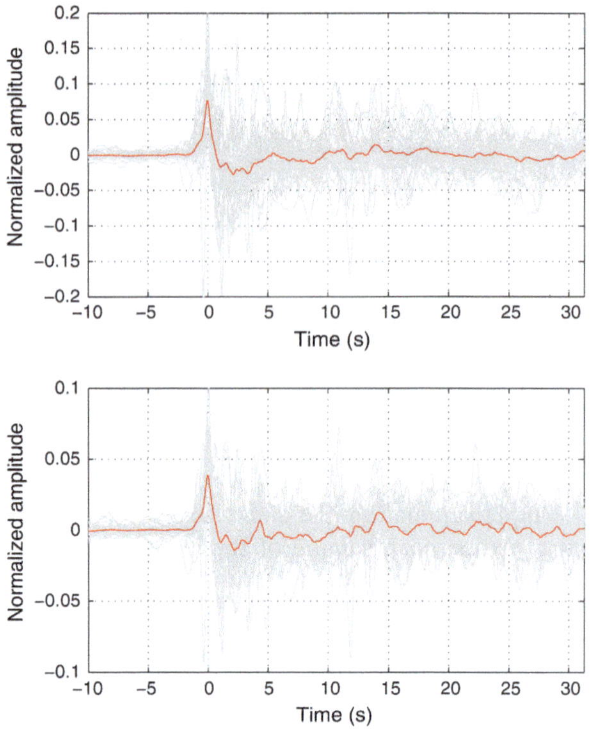

Fig. 4.12 Stack of P arrivals for receiver function analysis for event is shown in Fig. 4.11. A total of 44 events were used with backazimuths between 290° and 320° and with ray parameters between 0.04 and 0.045 s km^{-1}. Seismograms were cut for the same time window, normalized to equal energy, and rotated to radial and tangential components

4.6.3 Results

Transdimensional inversion was carried out allowing between 2 and 60 layers. As noted above, each layer is either described by one or two parameters. An a priori constrain for minimum and maximum velocity value and anisotropy in each layer was applied. The algorithm was implemented for parallel computers, providing a thorough search of the model space, with an ensemble solution made of about 10^6 different Earth models. The posterior distribution is approximated from the distribution (i.e., the histogram) of the ensemble of models in the solution (Fig. 4.13). The solution is thus given by an ensemble of 1D models with variable number of layers, thicknesses, and elastic parameters. In order to visualize the final ensemble, the collected models can be projected into a number of physical spaces that are used for interpretation.

For example, Fig. 4.13a shows the marginal distribution for S wave velocities as a function of depth, simply constructed from the density plot of the ensemble of models in the solution. Here, a number of expected lithospheric discontinuities have been imaged, such as the mid-lithospheric discontinuity at 90 km and a sharp lithosphere–asthenosphere boundary at 200 km. But interestingly, we also note a sharp positive velocity change at 150 km, which establishes the base of an

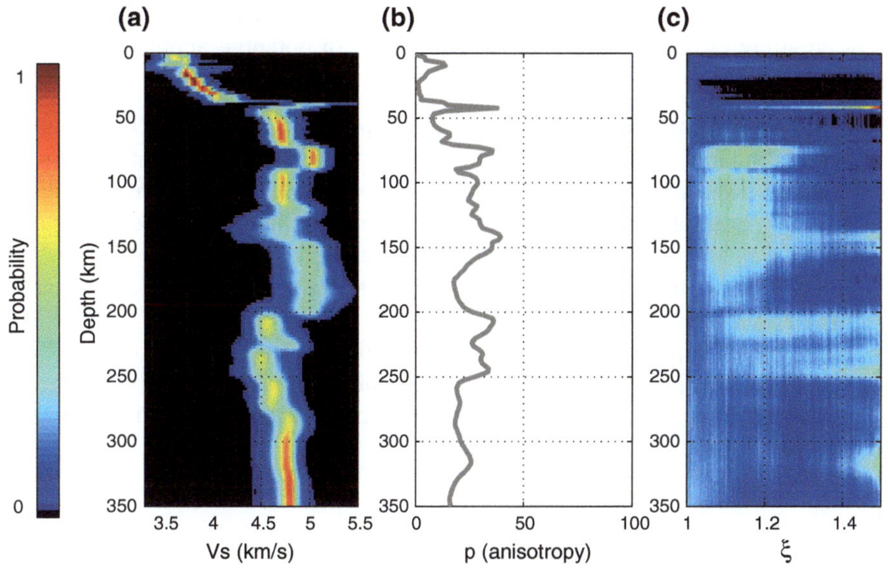

Fig. 4.13 Joint inversion of converted body waves in Fig. 4.12 and of the tomographic model in Fig. 4.10. *Left* Probability distribution for V_S. *Middle* Probability of having an anisotropic layer. *Right* Probability for ξ

intra-lithospheric low-velocity zone between 90 and 150 km. This low-velocity zone is clearly visible in Chen et al. (2007) who inverted Rayleigh wave phase-velocity observations (20 s $< T <$ 142 s) measured by the Yellowknife array. This feature is also observed in the regional full waveform tomographic model by Yuan and Romanowicz (2010). This low-velocity zone can be interpreted as a piece of archean subducted slab, stacked vertically over another archean block. For a discussion about upper mantel seismic discontibuities, see Schmerr (2015).

Since here we are interested in the relative contribution of layering to the observed anisotropy in SEMum2, we can look at the probability at each depth to have intrinsic anisotropy (i.e., $\xi \neq 1$). For each model, each layer is either isotropic or anisotropic. Therefore, at each depth, one can count the ratio of isotropic to anisotropic layers in the ensemble of models. This is shown in Fig. 4.13b. When no information is brought by the data (prior distribution), the probability to have anisotropy is 50 %. However here, the probability is lower and around 20 % across much of the depth profile. This implies that intrinsic anisotropy is not required to fit the smooth tomographic profile. In other words, the discontinuities required to fit the converted body waves may be enough to explain the anisotropy in the tomographic model. We view this as an important result, indicating that radial anisotropy (at least under cratons) should not be directly interpreted in tomographic models.

The third panel in Fig. 4.13c shows the probability distribution for ξ at each depth. Note that, for a given depth, around 80 % of models are isotropic with $\xi = 1$, and this distribution only represents the level of anisotropy in the 20 % remaining models.

Note that these are only preliminary results. This study is only a proof of concept, or study of feasibility, which opens a range of potential applications. Let us acknowledge a number limitations:

1. Here, we assume that smooth tomographic profiles are the true Earth that has been homogenized. Although this has been numerically demonstrated on 1D synthetic tests for VTI models (Capdeville et al. 2013), it may not always be true. For example, the effect of poor data sampling and hence model regularization may act as a smoothing operator not accounted for here.
2. Little is known about uncertainties in tomographic models, which are crucial in the context of a Bayesian formulation.
3. Here, we only place ourselves in the case of VTI profile. No azimuthal anisotropy is considered. Furthermore, it is well known that lateral heterogeneities (e.g., Moho topography) may also produce apparent radial anisotropy. This case has not been considered here.

4.7 Conclusion

Global tomographic imaging is an inverse problem where different types of observables (e.g., surface waves, body waves) are used to constrain different types of parameters (e.g., P and S wave velocity, anisotropy, density). Different measurements have different sensitivities, and all parameters are not equally well resolved. Some parameters present strong trade-offs. Furthermore, the elastic properties to be constrained are scale dependent. These issues have led seismologists to simplify the inverse problem and to invert separately different observables, different frequency bands, and with different spatial and physical parameterizations. Thus, different classes of models with different resolving power have been published, which are sometimes difficult to reconcile.

Recent theoretical developments as well as increased availability of computational power have allowed the emergence of full waveform inversion, where the full wave-field (and its derivatives with respect to the model) is exactly computed with purely numerical methods. Inverting the full wave-field in the time domain enables us to combine body and surface waves in the same inversion scheme. The recent local, regional, and global applications of full waveform tomography reveal indeed an improved resolution. For example, the amplitudes of the imaged anomalies increase, and more small-scale features are constrained. However, a remaining challenge is computational cost. One way to keep reasonable the time of computations is to limit the frequency content of waveforms and only invert the long periods. As a result, the resolving power of full waveform tomography is mainly determined by the minimum period or minimum wavelength.

Elastic properties are scale dependent, and hence, the small-scale heterogeneities that are not resolved in tomographic models are mapped into large-scale structure. This mapping is nonlinear, which makes large-scale structure difficult to interpret. In this study, we have focused on vertical transversely isotropic (VTI) models, where small-scale heterogeneities in S wave velocity are mapped into large-scale radial anisotropy.

A method has been proposed to “downscale” smooth tomographic models. Given the laws of homogenization which link a medium to its effective long-wave equivalent, we are able to explore the space of possible small-scale models that are equivalent to a given tomographic profile. We cast this inverse problem in a Bayesian formulation, which enables us to probabilistically quantify the trade-off between heterogeneity (in our case layering) and anisotropy.

We have applied this method to a tomographic profile of the North American craton, and added short-period information from receiver functions, to help locate the depth of discontinuities. This allows us to investigate the nature and history of the craton. But more importantly, we have shown that a large part of anisotropy present in the tomographic model may be due to unmapped discontinuities. Therefore, we conclude that one should not interpret radial anisotropy in tomographic models only in terms of geodynamics, e.g., mantle flow. The inferred radial anisotropy contains valuable information about the earth, but one has to keep in mind that this is only apparent anisotropy. It may only be interpreted when associated with higher frequency information, or with additional information from other disciplines (geology, mineral physics, or geodynamics).

Here, we have assumed that the observed radial anisotropy is either due to layering or intrinsic radial anisotropy. We recognize that this a strong approximation. Indeed, both lateral heterogeneities and intrinsic azimuthal anisotropy may also contribute to the observed radial anisotropy. Although these effects are not considered here, they are the subject of current work. For example, we expect observations of azimuthal anisotropy to provide important additional constraints: (1) These observations cannot be explained by horizontal layering; (2) they may sometimes be due to LPO with a tilted axis of symmetry, in which case part of the observed radial anisotropy may still be intrinsic.

Acknowledgments We thank Andreas Fichtner and Lapo Boschi for providing constructive reviews. This project was supported by the French-Berkeley Fund (FBF). Thomas Bodin wishes to acknowledge support from the Miller Institute for Basic Research at the University of California, Berkeley. Y. Capdeville was partly supported by the French ANR blanche “mm” (grant ANR-10-BLAN-613 MEME). Computations were performed on the CCIPL computer “Erdre.” Computational resources were also provided by the Institutional Computing program of LANL. We are grateful to Cheng Cheng and Scott French for providing figures. We also thank Scott French for providing model SEMum2. We are also very grateful to Heather Ford and Karen Fischer for providing teleseismic P waveforms recorded at station YKW3.

References

Abt D, Fischer K, French S, Ford H, Yuan H, Romanowicz B (2010) North american lithospheric discontinuity structure imaged by PS and SP receiver functions. J Geophys Res 115(10.1029)

Aki K (1982) Scattering and attenuation. Bull Seismol Soc Am 72(6B):S319–S330

Aki K, Christoffersson A, Husebye ES (1977) Determination of the three-dimensional seismic structure of the lithosphere. J Geophys Res 82(2):277–296

Allaire G (1992) Homogenization and two-scale convergence. SIAM J Math Anal 23:1482–1518
Anderson DL (1961) Elastic wave propagation in layered anisotropic media. J Geophys Res 66(9):2953–2963
Anderson DL, Dziewonski AM (1982) Upper mantle anisotropy: evidence from free oscillations. Geophys J Int 69(2):383–404
Auriault J-L, Sanchez-Palencia E (1977) Étude du comportement macroscopique d'un milieu poreux saturé déformable. J Mécanique 16(4):575–603
Auriault J-L, Borne L, Chambon R (1985) Dynamics of porous saturated media, checking of the generalized law of Darcy. J Acoust Soc Am 77:1641
Babuška V, Cara M (1991) Seismic anisotropy in the earth, vol 10. Springer, Berlin
Backus G, Gilbert F (1968) The resolving power of gross earth data. Geophys J Roy Astron Soc 16(2):169–205
Backus GE (1962) Long-wave elastic anisotropy produced by horizontal layering. J Geophys Res 67(11):4427–4440
Bayes T (1763) An essay towards solving a problem in the doctrine of chances, C. Davis, Printer to the Royal Society of London
Bensen G, Ritzwoller M, Yang Y (2009) A 3-d shear velocity model of the crust and uppermost mantle beneath the united states from ambient seismic noise. Geophys J Int 177(3):1177–1196
Bensoussan A, Lions J-L, Papanicolaou G (1978) Asymptotic analysis of periodic structures, North Holland. AMS Chelsea Publishing, Providence
Bercovici D, Karato S-I (2003) Whole-mantle convection and the transition-zone water filter. Nature 425(6953):39–44
Bijwaard H, Spakman W, Engdahl E (1998) Closing the gap between regional and global travel time tomography. J Geophys Res 103(B12):30055
Bodin T, Maupin V (2008) Resolution potential of surface wave phase velocity measurements at small arrays. Geophys J Int 172(2):698–706
Bodin T, Salmon M, Kennett B, Sambridge M (2012a) Probabilistic surface reconstruction from multiple data-sets-an example for the Australian moho. J Geophys Res 117(B10)
Bodin T, Sambridge M, Tkalčić H, Arroucau P, Gallagher K, Rawlinson N (2012b) Transdimensional inversion of receiver functions and surface wave dispersion. J Geophys Res 117:B02301
Bodin T, Yuan H, Romanowicz B (2014) Inversion of receiver functions without deconvolution application to the Indian craton. Geophys J Int 196(2):1025–1033
Bonnet M, Constantinescu A (2005) Inverse problems in elasticity. Inverse Prob 21(2):R1
Bostock M (1998) Mantle stratigraphy and evolution of the slave province. J Geophys Res Solid Earth (1978–2012) 103(B9):21183–21200
Bostock M (2014) Teleseismic body-wave scattering and receiver-side structure. Treatise Geophys 1:219–246
Box G, Tiao G (1973) Bayesian Inference in Statistical Inference. Addison-Wesley, Boston
Bozdağ E, Trampert J (2008) On crustal corrections in surface wave tomography. Geophys J Int 172(3):1066–1082
Burdick L, Langston C (1977) Modeling crustal structure through the use of converted phases in teleseismic body-wave forms. Bull Seismol Soc Am 67(3):677–691
Burdick S, Li C, Martynov V, Cox T, Eakins J, Mulder T, Astiz L, Vernon FL, Pavlis GL, van der Hilst RD (2008) Upper mantle heterogeneity beneath North America from travel time tomography with global and USArray transportable array data. Seismol Res Lett 79(3):384–392
Capdeville Y, Marigo J-J (2007) Second order homogenization of the elastic wave equation for non-periodic layered media. Geophys J Int 170(2):823–838
Capdeville Y, Cance P (2015) Residual homogenization for elastic wave propagation in complex media.Geophys J Int 200(2):984–997
Capdeville Y, To A, Romanowicz B (2003) Coupling spectral elements and modes in a spherical earth: an extension to the sandwich case. Geophys J Int 154(1):44–57

Capdeville Y, Gung Y, Romanowicz B (2005) Towards global earth tomography using the spectral element method: a technique based on source stacking. Geophys J Int 162(2):541–554

Capdeville Y, Guillot L, Marigo J-J (2010a) 1-D non-periodic homogenization for the seismic wave equation. Geophys J Int 181(2):897–910

Capdeville Y, Guillot L, Marigo J-J (2010b) 2-D non-periodic homogenization to upscale elastic media for p–sv waves. Geophys J Int 182(2):903–922

Capdeville Y, Zhao M, Cupillard P (2015) Fast fourier homogenization for elastic wave propagation in complex media. Wave Motion

Capdeville Y, Stutzmann É, Wang N, Montagner J-P (2013) Residual homogenization for seismic forward and inverse problems in layered media. Geophys J Int 194(1):470–487

Carcione JM (1994) The wave equation in generalized coordinates. Geophysics 59(12):1911–1919

Chapman C (2004) Fundamentals of seismic wave propagation. Cambridge University Press, Cambridge

Chen C-W, Rondenay S, Weeraratne DS, Snyder DB (2007) New constraints on the upper mantle structure of the slave craton from rayleigh wave inversion. Geophys Res Lett 34(10):L10301

Chen C-W, Rondenay S, Evans RL, Snyder DB (2009) Geophysical detection of relict metasomatism from an Archean (3.5 ga) subduction zone. Science 326(5956):1089–1091

Cheng C, Allen RM, Porritt RW, Ballmer MD (2015) Seismic constraints on a double-layered asymmetric whole-mantle plume beneath Hawaii. Hawaiian volcanoes: from source to surface. pp 19–34

Choblet G, Husson L, Bodin T (2014) Probabilistic surface reconstruction of coastal sea level rise during the twentieth century 2014. J Geophys Res (Solid Earth) 119:9206–9236. doi:10.1002/2014JB011639

Colli L, Fichtner A, Bunge H-P (2013) Full waveform tomography of the upper mantle in the south Atlantic region: imaging a westward fluxing shallow asthenosphere? Tectonophysics 604:26–40

Cook FA, Velden AJ, Hall KW, Roberts BJ (1999) Frozen subduction in Canada's Northwest Territories: lithoprobe deep lithospheric reflection profiling of the western Canadian shield. Tectonics 18(1):1–24

Crampin S (1981) A review of wave motion in anisotropic and cracked elastic-media. Wave Motion 3(4):343–391

Crampin S, Booth DC (1985) Shear-wave polarizations near the north Anatolian fault–II. Interpretation in terms of crack-induced anisotropy. Geophys J Int 83(1):75–92

Dahlen F, Hung S-H, Nolet G (2000) Fréchet kernels for finite-frequency traveltimes–I. Theory. Geophys J Int 141(1):157–174

Debayle E, Kennett B (2000) Anisotropy in the Australasian upper mantle from love and rayleigh waveform inversion. Earth Planet Sci Lett 184(1):339–351

Debayle E, Ricard Y (2012) A global shear velocity model of the upper mantle from fundamental and higher rayleigh mode measurements. J Geophys Res Solid Earth (1978–2012) 117(B10)

Dettmer J, Dosso S, Holland C (2010) Trans-dimensional geoacoustic inversion. J Acoust Soc Am 128:3393

Domingos P (1999) The role of Occam's razor in knowledge discovery. Data Min Knowl Disc 3(4):409–425

Dziewonski A, Anderson D (1981) Preliminary reference earth model. Phys Earth Planet Inter 25(4):297–356

Ekström G, Dziewonski AM (1998) The unique anisotropy of the pacific upper mantle. Nature 394(6689):168–172

Endrun B, Meier T, Lebedev S, Bohnhoff M, Stavrakakis G, Harjes H-P (2008) S velocity structure and radial anisotropy in the aegean region from surface wave dispersion. Geophys J Int 174(2):593–616

Ferreira A, Woodhouse J, Visser K, Trampert J (2010) On the robustness of global radially anisotropic surface wave tomography. J Geophys Res Solid Earth (1978–2012) 115(B4)

Fichtner A, Igel H (2008) Efficient numerical surface wave propagation through the optimization of discrete crustal models a technique based on non-linear dispersion curve matching (dcm). Geophys J Int 173(2):519–533

Fichtner A, Trampert J (2011) Resolution analysis in full waveform inversion. Geophys J Int 187(3):1604–1624
Fichtner A, Bunge H-P, Igel H (2006) The adjoint method in seismology: I. Theory. Phys Earth Planet Inter 157(1):86–104
Fichtner A, Kennett BL, Igel H, Bunge H-P (2009) Full seismic waveform tomography for upper-mantle structure in the australasian region using adjoint methods. Geophys J Int 179(3):1703–1725
Fichtner A, Kennett BL, Igel H, Bunge H-P (2010) Full waveform tomography for radially anisotropic structure: new insights into present and past states of the australasian upper mantle. Earth Planet Sci Lett 290(3):270–280
Fichtner A, Kennett BL, Trampert J (2013a) Separating intrinsic and apparent anisotropy. Phys Earth Planet Inter 219:11–20
Fichtner A, Trampert J, Cupillard P, Saygin E, Taymaz T, Capdeville Y, Villaseñor A (2013b) Multiscale full waveform inversion. Geophys J Int 194(1):534–556. doi:10.1093/gji/ggt118
Ford HA, Fischer KM, Abt DL, Rychert CA, Elkins-Tanton LT (2010) The lithosphere–asthenosphere boundary and cratonic lithospheric layering beneath australia from Sp wave imaging. Earth Planet Sci Lett 300(3):299–310
French S, Lekic V, Romanowicz B (2013) Waveform tomography reveals channeled flow at the base of the oceanic asthenosphere. Science 342(6155):227–230
Friederich W, Huang Z-X (1996) Evidence for upper mantle anisotropy beneath southern Germany from love and rayleigh wave dispersion. Geophys Res Lett 23(10):1135–1138
Fukao Y, Obayashi M (2013) Subducted slabs stagnant above, penetrating through, and trapped below the 660 km discontinuity. J Geophys Res Solid Earth 118(11):5920–5938
Fukao Y, Widiyantoro S, Obayashi M (2001) Stagnant slabs in the upper and lower mantle transition region. Rev Geophys 39(3):291–323
Geyer C, Møller J (1994) Simulation procedures and likelihood inference for spatial point processes. Scand J Stat 21(4):359–373
Gilbert F, Dziewonski A (1975) An application of normal mode theory to the retrieval of structural parameters and source mechanisms for seismic spectra. Philos Trans R Soc London A 278:187–269
Green P (1995) Reversible jump MCMC computation and Bayesian model selection. Biometrika 82:711–732
Green P (2003) Trans-dimensional Markov chain Monte Carlo. Highly Struct Stoch Syst 27:179–198
Guillot L, Capdeville Y, Marigo J-J (2010) 2-D non-periodic homogenization of the elastic wave equation: Sh case. Geophys J Int 182(3):1438–1454
Gung Y, Panning M, Romanowicz B (2003) Global anisotropy and the thickness of continents. Nature 422(6933):707–711
Hashin Z, Shtrikman S (1963) A variational approach to the elastic behavior of multiphase materials. J Mech Phys Solids 11:127–140
Hastings W (1970) Monte Carlo simulation methods using Markov chains and their applications. Biometrika 57:97–109
Helbig K, Thomsen L (2005) 75-plus years of anisotropy in exploration and reservoir seismics: a historical review of concepts and methods Geophysics 70(6):9ND–23ND
Hess H (1964) Seismic anisotropy of the uppermost mantle under oceans. Nature 203:629–631
Hopper E, Ford HA, Fischer KM, Lekic V, Fouch MJ (2014) The lithosphere–asthenosphere boundary and the tectonic and magmatic history of the northwestern united states. Earth Planet Sci Lett 402:69–81
Houser C, Masters G, Shearer P, Laske G (2008) Shear and compressional velocity models of the mantle from cluster analysis of long-period waveforms. Geophys J Int 174(1):195–212
Iaffaldano G, Bodin T, Sambridge M (2012) Reconstructing plate-motion changes in the presence of finite-rotations noise. Nat Commun 3:1048
Iaffaldano G, Bodin T, Sambridge M (2013) Slow-downs and speed-ups of India–Eurasia convergence since: data-noise, uncertainties and dynamic implications. Earth Planet Sci Lett 367:146–156

Kárason H, Van Der Hilst RD (2000) Constraints on mantle convection from seismic tomography. Hist Dyn Glob Plate Mot 277–288

Kawakatsu H, Kumar P, Takei Y, Shinohara M, Kanazawa T, Araki E, Suyehiro K (2009) Seismic evidence for sharp lithosphere-asthenosphere boundaries of oceanic plates. Science 324(5926):499–502

Kelly K, Ward R, Treitel S, Alford R (1976) Synthetic seismograms: a finite-difference approach. Geophysics 41(1):2–27

Kennett B (1995) Approximations for surface-wave propagation in laterally varying media. Geophys J Int 122(2):470–478

Kennett B, Nolet G (1990) The interaction of the S-wavefield with upper mantle heterogeneity. Geophys J Int 101(3):751–762

Khan A, Boschi L, Connolly J (2011) Mapping the earth's thermochemical and anisotropic structure using global surface wave data. J Geophys Res Solid Earth (1978–2012) 116(B1)

Kind R, Yuan X, Kumar P (2012) Seismic receiver functions and the lithosphere–asthenosphere boundary. Tectonophysics 536:25–43

Komatitsch D, Tromp J (1999) Introduction to the spectral element method for three-dimensional seismic wave propagation. Geophys J Int 139(3):806–822

Komatitsch D, Tromp J (2002) Spectral-element simulations of global seismic wave propagation—I. Validation. Geophys J Int 149(2):390–412

Komatitsch D, Vilotte J-P (1998) The spectral element method: an efficient tool to simulate the seismic response of 2d and 3d geological structures. Bull Seismol Soc Am 88(2):368–392

Komatitsch D, Vinnik LP, Chevrot S (2010) Shdiff-svdiff splitting in an isotropic earth. J Geophys Res 115(B7):B07312

Kustowski B, Ekström G, Dziewoński A (2008) Anisotropic shear-wave velocity structure of the earth's mantle: a global model. J Geophys Res Solid Earth (1978–2012) 113(B6)

Langston C (1979) Structure under Mount Rainier, Washington, inferred from teleseismic body waves. J Geophys Res 84(B9):4749–4762

Lebedev S, Van Der Hilst RD (2008) Global upper-mantle tomography with the automated multimode inversion of surface and s-wave forms. Geophys J Int 173(2):505–518

Lebedev S, Boonen J, Trampert J (2009) Seismic structure of Precambrian lithosphere: new constraints from broad-band surface-wave dispersion. Lithos 109(1):96–111

Lekić V, Romanowicz B (2011) Inferring upper-mantle structure by full waveform tomography with the spectral element method. Geophys J Int 185(2):799–831

Lekić V, Panning M, Romanowicz B (2010) A simple method for improving crustal corrections in waveform tomography. Geophys J Int 182(1):265–278

Lekic V, Cottaar S, Dziewonski A, Romanowicz B (2012) Cluster analysis of global lower mantle tomography: a new class of structure and implications for chemical heterogeneity. Earth Planet Sci Lett 357:68–77

Levshin A, Ratnikova L (1984) Apparent anisotropy in inhomogeneous media. Geophys J Int 76(1):65–69

Li C, van der Hilst RD, Engdahl ER, Burdick S (2008) A new global model for P wave speed variations in earth's mantle. Geochem Geophys Geosyst 9(5)

Li X-D, Romanowicz B (1995) Comparison of global waveform inversions with and without considering cross-branch modal coupling. Geophys J Int 121(3):695–709

Li X-D, Romanowicz B (1996) Global mantle shear velocity model developed using nonlinear asymptotic coupling theory. J Geophys Res Solid Earth (1978–2012) 101(B10):22245–22272

Li X-D, Tanimoto T (1993) Waveforms of long-period body waves in a slightly aspherical earth model. Geophys J Int 112(1):92–102

Liu Q, Gu Y (2012) Seismic imaging: from classical to adjoint tomography. Tectonophysics 566:31–66

Love A (1927) The mathematical theory of elasticity. CRC Press, Boca Raton

Mainprice D (2007) Seismic anisotropy of the deep earth from a mineral and rock 1022 physics perspective. In: Schubert G (ed) Treatise in geophysics, vol 2, pp 437–492

Malinverno A (2002) Parsimonious Bayesian Markov chain Monte Carlo inversion in a nonlinear geophysical problem. Geophys J Int 151(3):675–688

Maupin V (2001) A multiple-scattering scheme for modelling surface wave propagation in isotropic and anisotropic three-dimensional structures. Geophys J Int 146(2):332–348

Maupin V, Park J (2014) Theory and observations—wave propagation in anisotropic media. Seismol Struct Earth Treatise Geophys 1:289–321

Mégnin C, Romanowicz B (2000) The three-dimensional shear velocity structure of the mantle from the inversion of body, surface and higher-mode waveforms. Geophys J Int 143(3):709–728

Metropolis N et al (1953) Equations of state calculations by fast computational machine. J Chem Phys 21(6):1087–1091

Montagner J (2014) Upper mantle structure: global isotropic and anisotropic tomography. Treatise Geophys 1:559–590

Montagner J-P (1994) Can seismology tell us anything about convection in the mantle? Rev Geophys 32(2):115–137

Montagner J-P, Anderson DL (1989) Petrological constraints on seismic anisotropy. Phys Earth Planet Inter 54(1):82–105

Montagner J-P, Guillot L (2002) Seismic anisotropy and global geodynamics. Rev Mineral Geochem 51(1):353–385

Montagner J-P, Jobert N (1988) Vectorial tomography—II. Application to the Indian ocean. Geophys J Int 94(2):309–344

Montagner J-P, Nataf H-C (1988) Vectorial tomography—I. Theory. Geophys J 94(2):295–307

Murat F, Tartar L (1985) Calcul des variations et homogénéisation. In: Homogenization methods: theory and applications in physics (Bréau-sans-Nappe, 1983), vol 57 of Collect. Dir. Études Rech. Élec. France, Eyrolles, Paris, pp 319–369

Muyzert E, Paulssen H, Snieder R (1999) A seismic cross-section through the east European continent. Geophys J Int 136(3):695–704

Nachman AI (1988) Reconstructions from boundary measurements. Ann Math 128(3):531–576

Nakamura G, Uhlmann G (1994) Global uniqueness for an inverse boundary problem arising in elasticity. Inventiones mathematicae 118(1):457–474

Nolet G (1990) Partitioned waveform inversion and two-dimensional structure under the network of autonomously recording seismographs. J Geophys Res Solid Earth (1978–2012) 95(B6):8499–8512

Obrebski M, Allen RM, Pollitz F, Hung S-H (2011) Lithosphere–asthenosphere interaction beneath the western united states from the joint inversion of body-wave traveltimes and surface-wave phase velocities. Geophys J Int 185(2):1003–1021

Panning M, Romanowicz B (2006) A three-dimensional radially anisotropic model of shear velocity in the whole mantle. Geophys J Int 167(1):361–379

Panning M, Lekić V, Romanowicz B (2010) Importance of crustal corrections in the development of a new global model of radial anisotropy. J Geophys Res Solid Earth (1978–2012) 115(B12)

Park M, Odom RI (1999) The effect of stochastic rough interfaces on coupled-mode elastic waves. Geophys J Int 136(1):123–143

Pilia S, Rawlinson N, Cayley RA, Bodin T, Musgrave R, Reading AM, Direen NG, Young MK (2015) Evidence of micro-continent entrainement during crustal accretion. Scientific Reports, NPG, 5:8218. doi:10.1038/srep08218

Postma G (1955) Wave propagation in a stratified medium. Geophysics 20(4):780–806

Priolo E, Carcione JM, Seriani G (1994) Numerical simulation of interface waves by high-order spectral modeling techniques. J Acoust Soc Am 95(2):681–693

Rawlinson N, Sambridge M (2003) Seismic traveltime tomography of the crust and lithosphere. Adv Geophys 46:81–198

Rawlinson N, Kennett BLN, Salmon M, Glen RA (2015) Origin of Lateral Heterogeneities in the Upper Mantle Beneath South-east Australia from Seismic Tomography, this volume

Ray A, Key K (2012) Bayesian inversion of marine CSEM data with a trans-dimensional self parametrizing algorithm. Geophys J Int 191:1135–1151

Raykova RB, Nikolova SB (2003) Anisotropy in the earth's crust and uppermost mantle in southeastern Europe obtained from rayleigh and love surface waves. J Appl Geophys 54(3):247–256
Ricard Y, Durand S, Montagner J-P, Chambat F (2014) Is there seismic attenuation in the mantle? Earth Planet Sci Lett 388:257–264
Richards PG, Menke W (1983) The apparent attenuation of a scattering medium. Bull Seismol Soc Am 73(4):1005–1021
Rickers F, Fichtner A, Trampert J (2013) The Iceland–Jan Mayen plume system and its impact on mantle dynamics in the North Atlantic region: evidence from full-waveform inversion. Earth Planet Sci Lett 367:39–51
Ritsema J, van Heijst HJ, Woodhouse JH (1999) Complex shear wave velocity structure imaged beneath Africa and iceland. Science 286(5446):1925–1928
Ritsema J, Deuss A, Van Heijst H, Woodhouse J (2011) S40rts: a degree-40 shear-velocity model for the mantle from new rayleigh wave dispersion, teleseismic traveltime and normal-mode splitting function measurements. Geophys J Int 184(3):1223–1236
Romanowicz B (1987) Multiplet-multiplet coupling due to lateral heterogeneity: asymptotic effects on the amplitude and frequency of the earth's normal modes. Geophys J Roy Astron Soc 90(1):75–100
Romanowicz B (2002) Inversion of surface waves: a review. Int Geophys 81A:149–173
Romanowicz B (2003) Global mantle tomography: progress status in the past 10 years. Ann Rev Earth Planet Sci 31(1):303–328
Romanowicz BA, Panning MP, Gung Y, Capdeville Y (2008) On the computation of long period seismograms in a 3-d earth using normal mode based approximations. Geophys J Int 175(2):520–536
Sambridge M, Bodin T, Gallagher K, Tkalčić H (2013) Transdimensional inference in the geosciences. Philos Trans R Soc A Math Phys Eng Sci 371(1984)
Sanchez-Palencia E (1980) Non homogeneous media and vibration theory, no. 127 in Lecture Notes in Physics. Springer, Berlin
Schaeffer A, Lebedev S (2013) Global shear speed structure of the upper mantle and transition zone. Geophys J Int 194(1):417–449
Schaeffer AJ, Lebedev S (2015) Global Heterogeneity of the Lithosphere and Underlying Mantle: A Seismological Appraisal Based on Multimode Surface-Wave Dispersion Analysis, Shear-Velocity Tomography, and Tectonic Regionalization, this volume
Schmerr N (2015) Imaging Mantle Heterogeneity with Upper Mantle Seismic Discontinuities, this volume
Seriani G, Priolo E (1994) Spectral element method for acoustic wave simulation in heterogeneous media. Finite Elem Anal Des 16(3):337–348
Shearer P (1991) Imaging global body wave phases by stacking long-period seismograms. J Geophys Res 96(B12):20353–20364
Sheng P (1990) Effective-medium theory of sedimentary rocks. Phys Rev B 41(7):4507
Sigloch K, Mihalynuk MG (2013) Intra-oceanic subduction shaped the assembly of cordilleran North America. Nature 496(7443):50–56
Sisson S (2005) Transdimensional Markov Chains: a decade of progress and future perspectives. J Am Stat Assoc 100(471):1077–1090
Sivia D (1996) Data analysis: a Bayesian tutorial. Oxford University Press, USA
Tape C, Liu Q, Maggi A, Tromp J (2010) Seismic tomography of the southern california crust based on spectral-element and adjoint methods. Geophys J Int 180(1):433–462
Tarantola A (2005) Inverse problem theory and methods for model parameter estimation. Society for Industrial Mathematics
Tessmer E, Kosloff D, Behle A (1992) Elastic wave propagation simulation in the presence of surface topography. Geophys J Int 108(2):621–632
Thomson WT (1950) Transmission of elastic waves through a stratified solid medium. J Appl Phys 21:89

Tkalčić H, Young M, Bodin T, Ngo S, Sambridge M (2013) The shuffling rotation of the earth/'s inner core revealed by earthquake doublets. Nat Geosci 6:497–502
Trampert J, Woodhouse JH (2003) Global anisotropic phase velocity maps for fundamental mode surface waves between 40 and 150 s. Geophys J Int 154(1):154–165
Tromp J, Tape C, Liu Q (2005) Seismic tomography, adjoint methods, time reversal and banana-doughnut kernels. Geophys J Int 160(1):195–216
Van Der Lee S, Frederiksen A (2005) Surface wave tomography applied to the north american upper mantle. Geophys Monogr Ser 157:67–80
Van der Lee S, Nolet G (1997) Upper mantle s velocity structure of North America. J Geophys Res 102(B10):22815–22822
Vinnik L (1977) Detection of waves converted from P to SV in the mantle. Phys Earth Planet Inter 15(1):39–45
Vinnik LP, Farra V, Romanowicz B (1989) Azimuthal anisotropy in the earth from observations of SKS at Geoscope and NARS broadband stations. Bull Seismol Soc Am 79(5):1542–1558
Virieux J (1986) P-SV wave propagation in heterogeneous media: velocity-stress finite-difference method. Geophysics 51(4):889–901
Virieux J, Operto S (2009) An overview of full-waveform inversion in exploration geophysics. Geophysics 74(6):WCC1–WCC26
Wang N, Montagner J-P, Fichtner A, Capdeville Y (2013) Intrinsic versus extrinsic seismic anisotropy: the radial anisotropy in reference earth models. Geophys Res Lett 40(16):4284–4288
Woodhouse JH, Dziewonski AM (1984) Mapping the upper mantle: three-dimensional modeling of earth structure by inversion of seismic waveforms. J Geophys Res Solid Earth (1978–2012) 89(B7):5953–5986
Young MK, Rawlinson N, Bodin T (2013) Transdimensional inversion of ambient seismic noise for 3d shear velocity structure of the tasmanian crust. Geophysics 78(3):WB49–WB62
Yuan H, Romanowicz B (2010) Lithospheric layering in the North American craton. Nature 466(7310):1063–1068
Yuan X, Kind R, Li X, Wang R (2006) The s receiver functions: synthetics and data example. Geophys J Int 165(2):555–564
Zhu H, Tromp J (2013) Mapping tectonic deformation in the crust and upper mantle beneath Europe and the north atlantic ocean. Science 341(6148):871–875
Zhu H, Bozda E, Peter D, Tromp J (2012) Structure of the European upper mantle revealed by adjoint tomography. Nat Geosci 5(7):493–498

Chapter 5
Relationships Between Seismic Wave-Speed, Density, and Electrical Conductivity Beneath Australia from Seismology, Mineralogy, and Laboratory-Based Conductivity Profiles

A. Khan, S. Koch, T.J. Shankland, A. Zunino and J.A.D. Connolly

Abstract We present maps of the three-dimensional density (ρ), electrical conductivity (σ), and shear-wave speed (V_S) structure of the mantle beneath Australia and surrounding ocean in the depth range of 100–800 km. These maps derived from stochastic inversion of seismic surface-wave dispersion data, thermodynamic modeling of mantle mineral phase equilibria, and laboratory-based conductivity models. Because composition and temperature act as fundamental parameters, we obtain naturally scaled maps of shear-wave speed, density, and electrical conductivity that depend only on composition, physical conditions (pressure and temperature), and laboratory measurements of the conductivity of anhydrous mantle minerals. The maps show that in the upper mantle ρ, σ and V_S follow the continental-tectonic division that separates the older central and western parts of Australia from the younger eastern part. The lithosphere beneath the central and western cratonic areas appears to be relatively cold and Fe-depleted, and this is reflected in fast shear-wave speeds, high densities, and low conductivities. In contrast, the lithosphere underneath younger regions is

Electronic supplementary material The online version of this chapter (doi:10.1007/978-3-319-15627-9_5) contains supplementary material, which is available to authorized users.

A. Khan (✉) · S. Koch
Institute of Geophysics, ETH Zürich, Zürich, Switzerland
e-mail: amir.khan@erdw.ethz.ch

T.J. Shankland
Geophysics Group, Los Alamos National Laboratory, Los Alamos, USA

A. Zunino
Niels Bohr Institute, University of Copenhagen, Copenhagen, Denmark

J.A.D. Connolly
Institute of Geochemistry and Petrology, ETH Zürich, Zürich, Switzerland

© Springer International Publishing Switzerland 2015
A. Khan and F. Deschamps (eds.), *The Earth's Heterogeneous Mantle*,
Springer Geophysics, DOI 10.1007/978-3-319-15627-9_5

relatively hot and enriched with Fe, which is manifested in slow shear-wave speeds, low densities, and high conductivities. This trend appears to continue to depths well below 300 km. The slow-fast shear-wave speed distribution found here is also observed in independent seismic tomographic models of the Australian region, whereas the coupled slow-fast shear-wave speed, low-high density, and high-low electrical conductivity distribution has not been observed previously. Toward the bottom of the upper mantle at 400 km depth marking the olivine → wadsleyite transformation (the "410-km" seismic discontinuity), the correlation between V_S, ρ, and σ weakens. In the transition zone, V_S, ρ, and σ are much less correlated indicating a significant compositional contribution to lateral heterogeneity. In particular, in the lower transition zone, σ and ρ appear to be governed mostly by variations in Fe/(Fe + Mg), whereas lateral variations in V_S result from changes in (Mg + Fe)/Si and not, as observed in the upper mantle, from temperature variations. Lower mantle lateral variations in thermo-chemical parameters appear to smooth out, which suggests a generally homogeneous lower mantle in agreement with seismic tomographic images of the lower mantle. As a test of the regional surface-wave-based conductivity model, we computed magnetic fields of 24 h S_q variations and compared these to observations. The comparison shows that while our predicted conductivity model improves the fit to observations relative to a one-dimensional model, amplitudes of the computed conductivity anomalies appear not to be large enough to enable these to be discriminated at present.

Keywords Electrical conductivity · Seismic wave-speed · Tomography · Phase equilibria · Surface waves · Electromagnetic sounding · Mantle composition · Mantle temperatures

5.1 Introduction

Large-scale features of elastic properties from seismic tomography are reasonably well resolved and show strong correlation with surface-tectonic features (e.g., Schaeffer and Lebedev 2015; Lekic and Romanowicz 2012; Rawlinson et al. 2015), but smaller scale variations appear to be less well resolved and to have more complex thermal and chemical origins (e.g., Trampert and van der Hilst 2005). Mantle convection simulations favor a heterogeneous mantle made of a mechanical mixture of basalt and harzburgite (e.g. Christensen and Hofmann 1994; Xie and Tackley 2004), which are the products of partial melting at mid-ocean ridges. While such models produce a natural metric for understanding the generation and continued renewal of a laterally heterogeneous mantle through subduction of differentiated oceanic lithosphere (e.g., Xu et al. 2008), it is not yet clear whether such mixture models can account for the complexity observed across a wide range of geophysical data (e.g., Helffrich and Wood 2001; Schmerr 2015; Kawai and Tsuchiya 2015).

Detailed maps of compressional and shear-wave velocity models exist, but there are fewer constraints on mantle density structure (e.g., Kennett 1998; Ishii and Tromp 2004) because seismic data underlying tomographic models are relatively insensitive to density (e.g., Resovsky and Ritzwoller 1999; Romanowicz 2001; Kuo and Ramanowicz 2002). Yet, geodynamical modeling expresses internal dynamics driven by lateral density variations (e.g., Forte and Perry 2000; Deschamps et al. 2001), and improved knowledge of mantle density structure could prove an important additional constraint.

Complementary means to obtain structural information can come from studies that relate elastic properties to chemical composition and crystal structure to constrain systematic relations between seismic wave-speed and density in mantle minerals (e.g., Birch 1961; Shankland 1972). These systematics can subsequently be compared to density, P-, and S-wave speed profiles that are deduced from geophysical studies or even be applied in an inverse sense to obtain compositional information (e.g., Shankland 1977). Other means rely on transport properties, e.g., electrical and thermal conductivities, and viscosity, (e.g., Shankland 1981; Poirier 2000). Electrical conductivity, for example, is more sensitive to composition and temperature than in elasticity (e.g., Xu et al. 2000; Dobson and Brodholt 2000; Khan et al. 2006; Verhoeven et al. 2009). However, the perceived advantage of studies based on electrical conductivity is diminished by a smaller number of world-wide geomagnetic observatories in comparison with the present-day global network of seismic stations (e.g., Kuvshinov 2012).

Global and semi-global three-dimensional (3D) conductivity images of the mantle are most reliable in imaging transition zone and outermost lower mantle (e.g., Utada et al. 2009; Kelbert et al. 2009; Tarits and Mandea 2010; Shimizu et al. 2010; Semenov and Kuvshinov 2012), where they show relatively low lateral and radial resolution in comparison with their seismic counterparts. Resolution aside, current global-scale conductivity, and seismic tomography maps appear to have few features in common (for a comparison see, e.g., Semenov and Kuvshinov 2012). Fast and slow velocity anomalies that are usually interpreted as evidence of cold and hot mantle conditions, respectively, do not always correlate with low and high conductivities as might be expected from the observation that conductivity has a strong temperature dependence following an Arrhenius relation (e.g., Poirier 2000). On a more fundamental level, transport and elastic properties of minerals are different phenomena and per se need not be correlated. However, if temperature is the dominant source of mantle velocity anomalies as frequently argued (e.g., Goes et al. 2000), then we might expect a minimum degree of correlation. Such a correlation is not usually observed for reasons yet to be determined. On the other hand, if composition in the form of variations in major element chemistry is responsible for a significant part of the signal, then we might not observe significant correlations between the two physical properties. Hence, these features could provide potential means to unravel compositional contributions to lateral variations in mantle properties.

To examine relations between elastic properties and electrical conductivity, we link elasticity, density, and electrical conductivity through thermodynamic modeling of mantle minerals (e.g., Khan et al. 2006; Verhoeven et al. 2009). Using a self-consistent scheme (Gibbs free-energy minimization) to calculate phase equilibria, followed by equation of state modeling, we compute properties that depend only on physical conditions (pressure and temperature) and on chemical composition (e.g., Connolly 2005; Ricard et al. 2005; Stixrude and Lithgow-Bertelloni 2005; Piazzoni et al. 2007). Thus, the use of free-energy minimization methods results in physically realistic models that include natural scaling between the various properties.

We first consider shear-wave velocities of the mantle beneath continental Australia to a depth of 800 km that are derived from stochasticinversion of surface-wave dispersion data as described in Khan et al. (2013). In the inversion, we map mantle temperatures and compositions from elastic responses. Having these properties together with independent laboratory conductivity measurements of different mineralogies and temperatures permit calculating mantle electrical conductivities independently of elastic properties at depth. From these conductivity maps, it is possible to calculate an equivalent conductivity response. Hence, this approach provides a natural basis for analyzing different data sets and for addressing fundamental issues related to the structure and constitution of the Earth (e.g., Khan et al. 2006, 2013; Verhoeven et al. 2009; Fullea et al. 2009, 2011, 2012; Afonso et al. 2013a, b; Jones et al. 2013; Drilleau et al. 2013; Kuskov et al. 2014).

Joint inversion of seismic, gravity, and electromagnetic sounding data would be a preferable approach; owing to the complexity and the computational resources required for solving the joint problem, it is beyond the scope of this study and will be considered in future studies. However, we would like to emphasize that this procedure, while based on inversion of seismic data, does not imply that the derived conductivity structure simply or tautologically follows that dictated by seismic data because independent chemical, mineralogical, and laboratory conductivity databases are interposed. Finally, we note that because the main purpose of this study is to examine relative behaviors of elastic and transport properties, we do not consider the inverse problem of constraining composition and temperature from seismic data per se in any detail here, but consider this part a *fait accompli* with details supplied in the previous treatment (Khan et al. 2013).

As the main focus in this study centers on mantle compositional (major elements) and thermal variations, we assume anhydrous and subsolidus conditions, i.e., we consider the mantle to be dry and melt-free. We make such a simplifying assumption out of necessity as (1) thermodynamic data for modeling shear-wave speed for water- and melt-bearing phases are uncertain and (2) several studies have showed that an anhydrous mantle is not inconsistent with conductivity profiles obtained from global electromagnetic sounding data (e.g., Manthilake et al. 2009; Khan and Shankland 2012). Moreover, although available data have been used to construct parameterized approaches to modeling the effect of water and melt on shear-wave speed and conductivity (e.g., Laske et al. 2011; Goes et al. 2012), these cannot be modeled self-consistently in line with the main approach of this study. It is left to future studies to consider this and contributions arising from other effects such as the presence of melt (e.g., Shankland et al. 1981; Park and Ducea 2003; Toffelmier and Tyburczy 2007; Khan and Shankland 2012; Koyama et al. 2006, 2013; Karato 2011; Pommier 2014).

5.2 Data and Methods

5.2.1 The Forward Problem

The solution to the forward problem, i.e., estimation of surface-wave dispersion data (d_{seis}) and magnetic field variations (d_{mag}) from composition c, temperature T, and a laboratory conductivity database (d_σ) can be summarized as follows:

$$\begin{array}{ccccccc} c,T & \overset{g_1}{\rightarrow} & M & \overset{g_2}{\rightarrow} & \rho, V_S, V_P & \overset{g_4}{\rightarrow} & d_{seis} \\ & & & \overset{g_3}{\searrow} & & & \\ & & d_\sigma & \overset{g_3}{\rightarrow} & \sigma & \overset{g_5}{\rightarrow} & d_{mag} \end{array}$$

where M indicates equilibrium modal mineralogy (all parameters are implicitly assumed to be functions of radius) and the various g's embody physical laws in the form of thermodynamic modeling (g_1), equation of state modeling (g_2), bulk rock conductivity estimation (g_3), surface-wave dispersioncalculations (g_4), and computation of magnetic field variations (g_5). Here, we do not consider the part related to g_4 for reasons outlined earlier. From the above forward scheme, we observe that electrical conductivity is not immediately available from free-energy minimization, but has to be computed using separate laboratory-based conductivity data.

5.2.2 Parameterization

As outlined in more detail in Khan et al. (2013), we parameterized the model laterally in terms of 5° × 5° pixels and beneath the center of each pixel radially by a number of layers with a fixed node-spacing in the mantle of 50–200 km for temperature and composition (Fig. 5.1). The crust is parameterized using purely physical properties that include P- and S-wave speeds, density, conductivity, and depth to crust-mantle interface. With these parameters assigned, we performed Gibbs free-energy minimization (to be described below) and computed modal mineralogy and mantle physical properties on a denser fixed radial grid with node spacing in the range of 10–50 km.

5.2.3 Thermodynamic Modeling

Application of free-energy minimization methods in geophysics is not new (e.g., Saxena and Eriksson 1983; Wood and Holloway 1984; Kuskov and Panferov 1991; Sobolev and Babeyko 1994; Bina 1998; Fabrichnaya 1999), but it has been limited by the extent of early thermodynamic databases. Here, we use the

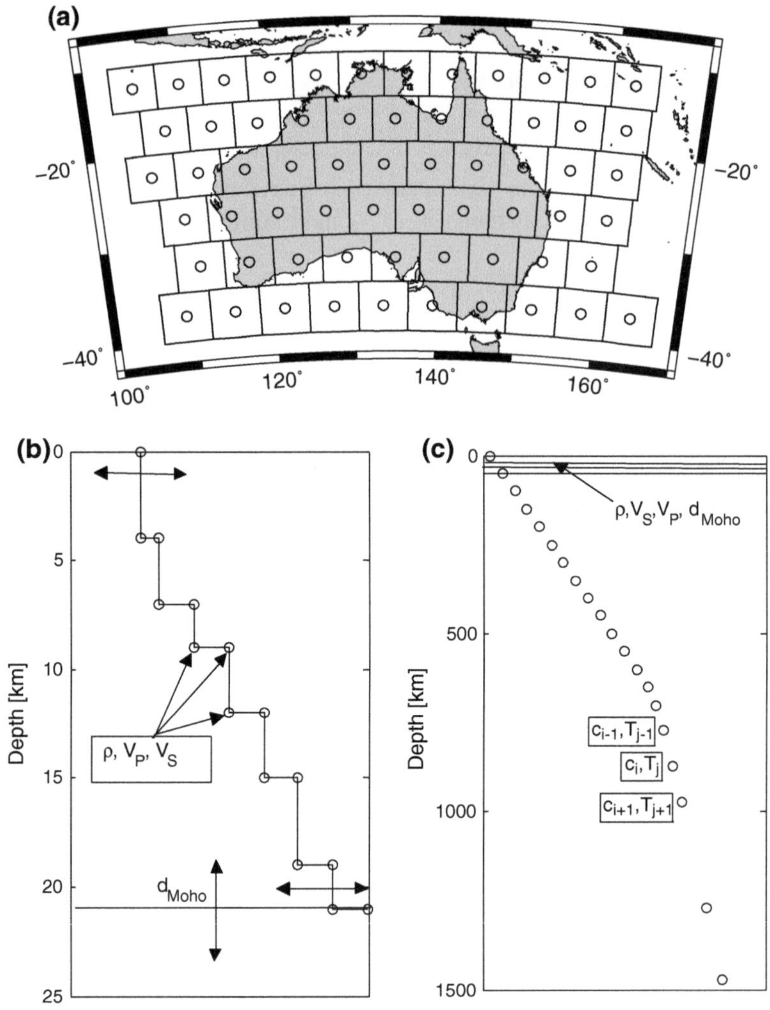

Fig. 5.1 Model parameterization: lateral grid spacing is 5° × 5° (**a**) and radially model parameterization consists of layers (**b–c**). The crust (**b**) is delineated by density (ρ), P- and S-wave speeds (V_P, V_S), and depth to crust-mantle interface (d_{Moho}). Mantle layers (**c**) are parameterized by composition (c) and temperature (T). Circles in both plots denote location of a set of prefixed depth nodes (corresponding to indices i and j in plot **c**), except in the crust where depth to crust–mantle interface is variable. Modified from Khan et al. (2013)

free-energy minimization strategy described by Connolly (2009) to predict rock mineralogy, elastic moduli, and density as a function of pressure, temperature, and bulk composition. For this purpose, we employ the thermodynamic formulation of Stixrude and Lithgow-Bertelloni (2005) with parameters as in Stixrude and Lithgow-Bertelloni (2011). Possible mantle compositions are explored within the Na_2O–CaO–FeO–MgO–Al_2O_3–SiO_2 (NCFMAS) system, which accounts

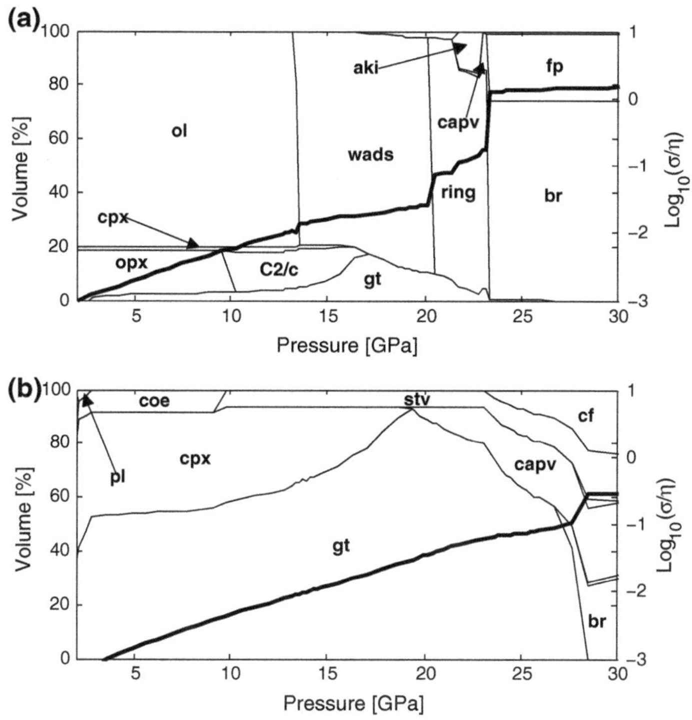

Fig. 5.2 Variations in phase proportions and physical properties in the upper mantle, transition zone,andoutermost lower mantle (0–30 GPa, 0–750 km depth). Aggregate rock conductivity (*bold black lines*) and phase equilibria are calculated for two different mantle compositions in the NCFMAS system (comprising oxides of the elements Na_2O–CaO–FeO–MgO–Al_2O_3–SiO_2): harzburgite (**a**) and MORB (**b**) along the mantle adiabat of Brown and Shankland (1981) (see Fig. 5.3). Phases are: olivine (ol), orthopyroxene (opx), clinopyroxene (cpx), pl (plagioclase), coe (coesite), stishovite (stv), high-pressure Mg-rich cpx (C2/c), garnet (gt), wadsleyite (wad), ringwoodite (ring), akimotoite (aki), calcium silicate perovskite (capv), ferropericlase (fp),bridgemanite (br), and calcium ferrite (cf)

for ~99 % of the mass of Earth's mantle (e.g., Irifune 1994). Because the equilibrium assumption is dubious at low temperature (e.g., Wood and Holloway 1984), for models that require rock properties at temperatures below 800 K, the stable mineralogy is first calculated at 800 K and its physical properties are then computed at the temperature of interest. Bulk rock elastic moduli are estimated by Voigt-Reuss-Hill (VRH) averaging. The pressure profile is obtained by integrating the load from the surface (boundary condition $p = 10^5$ Pa). Examples of equilibrium mineralogy and bulk rock conductivity for two different compositions along the mantle adiabat of Brown and Shankland (1981) are shown in Fig. 5.2. The influence of phase equilibria in these examples representing end-member compositions is apparent from the discontinuities in conductivity associated with mineral phase transformations. The difference in geophysical response incurred by the two conductivity profiles shown in Fig. 5.2 is easily detectable with an appropriate set of electromagnetic response functions.

Uncertainties associated with the thermodynamic data are notoriously difficult to assess because of the nonlinearity of the free-energy minimization problem and correlations among the various thermodynamic parameters. A preliminary Monte Carlo analysis of this uncertainty indicates uncertainties in density, P-, and S-wave speeds on the order of <0.1, <0.5, and <1 %, respectively (unpublished data). Alternative means of investigating this uncertainty are illustrated in the applications of Afonso et al. (2013a) and Kuskov et al. (2014) that employ similar approaches to the method outlined here. In a methodological study of determining thermo-chemical mantle structure from a diverse set of geophysical data, Afonso et al. (2013a) assess the uncertainty by comparing predicted and observed modal compositions for a set of xenoliths and generally find that differences between computed and observed values are within experimental error. Differences in physical properties between computed and laboratory measurements are also found to be negligible. In a related study, Kuskov et al. (2014) map the thermo-chemical structure of the lithospheric mantle beneath the Siberian craton using seismic velocity profiles and find that temperatures and depth to the lithosphere–asthenosphere boundary can only be determined to within an accuracy of ± 100 °C and ± 30 km, respectively, given the uncertainties in the thermodynamic parameters and modeled velocities.

5.2.4 Laboratory Electrical Conductivity Data

The conductivity data employed in this study are detailed in Khan and Shankland (2012) and shown in Fig. 5.3. The conductivity database comprises the minerals olivine (ol), orthopyroxene (opx), clinopyroxene (cpx), garnet (gt), wadsleyite (wads), ringwoodite (ring), ferropericlase (fp), and bridgemanite (br). This data set allows us to model mineral conductivities as functions of major element composition, temperature, and pressure. In order to construct a bulk conductivity profile from single mineral conductivities, we make use of a self-consistent solution based on effective medium theory (Landauer 1952; Berryman 1995) for conducting composites (for more discussion see, e.g., Xu et al. 2000; Khan and Shankland 2012)

$$\sum_{i=1}^{N} x_i \left[\frac{\sigma_i - \sigma_{sc}}{\sigma_i + 2\sigma_{sc}} \right] = 0, \tag{5.1}$$

where x_i is volume fraction of mineral i, N the total number of minerals, and σ_{sc} represents the self-consistent solution that has to be determined iteratively to satisfy Eq. 5.1, while bounded by the Hashin-Shtrikman bounds. The latter are the narrowest bounds that exist for a multiphase system in the absence of information about the geometrical arrangement of the constituent phases (Hashin and

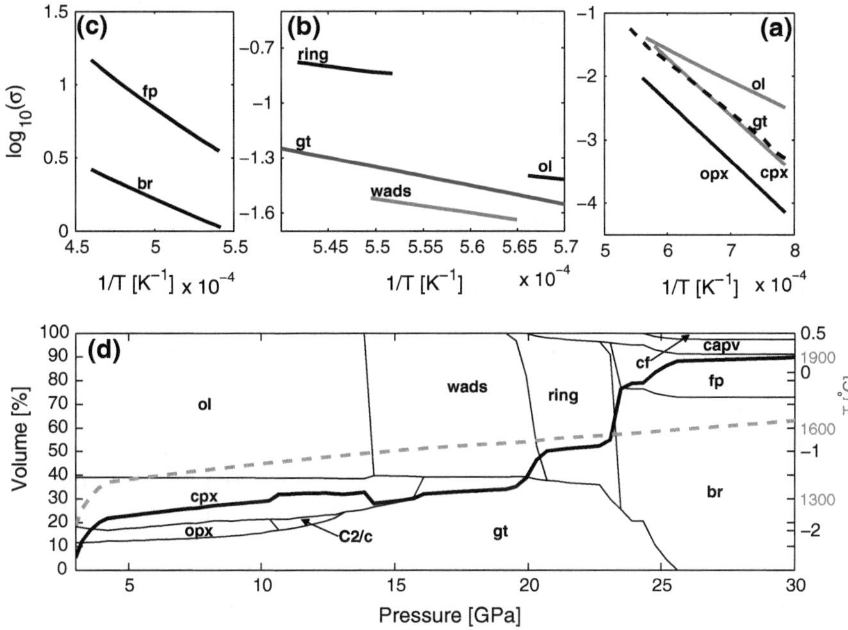

Fig. 5.3 Summary of mineral electrical conductivities measured in the laboratory as a function of inverse temperature for **a** upper mantle, **b** transition zone, and **c** upper part of lower mantle. **d** Variations in mineral phase proportions and laboratory-based bulk conductivity profile. For the example shown here bulk conductivity profile and mineral modes were computed using a homogeneous adiabatic pyrolitic and anhydrous mantle as a function of pressure (depth). The *solid black* and *dotted green lines* in plot **d** show bulk conductivity and the adiabat of Brown and Shankland (1981), respectively. For phase names, see Fig. 5.2; the main text contains further discussion. Modified from Khan et al. (2011)

Shtrikman 1962; Watt et al. 1976). Phases present at levels <10 vol% (e.g., coesite, stishovite, and akimotoite) are not considered as these have no effect on bulk conductivity (e.g., Khan et al. 2011).

To illustrate the methodology, we computed modal mineralogy and corresponding self-consistently determined bulk conductivity profile for a homogeneous adiabatic mantle made of pyrolite (Fig. 5.3). The bulk conductivity profile reveals a number of interesting features observed previously that include an almost vanishing "410-km" discontinuity (olivine → wadsleyite), and relatively strong "520-km" and "660-km" discontinuities (wadsleyite → ringwoodite and ringwoodite → bridgemanite + ferriperoclase), respectively. As a consequence, absence of a strong "410" discontinuity in conductivity is likely to be a prominent feature of conductivity images relative to those showing shear-wave speed. Variations in V_S are expected to be strong across the "410." In the case of the "660," a relatively strong compositional dependence through variations in Fe content is possible.

5.3 Results and Discussion

Maps of mean mantle shear-wave speed [these are anelastically corrected using the approach outlined in Khan et al. (2013)], electrical conductivity, density, mineralogy, and thermochemical variations derived from inversion of surface-wave data are shown in Fig. 5.4. Mean bulk compositions are indicated as ratios of the three major elements Mg, Fe, and Si derived from the oxide models c; the remaining elements are less well determined. The mean model was computed from the 10^4 models that were sampled and considered in the analysis of Khan et al. (2013). Note that variations are displayed relative to mean reference models that are computed from lateral averages of all sampled models across all pixels. The mean reference profiles are shown in Fig. 5.5. Moreover, 1D marginal posterior distributions for the parameters displayed in Fig. 5.4 (for a single location in central Australia) are depicted in Fig. 5.6. With regard to the mean model displayed here, we would like to note that any model picked randomly from the posterior probability distribution is as representative as the mean or median model, if not more so, since the latter models are unlikely to be sampled and therefore constitute a poor representation of the solution (small posterior probability).

Rather than compute uncertainty measures such as credible intervals from the suite of sampled models (see Khan et al. 2013), we depict model parameter uncertainty in the form of a movie sequence that displays 250 models taken randomly from the posterior distribution (see online supporting material). The movie provide insight into overall model variability. The features that are well resolved will tend to be more stable across the maps and models, whereas less well-resolved features appear more unstable. Variations in mantle temperature and physical properties, for example, are seen to be remarkably stable, particularly throughout the upper mantle. Compositional variations, on the other hand, only appear to be stable down to 200 km depth. However, a closer look actually reveals that the relative geographical distribution of compositional variations is stable. As an aid in interpreting the maps, we have computed correlations between the various physical and thermochemical parameters as a function of depth (Figs. 5.7 and 5.8). Note that correlations are shown as distributions that are computed from 10^4 models, i.e., from all samples considered in Khan et al. (2013), and indicate the probability for observing a correlation coefficient in a particular range. The more peaked the correlation coefficient is, the better the parameter is resolved. It must also be mentioned that the correlations in and around the transition zone do not necessarily coincide with the depths at which mineralogical phase transitions occur.

5.3.1 Correlated Lateral Variations of Physical Properties

Figures 5.7 and 5.8 provide a number of interesting interrelationships as lateral correlations between physical properties, chemistry, mineralogy, and temperature.

The distributions indicate variations with lateral distances for various depth ranges indicated by color coding. Thus, lateral changes are isobaric, i.e., independent of phase changes with pressure. We emphasize the significance only of distributions outside the range −0.5 to 0.5 where statistics can be inconclusive. This section interprets physical connections between different variables, and the following section treats their geophysical variations.

The top row of Fig. 5.7 displays temperature effects and has the clearest correlations, either strongly positive or negative. As expected, density ρ decreases with T as a consequence of thermal expansion. Velocities also decrease with T as measured in laboratory measurements, a consequence of weakening interatomic bonds (Wang 1970). The positive conductivity variation with T reflects the exponential increase of σ with T in semiconductors and thus is a right–left mirror image of the other two figures. With increasing depth, lateral thermal effects become less correlated, produce fewer lateral changes of physical properties, and require other explanations.

In the second row, density increases with replacement of Mg by the heavier element Fe in crystal structures; within a given crystal structure, this replacement leads to decreased shear-wave speed (e.g., Birch 1961; Shankland 1972) as in Fig. 5.7e. Electrical conductivity within different silicate mineralogies commonly increases with Fe content (e.g., Hirsch et al. 1993; Yoshino et al. 2012), and this is seen in Fig. 5.7f.

In the third row, to the degree that increasing (Mg + Fe)/Si represents increase of ρ as olivine increases with respect to pyroxene, there can be lateral variations of these two minerals in the upper mantle. In the lower mantle, an increase of fp (Mg, Fe)O with respect to br (Mg, Fe)SiO_3 would increase (Mg + Fe)/Si but decrease density because of the lower density of fp with respect to br (Ricolleau et al. 2009). V_S distributions resemble those of ρ, which suggests the same underlying causes (relative quantities of ol, px; br, fp) for their lateral variations. σ shows something like a mirror image; the higher conductivity of fp relative to br (Yoshino 2010) could lead to a positive correlation with (Mg + Fe)/Si. However, it is not clear why there should be a negative correlation involving ol and px contents (Fig. 5.7i). None of the properties in this row resembles the T-effects of the first and second rows, and this argues for complex changes in mineralogy. Note that (Mg + Fe)/Si from 1 to 2 corresponds to a bulk composition from pyroxene to olivine stoichiometry, i.e., in the ranges of peridotites. The correlations of Fig. 5.7j reflect the tight linkages created by the forward operations g_2 and g_4. The final column of Fig. 5.4 maps upper mantle mineral fractions px/(ol + px) and lower mantle fp/(br + fp). In both cases, the maps show good correlation between bulk chemistry and mineralogy. Because these minerals volumetrically dominate their respective depth ranges, they are of the most interest. Minor minerals are included in volumetric proportions for density and for elastic wave velocities in the VRH averaging scheme. However, because they are unlikely to form interconnected phases, their contributions to electrical conductivity are less significant and therefore more easily neglected.

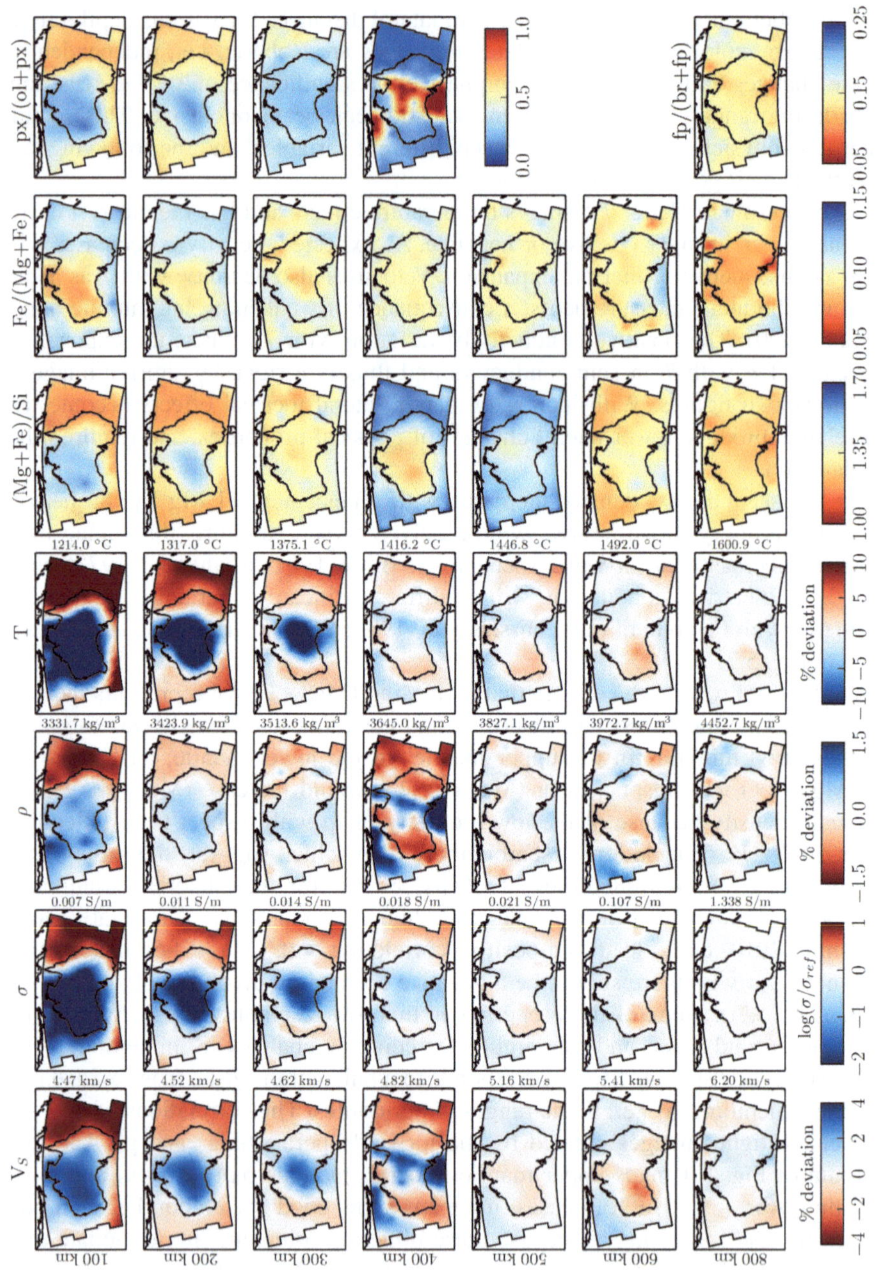

px/(ol+px)
fp/(br+fp)
Fe/(Mg+Fe)
(Mg+Fe)/Si
T
ρ
σ
V_S
0.0 0.5 1.0
0.05 0.15 0.25
0.05 0.10 0.15
1.00 1.35 1.70
% deviation
−10 −5 0 5 10
−1.5 0.0 1.5
log(σ/σ_ref)
−2 −1 0 1
−4 −2 0 2 4
1214.0 °C
1317.0 °C
1375.1 °C
1416.2 °C
1446.8 °C
1492.0 °C
1600.9 °C
3331.7 kg/m³
3423.9 kg/m³
3513.6 kg/m³
3645.0 kg/m³
3827.1 kg/m³
3972.7 kg/m³
4452.7 kg/m³
0.007 S/m
0.011 S/m
0.014 S/m
0.018 S/m
0.021 S/m
0.107 S/m
1.338 S/m
4.47 km/s
4.52 km/s
4.62 km/s
4.82 km/s
5.16 km/s
5.41 km/s
6.20 km/s
100 km
200 km
300 km
400 km
500 km
600 km
800 km

Fig. 5.4 Maps of mean mantle thermochemical anomalies and variations in physical properties beneath Australia. Isotropic mantle shear-wave velocity (first column), electrical conductivity(second column), density (third column), temperature (fourth column), (Mg + Fe)/Si (fifth column, atomic fraction), Fe/(Fe + Mg) (sixth column, atomic fraction), and upper and lower mantle mineral ratios px/(ol + px) and fp/(br + fp) (seventh column, atomic fraction). Shear-wave speed, density, and temperature are given in % deviations from a mean model (Fig. 5.5), respectively. Electrical conductivity is relative to a reference electrical conductivity profile (Fig. 5.5). Mean reference values for all properties are indicated on the *right side* of each panel. Note that *color bars* are inverted for shear-wave speed and conductivity so that fast (slow) velocity anomalies correspond to low (high) conductivities

Figure 5.8 presents another picture in which thermochemical parameters are implicit, and it thus displays lateral interrelationships between physical properties. The overall picture behaves somewhat like the *T*-effects in the first row of Fig. 5.7; there is declining significance of *T* variation with depth. Relations in the first figure represent velocity–density systematics (e.g., Birch 1961; Shankland 1977). The most positive correlations characterize the case where density increases in different mineral phases having minimal change of chemical composition as represented by mean atomic weight $\bar{m}$. Mean atomic weight equals molecular weight of a compound divided by the number of atoms in its chemical formula and mostly reflects Fe enrichment. For instance, in the olivine series $((Mg_{1-x}, Fe_x)_2SiO_4)$ $\bar{m}$ varies with x from 20.1 to 29.1. Negative correlations are strongest in a given crystal phase for increasing iron content. It appears that the wide range of lateral correlations in Fig. 5.8a could represent both effects beneath the Australian continent.

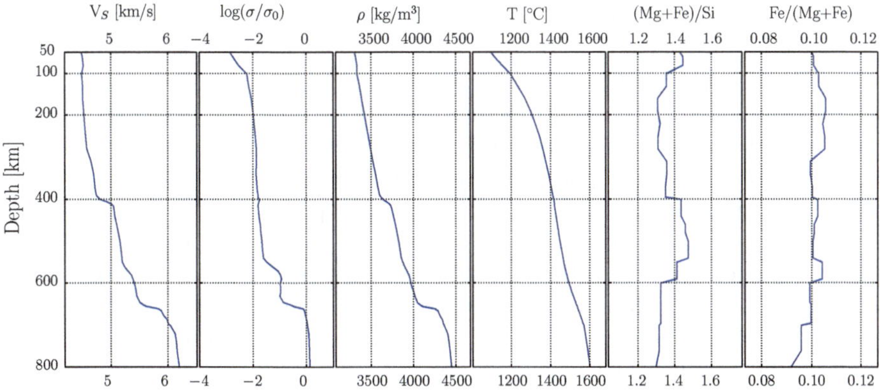

Fig. 5.5 Mean reference profiles (*blue*) showing radial variations in shear-wave speed (V_S), electrical conductivity $\log(\sigma/\sigma_0)$, density (ρ), temperature (T), and composition (Mg + Fe)/Si and Fe/(Mg + Fe) in atomic fractions, respectively. The mean profiles were constructed from lateral averages of each property at every depth node. In the conductivity plot $\sigma_0 = 1$ S/m

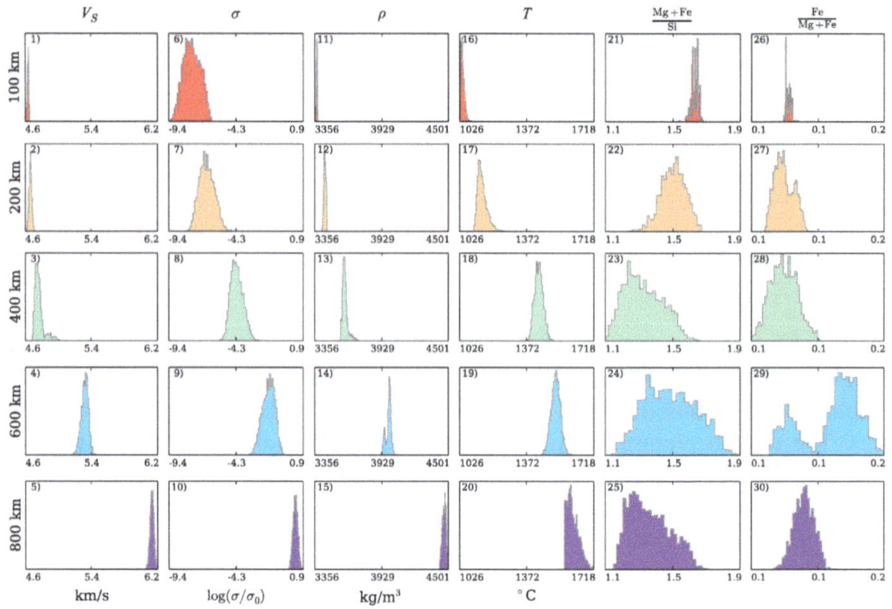

Fig. 5.6 One-dimensional marginal distributions for the model parameters are shown in Fig. 5.5 for a specific location in the center of the Australian continent (120°E, 27°S). Compositional distributions (Mg + Fe/Si and Fe/Mg + Fe) are in atomic fractions. In the conductivity plot $\sigma_0 = 1$ S/m

5.3.2 Upper Mantle Structure

There is a good correlation in the upper mantle where shear-wave, density, and conductivity anomaliesoutline the continental boundaries of the region. Lateral velocity variations are commonly observed in seismic tomography images of the region (e.g., Fishwick and Rawlinson 2012; Kennett et al. 2012; Rawlinson et al. 2015). The continental regime has a distinct thermochemical signature and is divided into slow (corresponding to "hot" with relatively low (Mg + Fe)/Si and high Fe/(Mg + Fe) values) and fast (corresponding to "cold" with relatively high (Mg + Fe)/Si and low Fe/(Mg + Fe) values) patterns that extend well into the upper mantle.

Fe/(Mg + Fe) is of order 0.1 ± 0.03 in mantle minerals; higher ratios yield higher conductivities (e.g., Hirsch et al. 1993) principally in the polaron reaction $Fe^{2+} \leftrightarrows Fe^{2+} + e^{-}$. Moreover, low upper mantle conductivities clearly correlate with fast velocity and positive density anomaliesover the older western and central parts of the continent, whereas high conductivity anomalies correlate with slow velocity and negative density variations over the younger eastern and oceanic lithospheric regions. This conductivity pattern has not been resolved previously; to first order, it appears to be governed by temperature variations through the strong T-dependence of σ.

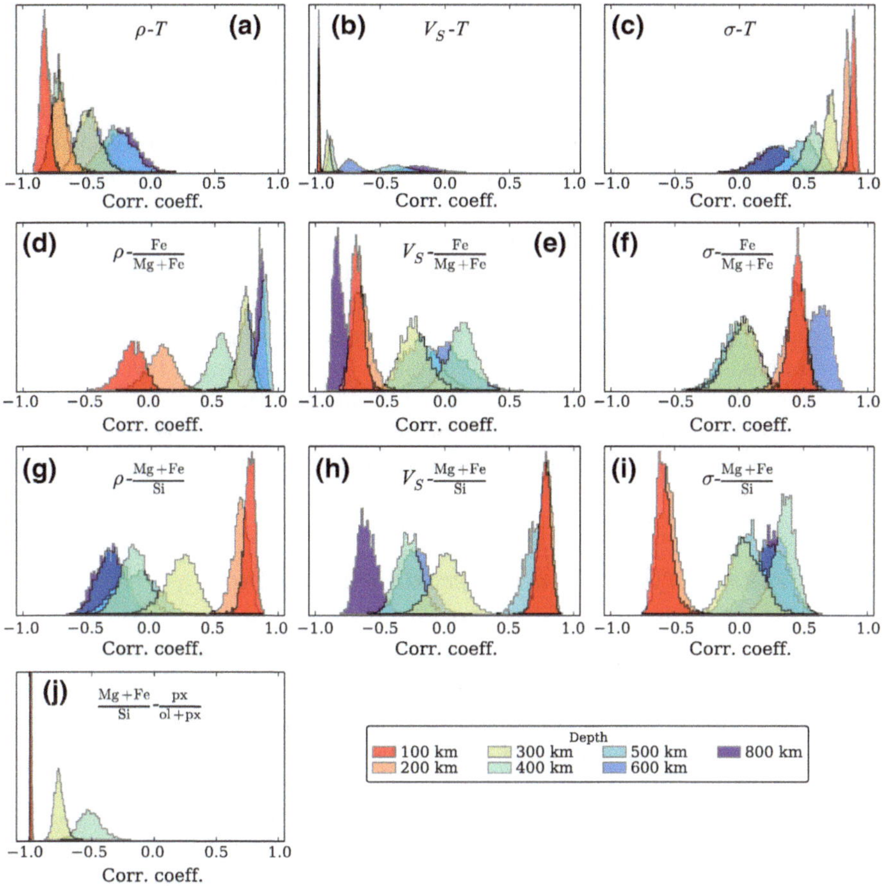

Fig. 5.7 Correlation between physical and thermochemical parameters as functions of depth. Distributions are normalized to unity for the highest peak in each plot, i.e., peak heights/widths are independent between the various properties shown (e.g., $V_S - T$ and V_S − Fe/Mg + Fe)

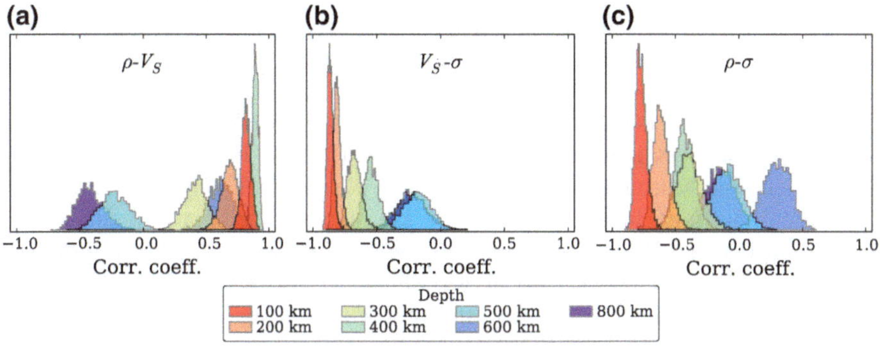

Fig. 5.8 Correlation between physical parameters as functions of depth. Distributions are normalized to unity for the highest peak, i.e., peak heights/widths are independent between the various properties shown (e.g., $V_S - T$ and V_S − Fe/Mg + Fe)

This observation is strongly supported by the distribution of correlation coefficients (Fig. 5.7). The correlation pairs $V_S - T$, $\rho - T$, and $\sigma - T$ are very peaked in the depth range 100–300 km relative to, e.g., Fe/(Mg + Fe) and (Mg + Fe)/Si, although compositional variations do appear to be important. In particular, we observe ρ to correlate positively with (Mg + Fe)/Si and V_S and σ to correlate with both Fe/(Mg + Fe) and (Mg + Fe)/Si. The absolute values of the correlation coefficients are as expected in the case of temperature, with negative correlations existing for $V_S - T$ and $\rho - T$ and positive correlations for $\sigma - T$.

Toward the bottom of the upper mantle, thermal effects become progressively less important, giving way to more complex compositionally-dependent structures. This is most apparent in the case of ρ at 300 km depth, which reveals a relatively strong compositional signature that correlates positively with Fe/(Mg + Fe). In contrast, V_S and σ are largely controlled by lateral thermal effects (Fig. 5.7b, c). For additional discussion of upper mantle structure see the contributions by Rawlinson et al. (2015), Schmerr (2015), and Katsura and Yoshino (2015).

5.3.3 Transition Zone and Lower Mantle Structure

Within the transition zone (400–600 km depth), lateral velocity, density, and conductivity variations are reflected in correlations between the various parameters. Temperature has less influence compared with chemical/mineral contributions. Lateral changes of ρ correlate strongly with Fe/(Mg + Fe) in particular, a result of substituting the heavier Fe atom for Mg. However, transition zone density correlates poorly with oxide (mineral) content, and V_S behaves similarly, which would argue for incoherent lateral variation in these properties. V-ρ systematics Fig. 5.8a has more structure with more lateral changes still in the range close to zero correlations. This also holds for electrical conductivity except for the 600 km regime. Here, the Fe/(Mg + Fe)-dependence at 600 km probably relates to conductivity dependence on Fe in transition-zone minerals, particularly in ringwoodite (Yoshino and Katsura 2013). The lower transition zone beneath the main continent appears to have a different style of lateral variation with slow velocity, negative density, and weak-to-positive conductivity relations. Overall, transition zone displays a general decoupling of structure relative to the upper mantle which is driven by an increase in (Mg + Fe)/Si with depth, (Fig. 5.5). The decoupling appears uncorrelated with Fe/(Mg + Fe), while thermal anomaliesare smoothed out and less dominant relative to the upper mantle. For additional discussion of transition zone structure see the contribution by Kawai and Tsuchiya (2015).

In the outermost lower mantle (800 km depth), variations in all properties diminish in amplitude, in agreement with a lower mantle that is governed by small-amplitude thermochemical variations (e.g., Jackson 1998). Figure 5.7 reveals that both ρ and V_S correlate strongly with Fe/(Mg + Fe), although with opposite signs, resulting in a negative correlation between ρ and V_S in the outermost lower mantle. V_S also shows significant negative correlation with (Mg + Fe)/

Si. The weak negative correlation in Fig. 5.8a suggests lateral Fe enrichment in the lower mantle (in contrast to lateral mineralogical variation in the upper mantle). The small-scale features found here suggest a relatively homogeneous lower mantle. This observation is in line with what is inferred from global seismic tomography that also shows low-amplitude velocity variations in the lower mantle (e.g., Kustowski et al. 2008).

5.3.4 Comparison with Other Studies

The shear wave-speed variations found here in the upper mantle have, as discussed in more detail in Khan et al. (2013), also been reported elsewhere (see the Australian upper mantle seismic tomography models of, e.g., Fichtner et al. (2010), Fishwick and Rawlinson (2012), Kennett et al. (2012). Most of these models (not shown for brevity) are sensitive to about 300 km. Their prominent features to this depth include separation of structure into tectonic regimes that are slow (corresponding to our relatively "hot," low Fe/(Mg + Fe) and (Mg + Fe)/Si) or fast (relatively "cold," high Fe/(Mg + Fe) and (Mg + Fe)/Si). Details among the various models nonetheless differ as a result of differences in methodology, data, and parameterization. However, the overall level of agreement with other regional models is encouraging and is considered evidence in support of the joint thermodynamic analysis performed here.

Several studies have tried to unravel the physical causes of the observed velocity variations that are seen in seismic tomography images (e.g., Masters et al. 2000; Saltzer et al. 2001; Deschamps et al. 2002; Deschamps and Trampert 2003; Resovsky and Trampert 2003; Simmons et al. 2009; Afonso et al. 2010). For this purpose, the ratio of relative changes in shear (V_S) and compressional wave velocities (V_P), defined as $R_{P/S} = \partial \ln V_S / \partial \ln V_P$, has been computed directly from seismic *P*- and *S*-wave tomography models. An increase in $R_{P/S}$ with depth is thought to indicate an increased compositional contribution to the observed structural variations. Generally, these studies find that the mean value of $R_{P/S}$ increases radially, although important, but less-defined, lateral variations also exist. However, as is often the case independent overlapping information on *P*- and S-wave velocities is not available, and even in cases where it is, Deschamps and Trampert (2003) have shown that $R_{P/S}$ is only able to provide qualitative indication of chemical variations. Deschamps and Trampert (2003), nonetheless conclude from the distribution of $R_{P/S}$ and its dispersion that the cause of P- and S-wave speed anomalies is not only temperature, but it likely bears a compositional component. However, separation of thermal and chemical effects by seismic wave-speeds alone is difficult and, as already mentioned, is further complicated by the relative insensitivity of seismic wave-speeds to the density contrasts that must ultimately drive mantle convection.

This contrasts with the approach here where composition and temperature act as fundamental parameters, that, when connected to geophysical data via material properties through self-consistent thermodynamic computations of mineral phase equilibria provides a natural means of determining all properties and their variations simultaneously. This obviates the need for unknown or *ad hoc* scaling relationships between the various physical parameters as typically invoked in seismic tomography. The advantage of our scheme is that geophysical/petrological knowledge is implicitly involved, so that more physically realistic models that depend only on the specific composition, temperature, and pressure conditions are produced. For example, electrical conductivity can enhance pictures of mineral composition and density in mantle models.

Limited comparisons for upper mantle conductivity models of the region are available. A conductivity-depth profile based on the Australian hemisphere model of solar quiet (S_q) daily variations was obtained by Campbell et al. (1998). Conductivities were found to range from 0.02 to 0.04 S/m at 200 km depth to 0.1–0.15 S/m at 400 km depth in general agreement with present findings. Comparison with the global 3D conductivity models of Kelbert et al. (2009) and Semenov and Kuvshinov (2012) (not shown for brevity) is difficult given the low lateral resolution of their models on continental scales. Their models, valid in the depth range 400–1600 km, generally differ from each other by one log-unit in conductivity and appear to have few features in common with this work, although a somewhat higher conductivity anomaly over eastern Australia relative to the rest of the continent is discernable in both studies. This anomaly has also been observed recently by Koyama et al. (2014). Conductivities across the continent range from ~0.1 to ~1 S/m in the transition zone and increase to 1–3.5 S/m below. Anomaly patterns aside, these conductivity ranges essentially bracket those found here where conductivities range from ~0.05 to 1 S/m in the transition zone and ~1 to 3 S/m in the outermost lower mantle. For comparison, conductivities computed from laboratory data assuming a "standard" mantle adiabat and a uniform mantle composed of "dry" pyrolite lies in the range 0.1–0.3 S/m. Comparison of our Australian conductivity model with the model of Semenov and Kuvshinov (2012) shows some agreement, particularly in the lower mantle where both models appear relatively uniform. In the depth range 700–800 km, our model and Semenov and Kuvshinov's model suggest conductivities around 1–2 and 1–3 S/m, respectively, whereas the models of Kelbert et al. and Tarits and Mandea (2010) have larger variations that range from ~0.5 to 5 S/m to ~0.1 to 10 S/m, respectively. Such large conductivity variations can be difficult to explain in the apparent absence of unusually strong thermochemical anomalies associated with water and/or melt (Yoshino and Katsura 2013).

5.4 Testing the Conductivity Model Against Observations

To test this Australian conductivity model, we computed magnetic fields of 24 h S_q variations using the S3D method described in Koch and Kuvshinov (2013) and compared these to observed magnetic fields acquired from an Australia-wide

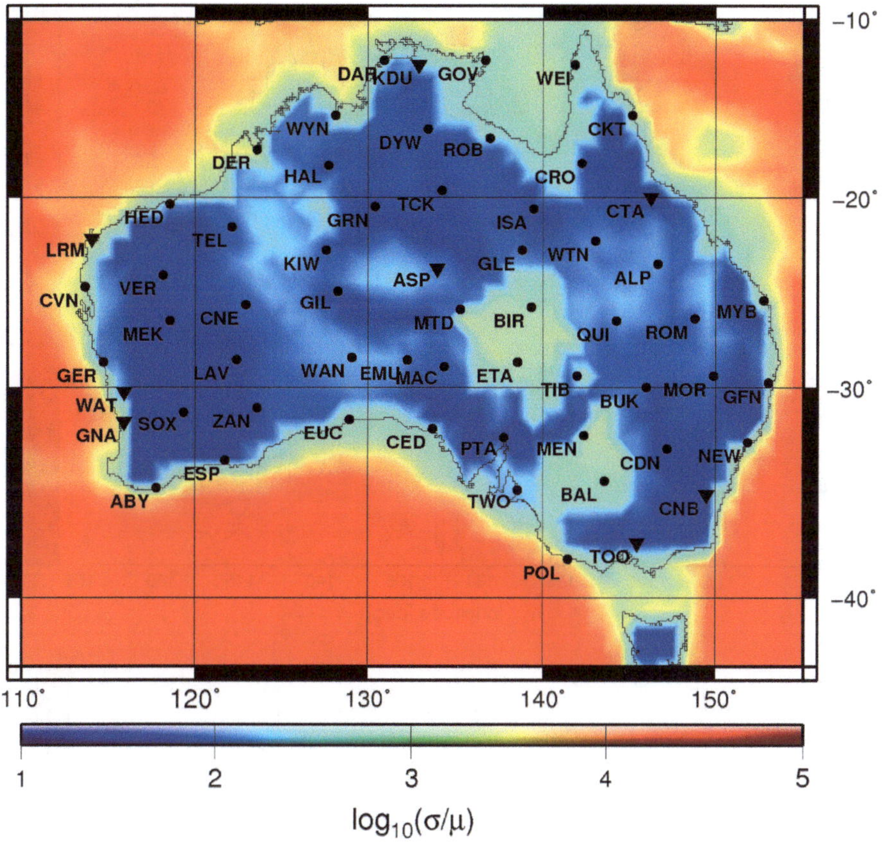

Fig. 5.9 Non-uniform surface conductance map (Manoj et al. 2006) used for magnetic field computations. Location of geomagnetic stations are also indicated: 53 non-permanent vector magnetometer (*dots*) of the Australia-wide array of geomagnetic stations (AWAGS) alongside with the eight permanent magnetic observatories (*inverted triangles*) of which four (CNB, CTA, GNA, and LRM) were part of AWAGS. $\mu = 1$ S

array of 57 3-component magnetometers (AWAGS) that operated for a period of 8 months (e.g., Chamalaun and Barton 1993). Station distribution and the currently employed surface conductance model are shown in Fig. 5.9. Observed and computed magnetic fields for all AWAGS stations based on our Australian model and a purely radial conductivity model (mean of the 3D model) are compared in Fig. 5.10. The misfit between models is computed from

$$B_{\text{mis}}(\theta,\phi) = \frac{1}{24}\sum_{i=1}^{24} B_i^{\text{obs}}(\theta,\phi) - B_i^{\text{mod}}(\theta,\phi), \tag{5.2}$$

where $B_i^{\text{obs}}(\theta,\phi)$ and $B_i^{\text{mod}}(\theta,\phi)$ represent observed and computed magnetic fields, respectively, at a given observatory.

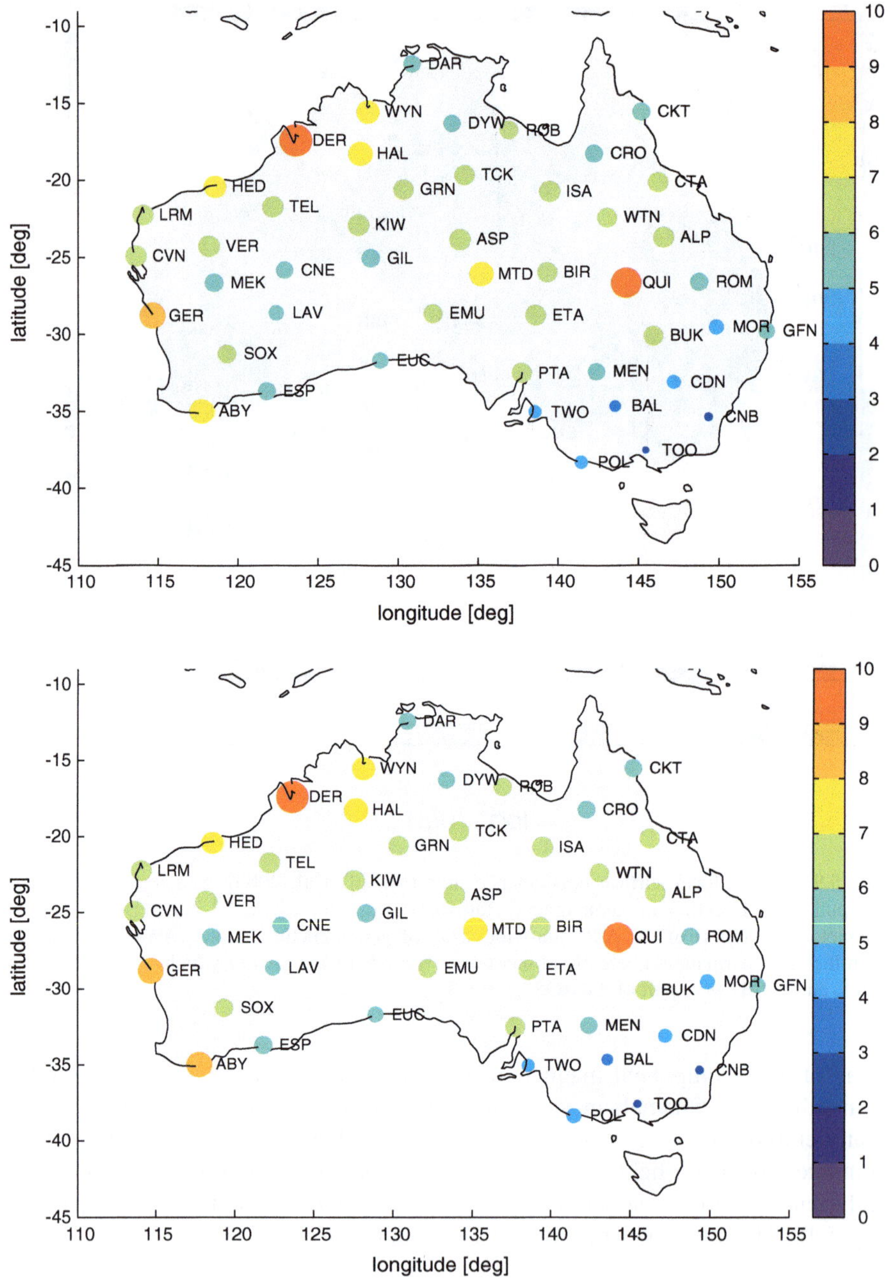

Fig. 5.10 Comparison between observed and computed horizontal magnetic fields across the Australia-wide array of geomagnetic stations (AWAGS) for the three-dimensional (3D) conductivity model is shown in Fig. 5.4 (**a**) and a radial one-dimensional conductivity model (mean of 3D model) (**b**). Both models are overlain by the non-uniform surface conductance map shown in Fig. 5.9. Color bar on the right side of each plot shows size of residual over 24 h as computed from Eq. 5.2 and is given in [nT]. Color coding as follows: *blue* (*small*) and *red* (*big*) *circles* designate small and large residuals, respectively

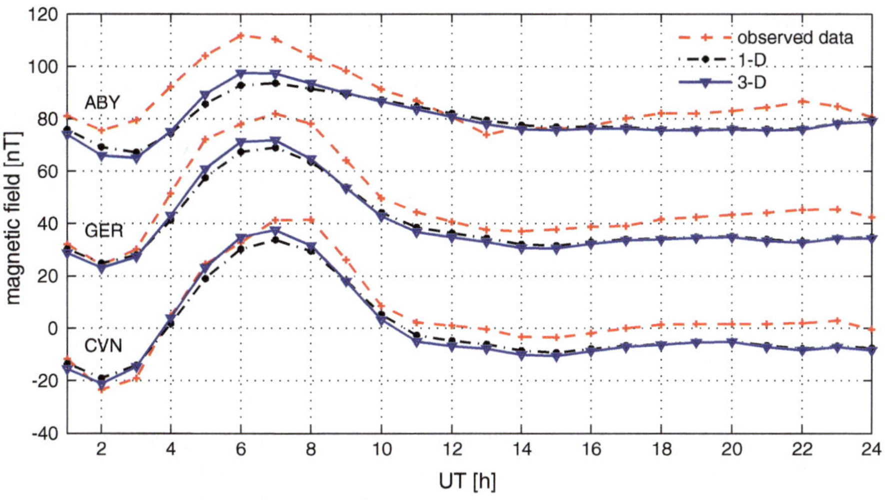

Fig. 5.11 Comparison of computed and observed horizonal-component magnetic fields of 24 h S_q-variations (May 15, 1990) at three representative observatories of the Australia-wide array of geomagnetic stations (AWAGS). *Dot-dashed curves* (labeled *1D*) designate synthetic data at observatory positions obtained using the surface conductance map shown in Fig. 5.9 and a one-dimensional (*1D*) conductivity background section constructed as the mean of the three-dimensional (*3D*) model. Solid curves (labeled *3D*) designate synthetic data at observatory positions obtained employing the same surface conductance shell map and the 3D upper mantle conductivity model (Fig. 5.4). Observed data are shown as *dashed lines*. In order to determine the S_q source that drives the forward modeling, we employed the S3D method (Koch and Kuvshinov 2013) for analysis of a magnetically very quiet day with the typical symmetric characteristic of the northern and southern S_q vortices

Examples of observed and computed magnetic field variations over a full 24 h-period at three stations are shown in Fig. 5.11; although misfits appear to be similar across the continent, differences between computed and observed magnetic fields are nonetheless observed to be present. The 3D conductivity model provides a better fit to data than the purely radial conductivity model (both overlain by the non-uniform surface conductance model shown in Fig. 5.9). The level of improvement relative to the purely radial model suggests that the lateral conductivity variations found here are supported by data, although further improvement in data fits probably calls for stronger regional variations than seen presently (e.g., Koyama et al. 2014). In this context, 3D global conductivity models (e.g., Kelbert et al. 2009; Tarits and Mandea 2010; Semenov and Kuvshinov 2012) show anomalies that vary several orders of magnitude across regions spanning continents and oceans. This difference between field- and laboratory-derived conductivities has presented long-standing difficulties.

An explanation for the discrepancy between conductivities determined in the laboratory and those derived from long-period EM-induction data could be due to inadequate or incomplete modeling of the magnetospheric source. Thus, if a complex source is modeled using a simple source structure such as P_0^1 (implicit in the

C-response concept), for example, then a large part of the modeled signal could be regarded as emanating entirely from the mantle. Such an effect could potentially lead to overestimation of mantle conductivity. This hypothesis is currently being investigated by Püthe et al. (2015) who are studying the use and implications of a set of new transfer functions that account for complex source structures when inverting global EM-induction data. Indeed, Püthe et al. (2015) find that inaccurate description of source structure leads to erroneous estimations of mantle conductivity.

5.5 Conclusions

The connection between geophysical observables, physical rock properties (seismic wave-speeds, density, and electrical conductivity), and thermo-chemistry is contained in the use of a self-consistent thermodynamic modeling scheme of mantle mineral phase equilibria that depends only on composition, temperature, and pressure. The great advantage of this approach is that it inserts geophysical/petrological knowledge of, e.g., discontinuities that straightforward inversions would not necessarily be able to resolve even though they have to be present. In this manner, we produce profiles of physical properties to obtain models of mantle conditions that simultaneously combine features of both laboratory and geophysical data.

In this study, we have examined the relative behavior between various mantle physical properties (elasticity, density, and electrical conductivity) that derive from stochastic inversion of seismic data. We have presented maps of the three-dimensional density, electrical conductivity, and shear-wave speed of the mantle beneath Australia and surrounding ocean in the depth range 100–800 km. The inversion mapped mantle temperatures and compositions, which, when combined with independent laboratory conductivity measurements of different mineralogies and temperatures, allowed us to compute mantle electrical conductivities at depth. Thus, although the conductivity maps obtained here are not constrained by data that are directly sensitive to conductivity (e.g., electromagnetic sounding data or magnetic data), we have calculated equivalent magnetic response data for a model and compared these to observations.

For Australia and its surroundings, we have shown from a combined analysis of seismic surface-wave data, phase-equilibrium computations, and laboratory-measured electrical conductivities that in the upper mantle seismic wave-speed, density, and conductivity anomalies appear to follow continental boundaries. Low conductivities correlate with the old stable central and western parts of the continent (relatively cold and seismically fast and dense), whereas high conductivity anomalies correlate with younger continental regions and oceanic lithosphere (relatively hot and seismically slow and less dense). Contributions to variations in structure are to a large extent temperature-controlled, although composition does appear to play a non-negligible role. Toward the bottom of the upper mantle and within the transition-zone lateral shear-wave speed, density, and conductivity variations appear to correlate less in comparison with the upper mantle, which suggests a compositional signature in observed anomaly patterns. Apart from the strong changes in properties that occur at seismic discontinuities, which are due to variations in thermochemically induced phase transformation of olivine to wadsleyite, transition-zone shear-wave speed, density, and conductivity maps are relatively smooth—a distinct feature of many seismic tomography images. In the lower mantle, compositional variations seem to govern relative behaviors of shear-wave speeds and conductivity to a greater extent than in the shallower mantle. There is additional evidence for bulk compositional variations between upper and lower mantle with the transition zone possibly acting as an intermediary layer.

Finally, the 3D regional conductivity model presented here has been tested against observed magnetic fields based on S_q-variations from an Australia-wide array of geomagnetic stations and found to provide an adequate fit. As a result, the present model should be a good choice as a starting model for 3D electromagnetic inversions of the Australian region. Validation of the proposed conductivity model will have to await solution of the inverse problem, which will be the focus of future studies.

Acknowledgements We wish to thank J.C. Afonso and an anonymous reviewer for helpful comments as well as T. Koyama, C. Püthe, and A. Kuvshinov for informed discussions. TJS thanks the Office of Basic Energy Sciences, U.S. Department of Energy for support. Numerical computations were performed on the ETH cluster Brutus.

References

Afonso JC, Ranalli G, Fernandez M, Griffin WL, O'Reilly SY, Faul U (2010) On the Vp/Vs—Mg# correlation in mantle peridotites: implications for the identification of thermal and compositional anomalies in the upper mantle. Earth Planet Sci Lett 289:606

Afonso JC, Fullea J, Griffin WL, Yang Y, Jones AG, Connolly JAD, O'Reilly SY (2013a) 3D multi-observable probabilistic inversion for the compositional and thermal structure of the lithosphere and upper mantle I: a priori information and geophysical observables. J Geophys Res. doi:10.1002/jgrb.50124

Afonso JC, Fullea J, Yang Y, Connolly JAD, Jones AG (2013b) 3D multi-observable probabilistic inversion for the compositional and thermal structure of the lithosphere and upper mantle II: general methodology and resolution analysis. J Geophys Res. doi:10.1002/jgrb.50123

Berryman JG (1995) Mixture theories for rock properties. In: Ahrens TJ (ed) American geophysical union handbook of physical constants. AGU, New York, p 205

Bina CR (1998) Free energy minimization by simulated annealing with applications to lithospheric slabs and mantle plumes. Pure Appl Geophys 151:605

Birch F (1961) The velocity of compressional waves in rocks to 10 kilobars, 2. J Geophys Res 66:2199

Brown JM, Shankland TJ (1981) Thermodynamic parameters in the earth as determined from seismic profiles. Geophys J Int 66:579

Campbell WH, Barton CE, Chamalaun FH, Webb W (1998) Quiet-day ionospheric currents and their applications to upper mantle conductivity in Australia. Earth Planet Space 50:347

Chamalaun FH, Barton CE (1993) Electromagnetic induction in the Australian crust: results from the Australia-wide array of geomagnetic stations. Explor Geophys 24:179

Christensen UR, Hofmann AW (1994) Segregation of subducted oceanic crust in the convecting mantle. J Geophys Res 99(19):867. doi:10.1029/93JB03403
Connolly JAD (2005) Computation of phase equilibria by linear programming: a tool for geodynamic modeling and an application to subduction zone decarbonation. Earth Planet Sci Lett 236, doi:10.1016/j.epsl.2005.04.033
Connolly JAD (2009) The geodynamic equation of state: what and how. Geophys Geochem Geosys 10:Q10014. doi:10.1029/2009GC002540
Deschamps F, Trampert J (2003) Mantle tomography and its relation to temperature and composition. Phys Earth Planet Int 140:277
Deschamps F, Snieder R, Trampert J (2001) The relative density-to-shear velocity scaling in the uppermost mantle. Phys Earth Planet Int 124:193
Deschamps F, Trampert J, Snieder R (2002) Anomalies of temperature and iron in the uppermost mantle inferred from gravity data and tomographic models. Phys Earth Planet Int 129:245
Dobson DP, Brodholt JP (2000) The electrical conductivity and thermal profile of the Earth's mid-mantle. Geophys Res Lett 27:2325
Drilleau M, Beucler E, Mocquet A, Verhoeven O, Moebs G, Burgos G, Montagner J-P, Vacher P (2013) A Bayesian approach to infer radial models of temperature and anisotropy in the transition zone from surface wave data. Geophys J Int. doi:10.1093/gji/ggt284
Fabrichnaya OB (1999) The phase in the FeO–MgO–$Al2O_3$–SiO_2 system: assessment of thermodynamic properties and phase equilibria at pressures up to 30 GPa. Calphad 23:19
Fichtner A, Kennett BLN, Igel H, Bunge H-P (2010) Full waveform tomography for radially anisotropic structure: new insights into present and past states of the Australasian upper mantle. Earth Planet Sci Lett 290, doi:10.1016/j.epsl.2009.12.003
Fishwick S, Rawlinson N (2012) 3-D structure of the Australian lithosphere from evolving seismic datasets. Austr J Earth Sci 59:809
Forte AM, Perry HC (2000) Geodynamic evidence for a chemically depleted continental tectosphere. Science 290:1940
Fullea J, Afonso JC, Connolly JAD, Fernandez M, Garcia-Castellanos D, Zeyen H (2009) LitMod3D: an interactive 3D software to model the thermal, compositional, density, rheological, and seismological structure of the lithosphere and sublithospheric upper mantle. Geochem Geophys Geosyst 10:Q08019. doi:10.1029/2009GC002391
Fullea J, Muller MR, Jones AG (2011) Electrical conductivity of continental lithospheric mantle from integrated geophysical and petrological modeling: application to the Kaapvaal Craton and Rehoboth Terrane, southern Africa. J Geophys Res 116:B10202. doi:10.1029/2011JB008544
Fullea J, Lebedev S, Agius MR, Jones AG, Afonso JC (2012) Lithospheric structure in the Baikal central Mongolia region from integrated geophysical-petrological inversion of surface-wave data and topographic elevation. Geochem Geophys Geosyst 13, Q0AK09. doi:10.1029/2012GC004138
Goes S, Govers R, Vacher P (2000) Shallow mantle temperatures under Europe from P and S wave tomography. J Geophys Res 105:11153
Goes S, Armitage J, Harmon N, Smith H, Huismans R et al (2012) Low seismic velocities below mid-ocean ridges: attenuation versus melt retention. J Geophys Res 117. doi:10.1029/2012JB009637
Hashin Z, Shtrikman S (1962) A variational approach to the theory of effective magnetic permeability of multiphase materials. J Appl Phys 33:3125
Helffrich GR, Wood BJ (2001) The Earth's Mantle. Nature 412–501
Hirsch LM, Shankland TJ, Duba AG (1993) Electrical conduction and polaron mobility in Fe-bearing olivine. Geophys J Int 114:36
Irifune T (1994) Absence of an aluminous phase in the upper part of the Earths lower mantle. Nature 370:131
Ishii M, Tromp J (2004) Constraining large-scale mantle heterogeneity using mantle and inner-core sensitive normal modes. Phys Earth Planet Inter 146:113

Jackson I (1998) Elasticity, composition and temperature of the Earths lower mantle: a reappraisal. Geophys J Int 134:291

Jones AG, Afonso JC, Fullea J, Salajegheh F (2013) The lithosphere-asthenosphere system beneath Ireland from integrated geophysical-petrological modeling—I: observations, 1D and 2D hypothesis testing and modeling. Lithos (in press)

Karato S-I (2011) Water distribution across the mantle transition zone and its implications for global material circulation. Earth Planet Sci Lett 301. doi:10.1016/j.epsl.2010.1.038

Katsura T, Yoshino T (2015) Heterogeneity of electrical conductivity in the oceanic upper mantle, this volume

Kawai K, Tsuchiya T (2015) Elasticity of continental crust around the mantle transition zone, this volume

Kelbert A, Schultz A, Egbert G (2009) Global electromagnetic induction constraints on transition-zone water content variations. Nature 460. doi:10.1038/nature08257

Kennett BLN (1998) On the density distribution within the Earth. Geophys J Int 132:374

Kennett BLN, Fichtner A, Fishwick S, Yoshizawa K (2012) Australian seismological reference model (AuSREM): Mantle component. Geophys J Int 192. doi:10.1093/gji/ggs065

Khan A, Shankland TJ (2012) A geophysical perspective on mantle water content and melting: inverting electromagnetic sounding data using laboratory-based electrical conductivity profiles. Earth Planet Sci Lett 317–318. doi:10.1016/j.epsl.2011.11.031

Khan A, Connolly JAD, Olsen N (2006) Constraining the composition and thermal state of the mantle beneath Europe from inversion of long-period electromagnetic sounding data. J Geophys Res 111:B10102. doi:10.1029/2006JB004270

Khan A, Kuvshinov A, Semenov A (2011) On the heterogeneous electrical conductivity structure of the Earth's mantle with implications for transition zone water content. J Geophys Res 116(B01103):2011. doi:10.1029/2010JB007458

Khan A, Zunino A, Deschamps F (2013) Upper mantle compositional variations and discontinuity topography imaged beneath Australia from Bayesian inversion of surface-wave phase velocities and thermochemical modeling. J Geophys Res 118. doi:10.1002/jgrb.50304

Koch S, Kuvshinov A (2013) Global 3-D EM inversion of Sq-variations based on simultaneous source and conductivity determination. A concept validation and resolution studies. Geophys J Int doi:10.1093/gji/ggt227

Koyama T, Shimizu H, Utada H, Ichiki M, Ohtani E, Hae R (2006) Water content in the mantle transition zone beneath the North Pacific derived from the electrical conductivity anomaly. In: Jacobsen S, van der Lee S (eds) Earth's deep water cycle, vol 168. AGU, Washington, DC, p 171 (Geophys Monogr Ser)

Koyama T, Khan A, Kuvshinov A (2014) Three-dimensional electrical conductivity structure beneath Australia from inversion of geomagnetic observatory data: evidence for lateral variations in transition-zone temperature, water content, and melt. Geophys J Int. doi:10.1093/gji/ggt455

Kuo C, Romanowicz B (2002) On the resolution of density anomalies in the Earth's mantle using spectral fitting of normal mode data. Geophys J Int 150:162

Kuskov OL, Panferov AB (1991) Phase diagrams of the FeO–MgO–SiO_2 system and the structure of the mantle discontinuities. Phys Chem Minerals 17:642

Kuskov OL, Kronrod VA, Prokofyev AA, Pavlenkova NI (2014) Thermo-chemical structure of the lithospheric mantle underneath the Siberian craton inferred from long-range seismic profiles. Tectonophysics 615616:154. doi:10.1016/j.tecto.2014.01.006

Kustowski B, Ekström G, Dziewonski AM (2008) Anisotropic shear-wave velocity structure of the Earth's mantle: a global model. J Geophys Res 113:B06306. doi:10.1029/2007JB005169

Kuvshinov A (2012) Deep electromagnetic studies from land, sea and space. Progress status in the past 10 years. Surv Geophys 33. doi:10.1007/s10712-011-9118-2

Landauer R (1952) The electrical resistance of binary metallic mixtures. J Appl Phys 23:779

Laske G, Markee A, Orcutt JA, Wolfe CJ, Collins JA, Solomon SC, Detrick RS, Bercovici D, Hauri EH (2011) Asymmetric shallow mantle structure beneath the Hawaiian Swell—evidence from Rayleigh waves recorded by the PLUME network. Geophys J Int 187(1725):2011

Lekic V, Romanowicz B (2012) Tectonic regionalization without a priori information: a cluster analysis of upper mantle tomography. Earth Planet. Sci Lett 308, doi:10.1016/j.epsl.2011.05.050

Manoj C, Kuvshinov A, Maus S, Lühr H (2006) Ocean circulation generated magnetic signals. Earth Planet Space 58:429

Manthilake M, Matsuzaki T, Yoshino T, Yamashita S, Ito E, Katsura T (2009) Electrical conductivity of wadsleyite as a function of temperature and water content. Phys Earth Planet Int. doi:10.1016/j.pepi.2008.06.001

Masters G, Laske G, Bolton H, Dziewonski AM (2000) The relative behaviour of shear velocity, bulk sound speed, and compressional velocity in the mantle: Implications for chemical and thermal structure. In: Karato et al. (eds) Earths deep interior: mineral physics and tomography from the atlantic to the global scale, vol 117. AGU, Washington, DC, p 63 (Geophys Monogr Ser)

Park SK, Ducea MN (2003) Can in situ measurements of mantle electrical conductivity be used to infer properties of partial melts? J Geophys Res 108:2270. doi:10.1029/2002JB001899

Piazzoni AS, Steinle-Neumann G, Bunge HP, Dolejs D (2007) A mineralogical model for density and elasticity of the Earths mantle. Geochem Geophys Geosyst 8:Q11010. doi:10.1029/2007 GC001697

Poirier JP (2000) Introduction to the physics of the Earth's interior, 2nd edn. Cambridge University Press, Cambridge, p 312

Pommier A (2014) Interpretation of magnetotelluric results using laboratory measurements. Surv Geophys 35. doi:10.1007/s10712-013-9226-2

Püthe C, Kuvshinov A, Olsen N (2015) Handling complex source structures in global EM induction studies: from C-responses to new arrays of transfer functions. Geophys J Int 201, doi: 10.1093/gji/ggv021

Rawlinson N, Kennett BLN, Salmon M, Glen RA (2015) Origin of lateral heterogeneities in the upper mantle beneath south-east australia from seismic tomography, this volume

Resovsky JS, Ritzwoller MH (1999) Regularization uncertainty in density models estimated from normal mode data. Geophys Res Lett 26:2319

Resovsky J, Trampert J (2003) Using probabilistic seismic tomography to test mantle velocity-density relationships. Earth Planet Sci Lett 215:121

Ricard Y, Mattern E, Matas J (2005) Synthetic tomographic images of slabs from mineral physics. In: Hilst RVD, Bass JD, Matas J, Trampert J (eds) Earths deep Mantle: structure, composition, and evolution. AGU, Washington, DC, pp 283–300

Ricolleau et al (2009) Density profile of pyrolite under the lower mantle conditions. Geophys Res Lett 36:L06302. doi:10.1029/2008GL036759

Romanowicz B (2001) Can we resolve 3D density heterogeneity in the lower mantle? Geophys Res Lett 28:1107

Saltzer RL, van der Hilst RH, Karason H (2001) Comparing P and S wave heterogeneity in the mantle. Geophys Res Lett 28:1335

Saxena SK, Eriksson G (1983) Theoretical computation of mineral assemblages in pyrolite and lherzolite. J Petrol 24:538

Schaeffer AJ, Lebedev S (2015) Global heterogeneity of the lithosphere and underlying mantle: a seismological appraisal based on multimode surface-wave dispersion analysis, shear-velocity tomography, and tectonic regionalization, this volume

Schmerr N (2015) Imaging mantle heterogeneity with upper mantle seismic discontinuities, this volume

Semenov A, Kuvshinov A (2012) Global 3-D imaging of mantle electrical conductivity based on inversion of observatory C-responses—II. Data analysis and results. Geophys. J Int. doi:10.1111/j.1365-246X.2012.05665.x

Shankland TJ (1972) Velocity-density systematics: derivation from Debye theory and the effect of ionic size. J Geophys Res 77:3750

Shankland TJ (1977) Elastic properties, chemical composition, and crystal structure of minerals. Surv Geophys 3:89

Shankland TJ (1981) Electrical conduction in Mantle materials. In: O'Connell RJ, Fyfe WS (eds) Evolution of the Earth, American Geophysical Union, vol 5. Washington, DC. p 256, doi: 10.1029/GD005p0256

Shankland TJ, O'Connell RJ, Waff HS (1981) Geophysical constraints on partial melt in the upper mantle. Rev Geophys Space Phys 19:394

Shimizu H, Utada H, Baba K, Koyama T, Obayashi M, Fuka Y (2010) Three-dimensional imaging of electrical conductivity in the mantle transition zone beneath the North Pacific Ocean by a semi-global induction study. Phys Earth Planet Int 183. doi:10.1016/j.pepi.2010.01.010

Simmons NA, Forte AM, Grand SP (2009) Joint seismic, geodynamic and mineral physical constraints on three-dimensional mantle heterogeneity: Implications for the relative importance of thermal versus compositional heterogeneity. Geophys J Int 177:1284

Sobolev SV, Babeyko AY (1994) Modeling of mineralogical composition, density and elastic-wave velocities in anhydrous magmatic rocks. Surv Geophy 15:515

Stixrude L, Lithgow-Bertelloni C (2005) Thermodynamics of mantle minerals I. Physical properties. Geophys J Int 162:610

Stixrude L, Lithgow-Bertelloni C (2011) Thermodynamics of mantle minerals II. Phase equilibria. Geophys J Int 184:1180

Tarits P, Mandea M (2010) The heterogeneous electrical conductivity structure of the lower mantle. Phys Earth Planet Int 183. doi:10.1016/j.pepi.2010.08.002

Toffelmier DA, Tyburczy JA (2007) Electromagnetic detection of a 410-km-deep melt layer in the Southwestern United States. Nature 447. doi:10.1038/nature05922

Trampert J, Van der Hilst RD (2005) Towards a quantitative interpretation of global seismic tomography. In: Van der Hilst RD, Bass JD, Matas J, Trampert J (eds) Earth's Deep Interior: Structure, Composition, and Evolution, Geophysical Monograph 160. American Geophysical Union, Washington, p 47–62

Utada, H., Koyama, T., Obayashi, M., Fukao, Y., 2009. A joint interpretation of electromagnetic and seismic tomography models suggests the mantle transition zone below Europe is dry, Earth Planet Sci Lett. doi:10.1016/j.epsl.2009.02.027

Verhoeven O et al (2009) Constraints on thermal state and composition of the Earth's lower mantle from electromagnetic impedances and seismic data. J Geophys Res 114:B03302. doi:10.1029/2008JB005678

Wang CY (1970) Density and constitution of the mantle. J Geophys Res 75:3264

Watt JP, Davies GF, O'Connell RJ (1976) The elastic properties of composite materials. Rev Geophys Space Phys 14:541

Wood BJ, Holloway JR (1984) A thermodynamic model for subsolidus equilibria in the system $CaO–MgO–Al_2O_3–SiO_2$. Geochim Cosmochim Acta 66:159

Xie S, Tackley PJ (2004) Evolution of helium and argon isotopes in a convecting mantle. Phys Earth Planet Inter 146:417. doi:10.1016/j.pepi.2004.04.003

Xu Y, Shankland TJ, Poe BT (2000) Laboratory-based electrical conductivity in the Earth's mantle. J Geophys Res 108:2314

Yoshino T (2010) Laboratory electrical conductivity measurement of mantle minerals. Surv Geophys 31:163206. doi:10.1007/s10712-009-9084-0

Yoshino T, Katsura T (2013) Electrical conductivity of mantle minerals: role of water in conductivity anomalies. Ann Rev Earth Planet Sci 41. doi:10.1146/annurev-earth-050212-124022

Yoshino T et al (2012) Effect of temperature, pressure and iron content on the electrical conductivity of olivine and its high-pressure polymorphs. J Geophys Res 117:B08205. doi:10.1029/2011JB008774

Chapter 6
Heterogeneity of Electrical Conductivity in the Oceanic Upper Mantle

Tomoo Katsura and Takashi Yoshino

Abstract We discuss conductivity heterogeneities of the oceanic upper mantle using available experimental data. The activation energy of the polaron conduction in olivine, wadsleyite, and ringwoodite is similar, at 1.4–1.6 eV. The ionic conduction is significant in olivine, but not in wadsleyite and ringwoodite. Its activation energy is much larger than that of the small polaron conduction (7–12 eV). The proton conductions have a smaller activation energy than the small polaron conduction (less than 1 eV) and are negligible at high temperatures in the depleted MORB mantle. The effects of the secondary minerals are negligible. No significant conductivity jump is associated with the olivine–wadsleyite transition. Volatile components greatly increase conductivity of basaltic melt. The anisotropy in both intrinsic and proton conditions in olivine is small. Sheared, partially molten peridotite can show conductivity anisotropy. The high conductivity below mid-oceanic ridges could be caused by partial melting. Conductivity at several locations suggests a melt fraction of the order of 0.1 vol.%, whereas that under the East Pacific Rise at 9°N suggests one of 15 vol.%. Lithosphere has low conductivity, which should be primarily due to its low temperature. However, the conductivity is too high, judging from the temperature structure and intrinsic conduction of olivine. The circulation of water does not provide enough explanation. The high-conductivity layer at the top of the asthenosphere is not a ubiquitous feature of the mantle—it is relatively limited to regions under young plates. The associated conductivity anisotropy suggests its partial melting origin. The conductivity in the mantle transition zone can be explained by the intrinsic conduction of wadsleyite and ringwoodite. Estimations of water content in the transition zone are largely affected by the uncertainty of geophysical modeling. The MT studies do not detect mantle

T. Katsura (✉)
Bayerisches Geoinstitut, University of Bayreuth, 95440 Bayreuth, Germany
e-mail: tomo.katsura@uni-bayreuth.de

T. Yoshino
Institute for Study of the Earth's Interior, Okayama University,
Misasa 612-0193, Japan

© Springer International Publishing Switzerland 2015
A. Khan and F. Deschamps (eds.), *The Earth's Heterogeneous Mantle*,
Springer Geophysics, DOI 10.1007/978-3-319-15627-9_6

plumes, although the seismic studies show the presence of low-velocity zones. The conductivity anomalies, whose origins are not understood, are observed under the southern Philippine Sea and the broad region north of Hawaii.

Keywords Electrical conductivity · Small polaron conduction · Ionic conduction · Proton conduction · Magnetotellurics · Olivine · Wadsleyite · Ringwoodite · Mantle heterogeneity

6.1 Introduction

Electrical conductivity is a physical parameter that can be estimated from the earth's surface by means of the magnetotelluric (MT) and geomagnetic deep soundings (GDS) method. Unfortunately, however, spatial resolutions of MT studies are by far worse than those of seismic studies (cf. Fukao et al. 2004). Additionally, although seismic studies can detect variation in elasticity of the order of 0.1 %, variation by factors is necessary to detect conductivity anomalies by the MT and GDS methods. Nevertheless, electrical conductivity can provide valuable information about the earth's depths.

Fortunately, conductivity of minerals can vary by orders of magnitude. One reason for the importance of conductivity studies is this large variation: A large variation in conductivity allows the detection of a conductive phase with a small volume fraction. Furthermore, some chemical variations that are difficult to detect with a seismic study can be detected: Those in defect structures and amounts of impurities produce conductivity variation, whereas those in chemical bonding between major constituent atoms could be caused by elastic variation. As Jacobsen (2006) suggested, although elastic moduli decrease with increasing water content at a given pressure, the effects of water on the elastic moduli become smaller with increasing pressure because of the larger pressure dependence of elastic moduli with the increasing water content. On the other hand, many workers reported that water incorporation in nominally anhydrous minerals largely increases their conductivity (cf. Yoshino and Katsura 2013). Thus, conductivity studies should provide different and independent information about the earth's interior to that from seismic studies.

Another point that we should mention is that regions that can be studied with electrical methods are different from those that can be seismically studied. The distribution of seismicity is quite heterogeneous in the earth, whereas the source signals of the MT and GDS observations are the variations in the electromagnetic field from space, available anywhere on earth. The electromagnetic waves from space penetrate the mantle vertically, allowing us to study conductivity structures under observatories. Seismic structures are obtained for the regions between observatories and large earthquakes.

In this article, we try to understand the conductivity variation in the oceanic upper mantle. Because the oceans cover 70 % of the earth's surface, the oceanic mantle should be representative of the earth's mantle. We focus on the upper mantle and do not consider the lower mantle, although results of observational and experimental studies for the lower mantle conductivity recently exist (cf. Velimsky 2010; Ohta et al. 2010). We first review results of laboratory conductivity measurements to discuss possible reasons for the creation of conductivity heterogeneity in the mantle. We then discuss geophysical observations of conductivity heterogeneity in the oceanic upper mantle, using the experimental results discussed in the first part. Since the authors have been studying electrical conductivity of geomaterials (cf. Fuji-ta et al. 2004, 2007; Katsura et al. 1998, 2007; Manthilake et al. 2009; Yoshino and Katsura 2009, 2012, 2013; Yoshino et al. 2003, 2004, 2006, 2008a, b, c, 2009a, b, 2010, 2011; 2012a, b; Zhang et al. 2012, 2014), but are no MT and GDS experts, we have for the purposes of this article critically evaluated and selected experimental studies based on our experience. Of course, we are aware of a number of problems with MT and GDS studies, namely (1) strong model dependency (Shimizu et al. 2010a) and (2) the effects of surrounding geometry and conductivity structures (Heinson and Constable 1992). On the other hand, we have less MT and GDS study results that have been clearly disputed in literature.

Throughout this article, we assume that the mantle consists of two kinds of materials: the depleted MORB mantle (DMM) and the enriched mantle (EM). The DMM occupies the majority of the upper mantle, whereas the EM is associated with mantle plumes. As Workman et al. (2006) argued, the EM contains several times more volatile components than the DMM. According to Hirschmann (2010), we assume that the DMM contains 100 and 60 ppm of H_2O and CO_2, respectively, whereas the EM contains 300–900 and 120–1800 ppm of H_2O and CO_2, respectively.

6.2 Laboratory Conductivity Measurements of Upper Mantle Constituents

6.2.1 Mathematical Expression of Conductivity

Electrical conduction in the upper mantle minerals is a thermal activation process. For this reason, the temperature and pressure dependence of conductivity of minerals is assumed to follow the Arrhenius relation.

$$\sigma = \sigma_0 \exp\left[-\frac{H_a}{RT}\right] = \sigma_0 \exp\left[-\frac{E_a + PV_a}{RT}\right] \tag{6.1}$$

where σ_0 is the pre-exponential term, H_a is the activation enthalpy, E_a is the activation energy, V_a is the activation volume, R is the gas constant, T is the temperature, and P is the pressure. The pre-exponential term σ_0 expresses the population of charge. The activation energy E_a expresses the energy barrier for the charge transfer at ambient pressure—the large E_a means that conductivity rapidly decreases with decreasing temperature. The activation enthalpy H_a expresses the high-pressure correction of E_a. The pressure dependence is considered to be linear with the constant activation volume V_a. The positive V_a means that conductivity has negative pressure dependence. Note that, although the population of charge could vary with pressure, it is usually assumed that only the energy barrier varies with pressure. The frequently used units of H_a, E_a, V_a, P, and T are kJ/mol, kJ/mol, cm^3, GPa, and K, respectively. The unit eV is also frequently used for E_a. In this case, the gas constant should be replaced by the Boltzmann constant k. If more than two mechanisms are considered to contribute to the bulk conductivity, the conductivity is expressed by the sum of Arrhenius equations.

Electrical conduction in liquid phases such as aqueous fluids, silicate melts, and carbonate melts is also a thermal activation process. Due to the complexity of electrical conduction in liquid phases, there is no widely accepted formula for the electrical conduction of liquid phases. In some studies, the obtained temperature dependence of electrical conductivity is well expressed using a single Arrhenius relation (Eq. 6.1) (cf. Waff and Tyburczy 1983; Gaillard 2004). Other studies find that the temperature dependence of conductivity largely deviates from the Arrhenius relations (cf. Ni et al. 2011). Ni et al. (2011) applied the Vogel–Tammann–Fulcher (VTF) equation (Angel and Smith 1982) originally developed to express the temperature dependence of viscosity in the expression of conductivity of liquid phases, because both electrical conduction and viscosity are related to the body motion of ions in liquid phases. The simple formula for the VTF equation is:

$$\sigma = A \exp\left[-\frac{B}{T - T_0}\right] \tag{6.2}$$

where A, B, and T_0 are empirical parameters. Conductivity in this expression decreases with decreasing temperature less strongly than in the Arrhenius equation, especially at lower temperatures.

6.2.2 *Intrinsic Conduction of Olivine*

Since olivine is the most abundant mineral in the upper mantle, its electrical properties have been studied extensively. Most efforts to understand the intrinsic conduction were made until 2000. Intrinsic conduction in this paper refers to electrical conduction by major chemical components. Since olivine is a ferromagnesian mineral, hopping of electron holes between ferrous and ferric iron, often called the small polaronconduction (cf. Wanamaker and Duba 1993), should be the most essential intrinsic conduction mechanism. Another mechanism is the migration of Mg vacancies, called ionic conduction in this chapter. Schock et al. (1989) demonstrated the importance of this mechanism at high temperatures. Separation of these two mechanisms is often difficult because ionic conduction becomes visible only in a small range of inverse temperatures in an Arrhenius plot. Therefore, we treat these two mechanisms as one intrinsic mechanism in most of this paper.

The following is a summary of properties of the intrinsic conduction in olivine.

1. Small polaron conduction is dominant up to 1500 K. Above this temperature, ionic conduction gradually becomes significant (Constable 2006).
2. The activation energy of small polaron conduction is about 1.5–1.7 eV (Constable and Duba 1990; Shankland and Duba 1990; Constable et al. 1992; Xu et al. 1998; Yoshino et al. 2009a, b). These values are smaller than the ionic conduction but larger than that of the proton conduction mentioned later. The small polaron conduction decreases significantly at low temperatures.
3. The activation energy of ionic conduction is very high. Due to difficulty in separating the ionic conduction from the small polaron conduction, its activation energy is not clear. However, it is reported to be from 7 to 12 eV (Constable et al. 1992; Constable and Duba 1990).
4. The absolute values of conductivity increase with increasing Fe content (decreasing Mg# = 100 Mg/(Mg + Fe)) at the same oxygen buffer (Omura et al. 1989; Yoshino et al. 2012). Although the degree of increase in conductivity with Fe content varies with temperature because of the Fe content dependence of activation energy mentioned below, conductivity of olivine with Mg# = 80 composition is half an order of magnitude higher than that with Mg# = 90 composition at upper mantle temperatures.
5. The activation energy decreases with increasing Fe content (decreasing Mg#) (Omura et al. 1989; Yoshino et al. 2012). The activation energy decreases by 0.2 eV when the Mg# changes from 90 to 80.
6. Conductivity increases proportionally to 0.18 the power of oxygen fugacity (Wanamaker and Duba 1993; Duba and Constable 1993). Namely, the oxidation state slightly affects conductivity.
7. Silica activity slightly affects conductivity. Wanamaker and Duba (1993) showed that a pyroxene-buffered sample has up to 0.2 log unit lower conductivity than a self-buffered sample.
8. Conductivity in the [001] direction is by a factor of about two higher than that in the [100] and [010] directions at temperatures of the upper mantle (Schock et al. 1989; Shankland and Duba 1990; Du Frane et al. 2005). Although the conductivity anisotropy in olivine is small, the degree of anisotropy increases with decreasing temperature.
9. Olivine conductivity has positive pressure dependence, namely negative activation volume (Yoshino et al. 2012b). The absolute values of activation volume decrease with increasing Mg#, and the pressure dependence of the intrinsic conduction of the mantle olivine is very small.

In this paper, we use the Constable's (2006) SO3 model as a reference for the intrinsic conductivity of mantle olivine with the assumption that the mantle oxygen fugacity is near the QFM buffer (Fig. 6.1).

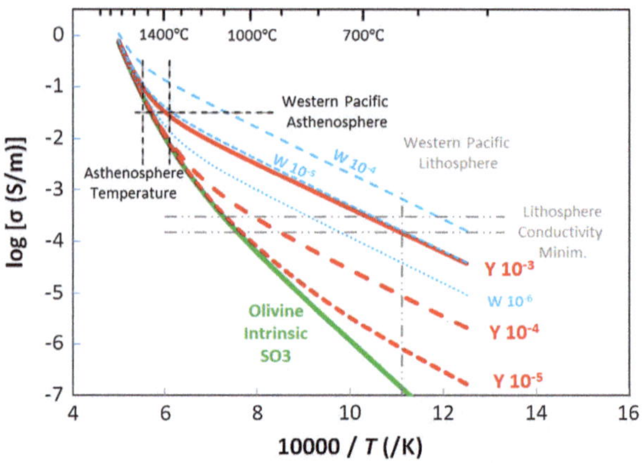

Fig. 6.1 Electrical conductivity of olivine. The *green curve* denotes a reference intrinsic conduction model of mantle olivine "SO3" proposed by (Constable (2006)). The *red* and *blue curves* denote the sum of the intrinsic and proton conduction with different water content, where the proton conduction is based on (Yoshino et al. (2009a, b) and Wang et al. (2006). The numbers next to these *curves* indicate the weight fraction of water in olivine, namely 10^{-3} or 0.1 wt% of water. The contribution of proton conduction is prominent at low temperatures but becomes negligible at temperatures corresponding to the asthenosphere. The two *vertical black broken lines* denote the range of temperature in the asthenosphere down to 400 km depth according to Katsura et al. (2010). The *horizontal black broken line* denotes conductivity of asthenosphere under the Western Pacific to the east of Izu Islands at 250–300 km depth reported by Baba et al. (2010). The *vertical gray line denotes* the possible temperature at a depth of 50 km in this region. The *horizontal gray broken lines* denote the minimum conductivity in this region

6.2.3 Proton Conduction of Olivine

Proton conduction is an electrical charge transfer by the migration of H^+. Since olivine can incorporate water under pressure in spite of its nominally anhydrous formula, the proton conduction could be significant as the dominant conduction mechanism in the deep mantle (Karato 1990). Subsequent to this finding, proton conduction has received attention from many scientists in this field (cf. Constable 1993; Evans et al. 1999; Baba 2005). Before proton conduction in olivine was actually measured, it was estimated using hydrogen diffusion data in olivine provided by Mackwell and Kohlstedt (1990) and Kohlstedt and Mackwell (1998). This estimation predicted more than one order of magnitude higher conductivity than the intrinsic conduction and more than one order of magnitude higher anisotropy (cf. Karato 1990; Evans et al. 2005).

In the late 2000s, three groups reported proton conduction of olivine (Wang et al. 2006; Yoshino et al. 2006, 2009a, b; Poe et al. 2010). Although the results of Poe et al. (2010) agree with those of Yoshino et al. (2009a, b) at relatively low water contents, Wang et al. (2006) report significantly higher proton conduction at the same water content than those of the other two groups. There are a few indicators of the reliability of the Yoshino et al. (2009a, b) results. Firstly, Yoshino et al.

(2006), (2009a, b) and Yoshino and Katsura (2012) demonstrated that water in doped olivine is released at temperatures above 1000 K. Above 1000 K, olivine conductivity is masked by a fluid phase and therefore, it is impossible to measure the conductivity of anhydrous olivine. Nevertheless, Wang et al. (2006) measured conductivity at temperatures up to 1273 K. Secondly, Yoshino et al. (2006) observed changes in conductivity, with water content down to the conditions where the effects of proton conduction are almost masked by the intrinsic conduction. Wang et al. (2006) measured proton conduction only when conductivity is more than one order of magnitude higher than the supposed intrinsic conduction. This group has actually never shown measurements of the intrinsic conduction of olivine. Thirdly, the impedance spectra produced by Yoshino et al. (2009a, b) showed almost perfect semi-arcs with a tail, which indicates that only one conduction mechanism is responsible for conductivity, although there is some reaction with the electrode. Although Wang et al. (2006) did not show any impedance spectra, impedance spectra of water-doped, nominally anhydrous minerals given by this group usually show strong distortion, further discussed by Yoshino and Katsura (2012). Since Yoshino and Katsura (2013) discussed all these points in detail, the proton conduction in olivine discussed in this article is based on Yoshino et al. (2006, 2009a, b) .

The following are summaries of the proton conductivity of olivine from Yoshino et al. (2006, 2009a, b).

1. Proton conduction increases with increasing water content. Although there is an additional effect on the activation energy mentioned below, the magnitude of proton conduction is essentially proportional to the water content.
2. The activation energy decreases with increasing water content, linearly with one-third power of water content: $\sigma_{\mathrm{P}} = \sigma_{\mathrm{P}_0}\exp\left[-\frac{E_0-\alpha C_{\mathrm{H_2O}}^{1/3}}{kT}\right]$, where σ_{P} is the proton conduction, σ_{P_0} is the pre-exponential term of the proton conduction, E_{P_0} is the activation energy of the proton conduction at zero water, $C_{\mathrm{H_2O}}$ is the water content, k is the Boltzmann factor, T is the absolute temperature, and α is the constant. The activation energy of the proton conduction at zero water is $E_{\mathrm{P}_0} = 0.92$ eV. Thus, the activation energies of proton conduction are lower than those of intrinsic conduction (1.4–12 eV). Consequently, proton conduction dominates at low temperatures. At asthenospheric temperatures, the contribution of proton conduction is very small.
3. Although a large anisotropy is expected from the hydrogen diffusion data (cf. Evans et al. 2005), the proton conduction in olivine is essentially isotropic at asthenospheric temperatures (Yoshino et al. 2006; Yang 2012) (Fig. 6.2). The anisotropy of proton conduction is actually even smaller than that of the intrinsic conduction. Although the hydrogen diffusion experiments (Mackwell and Kohlstedt 1990; Kohlstedt and Mackwell 1998) did not report a decrease in anisotropy with increasing temperature, this is due to their large experimental uncertainty.
4. There is no report about the activation volume of proton conduction of olivine. Since the ionic volume of protons is small, the pressure effect on proton conduction should be small, i.e., the activation energy is nearly zero.

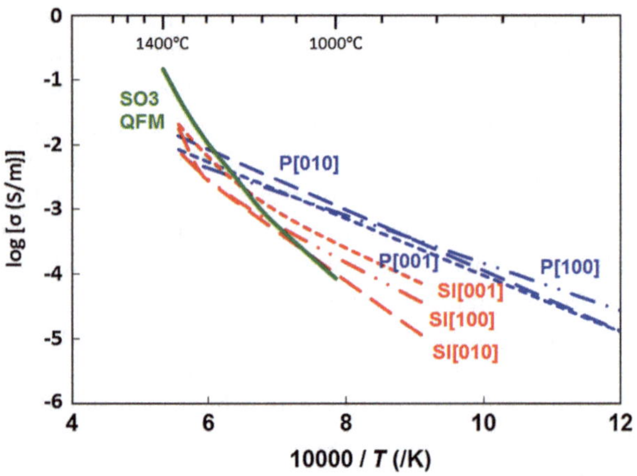

Fig. 6.2 Conductivity anisotropy of the intrinsic (SI, *red*) and proton (P, *blue*) conductions of olivine. The original data are from Constable et al. (1992) (intrinsic conduction) and Yoshino et al. (2006) (proton conduction). The anisotropy in proton conduction is negligible at high temperatures between 1000 and 1700 K and even smaller than that in intrinsic conduction

In comparison, the following are summaries of proton conduction from Wang et al. (2006) (Fig. 6.1).

1. Proton conduction increases proportionally to the power of 0.62 of the water content.
2. The activation energy is independent of water content and is 0.92 eV.
3. The magnitudes of the proton conductions are about two orders of magnitude higher than those reported by Yoshino et al. (2006).

6.2.4 Olivine High-Pressure Polymorphs

Olivine transforms to wadsleyite and then to ringwoodite in the upper mantle. The conductivity of wadsleyite and ringwoodite has been studied extensively compared to olivine because their conductivity is thought to govern the conductivity of the transition zone (cf. Yoshino and Katsura 2013). Since wadsleyite and ringwoodite incorporate even higher amounts of water than olivine (Keppler and Bolfan-Casanova 2006), the proton conduction of these minerals was considered to be important in understanding the transition zone conductivity as well as the intrinsic conductivity. We note that there are huge discrepancies in the magnitude of proton conduction of wadsleyite and ringwoodite, as well as that of olivine, among different data sets (Huang et al. 2005; Yoshino et al. 2008a, b, c; Dai and Karato 2009; Karato and Dai 2009; Romano et al. 2009; Yoshino and Katsura 2009, 2013). Although the conclusion is still indefinite, we will adopt the results from Yoshino et al. (2008a, b, c), Manthilake et al. (2009), and Yoshino and Katsura (2012) in the present chapter, as explained in Yoshino and Katsura (2013).

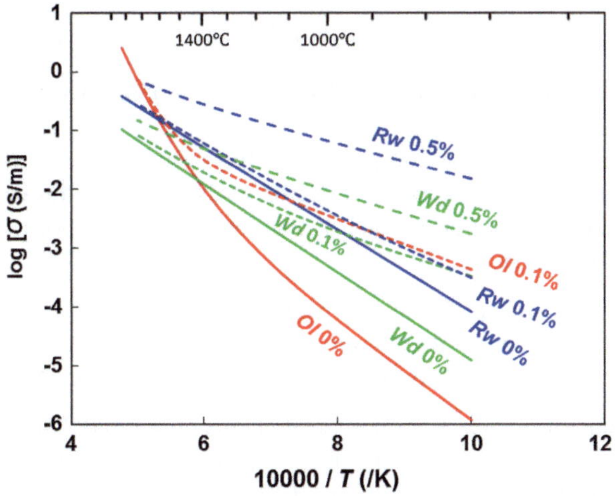

Fig. 6.3 Comparison of the intrinsic and total (intrinsic + proton) conductions of olivine (*red*), wadsleyite (*green*), and ringwoodite (*blue*). The *solid, short-broken,* and *long-broken curves* denote conductivity of these minerals with water contents of 0, 0.1, and 0.5 wt%. The original data are from Yoshino et al. (2008a, b, c, 2009a, b). The intrinsic conduction increases with the high-pressure phase transition from olivine to wadsleyite and then to ringwoodite. The differences in conductivity among the three phases become smaller with the incorporation of 0.1 % of water. The contribution of the proton conduction of ringwoodite becomes significant above this water content

The following is a summary of the conductivity of wadsleyite and ringwoodite (Fig. 6.3).

1. The conductivity of these minerals without water increases with increasing iron content (Yoshino and Katsura 2009; Yoshino et al. 2012b). Therefore, the small polaronconduction should be the essential conduction mechanism.
2. There is no clear evidence for dominance of ionic conduction at high temperatures (cf. Xu et al. 1998; Yoshino et al. 2008a, b, c).
3. The activation energy of the small polaron conduction of wadsleyite and ringwoodite is similar to or slightly smaller than that of olivine, namely 1.5 and 1.4 eV, respectively (Yoshino and Katsura 2013).
4. The small polaron conductivity of wadsleyite is by 0.5 log unit higher than that of olivine. That of ringwoodite is by 0.7 log unit higher than that of wadsleyite (Yoshino and Katsura 2013).
5. Due to ionic conduction in olivine but not in wadsleyite and ringwoodite, the intrinsic conductivity of wadsleyite and ringwoodite is smaller than that of olivine at temperatures in the asthenosphere and transition zone (see the "*Ol* 0 %" curve in Fig. 6.3).
6. The magnitude of proton conduction in ringwoodite is much larger than that in olivine and wadsleyite. That in wadsleyite is slightly lower than that in olivine (Yoshino and Katsura 2013).
7. The activation energy of proton conduction in wadsleyite and ringwoodite decreases with increasing water content, linearly with one-third power of water

content as in olivine: $\sigma_p = \sigma_{p_0} \exp\left[-\frac{E_0 - \alpha C_{H_2O}^{1/3}}{kT}\right]$. The α values of olivine and wadsleyite are similar (0.16 and 0.20, respectively) but that of ringwoodite is 3 times larger (0.67).

8. As is assumed for olivine, the pressure effects on conduction should be negligible. Additionally, the stability fields of wadsleyite and ringwoodite are very narrow in pressure; therefore, their activation volumes do not have significance in terms of geophysics.

Following are summaries of proton conduction from Huang et al. (2005) and Dai and Karato (2009). The general tendencies to discrepancies in wadsleyite and ringwoodite are similar to those in olivine between Yoshino et al. (2009a, b) and Wang et al. (2006).

1. Proton conduction increases proportionally to the power of 0.7 of the water content.
2. The activation energyis independent of water content and is 0.90 and 0.67 eV in wadsleyite and ringwoodite, respectively.
3. The magnitudes of the proton conductions are about two orders of magnitude higher than those reported by Yoshino et al. (2006).

Figure 6.3 shows total conduction of olivine with 0 and 0.1 % of water, and wadsleyite and ringwoodite with 0, 0.1, and 0.5 % of water. Since ionic conduction is significant in olivine but not in wadsleyite and ringwoodite, the intrinsic conductivity of olivine exceeds that of wadsleyite at the temperature at 410 km depth (1810 ± 20 K) in the present model. It is not clear whether this feature is the real case or not. Many studies measured intrinsic conductivity of wadsleyite only up to 1470 K (Xu et al. 1998; Dai and Karato 2009), which cannot provide any information about the ionic conduction. Yoshino and Katsura (2012) measured wadsleyite conductivity up to 2000 K, and their data show some possible contribution by ionic conduction. The number of data points in this paper is, however, insufficient to estimate the contribution by ionic conduction by numerically separating it from the small polaron conduction. A detailed conductivity measurement for wadsleyite at even higher temperature is necessary to understand the ionic conduction of wadsleyite.

After Xu et al. (1998) claimed a large conductivity jump associated with the olivine–wadsleyite transition, some geophysical studies artificially set a conductivity jump at 400 km depth by releasing the smoothing conditions (cf. Utada et al. 2003). However, as Huang et al. (2005) later demonstrated, the wadsleyite sample of Xu et al. (1998) contained 0.07 % of water, and therefore, their data cannot be used to discuss the intrinsic conduction of wadsleyite. Hence, it is unnecessary to assume a large conductivity jump at a depth of 400 km in MT modeling based on the current experimental data.

6.2.5 Secondary Minerals

Olivine is not the only constituent mineral in the upper mantle. Garnet, Ca-rich pyroxene, and Ca-poor pyroxene are also petrologically important constituent minerals. Their conductivity has been measured less extensively than olivine, and the reliability of the measurement is not high. One reason for the confusing results is their complex compositions. In the case of olivine, we can reasonably approximate its composition by a binary solution Mg_2SiO_4–Fe_2SiO_4. On the other hand, at least four components, e.g., $MgSiO_3$–$FeSiO_3$–$CaSiO_3$–Al_2O_3, are necessary to describe the compositions of pyroxene and garnet. They can also contain large amounts of ferric iron. Accordingly, their conductivity should vary largely with a Fe^{3+}/Fe^{2+} ratio. Although some efforts have been made to estimate the bulk rock conductivity from each mineral conductivity based on the Hashin–Shtrikman bounds, there is no experimental data showing that secondary minerals affect the bulk rock conductivity of peridotite (cf. Jones et al. 2009). Actually, Duba and Constable (1993) report that the conductivity of lherzolite with 65 % of olivine modal composition is identical to that of olivine. Olivine forms a network of conductivity in peridotite, which masks the effects of other minerals. The effect of these minerals on the bulk conductivity in the mantle should therefore be negligible. Hence, we ignore the effects of the other minerals on mantle conductivity in this paper.

6.2.6 Liquid Phases

In contrast to the secondary solid minerals, liquid phases could affect whole-rock conductivity. Waff and Bulau (1979) showed dihedral angles of 30–50° for the basalt–olivine systems. Yoshino et al. (2009a, b) reported that dihedral angles in the forsterite–H_2O system decrease with pressure and become zero above 7 GPa. These reports suggest that liquid phases form an interconnected network in the deep upper mantle even though their fractions are small. Moreover, the surface energy at grain edges formed by three minerals must be higher than that of grain surfaces formed by two minerals. Therefore, it is expected that liquid phases, even if their amounts are trace, should affect conductivity paths in the peridotite matrix.

The primary candidate for a liquid phase in the upper mantle is basaltic melt. Some studies measure electrical conductivity of basaltic melt (cf. Tyburczy and Waff 1983; Ni et al. 2011) (Fig. 6.3). The results suggest that the conductivity of dry basaltic melt is 0.5–0.9 log S/m at the temperature at the top of the oceanic asthenosphere (Fig. 6.4) and slightly increases with increasing temperature. The conductivity slightly decreases with increasing pressure up to 0.5–1.5 GPa, but it becomes almost constant above 1.5 GPa (Tyburczy and Waff 1983). Conductivity of dry basaltic melt is higher than that of the intrinsic conduction of olivine by about three orders of magnitude. Therefore, when the basaltic melt fraction is more than 1 %, the bulk conductivity is primarily a function of conductivity and the fraction of basaltic melt. Yoshino et al. (2010) studied the logarithmic relations between the bulk conductivity and basaltic melt fraction and obtained a basaltic melt fraction exponent of 0.89 (Fig. 6.5). Thus, the bulk conductivity is approximately proportional to the basaltic melt fraction.

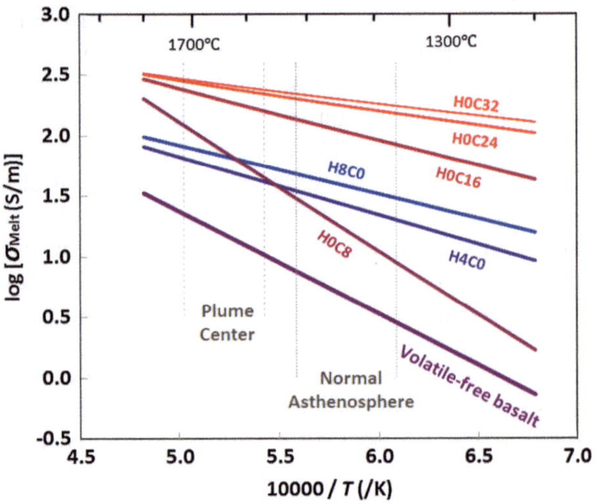

Fig. 6.4 Conductivity of basaltic melt with various H_2O and CO_2 contents by Sifre et al. (2014). The marking letters "H_C_" denote H_2O and CO_2 contents in the melt: The numbers after H and C indicate H_2O and CO_2 contents in wt%. The two pairs of *vertical bracketing lines* denote the temperature ranges of the normal asthenosphere based on Katsura et al. (2010) and the plume center based on the seismic wave anomalies reported by Montelli et al. (2006), and the relations between the temperature and shear-wave velocity reported by Faul (1997)

Recently, some groups studied conductivity anisotropy caused by small amounts of basaltic melt under shear (Caricchi et al. 2011; Zhang et al. 2014). Caricchi et al. (2011) measured the conductivity of natural peridotite with 4 or 8 % of basaltic melt in a torsion experiment. They found that the bulk conductivity normal to the shear is smaller than that under static conditions by one order of magnitude. Zhang et al. (2014) measured the conductivity of natural and model peridotite with 2 % of basaltic melt in the normal and parallel to shear on a shear plane simultaneously. They reported that the bulk conductivity in the normal and parallel shear directions is equivalent in the beginning of shear deformation, whereas that in the shear direction increased by one order of magnitude with increasing shear strain. Micro-texture analysis demonstrated that melt tubes were aligned in the shear direction, which should account for the high bulk conductivity in the shear direction. Since anisotropy of olivine does not cause conductivity anisotropy by more than a factor of three as previously mentioned, high-conductivity anisotropy should indicate the presence of a small fraction of liquid phases (Fig. 6.6).

H_2O is the most important volatile component in the upper mantle. Gaillard (2004) observed that conductivity of silicate melt (obsidian) increases in association with the addition of water. Ni et al. (2011) measured the conductivity of hydrous basaltic melt as a function of temperature and water content (Fig. 6.4). Hirschmann (2010) suggested that water content in the basaltic melt in the DMM and EM could be up to 2 and 6 wt%, respectively, at low degree of partial melting.

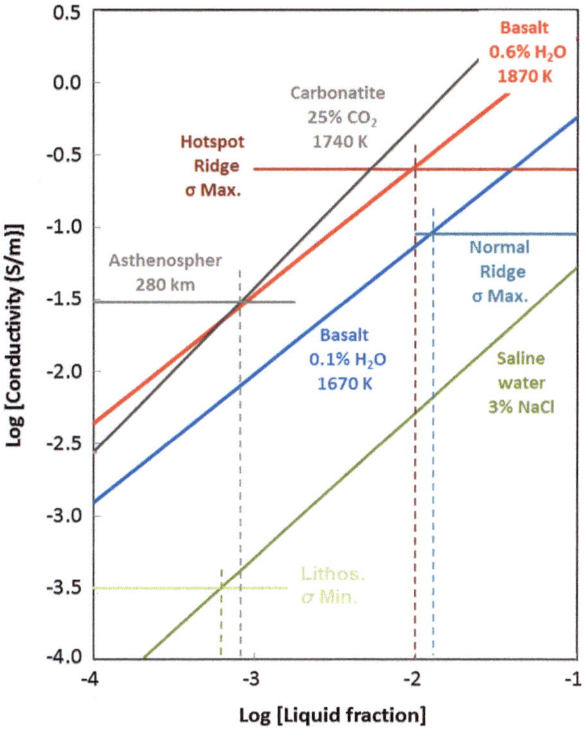

Fig. 6.5 Bulk conductivity of rocks against the fractions of liquid phases. *Green* saline water with 3 % of NaCl in quartzite (Shimojuku et al. 2014). *Blue* basalt with 0.1 % of H_2O in peridotite at 1670 K. *Red* basalt with 0.6 % of H_2O in peridotite at 1870 K based on the conductivity of hydrous basalt by Ni et al. (2011) and melt fraction exponent by Yoshino et al. (2010). *Gray* carbonatite with 25 % of CO_2 at 1740 K (Yoshino et al. 2010). The *dashed lines* denote the estimated fractions of the liquid phases at various tectonic settings

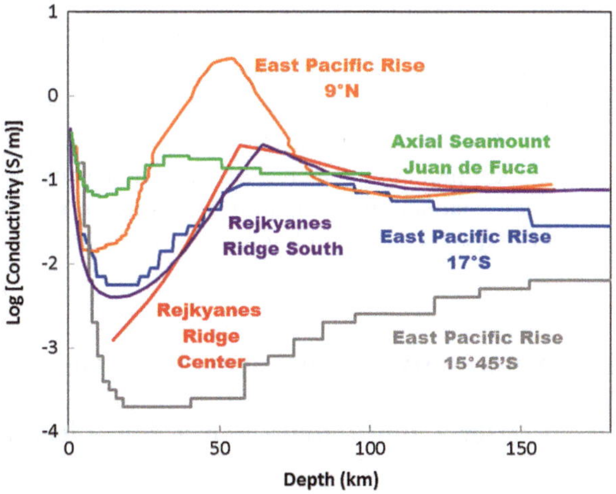

Fig. 6.6 One-dimensional conductivity beneath mid-oceanic ridges. *Blue and Gray* East Pacific Rise at 17°S and 15°45'S, respectively (Baba et al. 2006a, b). *Red* and *violet* Reykyanes Ridge (Heinson et al. 2000). *Green* Axial Seamount on the Juan de Fuca Ridge (Heinson et al. 1996). The intrinsic conductivity of olivine is in the order of 10^{-4} S/m and below the conductivity range shown in this figure

The 6 wt% of water increases the conductivity of basaltic melt by one order of magnitude.

CO_2 is another important volatile component in the mantle. Gaillard et al. (2008) demonstrated that molten carbonate has a conductivity of more than

3×10^2 S/m, which is by 2.5–3.5 orders of magnitude higher than that of basaltic melt. Sifre et al. (2014) modeled the effects of CO_2 and H_2O on the conductivity of basaltic melt by assuming the compensation law (Tyburczy and Waff 1983) and parallel conductive processes of hydrous basaltic melt and carbonate (Fig. 6.4). The effects of H_2O are large at low water content, but saturated at high water contents. The effects of CO_2 are smaller than H_2O below 10 wt%. However, they become significant at a CO_2 content above 8 %, exceeding the effects of water. These effects also seem saturated above 20 wt%. Proton can be incorporated in olivine and pyroxene, whereas carbon cannot be incorporated in silicate minerals. Consequently, the CO_2 content in the incipient melts can reach about 50 %, whereas that of H_2O should be limited to up to 7 % (Sifre et al. 2014). Therefore, the effects of CO_2 are more significant than those of H_2O in the majority of the asthenosphere, where the degree of partial melting should be less than 0.3 % (Hirschmann 2010). The relation of the bulk conductivity to the carbonatite melt fraction was studied by Yoshino et al. (2010), who reported a melt fraction exponent of 1.12 (Fig. 6.5).

Aqueous fluids are another possible liquid phase in the earth. However, silicate melts and fluid phases become miscible at high pressures. Mibe et al. (2011) demonstrated that the second critical point of the basalt + H_2O system is located at 3.4 GPa and 1100 K (Fig. 6.5). Therefore, no aqueous fluid phase is present below 100 km depth. In addition, the miscibility gap between a fluid phase and silicate melt should close at higher temperatures even at lower pressures. Mibe et al. (2011) observed that the miscibility gap closes at a temperature of 1600 K at a pressure of 1.8 GPa. Therefore, conductivity of fluid phases is important only at shallower parts of the mantle.

Although there is no measurement for the conductivity of aqueous fluid coexisting with peridotite, Shimojuku et al. (2014) measured the conductivity of brine-bearing quartzite at temperatures of 800–1100 K and a pressure of 1 GPa. They showed that the fluid conductivity is almost independent of temperature and the bulk rock conductivity increases with increasing salinity and fluid fraction. The bulk rock conductivity with a fluid phase with a constant salinity near that of seawater (3 wt%) is shown in Fig. 6.4.

6.3 Conductivity Heterogeneity in the Upper Mantle

6.3.1 High Conductivity Below Mid-Oceanic Ridges

Large amounts of basalt erupt at mid-oceanic ridges and huge partial melting regions are therefore expected under mid-oceanic ridges. Forsyth et al. (1998) estimated a minimum melt of 1–2 % down to 100 km depth under the East Pacific Rise,based on the seismic low-velocity anomaly. It is therefore expected that mid-oceanic ridge magmatism should cause a high-conductivity anomaly under ridges. There have actually been a number of reports on high-conductivity regions

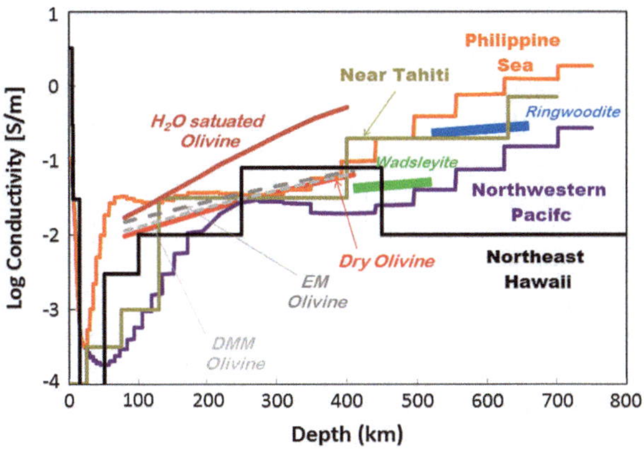

Fig. 6.7 One-dimensional conductivity structures under the Philippine Sea (*orange*) (Baba et al. 2010) and the northwestern Pacific near the Izu Islands (*violet*) (Baba et al. 2010) and near Tahiti (*olive*) (Nolasco et al. 1998). The conductivity of olivine under dry and water-saturated conditions is shown in *red* and *brown*. The conductivity of olivine with the DMM and EM compositions is denoted by *broken curves* in *dark* and *light gray*, respectively. The conductivity of dry wadsleyite and ringwoodite is shown in *green* and *blue*

under mid-oceanic ridges (Heinson and Lilley 1993; Heinson et al. 1993, 1996, 2000; Constable et al. 1997; Sinha et al. 1997; Evans et al. 1999, 2005; Baba et al. 2006a, b; Key et al. 2013) (Fig. 6.7).

Baba et al. (2006a) studied the two-dimensional conductivity structure under the East Pacific Rise at 17°S and showed a high-conductivity region of 9×10^{-2} S/m, 40 km in width, and vertically elongated from 50 to 120 km in depth immediately beneath the ridge axis. On the other hand, Baba et al. (2006b) showed no high-conductivity region under the East Pacific Rise at 15° 45′S: The conductivity at 50–80 km depth just beneath the axis is about $1\text{–}3 \times 10^{-3}$ S/m. Key et al. (2013) illustrated the very detailed conductivity structure under the East Pacific Rise at 9°N. Their model shows that conductivity at the center of the high-conductivity region reaches 2 S/m. The regions with conductivity up to 2×10^{-1} S/m extend vertically from 20 to 80 km in depth and horizontally over 60 km. Thus, the presence of high-conductivity regions is very heterogeneous, even beneath the same ridge, the East Pacific Rise.

The adiabatic temperature at 50–80 km depth is about 1630–1680 K (Katsura et al. 2010). The intrinsic olivine conductivity is $3\text{–}5 \times 10^{-3}$ S/m at these temperatures. These values explain conductivity beneath the East Pacific Rise at 15° 45′S but are far lower than the observed values at 17°S and 9°N. Therefore, the high conductivity at these regions should be explained by the MORB magmatism. The N-MORB (normal MORB) contains about 0.1 % of H_2O (cf. Sobolev and Chaussidon 1996), and such basaltic melt should have a conductivity of 20 S/m. Conductivities of 9×10^{-2} and 2 S/m suggest 0.2 and 10 vol.% of melt fractions beneath the East Pacific Rise at 17°S and 9°N, respectively, based on the melt fraction exponent of 0.89 for basaltic melt from Yoshino et al. (2010) (Fig. 6.4).

This melt fraction of 15 vol.% exceeds the threshold fraction for melt segregation (2–3 vol.%) (Faul 1997). The melt should be successively supplied and rise in this region. On the other hand, Forthys et al. (1998) estimate the melt fraction at 17°S to be 1–2 vol.%. Thus, the melt fraction based on the conductivity is much lower than that based on the seismic velocity anomaly.

Heinson et al. (1996) presented a one-dimensional conductivity structure under Axial Seamount on the Juan de Fuca Ridge. They showed a conductivity maximum of 2×10^{-1} S/m at a depth of 30–40 km. Heinson et al. (2000) studied the conductivity structures under the spreading center of the Reykjanes Ridge, the northern section of the mid-Atlantic Ridge. They showed conductivity maximum of 3×10^{-1} S/m at a depth of 60 km. These locations are known as the overlap of the ridge and a hot spot. As previously mentioned, the EM contains much higher volatile components than the DMM. Hung et al. (2004) studied the seismic structures beneath Iceland. From the velocity anomalies, they estimated that the temperature anomaly in this region is 150–200 K higher than the surrounding mantle. Nicholas et al. (2002) reported that the H_2O content in basaltic melt from the Reykjanes Ridge is up to 0.6 wt%. Such high water content would increase the melt conductivity, which should be enhanced by these high temperatures and high-volatile contents, and is estimated to be 40–60 S/m. From these values, the melt fraction is estimated to be 0.5–1 vol.%.

6.3.2 Lithosphere

As mentioned earlier on, the conductivity of minerals and liquids is a thermally activated process, and the mantle should have low conductivity at low-temperature regions. We emphasize that the Arrhenius relation (6.1) indicates that the magnitude of the conductivity anomaly by a low-temperature anomaly is more significant than that by a high-temperature one, if the magnitudes of the temperature anomalies are comparable.

Conductivity models under oceans depict low-conductivity layers at shallow regions. For example, Baba et al. (2010) showed conductivity minimums of 3×10^{-4} and 1.5×10^{-4} S/m at 20 and 50 km depth under the northern Philippine Sea and the Pacific Ocean to the east of the Izu Islands (Fig. 6.7). Conductivity in these regions monotonically increases with increasing depth below these depths. The Philippine Sea plate was formed 60 million years ago, and the Western Pacific plate of the studied area formed 125–150 million years ago. The temperatures at 20 and 50 km depth under the Philippine Sea and Western Pacific plates can be estimated at 600–1100 K and 900 K, respectively, based on the cooling model of the ocean plates. If the intrinsic conduction of olivine is the dominant conduction mechanism in such regions, the olivine conductivity should be lower than 1×10^{-5} S/m (Fig. 6.1). Therefore, the observed values are by more than one order of magnitude higher than expected from the intrinsic conductivity of olivine. Heinson and Constable (1992) had already indicated that the observed conductivity in the lithosphere is significantly higher than that expected from the intrinsic conductivity of olivine. They argued that such discrepancy could be due to overestimation by geophysical modeling because of the coastal effect, for example.

Should a more conductive lithosphere than expected from the intrinsic conductivity of olivine not be due to analysis problems in geophysical modeling, it could be caused by proton conduction. However, proton conduction also has positive temperature dependence. In order to produce proton conduction of 1×10^{-4} S/m at 1000 K, about 1000 ppm of water is necessary, according to data from Yoshino et al. (2009a, b) (Fig. 6.1). However, the solubility of water at 1 GPa (30 km depth) is 100–200 ppm (cf. Keppler and Bolfan-Casanova 2006). Thus, proton conduction is insufficient for explaining the magnitude of the lithospheric conductivity.

Heinson and Constable (1992) argue that high conductivity at the shallowest depth could reflect circulation of saline water and the decrease in conductivity with depth in spite of a temperature increase should be due to crack closure with increased lithostatic pressure. By assuming the comparable salinity with seawater (3 %), the conductivity of 1–3 $\times 10^{-4}$ S/m can be achieved with 0.02–0.03 vol.% of the fluid phase (Fig. 6.5). However, the conductivity increases from the minimum 20 and 50 km depth to the deeper regions (Baba et al. 2010) which cannot be explained by this hypothesis because the conductivity of saline water is almost independent of temperature and a fluid fraction increase with increasing depth can hardly be considered.

6.3.3 Asthenosphere

Several studies in the 1970s and 1980s (cf. Larsen 1975; Filloux 1977; Oldenburg 1981) report "high-conductivity layers" in the oceanic asthenosphere. "High-conductivity layer" here is defined as a horizontal layer with maximum conductivity in the vertical direction. Since temperatures increase with increasing depth in the upper mantle, except for subduction zones, and conductivities of mantle minerals have positive temperature dependence but negligible pressure dependence, we specifically need to account for the high-conductivity layer. Heinson and Constable (1992), however, re-examined Oldenburg's (1981) analysis and concluded that the presence of the high-conductivity layer proposed by this study could be true but not robust enough. In addition, the majority of reported high conductivity in the oceanic asthenosphere should originate in the mid-oceanic ridge magmatism discussed above.

Nevertheless, we currently have examples showing high-conductivity layers at the top of the asthenosphere. Baba et al. (2006a) studied the conductivity structure in the region 150 km west to 350 km east of the East Pacific Rise at 17°S, down to 200 km depth, and defined a high-conductivity layer with a maximum of 9×10^{-2} S/m from 50 to 350 km east at 70–120 km depth (Fig. 6.8). Baba et al. (2006a)

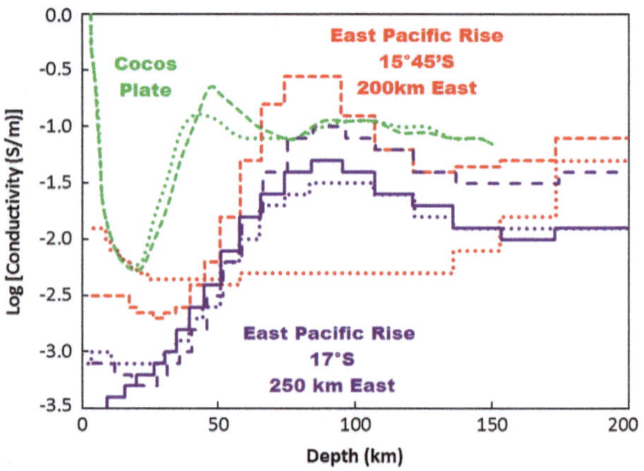

Fig. 6.8 The conductivity structures at the top of the asthenosphere under young plates. *Violet* East Pacific Rise at 17°S (Baba et al. 2006a). *Red* East Pacific Rise at 15° 45′S (Baba et al. 2006b). *Green* Cocos plate (Naif et al. 2013). The *solid line* denotes an isotropic structure. The *broken* and *dotted lines* denote anisotropic conductivity structures in the direction parallel and normal to the plate motion, respectively

and Evans et al. (2005) reported conductivity anisotropy in this high-conductivity layer, in which conductivity is by one order of magnitude higher in the direction parallel to the plate motion than in the normal direction. Naif et al. (2013) showed a conductivity structure in the 23–24-million-year-old region of the Cocos plate down to 150 km depth. They presented an anisotropic conductivity model with a maximum conductivity of 2×10^{-1} S/m at 50 km depth in the direction of the plate motion but almost no maximum in the other direction. Baba et al. (2006b) also illustrated an anisotropic conductivity model near the East Pacific Rise at 15° 45′S showing a high-conductivity layer of 3×10^{-1} S/m from 100 km east of the East Pacific Rise at 80–100 km depth in the direction of the plate motion. Although these regions are located near mid-oceanic ridges, Baba et al. (2010) studied conductivity structure under the northern Philippine Sea at depths down to 750 km. The conductivity in this region increases largely to reach 3×10^{-2} S/m from 20 to 70 km depth and slightly decreases by at most 1×10^{-2} S/m at 120 km depth (Fig. 6.7).

Nolasco et al. (1998) reported the conductivity structure under Tahiti, whose lithospheric age is 70 million years (Fig. 6.7). Their profile starts to increase from 1×10^{-4} S/m at the surface to 3×10^{-2} S/m at 130 km depth. The conductivity stays constant from this depth to 400 km depth. Thus, there is no high-conductivity layer in this region. Lizarralde et al. (1995) studied a conductivity structure between Hawaiiand California at depths from 150 to 1000 km. Their model reveals a conductive zone (0.05–0.1 S/m) between 250 and 450 km depth and a positive gradient in regions deeper than 500 km (Fig. 6.7). Baba et al. (2010) explained that conductivity increases to reach 3×10^{-2} S/m between 50 and 250–300 km depth and slightly decreases at deeper parts of the upper mantle, under the Pacific Ocean to the east of the Izu Islands. These two models show no high-conductivity layer at the top of the asthenosphere, but some high-conductivity regions at deeper levels.

In summary, the high-conductivity layers with a conductivity maximum exist at the top of the asthenosphere under relatively young plates, with associated conductivity anisotropy. Such layers do not exist at the top of the asthenosphere under mature plates, and high-conductivity regions without a clear maximum may exist in deeper regions (150–400 km depth).

The high-conductivity layers were often interpreted in view of the proton conduction of olivine (Karato 1990; Evans et al. 2005). However, Yoshino et al. (2006, 2009a, b) argued that the magnitudes of the proton conduction are not enough to explain the high-conductivity layers at the top of the asthenosphere, because of the limited water solubility in olivine at low pressures. The temperature at 100 km depth along the adiabatic geotherm is 1670 ± 20 K, and the olivine intrinsic conduction is 4×10^{-3} S/m. The maximum water content in the 3 GPa is 600 ppm, according to Mosenfelder et al. (2006), which gives a proton conduction of 4×10^{-3} S/m. Yoshino et al. (2006) also argued that the anisotropy of the proton conduction is negligibly small at temperatures corresponding to the top of the asthenosphere. Thus, the high-conductivity layers associated with conductivity anisotropy at the top of the asthenosphere cannot be explained by the proton conduction, although the magnitude of the high-conductivity layers can be explained using the data set from Wang et al. (2006).

Partial melting is an alternative interpretation of the high-conductivity layers. Although it is usually difficult to expect partial melting in regions deeper than 80 km in anhydrous peridotite, the trace amounts of volatiles, H_2O, and CO_2 should produce partial melting down to 180–200 km depth even in a DMM composition (Hirschmann 2010; Dasgupta et al. 2013). Sifre et al. (2014) argued that the high-conductivity layers should originate in trace partial melting induced by trace amounts of H_2O and CO_2. Sifre et al. (2014) assumed the H_2O and CO_2 contents of 200 and 100–500 ppm, respectively, for the DMM mantle and interpreted the high-conductivity layers. However, their assumed CO_2 contents are too high in comparison with the commonly accepted values (cf. Workman et al. 2006). Here, we have to estimate the bulk conductivity of peridotite with incipient melting due to 100 ppm of H_2O and 50 ppm of CO_2. The melt fraction at 100 km depth is about 0.1 wt%, according to Dasgupta et al. (2013). Sifre et al. (2014) analysis suggests that the CO_2 and H_2O contents in the 0.1 % partial melt should be 1.3 and 6.5 %, respectively, and its conductivity should be 3.2×10^1 S/m. Therefore, the bulk conductivity would be 3×10^{-2} S/m. Therefore, the partial melting hypothesis is also insufficient to explain the high-conductivity layers near mid-oceanic ridges if we assume the above volatile contents in the DMM.

As mentioned above, the high-conductivity layer shows conductivity anisotropy (Evans et al. 2005; Baba et al. 2006a, b; Naif et al. 2013). Since conductivity anisotropies of intrinsic and proton conductions of olivine are both small, the anisotropy in the high-conductivity layer cannot be explained with the conductivity of olivine. On the other hand, Zhang et al. (2014) demonstrated that partially melted peridotite can produce conductivity anisotropy between the normal and parallel to shear directions on a shear plane. Therefore, the presence of conductivity anisotropy implies the presence of partial melting in these regions. Since the magnitude of partially molten peridotite at the top of the asthenosphere with 100 and 50 ppm of H_2O and CO_2, respectively, is much lower than the observation, the actual volatile contents in the DMM could be higher than these values and might be 200 and 100–500 ppm, respectively, as Sifre et al. (2014) assumed.

6.3.4 Vertical Conductivity Structure in the Deep Upper Mantle

Xu et al. (1998) claimed a conductivity jump by more than one order of magnitude for the olivine–wadsleyite transition and essentially no jump for the wadsleyite and ringwoodite transition. Although smoothing conditions are included in building a conductivity model from MT data, these were not applied at a 400 km depth to create a conductivity jump for the olivine–wadsleyite transition in order to construct a large-scale model (Utada et al. 2003). The current experimental data, however, suggest no large conductivity jump associated with the olivine–wadsleyite transition in intrinsic conduction or even imply a decrease in conductivity for this transition. It seems that Xu et al. (1998) obtained the high conductivity of wadsleyite and ringwoodite due to proton conduction (Huang et al. 2005). The decrease in conductivity for the olivine–wadsleyite transition is due to the ionic conduction in olivine but not in wadsleyite. It is not clear whether the absence of ionic conduction in wadsleyite is due to the limited experimental technology in laboratory conductivity measurement (Fig. 6.9).

Figure 6.10 shows the most recent semi-global 1D conductivity models with and without the application of smoothing conditions at 400, 550, and 650 km depths under the northern Pacific Ocean reported by Shimizu et al. (2010a, b).

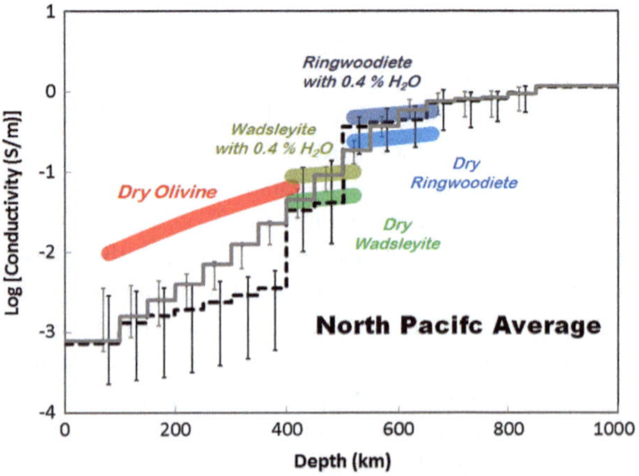

Fig. 6.9 The average one-dimensional conductivity structure under the North Pacific Ocean reported by Shimizu et al. (2010a). This model describes three conductivity jumps at 400, 500, and 650 km depths. The conductivities of wadsleyite and ringwoodite with 0 and 0.4 wt% of water are shown based on data by (Yoshino et al. (2008a, b, c)). The conductivity of dry olivine is also shown (Yoshino et al. 2009a, b)

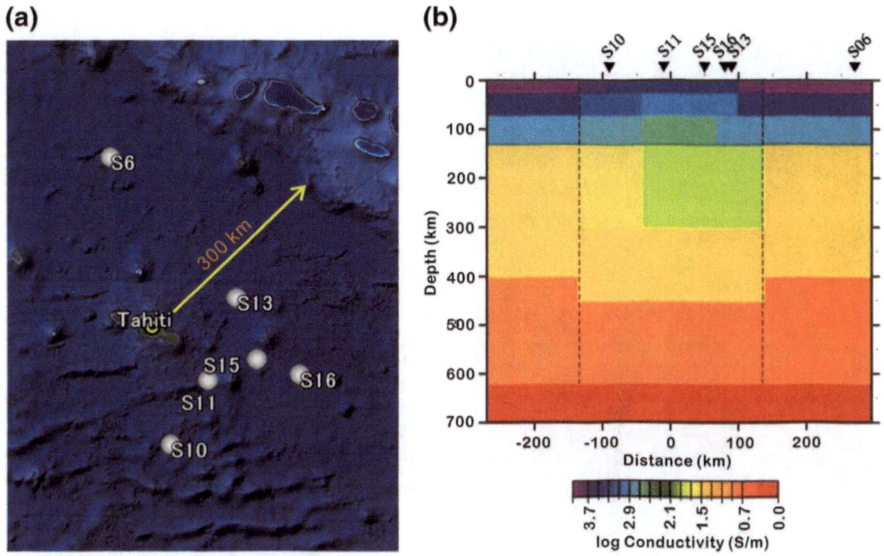

Fig. 6.10 The 2D conductivity structure near Tahiti (modified from Nolasco et al. 1998). **a** The locality of the observatories (labeled S6, S10, S1, S13, S15, and S16). The size of the Hawaiian plume, marked by an *arrow*, is 300 km (Montelli et al. 2006). **b** The two-dimensional conductivity structure approximately from NNW to SSE according to Nolasco et al. (1998). The locations of the observatories are marked on the figure. The conductivity is higher than the surrounding mantle in the region that is shallower than 130 km depth, whereas it is identical or even lower in the deeper regions

The conductivity of dry olivine, wadsleyite and ringwoodite, and wadsleyite and ringwoodite with 0.4 % of water is shown for comparison. The conductivity in the upper mantleabove 400 km in these models is significantly lower than that of intrinsic conduction in olivine. However, Shimizu et al. (2010a) used the long period range of MT signals (0.5–113 days), and therefore, the conductivity above 400 km depth was not well constrained (Baba et al. 2010). The conductivity in the transition zone in these models can be explained by intrinsic conduction in wadsleyite and ringwoodite. Because large amounts of water can be incorporated in these minerals, many attempts have been made to estimate water contents in the mantle transition zone by comparing the geophysical models and laboratory measurements (Huang et al. 2005; Yoshino et al. 2008a, b, c; Manthilake et al. 2009; Dai and Karato 2009; Romano et al. 2009; Yoshino and Katsura 2012). However, Fig. 6.10 demonstrates that the conductivity values of the transition zone depend on the modeling assumption. The uncertainty of each model is also large, and it is therefore found that estimation of the abundance of water in the transition zone in this way is not yet meaningful.

6.3.5 Local Conductivity Structures Under Hot Spots

Nolasco et al. (1998) studied conductivity structures under the Tahiti hot spot (Fig. 6.11). They showed about half an order of magnitude higher conductivity

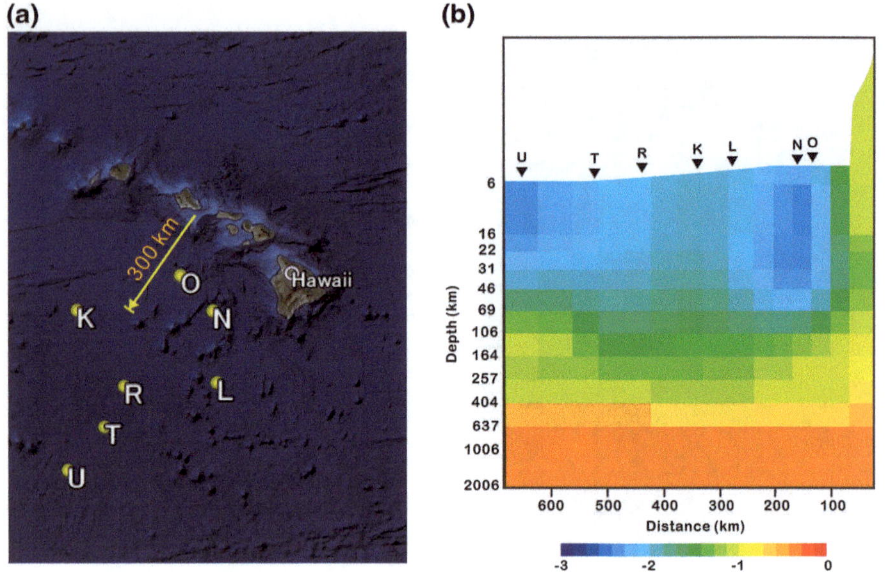

Fig. 6.11 The 2D conductivity structure southwest of Hawaii (modified from Constable and Heinson 2004). **a** The localities of observatories (labeled K, L, N, O, R, T, and U). The size of the Hawaiian plume, marked by an *arrow*, is 300 km (Montelli et al. 2006). **b** The two-dimensional conductivity structure approximately NE to SW, indicating the locations of observatories (Constable and Heinson 2004)

down to the depth of 130 km. However, deeper regions have equivalent or even lower conductivity than the surrounding regions. On the other hand, a seismic study shows clear, low-velocity anomalies from the depth of 300 km to the bottom of the mantle (Montelli et al. 2006). Constable and Heinson (2004) showed a two-dimensional conductivity model beneath the region southwest of Hawaii (Fig. 6.12). Their model also does not show conductivity anomalies in this region. Seismic tomography, however, shows a clear 1 % P-wave anomaly beneath Hawaii (Montelli et al. 2006).

The 1 % P-wave velocity anomaly would correspond to a temperature higher by 150 K. This temperature anomaly would increase the intrinsic conductivity of the main upper mantle minerals by half an order of magnitude. This high conductivity should be detectable by MT observations. In addition, geochemical studies show that sources of the ocean island basalt are enriched and have up to several factors higher volatile components (cf. Wallace 1998). As previously mentioned, high-volatile contents should always increase bulk conductivity by the proton conduction or melt network. For these reasons, the absence of high-conductivity zones under the hot spots cannot be explained consistently, based on available knowledge of mineral physics, seismology, and geochemistry.

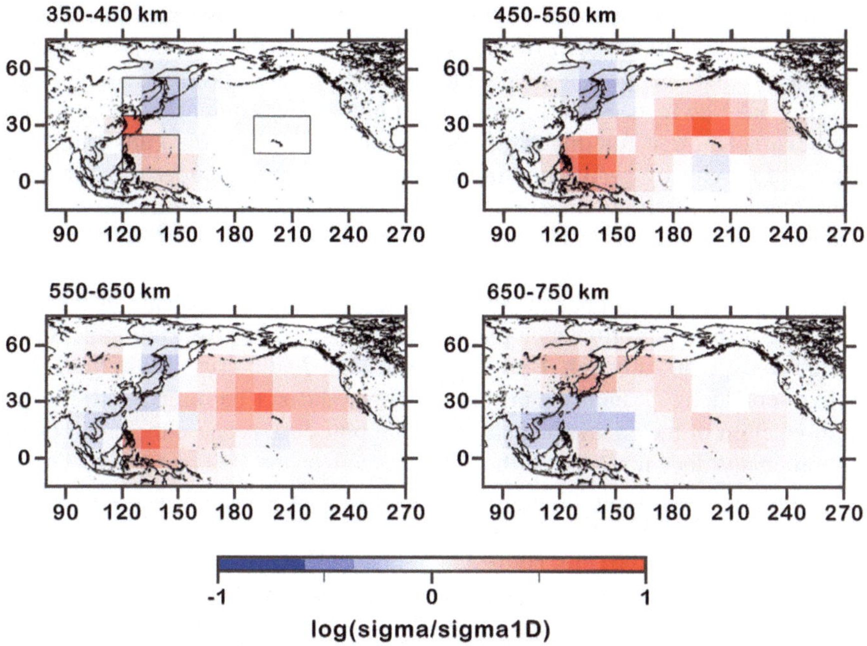

Fig. 6.12 Three-dimensional conductivity structure under the Pacific Ocean (modified from Shimizu et al. 2010b). The strong high-conductivity anomaly (*red*) is visible under the southern Philippine Sea. Another weak and broad high-conductivity anomaly is visible in the north of Hawaii

6.3.6 Conductivity Anomalies in the Transition Zone

Shimizu et al. (2010b) report a 3D conductivity structure from a depth of 350–850 km beneath the Pacific Ocean. They show a strong high-conductivity anomaly in the mantle transition zone under the southern Philippine Sea (Fig. 6.11). The center of the high-conductivity anomaly is located from the West Philippine Basin to the Caroline Basin. In contrast, the northern part of the Philippine Sea shows no conductivity anomaly. The maximum magnitude of high conductivity is about 0.8 log unit. The center of the high-conductivity anomaly is located at a depth of 450–550 km, becoming weaker at a depth of 550–650 km depth and turning into a low-conductivity anomaly in the top part of the lower mantle.

It is possible that this high-conductivity anomaly is caused by high-volatile contents in the region. If this high-conductivity anomaly was caused by the proton conduction of wadsleyite, the water contents in this region would be found to be 0.5 wt% in the upper part of the transition zone. Since wadsleyite and ringwoodite can contain water more than 2 wt% (Keppler and Bolfan-Casanova 2006), the amount of 0.5 wt% water in wadsleyite is possible. It is noteworthy that the petrological study on Mariana back-arc trough magma (Stolper and Newman 1994) suggests such high water content in these regions. Another explanation for this is partial melting. However, the enhancement of bulk conductivity by partial melting has not been studied under the conditions of the mantle transition zone, and we therefore cannot argue this possibility at present.

Although the high-conductivity anomaly can be explained by assuming larger amounts of volatile components, the possible high-volatile contents in the upper part of the mantle transition zone under the southern part of the Philippine Sea are difficult to explain. Although there are complex subduction systems in this region, the high-conductivity anomaly in this region should not be directly attributed to the subduction of the Pacific plate to the Philippine Sea, because the high-conductivity anomaly does not exist in the northern part of the Philippine Sea. There is no reason for a drastic difference in transported water between the northern and southern parts of the Philippine Sea.

Shimizu et al. (2010b) also show a high-conductivity anomaly under the region north of Hawaii. The depth range of this anomaly is from 450 to 650 km. The east–west dimension of the high-conductivity region extends over 6000 km. The global seismic tomography shows a low-velocity anomaly beneath Hawaii. However, the low-velocity anomaly is located just below or even west of Hawaii. Consequently, high-conductivity and low-velocity anomalies are not caused by the same mechanism. The low-velocity anomaly is caused by a temperature anomaly, whereas the high-conductivity anomaly is caused by a compositional anomaly. This difference results in variation of a specific scale between these two anomalies. The very small dimension of high-temperature regions should be detected by P-wave tomography but would be too small for the analysis of the semi-global 3D structure of conductivity.

6.3.7 Fine-Scale Conductivity Variation Under the Philippine Sea

As previously explained, the semi-global 3D structures in conductivity show the high-conductivity anomaly in the upper part of the mantle transition zone under the south Philippine Sea. In addition to this, a regional study shows other anomalous features under the Philippine Sea (Seama et al. 2007). The study comprised a two-dimensional conductivity structure (Fig. 6.12b) from the south of the Mariana Trough to the north of the Daito Ridge (under the yellow line in Fig. 6.12a). This showed a strong high-conductivity anomaly exceeding 1×10^{-1} S/m under one point east of the Ryukyu-Palau ridge between the Shikoku and Parece Vela basins (Point 4 in Fig. 6.12a) and weaker but still strong anomalies west of the northern part of the Ryukyu-Palau Ridge (Points 5 and 6). The peaks of these high-conductivity anomalies are located at a depth of 50–150 km. In contrast to these high-conductivity anomalies, the study found no high-conductivity anomaly under the Mariana Trough. Matsuno et al. (2010) also reported an absence of high-conductivity anomalies beneath the Mariana Trough (under the pink line in Fig. 6.12a).

The regions in which the high-conductivity anomalies are found were active in the past (40–60 million years ago). The region of no conductivity anomalies is located in the currently active volcanic zone. A seismic study shows a very strong S-wave velocity anomaly up to −8 % in this region (Isse et al. 2009) (Fig. 6.12c). It also shows that the low-velocity anomaly becomes weaker and deeper with increasing plate age from 0 to 100 million years. Actually, the region that shows the strongest S-wave velocity anomaly is under the Mariana Trough. The 8 % S-wave anomaly may indicate a temperature anomaly of 200 K, if we attribute the S-wave anomaly to a temperature anomaly only. This temperature anomaly should raise conductivity by 0.6 orders of magnitude, assuming a base temperature of 1660 K at 80 km depth. Thus, the absence of a high-conductivity layer is contrary to our expectation from the tectonic setting and seismic observation.

One possible explanation is that while volatile components or melts are successively supplied from the deeper part of the mantle to the upper mantle, the currently active zone has been releasing the volatile components. As a result, the non-active regions become volatile rich, whereas the active region becomes volatile poor. However, we note that the maximum water content at 100 km depth is only 500 ppm (cf. Keppler and Bolfan-Casanova 2006). The magnitude of proton conduction by this amount of water is negligible in comparison with the intrinsic conduction. The high-conductivity anomaly could be explained by the presence of partial melt. The Mariana Trough basalt contains 2 % of water (Stolper and Newman 1994), which suggests melt conductivity of 10 S/m. Therefore, the conductivity anomaly exceeding 1×10^{-1} S/m should indicate a melt fraction more than 1 %, which should cause some degree of low shear-wave anomaly. Actually, Isse et al. (2009) showed some low shear-wave anomaly under the Points 4, 5, and 6 in the Seama et al. (2007) study. This hypothesis could thus explain the high-conductivity anomaly under the Kyushu-Palau Ridge. There is yet no persuasive explanation for the low-conductivity anomaly under the Mariana Trough (Fig. 6.13).

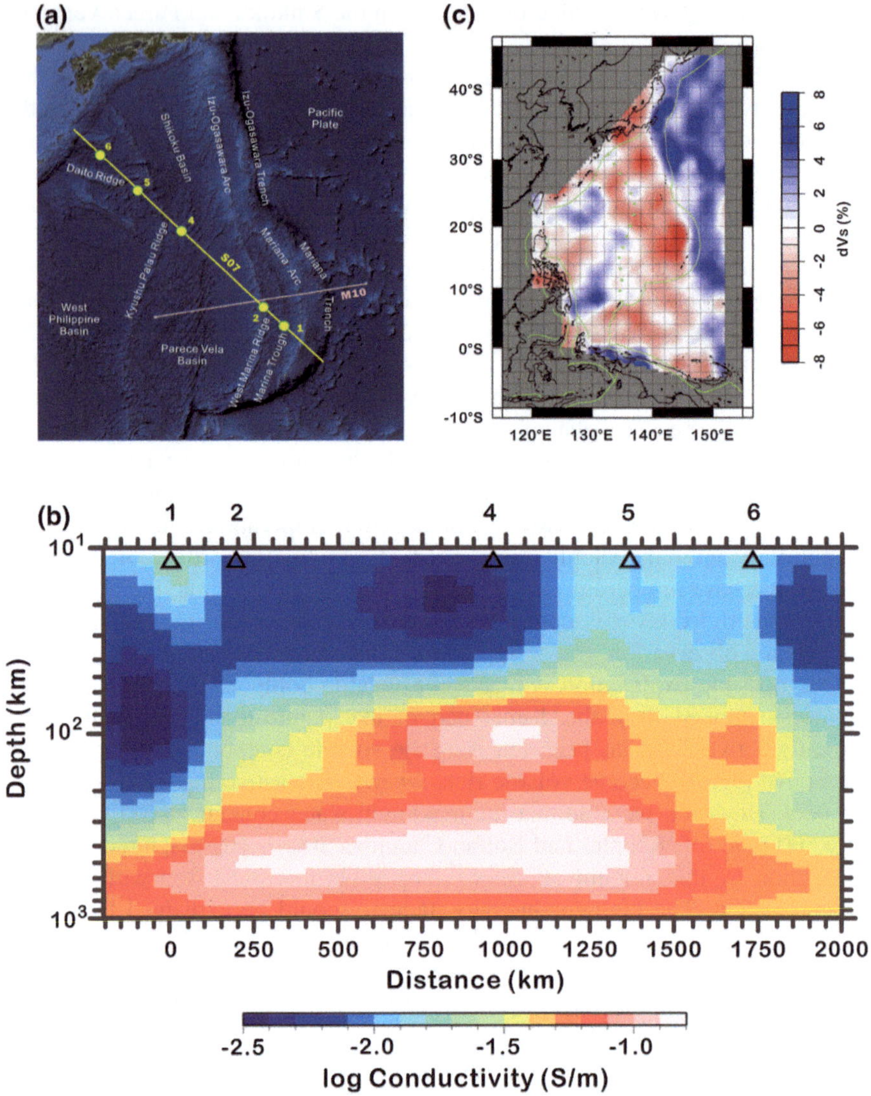

Fig. 6.13 Comparison of conductivity and seismic anomalies under the Philippine Sea. **a** Location of the observation sites in Seama et al. (2007). **b** 2D conductivity structure along the observation line S07 shown in **a**. The region under the Mariana Trough (OBEM1) at 40–200 km depth has low conductivity (modified from Seama et al. 2007). **c** The shear-wave anomaly under the Philippine Sea at 80 km depth (Isse et al. 2009). The region under the Mariana Trough has the low shear-wave anomaly, suggesting high temperatures (modified from Isse et al. 2009)

6.4 Concluding Remarks

We have reviewed the results of laboratory experiments about the conductivity of upper mantle minerals and representative MT observations under oceans. Conductivity in the upper mantle varies by more than three orders of magnitude. We have the following qualitative explanation for the following results obtained by the MT method.

1. The high conductivity just below the mid-oceanic ridges is mainly due to partial melting and partly due to high temperature.
2. The very low conductivity in the lithosphere is due to low temperatures.
3. The gradually increasing conductivity in the asthenosphere is due to the temperature increases in the adiabatic geotherm.
4. The high-conductivity layers are a feature of the asthenosphere under young plates. They associate conductivity anisotropy, which implies its partial melting origin.
5. There should be no large conductivity jump at the 410 km depth.
6. The high-conductivity anomaly in the mantle transition zone under the south of the Philippine Sea may be due to high water content.
7. The high-conductivity anomaly in the mantle transition zone north of Hawaii may be due to the volatile components of plume source.

The mantle structures under oceans are expected to be simpler than those under continents. Therefore, we expect that we can obtain quantitative and robust explanations. Nevertheless, the above explanations have the following serious problems.

1. The conductivity in the uppermost part of the lithosphere is too high by far, considering its low temperatures based on the ocean floor cooling model.
2. The magnitude of the high-conductivity layers at the top of the asthenosphere is difficult to explain based on the usually assumed volatile contents in the DMM.
3. The high-temperature anomalies inferred from the seismic tomography indicate no or little effect on the conductivity structures.
4. The local conductivity anomalies disagree with those expected from the tectonic setting.
5. In general, the conductivity in the upper mantle is higher than that expected from the experimental results.

Acknowledgement The author acknowledges S. Baba and T. Kogiso for their helpful discussion.

References

Angell CA, Smith DL (1982) Test of the entropy basis of the vogel-tammann-fulcher equation—dielectric-relaxation of polyalcohols near tg. J Phys Chem 86(19):3845–3852

Baba K (2005) Electrical structure in marine tectonic settings. Surv Geophys 26(6):701–731

Baba K, Chave AD, Evans RL, Hirth G, Mackie RL (2006a) Mantle dynamics beneath the East Pacific Rise at 17 degrees S: Insights from the Mantle Electromagnetic and Tomography (MELT) experiment. J Geophys Res-Solid Earth 111(B2). doi:10.1029/2004JB003598

Baba K, Tarits P, Chave AD, Evans RL, Hirth G, Mackie RL (2006b) Electrical structure beneath the northern MELT line on the East Pacific Rise at 15° 45′S. Geophys Res Lett 33(22). doi:10.1029/2006gl027528

Baba K, Utada H, Goto T, Kasaya T, Shimizu H, Tada N (2010) Electrical conductivity imaging of the Philippine Sea upper mantle using seafloor magnetotelluric data. Phys Earth Planet Inter 183(1–2):44–62

Caricchi L, Gaillard F, Mecklenburgh J, Trong EL (2011) Experimental determination of electrical conductivity during deformation of melt-bearing olivine aggregates: implications for electrical anisotropy in the oceanic low velocity zone. Earth Planet Sci Lett 302(1–2):81–94. doi:10.1016/j.epsl.2010.11.041

Constable S (1993) Conduction by mantle hydrogen. Nature 362(6422):704

Constable S (2006) SEO3: a new model of olivine electrical conductivity. Geophys J Int 166(1):435–437

Constable S, Duba A (1990) Electrical conductivity of olivine, a dunite, and the mantle. J Geophys Res 95(B5):6967–6978

Constable S, Heinson G (2004) Hawaiian hot-spot swell structure from seafloor MT sounding. Tectonophysics 389(1–2):111–124

Constable S, Shankland TJ, Duba A (1992) The electrical conductivity of an isotropic olivine mantle. J Geophys Res 97(B3):3397–3404

Constable SC, Heinson GS, Anderson G, White A (1997) Seafloor electromagnetic measurements above axial seamount, juan de fuca ridge. J Geomagn Geoelectr 49(11–12):1327–1342

Dai L, Karato SI (2009) Electrical conductivity of wadsleyite at high temperatures and high pressures. Earth Planet Sci Lett 287:277–283

Dasgupta R, Mallik A, Tsuno K, Withers AC, Hirth G, Hirschmann MM (2013) Carbon-dioxide-rich silicate melt in the Earth's upper mantle. Nature 493(7431):211–215. doi:10.1038/nature11731

Du Frane WL, Roberts JJ, Toffelmier DA, Tyburczy JA (2005) Anisotropy of electrical conductivity in dry olivine. Geophys Res Lett 32(24):L24315

Duba A, Constable S (1993) The electrical conductivity of lherzolite. J Geophys Res Solid Earth 98(B7):11885–11899

Evans RL et al (1999) Asymmetric electrical structure in the mantle beneath the East Pacific rise at 17° S. Science 286(5440):752–756 (Artn l24315)

Evans RL, Hirth G, Baba K, Forsyth D, Chave A, Mackie R (2005) Geophysical evidence from the MELT area for compositional controls on oceanic plates. Nature 437(7056):249–252

Faul UH (1997) Permeability of partially molten upper mantle rocks from experiments and percolation theory. J Geophys Res-Solid Earth 102(B5):10299–10311. doi:10.1029/96jb03460

Filloux JH (1977) Ocean-floor magnetotelluric sounding over north-central Pacific. Nature 269:297–301

Forsyth DW et al (1998) Imaging the deep seismic structure beneath a mid-ocean ridge: the MELT experiment. Science 280(5367):1215–1218

Fuji-ta K, Katsura T, Tainosho Y (2004) Electrical conductivity measurement of granulite under mid- to lower crustal pressure-temperature conditions. Geophys J Int 157:79–86

Fuji-ta K, Katsura T, Matsuzaki T, Ichiki M, Kobayashi T (2007) Electrical conductivity measurement of gneiss under mid- to lower crustal P-T conditions. Tectonophysics 434:93–101

Fukao Y, Koyama T, Obayashi M, Utada H (2004) Trans-Pacific temperature field in the mantle transition region derived from seismic and electromagnetic tomography. Earth Planet Sci Lett 217(3–4):425–434

Gaillard F (2004) Laboratory measurements of electrical conductivity of hydrous and dry silicic melts under pressure. Earth Planet Sci Lett 218(1–2):215–228

Gaillard F, Malki M, Iacono-Marziano G, Pichavant M, Scaillet B (2008) Carbonatite melts and electrical conductivity in the asthenosphere. Science 322(5906):1363–1365

Heinson G, Constable SH (1992) The electrical conductivity of the oceanic upper mantle. Geophys J Int 110(1):159–179

Heinson GS, Lilley FEM (1993) An application of thin-sheet electromagnetic modelling to the Tasman Sea. Phys Earth Planet Inter 81(1–4):231–251

Heinson GS, White A, Law LK, Hamano Y, Utada H, Yukutake T, Segawa J, Toh H (1993) EMRIDGE: the electromagnetic investigation of the Juan de Fuca Ridge. Mar Geophys Res 15(2):77–100. doi:10.1007/bf01204130

Heinson G, Constable S, White A (1996) Seafloor magnetotelluric sounding above axial seamount. Geophys Res Lett 23(17):2275–2278

Heinson G, Constable S, White A (2000) Episodic melt transport at mid-ocean ridges inferred from magnetotelluric sounding. Geophys Res Lett 27(15):2317–2320

Hirschmann MM (2010) Partial melt in the oceanic low velocity zone. Phys Earth Planet Inter 179(1–2):60–71

Huang XG, Xu YS, Karato SI (2005) Water content in the transition zone from electrical conductivity of wadsleyite and ringwoodite. Nature 434:746–749

Hung SH, Shen Y, Chiao LY (2004) Imaging seismic velocity structure beneath the Iceland hot spot: a finite frequency approach. J Geophys Res-Solid Earth 109(B8), B08305. doi: 10.1029/2003jb002889

Isse T et al (2009) Seismic structure of the upper mantle beneath the Philippine Sea from seafloor and land observation: Implications for mantle convection and magma genesis in the Izu-Bonin-Mariana subduction zone. Earth Planet Sci Lett 278(1–2):107–119

Jacobsen SD (2006) Effect of water on the equation of state of nominally anhydrous minerals. In: Keppler H, Smith JR (eds) Water in nominally anhydrous minerals. Geochemical Society, Mineralogical Society of America, USA, pp 321–342

Jones AG, Evans RL, Eaton DW (2009) Velocity-conductivity relationships for mantle mineral assemblages in Archean cratonic lithosphere based on a review of laboratory data and Hashin-Shtrikman extremal bounds. Lithos 109(1–2):131–143. doi:10.1016/j.lithos.2008.10.014

Karato S (1990) The role of hydrogen in the electrical conductivity of the upper mantle. Nature 347(6290):272–273

Karato S, Dai LD (2009) Comments on "Electrical conductivity of wadsleyite as a function of temperature and water content" by Manthilake et al. Phys Earth Planet Inter 174:19–21

Katsura T, Sato K, Ito E (1998) Electrical conductivity of silicate perovskite at lower-mantle conditions. Nature 395:493–495

Katsura T, Yokoshi S, Kawabe K, Shatskiy A, Okube M, Fukui H, Ito E, Nozawa A, Funakoshi K-I (2007) Pressure dependence of electrical conductivity of (Mg, Fe)SiO3 ilmenite. Phys Chem Miner 34:249–255

Katsura T, Yoneda A, Yamazaki D, Yoshino T, Ito E (2010) Adiabatic temperature profile in the mantle.Phys. Earth Planet. Inter., 183(1–2):212–218. doi:10.1016/j.pepi.2010.07.001

Keppler H, Bolfan-Casanova N (2006) Thermodynamics of water solubility and partitioning. In: Keppler H, Smith JR (eds) Water in nominally anhydrous minerals. Geochemical Society, Mineralogical Society of America, USA, pp 193–230

Key K, Constable S, Liu LJ, Pommier A (2013) Electrical image of passive mantle upwelling beneath the northern East Pacific Rise. Nature 495(7442):499–502. doi:10.1038/nature11932

Kohlstedt DL, Mackwell SJ (1998) Diffusion of hydrogen and intrinsic point defects in olivine. Zeitschrift Fur Physikalische Chemie-Int J Res Phys Chem Chem Phys 207:147–162

Larsen JC (1975) Low-frequency (0.1-6.0 CPD) Electromagnetic study of deep mantle electrical-conductivity beneath Hawaiian-Islands. Geophys J Roy Astron Soc 43(1):17–46

Lizarralde D, Chave A, Hirth G, Schultz A (1995) Northeastern pacific mantle conductivity profile from long-period magnetotelluric sounding using Hawaii-to-California submarine cable data. J Geophys Res-Solid Earth 100(B9):17837–17854

Mackwell SJ, Kohlstedt DL (1990) Diffusion of hydrogen in olivine: implications for water in the mantle. J Geophys Res 95(B4):5079–5088

Manthilake M, Matsuzaki T, Yoshino T, Yamashita S, Ito E, Katsura T (2009) Electrical conductivity of wadsleyite as a function of temperature and water content. Phys Earth Planet Inter 174(1–4):10–18. doi:10.1016/j.pepi.2008.06.001

Matsuno T et al (2010) Upper mantle electrical resistivity structure beneath the central Mariana subduction system. Geochem Geophys Geosyst 11:Q09003

Mibe K, Kawamoto T, Matsukage KN, Fei Y, Ono S (2011) Slab melting versus slab dehydration in subduction-zone magmatism. Proc Natl Acad Sci USA 108(20):8177–8182

Montelli R, Nolet G, Dahlen FA, Masters G (2006) A catalogue of deep mantle plumes: New results from finite-frequency tomography. Geochem Geophys Geosyst 7(11):Q11007

Mosenfelder JL, Deligne NI, Asimow PD, Rossman GR (2006) Hydrogen incorporation in olivine from 2-12 GPa. Am Miner 91(2–3):285–294. doi:10.2138/am.2006.1943

Naif S, Key K, Constable S, Evans RL (2013) Melt-rich channel observed at the lithosphereasthenosphereboundary. Nature 495(7441):356–359. doi:10.1038/nature11939

Ni HW, Keppler H, Behrens H (2011) Electrical conductivity of hydrous basaltic melts: implications for partial melting in the upper mantle. Contrib Mineral Petrol 162(3):637–650

Nichols ARL, Carroll MR, Hoskuldsson A (2002) Is the Iceland hot spot also wet? Evidence from the water contents of undegassed submarine and subglacial pillow basalts. Earth Planet Sci Lett 202(1):77–87. doi:10.1016/s0012-821x(02)00758-6

Nolasco R, Tarits P, Filloux JH, Chave AD (1998) Magnetotelluric imaging of the Society Islands hotspot. J Geophys Res-Solid Earth 103(B12):30287–30309

Ohta K, Hirose K, Ichiki M, Shimizu K, Sata N, Ohishi Y (2010) Electrical conductivities of pyrolitic mantle and MORB materials up to the lowermost mantle conditions. Earth Planet Sci Lett 289(3–4):497–502

Oldenburg DW (1981) Conductivity structure of oceanic upper mantle beneath the Pacific plate. Geophys J Royal Astron Soc 65:359–394

Omura K, Kurita K, Kumazawa M (1989) Experimental study of pressure dependence of electrical conductivity of olivine at high temperatures. Phys Earth Planet Inter 57(3–4):291–303

Poe BT, Romano C, Nestola F, Smyth JR (2010) Electrical conductivity anisotropy of dry and hydrous olivine at 8 GPa. Phys Earth Planet Inter 181(3–4):103–111

Romano C, Poe BT, Tyburczy J, Nestola F (2009) Electrical conductivity of hydrous wadsleyite. Eur J Mineral 21:615–622

Schock RN, Duba AG, Shankland TJ (1989) Electrical-conduction in olivine. J Geophys Res-Solid Earth Planet 94(B5):5829–5839

Seama N, Baba K, Utada H, Toh H, Tada N, Ichiki M, Matsuno T (2007) 1-D electrical conductivity structure beneath the Philippine Sea: results from an ocean bottom magnetotelluric survey. Phys Earth Planet Inter 162(1–2):2–12

Shankland TJ, Duba AG (1990) Standard electrical conductivity of isotropic, homogeneous olivine in the temperature range 1200°–1500°C. Geophys J Int 103:25–31

Shimizu H, Koyama T, Baba K, Utada H (2010a) Revised 1-D mantle electrical conductivity structure beneath the north Pacific. Geophys J Int 180(3):1030–1048

Shimizu H, Utada H, Baba K, Koyama T, Obayashi M, Fukao Y (2010b) Three-dimensional imaging of electrical conductivity in the mantle transition zone beneath the North Pacific Ocean by a semi-global induction study. Phys Earth Planet Inter 183(1–2):252–269

Shimojuku A, Yoshino T, Yamazaki D (2014) Electrical conductivity of brine-bearing quartzite at 1 GPa: implications for fluid content and salinity of the crust. Earth Planet Space 66:9. doi:10.1186/1880-5981-66-2

Sifre D, Gardes E, Massuyeau M, Hashim L, Hier-Majumder S, Gaillard F (2014) Electrical conductivity during incipient melting in the oceanic low-velocity zone. Nature 509(7498):81–85. doi:10.1038/nature13245

Sinha MC, Navin DA, Macgregor LM, Constable S, Peirce C, White A, Heinson G, Inglis MA (1997) Evidence for accumulated melt beneath the slow-spreading Mid-Atlantic Ridge. Philos Trans R Soc A Math Phys Eng Sci 355(1723):233–253

Sobolev AV, Chaussidon M (1996) H_2O concentrations in primary melts from suprasubduction zones and mid-ocean ridges: implications for water storage and recycling in the mantle. Earth Planet Sci Lett 137:45–55

Stolper E, Newman S (1994) The role of water in the petrogenesis of Mariana trough magmas. Earth Planet Sci Lett 121(3–4):293–325

Tyburczy JA, Waff HS (1983) Electrical conductivity of molten basalt and andesite to 25 kilobars pressure: geophysical significance and implications for charge transport and melt structure. J Geophys Res 88(B3):2413–2430

Utada H, Koyama T, Shimizu H, Chave AD (2003) A semi-global reference model for electrical conductivity in the mid-mantle beneath the north Pacific region. Geophys Res Lett 30(4), 1194. doi:10.1029/2002GL016092

Velimsky J (2010) Electrical conductivity in the lower mantle: constraints from CHAMP satellite data by time-domain EM induction modelling. Phys Earth Planet Inter 180(3–4):111–117

Waff HS, Bulau JR (1979) Equilibrium fluid distribution in an ultramafic partial melt under hydrostatic stress conditions. J Geophys Res 84(NB11):6109–6114

Wallace PJ (1998) Water and partial melting in mantle plumes: Inferences from the dissolved H2O concentrations of Hawaiian basaltic magmas. Geophys Res Lett 25(19):3639–3642. doi:10.1029/98gl02805

Wanamaker BJ, Duba AG (1993) Electrical conductivity of San Carlos olivine along [100] under oxygen- and pyroxene-buffered conditions and implications for defect equilibria. J Geophys Res 98(B1):489–500

Wang D-J, Mookherjee M, Xu YS, Karato S (2006) The effect of water on the electrical conductivity of olivine. Nature 443(7114):977–980

Workman RK, Hauri E, Hart SR, Wang J, Blusztajn J (2006) Volatile and trace elements in basaltic glasses from Samoa: implications for water distribution in the mantle. Earth Planet Sci Lett 241(3–4):932–951. doi:10.1016/j.epsl.2005.10.028

Xu Y-S, Poe BT, Shankland TJ, Rubie DC (1998) Electrical conductivity of olivine, wadsleyite, and ringwoodite under upper-mantle conditions. Science 280(5368):1415–1418

Yang X-Z (2012) Orientation-related electrical conductivity of hydrous olivine, clinopyroxene and plagioclase and implications for the structure of the lower continental crust and uppermost mantle. Earth Planet Sci Lett 317:241–250

Yoshino T, Katsura T (2009) Effect of iron content on electrical conductivity of ringwoodite, with implications for electrical structure in the transition zone. Phys Earth Planet Inter 174(1–4):3–9

Yoshino T, Katsura T (2012) Re-evaluation of electrical conductivity of anhydrous and hydrous wadsleyite. Earth Planet Sci Lett 337:56–67

Yoshino T, Katsura T (2013) Electrical conductivity of mantle minerals: role of water in conductivity anomalies. In: Jeanloz R (ed) Annual review of earth and planetary sciences, vol 41. Annual Reviews, Palo Alto, p 605

Yoshino T, Walter MJ, Katsura T (2003) Core formation in planetesimals triggered by permeable flow. Nature 422:154–157

Yoshino T, Walter MJ, Katsura T (2004) Connectivity of molten Fe alloy in peridotite based on in situ electrical conductivity measurements: implications for core formation in terrestrial planets. Earth Planet Sci Lett 222:625–643

Yoshino T, Matsuzaki T, Yamashita S, Katsura T (2006) Hydrous olivine unable to account for conductivity anomaly at the top of the asthenosphere. Nature 443(7114):973–976

Yoshino T, Manthilake G, Matsuzaki T, Katsura T (2008a) Dry mantle transition zone inferred from the conductivity of wadsleyite and ringwoodite. Nature 451:326–329

Yoshino T, Nishi M, Matsuzaki T, Yamazaki D, Katsura T (2008b) Electrical conductivity of majorite garnet and its implications for electrical structure in the mantle transition zone. Phys Earth Planet Inter 170:193–200

Yoshino T, Yamazaki D, Ito E, Katsura T (2008c) No interconnection of ferro-periclase in post-spinel phase inferred from conductivity measurement. Geophys Res Lett 35:5

Yoshino T, Matsuzaki T, Shatskiy A, Katsura T (2009a) The effect of water on the electrical conductivity of olivine aggregates and its implications for the electrical structure of the upper mantle. Earth Planet Sci Lett 288(1–2):291–300

Yoshino T, Yamazaki D, Mibe K (2009b) Well-wetted olivine grain boundaries in partially molten peridotite in the asthenosphere. Earth Planet Sci Lett 283(1–4):167–173

Yoshino T, Laumonier M, McIsaac E, Katsura T (2010) Electrical conductivity of basaltic and carbonatite melt-bearing peridotites at high pressures: Implications for melt distribution and melt fraction in the upper mantle. Earth Planet Sci Lett 295(3–4):593–602. doi:10.1016/j.epsl.2010.04.050

Yoshino T, Ito E, Katsura T, Yamazaki D, Shan SM, Guo XZ, Nishi M, Higo Y, Funakoshi K (2011) Effect of iron content on electrical conductivity of ferropericlase with implications for the spin transition pressure. J Geophys Res-Solid Earth 116, B04202. doi:10.1029/2010jb007801

Yoshino T, McIsaac E, Laumonier M, Katsura T (2012a) Electrical conductivity of partial molten carbonate peridotite. Phys Earth Planet Inter 194:1–9

Yoshino T, Shimojuku A, Shan S-M, Guo X-Z, Yamazaki D, Ito E, Higo Y, Funakoshi K (2012b) Effect of temperature, pressure and iron content on the electrical conductivity of olivine and its high-pressure polymorphs. J Geophys Res-Solid Earth 117, B08205. doi:10.1029/2011jb008774

Zhang BH, Yoshino T, Wu XP, Matsuzaki T, Shan SM, Katsura T (2012) Electrical conductivity of enstatite as a function of water content: Implications for the electrical structure in the upper mantle. Earth Planet Sci Lett 357:11–20

Zhang B-H, Yoshino T, Yamazakil D, Manthilake G, Katsura T (2014) Electrical in partially molten peridotite under shear deformation, Earth Planet Sci Lett 405:98–109

Chapter 7
Melt Pockets and Spongy Clinopyroxenes in Mantle Xenoliths from the Plio-Quaternary Al Ghab Volcanic Field, NW Syria: Implications for the Metasomatic Evolution of the Lithosphere

George S.-K. Ma, Kuo-Lung Wang, John Malpas, Yoshiyuki Iizuka, Costas Xenophontos, Abdulsalam A. Turkmani, Gavin H.-N. Chan, Tadashi Usuki and Queenie H.-S. Chan

Monograph: Mantle heterogeneities (Ed. F. Deschamps).

Electronic supplementary material The online version of this article (doi:10.1007/978-3-319-15627-9_7) contains supplementary material, which is available to authorized users.

G.S.-K. Ma (✉) · K.-L. Wang · Y. Iizuka · T. Usuki
Institute of Earth Sciences, Academia Sinica, Taipei 11529, Taiwan
e-mail: georgema@graduate.hku.hk

G.S.-K. Ma
Dragon Mining Consulting, Unit 1701, Grand Millennium Plaza,
181 Queen's Road Central, Hong Kong, Hong Kong

G.S.-K. Ma · J. Malpas · C. Xenophontos · Q.H.-S. Chan
Department of Earth Sciences, The University of Hong Kong,
Pokfulam Road, Pokfulam, Hong Kong

A.A. Turkmani
General Establishment of Geology and Mineral Resources,
Ministry of Petroleum and Mineral Resources, Damascus, Syria

A.A. Turkmani
Al Badia Cement Company (JSC), Damascus, Syria

G.H.-N. Chan
Department of Earth Sciences, Oxford University, Parks Road, Oxford OX1 3PR, UK

G.H.-N. Chan
SRK Consulting, A1, 11/F, One Capital Place, 18 Luard Road, Wanchai, Hong Kong

Q.H.-S. Chan
ARES, NASA Johnson Space Center, Houston TX 77058, USA

© Springer International Publishing Switzerland 2015

A. Khan and F. Deschamps (eds.), *The Earth's Heterogeneous Mantle*,
Springer Geophysics, DOI 10.1007/978-3-319-15627-9_7

Abstract Spongy minerals, especially clinopyroxenes, and fine-grained, often glass-bearing mineral assemblages, commonly referred to as melt pockets, occur in many mantle xenolith suites worldwide, but their origins remain far from being clearly understood. We describe a suite of spongy clinopyroxene- and melt pockets-bearing peridotite xenoliths from the Plio-Quaternary volcanic field in the Al Ghab Depression, a pull-apart basin of the Dead Sea Fault System in northwestern Syria. The melt pockets comprise fine-grained olivines, clinopyroxenes, spinels and feldspars ± amphiboles ± glasses. Petrography and major and trace element mineral chemistry suggest that the xenoliths have experienced at least two stages of metasomatism with the formation of the melt pockets being associated with the latest event. The first metasomatic episode featuring LREE, Na, Th, U enrichment and relative Ti and Zr depletion in the cores of primary clinopyroxenes involved metasomatism by a low-silica, CO_2-rich agent. The second metasomatic episode was associated with the development of melt pockets which evolved from decompressional and perhaps metasomatism-induced incongruent melting of clinopyroxene ± spinel. The spongy clinopyroxenes that occur as coronas around clear, primary clinopyroxenes represent a transitional stage of the partially melted crystals. The incongruent melting produced a liquid that evolved within the melt pockets and eventually migrated out to form amphiboles and micas elsewhere in the lithosphere. Albeit with some uncertainty, geothermobarometric estimations reveal significant, systematic differences in the equilibration pressures between the primary minerals (0.8–1.4 GPa), and spongy and melt-pocket minerals (0.7–0.9 GPa), lending good support for the decompressional origins of the spongy clinopyroxenes and melt pockets. It is interpreted that decompression resulted from local transtension associated with the development of the Al Ghab pull-apart basin, a step-over zone of the Dead Sea Fault System, in Plio-Quaternary time.

Keywords Mantle metasomatism · Spongy texture · Thermobarometry · Lithospheric mantle · Melt pocket

7.1 Introduction

Mantle xenoliths can be of value in providing first-hand information about the history of the deep lithosphere from which they are derived. Their textures, mineral composition and zonation, for instance, can provide insight to the thermal and compositional evolution of the lithosphere. This is of particular importance for studying the mantle of continental lithosphere, which is generally thought to have experienced multi-stage melting and fluid/melt–rock interaction events (Menzies 1983; Downes 2001; Dawson 2002; Ackerman et al. 2007; Griffin et al. 2009; O'Reilly and Griffin 2013; Tang et al. 2013). In recent decades, a plethora of studies have described different styles of fluid/melt–rock interaction as a result of infiltration or percolation of mafic/ultramafic silicate melts, carbonate-rich melts and/or C–O–H-rich fluids (Dawson 1984; Downes 2001; Luth 2003; O'Reilly and Griffin 2013). These studies, plus experimental work, have led to the establishment of many indicators, largely based on the trace element systematics of clinopyroxenes, to gauge the effects of the various styles of metasomatism (e.g. Yaxley et al. 1998; Coltorti et al. 1999).

In many xenolith suites such as those from the Kenya rift, western Qinling, central China, the Lashaine volcano, northern Tanzania, the Bakony–Balaton Highland volcanic field, western Hungary, the Tariat Depression and Dariganga lava plateau, Mongolia, the Harrat Uwayrid, Saudi Arabia and Jabel El Arab (Harrat Ash Shamah), and southern Syria (Ionov et al. 1994; Bali et al. 2002, 2008; Dawson 2002; Kaeser et al. 2007; Kaliwoda et al. 2007; Ismail et al. 2008; Su et al. 2010a, 2011, 2012), it is clear from the presence of spongy (spongy cellular or sieve)-textured minerals and fine-grained secondary minerals such as clinopyroxene, olivine and spinel, often associated with glasses, that the mantle has undergone some chemical or pressure–temperature (*P–T*) modification. However, the exact origin of these "quenched" reactive products is yet far from clear, as is whether or not an agent of metasomatism is required.

To decipher the nature of metasomatic reactions in the mantle, we present a mineral chemical study, integrated with textural constraints, of a little known peridotite xenolith suite from the Al Ghab Depression, a pull-apart basin associated with the northern Dead Sea Fault System, in NW Syria. The xenoliths contain texturally and compositionally variable phases, with clear signs of metasomatism such as the presence of amphiboles and light rare earth element (LREE)-enriched high-Na clinopyroxenes which are often spongy textured, along with some spinels. Associated with these spongy minerals are fine-grained patches, hereafter melt pockets, comprising secondary low-Na clinopyroxenes, high-Ca olivines, tiny euhedral spinels, interstitial/poikilitic feldspars and occasionally glass and amphiboles. We will demonstrate that the spongy minerals and melt pockets are genetically related and that they developed in the peridotite which had already experienced an earlier stage of metasomatic alteration. These spongy minerals and melt pockets are thought to have formed during times of local lithospheric extension associated with the evolution of the northern Dead Sea Fault System.

7.2 Geology and Samples

Previous structural and geophysical investigations have provided a fair understanding of the timing and mechanism of the development of the Al Ghab Depression and the northern Dead Sea Fault System in Syria (Trifonov et al. 1991; Domas 1994; Devyatkin et al. 1997; Kopp et al. 1999; Brew et al. 2001b). It has been proposed that the depression formed as a pull-apart basin in a complex left-lateral step-over zone of the fault (Fig. 7.1) in Plio-Quaternary time (Brew et al. 2001b). Seismic refraction experiments suggest that the Moho depth beneath the Aleppo Plateau is ~28–31 km (Brew et al. 2001a), but that beneath the Al Ghab Depression is likely shallower. High-resolution mapping of the lithosphere–asthenosphere boundary (LAB) across Al Ghab and the Aleppo Plateau is not available, but S-wave receiver functions-based regional surveys suggest a regional LAB depth of ~100 km (Angus et al. 2006).

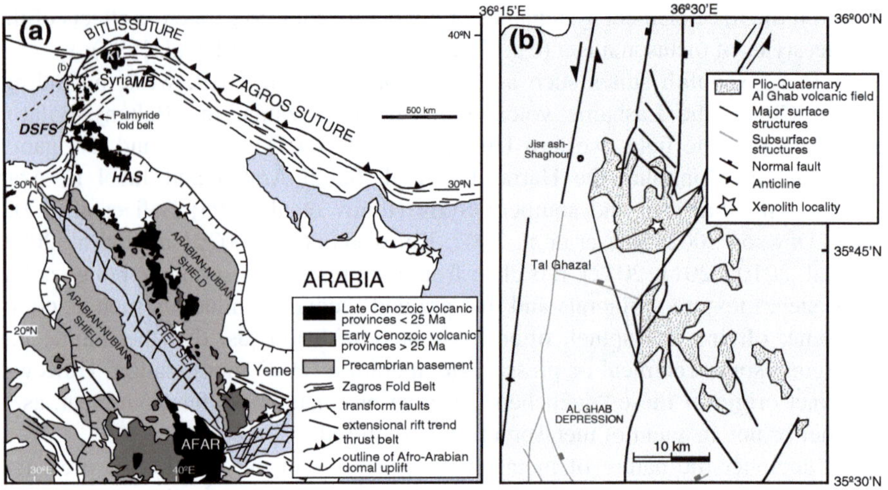

Fig. 7.1 Geographic locations of Cenozoic volcanism in (**a**) Arabia (modified after Davidson et al. 1994), with highlights of xenolith localities, and (**b**) the Al Ghab Depression, northwestern Syria. *Dashed box* in (**a**) marks the area shown in (**b**). The xenolith occurrence in Tal Ghazal is shown by a star symbol in (**b**). *DSFS* Dead Sea Fault System; *HAS* Harrat Ash Shamah; *KV* Karacadağ volcano; *MB* Mesopotamian Basin

The Al Ghab volcanic field comprises some 40 cinder and lava cones within an area of 600 km^2, and the volcanic rocks are mainly basanites with subordinate amounts of hawaiites and alkali basalts (Ma et al. 2011a). The volcanic rocks have been dated by K–Ar and ^{40}Ar/^{39} Ar methods at 4.0–1.1 Ma, with the majority younger than 2 Ma (Sharkov et al. 1994, 1998; Krienitz et al. 2009; Searle et al. 2010). Among all, Tal Ghazal (Fig. 7.1), a Late Pliocene cinder cone, is one of the best exposed xenolith-bearing cones in northern Al Ghab. The lava flows overly the scoriae, and their amount at Tal Ghazal is small compared with the scoriae. More than 100 xenolith samples were investigated by Turkmani and Al-Shara'a (Turkmani and Al-Shara'a 2004) who identified the majority (80 %) of xenoliths as being green Cr-diopside-bearing spinel lherzolites; no garnet-bearing samples have been found. We have examined many of the samples provided by these authors, together with our own collections that were taken from Tal Ghazal and its vicinity (within a 1 km distance of the cone). Overall, the specimens include (1) a majority of green Cr-diopside-bearing spinel lherzolites and harzburgites which are in some places amphibole-bearing and (2) a subordinate amount of spinel-bearing pyroxenites, hornblendites and megacrysts (e.g. kaersutite and sanidine). In the present study, a focus is placed on the melt-pocket bearing and spongy-mineral bearing peridotites. To better understand the origin of the secondary minerals/mineral assemblages, one melt-pocket free sample, G79, is included in the study suite, making up a total of ten samples.

7.3 Petrography

The xenoliths are rounded to discoidal in shape, and the dimensions are between a few cm and 20 cm (typically <10 cm). There are at least two generations of mineral assemblages; one being "primary" that forms the bulk of the peridotite and another being "secondary" in the melt pockets that are clearly formed later than the host peridotite (Fig. 7.2). As will be discussed below, the primary assemblage represents a combination of texturally equilibrated olivines, orthopyroxenes, spinels and occasionally zoned Cr-diopsides that have been affected by a metasomatic overprint. Modal metasomatic phases, including amphiboles, apatites and micas, are also present in some samples, but we will demonstrate that they are younger than the restite olivines and orthopyroxenes, and most likely related to the melt pockets. In this paper, we

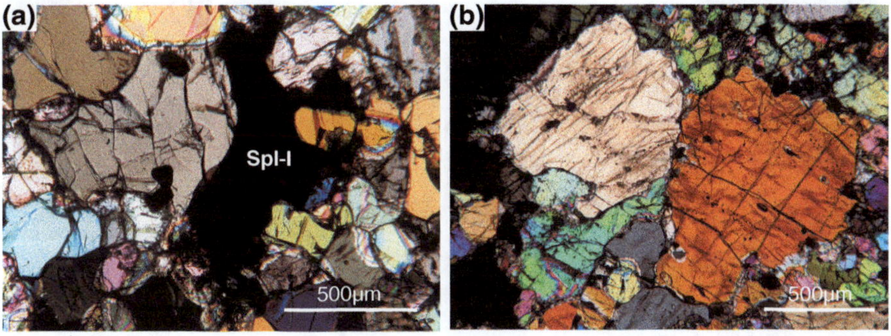

Fig. 7.2 Photomicrographs of thin sections in cross-polarised light (a–d), back-scattered electron micrographs (e, f, h–q), a calcium compositional zonation map (g) and an optical image of the polished surface of specimen (r) for the Al Ghab peridotite xenoliths. (a) Typical protogranular texture in the non-melt-pocket-bearing sample G79. Note the vermicular form of primary spinel (Spl-I). (b) Protogranular and somewhat porphyroclastic texture in sample G10. Spongy (Cpx-Spo; c, d, f–j) and non-spongy (Cpx-I; e) clinopyroxene. Note the close association of Cpx-Spo with melt pockets (c, d, k and m). The development of the spongy texture appears to be more mature when the clinopyroxene is adjacent or close to the melt pockets. Apparently, such clinopyroxene breaks down to form fine-grained melt-pocket clinopyroxene (Cpx-P), as relics of Cpx-I are sometimes observed within the melt pockets. Many of these relics and Cpx-P within single melt pockets are in optical continuity (see main text). In sample G47 (i and j), Cpx-I is clearly melting to give Cpx-Spo and a melt (now glass) that is propagating outwards. (c, d, k–q) Typical (dirty) appearance of melt pockets showing fine-grained aggregates of secondary clinopyroxene (Cpx-P), olivine (Ol-P), spinel (Spl-P) and feldspar (Flsp) in G10 (c, q), G47 (d, l, m, o), S52-05A (k, n) and G72 (p). Note that Ol-P is usually subhedral to euhedral, whereas Cpx-P varies from subhedral to anhedral and is usually spongy. Spl-P (l–p) is tiny and euhedral, and sometimes somewhat skeletal with overgrowths of discontinuous Spl-P chains ("atolls"), commonly enclosed in euhedral Ol-P and interstitial Flsp. By contrast, relic Spl-I (q and to a lesser extent l) in the centre of some melt pockets is rounded and optically less heterogeneous. The euhedral Spl-P often forms a chain ("necklace"), partially included in euhedral overgrowths of olivine along the rims of melt pockets. In (p), the interstitial, matrix material in the melt pocket is glass, instead of Flsp as commonly seen in others. Many melt pockets, as exemplified in (r), form discontinuous chains aligned parallel to the deformation fabrics of the rocks [see also (q)]. Samples: (a, e) G79; (b, c, q) G10; (d, f, g, i, j, l, m, o) G47; (h, k, n, r) S52-05A; (p) G72

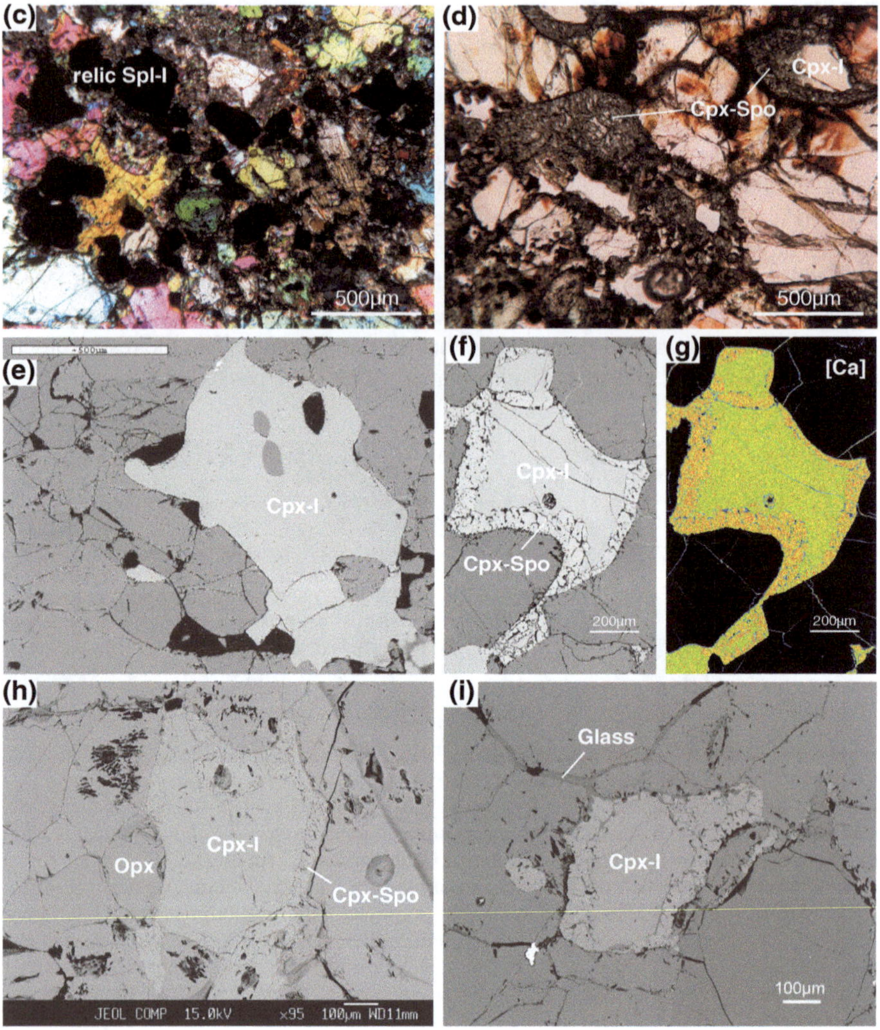

Fig. 7.2 (continued)

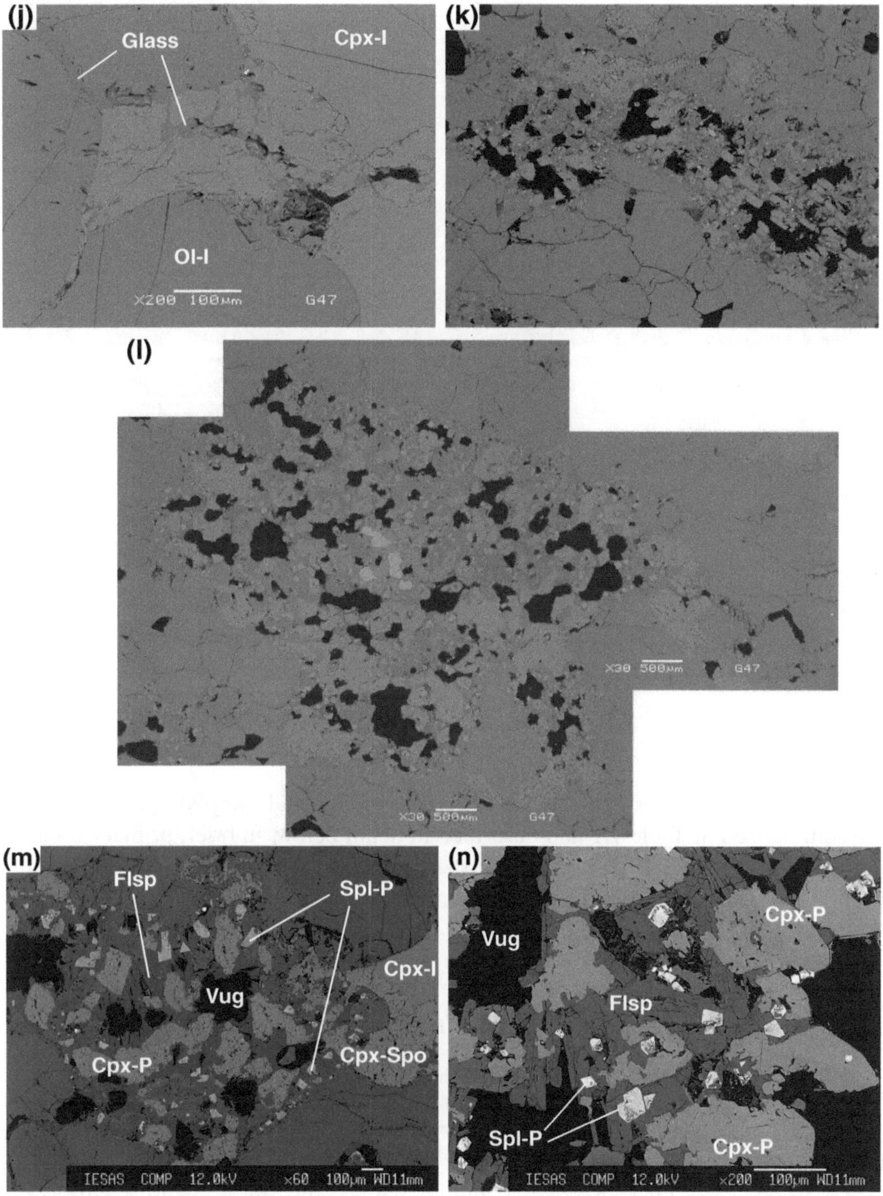

Fig. 7.2 (continued)

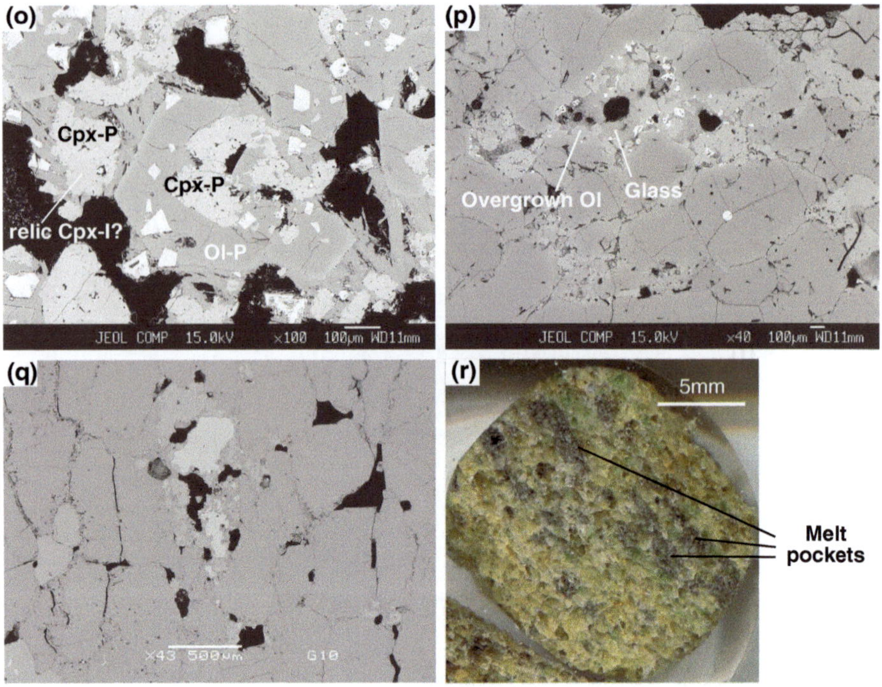

Fig. 7.2 (continued)

label the minerals according to their textural positions and morphologies. The labels are summarised in Table 7.1 and described in detail below. In brief, primary minerals are suffixed "I" (e.g. Ol-I) and melt-pocket minerals "P" (e.g. Cpx-P). The spongy portions (usually the rims) of the primary minerals are suffixed "Spo" (e.g. Cpx-Spo). Secondary Ti-rich rims mantling/replacing Cpx-I are suffixed "TR" (i.e. Cpx-TR).

7.3.1 Petrography of the Peridotites

The peridotite samples exhibit a tabular to protogranular texture, and their general characteristics are given in Table 7.2. Deformation is common and marked by strained and kink-banded Ol-I and Opx-I. Subgrain rotation recrystallisation is common in the Ol-I and Opx-I, though apparently more easily developed in larger porphyroclasts, having subgrains or neoblasts formed around larger grains. In many places, the Opx-I does not show the same degree of deformation as the Ol-I, despite their similar size (typically 0.5–2 mm). Ol-I is colourless, and its margins often have euhedral projections into the melt pockets, indicating that Ol-I is a good nucleation site for overgrowth. Opx-I is beige or colourless and sometimes contains abundant fluid inclusion trails. Unlike Cpx-I, Opx-I appears to be a stable phase not showing any spongy rims. Only rarely, usually adjacent to melt pockets, does it break down to form a narrow region (a few to tens of microns

Table 7.1 Abbreviations for the various minerals identified in the Al Ghab peridotite xenoliths

Abbreviation	Description	Petrography	Key chemical features
Ol-I	Primary olivine	Commonly strained and exhibit kink banding	Ca poor; high Fo
Opx-I	Primary orthopyroxene		High Mg#
Cpx-I	Primary clinopyroxene (spongy-free portion)	Commonly strain-free	Na- and Al-rich
Spl-I	Primary spinel	Rounded–holly leaf shaped or rarely rounded inclusions in Ol-I and Opx-I	
Cpx-Spo	Spongy (sieve-textured) portion of primary clinopyroxene	Empty vug or channel-rich portion, usually coronas or sometimes entire grains, of otherwise Cpx-I or Spl-I; Cpx-Spo never develops when in contact with Opx-I	Na and Al poorer, Ca richer and Cr# higher than Cpx-I
Spl-Spo	Spongy (sieve-textured) portion of primary spinel		Cr# higher than Spl-I
Ol-P	Melt pocket olivine	Subhedral–euhedral; embayed; often enclose Spl-P; optically continuous with adjacent Ol-P	Ca-rich
Cpx-P	Melt pocket clinopyroxene	Subhedral–euhedral; embayed; optically continuous with adjacent Ol-P; strongly zoned; worm-like when associated with decomposing Opx-I	Na poorer (similar to Cpx-Spo) and variable but generally Al richer than Cpx-I
Spl-P	Melt pocket spinel	Euhedral; tiny; some with discontinuous chains of Cr-rich Spl overgrowths	
Flsp	Melt pocket feldspar	Interstitial; anhedral or as microlaths	Wide compositional range from labradorite to sanidine in individual xenolith samples and melt pockets

in width) of orthopyroxene + clinopyroxene ± spinel symplectites that gradate towards the melt-pocket phases.

Cpx-I, typically <1.5 mm, is generally smaller than Ol-I and Opx-I and shows the typical green colour of Cr-diopside. Orthopyroxene exsolution lamellae are present in Cpx-I cores of some samples (e.g. G72). Except for sample G79, Cpx-I always shows a spongy texture along the rims, on the margins of fractures, or in places the entire grains (Cpx-Spo). The spongy texture is accompanied by entrapment of irregularly shaped vugs and appears to be better developed (thicker spongy rims or in many instances spongy cores) in Cpx closer to the melt pockets. In well-developed Cpx-Spo, the vugs may become networked channels (Fig. 7.2f, i). In most cases, the vugs are empty (Fig. 7.2f), but rarely, glasses and feldspars are found within the vugs (Fig. 7.2i, j). Where occurring within or in direct contact with the melt pockets, Cpx-Spo is strongly embayed to give an anhedral boundary and associated with Flsp, Spl-P, Ol-P and Cpx-P (Fig. 7.2d, k). The latter, Cpx-P, sometimes has the same crystallographic orientation as the adjacent decomposing

Table 7.2 Modal proportions of the peridotite xenoliths from Al Ghab volcanic field, NW Syria (in vol.%)

Sample	Rock type	Texture	Ol	Cpx	Opx	Spl	Amph	Mica	Apt
G10	Harzburgite	tabular–protogranular	88	3	6	1	3		tr
G47	Lherzolite	tabular–protogranular	67	11	22				
G49	Lherzolite	protogranular	69	10	19	1	1	tr	
G64	Lherzolite	tabular	67	6	25	2	tr		
G72	Lherzolite	tabular–protogranular	70	16	11	3			
G79	Lherzolite	protogranular	64	12	22	3			
S32-05	Lherzolite	tabular–porphyroclastic	80	6	12	2			
S52-05A	Lherzolite	tabular–porphyroclastic	75	10	15				
S52-05D	Lherzolite	tabular–porphyroclastic	63	14	21	3	tr	tr	
S52-05GI	Harzburgite	protogranular–porphyroclastic	69	3	18	1	9		

tr, trace

Cpx-Spo. Of note, Cpx-Spo rims never develop in direct contact with Opx-I (Fig. 7.2h), an observation also made in the peridotite xenoliths from Mongolia (Ionov et al. 1994). Where Opx-I is missing from the immediate surroundings of Cpx, incomplete Cpx-Spo (portions of clear rims) sometimes occurs against Ol-I.

Rims of Cpx-I and Cpx-Spo are in places replaced by a Ti-rich variety of Cpx (Cpx-TR) in the amphibole-bearing samples, and the replacement is usually accompanied by an increase in Al and a decrease in Mg# in the pyroxene. Progressive replacement appears to have converted some Cpx-TR to amphibole.

Spl-I (Cr-spinel) is reddish brown or yellowish brown and is rounded or holly leaf shaped (Fig. 7.2a), interstitial among Ol-I, Opx-I and Cpx-I, although some occurs as rounded inclusions in Ol-I and Opx-I. It is found as small crystals, usually <1 mm, and generally occupies no more than 2 vol.% in the rocks. It is absent from rocks that have high proportions of (large) melt pockets (>10 vol.%; e.g. G47 and S52-05a), and relics of Spl-I are in places found in the centres of the melt pockets (Fig. 7.2c, q), suggesting its consumption by the melt pockets. Similar to Cpx-Spo, Spl-Spo develops at rims on Spl-I, accompanied by a change in colour from lighter brown to darker brown from the core to the rim, reflecting also a change of the chemistry. However, the rims are always too thin or too "spongy" to obtain valuable electron microprobe analyses.

Small amounts of volatile-bearing phases, such as amphibole, mica and apatite, which are usually attributed to metasomatic origins (e.g. Menzies 1983), are present in most of the samples (except G79). The amphiboles are usually Ti-pargasite and occur as an interstitial phase in samples G10, G49, G64, S52-05D and S52-05GI. The latter sample, S52-05GI, also contains ~0.1–0.2-mm-thick veinlets of amphiboles and clinopyroxenes. The interstitial amphiboles in these samples are generally associated with (usually surrounding) Spl-I, but independent amphiboles are also common. In rare instances (e.g. G49), they occur as exsolution lamellae in Cpx-I which otherwise shows no exsolution or orthopyroxene lamellae. As described above, some amphiboles in G10, G49 and S52-05GI apparently replace Cpx-I, Cpx-Spo and Cpx-P that appear to have been preceded by conversion to Ti–Al-rich clinopyroxene rims (Cpx-TR; see below). The micas are altered to a mixture of biotite and phlogopite and are found only in G49 and S52-05D as an interstitial phase. The apatite is also rare and is only found in G10 interstitially between Cpx-I and Opx-I.

Melt-pocket free sample G79 has a protogranular texture and does not contain any hydrous minerals (Fig. 7.2a). Local recrystallisation, i.e. neoblast formation, is evidenced by aggregates of polygonal grains. Its Ol-I (<2 mm), Opx-I (<2 mm), Cpx-I (<0.5 mm) and Spl-I (<1 mm) appear to be texturally equilibrated, although one Spl-I grain near the xenolith margin shows a darker Spl-Spo rim. Cpx-I is green and always clear, without spongy rims.

7.3.2 Petrography of the Melt-Pockets

The shape of the melt pockets varies from rounded to, more commonly, elongated (Fig. 7.2k, q, r). Most of them are isolated from each other, but some are networked with adjacent pockets (Fig. 7.2r). Elongated melt pockets often form chain-like clusters parallel to the deformation fabrics of the rocks (Fig. 7.2q, r). None of these clusters nor individual melt pockets appear to be connected with the host basalts. Albeit uncommon, in places—<1 mm from the host basalts—where the host magma infiltration occurs, the xenoliths' primary minerals are strongly resorbed. The infiltrated reaction zones are easily distinguished from the xenoliths' melt pockets in that the olivine phenocrysts of the former are iddingsitised, a common feature of the olivine phenocrysts in the host magma (Ma et al. 2011a).

The boundaries of the melt pockets against the peridotite hosts are gradational, with the primary olivines corroded or overgrown with euhedral Ol-P (Fig. 7.2p). Although not common, the rims of some primary orthopyroxenes against the melt pockets are destabilised as symplectites of clinopyroxenes + olivines ± spinels. Where they occur, primary clinopyroxenes at melt-pocket boundaries are highly embayed and spongy (Fig. 7.2d, k).

The melt pockets are "vesicular", with up to 15 vol.% rounded or irregular-shaped empty vugs. The vugs must have once been filled with fluids; otherwise, the pressure at depth would have collapsed them. The melt-pocket mineral assemblage comprises fine-grained olivines, clinopyroxenes and spinels (Ol-P, Cpx-P and Spl-P, respectively), in subhedral to euhedral habit (Fig. 7.2k, o), associated with interstitial feldspars (Flsp; Fig. 7.2m–o). Despite its ubiquity in the melt pockets, Flsp never occurs as a primary phase. Glasses (Fig. 7.2p) and amphiboles are uncommon, and where present, occupy interstitial positions. The latter also appear to replace Cpx-P on rare occasions. In most samples in particular G10, vermicular relic Spl-I is a common phase in the centre of the melt pockets. However, in samples G47 and S52-05A, Spl-I is absent both as a primary phase and as relic grain in the melt pockets. The melt pockets in the latter two samples are generally larger than those in G10 and other samples.

In many instances, the melt pockets are rimmed by a "necklace" of Spl-P on the surrounding Ol-I that now is overgrown with euhedral Ol-P (Fig. 7.2k, p). Individual Spl-P is also common in interstitial Flsp. Some Spl-P is zoned with Cr-rich rims or may have overgrowths of discontinuous Spl-P chains ("atolls"; Fig. 7.2n, o) that are usually Cr-richer (Mg-chromites). Individual Ol-P and Cpx-P, respectively, within a single pocket are usually at identical crystallographic orientations (i.e. they are optically continuous), indicating that they have recrystallised from or overgrown on strongly resorbed, formerly, larger gains. The petrographic observations suggest that the melt pockets are later features, having formed after clinopyroxenes and spinels, with little evidence for textural re-equilibration with the host peridotite, indicating their transient nature. Overall, all the primary phases have been corroded at the melt-pocket margins, and their susceptibility to corrosion appears to be Cpx-I > Spl-I > Ol-P ~ Opx-I. It appears that there is an intimate relationship among breakdown of Cpx-I and Spl-I, development of Cpx-Spo and Spl-Spo and formation of the melt pockets.

7.4 Compositions of Minerals and Glasses

Representative electron microprobe analyses of olivines, spinels, orthopyroxenes, clinopyroxenes, amphiboles, micas and glasses are given in Tables 7.3, 7.4, 7.5, 7.6 and 7.7 and in the online supplementary dataset.

7.4.1 Mineral Chemistry of the Peridotite and Melt-Pockets

Ol-I and Ol-P are readily distinguishable by their CaO and Fo contents (Fig. 7.3a; Table 7.3). Most Ol-I has CaO <~0.08 wt% and Fo in the range of ~87.5–90, and most Ol-P has significantly higher CaO between 0.05 and 0.35 wt% (up to 0.7 wt%) in the Fo range of 87.5–93.

Most Spl-I is Cr-spinel with 14–39 wt% Cr_2O_3 and Spl-P is Cr-spinel or Mg-chromite with 13–62 wt% Cr_2O_3. Their compositions (Table 7.4) with respect to Mg# and Cr# are shown in Fig. 7.3b, displaying a broad negative correlation. Spl-I has a narrower range of Mg# (66–84) and Cr# (15–49) than Spl-P (41–96 and 14–84, respectively), although there is considerable overlap. Two of the Spl-P analyses from G72 have a distinct composition, characterised by an elevated magnetite component (33 wt% FeO_T).

The rims of some Opx-I in two amphibole-bearing samples G49 and S52GI are enriched in Fe, with Mg# as low as 88 and 87.2, respectively; otherwise, Mg# of the majority of Opx-I is in the range of 89.5–91.5 (Table 7.5). Al_2O_3 is variable from 1.7 to 4.7 wt% and is mainly a result of mineral zonation (see below).

Compositional variation between different classes of clinopyroxenes is shown in Fig. 7.3c–e, portraying a broadly negative correlation between TiO_2 and Mg#, with the highest TiO_2 values being those of the secondary clinopyroxenes. In general, Cpx-I is Cr-diopside with the majority having Mg# in the range of ~89–92, Al_2O_3 of ~4–6 wt%, CaO of ~18.5–20.5 wt%, Na_2O > ~1.5 wt% and

Table 7.3 Representative chemical compositions of olivines in Al Ghab peridotite xenoliths and melt pockets, NW Syria (in wt%)

Sample	Analysis#	Mineral	SiO_2	TiO_2	Al_2O_3	Cr_2O_3	FeO	MnO	NiO	MgO	CaO	Na_2O	K_2O	Total	Fo	Comment
G10	G10_GZ18	Ol-I	41.09	0.05	0.01	0.07	9.19	0.15	0.37	49.92	0.04	n.d.	n.d.	100.9	90.6	
G10	G10_S-72	Ol-I	41.31	n.d.	n.d.	n.d.	9.25	0.21	0.46	49.68	0.01	n.d.	n.d.	100.9	90.5	
G10	G10_T-27	Ol-I	40.85	0.08	0.02	0.07	8.79	0.09	0.14	49.70	0.21	0.01	n.d.	100.0	91.0	
G47	G47_S-47	Ol-I	40.69	n.d.	0.02	0.02	8.35	0.17	0.51	49.73	0.12	n.d.	n.d.	99.6	91.4	
G47	G47_T-4	Ol-P	40.63	0.08	0.05	n.d.	9.40	0.17	0.43	49.08	0.07	n.d.	n.d.	99.9	90.3	
G47	G47_GZ10	Ol-I/P	41.96	0.04	0.20	0.05	6.95	0.04	0.51	50.21	0.23	0.02	n.d.	100.2	92.8	
G49	G49_T-1	Ol-I	40.56	n.d.	n.d.	0.01	10.81	0.14	0.28	48.84	0.07	n.d.	0.03	100.7	89.0	
G64	G64_S-80	Ol-P	40.63	n.d.	0.02	n.d.	9.63	0.09	0.19	49.92	0.06	n.d.	0.02	100.6	90.2	
G72	G72_T-51	Ol-I	40.65	n.d.	n.d.	n.d.	9.60	0.20	0.26	49.28	0.05	n.d.	0.03	100.1	90.2	
G72	G72_T-75	Ol-I	40.93	n.d.	0.13	0.10	10.61	0.15	0.12	48.62	0.15	0.05	n.d.	100.9	89.1	
G79	G79_GZ38	Ol-I	41.41	n.d.	n.d.	n.d.	9.61	0.10	0.42	48.63	0.04	n.d.	n.d.	100.2	90.0	
G79	G79_GZ50	Ol-P	41.02	n.d.	0.01	0.03	9.56	0.14	0.33	48.09	0.06	0.04	n.d.	99.3	90.0	rim of G79_GZ38
S32-05	S32_GZ56	Ol-I/P	41.19	n.d.	0.01	0.03	9.54	0.18	0.42	48.84	0.05	n.d.	n.d.	100.3	90.1	
S52-05A	S52A_A-22	Ol-P	40.85	n.d.	0.02	n.d.	8.85	0.12	0.26	50.17	0.18	0.02	n.d.	100.5	91.0	
S52-05A	S52A_C2-80	Ol-I	40.88	0.01	0.12	0.18	8.30	0.08	0.29	49.57	0.20	0.01	0.01	99.6	91.4	
S52-05A	S52A_D-96	Ol-I	41.05	n.d.	n.d.	0.02	10.09	0.11	0.10	49.17	0.03	n.d.	n.d.	100.6	89.7	
S52-05D	S52D_GZ09	Ol-I	41.25	n.d.	0.04	0.04	10.56	0.19	0.45	47.64	0.06	0.03	n.d.	100.3	88.9	
S52-05GI	S52GI_GZ10	Ol-I	41.35	n.d.	0.02	n.d.	10.82	0.14	0.36	47.64	0.06	n.d.	n.d.	100.4	88.7	

Table 7.4 Representative chemical compositions of spinels in Al Ghab peridotitexenoliths and melt pockets, NW Syria (in wt%)

Sample	Analysis#	Textural class	TiO_2	Al_2O_3	Cr_2O_3	Fe_2O_3[a]	FeO	NiO	MnO	MgO	Mg#	Cr#	Comments
G10	G10_S-65	Spl-P	0.95	28.70	40.26	2.00	12.90	0.00	0.38	16.03	68.9	48.5	
G47	G47_GZ22	Spl-P	0.08	49.87	17.84	1.80	11.60	0.60	0.19	17.99	73.4	19.4	core
G47	G47_GZ23	Spl-P	0.16	51.40	15.71	2.80	11.54	0.49	0.17	18.41	74.0	17.0	rim of G47_GZ22
G47	G47_S-32	Spl-P	0.16	53.21	15.36	1.46	9.25	0.22	0.21	20.20	79.6	16.2	
G49	G49_U-70	Spl-I	0.46	48.86	19.07	2.19	10.57	0.25	0.07	19.28	76.5	20.7	
G64	G64_GZ39	Spl-I	0.19	44.67	24.39	1.03	10.96	0.26	0.08	18.21	74.8	26.8	
G64	G64_S-61	SPN	0.69	45.70	21.04	2.60	9.86	0.22	0.10	19.29	77.7	23.6	
G72	G72_GZ57	Spl-I	0.09	51.11	16.24	2.10	10.17	0.38	0.12	19.15	77.1	17.6	core
G72	G72_GZ58	Spl-I	0.01	49.95	16.11	3.39	10.38	0.24	0.19	18.78	76.3	17.8	rim of G72_GZ57
G72	G72_GZ25	Spl-P	1.90	33.92	17.68	13.87	20.69	0.22	0.34	11.59	50.0	25.9	
G72	G72_GZ46	Spl-P	0.17	52.95	13.29	2.92	9.97	0.32	0.20	19.48	77.7	14.4	
G79	G79_GZ55	Spl-I	0.19	52.81	13.84	3.16	8.22	0.33	0.08	20.70	81.8	15.0	core
G79	G79_GZ56	Spl-I	0.17	53.20	13.59	3.62	8.25	0.39	0.10	20.82	81.8	14.6	rim of G79_GZ55
S32-05	S32_GZ30	Spl-I	0.80	32.23	34.54	3.40	14.24	0.22	0.26	15.31	65.7	41.8	core
S32-05	S32_GZ31	Spl-I	0.85	33.54	32.07	4.43	13.75	0.22	0.22	15.78	67.2	39.1	rim of S32_GZ30
S32-05	S32_GZ62	Spl-P	0.54	37.70	32.40	0.71	11.58	0.22	0.17	17.50	72.9	36.6	
S52-05A	S52-a_A-7	Spl-p	0.29	7.79	62.13	1.38	20.74	0.15	0.55	8.01	40.8	84.3	
S52-05A	S52-a_B3-43	Spl-P	1.18	31.28	33.30	4.77	13.49	0.22	0.37	15.70	67.5	41.7	
S52-05A	S52-a_B3-33	Spl-P	0.18	43.60	24.68	2.46	11.51	0.34	0.28	17.78	73.4	27.5	
S52-05D	S52d_GZ57	Spl-P	1.85	33.61	28.65	6.07	13.31	0.32	0.15	16.53	68.9	36.4	
S52-05GI	S52gI_GZ22	Spl-I	0.40	34.69	29.51	6.13	13.20	0.16	0.16	15.77	68.1	36.3	

[a]Estimated assuming perfect stoichiometry

Table 7.5 Representative chemical compositions of orthopyroxenes in Al Ghab peridotite xenoliths and melt pockets, NW Syria (in wt%)

Sample	Analysis#	SiO_2	TiO_2	Al_2O_3	FeO	MnO	MgO	CaO	Na_2O	K_2O	Cr_2O_3	Total	Mg#	En	Fs	Wo	Comment
G10	G10_Line 16 G2	57.39	0.06	2.05	5.58	0.16	34.33	0.80	0.11	n.d.	0.53	101.0	91.6	90.0	8.5	1.5	core
G10	G10_Line 1 G2	56.58	0.13	2.52	5.60	0.17	34.34	0.78	0.16	n.d.	0.61	100.9	91.6	90.0	8.5	1.5	rim
G47	G47_Line 13 G8	56.70	0.04	2.76	5.97	0.09	34.04	0.75	0.11	n.d.	0.38	100.8	91.0	89.6	8.9	1.4	core
G47	G47_Line 1 G8	56.39	0.01	3.05	6.19	0.19	34.12	0.72	0.17	n.d.	0.34	101.2	90.8	89.3	9.4	1.3	rim
G49	G49_Line 13 G11	56.75	0.06	1.52	6.83	0.13	33.65	0.72	0.08	n.d.	0.37	100.1	89.8	88.4	10.3	1.4	core
G49	G49_Line 1 G11	55.08	0.19	3.59	7.18	0.12	32.79	0.76	0.07	n.d.	0.49	100.3	89.1	87.6	10.9	1.5	rim
G64	G64_Line 9 G1	55.25	n.d.	3.62	5.95	0.09	33.69	0.71	0.17	n.d.	0.38	99.9	91.0	89.6	9.0	1.4	core
G64	G64_Line 15 G1	54.64	0.08	3.61	6.21	0.09	33.63	0.70	0.14	n.d.	0.51	99.6	90.6	89.3	9.4	1.3	rim
G72	G72_Line 5 G10	55.13	n.d.	3.73	5.95	0.15	33.15	0.61	0.11	n.d.	0.29	99.1	90.8	89.6	9.3	1.2	core
G72	G72_Line 1 G10	55.41	n.d.	4.01	6.10	0.17	33.05	0.73	0.12	n.d.	0.33	99.9	90.6	89.1	9.5	1.4	rim
G79	G79_GZ05	56.14	n.d.	3.81	6.06	0.07	32.43	0.67	0.08	0.01	0.41	99.7	90.5	89.2	9.5	1.3	core
G79	G79_GZ02	56.04	0.13	4.06	6.31	0.13	32.18	0.78	0.11	n.d.	0.32	100.0	90.1	88.5	9.9	1.5	rim
S32-05	S32_Line 7 G3	56.12	0.10	2.50	6.25	0.25	34.07	0.85	0.13	n.d.	0.33	100.6	90.7	88.9	9.5	1.6	core
S32-05	S32_Line 2 G3	55.62	0.15	2.87	6.37	0.08	33.93	0.79	0.07	n.d.	0.57	100.4	90.5	89.0	9.5	1.5	rim
S52-05A	S52a_GZ39	56.27	n.d.	2.66	6.48	0.18	32.67	0.66	0.14	n.d.	0.39	99.4	90.0	88.6	10.1	1.3	core
S52-05A	S52a_GZ31	55.96	n.d.	2.59	6.58	0.20	32.56	0.66	0.13	0.02	0.40	99.1	89.8	88.4	10.3	1.3	rim
S52-05D	S52d_GZ24	56.33	0.05	2.70	6.43	0.16	32.39	0.79	0.20	0.02	0.50	99.6	90.0	88.4	10.1	1.6	core
S52-05D	S52d_GZ17	56.35	0.24	3.30	6.66	0.21	31.69	0.89	0.19	n.d.	0.62	100.2	89.5	87.6	10.6	1.8	rim
S52-05GI	S52GI_Line 10 G9	55.63	0.20	3.25	7.03	0.13	32.65	0.83	0.14	n.d.	0.46	100.3	89.2	87.6	10.8	1.6	core
S52-05GI	S52GI_Line 1 G9	55.95	0.17	3.25	6.86	0.10	32.63	0.80	0.16	n.d.	0.41	100.3	89.4	87.9	10.5	1.6	rim

Table 7.6 Representative chemical compositions of clinopyroxenes in Al Ghab peridotite xenoliths and melt pockets, NW Syria (in wt%)

Sample	Analysis#	Textural Class	SiO_2	TiO_2	Al_2O_3	FeO	MnO	MgO	CaO	Na_2O	K_2O	Cr_2O_3	NiO	Total	Mg#	Wo	En	Fs	Al(IV)	AL(VI)	AlIV/AlVI	Comment
G10	G10_Line 17 G8	Cpx-TR	48.63	1.88	6.86	2.68	0.11	15.05	20.83	1.28	n.d.	1.98	n.d.	99.3	90.9	47.4	47.6	5.0	0.208	0.089	2.330	rim
G10	G10_GZ41	Cpx-I	52.77	0.47	4.90	2.57	0.08	15.21	19.36	2.10	n.d.	1.89	0.08	99.4	91.3	45.5	49.7	4.9	0.080	0.130	0.610	core
G10	G10_GZ33	Cpx-Spo	49.92	1.44	5.54	2.77	0.09	16.18	19.90	1.08	n.d.	2.14	0.09	99.1	91.2	44.6	50.4	5.0	0.166	0.074	2.229	rim
G10	G10_GZ89	Cpx-P	52.32	0.34	4.32	2.89	0.11	15.63	19.62	1.90	n.d.	1.81	0.15	99.1	90.6	44.9	49.8	5.4	0.083	0.104	0.797	core
G10	G10_GZ90	Cpx-P	47.51	1.75	6.25	4.93	0.11	13.73	23.12	0.60	0.02	0.82	0.11	98.9	83.3	50.1	41.4	8.5	0.216	0.060	3.579	rim of G10_GZ89
G47	G47_GZ20	Cpx-P	52.96	0.08	3.39	2.47	0.07	16.72	22.18	0.58	0.01	1.65	0.09	100.2	92.3	46.8	49.1	4.2	0.079	0.065	1.212	core
G47	G47_GZ53	Cpx-I	53.32	0.14	4.51	2.73	0.15	16.17	19.32	1.89	n.d.	1.18	0.02	99.4	91.3	43.8	51.1	5.1	0.064	0.129	0.501	core
G47	G47_GZ54	Cpx-Spo	53.35	n.d.	2.67	2.83	0.14	17.41	21.70	0.59	0.01	1.42	0.09	100.2	91.6	45.0	50.2	4.8	0.065	0.049	1.322	rim of G47_GZ53
G49	G49_GZ21	Cpx-P	52.68	0.62	5.51	3.18	0.11	15.76	19.49	1.65	n.d.	0.97	0.06	100.0	89.8	44.3	49.9	5.8	0.095	0.139	0.685	core
G49	G49_GZ02	Cpx-TR	50.53	0.83	6.63	3.37	0.17	14.88	20.71	1.74	0.01	1.22	0.05	100.1	88.7	46.9	46.9	6.3	0.157	0.128	1.233	rim
G49	G49_Line 1 G1	Cpx-Spo	51.71	0.97	3.49	2.94	0.12	15.41	22.34	0.81	n.d.	1.51	n.d.	99.3	90.3	48.4	46.4	5.2	0.097	0.054	1.800	rim
G49	G49_GZ30	Cpx-I	52.43	0.70	5.15	3.29	0.16	15.77	18.95	1.88	n.d.	1.54	0.02	99.9	89.5	43.5	50.3	6.2	0.098	0.122	0.802	core
G64	G64_GZ22	Cpx-I	51.79	0.35	5.93	3.17	0.06	15.08	20.12	2.17	0.01	1.18	0.02	99.9	89.5	46.1	48.1	5.8	0.115	0.139	0.824	core
G64	G64_T-9	Cpx-Spo	51.79	0.32	3.78	3.18	0.08	16.72	21.78	0.46	n.d.	1.53	0.09	99.7	90.4	45.8	48.9	5.3	0.108	0.055	1.982	rim
G64	G64_GZ35	Cpx-P	54.11	0.14	5.33	3.08	0.12	15.16	18.93	2.02	0.02	1.06	0.12	100.1	89.8	44.5	49.6	5.9	0.053	0.173	0.305	core
G64	G64_Line 1 g3	Cpx-TR	49.83	1.25	7.93	3.12	0.07	15.57	19.72	1.31	n.d.	1.73	n.d.	100.5	89.9	45.0	49.4	5.7	0.197	0.141	1.401	rim
G72	G72_GZ20	Cpx-P	49.84	1.09	6.68	5.79	0.11	13.45	20.66	1.17	n.d.	0.67	0.04	99.5	80.5	47.0	42.6	10.5	0.157	0.135	1.163	core
G72	G72_GZ60	Cpx-I	52.26	0.14	5.46	2.97	0.08	15.39	20.28	1.85	n.d.	0.79	0.08	99.3	90.2	46.0	48.6	5.4	0.092	0.142	0.650	core
G72	G72_GZ28	Cpx-Spo	53.24	0.18	2.96	2.57	0.04	16.16	23.40	0.38	n.d.	0.96	0.07	100.0	91.8	48.8	46.9	4.3	0.063	0.064	0.978	rim
G72	G72_Line 1 G3	Cpx-TR	49.58	0.93	6.04	4.38	0.08	15.25	20.36	1.20	n.d.	0.79	n.d.	98.6	86.1	45.2	47.1	7.7	0.159	0.105	1.509	rim
G79	G79_GZ19	Cpx-I	52.58	0.16	4.90	2.75	0.08	15.89	21.62	0.88	0.01	0.49	0.05	99.4	91.2	47.1	48.1	4.8	0.085	0.125	0.679	rim of G79_GZ22
G79	G79_GZ22	Cpx-I	53.34	0.22	4.65	2.77	0.13	15.89	21.65	0.89	0.02	0.61	0.03	100.2	91.1	47.0	48.1	4.9	0.074	0.124	0.595	core
S32-05	S32_GZ19	Cpx-Spo	52.81	0.63	3.40	2.85	0.12	15.51	22.80	0.66	n.d.	1.64	0.01	100.4	90.7	48.8	46.2	5.0	0.082	0.064	1.286	rim of S32_GZ26

(continued)

Table 7.6 (continued)

S32-05	S32_GZ26	Cpx-I	53.24	0.46	4.89	2.99	0.12	15.22	20.25	1.75	n.d.	1.75	0.07	100.7	90.1	46.2	48.3	5.5	0.083	0.125	0.665	core
S32-05	S32_GZ29	Cpx-I	52.71	0.40	4.85	3.22	0.11	15.45	20.40	1.79	0.01	1.74	0.07	100.7	89.5	45.8	48.3	5.8	0.096	0.110	0.873	rim of S32_GZ26
S32-05	S32_50	Cpx-P	52.15	0.52	3.99	3.84	n.d.	15.72	20.49	0.94	n.d.	1.76	0.11	99.5	87.9	45.2	48.2	6.6	0.090	0.082	1.090	core
S52-05A	S52a_GZ25	Cpx-I	53.79	0.04	5.05	2.98	0.12	14.92	19.69	2.23	0.01	1.25	0.12	100.2	89.9	45.9	48.4	5.6	0.058	0.156	0.374	core
S52-05A	S52a_GZ26	Cpx-Spo	53.09	n.d.	2.75	3.17	0.12	16.69	21.55	0.77	0.02	1.50	0.05	99.7	90.4	45.5	49.1	5.4	0.062	0.057	1.085	rim of S52a_GZ25
S52-05A	S52a_GZ29	Cpx-P	53.47	0.35	3.51	2.75	0.15	16.73	22.01	0.58	n.d.	1.32	n.d.	100.8	91.6	46.3	49.0	4.7	0.076	0.073	1.036	core
S52-05A	S52a_A-19	Cpx-Spo	52.25	0.04	2.66	2.42	n.d.	17.29	21.57	0.78	0.01	1.72	0.02	98.7	92.7	45.4	50.6	4.0	0.079	0.036	2.162	rim
S52-05D	S52d_GZ48	Cpx-I	53.34	0.60	4.97	3.48	0.07	15.25	18.67	2.10	0.01	1.47	n.d.	100.0	88.7	43.8	49.7	6.5	0.070	0.142	0.495	rim
S52-05D	S52d_GZ61	Cpx-P	51.41	0.84	4.72	3.89	0.14	16.52	19.98	0.93	n.d.	1.38	0.02	99.8	88.3	43.3	49.8	6.8	0.124	0.079	1.557	core
S52-05GI	S52gI_GZ32	Cpx-I	52.60	0.20	5.54	3.58	0.06	15.28	19.76	2.01	n.d.	1.01	0.15	100.2	88.4	45.1	48.5	6.5	0.094	0.143	0.655	core

Table 7.7 Representative chemical compositions of feldspars, glasses, amphiboles and micas in Al Ghab peridotite xenoliths and melt pockets, NW Syria (in wt%)

Sample	Analysis#	SiO_2	TiO_2	Al_2O_3	FeO	MnO	MgO	CaO	Na_2O	K_2O	Cr_2O_3	NiO	H2O	Total	An	Ab	Or	Comment
Feldspar																		
G47	G47_GZ19	61.99	0.12	23.67	0.22	n.d.	0.03	4.64	7.97	1.79	n.d.	n.d.		100.4	21.9	68.1	10.0	interstitial
G47	G47_T-38	64.72	0.02	18.90	0.69	0.04	0.15	0.29	6.01	7.28	0.02	n.d.		98.1	1.5	54.8	43.7	interstitial
G64	G64_GZ38	61.48	0.75	23.10	0.24	0.02	0.03	4.73	8.16	1.36	n.d.	0.02		99.9	22.4	69.9	7.7	interstitial
S52-05A	S52A_D-82	65.48	0.08	20.02	0.46	n.d.	0.12	0.93	8.15	3.75	n.d.	n.d.		99.0	4.6	73.2	22.2	interstitial
S52-05A	S52a_GZ76	55.87	0.03	27.44	0.29	n.d.	0.05	9.80	6.03	0.57	0.05	n.d.		100.1	45.8	51.0	3.2	interstitial
S52-05A	S52A_A-27	64.82	0.07	20.14	0.29	0.15	0.09	1.26	8.45	3.25	0.05	0.11		98.7	6.2	74.9	18.9	interstitial
Glass																		
G47	G47_U-2	60.35	0.35	23.06	1.26	0.01	1.30	4.74	7.48	1.27	n.d.	0.08		99.9	–	–	–	vein
G47	G47_T-9	61.55	0.44	21.15	1.24	n.d.	3.46	2.93	6.51	2.05	0.08	0.04		99.4	–	–	–	vein
G72	G72_GZ26	52.81	1.67	21.40	5.77	0.09	1.81	5.58	7.31	2.67	n.d.	0.03		99.1	–	–	–	interstitial
G72	G72_GZ27	54.49	1.10	21.73	4.61	0.09	2.01	5.29	6.56	3.18	0.04	0.02		99.1	–	–	–	interstitial
Amphibole															Mg#			
G10	G10_GZ51	44.26	2.20	12.26	3.61	0.11	17.60	10.40	3.26	1.59	2.54	0.10	2.12	100.0	89.7			interstitial
G49	G49_GZ13	41.95	2.86	13.66	4.66	0.07	16.54	10.91	3.55	1.44	1.52	0.21	2.08	99.4	86.4			interstitial
G49	G49_GZ17	41.83	3.05	14.07	4.34	0.08	16.26	11.04	3.49	1.80	1.31	0.06	2.07	99.4	87.0			interstitial
G64	G64_GZ27	43.16	3.52	14.18	4.08	0.13	16.94	10.85	3.25	1.34	0.78	0.11	2.12	100.5	88.1			interstitial
G64	G64_GZ28	43.37	3.23	14.12	4.32	0.05	17.04	10.68	3.14	1.23	0.91	0.17	2.13	100.4	87.5			interstitial
S52-05GI	S52gI_GZ36	43.26	3.08	13.23	4.92	0.07	16.49	10.72	3.18	1.83	0.12	0.08	2.08	99.0	85.7			vein
S52-05GI	S52gI_GZ37	43.57	2.82	12.96	5.01	0.05	16.57	10.72	3.23	1.83	0.41	0.05	2.08	99.3	85.5			vein
S52-05GI	S52gI_GZ71	42.78	1.84	13.68	5.14	0.08	16.32	10.59	3.23	1.74	1.96	0.11	2.09	99.6	85.0			interstitial
Mica																		
G49	G49_U-112	37.82	5.10	16.54	5.31	0.08	19.14	0.02	0.97	7.71	1.34	0.19	4.18	98.4	86.5			interstitial
G49	G49_U-113	36.56	4.56	16.74	4.98	n.d.	20.19	0.07	0.88	8.02	1.26	0.41	4.13	97.8	87.8			interstitial

TiO_2 < ~1 wt% (Table 7.6). Most of the secondary clinopyroxenes (Cpx-Spo, Cpx-TR and Cpx-P) have lower Na_2O, < ~1.5 wt% (Table 7.6; Fig. 7.3e). Compared to Cpx-I, Cpx-Spo tends to have lower Al_2O_3, higher CaO with variable Cr# that extends to as high as ~56, and Cpx-TR tends to have higher Al_2O_3, lower CaO with variable Mg# that extends to as low as 83. A large number of Cpx-Spo cluster at high values of Mg# between 90 and 93, although some have high TiO_2 (up to 2 wt%). The compositions of Cpx-P encompass nearly all those of Cpx-I, Cpx-Spo and Cpx-TR in terms of Al_2O_3, CaO, TiO_2, Mg# and Cr# (Fig. 7.3c–e). Note that Cpx-I from the modally unmetasomatised sample G79 has the lowest Na_2O

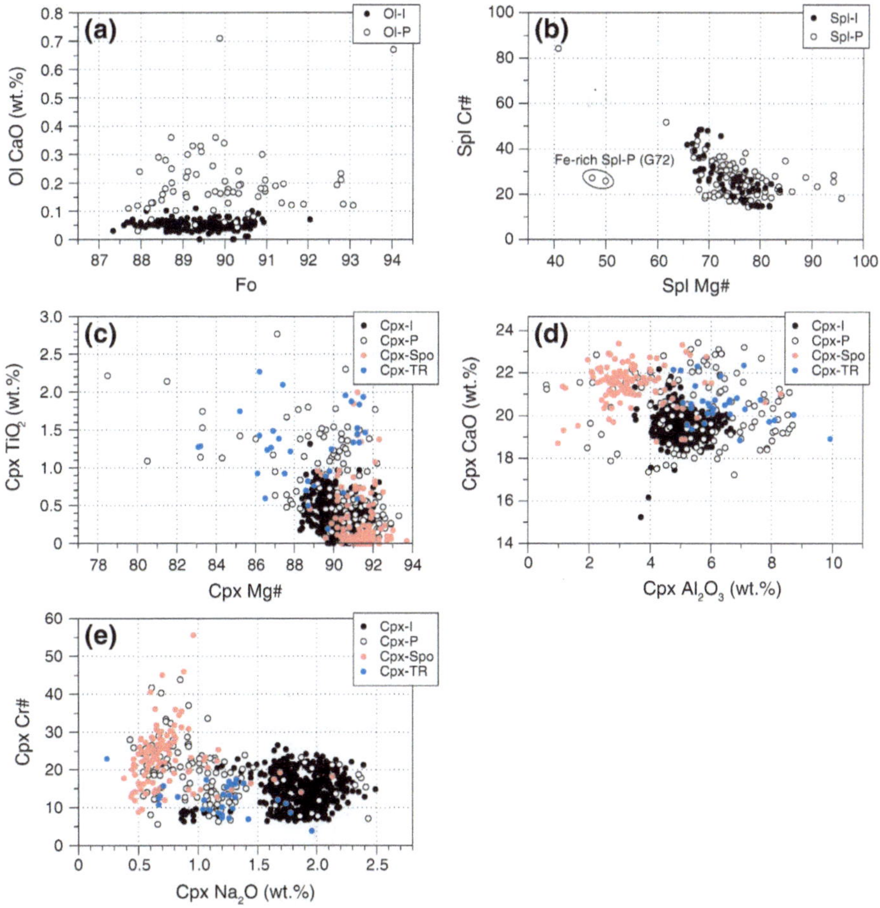

Fig. 7.3 Plots of (**a**) CaO versus Fo (mol%) for olivines, (**b**) Cr# versus Mg# for spinels, (**c**) TiO_2 versus Mg#, (**d**) CaO versus Al_2O_3, and (**e**) Cr# versus Na_2O for clinopyroxenes from Al Ghab peridotite xenoliths and melt pockets. Abbreviations of different textural types of minerals are given in Table 7.1 and described in text

of 0.8–1.3 wt% among all Cpx-I, suggesting that other samples may have been affected by a flux of Na prior to the development of Cpx-Spo, Cpx-TR and Cpx-P.

Amphiboles (Table 7.7) are titanian pargasites. Mg# and TiO_2 of the interstitial variety are in the ranges of 83.2–90.1 and 1.6–3.5 wt%, covering those of the metasomatic vein amphiboles from sample S52GI (85.1–87.6, 2.3–3.5 wt%, respectively). The interstitial amphiboles, often associated with Cr-spinels, are enriched in Cr_2O_3 (0.8–2.6 wt%), whereas the vein amphiboles are poorer in Cr_2O_3 (0.1–0.8 wt%).

Micas from S52D are altered; no representative EMP data can be obtained. The few analyses (Table 7.7) from G49 are titaniferous phlogopites with 4.6–5.1 wt% TiO_2 and 1.3 wt% Cr_2O_3 in the Mg# range of 86–88.

Feldspars from the melt-pockets are compositionally variable, ranging from andesine to Na-sanidine (Fig. 7.4; Table 7.7). The variability occurs both across and within individual melt-pockets of the same samples, and no spatial–compositional relationship is observed at the thin section scale.

Glasses (Table 7.7) are found only in melt pockets in sample G72 and as veinlets propagating from Cpx-Spo in G47 (Fig. 7.5). Both types are enriched in SiO_2 (52.8–61.5 wt%), Al_2O_3 (21.1–24.0 wt%), Na_2O (6.0–7.5 wt%) and K_2O (0.6–3.2 wt%), with high Na_2O/K_2O, typical of glasses from mantle xenoliths worldwide (Schiano and Clocchiatti 1994) and partly overlapping the field defined by mantle xenolith glasses from southern Syria (Ismail et al. 2008) (Fig. 7.5). The melt-pocket glasses from G72 differ from the veinlet type in G47 that they have lower SiO_2 (Fig. 7.5a), higher CaO (Fig. 7.5b) and TiO_2 (0.6–1.7 wt%; 0.3–0.5 wt% for G47). No Mg-rich glasses are found (0.9–2.8 wt% for G72 and 1.3–3.5 wt% for G47).

Fig. 7.4 Compositions of feldspars present in the melt-pockets of the Al Ghab peridotite xenoliths

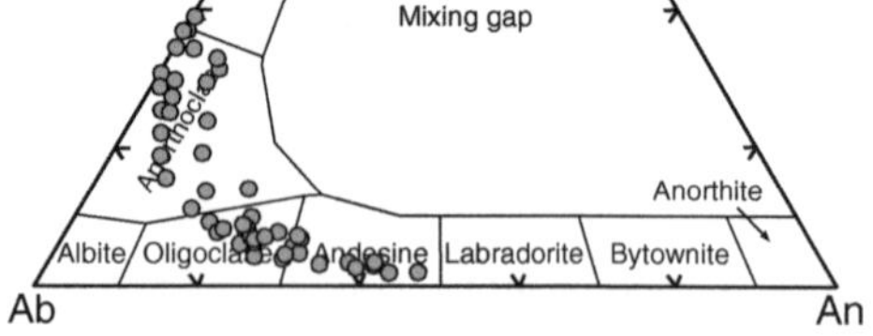

Fig. 7.5 (**a**) Total alkalis ($Na_2O + K_2O$) versus SiO_2 (TAS) diagram of Le Maitre (2002), and plots of (**b**) CaO versus SiO_2 and (**c**) Al_2O_3 versus Na_2O/K_2O for glasses from melt pockets (G72) and veinlets (G47) from the Al Ghab peridotite xenoliths (all in wt%). Also shown for comparison in (**a**) and (**b**) are the compositional ranges of glasses from Tel Tannoun peridotite xenoliths, Jabel El Arab, southern Syria (Ismail et al. 2008)

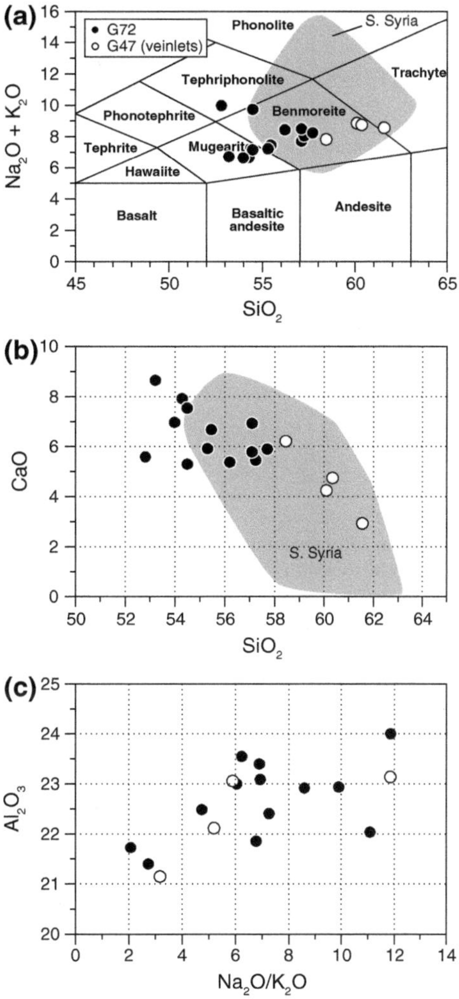

7.4.2 Major Element Zonation in the Primary Pyroxenes

Investigation of primary pyroxene zonation was conducted by electron microprobe traverses across, and mapping over, single grains. A focus was placed on the primary clinopyroxenes and their spongy rims (Cpx-Spo). Representative zonation patterns are highlighted in Figs. 7.6 and 7.7, and the representative core–rim values are given in Tables 7.5 and 7.6, and complete data of the traverses are given as online supplementary material.

Albeit subtle in some cases, most Opx-I is zoned in terms of Al, with a low-concentration plateau or trough in the core, gently increasing towards the rims

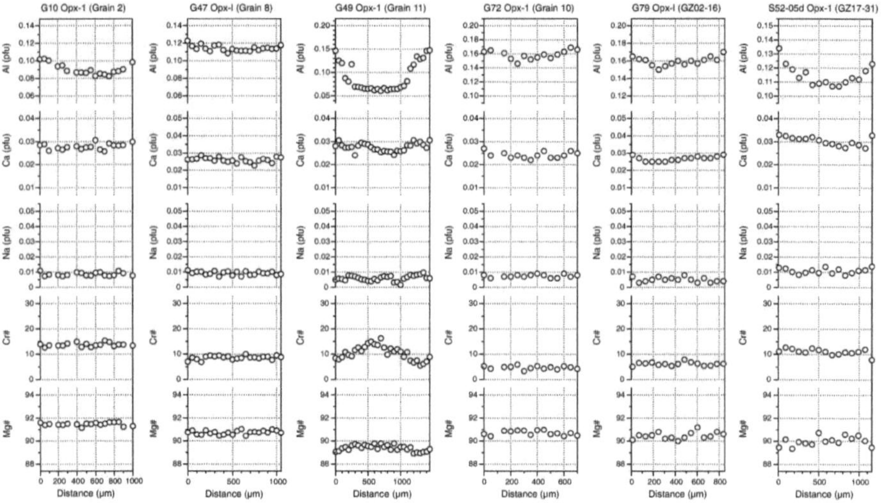

Fig. 7.6 Representative chemical zoning patterns of primary orthopyroxenes from Al Ghab peridotite xenoliths

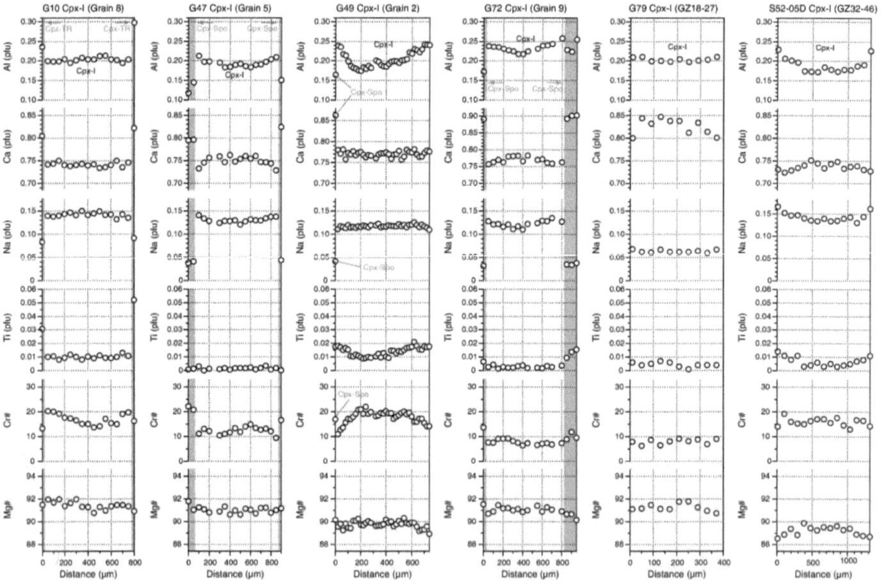

Fig. 7.7 Representative chemical zoning patterns of primary clinopyroxenes from Al Ghab peridotite xenoliths

(Fig. 7.6). Ca is less zoned, and when present, the zonation mimics that of Al. Other elements and Mg# are generally homogeneous.

Clinopyroxene zonation (Fig. 7.7) is usually more complex, revealing two (or three) stages of modification, with the later ones being formation of Cpx-Spo and Cpx-TR. Analogous to Opx-I, clear, non-spongy portions of Cpx-I show varying degrees of Al zonation, increasing from cores to rims. Ca patterns are either flat or slightly decreasing rimwards. Other elements and Mg# are essentially homogeneous. Subsequently, two other types of zonation overprinted this type of Al-dominated zonation, and they must have occurred very recently judging from the sharp, steplike zonation in the rims. One of such late modifications (Cpx-Spo) is accompanied by the development of a spongy texture, where the modified rims become Al- and Na-poor, and Ca- (e.g. Figure 7.7) and Cr-rich. On occasion, accompanying these are slight increases in Ti and Mg#. The other type of modification (Cpx-TR) also features increasing Ca and decreasing Na in rims, but Al and Ti are much higher, sometimes with Mg# decreasing rimwards (Fig. 7.7).

7.4.3 Trace Elements in the Peridotite and Melt-Pocket Phases

Laser ablation ICP-MS analyses have been made of clinopyroxenes in different textural positions (Cpx-I, Cpx-Spo and Cpx-P) and in a few cases of amphiboles, feldspars and glasses. For laser ablation, meticulous care has been taken to choose surfaces of Cpx-Spo, Cpx-P and Flsp that are apparently inclusion free. Despite such effort, chemical heterogeneity is observed on occasion during laser drilling. In most case, the heterogeneities, most severely in Cpx-Spo, can be picked up by a sharp increase in Ba counts, presumably reflecting sampling of Ba-rich fluid/melt inclusions in the minerals by the laser beam. In all such instances, only the signal intervals prior to the heterogeneous signals are used and regarded as representative of the minerals themselves.

Results of the LA-ICPMS analysis are given in Tables 7.8 and 7.9 and in online supplementary files and shown in chondrite-normalised REE and primitive mantle-normalised multi-element plots in Figs. 7.8 and 7.9. Normalisation values are those of Sun and McDonough (1989) and McDonough and Sun (1995), respectively. In all clinopyroxenes, amphiboles, feldspars and glasses, regardless of their textural positions, LREE is always enriched relative to HREE, except Cpx-I from sample G79. Details of the trace element characteristics are described below.

7.4.3.1 Clinopyroxenes

On the chondrite-normalised plots, most analyses of Cpx-I are characterised by a smooth HREE to LREE increasing concave downward pattern that peaks at Ce–Pr or flattens out at La–Ce, whereas their spongy rims (Cpx-Spo) and Cpx-P are

Table 7.8 Representative trace element compositions (ppm) of clinopyroxene in Al Ghab peridotite xenoliths and melt pockets, NW Syria, obtained by LA-ICPMS

Sample	G47	G47	G47	G49	G49	G49	G49	G72	G72	G72	G79	S52-05A	S52-05A	S52-05A
Analysis#	G47_GZ86	G47_GZ54	G47_GZ15	G49_GZ02	G49_GZ30	G49_GZ21	G49_GZ22	G72_GZ60	G72_GZ28	G72_GZ45	G79_GZ21-22-23	S52a_GZ64	S52a_GZ65	S52a_GZ77
Textural Class	Cpx-I	Cpx-Spo	Cpx-P	Cpx-TR	Cpx-I	Cpx-P	Cpx-P	Cpx-I	Cpx-Spo	Cpx-P	Cpx-I	Cpx-I	Cpx-Spo	Cpx-P
CaO (wt%)	19.5	21.7	22.9	20.6	19	19.5	23.1	20	23.4	21.1	21.6	19.6	22.2	22.8
Sc	71.6	79.3	116	53.7	51.9	54	91	73.3	87.1	125	69.4	77.6	83.3	125
Ti	349	391	2872	6654	4760	5665	22421	579	953	4376	1864	414	1181	2080
V	224	235	489	260	177	239	349	204	274	500	277	241	303	361
Rb	0.07	<0.861	0.243	<0.0813	<0.0678	0.258	0.22	<0.109	0.516	<0.153	<0.0282	<0.177	<0.307	<0.140
Sr	415	160	148	162	168	169	221	417	527	184	9.54	425	151	139
Y	15.7	18.1	28.6	18.5	17.5	16	21.6	23.2	30.2	40.7	11.7	19.4	22.4	26.5
Zr	5.24	3.04	5.65	183	226	171	140	94.2	91.5	51.4	2.27	6.65	11.7	6.15
Nb	0.217	0.0713	1.03	0.349	0.241	0.907	6.24	0.855	7.07	2.95	0.0725	0.162	0.318	0.373
Cs	<0.0195	<0.0463	<0.0303	<0.0271	<0.0297	0.127	<0.554	<0.0541	<0.115	<0.0568	<0.0114	1.46	<0.129	<0.0639
Ba	<0.154	<0.294	3.09	<0.102	<0.119	3.76	3.94	<0.227	46.6	<0.337	<0.0675	0.974	1.22	1.42
La	20.5	10.4	8.13	6.7	5.89	6.49	9.56	17	22.7	7.78	0.487	16.2	6.72	4.28
Ce	50.4	35.1	31.3	21.5	19	20.8	31.6	49.4	63.9	31.8	1.15	51.4	30.7	21.4
Pr	6.21	5.42	5.61	3.51	3.19	3.43	5.36	6.95	8.78	5.98	0.158	7.29	5.9	4.51
Nd	25.4	24	30.7	18.7	18	17.8	29.3	31.3	40.6	35.3	0.958	32.9	33	27
Sm	4.92	5.28	7.58	5.43	5.23	5.36	7.84	7.41	8.84	10.5	0.608	6.59	7.85	8.33
Eu	1.52	1.54	2.7	1.95	1.81	1.59	2.71	2.58	2.92	3.9	0.266	2.21	2.63	2.82
Gd	3.83	4.4	7.3	5.6	5.47	4.61	7.69	6.19	7.81	11.5	1.2	5.9	6.79	8.3
Tb	0.534	0.596	1.06	0.753	0.764	0.681	1.01	0.82	1.13	1.56	0.25	0.676	0.928	1.14
Dy	3.12	4.11	6.24	4.6	4.37	3.83	5.62	4.57	6.57	9.1	1.95	4.21	5.16	5.98
Ho	0.611	0.641	0.954	0.747	0.726	0.691	0.903	0.927	1.16	1.68	0.447	0.731	0.892	1.02
Er	1.53	1.74	2.63	1.84	1.64	1.58	1.81	2.16	2.92	4.08	1.34	1.87	2.06	2.71

(continued)

Table 7.8 (continued)

Sample	G47	G47	G47	G49	G49	G49	G49	G72	G72	G72	G79	S52-05A	S52-05A	S52-05A
Tm	0.211	0.247	0.319	0.221	0.2	0.189	0.249	0.316	0.403	0.465	0.195	0.265	0.28	0.315
Yb	1.34	1.65	1.83	1.33	1.14	1.29	1.19	1.88	2.57	3.09	1.37	1.71	1.66	1.71
Lu	0.193	0.229	0.285	0.165	0.16	0.148	0.19	0.242	0.359	0.371	0.189	0.207	0.254	0.247
Hf	0.163	0.153	0.156	5.56	7.69	4.35	5.32	0.399	0.295	0.249	0.26	0.198	0.371	0.182
Ta	<0.0116	<0.0216	0.0274	0.0951	0.115	0.124	0.455	0.0725	0.124	0.215	0.00854	<0.0193	<0.0394	<0.0158
Pb	0.727	<0.150	0.288	0.212	0.159	0.172	0.689	0.634	1.11	0.176	0.239	0.345	<0.146	<0.0813
Th	2.51	0.254	0.127	0.375	0.278	0.339	0.206	0.833	1.33	0.0945	0.0326	0.562	0.0442	0.02
U	0.767	0.0352	0.0331	0.139	0.0914	0.104	0.0978	0.243	0.37	0.025	<2.60	0.182	0.0127	0.00941
$(La/Pr)_N$	1.29	0.75	0.57	0.75	0.72	0.74	0.70	0.96	1.01	0.51	1.21	0.87	0.45	0.37
$(Sm/Yb)_N$	3.99	3.48	4.50	4.43	4.98	4.51	7.16	4.28	3.74	3.69	0.48	4.19	5.14	5.29
$(La/Yb)_N$	10.41	4.29	3.02	3.43	3.52	3.42	5.47	6.15	6.01	1.71	0.24	6.45	2.76	1.70
Ti/Eu	230	254	1064	3413	2630	3563	8274	224	326	1122	7009	187	449	738

Table 7.9 Representative trace element compositions (ppm) of feldspar, glass and amphibole in Al Ghab peridotite xenoliths and melt pockets, NW Syria, obtained by LA-ICPMS

Sample	G47	G64	S52-05A	G72	G72	G72	G64	G10	G49	S52-05GI	S52-05GI	S52-05GI
Analysis#	G47_GZ19	G64_GZ38	S52a_GZ82	G72_GZ26	G72_GZ27	G72_GZ43	G64_GZ27	G10_GZ51	G49_GZ13-14	S52gI_GZ71	S52gI_GZ36	S52gI_GZ37
Mineral	Flsp	Flsp	Flsp	Glass	Glass	Glass	Amph (interstitial)	Amph (interstitial)	Amph (interstitial)	Amph (interstitial)	Amph (vein)	Amph (vein)
CaO	4.64	4.73	9.58	5.58	5.29	6.93	10.9	10.4	10.9	10.6	10.7	10.7
Ti	1667	5773	1145	10251	11211	4616	17565	18884	23201	15707	30754	29376
Sc	4.92	<5.58	1.59	1.72	1.74	15.5	69.2	52.0	39.1	40.6	27.5	32.0
V	114	50.8	19.5	56.4	60.3	204	575	343	344	340	300	317
Rb	2.1	3.31	2.29	14.1	19	3.71	2.09	5.86	16.7	8.58	7.32	8.07
Sr	1700	1370	2790	4600	2840	2030	918	561	471	501	667	641
Y	6.74	6.4	4.14	15.3	17.4	25.8	23.8	18.1	18.3	14.4	15.6	16.5
Zr	8.84	156	4.77	206	223	129	123	76.2	166	106	113	115
Nb	32.1	77.4	12.9	128	139	159	93.0	52.2	22.7	48.1	45.7	46.3
Cs	<0.0481	<0.618	<0.123	0.19	0.241	<0.183	<0.489	<0.0398	<0.0346	<0.141	<0.0754	<0.161
Ba	758	461	513	2400	1110	1180	231	145	264	169	194	193
La	33.9	26.4	28.5	66.5	74	59.2	12.0	13.5	8.75	9.30	13.6	13.0
Ce	63.6	62.6	56	120	133	126	45.1	38.2	26.2	30.0	40.2	39.4
Pr	6.18	8.27	5.49	11.8	12.9	14.6	6.80	5.61	3.91	4.51	5.61	5.96
Nd	21	24.2	19.2	39.7	43.2	52.8	35.5	27.2	20.6	21.2	28.0	28.2
Sm	2.71	5.15	2.09	6.28	6.84	9.8	9.51	6.48	5.62	5.24	6.26	5.87
Eu	1.18	2.06	3.69	2.3	2.2	3.02	3.63	2.18	1.96	1.72	2.35	2.34
Gd	1.73	3.67	1.59	4.83	5.01	6.23	6.12	5.80	5.45	5.15	5.81	5.44
Tb	0.213	<0.309	0.15	0.513	0.686	0.747	1.29	0.78	0.76	0.59	0.80	0.76
Dy	1.19	1.88	0.594	3.12	3.8	4.97	4.56	4.40	4.35	3.32	3.90	3.70
Ho	0.217	0.171	0.113	0.585	0.655	0.905	0.75	0.73	0.74	0.52	0.65	0.61
Er	0.609	<1.33	0.343	1.49	1.66	2.5	2.22	1.70	1.79	1.17	1.50	1.48
Tm	0.0957	<0.407	0.0927	0.155	0.248	0.419	<0.409	0.21	0.20	0.13	0.17	0.15

(continued)

Table 7.9 (continued)

Sample	G47	G64	S52-05A	G72	G72	G72	G64	G10	G49	S52-05GI	S52-05GI	S52-05GI
Yb	0.535	<0.875	0.391	1.03	1.31	2.6	1.14	1.22	1.22	0.96	1.01	1.04
Lu	0.0904	<0.231	0.0432	0.161	0.159	0.369	0.29	0.14	0.15	0.12	0.13	0.14
Hf	0.152	3.74	0.152	2.93	3.41	0.648	1.81	1.82	3.77	2.54	3.30	3.29
Ta	0.167	5.66	0.0758	6.12	6.63	2.89	5.13	2.56	1.67	3.59	2.79	2.70
Pb	2.82	12.7	0.667	8.84	7.85	3.73	3.89	0.75	0.61	0.37	0.29	<0.292
Th	2.77	0.59	0.364	7.55	8.47	3.75	0.13	0.36	0.32	0.22	0.29	0.21
U	0.87	0.163	0.11	2.62	2.88	1.26	0.06	1.06	0.10	0.08	0.07	0.07
$(La/Pr)_N$	2.15	1.25	2.03	2.21	2.25	1.59	0.69	0.94	0.88	0.81	0.95	0.85
$(Sm/Yb)_N$	5.50	–	5.81	6.62	5.67	4.09	9.06	5.77	5.00	5.91	6.73	6.13
$(La/Yb)_N$	43.12	–	49.61	43.94	38.44	15.50	7.16	7.53	4.88	6.57	9.16	8.51
Ti/Eu	1412	2803	310	4457	5096	1529	4839	8663	11837	9132	13087	12554

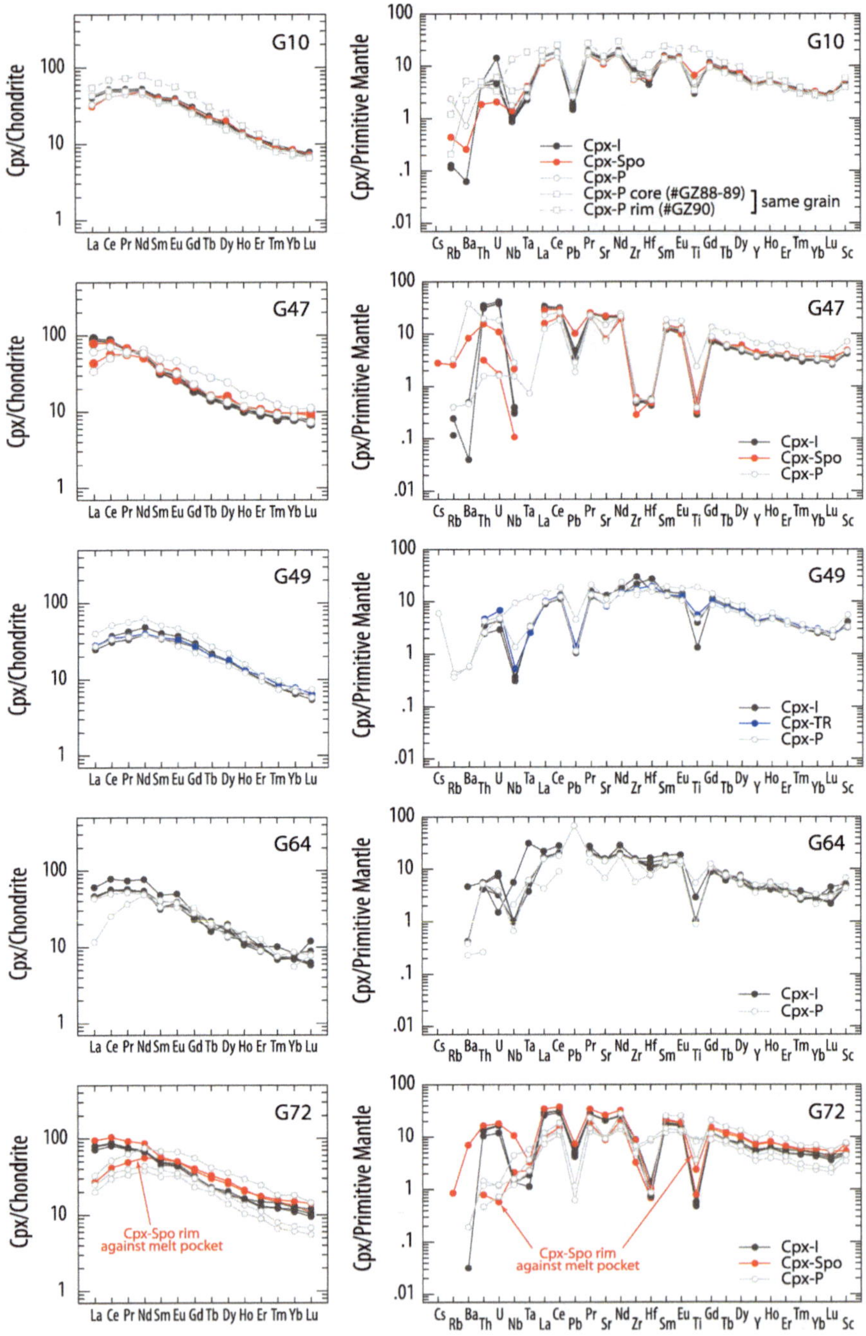

Fig. 7.8 Chondrite and primitive mantle normalised rare earth element (REE) and multi-trace element diagrams for clinopyroxenes from Al Ghab peridotite xenoliths and melt-pockets. Normalising values are from Sun and McDonough (1989) and McDonough and Sun (1995), respectively

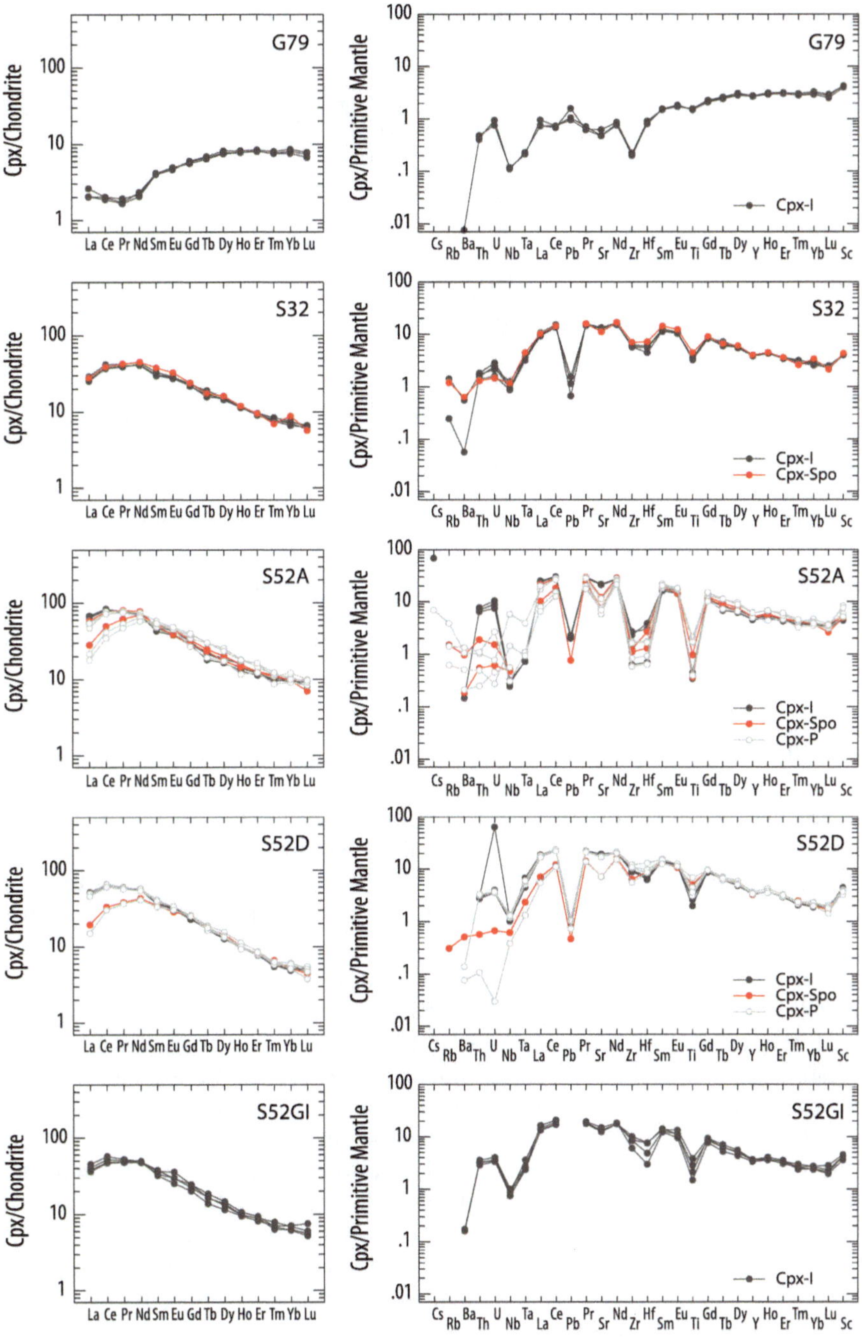

Fig. 7.8 (continued)

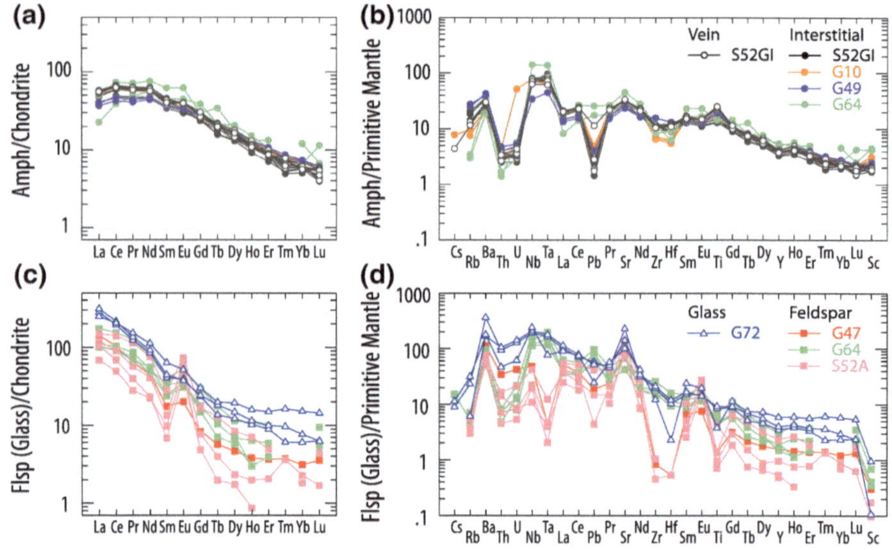

Fig. 7.9 Chondrite and primitive mantle normalised rare earth element (REE) and multi-trace element diagrams for (**a**) amphiboles, (**b**) glasses and feldspars from Al Ghab peridotite xenoliths and melt-pockets. Normalising values are from Sun and McDonough (1989) and McDonough and Sun (1995), respectively

generally characterised by a stronger concave downward pattern with a maximum at Pr–Nd or Nd–Sm (Fig. 7.8). Few Cpx-Spo and Cpx-P exhibit the high-La–Ce pattern that is typical of Cpx-I. On a primitive mantle-normalised basis, Cpx-I shows pronounced positive Th–U anomalies and strong relative depletions for LILE (Rb, Ba and Pb) and HFSE (Nb, Zr–Hf, Ti), although often times Rb and Ba (Cs always) are below the detection limits. Where data are available, Ba is always depleted relative to Rb.

With the exception of G72 where the REE abundances of Cpx-P are highly variable, most Cpx-P shows similar HREE features to Cpx-I, and their HFSE negative anomalies are in general less pronounced or sometimes absent (for Nb). Negative Sr anomalies are more pronounced, and La/Yb, La/Sm and La/Ce are lower in Cpx-P. The most significant distinction lies at the concentrations of LILE and Th–U—the characteristic positive Th–U anomaly in Cpx-I becomes less pronounced or even absent in Cpx-P which is also accompanied by enrichment of LILE despite Ba/Rb ratios being highly variable. Most Cpx-Spo shares a considerable trace element similarity with Cpx-P, and a few analyses intermediate between that of Cpx-P and Cpx-I, indicating some intimate genetic relationship among Cpx-I, Cpx-Spo and Cpx-P. This is justified, for instance by the Cpx-Spo analyses of G72, where the Cpx-Spo in contact with melt-pocket glasses has a trace element characteristic similar to Cpx-P and that away from any melt pocket has a trace element characteristic more akin to Cpx-I (Fig. 7.8).

Whereas the preceding descriptions pertain to most clinopyroxenes, there is an exception. Cpx-I from G79 is LREE-depleted. The HREE are almost constant on the chondrite-normalised diagram, decreasing at MREE and flattening out at LREE, or showing a small trough at Pr. All of the LILE and HFSE are depleted relative to HREE, and the positive Th–U and negative Nb anomalies are similar to other LREE-enriched Cpx-I. Zr is depleted with respect to Nd, but the depletion is not accompanied by depletion of Hf as seen in the LREE-enriched Cpx-I.

7.4.3.2 Amphiboles

Amphiboles display typical concave downward patterns with maxima at (Pr)–Nd–Sm on the chondrite-normalised REE diagrams (Fig. 7.9a), similar to those of Cpx-P. They exhibit extreme enrichments in LILE, with primitive mantle normalised spider patterns (Fig. 7.9b) almost reciprocal to those of Cpx-I (cf. Figure 7.8). They show strong positive Ba and Nb–Ta and weak positive Sr and Ti anomalies, and strong negative Th–(U) and Pb and weak Zr–Hf (except G49) negative anomalies, reflecting typical amphibole–melt partition coefficients (LaTourrette et al. 1995; Tiepolo et al. 2007; Pilet et al. 2011). Amphiboles from sample G64 are exceptional in that they show no salient Pb and Ti anomalies.

7.4.3.3 Feldspars

Feldspars are enriched in LREE relative to HREE, with variable positive Eu anomalies on chondrite-normalised REE diagrams (Fig. 7.9c). On primitive mantle-normalised multi-element diagrams (Fig. 7.9d), they show enrichments in LILE relative to HREE and most HFSE except Nb and exhibit positive Ba, Sr and Sm–Eu anomalies and negative Rb, Th–U, Zr–Hf and Ti anomalies. Zr–Hf anomalies are variable, being strongly negative in samples G47 and S52A, but absent in G64. Compared to the clinopyroxenes, all of the feldspars have higher concentrations of LILE, Th, U, Nb, La and Sr, and lower MREE and HREE.

7.4.3.4 Glasses

Glasses occur only in sample G72. The three analyses have a straight, steep positive slope from LREE to HREE on the chondrite-normalised REE diagrams (Fig. 7.9c). Their chondrite-normalised $(La/Yb)_N$ is 16.3–46.5, much higher than in the clinopyroxenes from the same sample (2.1–6.7). Their primitive mantle-normalised spider patterns are concave downward with a maximum at Nb or Ba (Fig. 7.9d). Strontium is enriched, while Th–U and Cs–Rb are depleted relative to their neighbouring elements, but note that all of the LILE and HFSE are enriched relative to the primitive mantle and the coexisting clinopyroxenes.

7.5 Metasomatic Evolution of the Lithospheric Mantle Beneath Al Ghab

7.5.1 Metasomatism in the Peridotites

It is generally understood that continental lithospheric mantle undergoes multiple episodes of depletion and refertilisation (metasomatic) events, which continuously overprint former events (Menzies 1983; Downes 2001; Dawson 2002; Ackerman et al. 2007; Griffin et al. 2009; O'Reilly and Griffin 2013; Tang et al. 2013). Of particular interest is the latter, mantle metasomatism, that has dominated the recent histories of many intraplate xenolith suites, leaving little trace of former partial melting records (Kaeser et al. 2007; Su et al. 2010b). Since the pioneer work of Dawson (1980, 1984) and others (Harte 1983; Menzies, 1983), mantle metasomatism has been considered in two categories: modal—one that introduces new phases, usually amphiboles and micas, into the peridotites—and cryptic—one that affects only the chemistry of the primary minerals. The metasomatic agents that react with the peridotites and bring into these changes may range from those typical of low-degree silicate partial melts to those typical of carbonate-rich (or even carbonatite) melts (Yaxley et al. 1998; Coltorti et al. 1999, 2000; Downes 2001; Powell et al. 2004; Powell and O'Reilly 2007), although some recent studies have tried to identify a genetic link between the two agents (Ionov et al. 2002a, 2002b; Bodinier et al. 2004).

Amphiboles, micas and apatites are generally considered as typical metasomatic minerals (Menzies 1983; Dawson 1984) and are prone to break down as a result of *P–T* or even chemical modifications (Stosch and Seck 1980; Mengel and Green 1989; Kaeser et al. 2007). The occurrence of these minerals in interstitial positions in some of the Al Ghab peridotite xenoliths suggests relatively recent metasomatism that brought in substantial volatiles and trace elements. The LREE enrichment Cpx-I in many of xenoliths, however, must have occurred earlier than this amphibole-forming metasomatism, judging from petrographic observations that amphiboles replace LREE-enriched Cpx-I.

The Al Ghab LREE-enriched Cpx-I exhibits positive anomalies of Ba, Th, U and Sr and negative anomalies of Ti, Zr and Hf on the primitive mantle-normalised multi-element diagrams, features that have been ascribed to carbonate metasomatism for many xenolith suites worldwide (Yaxley et al. 1998; Coltorti et al. 1999; Powell et al. 2004; Powell and O'Reilly 2007; Su et al. 2010b). The contrasting styles of silicate and carbonate metasomatism are perhaps best distinguished by a plot of $(La/Yb)_N$ versus Ti/Eu or similar analogues developed by Rudnick et al. (1993), Yaxley et al. (1998) and Coltorti et al. (1999), on the basis of the concept that carbonatites are LREE–MREE-rich and Ti-poor, with Ti/Eu much lower than the primitive mantle value of ~7000–8000 (McDonough and Sun 1995; Palme and O'Neill 2003). The Cpx-I of the Al Ghab xenoliths shows affinities to carbonate metasomatism as already proposed, and the only LREE-depleted sample G79 exhibits primitive mantle-like $(La/Yb)_N$ and Ti/Eu (Fig. 7.10).

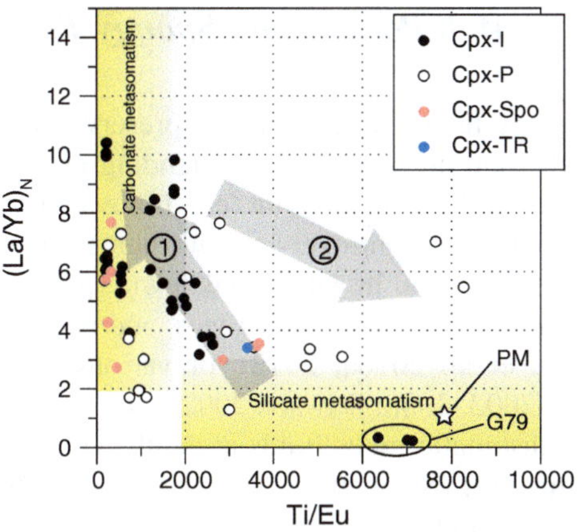

Fig. 7.10 Plot of $(La/Yb)_N$ versus Ti/Eu for clinopyroxenes from Al Ghab peridotite xenoliths and melt pockets. The shaded fields of carbonate and silicate metasomatism are modified after Coltorti et al. (1999). Numbered arrows indicate inferred metasomatic evolution of the Al Ghab lithospheric mantle recorded in the clinopyroxenes, i.e. carbonate metasomatism followed by silicate metasomatism. Note that the apparent clustering of data points at lower Ti/Eu may partly reflect a sampling bias that most of LA-ICPMS analysis was made in the cores of Cpx-I, while the silicate metasomatism, as marked by Ti enrichment (see main text), is more prominently observed in the rims of the clinopyroxenes (Cpx-TR) and Cpx-P

The carbonate metasomatism that affected the Al Ghab lithospheric mantle behaves very much more in a cryptic manner than modal, presumably as a result of reaction under a very low melt–rock ratio (Bodinier et al. 1990; Rudnick et al. 1993) which is permitted by the low viscosity of CO_2-laden fluids. Despite the presence of accessory apatite that may have formed during the carbonate metasomatism, no discrete carbonates are observed in the Al Ghab xenoliths. Most, if not all, of the amphiboles as well as the closely associated micas appear to have formed in a later, distinct metasomatic event, judging from petrographic observations and the chemistry of the amphiboles. The key arguments are as follows: (1) the Mg# of the amphiboles (interstitial: 83–90; vein: 85–88) ranges from those typical of the coexisting Cpx-I (~90) to notably lower than the Cpx-I which are in the majority; (2) petrographic evidence and major element zonation of Cpx-I in some samples suggest successive modification of Cpx-I rims to high-Ti, high-Al and low-Mg# varieties and subsequently replacement of such rims by amphiboles; (3) the amphiboles and micas are Ti-rich, contrary to the low-Ti nature of carbonate metasomatism. These features indicate that the amphibole-forming metasomatism post-dated the earlier carbonate metasomatism that was responsible for the trace element enrichment in the Cpx-I.

It is important to note that similar amphiboles and high-Ti–Al Cpx rims also occur in the melt pockets. In the following sections, we will examine the origin of the melt pockets and their relationships with the second stage, amphibole-forming metasomatism.

7.5.2 Origin of the Melt Pockets and Spongy Textures

Melt pockets and spongy textured minerals are preserved reacting materials from the lithospheric mantle and are keys to understanding processes (i.e. melting and metasomatic) operating in the upper mantle (e.g. Ionov et al. 1994; Dawson 2002). The origin of melt pockets is controversial with numerous discussions revolving around whether they represent in situ breakdown (partial melting) of mantle minerals, typically amphiboles, clinopyroxenes and/or orthopyroxene (Francis, 1976; Shaw 2009; Su et al. 2010a, 2012), or various kinds of melt–peridotite interaction (Bali et al. 2002, 2008; Dawson 2002; Su et al. 2010a, 2012). The latter concerns whether metasomatic melts/fluids at depths (Ionov et al. 1994; Su et al. 2010a) or the host magmas (Shaw et al. 2006; Shaw and Dingwell 2008) are the agents inciting the dissolution.

Likewise, the origin of the spongy textures in clinopyroxene, spinel and occasionally orthopyroxene is far from well understood. Diverse opinions have been held on the textures, attributing them to partial dissolution in response to heating,decompression, mantle metasomatism or transport-related metasomatism (Ionov et al. 1994; Carpenter et al. 2002; Bonadiman et al. 2005; Shaw et al. 2006; Perinelli et al. 2008; Shaw and Dingwell 2008; Su et al. 2011, 2012). In spite of the similar mechanisms proposed, melt-pockets and spongy textures are not often considered together and few authors (e.g. Ionov et al. 1994; Su et al. 2012) have addressed the possible genetic link between the two. Ionov et al. (1994) proposed that the melt-pockets in the peridotite xenoliths from Mongolia formed after dissolution of clinopyroxene and spinel, whereas Su et al. (2012) proposed that those from western Qinling, central China formed after orthopyroxene. The melt-pockets in our Al Ghab xenolith suite are very similar in terms of the texture and mineral assemblage to the melt pockets in these cases. Major differences from the western Qinling case, however, are in the spongy minerals; orthopyroxene appears to be a stable phase in the Syrian xenoliths. Instead, textural evidence suggests that the clinopyroxene and spinel are breaking down to form the melt pockets, similar to the Mongolian samples (Ionov et al. 1994). What remains unclear yet is what triggered dissolution or partial melting of the clinopyroxene and spinel, and then formation of the melt pockets, and how they are related to the amphibole-forming metasomatism as discussed in the preceding sections. We now explore the petrological and geochemical aspects of the melt pockets and spongy minerals.

7.5.2.1 Infiltration of Host Basalts?

Several experimental studies have in early years attempted to quantify the rate at which magma can penetrate into a melt-free rock (Watson 1982; Riley and Kohlstedt 1991; Daines and Kohlstedt 1993), and the results in general suggest infiltration of ~10 mm in ~100 days within which most xenoliths are transported to the surface (e.g. Demouchy et al. 2006). Although the effect of decompression (during transport in the magma) on the infiltration rate is not well understood, the short penetration distance compared to the size of the Al Ghab mantle xenoliths (fist size) and the fact that the melt pockets and spongy minerals do not concentrate at the rims of the xenoliths strongly argues against a host magma infiltration model for the Al Ghab xenoliths. Furthermore, the clinopyroxenes from the Al Ghab host basalts have Cr# <80 and Mg# <80, significantly different from the melt-pocket clinopyroxenes in the peridotite xenoliths.

7.5.2.2 Decompression-Induced Incongruent Melting of Primary Clinopyroxenes

The occurrence of newly crystallised Ol-P that is more forsteritic than Ol-I and which is intimately associated with precipitation of Spl-P is significant. Similar associations have been reported in melt-pockets in peridotite xenoliths from Mongolia (Ionov et al. 1994) and northern Tanzania (Dawson 2002). In these reports, some agreement has been reached on the generation of high-Fo olivines (and porphyritic clinopyroxenes) in the melt pockets through fractional crystallisation of melts derived from incongruent partial melting of primary clinopyroxene and spinel to produce residual chromite, phenocrysts of olivine and clinopyroxene of a new composition, and melts of variable composition that are now quenched as glass (Ionov et al. 1994; Dawson 2002). In the Tanzanian case, chromite occurs as a liquidus phase followed by olivine and clinopyroxene in the melt pockets (Dawson 2002), a crystallisation sequence also true for the melt pockets in the Syrian xenoliths except that feldspar is present in the Syrian melt pockets and glass not very common. In the Syrian case, the Spl-P occurs as small, euhedral grains enclosed by later crystallising Ol-P and Flsp, but less commonly by Cpx-P, suggesting that at least some of the clinopyroxenes are present before crystallisation of Spl-P. These "early" clinopyroxenes are presumably relics of dissolving primary clinopyroxenes, and this is supported by the zoned textures of the Cpx-P which in places shows cores typical of Cpx-I compositions. Nevertheless, the textural evidence lends strong support for a partial dissolution followed by crystallisation origin of the melt pockets, and the high CaO content of Ol-P is clearly a feature of magmatic crystallisation (Thompson and Gibson 2000). A possible explanation for the crystallisation of high-Fo Ol-P is that these olivines formed at times of transient Fe depletion as a result of preferential Fe partitioning into liquidus Cr-spinel (Spl-P) (Dawson 2002), which may indeed be a reflection of crystallisation under oxidising conditions.

Assuming a single process that produced the Cpx-Spo and Spl-Spo and subsequently the melt pockets in the Al Ghab xenoliths, the presence of Flsp in the melt pockets as well as in the vugs of Cpx-Spo, but not the host peridotites, strongly argues for a shallower origin ($\leq$30 km deep) for this process. Cpx-Spo in this study is characterised by lower Na and Al, higher Ca and Cr# than the Cpx-I of the same samples, and Cpx-P also shares similar characteristics, but its Al usually extends to higher values. Increases in Cr# and decreases in Al in clinopyroxenes and spinels have frequently been shown in experimental and natural samples to be indicators of partial melting, and thus, the Cpx-Spo can be considered as partially melted, residual portions of primary clinopyroxenes.

Chemical evidence for decompressional melting lies in the Na contents of the clinopyroxenes, given that jadeite component in clinopyroxenes is a function of pressure. In Thetford Hill, Vermont, sodian augite phenocrysts in a limburgite are transformed to non-sodian augites, in response to a drastic reduction of geostatic pressures during intrusion (Hibbard and Sjoberg 1994). Likewise, the drop of Na in Cpx-P and Cpx-Spo in this study and others most likely reflects the low $D_{Na}^{Cpx/melt}$ at low pressures where the exchange of Na for Ca, Fe and Mg in the M2 site of the pyroxene structure is favoured.

The relative proportions of Al in the tetrahedral (Al^{IV}) and octahedral (Al^{VI}) sites of clinopyroxenes may also provide some useful pressure information. The Al^{IV}/Al^{VI} ratio in clinopyroxenes is known to reflect in part the equilibrium pressures, showing an inverse relationship with pressures and Si (Aoki and Kushiro 1968; Wass 1979). Both Cpx-Spo and Cpx-P in the Al Ghab xenoliths exhibit higher Al^{IV}/Al^{VI} and lower Na than Cpx-I (Fig. 7.11; Table 7.6), testifying to their lower pressure origin. Note that individual Cpx-P with identical crystallographic orientation suggests a replacement origin for Cpx-P, rather than crystallisation from a melt in an open space. The differences in the Al abundances of Cpx-P and Cpx-Spo are also consistent with this origin. Whereas the low Al in Cpx-Spo indicates residual portions after partial melting, the variable and generally high Al in Cpx-P may reflect a more advanced state of clinopyroxene breakdown in which the recrystallised clinopyroxenes (now Cpx-P) or overgrowths on relic Cpx-I have been variably re-equilibrated with the Al-rich partial melt thus derived. This is evidenced in some Cpx-P grains in which Al-rich rims are observed and in the only Cpx-P rim trace element data (Analyses# GZ88–89 and GZ90 of G10; Fig. 7.8; Table 7.6) which shows substantially higher LREE than in the core. Note that this rim has somewhat lower Mg# (83.3) than the core (90.6), indicating some degree of fractional crystallisation within the melt pocket.

7.5.2.3 Formation of the Melt Pockets in an Open System

The melt pockets examined in the present study comprise microphenocrysts of Ol-P, Cpx-P, Spl-P and interstitial Flsp, with variable, usually absent or minor amounts, of amphiboles and glasses. The dominance of Flsp over glasses suggests that the melt pockets formed through prolonged durations at depth, rather than quenching at the surface. However, the lack of textural and chemical equilibrium of the melt pockets with the surrounding primary minerals provides strong evidence that the melt pockets must be short-lived and likely formed shortly before the eruption. The ubiquitous presence of empty vugs in the melt pockets also indicates that volatiles were a significant component in the crystallizing melt in the pockets.

The significance of volatiles in the melt pockets is difficult to reconcile with a simple incongruent melting of Cpx-I and Spl-I which are both nominally

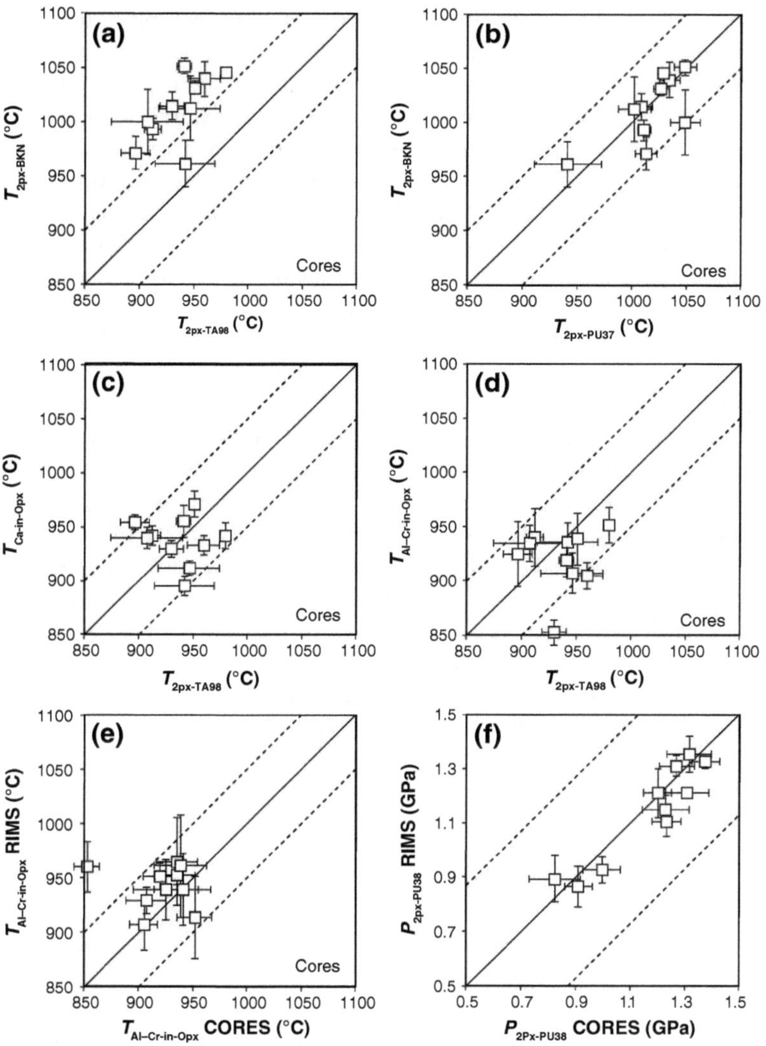

Fig. 7.11 Comparisons of results obtained by various thermometers and barometers listed in Table 7.11. Error bars show standard deviations of the averages. Typical precisions of ±50 °C and ±0.37 GPa for the thermometers and Putirka's (2008) two-pyroxene barometer are shown as dashed lines. $T_{2Px\text{-}BKN}$, two-pyroxene thermometer of Brey and Köhler (1990); $T_{2Px\text{-}TA98}$, two-pyroxene thermometer of Taylor (1998); $T_{2Px\text{-}PU37}$, two-pyroxene thermometer of Putirka's (2008) Eq. 37; $T_{Ca\text{-}in\text{-}Opx}$, Ca-in-orthopyroxene thermometer of Brey and Köhler (1990) modified by Nimis and Grütter (2010); $T_{Al\text{–}Cr\text{-}in\text{-}Opx}$, Cr- and Al-in-orthopyroxene thermometer of Witt-Eickschen and Seck (1991); $P_{2Px\text{-}PU38}$, two-pyroxene barometer of Putirka's (2008) Eq. 38

anhydrous. This suggests that either the partial melting took place not in a closed system or that an additional hydrous phase was involved in the incongruent melting. The observations that the Cpx-P and Ol-P often show embayed boundaries and Spl-P shows "atolls" of overgrowths lend good textural support to the former. The compositions of the melt-pocket feldspars also reveal signs of open-system crystallisation. The wide compositional range from An_{46} to Or_{50} and high Or contents of some of the feldspars demand extensive fractional crystallisation within the melt pockets, which is unrealistic given the small size of the melt pockets and the K-poor nature of the primary phases in the peridotites. More likely, K has been added through a fluid or melt phase.

To further constrain the role of any external melt/fluid in the formation of the melt pockets, we have calculated the bulk compositions of three representative melt pockets from sample S52-05A using the mode estimated from image analysis software and average mineral compositions of the individual melt pockets. These three melt pockets are so chosen that no large relic minerals (Cpx-I and Spl-I) are now present in the pockets, representing a more mature stage of incongruent melting/reaction and, as such, reducing the uncertainty of how much relic mineral should be included in the mode estimation. The results of the calculated bulk compositions are provided in Table 7.10. They have very high Mg# (~89), implying that direct melt impregnation cannot be the case because such Mg# (~89) is too high to be of a melt and the Fo_{90} Ol phenocrysts would not have been in equilibrium with such a melt. The high Mg# must reflect the breakdown of mantle phases. The low Ti contents calculated further confirm that the pre-existing

Table 7.10 Estimated chemical (wt%) and modal (vol. %) compositions of representative melt pockets from S52-05A

Melt Pocket	C	F	H
SiO_2	46.79	47.77	51.53
TiO_2	0.21	0.20	0.13
Al_2O_3	11.39	10.73	8.09
FeO	4.63	4.34	4.61
MnO	0.10	0.04	0.12
MgO	22.09	19.99	23.00
CaO	9.27	10.73	6.12
Na_2O	2.17	2.39	2.99
K_2O	0.52	0.22	1.43
Cr_2O_3	2.28	2.62	1.02
NiO	0.14	0.14	0.10
Total	99.6	99.2	99.1
Mg# (mol%)	89.5	89.1	89.9
Cpx-P (%)	34	43.5	28
Ol-P (%)	31	24.7	36
Spl-P (%)	6	5.2	1
Flsp (%)	29	26.5	35

amphiboles and micas (Ti-rich varieties in the Al Ghab xenoliths) are not the major reacting phases. Indeed, the calculated bulks, with high Si, Mg, Ca, Al and Cr, resemble mixtures of Cpx-I and Spl-I. The retention of much of Cpx-I + Spl-I signature in the melt pockets and enrichment of K point to a flush of K-laden fluid/melt at a low fluid/melt–rock ratio. We seek trace element evidence in the following.

Compared with the Cpx-I, most of the Cpx-P and Cpx-Spo display lower LREE and Zr/Hf and higher Cs, Rb, Ba, Nb, Ce/Pb and Ti/Eu (Fig. 7.8) at similar or subtly lower Th and U. Although partial melting may account for the lowering of LREE and Zr/Hf, given $D_{LREE/HREE}{}^{Cpx/melt}$ and $D_{Zr/Hf}{}^{Cpx/melt} < 1$ (David et al. 2000; Weyer et al. 2003; Adam and Green 2006), the other features argue against a simple partial melting model; for instance, the LILEs are known to be highly incompatible in clinopyroxenes (e.g. Adam and Green 2006). Likewise, porous-flowing melts or fluids in the lithosphere may interact and modify the lithospheric mantle in a chromatographic manner, thereby implanting a low LREE/MREE and high MREE/HREE feature to the peridotite wall rock (Vernières et al. 1997), but additional mechanisms are still needed to explain the LILE and HFSE of the Al Ghab Cpx-P and Cpx-Spo.

Instead, low LREE and Zr/Hf and high Cs, Rb, Ba, Nb, Ce/Pb and Ti/Eu as seen in the Cpx-P and Cpx-Spo have been used in many other xenolith suites and massif peridotite provinces to indicate silicate metasomatism (Johnson et al. 1996; Coltorti et al. 1999; Downes 2001; Powell and O'Reilly 2007; Ismail et al. 2008). Specifically, breakdown of amphiboles may provide the necessary chemistry, i.e. high LILE/Th–U, Ce/Pb, Ti/Eu and moderate LREE/HREE and Zr/Hf, as well as volatiles for the melt, which by consideration of previous experimental and empirical studies would have a silica-undersaturated basalt affinity (Pilet et al. 2008; Ma et al. 2011a, b, 2013), to metasomatise the lithospheric mantle. Both textural and chemical evidence suggests that such amphiboles are not those presently observed in the xenoliths and that the metasomatism likely occurred simultaneously with, or presumably induced, breakdown of Cpx-I and formation of Cpx-Spo and melt pockets. This is evidenced by the Cpx-P of the amphibole-bearing sample G10, in which the rim of the Cpx-P (analysis# G10_GZ90) shows elevated Al, Ti and Ca at the expense of Si, and a trace element spider pattern intermediate between its core (analysis# G10_GZ88-89) and the amphibole-replaced Cpx-I in the rocks (Table 7.6 and Fig. 7.8; see also Fig. 7.9a, b). This rim also shows a characteristic shift of peak from La–Ce (core) to Sm–Nd (rim) in the chondrite-normalised REE plot, more typical of amphibole. It is interpreted that this Cpx-P was being replaced by amphibole through interaction with an amphibole-derived melt/fluid, the same liquid that incited incongruent melting of Cpx-I and Spl-I and replaced Cpx-I by amphiboles. It is interpreted that this melt was initially not particularly enriched in Ti. It reacted with the Cpx-I and Spl-I and incongruently crystallised Spl-P, Ol-P, Cpx-P and perhaps also some Flsp becoming enriched in incompatible elements such as Ti, LREE, Nb and Rb. Eventually, the melt was saturated with, and started crystallising, Ti-amphiboles in the melt pockets and elsewhere upon migration.

7.6 Thermobarometric Constraints on the Evolution of the Lithospheric Mantle Beneath Al Ghab

7.6.1 Application of Geothermobarometers

A number of experimentally and empirically calibrated thermometers and barometers are applicable to the upper mantle system (Brey and Köhler 1990; Brey et al. 1990; Taylor 1998; Smith 1999; Nimis and Grütter 2010), where *P* estimation almost exclusively relies on the Al exchange between garnet and orthopyroxene (e.g. Taylor 1998). In garnet-free rocks, i.e. spinel peridotites, complementary *P–T* constraints are difficult, as Al in pyroxenes becomes more sensitive to *T* than *P* (Gasparik 1984). However, in recent years, Putirka (2008) has calibrated new thermometers and barometers for as high as ~2000 °C and ~7 GPa, encompassing both the spinel- and garnet-facies peridotite stability fields. More importantly, the author also developed a method using the iterative function in Microsoft Excel to solve the *P* and *T* simultaneously from two equations, opening a new window to more accurate geothermobarometry. This iterative method is adopted here to estimate the equilibrium *P–T* of the primary peridotites and the melt-pockets. The results estimated using a combination of En–Di exchange two-pyroxene thermometers ($T_{\mathrm{2Px\text{-}BKN}}$: Brey and Köhler 1990; $T_{\mathrm{2Px\text{-}TA98}}$: Taylor 1998), Ca-in-orthopyroxene thermometer ($T_{\mathrm{Ca\text{-}in\text{-}Opx}}$: Brey and Köhler 1990 modified by Nimis and Grütter 2010), Al–Cr-in-orthopyroxene thermometer ($T_{\mathrm{Al\text{–}Cr\text{-}in\text{-}Opx}}$: Witt-Eickschen and Seck 1991) and Putirka's (2008) *P–T* equations (his Eqs. 37 and 38: $T_{\mathrm{2Px\text{-}PU37}}$ and $P_{\mathrm{2Px\text{-}PU38}}$) which are given in Table 7.11. Only chemically equilibrated orthopyroxene–clinopyroxene pairs, which are those having an Fe–Mg exchange $[K_D(\mathrm{Fe–Mg})^{\mathrm{cpx-opx}}] = 1.09 \pm 0.14$, are used in the calculations. The use of multiple thermometers in conjunction with a consideration of mineral zonation patterns and element diffusion speeds can serve to evaluate the tempo-thermal evolution of the lithosphere. We emphasise that the *P* estimation for spinel-facies peridotites employing Putirka's (2008) iterative method still carries some uncertainty, but as will be demonstrated below, because the primary peridotite assemblages and the melt-pocket assemblages of the Al Ghab xenolith suite reveal a systematic difference in their equilibrium *P* intervals, the *P* estimation may still be appreciative. This *P* difference is more insightful than the absolute values of the *P* estimated.

7.6.2 *P–T* Characteristics Deduced from the Primary Peridotite Phases

As stated in the preceding section, Al in pyroxenes is a function of *T* and varies little with *P* in spinel peridotites (Gasparik 1984). Its slow diffusion (Sautter et al. 1988; Witt-Eickschen and Seck 1991) also makes it less prone to modification by short-lived thermal events such as the heating associated with the transporting magmas (contrast Ca in olivines) and thus enables evaluation of the lithosphere's thermal evolution by comparing the pyroxene core and rim differences in Al contents. In general, Al increases in both clinopyroxenes and orthopyroxenes with increasing T (Gasparik 1984; Witt-Eickschen and Seck 1991). Likewise, Ca or more precisely Ca/(1-Na) in pyroxenes is also sensitive to T changes and decreases in clinopyroxenes and increases in orthopyroxenes with increasing T (Brey and Köhler 1990).

Most of the samples reveal a former heating event before the development of Cpx-Spo rims, as indicated by their variable core-to-rim increases in Al contents in Opx-I and the clear portions of Cpx-I (Figs. 7.7 and 7.8), although the calculated temperatures from $T_{\text{Al–Cr-in-Opx}}$ show only subtle differences in the cores (853–952 °C) and rims (907–965 °C; Table 7.11; Fig. 7.11e). Concomitant Ca zonation patterns are often less pronounced or absent, a feature ascribed to the faster diffusion of Ca than Al. $T_{\text{Ca-in-Opx}}$ yields 895–971 °C for the cores (Fig. 7.11c) and 934–999 °C for the rims (Table 7.11). The two-pyroxene thermometers reveal no notable differences in the cores and rims (e.g. 896–980 and 900–1016 °C for $T_{\text{2Px-TA98}}$; Table 7.11), reflecting equilibration in terms of Na, Mg and Fe (Mg#) which are all fast diffusing elements.

Temperatures estimated from $T_{\text{2Px-BKN}}$ and $T_{\text{2Px-PU37}}$ are comparable (Fig. 7.11b), but systematically higher than those from $T_{\text{2Px-TA98}}$, $T_{\text{Al–Cr-in-Opx}}$ and $T_{\text{Ca-in-Opx}}$ (Figs. 7.11). Nimis and Grütter (2010) claim that $T_{\text{2Px-TA98}}$ is by far the most reliable thermometer capable of reproducing most experimental temperatures in a variety of simple and natural peridotitic systems between ~700 and ~1500 °C and that the $T_{\text{2Px-BKN}}$ tends to overestimate T for Na-rich clinopyroxenes (>0.05 atoms per 6-oxygen formula). The Cpx-I of the Al Ghab xenoliths is invariably Na-rich, and therefore, we prefer $T_{\text{2Px-TA98}}$.

Average pressures from $P_{\text{2Px-PU38}}$ show similar, narrow ranges for cores (0.8–1.4 GPa) and non-spongy rims (0.9–1.4 GPa; Fig. 7.11f). The P–T space constructed from clear cores using $T_{\text{2Px-TA98}}$ and $P_{\text{2Px-PU38}}$ is set out in Fig. 7.12, and the samples span the depth range of ~24–42 km, in general agreement with the geotherm derived from mantle and lower crustal xenoliths from southern Syria (Nasir and Safarjalani 2000).

The Al zonation patterns of clinopyroxene may be used to constrain the timescale of the thermal event(s). A model of radial diffusion in a sphere (Crank 1975) was employed for representative Cpx-I grains, and the modelling results are provided in Table 7.12. Albeit varying (2–20 m.y.), the results suggest heating in the order of millions of years prior to the eruptions, verifying that the heat source was not related to the magma that transported the xenoliths. Potentially, the heating was related to nearby Early–Middle Miocene volcanism on the Aleppo Plateau.

Table 7.11 Estimated equilibrium temperature (°C) and pressure (GPa) for the Al Ghab peridotite xenoliths and melt pockets, NW Syria

Sample	$T_{2Px\text{-}BKN}$		$T_{2Px\text{-}BKN}$		$T_{2Px\text{-}P37}$		$T_{Ca\text{-}in\text{-}Opx}$		$T_{Al\text{–}Ca\text{-}in\text{-}Opx}$		$P_{2Px\text{-}P38}$	
	cores	rims	cores	rims	cores	rims	cores	rims	cores	rims	cores	rims
G10	941 ± 7	951 ± 2	### ± 7	### ± 5	### ± 10	### ± 5	955 ± 15	954 ± 12	919 ± 15	951 ± 12	1.2 ± 0.1	1.1 ± 0.1
G47	959 ± 15	959 ± 33	### ± 16	### ± 30	### ± 9	### ± 16	933 ± 9	955 ± 7	905 ± 12	907 ± 23	1.3 ± 0.1	1.4 ± 0.1
G49	930 ± 11	933 ± 6	### ± 13	### ± 6	### ± 10	993 ± 14	930 ± 8	946 ± 12	853 ± 12	960 ± 23	1.0 ± 0.1	0.9 ± 0.1
G64	912 ± 8	928 ± 18	993 ± 10	### ± 18	### ± 6	### ± 7	942 ± 9	944 ± 5	940 ± 27	940 ± 33	1.3 ± 0.1	1.3 ± 0.1
G72	946 ± 29	957 ± 7	### ± 30	### ± 7	### ± 15	### ± 5	912 ± 7	943 ± 6	907 ± 18	929 ± 12	1.4 ± 0.1	1.3 ± 0.1
G79	942 ± 28	### ± 1	961 ± 21	### ± 2	941 ± 31	### ± 4	895 ± 9	954 ± 4	935 ± 19	965 ± 40	0.8 ± 0.1	0.9 ± 0.1
S32-05	896 ± 13	912 ± 1	971 ± 15	984.48 ± 4	### ± 10	### ± 8	954 ± 7	947 ± 9	925 ± 30	939 ± 28	0.9 ± 0.1	0.9 ± 0.1
S52-05A	907 ± 33	900 ± 11	### ± 30	993.00 ± 9	### ± 14	### ± 10	940 ± 10	934 ± 1	935 ± 17	953 ± 9	1.3 ± 0.1	1.2 ± 0.1
S52-05D	951 ± 5	916 ± 14	### ± 6	### ± 11	### ± 7	### ± 13	971 ± 12	999 ± 4	939 ± 24	961 ± 47	1.2 ± 0.1	1.2 ± 0.1
S52-05GI	980 ± 2	970 ± 11	### ± 3	### ± 9	### ± 4	### ± 8	942 ± 12	966 ± 13	952 ± 17	914 ± 38	1.2 ± 0.1	1.1 ± 0.1
G10 Ti-cpx rims		886 ± 39		992 ± 40		### ± 30						0.9 ± 0.1
G49 Ti-cpx rims		824 ± 29		873 ± 39		893 ± 31						0.6 ± 0.1
G72 Low-Na Cpx-Spo rims		910 ± 40		923 ± 41		937 ± 26						0.8 ± 0.1

(continued)

Table 7.11 (continued)

Sample	$T_{2Px\text{-}BKN}$		$T_{2Px\text{-}BKN}$		$T_{2Px\text{-}P37}$		$T_{Ca\text{-}in\text{-}Opx}$		$T_{Al\text{–}Ca\text{-}in\text{-}Opx}$		$P_{2Px\text{-}P38}$	
	cores	rims	cores	rims	cores	rims	cores	rims	cores	rims	cores	rims
S52-05A Cpx-Spo rim		954		968		984						0.7 ± 0.1
G72 Low-Na Cpx-P rims		826 ± 43		887 ± 50		940 ± 38						0.9 ± 0.1
S52-05A Cpx-P		981 ± 1		### ± 2		### ± 4						0.9 ± 0.1

The figures represent averages of multiple estimates and their standard deviation

$T_{2Px\text{-}BKN}$, two-pyroxene thermometer of Brey and Köhler (1990); $T_{2Px\text{-}TA98}$, two-pyroxene thermometer of Taylor (1998); $T_{2Px\text{-}PU37}$, two-pyroxene thermometer of Putirka's (2008) Eq. 37; $T_{Ca\text{-}in\text{-}Opx}$, Ca-in-orthopyroxene thermometer of Brey and Köhler (1990) modified by Nimis and Grütter (2010); $T_{Al\text{–}Cr\text{-}in\text{-}Opx}$, Cr- and Al-in-orthopyroxene thermometer of Witt-Eickschen and Seck (1991); $P_{2Px\text{-}PU38}$, two-pyroxene barometer of Putirka's (2008) Eq. 38

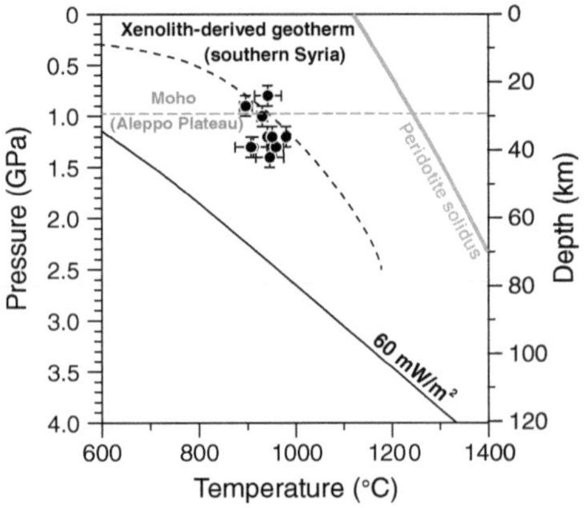

Fig. 7.12 *P–T* estimates for peridotite xenoliths from Al Ghab calculated using $T_{2Px\text{-}TA98}$ and $P_{2Px\text{-}PU38}$ (Table 7.11). Error bars show standard deviations of the averages. Shown from comparison are the two geotherms derived from *P–T* estimates for xenoliths from southern Syria (Nasir and Safarjalani 2000) and from surface heat flow of 60 mW/m^2 (Förster et al. 2010), a typical peridotite solidus (Hirschmann 2000), and the Moho depth for the nearby Aleppo Plateau (Brew et al. 2001a). Note that the Moho depth for Al Ghab, although lacking precise geological/geophysical estimation, is considered to be shallower than that of the Aleppo Plateau (see Sect. 7.2)

Table 7.12 Estimated diffusion time for Al zonation patterns in clinopyroxene based on the radial diffusion model

Sample	ID	Radius, *a* (μm)	Initial Al, C_0 (pfu)	Final (rim) Al, C_1 (pfu)	Best-fit t (m.y)
G47	Grain 5	425	0.185	0.210	7
G49	Grain 2[a]	180	0.174	0.239	2
G49	Grain 2[a]	540	0.174	0.241	20
G72	Grain 9	325	0.217	0.238	11
S52-05D	GZ32–46	652	0.172	0.229	18

Diffusion time was estimated using the radial diffusion equation of Crank (1975), assuming the mineral as a perfect sphere. Diffusion coefficient was 2.8×10^{-19} cm^2/s, calculated on the basis of Sautter and Harte (1990) and assuming a temperature of 1000 °C

[a]The Al profile is not symmetrical, and two different values of radius were used in the modelling

7.6.3 *P–T* Characteristics Deduced from the Secondary Phases

In the melt pockets, *P–T* estimates are less straightforward, owing to the absence of orthopyroxene and to the fact that equilibrium was barely reached judging from textural grounds. Nevertheless, rims of Opx-I occurring in the wall rock of the melt pockets may be used with Cpx-P to provide some constraints on the

P–T conditions of the melt pockets. Cpx-P in the melt-pocket centres is always out of equilibrium with cores and rims of wall rock Opx-I in terms of their Fe–Mg exchange, which becomes less disequilibrated or in a few cases equilibrated (in terms of Mg# exchange) for Cpx-P occurring close/adjacent to rims of wall rock Opx-I. Note that for many melt pockets, the equilibrium conditions cannot be evaluated in this way, as Opx-I is absent or Opx-I rim analyses not available. The combination of $T_{2Px\text{-}TA98}$–$P_{2Px\text{-}PU38}$ for the few equilibrated pairs gives 826 ± 43 °C and 0.9 ± 0.1 GPa for G72 and 981 ± 1 °C 0.9 ± 0.1 GPa for S52-05A.

Likewise, Cpx-Spo is often out of equilibrium with the Opx-I, but local equilibration with the rims of nearby Opx-I is reached in a few instances in the same samples G72 and S52-05A. Two-pyroxene temperatures and pressures ($T_{2Px\text{-}TA98}$–$P_{2Px\text{-}PU38}$) for these pairs are 910 ± 40 °C and 0.8 ± 0.1 GPa for G72, and 954 °C and 0.7 GPa for S52-05A, comparable in terms of P with those of the melt pockets.

Although most two-pyroxene-based barometers usually carry substantial uncertainties, the estimated P (equivalent to depths of 21–27 km) for Cpx-Spo and the melt pockets agree well with the occurrences of Flsp and indicate that the systematic P differences between the primary and secondary phases are significant. The P differences may be regarded as evidence for a rifting scenario as we will show below that the P drops, i.e. the development of the Cpx-Spo and melt pockets, did not result from, but preceded, entrainment of the xenoliths by the host magma. Given that the melt pockets are transient features, the rifting must be recent, most likely related to the development of the pull-apart Al Ghab Depression in the Plio-Quaternary, the time that the mantle xenoliths were erupted.

7.7 Synthesis and Geodynamic Implications

The occurrence of the latest Miocene–Quaternary Homs–Al Ghab volcanism agrees in time and space with the development of the Syrian segment of the northern Dead Sea Fault System which formed the Al Ghab pull-apart basin (Ma et al. 2011a). The Plio-Quaternary volcanism in Al Ghab carries a suite of peridotite xenoliths that contain equilibrated and unequilibrated phases and serve as a probe into the recent thermal and chemical evolution of the lithosphere.

The primary peridotite phases are derived from depths of 24–42 km (0.8–1.4 GPa) at 896–980 °C, estimated using the two-pyroxene geothermobarometer of Taylor (1998) and Putirka (2008). Aluminium, and to a lesser extent Ca, zonation in the primary clinopyroxenes and orthopyroxenes suggest subtle thermal perturbation in the recent past. It is as yet unclear whether this heating event was associated with the metasomatic events recorded in primary clinopyroxenes, metasomatic amphibole, and melt pockets, or not. Aluminium-in-clinopyroxene speedometry suggests heating on the order of several to tens of millions of years prior to eruption, consistent with the Miocene intraplate volcanic activity of the Aleppo

Plateau which peaked at ~19–18 and ~13–12 Ma (Ma et al. 2013), events that are thought to be associated with the Jordanian mantle plume arrival (Ma et al. 2013; see also Chang and Van der Lee 2011). In this scenario, one may argue that the Syrian xenoliths of this study represent fragments of a metasomatised continental lithosphere, instead of a cooled, recent plume head which would have otherwise resulted in "cooling" Al–Ca zonation profiles in the primary clinopyroxene of the xenoliths.

The Al Ghab xenolithsrecord two stages of metasomatism. The first is recorded in the cores of primary clinopyroxenes from most of the samples. They are LREE-enriched, with characteristic enrichment of Na, Th, U, and depletion of Ti and Zr that indicate carbonate metasomatism. As carbonatites are not reported in Syria and its vicinity, we are inclined to interpret the metasomatic agent as some sort of low-silica, CO_2-rich melt or fluid. The timing of metasomatism cannot be fully constrained, but as with the subtle thermal event, it most likely occurred prior to the formation of the Al Ghab pull-apart basin, because the metasomatism clearly pre-dated a decompressional event that formed the spongy rims in the clinopyroxenes.

The second stage of the metasomatism came along with the development of the spongy clinopyroxenes and spinels, melt pockets, amphiboles and micas. Textural and chemical evidence suggest that the melt pockets evolved from spongy clinopyroxenes and spinels through decompression and perhaps also metasomatic fluid/melt-induced incongruent melting of primary minerals. The reaction produced secondary Cr-spinels, olivines, clinopyroxenes, feldspars, and a melt enriched in Ti and other incompatible elements. This resultant melt formed amphiboles within the melt pockets and elsewhere in the peridotite upon migration. Geothermobarometry is based on limited pairs of melt-pocket clinopyroxenes–wall rock orthopyroxene rims and spongy clinopyroxene rims–nearby orthopyroxene rims yield equilibrium temperature and pressures of 826–981 °C and 0.7–0.9 GPa (21–27 km). Given that the melt pockets are transient features and must represent recent processes and that the dominance of feldspars over glasses in the melt-pockets implies a prolonged crystallisation process and thus a relatively slow decompressional path, the decompression is interpreted to be a local feature associated with the development of the Al Ghab pull-apart basin in Plio-Quaternary time. The decompression may have destabilised the pre-existing (Ti-poor?) amphiboles deep down in the lithospheric mantle, causing them to break down and release a melt, likely of silica-undersaturated basalt affinity (Pilet et al. 2008; Ma et al. 2011a, b, 2013), that migrated upwards and reacted with the clinopyroxenes and spinels, forming the melt pockets. Shortly afterwards, the peridotites were picked up by the magma that eventually formed the Plio-Quaternary Al Ghab volcanic field in NW Syria.

Acknowledgments Many thanks are due to Mei-fu Zhou for financing EPMA work at Guangzhou, Simon Leung, Lin-Li Chen, Hui-Ho Hsieh for assisting EPMA work at Guangzhou and IESAS, Mike Tubrett for assisting LA-ICPMS work at Memorial, editor Frédéric Deschamps for invitation and encouragement in preparing this manuscript and Evgenii Sharkov and an anonymous referee for reviewing the manuscript. This study would not have been possible without the support and permission for fieldwork by the General Establishment of Geology and Mineral Resources (GEGMR), Damascus, Syria. Morwan Al-Shara'a, Tamam Darouz and Nouh Wappy from GEGMR are gratefully acknowledged for their invaluable assistance and hospitality in Syria. This work was supported by research funds provided by HKU (grant numbers 200607176152 and 200807176091 to J.M., and a CRCG Travel Grant [2009] to G.S.-K.M.), Postdoctoral Fellow Program of Academia Sinica, Taipei (to G.S.-K.M.), and National Science Council, Taipei (to K.-L.W and Y.I. [grant# 101-2116-M-001-011 to Y.I.]). This is No. IESAS 1841 publication of Institute of Earth Sciences, Academia Sinica.

Appendix: Analytical Techniques

A1 Electron Probe Microanalysis

Quantitative mineral analysis was performed at the Institute of Earth Sciences, Academia Sinica, Taipei, using a JEOL JXA-8500F field emission electron probe microanalyser (EPMA), equipped with five wavelength dispersive spectrometers (WDSs) and a JEOL JXA-8900R thermal emission (W-filament) EPMA with four WDSs. Secondary- and back-scattered electron images were used to guide the analysis on target positions of minerals. A 2-μm defocused beam was operated for quantitative analysis of silicates and oxides at acceleration voltage of 12 kV and 15 kV with beam current of 6 and 12 nA in FE-EPMA and W-EPMA, respectively. The measured intensities were ZAF-corrected using both synthetic and natural materials as follows: diopside or wollastonite for Si, rutile for Ti, corundum for Al, chromium oxide for Cr, fayalite or hematite for Fe, tephroite or Mn-oxide for Mn, periclase for Mg, Ni-oxide for Ni, wollastonite for Ca, albite for Na, and adularia for K. Counting times for each element and both upper and lower baselines were 10 and 5 s, respectively.

To evaluate the general characteristics of crystal zoning and the relationship among minerals, elemental distribution mapping of selected minerals was performed at 15 kV and 25 nA with a 2-μm beam using the W-EPMA. X-ray intensities were counted for 0.025 s at the interval of 2 μm with the X-Y stage driving.

A subset of quantitative mineral analyses was acquired at Guangzhou Institute of Geochemistry, Chinese Academy of Sciences, using a JEOL JXA-8100 (W-EPMA) Superprobe equipped with five WDSs (~1 μm beam, 15 kV, 20 nA and 10–30 s counting time) and at Oxford University with a JEOL JXA8800R (W-EPMA) equipped with four WDSs. Analytical details employed at these laboratories are given in Ma et al. (2011a) and Chan et al. (2009), respectively. For mineral grains that were analysed at different laboratories and machines, the results agree very well.

A2 Laser Ablation Inductively Coupled Plasma-Mass Spectrometry

Concentrations of 31 elements, Al, Si, Sc, Ti, V, Rb, Sr, Y, Zr, Nb, Cs, Ba, La, Ce, Pr, Nd, Sm, Eu, Gd, Tb, Dy, Ho, Er, Tm, Yb, Lu, Hf, Ta, Pb, Th and U, of representative clinopyroxenes, amphiboles, glasses and feldspars, were determined at the MAF-IIC laboratories at Memorial University of Newfoundland (MUN) using laser ablation inductively coupled plasma-mass spectrometry (LA ICP-MS).

The analytical system is a Finnigan ELEMENT XR, a high-resolution double-focusing magnetic sector inductively coupled plasma-mass spectrometer (HR-ICPMS) coupled to a GEOLAS 193-nm excimer laser system. A helium flow rate of 1–1.1 l/min was used to carry ablated material to the ICP, with an additional 0.75 l/min argon make up argon gas added after the ablation cell. Depending on target grain size, a laser spot size of between 20 and 69 μm was used for analyses (mostly 40 μm). Laser energy was approximately 4 J/cm^2, and the laser repetition rate was 10 Hz. Time-resolved intensity data were acquired by peak jumping in a combination with pulse-counting and analogue modes, depending on signal strength, with one point measured per peak for masses. Concentrations were calibrated with the NIST 612 glass. CaO was the internal standard used to deal with differences in ablation yields and matrix effects between the unknown clams and the calibration materials (NIST glasses). The CaO concentrations of the unknowns were predetermined by electron microprobe. Approximately 40 s of gas background data were collected prior to each laser ablation of both standards and unknowns. During the course of analysis, Th/ThO^+ was always lower than 0.5 %.

The data acquisition methodology employed an analytical sequence of two analyses of the NIST 612 standard and one of BCR2G reference material with analyses of up to 14 unknown minerals, closing with a repetition of the same standards in reverse order. The BCR2G was treated as an unknown, and data were acquired to allow the monitoring of accuracy and precision of the dataset and the technique in general. The results of BCR2G are given in the supplementary files along with the MUN long-term BCR2G values.

Data were reduced using MUN's in-house CONVERT and LAMTRACE spreadsheet programs, which employ procedures described by Longerich et al. (1996). LAMTRACE allows for selection of representative signal intervals, background subtraction, internal standard correction for ablation yield differences, correction for instrument sensitivity drift during the analytical session, and conversion of count rates into concentrations by reference to the standards.

References

Ackerman L, Mahlen N, Jelinek E, Medaris G Jr, Ulrych J, Strnad L, Mihaljevic M (2007) Geochemistry and evolution of subcontinental lithospheric mantle in central Europe: evidence from peridotite xenoliths of the Kozkov volcano, Czech Republic. J Petrol 48:2235–2260

Adam J, Green TH (2006) Trace element partitioning between mica- and amphibole-bearing garnet lherzolite and hydrous basanitic melt: 1. Experimental results and the investigation of controls on partitioning behaviour. Contrib Miner Petrol 152:1–17

Angus DA, Wilson DC, Sandvol E, Ni JF (2006) Lithospheric structure of the Arabian and Eurasian collision zone in Eastern Turkey from S-wave receiver functions. Geophys J Int 166:1335–1346

Aoki K, Kushiro I (1968) Some clinopyroxenes from ultramafic inclusions in Dreiser Weiher, Eifel. Contrib Miner Petrol 18:326–337

Bali E, Szabó C, Vaselli O, Török K (2002) Significance of silicate melt pockets in upper mantle xenoliths from the Bakony-Balaton Highland Volcanic Field, Western Hungary. Lithos 61:79–102

Bali E, Zanetti A, Szabó C, Peate DW, Waight TE (2008) A micro-scale investigation of melt production and extraction in the upper mantle based on silicate melt pockets in ultramafic xenoliths from the Bakony-Balaton Highland Volcanic Field (Western Hungary). Contrib Miner Petrol 155:165–179

Bodinier J-L, Vasseur G, Vernieres J, Dupuy C, Fabries J (1990) Mechanisms of mantle metasomatism—geochemical evidence from the Lherz orogenic peridotite. J Petrol 31:597–628

Bodinier J-L, Menzies MA, Shimizu N, Frey FA, McPherson E (2004) Silicate, hydrous and carbonate metasomatism at Lherz, France: contemporaneous derivatives of silicate melt-harzburgite reaction. J Petrol 45:299–320

Bonadiman C, Beccaluva L, Coltorti M, Siena F (2005) Kimberlite-like metasomatism and 'garnet signature' in spinel-peridotite xenoliths from Sal, Cape Verde archipelago: relics of a subcontinental mantle domain within the Atlantic oceanic lithosphere? J Petrol 46:2465–2493

Brew G, Barazangi M, Al-Maleh AK, Sawaf T (2001a) Tectonic and geologic evolution of Syria. GeoArabia 6:573–616

Brew G, Lupa J, Barazangi M, Sawaf T, Al-Imam A, Zaza T (2001b) Structure and tectonic development of the Ghab basin and the Dead Sea fault system, Syria. J Geol Soc London 158:665–674

Brey GP, Köhler T (1990) Geothermobarometry in 4-phase lherzolites II. New thermobarometers, and practical assessment of existing thermobarometers. J Petrol 31:1353–1378

Brey GP, Kóhler T, Nickel KG (1990) Geothermobarometry in 4-phase lherzolites I. Experimental results from 10 to 60 Kb. J Petrol 31:1313–1352

Carpenter RL, Edgar AD, Thibault Y (2002) Origin of spongy textures in clinopyroxene and spinel from mantle xenoliths, Hessian Depression, Germany. Mineral Petrol 74:149–162

Chan GH-N, Waters DJ, Searle MP, Aitchison JC, Horstwood MSA, Crowley Q, Lo C-H, Chan JS-L (2009) Probing the basement of southern Tibet: evidence from crustal xenoliths entrained in a Miocene ultrapotassic dyke. J Geol Soc London 166:45–52

Chang S-J, Van der Lee S (2011) Mantle plumes and associated flow beneath Arabia and East Africa. Earth Planet Sci Lett 301:448–454

Coltorti M, Bonadiman C, Hinton RW, Siena F, Upton BGJ (1999) Carbonatite metasomatism of the oceanic upper mantle: evidence from clinopyroxenes and glasses in ultramafic xenoliths of Grande Comore, Indian Ocean. J Petrol 40:133–165

Coltorti M, Beccaluva L, Bonadiman C, Salvini L, Siena F (2000) Glasses in mantle xenoliths as geochemical indicators of metasomatic agents. Earth Planet Sci Lett 183:303–320

Crank J (1975) The Mathematics of Diffusion. Oxford University Press, London

Daines MJ, Kohlstedt DL (1993) A laboratory study of melt migration. Philos Trans Roy Soc London Ser A: Phys Eng Sci 342:43–52

David K, Schiano P, Allégre CJ (2000) Assessment of the Zr/Hf fractionation in oceanic basalts and continental materials during petrogenetic processes. Earth Planet Sci Lett 178:285–301

Davidson I, Al-Kadasi M, Al-Khirbash S, Al-Subbary AK, Baker J, Blakey S, Bosence D, Dart C, Heaton R, McClay K, Menzies M, Nichols G, Owen L, Yelland A (1994) Geological evolution of the southeastern Red Sea Rift margin, Republic of Yemen. Geol Soc Am Bull 106:1474–1493

Dawson JB (1980) Kimberlites and their xenoliths. Springer, New York

Dawson JB (1984) Contrasting types of mantle metasomatism? In: Kornprobst J (ed) Kimberlites; the mantle and crust-mantle relationships. Elsevier, Amsterdam, pp 289–294

Dawson JB (2002) Metasomatism and partial melting in upper-mantle peridotite xenoliths from the Lashaine volcano, northern Tanzania. J Petrol 43:1749–1777

Demouchy S, Jacobsen SD, Gaillard F, Stern CR (2006) Rapid magma ascent recorded by water diffusion profiles in mantle olivine. Geology 34:429–432

Devyatkin EV, Dodonov AE, Sharkov EV, Zykin VS, Simakova AN, Khatib K, Nseir H (1997) The El-Ghab Rift depression in Syria: its structure, stratigraphy, and history of development. Stratigr Geol Correl 5:362–374

Domas J (1994) The late Cenozoic of the Al Ghab Rift, NW Syria. Sbornik Geologickych Ved. Antropozoikum 21: 57–73

Downes H (2001) Formation and modification of the shallow sub-continental lithospheric mantle: a review of geochemical evidence from ultramafic xenolith suites and tectonically emplaced ultramafic massifs of Western and Central Europe. J Petrol 42:233–250

Förster H-J, Förster A, Oberhänsli R, Stromeyer D (2010) Lithospheric composition and thermal structure of the Arabian Shield in Jordan. Tectonophysics 481:29–37

Francis DM (1976) The origin of amphibole in lherzolite xenoliths from Nunivak Island, Alaska. J Petrol 17:357–378

Gasparik T (1984) Two-pyroxene thermobarometry with new experimental data in the system CaO-MgO-Al2O3-SiO2. Contrib Miner Petrol 87:87–97

Griffin WL, O'Reilly SY, Afonso JC, Begg GC (2009) The Composition and evolution of lithospheric mantle: a re-evaluation and its tectonic implications. J Petrol 50:1185–1204

Harte B (1983) Mantle peridotites and processes—the kimberlite sample. In: Hawkesworth CJ, Norry MJ (eds) Continental basalts and mantle xenoliths. Shiva, Nantwich, pp 46–91

Hibbard M, Sjoberg J (1994) Signs of incongruent melting of clinopyroxene in limburgite, Thetford Hill, Vermont. Can Mineral 32:307–317

Hirschmann MM (2000) Mantle solidus: experimental constraints and the effects of peridotite composition. Geochem Geophys Geosyst 1. doi:10.1029/2000GC000070

Ionov DA, Hofmann AW, Shimizu N (1994) Metasomatism-induced melting in mantle xenoliths from Mongolia. J Petrol 35:753–785

Ionov D, Bodinier J-L, Mukasa SB, Zanetti A (2002a) Mechanisms and sources of mantle metasomatism: major and trace element compositions of peridotite xenoliths from Spitsbergen in the context of numerical modelling. J Petrol 43:2219–2259

Ionov D, Mukasa SB, Bodinier J-L (2002b) Sr–Nd–Pb isotopic compositions of peridotite xenoliths from Spitsbergen: numerical modelling indicates Sr–Nd decoupling in the mantle by melt percolation metasomatism. J Petrol 43:2261–2278

Ismail M, Delpech G, Cottin JY, Grégoire M, Moine BN, Bilal A (2008) Petrological and geochemical constraints on the composition of the lithospheric mantle beneath Syrian rift, northern part of the Arabian plate. In: Coltorti M, Grégoire M (eds) Metasomatism in oceanic and continental lithospheric mantle, vol 293. Geological Society Special Publication. pp 223–251

Johnson KE, Davis AM, Bryndzia LT (1996) Contrasting styles of hydrous metasomatism in the upper mantle: an ion microprobe investigation. Geochim Cosmochim Acta 60:1367–1385

Kaeser B, Kalt A, Pettke T (2007) Crystallization and breakdown of metasomatic phases in graphite-bearing peridotite xenoliths from Marsabit (Kenya). J Petrol 48:1725–1760

Kaliwoda M, Altherr R, Meyer HP (2007) Composition and thermal evolution of the lithospheric mantle beneath the Harrat Uwayrid, eastern flank of the Red Sea rift (Saudi Arabia). Lithos 99:105–120

Kopp MP, Adzhamyan Z, Il'yas K, Fakiani F, Khafez A (1999) Mechanism of formation of the El Ghab wrench graben (Syria) and the Levant transform fault propagation. Geotectonics 33:408–422

Krienitz M-S, Haase KM, Mezger K, van den Bogaard P, Thiemann V, Shaikh-Mashail MA (2009) Tectonic events, continental intraplate volcanism, and mantle plume activity in northern Arabia: constraints from geochemistry and Ar-Ar dating of Syrian lavas. Geochem Geophys Geosyst 10. doi:10.1029/2008GC002254

LaTourrette T, Hervig RL, Holloway JR (1995) Trace element partitioning between amphibole, phlogopite, and basanite melt. Earth Planet Sci Lett 135:13–30

Le Maitre RW (2002) Igneous rocks: a classification and glossary of terms: recommendations of international union of geological sciences subcommission on the systematics of igneous rocks. Cambridge University Press, Cambridge

Longerich HP, Jackson SE, Günther D (1996) Laser ablation-inductively coupled plasma-mass spectrometric transient signal data acquisition and analyte concentration calculation. J Anal At Spectrom 11:899–904

Luth RW (2003) Mantle volatiles—distribution and consequences. In: Carlson RW (ed) The mantle and core. Treatise on geochemistry, vol 2. Elsevier-Pergamon, Oxford, pp 319–361

Ma GS-K, Malpas J, Xenophontos C, Chan GH-N (2011a) Petrogenesis of latest miocene-quaternary continental intraplate volcanism along the northern Dead Sea fault system (Al Ghab-Homs Volcanic Field), western Syria: evidence for lithosphere–asthenosphere interaction. J Petrol 52:401–430

Ma GS-K, Malpas J, Xenophontos C, Suzuki K, Lo C-H (2011b) Early cretaceous volcanism of the coastal ranges, NW Syria: magma genesis and regional dynamics. Lithos 126:290–306

Ma GS-K, Malpas J, Suzuki K, Lo C-H, Wang K-L, Iizuka Y, Xenophontos C (2013) Evolution and origin of the miocene intraplate basalts on the Aleppo Plateau, NW Syria. Chem Geol 335:149–171

McDonough WF, Sun SS (1995) The composition of the Earth. Chem Geol 120:223–253

Mengel K, Green DH (1989) Stability of amphibole and phlogopite in metasomatized peridotite under water-saturated and water-undersaturated conditions. In: Ross J (ed) Kimberlites and related rocks: proceedings of the fourth international kimberlite conference. Geological Society of Australia, Perth, pp 571–581

Menzies MA (1983) Mantle ultramafic xenoliths in alkaline magmas: evidence for mantle heterogeneity modified by magmatic activity. In: Hawkesworth CJ, Norry MJ (eds) Continental basalts and mantle xenoliths. Shiva, Nantwich, pp 92–110

Nasir S, Safarjalani A (2000) Lithospheric petrology beneath the northern part of the Arabian plate in Syria: evidence from xenoliths in alkali basalts. J Afr Earth Sci 30:149–168

Nimis P, Grütter H (2010) Internally consistent geothermometers for garnet peridotites and pyroxenites. Contrib Miner Petrol 159:411–427

O'Reilly SY, Griffin WL (2013) Mantle metasomatism. In: Harlov DE, Austrheim H (eds) Metasomatism and the chemical transformation of rock. Lecture notes in Earth system sciences. Springer, Berlin, pp 471–533

Palme H, O'Neill HSC (2003) Cosmochemical estimates of mantle composition. In: Carlson RW (ed) The mantle and core. Treatise on Geochemistry 2. Elsevier-Pergamon, Oxford, pp 1–38

Perinelli C, Orlando A, Conte AM, Armienti P, Borrini D, Faccini B, Misiti V (2008) Metasomatism induced by alkaline magma in the upper mantle of northern Victoria Land (Antarctica): an experimental approach. In: Coltorti M, Grégoire M (eds) Metasomatism in oceanic and continental lithospheric mantle, vol 293. Geological Society Special Publication, pp 223–251

Pilet S, Baker MB, Stolper EM (2008) Metasomatized lithosphere and the origin of alkaline lavas. Science 320:916–919

Pilet S, Baker MB, Müntener O, Stolper EM (2011) Monte Carlo simulations of metasomatic enrichment in the lithosphere and implications for the source of alkaline basalts. J Petrol 52:1415–1442

Powell W, O'Reilly SY (2007) Metasomatism and sulfide mobility in lithospheric mantle beneath eastern Australia: implications for mantle Re–Os chronology. Lithos 94:132–147

Powell W, Zhang M, O'Reilly SY, Tiepolo M (2004) Mantle amphibole trace-element and isotopic signatures trace multiple metasomatic episodes in lithospheric mantle, western Victoria, Australia. Lithos 75:141–171

Putirka K (2008) Thermometers and barometers for volcanic systems. In: Putirka K, Tepley F (eds) Minerals, inclusions and volcanic processes. Reviews in Mineralogy and Geochemistry, vol 69. Mineralogical Society of America, pp 61–120

Riley JGN, Kohlstedt DL (1991) Kinetics of melt migration in upper mantle-type rocks. Earth Planet Sci Lett 105:500–521

Rudnick RL, McDonough WF, Chappell BW (1993) Carbonatite metasomatism in the northern Tanzanian mantle: petrographic and geochemical characteristics. Earth Planet Sci Lett 114:463–475

Sautter V, Harte B (1990) Diffusion gradients in an eclogite xenolith from the Roberts Victor kimberlite pipe (2): kinetics and implications for petrogenesis. Contrib Miner Petrol 105:637–649

Sautter V, Jaoul O, Abel F (1988) Aluminum diffusion in diopside using the 27Al(ρ, ϒ)28Si nuclear reaction: preliminary results. Earth Planet Sci Lett 89:109–114

Schiano P, Clocchiatti R (1994) Worldwide occurrence of silica-rich melts in sub-continental and sub-oceanic mantle minerals. Nature 368:621–624

Searle MP, Chung S-L, Lo C-H (2010) Geological offsets and age constraints along the northern Dead Sea fault, Syria. J Geol Soc London 167:1001–1008

Sharkov EV, Chernyshev IV, Devyatkin EV, Dodonov AE, Ivanenko VV, Karpenko MI, Leonov YG, Novikov VM, Hanna S, Khatib K (1994) Geochronology of late Cenozoic basalts in western Syria. Petrology 2:439–448

Sharkov EV, Chernyshev IV, Devyatkin EV, Dodonov AE, Ivanenko VV, Karpenko MI, Leonov YG, Novikov VM, Hanna S, Khatib K (1998) New data on the geochronology of upper Cenozoic plateau basalts from the northeastern periphery of the Red Sea Rift area (northern Syria). Dokl Earth Sci 358:19–22

Shaw CSJ (2009) Textural development of amphibole during breakdown reactions in a synthetic peridotite. Lithos 110:215–228

Shaw CSJ, Dingwell DB (2008) Experimental peridotite-melt reaction at one atmosphere: a textural and chemical study. Contrib Miner Petrol 155:199–214

Shaw CSJ, Heidelbach F, Dingwell DB (2006) The origin of reaction textures in mantle peridotite xenoliths from Sal Island, Cape Verde: the case for "metasomatism" by the host lava. Contrib Miner Petrol 151:681–697

Smith D (1999) Temperatures and pressures of mineral equilibration in peridotite xenoliths: review, discussion, and implications. In: Fei Y, Bertka CM, Mysen BO (eds) Mantle petrology: field observations and high pressure experimentation: a tribute to Francis R. (Joe) Boyd, vol 6. Geochemical Society Special Publication, pp 171–188

Stosch HG, Seck HA (1980) Geochemistry and mineralogy of two spinel peridotite suites from Dreiser Weiher West Germany. Geochim Cosmochim Acta 44:457–470

Su B, Zhang H, Sakyi PA, Qin K, Liu P, Ying J, Tang Y, Malaviarachchi S, Xiao Y, Zhao X, Mao Q, Ma Y (2010a) Formation of melt pocket in mantle peridotite xenolith from western Qinling, Central China: partial melting and metasomatism. J Earth Sci 21:641–668

Su B-X, Zhang H-F, Sakyi PA, Ying J-F, Tang Y-J, Yang Y-H, Qin K-Z, Xiao Y, Zhao X-M (2010b) Compositionally stratified lithosphere and carbonatite metasomatism recorded in mantle xenoliths from the Western Qinling (Central China). Lithos 116:111–128

Su B-X, Zhang H-F, Sakyi P, Yang Y-H, Ying J-F, Tang Y-J, Qin K-Z, Xiao Y, Zhao X-M, Mao Q, Ma Y-G (2011) The origin of spongy texture in minerals of mantle xenoliths from the Western Qinling, central China. Contrib Miner Petrol 161:465–482

Su B-X, Zhang H-F, Yang Y-H, Sakyi PA, Ying J-F, Tang Y-J (2012) Breakdown of orthopyroxene contributing to melt pockets in mantle peridotite xenoliths from the Western Qinling, central China: constraints from in situ LA-ICP-MS mineral analyses. Mineral Petrol 104:225–247

Sun SS, McDonough WF (1989) Chemical and isotopic systematics of oceanic basalts: implications for mantle composition and processes. In: Saunders AD, Norry MJ (eds) Magmatism in the Ocean basins, vol 42. Geological Society Special Publication, pp 313–345

Tang Y-J, Zhang H-F, Ying J-F, Su B-X (2013) Widespread refertilization of cratonic and circum-cratonic lithospheric mantle. Earth-Sci Rev. doi:http://dx.doi.org/10.1016/j.earscirev.2013.01.004

Taylor WR (1998) An experimental test of some geothermometer and geobarometer formulations for upper mantle peridotites with application to the thermobarometry of fertile lherzolite and garnet websterite. Neues Jahrbuch Fur Mineralogie-Abhandlungen 172:381–408

Thompson RN, Gibson SA (2000) Transient high temperatures in mantle plume heads inferred from magnesian olivines in Phanerozoic picrites. Nature 407:502–506

Tiepolo M, Oberti R, Zanetti A, Vannucci R, Foley SF (2007) Trace-element partitioning between amphibole and silicate melt. In: Hawthorne FC, Oberti R, Della Ventura G, Mottana A (eds) Amphiboles: crystal chemistry, occurrence, and health issues, vol 67. Reviews in Mineralogy and Geochemistry, pp 417–451

Trifonov VG, Trubikhin VM, Adzhamyan Z, Dzhallad S, El' Khair Y, Ayed K (1991) Levant fault zone in northeast Syria. Geotectonics 25:145–154

Turkmani AS, Al-Shara'a M (2004) Mantle ultramafic xenoliths in alkaline magmas on the eastern part of Al-Ghab depression (Syria): evidence of shallow upper mantle, Ain Shams University. Earth Science Series 18. Middle East Research Center (MERC), pp 27–48

Vernières J, Godard M, Bodinier J-L (1997) A plate model for the simulation of trace element fractionation during partial melting and magma transport in the Earth's upper mantle. J Geophys Res 102:24771–24784

Wass SY (1979) Multiple origins of clinopyroxenes in alkali basaltic rocks. Lithos 12:115–132

Watson BW (1982) Melt infiltration and magma evolution. Geology 10:236–240

Weyer S, Münker C, Mezger K (2003) Nb/Ta, Zr/Hf and REE in the depleted mantle: implications for the differentiation history of the crust–mantle system. Earth Planet Sci Lett 205:309–324

Witt-Eickschen G, Seck HA (1991) Solubility of Ca and Al in orthopyroxene from spinel peridotite: an improved version of an empirical geothermermeter. Contrib Miner Petrol 106:431–439

Yaxley GM, Green DH, Kamenetsky V (1998) Carbonatite metasomatism in the southeastern Australian lithosphere. J Petrol 39:1917–1930

Chapter 8
Elasticity of Continental Crust Around the Mantle Transition Zone

Kenji Kawai and Taku Tsuchiya

Abstract The fate and amount of granitic materials subducted into the deep mantle are still under debate. The density, elastic property, and phase stability of granitic materials in the mantle pressures are key to clarifying them. Here, we modeled high-pressure properties of the granitic assemblage by using ab initio mineral physics data of grossular garnet, K-hollandite, jadeite, stishovite, and calcium ferrite (CF)-type phase. We find that the ongoing subducting granitic assemblage, such as sediments and average upper crust rocks, is much denser than pyrolite in the pressure range from 9 GPa (~270 km), at which coesite undergoes a phase transition to stishovite, to around 27 GPa (~740 km). Above this pressure, granitic material becomes less dense than a pyrolite. This indicates that the granitic assemblage becomes gravitationally stable at the base of the mantle transition zone (MTZ). Results suggest a possibility that the granitic materials could accumulate around the 740 km depth if carried into the depth deeper than 270 km and segregated at some depth. Comparison of the velocities between granitic and pyrolitic materials shows that granitic materials can produce substantial velocity anomalies in the MTZ and the uppermost lower mantle (LM). Seismic observations such as anomalously fast velocities, especially for the shear wave, around the 660-km discontinuity, the complexity of 660-km discontinuity, and the scatterers in the uppermost LM could be associated with the subducted granitic materials.

K. Kawai (✉)
Department of Earth Science, University of California, Santa Barbara, CA 93106, USA
e-mail: kenji@ea.c.u-tokyo.ac.jp

K. Kawai
Department of Earth and Planetary Sciences, Tokyo Institute of Technology, Ookayama 2-12-1, Meguro, Tokyo, 152-8551, Japan

T. Tsuchiya
Geodynamics Research Center, Ehime University, Bunkyo-cho 2-5, Matsuyama, Ehime, 790-8577, Japan

Present Address
K. Kawai
Department of Earth Science and Astronomy, Graduate School of Arts and Sciences, University of Tokyo, Komaba 3-8-1, Meguro, Tokyo, 153-8902, Japan

© Springer International Publishing Switzerland 2015

A. Khan and F. Deschamps (eds.), *The Earth's Heterogeneous Mantle*, Springer Geophysics, DOI 10.1007/978-3-319-15627-9_8

Keywords Elasticity · Granite · Phase transition · First principles calculation

8.1 Introduction

Earth is the only planet with granite and continent in the solar system (Campbell and Taylor 1983). Continental growth is essential in order to understand the evolution of the Earth's surface environment and the origin of life (Taylor and McLennan 1995; Maruyama et al. 2014; Tsuchiya et al. 2013). It has been considered that the granitic crust formed on the surface must have survived through the Earth's history because of its buoyancy. Against such a traditional view, ubiquitous subduction of continental crust was proposed by both geophysical and geological studies at convergent plate boundary, which suggest new concepts of tectonic erosion (Scholl and von Huene 2007) and arc subduction (Yamamoto et al. 2009). Moreover, recent geodynamic studies have suggested that the subduction channel between the subducted slab and mantle wedge could have a potential to carry a considerable amount of granitic material into the deep mantle due to a drag force from the subducting slab (Ichikawa et al. 2013a, 2015, this volume). Since the granitic materials possess a large amount of incompatible element and radiogenic nuclide (e.g., Hofmann 1997), the fate and amount of subducted granitic materials must be of importance in geochemistry and also in the terrestrial thermal history (Kawai et al. 2009, 2013).

We investigated the fate of subducted Archaean granitic material such as tonalite–trondhjemite–granodiorite (TTG) assuming a hypothetical TTG crust, whose mineral proportion has the molar fraction albite:quartz is 1:7 (Komabayashi et al. 2009), and found that the accumulated volume of subducted granitic material at the base of the mantle transition zone (MTZ) might amount to about six times the present volume of the continental crust, which could explain the anomalous seismic-wave velocities observed around 660 km depth (Kawai et al. 2013) although importance of effects of lateral heterogeneity was pointed out to interpret laterally homogeneous models of seismic velocity (Styles et al. 2011). On the other hand, ongoing subducting granitic material, such as sediment and average upper crust rocks, is relatively enriched in potassium (e.g., Martin et al. 2005). It is, therefore, important to understand the fate of ongoing subducting material because radiogenic heat source is supposed to affect the thermal evolution of the Earth. While the abovementioned assumption may be justified for TTG with potassium depleted, more realistic composition should be considered for ongoing subducting granitic material such as sediment and average upper crust rocks. Therefore, we have investigated elastic properties of other minerals such as pyrope, grossular garnet (Kawai and Tsuchiya 2012b, 2015), aluminous hexagonal phase (Kawai and Tsuchiya 2012a), and K-hollandite (Kawai and Tsuchiya 2013) to consider more realistic composition of continental crust.

In order to understand the fate of granitic materials in the mantle, detection of their seismological signature is one of the most effective ways, because seismic velocity jumps produced by phase transitions in granitic materials and velocity contrasts between granitic materials and their surrounding mantle could be expected to scatter and reflect seismic energy. For this purpose, it is required to elucidate the phase stability and elastic properties of granitic materials under high-pressure and high-temperature condition. Also, density profiles of granitic materials are required to consider their gravitational stability in the mantle. While Kawai et al. (2013) considered the subducted granitic crust in Archaean (TTG), we investigate the high-pressure phase stability, density, and elastic properties of the ongoing subducting granitic material such as sediment and average upper crust rocks in this study.

8.2 Modeling Density and Seismic Velocity for Sediment and Average Upper Crust Rock

We calculate phase stability and elastic properties by means of the ab initio density functional computation method in the pressure range from 9 GPa, where the coesite–stishovite phase transition occurs (Liu et al. 1996), to 60 GPa, corresponding to the mid-lower mantle (LM) pressure. We consider two mineral proportions proposed for granitic materials: sediments (Irifune et al. 1994) and average upper crust rocks (Wu et al. 2009). The mineral proportions in both composites are assumed to be in a volume fraction grossular garnet ($Ca_3Al_2Si_3O_{12}$):stishovite (SiO_2):K-hollandite ($KAlSi_3O_8$):jadeite ($NaAlSi_2O_6$) in the mantle transition zone (MTZ) of 3:3:3:1 and 2:3:2:3, respectively. We neglect other minor phases such as the CAS phase because of marginal volumetric proportions (Ishii et al. 2012). Solid-solution effects in minerals such as garnet and jadeite are also neglected in this study since these effects to seismic velocity would be relatively small. Although calculations are conducted at the static temperature (0 K), we consider the phase transition pressures along the normal geotherm (Brown and Shankland 1981) and the slab geotherm which is 500 K lower than the normal geotherm in order to clarify the effects of lateral temperature heterogeneity. First, the high-*P*–*T* phase relations in the granite constituent minerals are summarized (Fig. 8.1). We use reported phase relations: jadeite ⇔ CF-type phase + stishovite (Kawai and Tsuchiya 2012a), grossular garnet ⇔ Cor + Ca-pv (Akaogi et al. 2009; Kawai and Tsuchiya 2012b), stishovite ⇔ α-PbO_2 phase (Tsuchiya et al. 2004), and K-hollandite I ⇔ II (Nishiyama et al. 2005; Kawai and Tsuchiya 2012c). Since the high *P-T* dissociation boundary of grossular garnet and the phase transition boundary of K-hollandite have not been studied theoretically to date, we apply Clapeyron slopes determined experimentally (Akaogi et al. 2009; Nishiyama et al. 2005) to the static transition pressures obtained by theoretical studies (Kawai and Tsuchiya 2012b, 2013). They show that while the phase transition pressures of post-spinel, post-grossular, post-jadeite, and K-hollandite along a normal geotherm are 23.3, 23.6, 23.7, and 28.3 GPa, respectively, and those along a slab geotherm are 24.4, 23.6, 22.1, and 24.5 GPa, respectively (Fig. 8.1).

We employ published density and elastic constants of the $NaAlSi_2O_6$ system (jadeite and a CF-type phase) (Kawai and Tsuchiya 2010), SiO_2 stishovite (Karki et al. 2001), $Ca_3Al_2Si_3O_{12}$ (grossular garnet, corundum (Cor), and Ca-perovskite (pv))

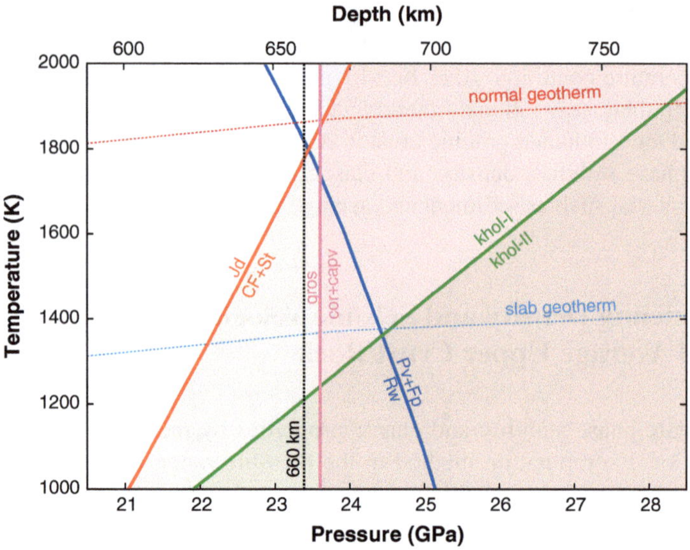

Fig. 8.1 Phase boundaries modified after Kawai and Tsuchiya (2012c). Phase boundaries of the post-jadeite, post-grossular, and K-hollandite phase transitions expected in the granitic materials at around 660 km depth with Clapeyron slopes of +2.8 MPa/K (Kawai and Tsuchiya 2012c), 0 MPa/K (Akaogi et al. 2009; Kawai and Tsuchiya 2012b), and +7 MPa/K (Nishiyama et al. 2005; Kawai and Tsuchiya 2013), respectively. A phase boundary of the post-spinel transition with a Clapeyron slope of −2.9 MPa/K at 1900 K (Yu et al. 2007), the normal mantle geotherm (Brown and Shankland 1981), and the cold slab geotherm 500 K colder than normal mantle geotherm are also potted. *Jd* jadeite, *CF* CF-type phase, *St* stishovite, *K-hol* K-hollandite, *gros* grossular garnet, *cor* corundum, *capv* calcium perovskite, *Rw* ringwoodite, *Pv* magnesium perovskite, *Pc* periclase

(Kawai and Tsuchiya 2012b), and the $KAlSi_3O_8$ systems (K-hollandite) (Kawai and Tsuchiya 2013) to determine the density variations for each composition with the phase transition pressures along the normal and slab geotherms (Fig. 8.2). +3.4 and +1.3 % density increases are estimated associated with the grossular garnet dissociation at 23.6 GPa and the jadeite dissociation at 23.7 GPa along the normal geotherm for the sedimentary composite (Irifune et al. 1994), and +2.3 and +3.8 % increases for the upper crust average composite (Wu et al. 2009), respectively. In contrast, +3.4 and +1.3 % density increases are estimated associated with the grossular garnet dissociation at 23.6 GPa and the jadeite dissociation at 22.1 GPa along the slab geotherm for the sedimentary composite, and +2.3 and +4.0 % increases for the upper crust average composite, respectively. The K-hollandite phase transition, as well as the post-stishovite transition, has no contribution to the discontinuous density change because it is a continuous second-order phase transition. Note that these values are calculated at the static temperature as mentioned, but they are likely expected to be insensitive to temperature.

We take the Voigt–Reuss–Hill averages of elastic constants and obtain the isotropic bulk and shear moduli for each mineral. And then, taking the Voigt–Reuss–Hill averages of bulk and shear moduli according to the volumetric

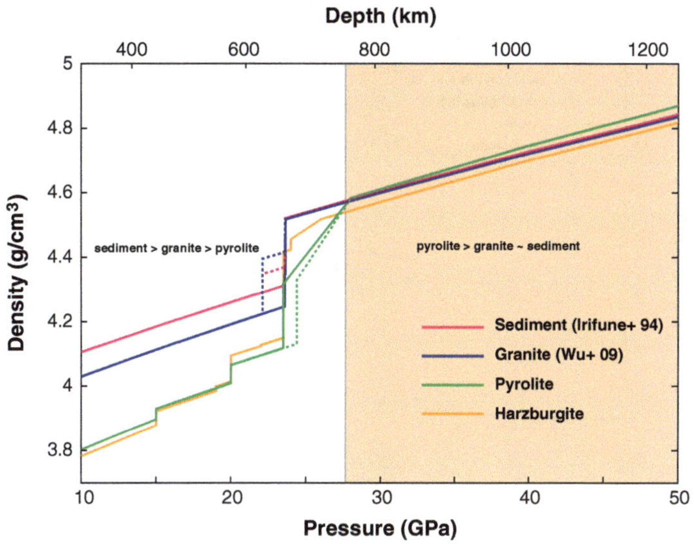

Fig. 8.2 Density profile calculated in the pressure range between 10 and 50 GPa at the static temperature. The phase transition pressures are shifted along the normal (*solid lines*) and slab (*dotted lines*) geotherms. The granitic materials are less dense than that of the pyrolitic composition over ~27 GPa in the LM. Although calculations are conducted at the static temperature (0 K), we consider the phase transition pressures along the normal geotherm (Brown and Shankland 1981) and the slab geotherm which is 500 K lower than the normal geotherm

proportions, we compute velocities for each assemblage (Fig. 8.3). The seismic velocity increases of +2.2 and +1.7 % for the P-velocity and +3.1 and +2.4 % for the S-velocity are obtained associated with the grossular garnet dissociation at 23.6 GPa and the jadeite dissociation at 23.7 GPa along the normal geotherm for the sedimentary composite (Irifune et al. 1994), and the increases of +1.4 and +5.1 % for the P-velocity and +2.0 and +7.0 % for the S-velocity for the upper crust average composite (Wu et al. 2009), respectively. Since the calculated pressures at which these dissociations occur are close to each other along the normal geotherm (23.6 and 23.7 GPa), those in the granitic materials yield a strong velocity discontinuity at a depth in the vicinity of the post-spinel phase transition (23.5 GPa) (solid lines in Fig. 8.3). On the other hand, when the temperature of granitic materials is deviated from the normal geotherm, two dissociations could occur separately due to the different Clapeyron slopes as shown in Fig. 8.1. Along the slab geotherm, the seismic velocity increases of +2.3 and +1.7 % for the P-velocity and +3.2 and +2.3 % for the S-velocity are obtained associated with the grossular garnet dissociation at 23.6 GPa and the jadeite dissociation at 22.1 GPa for the sedimentary composite, and the increases of +1.5 and +5.0 % for the P-velocity and +2.2 and +6.8 % for the S-velocity for the upper crust average composite, respectively (dashed lines in Fig. 8.3). The transitions of K-hollandite and stishovite at around 28 and 65 GPa along the normal geotherm

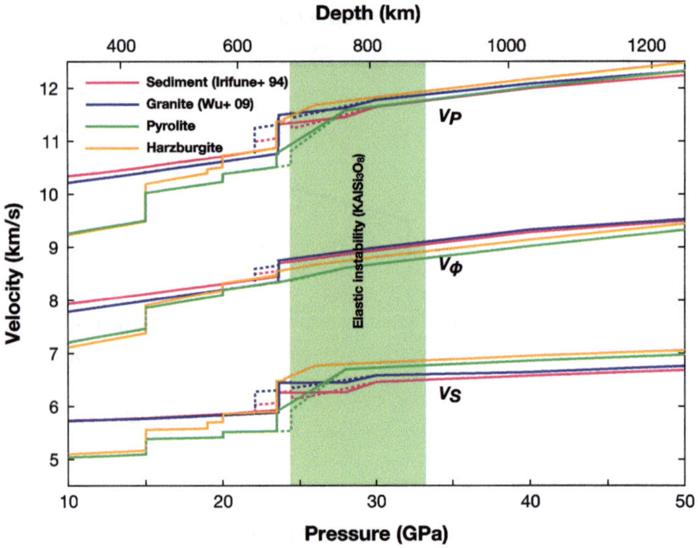

Fig. 8.3 Velocities for two granitic composites (Irifune et al. 1994; Wu et al. 2009), harzburgite and pyrolitic, at the static temperature. The phase transition pressures are shifted along the normal (*solid line*) and slab (*dotted line*) geotherms. The green shaded region indicates the pressure range for the ferroelastic anomaly associated with the phase transition between K-hollandite I and II. Although calculations are conducted at the static temperature (0 K), we consider the phase transition pressures along the normal geotherm (Brown and Shankland 1981) and the slab geotherm which is 500 K lower than the normal geotherm

and around 24 and 60 GPa along the slab geotherms, respectively, are known to accompany the elastic instabilities (Karki et al. 2001; Kawai and Tsuchiya 2013; Tsuchiya et al. 2004). Small anomalous pressure dependences of velocities seen especially in the P- and S-velocities in Fig. 8.3 are related to these instabilities.

8.3 Discussion and Geophysical Implications

8.3.1 Comparison with Elastic Properties for Ambient Mantle Material

The obtained density and seismic velocities of granitic materials are compared with those of pyrolite, a typical upper mantle composition model (e.g., Ringwood 1975). Major minerals in pyrolite are wadsleyite and majorite at 15−20 GPa (upper part of the MTZ), and ringwoodite and majorite at 20–23.5 GPa (lower part of the MTZ). Ringwoodite then dissociates into perovskite and ferropericlase at about 23.5 GPa. To model the pyrolitic assemblage, we approximate it by a composite of forsterite and majorite garnet including 12 and 7.5 mol% of Fe_2SiO_4 and $FeSiO_3$, respectively (e.g., Irifune 1987). The impurity effects such as iron for each mineral, that are usually substantial for density and shear modulus, are taken into account using those reported for perovskite (Tsuchiya and Tsuchiya 2006). As for the density, we consider only effects of mass change neglecting effects of volume change. This treatment might not be too rough at least for the Fe concentration considered here, but the effects should be calculated also for wadsleyite, ringwoodite, and majorite in the future. The partition coefficient between pv and fp is set to 0.3 (i.e., pv and fp include 7 and 18 mol% of $FeSiO_3$ and FeO, respectively) (Katsura and Ito 1996). The elastic constants for forsterite (da Silva et al. 1997), wadsleyite (Kiefer et al. 2001), ringwoodite (Kiefer et al. 1997), majorite (Li et al. 2007), and perovskite (Tsuchiya et al. 2004) are taken from literatures. The mineral proportions in pyrolite are assumed to be in a volume fraction spinel:majorite = 60:40 in the upper mantle and pv:fp = 80:20 in the lower mantle. We assume that the phase transition of majorite garnet occurs gradually from 23.5 to 28 GPa. Since the majorite–perovskite transition pressure varies depending strongly on the chemical composition (e.g., Irifune et al. 1996), contrary to the post-spinel phase boundary, we fix the transition pressure width irrespective of temperature. Also, we neglect Ca-perovskite in the LM because of its small amount with minor contributions. Resultant density variation of pyrolite, also shown in Fig. 8.2, indicates that the granitic materials are denser than pyrolite in the pressure range from 9 GPa (270 km) to 27 GPa (740 km). The seismic velocity jumps in pyrolite associated with the wadsleyite–ringwoodite (520 km) and the post-spinel (660 km) phase transitions are +1.4 and +2.4 % for the P-velocity, and +1.9 and +6.5 % for the S-velocity, respectively (Fig. 8.3). The magnitude of the velocity jump at the 660 km depth depends on how sharply majorite garnet dissociates (as mentioned, this reaction is assumed to occur gradually from 23.5 to 28 GPa in this study). The relative velocity difference between pyrolite and PREM (Dziewonski and Anderson 1981) in the MTZ seems to be relatively larger than that in the LM (Tsuchiya 2011). This is consistent with experimental results by Irifune et al. (2008), where the pyrolitic composition produces relatively slower velocities in the lower part of the MTZ.

8.3.2 Comparison with Elastic Properties for Harzburgite

We also computed density and velocities of harzburgite for comparison in the same matter as of pyrolite. Major minerals in harzburgite are wadsleyite and majorite at 15–19 GPa (upper part of the MTZ); ringwoodite, majorite, and stishovite at 19–22 GPa (middle part of the MTZ); and ringwoodite, majorite, and ilmenite at 22–24 GPa (lower part of the MTZ). Ringwoodite then dissociates into perovskite and ferropericlase at about 23.5 GPa. To model the harzburgite assemblage, we approximate it by a composite of forsterite and majorite garnet including 9 and 10 mol% of Fe_2SiO_4 and $FeSiO_3$, respectively (Irifune and Ringwood 1987). The mineral proportions in pyrolite are assumed to be in a volume fraction spinel:majorite = 80:20 in the upper mantle; spinel:majorite:stishovite = 88:10:2 in the middle MTZ;

spinel:majorite:ilmenite = 80:10:10 in the lower MTZ; and pv:fp = 80:20 in the lower mantle. We assume that the phase transition of majorite garnet occurs gradually from 23.5 to 26 GPa (Irifune and Ringwood 1987). The elastic constants for ilmenite are taken from da Silva et al. (1997) and Karki and Wentzcovitch (2002). Resultant density variation of harzburgite, also shown in Fig. 8.2, indicates that the harzburgite are denser than pyrolite in the pressure range from 19 GPa (580 km) to 25 GPa (700 km). This is consistent with experimental results by Irifune and Ringwood (1987) in the pressure range from 23.5 to 25 GPa. Although the density difference is very small, experimental study suggested that pyrolite is denser than the harzburgite in the pressure range from 19 to 23.5 GPa in contrast to our results. Another theoretical study also showed that harzburgite is slightly denser than pyrolite composite (Cobden et al. 2008), which is consistent with our study. This discrepancy implies that the density contrast between pyrolite and harzburgite in the lower MTZ is very small and dependent on small volume fraction of minerals and distribution of elements.

8.3.3 Velocity Contrast

Velocity contrasts are computed to highlight the differences between granitic and pyrolitic materials along the normal geotherm (Fig. 8.4). Velocity contrasts between pyrolite and basalt and between pyrolite and harzburgite at the LM pressures (Tsuchiya 2011) are also shown in Fig. 8.4 for comparison. The granitic materials have faster velocities than pyrolite after the coesite–stishovite phase transition of SiO_2 in granite over 9 GPa and even after the post-spinel transition of $(Mg,Fe)_2SiO_4$ in pyrolite. As the velocity contrast in the upper mantle condition is greatest, the granitic materials were supposed to be seismologically detected. This may be due to the size-dependent problem because the scale for granitic subduction channel above the subducted slab (less than 2 km) is smaller than wavelength of seismic waves. The velocity contrasts decrease with increasing pressure up to 23.5 GPa through the α–β and β–γ phase transitions of $(Mg,Fe)_2SiO_4$. Nevertheless, the P-, S-, and bulk sound velocities of the granitic materials remain +3, +7, and +2 % faster than those of pyrolite in the lower part of the MTZ. Even after the post-spinel phase transition, the velocities of the granitic materials are faster than those of pyrolite up to at least 25 GPa (700 km depth). This is due to the dissociation of jadeite and grossular garnet, although the upper boundary of the positive velocity contrast is closely related to the dissociation pressure width of majorite. Sharp negative anomalies of the velocity contrasts are seen at the bottom of the MTZ around 23.5 GPa. This is because three dissociation reactions of spinel, grossular, and jadeite occur in a narrow pressure range of 0.4 GPa (Fig. 8.1). Note that this anomaly strongly depends on the temperature condition, since the jadeite and spinel dissociations have positive and negative Clapeyron boundaries, respectively. In contrast, in the LM condition, at $P >$ ~30 GPa, the compressional and shear velocities of the granitic materials are about −1 and −4 % slower than those of pyrolitic materials, while the bulk sound velocity is +2 % faster,

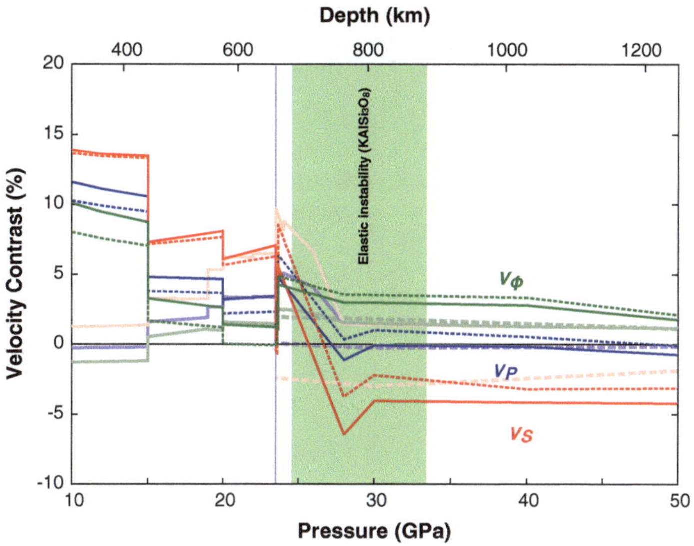

Fig. 8.4 Velocity contrasts relative to the pyrolitic composite ($(V_{\text{granite, harzburgite, sediment, or basalt}} - V_{\text{pyrolite}})/V_{\text{pyrolite}}$ (%) as a function of pressure. The phase transition pressures are shifted along the normal geotherm (Fig. 8.1). *Dark solid* and *dotted lines* indicate those for sediments (Irifune et al. 1994) and granite (Wu et al. 2009), respectively. *Light solid* and *dashed lines* indicate harzburgite (Irifune and Ringwood 1987) and basalt (Tsuchiya 2011), respectively

indicating that the subducted granitic materials can yield a strong anti-correlation between S and bulk sound velocity (Fig. 8.4). Since the granitic materials include K-hollandite and stishovite, which undergo ferroelastic phase transitions around 28 and 65 GPa, respectively (Kawai and Tsuchiya 2013; Tsuchiya et al. 2004), velocity anomalies, especially strong negative S-velocity contrasts, are produced at around these pressure ranges as also shown in Fig. 8.4. These behaviors could comport with observations of regional seismic discontinuities and scatterers in the uppermost LM (Kawakatsu and Niu 1994; Niu and Kawakatsu 1997; Kaneshima 2009) and in the mid-LM (Niu et al. 2003; Kaneshima and Helffrich 2009).

In contrast, along the slab geotherm, strong positive velocity contrasts are produced around the 660 km depth (Fig. 8.5). The velocity contrasts for the P-, S-, and bulk sound velocities reach about +9, +16, and +4 %, respectively. This could be consistent with evidence for the complexity around the 660-km discontinuity observed beneath North America and Izu–Bonin, where a cold subducted slab likely exists (Simmons and Gurrola 2000; Zhou et al. 2012) (Fig. 8.1). Except this depth range, the velocity contrasts are found insensitive to temperature. The density profiles calculated in this study suggest that if delaminated from the subducted slabs below 270 km, the granitic materials could accumulate at around 740 km depth due to their gravitational stability (Kawai et al. 2009, 2013). Since the granitic materials remain denser than the pyrolite until the completion of the majorite garnet dissociation in the pyrolite composition, they could penetrate into the LM accompanied with the slab subduction.

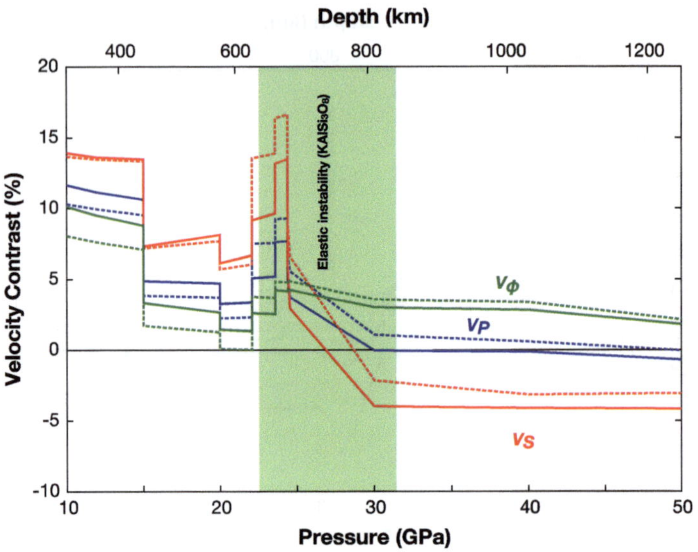

Fig. 8.5 Velocity contrasts from the pyrolitic composite as a function of pressure. The phase transition pressures are shifted along the slab geotherm (Fig. 8.1). *Solid* and *dotted lines* indicate the velocity contrasts for sediments (Irifune et al. 1994) and granite (Wu et al. 2009), respectively

8.3.4 Discussion

If only small volume fractions of continental material are being subducted, it may be difficult to detect them. Whether or not they are detectable depends on resolution of seismic data, i.e., wavelength of seismic waves and amount of seismic data available. If a considerable amount of granitic material accumulates, it may be detected by tomographic studies according to velocity contrasts shown in Figs. 8.4 and 8.5. If a small amount of granitic material is being subducted, it may be detected as seismic scatterers (e.g., Kaneshima 2009). As a new seismological technique such as (full-) waveform inversion has been developed and applied to data recently (e.g., Kawai et al. 2014; Konishi et al. 2014), the resolution in the deep mantle will be improved. Hence, velocity contrasts computed in this study will contribute to the interpretation of seismic images to consider chemical heterogeneities in the mantle in the future.

Whether or not continental crust stagnates at the base of the transition zone does not only depend on relative densities, but also depend on rheology, volume fractions, background mantle flow, grain size, and a whole load of other things. This will be a similar problem on the delamination of subducted oceanic crust (e.g., van Keken et al. 1996; Karato 1997). As for the delamination from a subducting slab above 270 km, Ichikawa et al. (2013a, 2015, this volume) conducted a numerical simulation of a granitic subduction channel based on the finite difference method and found that a sustainable thickness of the channel in the deep mantle is ~2–3 km and that its corresponding flux of continental materials integrated over the length of the current subduction zones is 2.2 km^3/year. This suggested that a considerable amount of granitic materials could be subducted below 270 km depth without delamination from the slab. The estimated total amount subducted to the deep mantle over 4 Gyr is found to be 10^{10} km^3. On the other hand, we have yet to study the other dynamical uncertainties related to the fate of continental material. It is essential to investigate whether or not continental material delaminates from the slab at the top of the lower mantle and whether granitic material is to rise back into the base of the transition zone. It depends on relative densities as well as rheology, volume fractions, and grain size. Further discussion should await these researches in the future.

Figures 8.4 and 8.5 both show a clear anti-correlation between V_φ and V_S in the LM. Although the basaltic materials similarly show the anti-correlation between V_φ and V_S in the LM (Tsuchiya 2011), the magnitude of the anti-correlation is much larger in the granitic materials than in the basaltic materials. This suggests that even smaller amount of the granitic materials can produce a large anti-correlation between V_φ and V_S in the LM. However, the basaltic oceanic crust is usually thought to be subducted in much larger amount than the granitic continental crust. These subducted crustal materials both are therefore consistent with a regional anti-correlation between V_φ and V_S seismologically observed in the LM (Niu et al. 2003). Recent studies have suggested that the anti-correlation between V_φ and V_S in the lower most mantle may be due to the presence of post-perovskite thermal effects (Wentzcovitch et al. 2006; Kawai and Tsuchiya 2009; Davies et al. 2012; Davies et al. 2015, this volume) or due to the wave propagation effects (Schuberth et al. 2012). These studies further pointed out that the depth, geographic location, extent, and existence of the anti-correlation strongly vary, depending on the model (Su and Dziewonski 1997; Kennett et al. 1998; Masters et al. 2000; Houser et al. 2008). However, the fact that they are also observed by normal-mode models (Ishii and Tromp 1999; Trampert et al. 2004; Mosca et al. 2012) indicates that they are not artifacts and favor a thermochemical origin. The detailed nature of these chemical heterogeneities is still debated (Deschamps et al. 2015, this volume). Deschamps et al. (2012) showed that an enrichment in recycled MORB would require very excess temperature and that an enrichment in iron is more likely. Although they do not account for thermal effects, our results (Figs. 8.4 and 8.5) indicate that these anti-correlations might be due to existence of continental materials.

Theoretical and experimental modeling both pointed out that the seismological models such as PREM have anomalously fast velocities around the bottom of the MTZ (~550–660 km) (Cobden et al. 2008; Irifune et al. 2008). Although this anomaly can be explained by the harzburgite-rich composition, the depth range where harzburgite is sufficiently denser than pyrolite composite is narrow (580–740 km). The subducted granitite material could be a candidate to explain the velocity anomalies around the 660-km discontinuity. The positive velocity contrasts produced by the granitic materials (Figs. 8.4 and 8.5) could explain this fast anomaly as suggested by Kawai et al. (2013).

If a considerable amount of granitic materials were accumulated at around 660 km, radioactive elements included could affect the thermal history in the mantle (Ichikawa et al. 2013a, b, 2015, this volume). Previous studies suggested that high concentrations of radioactive elements produce a temperature increase of ~200 K for 200 My (Senshu et al. 2009; Ichikawa et al. 2013a). This temperature increase will produce the upwelling plume due to the thermal instability, which might be related to the supercontinental cycle (Ichikawa et al. 2013b).

In this study, we investigated velocity contrasts between granitic and pyrolitic materials using their elasticity at static condition. The density contrast between the granitic material and the pyrolitic material will not be affected by the 1500–2000 K temperature increase in the mantle because their thermal expansivities are quite similar for many minerals in the depth range discussed in this study. On the other hand, velocities themselves may be more sensitive to temperature (e.g., Stixrude and Lithgow-Bertelloni 2005). However, as temperature dependences of many minerals in the depth range are similar, relative values such as velocity contrasts could be little influenced by a temperature increase of 2000 K. In addition, as shown in Figs. 8.4 and 8.5, the velocity contrasts of continental materials relative to pyrolite composition are positive in the depth range in the lowermost upper mantle. The mineral, which produces positive velocity anomaly for continental materials, is SiO_2 stishovite, while the surrounding ambient mantle is composed mainly of olivine (or spinel) and garnet. As minerals with high stiffness in general show weak temperature dependences on elasticity, the temperature dependence of stishovite is smaller than that of ambient mantle. At the temperature of 1500–2000 K in the mantle transition zone, the decrease of elastic velocity of ambient mantle relative to at static condition should be larger than that of continental materials. Accordingly, the velocity contrasts at ambient mantle condition could be even larger than those at static condition shown in this study. Hence, the conclusion in this chapter that granitic materials can produce substantial velocity anomalies in the mantle transition zone does not change.

8.4 Conclusions and Perspectives

In this study, we have calculated the elasticity of recycled granitic material in the deep mantle and found that the ongoing subducting granitic assemblage such as sediments and average upper crust rocks is much denser than pyrolite in the pressure range from 9 GPa (~270 km) to around 27 GPa (~740 km). Above this pressure, granitic material becomes less dense than pyrolite. This indicates that the granitic assemblage becomes gravitationally stable at the base of the MTZ. Results suggest a possibility that the granitic materials could accumulate around the 740 km depth if carried into the depth deeper than 270 km and segregated at some depth. Comparison of the velocities between granitic and pyrolitic materials shows that granitic materials can produce substantial velocity anomalies in the MTZ and the uppermost LM. Seismic observations such as anomalously fast velocities shown in global seismological models such as PREM, the regional complexity of 660-km discontinuity, and the scatterers in the LM could be related to the subducted granitic materials.

We investigated velocity contrasts between granitic and pyrolitic materials using their elasticity at static condition because even now it is challenging to compute elasticity for many minerals at high *P–T* condition. For example, previous studies on phase stability and elasticity of cubic Ca-perovskite (Capv) used small MD computational cell in order to reduce computational time (Li et al. 2006a, b; Adams and Oganov 2006). This led to insufficient convergence of computation as discussed in our recent paper (Kawai and Tsuchiya 2014). As elasticity of continental materials at mantle *P–T* condition is important, we are going to study elasticity of minerals composing granite at high *P–T* conditions in the future.

In this study, we did not consider effects of water and melting, which could affect elasticity of granitic material. We are going to study these effects on elasticity in the future. We also neglected the solid-solution effect to each mineral in granitic material such as sediment and average upper crust rock. In fact, as Ishii et al. (2012) recently reported that corundum is not stable but the assemblage of CF, Capv, St, and K-hol is stable for the continental crust materials in the lower mantle condition, the solid-solution effect for CF phase is important. Therefore, in order to obtain elastic properties for more realistic model for crustal material, the solid-solution effect in the granitic material in the upper and lower mantle condition will be studied in the future.

Acknowledgments We thank S. Yamamoto, H. Ichikawa, I. Jackson, and S. Maruyama for valuable discussion. We also thank J. Brodholt and an anonymous reviewer for constructive comments, which greatly improved our manuscript. This work was completed under the support in part of KAKENHI (Grant Nos. 23540560, 24840020, and 23224012).

References

Adams DJ, Oganov AR (2006) Ab initio molecular dynamics study of $CaSiO_3$ perovskite at P-T conditions on earth's lower mantle. Phys Rev B 73:184106

Akaogi M, Haraguchi M, Yaguchi M, Kojitani H (2009) High-pressure phase relations and thermodynamic properties of $CaAl_4Si_2O_{11}$ CAS phase. Phys Earth Planet Inter 173:1–6

Brown JM, Shankland TJ (1981) Thermodynamic parameters in the earth as determined from seismic profiles. Geophys J R Astr Soc 66:579–596

Campbell IH, Taylor SR (1983) No water, no granites—no oceans, no continents. Geophys Res Lett 10:1061–1064

Cobden L, Goes S, Cammarano F, Connolly JAD (2008) Thermochemical interpretation of one-dimensional seismic reference models for the upper mantle: evidence for bias due to heterogeneity. Geophys J Int 175:627–648

da Silva C, Stixrude L, Wentzcovitch RM (1997) Elastic constants and anisotropy of forsterite at high pressure. Geophys Res Lett 24:1963–1966

Davies DR, Goes S, Schuberth BSA, Bunge H-P, Ritsema J (2012) Reconciling dynamic and seismic models of earth's lower mantle: the dominant role of thermal heterogeneity. Earth Planet Sci Lett 353–354:253–269

Davies DR, Goes S, Lau HCP (2015) Thermally dominated deep mantle LLSVPs: a review. In: Khan A, Deschamps F (eds) The earth's heterogeneous mantle. Springer, Berlin

Deschamps F, Cobden L, Tackley PJ (2012) The primitive nature of large low shear-wave velocity provinces. Earth Planet Sci Lett 349–350:198–208

Deschamps F, Li Y, Tackley PJ (2015) Large-scale thermo-chemical structure of the deep mantle: observations and models. In Khan A, Deschamps F (eds) The earth's heterogeneous mantle, Springer, Berlin

Dziewonski AM, Anderson DL (1981) Preliminary reference earth model. Phys Earth Planet Int 25:297–356

Hofmann AW (1997) Mantle geochemistry: the message from oceanic volcanism. Nature 385:219–229

Houser C, Masters G, Shearer P, Laske G (2008) Shear and compressional velocity models of the mantle from cluster analysis of long-period waveforms. Geophys J Int 174:195–212

Ichikawa H, Kawai K, Yamamoto S, Kameyama M (2013a) Supply rate of continental materials to the deep mantle through subduction channels. Tectonophysics 592:46–52

Ichikawa H, Kameyama M, Kawai K (2013b) Mantle convection with continenal drift and heat source around the mantle transition zone. Gondwana Res 24:1080–1090

Ichikawa H, Kawai K, Yamamoto S, Kameyama M (2015) The effect of water on supply rate of continental materials to the deep mantle through subduction channels. In: Khan A, Deschamps F (eds) The earth's heterogeneous mantle, Springer, Berlin

Ishii M, Tromp J (1999) Normal-mode and free-air gravity constraints on lateral variations in velocity and density of earth's mantle. Science 285:1231–1236

Irifune T (1987) An experimental investigation of the pyroxene-garnet transformation in a pyrolite composition and its bearing on the constitution of the mantle. Phys Earth Planet Inter 45:324–336

Irifune T, Ringwood AE (1987) Phase transformations in a harzburgite composition to 26 GPa: implications for dynamical behavior of the subducting slab. Earth Planet Sci Lett 86:365–376

Irifune T, Higo Y, Inoue T, Kono Y, Ohfuji H, Funakoshi K (2008) Sound velocities of majorite garnet and the composition of the mantle transition zone. Nature 451:814–817

Irifune T, Ringwood AE, Hibberson WO (1994) Subduction of continental crust and terrigenous and pelagic sediments: an experimental study. Earth Planet Sci Lett 126:351–368

Irifune T, Koizumi T, Ando J (1996) An experimental study of the garnet– perovskite transformation in the system $MgSiO_3$–$Mg_3Al_2Si_3O_{12}$. Phys Earth Planet Inter 96:147–157

Ishii T, Kojitani H, Akaogi M (2012) High-pressure phase transitions and subduction behavior of continental crust at pressure-temperature conditions up to the upper part of the lower mantle. Earth Planet Sci Lett 357–358:31–41

Kaneshima S (2009) Seismic scatterers at the shallowest lower mantle beneath subducted slabs. Earth Planet Sci Lett 286:304–315

Kaneshima S, Helffrich G (2009) Lower mantle scattering profiles and fabric below Pacific subduction zones. Earth Planet Sci Lett 282:234–239

Karki BB, Wentzcovitch RM (2002) First-principles lattice dynamics and thermoplasticity of $MgSiO_3$ ilmenite at high pressure. J Geophys Res 107:2267. doi:10.1029/2001JB000702

Karki BB, Stixrude L, Wentzcovitch RM (2001) High-pressure elastic properties of major materials of earth's mantle from first principles. Rev Geophys 39:507–537

Katsura T, Ito E (1996) Determination of Fe-Mg partitioning between perovskite and magnesiowüstite. Geophys Res Lett 23:2005–2008

Karato S (1997) On the separation of crustal component from subducted oceanic lithosphere near the 660 km discontinuity. Phys Earth Planet Int 99:103–111

Kawakatsu H, Niu F (1994) Seismic evidence for a 920-km discontinuity in the mantle. Nature 371:301–305

Kawai K, Konishi K, Geller RJ, Fuji N (2014) Methods for inversion of body-wave waveforms for localized three-dimensional seismic structure and an application to D″ beneath Central America. Geophys J Int 197:495–524

Kawai K, Tsuchiya T (2009) Temperature profile in the lowermost mantle from seismological and mineral physics joint modeling. Proc Natl Acad Sci USA 106:22119–22123

Kawai K, Tsuchiya T (2010) Ab initio investigation of high-pressure phase relation and elasticity in the $NaAlSi_2O_6$ system. Geophys Res Lett 37:L17302. doi:10.1029/2010GL044310

Kawai K, Tsuchiya T (2012a) Phase stability and elastic properties of the NAL and CF phases in the $NaMg_2Al_5SiO_{12}$ system from first principles. Am Mineral 97:305–314

Kawai K, Tsuchiya T (2012b) First-principles investigations of elasticity and phase transition of grossular garnet. J Geophys Res 117:B02202. doi:10.1029/2011JB008529

Kawai K, Tsuchiya T (2012c) High-*P,T* phase relations in the $NaAlSi_2O_6$ system from first principles computation. Phys Chem Min 39:305–310

Kawai K, Tsuchiya T (2013) First principles study on the high-pressure phase transition and elasticity of $KAlSi_3O_8$ K-hollandite. Am Min 98:207–218

Kawai K, Tsuchiya T (2014) *P-V-T* equation of state of cubic $CaSiO_3$ perovskite from first principles computation. J Geophys Res Solid Earth 119:2801–2809

Kawai K, Tsuchiya T (2015) Elasticity and phase stability of pyrope garnet from ab initio computation. Phys Earth Planet Inter 240:125–131

Kawai K, Tsuchiya T, Tsuchiya J, Maruyama S (2009) Lost primordial continents. Gondwana Res 16:581–586

Kawai K, Yamamoto S, Tsuchiya T, Maruyama S (2013) The second continent: existence of granitic continental materials around the bottom of the mantle transition zone. Geosci Front 4:1–6

Kennett BLN, Widiyantoro S, van der Hilst RD (1998) Joint seismic tomography for bulk sound and shear wave speed in the earth's mantle. J Geophys Res 103:12469–12493

Kiefer B, Stixrude L, Hafner J, Kresse G (2001) Structure and elasticity of wadsleyite at high pressures. Am Min 86:1387–1395

Kiefer B, Stixrude L, Wentzcovitch RM (1997) Elastic constants and anisotropy of Mg_2SiO_4 spinal at high pressure. Geophys Res Lett 24:2841–2844

Komabayashi T, Maruyama S, Rino S (2009) A speculation on the structure of the D″ layer: the growth of anti-crust at the core-mantle boundary through the subduction history of the earth. Gondwana Res 15:342–353

Konishi K, Kawai K, Geller RJ, Fuji N (2014) Waveform inversion for localized 3-D seismic velocity structure in the lowermost mantle beneath the Western Pacific. Geophys J Int 199:1245–1267

Li L, Weidner DJ, Brodholt J, Price GD (2007) The effect of cation-ordering on the elastic properties of majorite: an ab initio study. Earth Planet Sci Lett 256:28–35

Li L, Weidner DJ, Brodholt J, Alfe D, Price GD, Caracas R, Wentzcovitch RM (2006a) Phase stability of $CaSiO_3$ perovskite at high pressure and temperature: insights from ab initio molecular dynamics. Phys Earth Planet Inter 155:260–268

Li L, Weidner DJ, Brodholt J, Alfe D, Price GD, Caracas R, Wentzcovitch RM (2006b) Elasticity of $CaSiO_3$ perovskite: implications for seismic discontinuities in the lower mantle. Phys Earth Planet Inter 155:249–259

Liu J, Topor L, Zhang J, Navrotsky A, Libermann RC (1996) Calorimetric study of the coesite-stishovite transformation and calculation of the phase boundary. Phys Chem Min 23:11–16

Martin H, Smithies RH, Rapp R, Moyen JF, Champion D (2005) An overview of adakite, tonalite-trondhjemite-granodiorite (TTG), and sanukitoid: relationships and some implications for crustal evolution. Lithos 79:1–24

Maruyama S, Sawaki Y, Ebisuzaki T, Ikoma M, Omori S, Komabayashi T (2014) Initiation of leaking earth: an ultimate trigger of Cambrian explosion. Gondwana Res 25:910–944

Mosca I, Cobden L, Deuss A, Ritsema J, Trampert J (2012) Seismic and mineralogical structures of the lower mantle from probabilistic tomography. J Geophys Res 117:B06304. doi:10.1029/2011JB008851

Nishiyama N, Rapp RP, Irifune T, Sanehira T, Yamazaki D, Funakoshi K (2005) Stability and P-V-T equation of state of $KAlSi_3O_8$-hollandite determined by in situ X-ray observations and implications for dynamics of subducted continental crust material. Phys Chem Min 32:627–637

Niu F, Kawakatsu H (1997) Depth variation of the mid-mantle seismic discontinuity. Geophys Res Lett 24:429–432

Niu F, Kawakatsu H, Fukao Y (2003) Seismic evidence for a chemical heterogeneity in the mid-mantle: a strong and slightly dipping seismic reflector beneath the Mariana subduction zone. J Geophys Res 108:2419. doi:10.1029/2002JB002384

Ringwood AE (1975) Composition and petrology of the earth's mantle. McGraw-Hill, New York

Scholl DW, von Huene R (2007) Crustal recycling at modern subduction zones applied to the past-issues of growth and preservation of continental basement crust, mantle geochemistry, and supercontinent reconstruction. Geol Soc Am Mem 200:9–32

Schuberth BSA, Zaroli C, Nolet G (2012) Synthetic seismograms for a synthetic earth: long-period P- and S-wave traveltime variations can be explained by temperature alone. Geophys J Int 200:1393–1412

Senshu H, Maruyama S, Rino S, Santosh M (2009) Role of tonalite-trodhjemite-granite (TTG) crust subduction on the mechanism of supercontinent breakup. Gondwana Res 15:433–442

Simmons NA, Gurrola H (2000) Multiple seismic discontinuities near the base of the transition zone in the earth's mantle. Nature 405:559–562

Styles E, Davies DR, Goes S (2011) Mapping spherical seismic into physical structure: biases from 3-D phase-transition and thermal boundary-layer heterogeneity. Geophys J Int 184:1371–1378

Su WJ, Dziewonski AM (1997) Simultaneous inversion for 3-D variations in shear and bulk velocity in the mantle. Phys Earth Planet Inter 100:135–156

Taylor SR, McLennan SM (1995) The geochemical evolution of the continental crust. Rev Geophys 33:241–265

Tsuchiya T (2011) Elasticity of subducted basaltic crust at the lower mantle pressures: Insights on the nature of deep mantle heterogeneity. Phys Earth Planet Inter 188:142–149

Tsuchiya T, Tsuchiya J (2006) Effect of impurity on the elasticity of perovskite and postperovskite: velocity contrast across the postperovskite transition in (Mg,Fe,Al)(Si,Al)O_3. Geophys Res Lett 33:L12S04. doi:10.1029/2006GL025706

Tsuchiya T, Kawai K, Maruyama S (2013) Expanding-contracting earth. Geosci Front 4:341–347

Tsuchiya T, Tsuchiya J, Umemoto K, Wentzcovitch RM (2004) Phase transition in $MgSiO_3$ perovskite in the earth's lower mantle. Earth Planet Sci Lett 224:241–248

Trampert J, Deschamps F, Resovsky JS, Yuen DA (2004) Probabilistic tomography maps significant chemical heterogeneities in the lower mantle. Science 306:853–856

van Keken PE, Karato S, Yuen DA (1996) Rheological control of oceanic crust separation in the transition zone. Geophys Res Lett 23:1821–1824

Wentzcovitch RM, Tsuchiya T, Tsuchiya J (2006) $MgSiO_3$ postperovskite at D″ condition. Proc Natl Acad Sci USA 103:543–546

Wu Y, Fei Y, Jin ZLX (2009) The fate of subducted upper continental crust: an experimental study. Earth Planet Sci Lett 282:275–284

Yamamoto S, Senshu H, Rino S, Omori S, Maruyama S (2009) Granite subduction: arc subduction, tectonic erosion and sediment subduction. Gondwana Res 15:443–453

Yu YG, Wentzcovitch RM, Tsuchiya T, Umemoto K, Weidner DJ (2007) First principles investigation of the postspinel transition in Mg2SiO4. Geophys Res Lett 34:L10306. doi:10.1029/2007GL029462

Zhou Y-Z, Yu X-W, Yang H, Zang S-X (2012) Multiplicity of the 660-km discontinuity beneath the Izu-Bonin area. Phys Earth Planet Inter 198–199:51–60

Chapter 9
Effect of Water on Subduction of Continental Materials to the Deep Earth

Hiroki Ichikawa, Kenji Kawai, Shinji Yamamoto and Masanori Kameyama

Abstract The flows in the subduction channels with wet mantle wedge are calculated by 1-D finite difference method with fine numerical resolutions. The water content largely affects the viscosity of the mantle wedge. Previous simulation result using dry rheology on the mantle wedge shows that viscosity of the subduction channels controls the process and that the sustainable thickness of the channel in the deep mantle is ~2–3 km. However, little is known about the effect of the water content in the mantle wedge on the subduction channels. Here, in order to estimate the supply rate of continental materials to the deep mantle with water-rich environment on the mantle wedge, we have conducted a numerical simulation of a subduction channel. The results show that the water content controls the flux of the continental materials especially when temperature of mantle wedge is high. Therefore, the water content of the mantle wedge can be more important in the ancient mantle because of its high temperature.

H. Ichikawa (✉) · M. Kameyama
Geodynamics Research Center, Ehime University, 2-5, Bunkyo-cho, Matsuyama, Ehime 790-8577, Japan
e-mail: h-ichikawa@sci.ehime-u.ac.jp

H. Ichikawa · K. Kawai
Earth-Life Science Institute, Tokyo Institute of Technology, Ookayama 2-12-1, Meguro, Tokyo 152-8550, Japan

K. Kawai
Department of Earth and Planetary Sciences, Tokyo Institute of Technology, Ookayama 2-12-1, Meguro, Tokyo 152-8551, Japan

S. Yamamoto
Department of Earth Science and Astronomy, Graduate School of Arts and Sciences, University of Tokyo, Komaba 3-8-1, Meguro, Tokyo 153-8902, Japan

Present Address
K. Kawai
Department of Earth Science and Astronomy, Graduate School of Arts and Sciences, University of Tokyo, Komaba 3-8-1, Meguro, Tokyo 153-8902, Japan

© Springer International Publishing Switzerland 2015
A. Khan and F. Deschamps (eds.), *The Earth's Heterogeneous Mantle*, Springer Geophysics, DOI 10.1007/978-3-319-15627-9_9

Keywords Subduction erosion · Sediment subduction · Subduction channel · Ultrahigh-pressure metamorphic rocks

9.1 Introduction

Evolution of life on the Earth is strongly related to the oceans and the continents, both of which are unique to the Earth (Campbell and Taylor 1983). Thermochemical evolution of the Earth's surface environment may be controlled from the continental materials, into which a large amount of incompatible and radiogenic elements are fractionated. In addition, granite is considered to be an optimal source of nutrients for life (Maruyama et al. 2013). Therefore, the growth of continents is most likely to play an essential role in the Earth's evolution.

According to recent geological studies, a significant amount of granitic crust is estimated to have been lost from the Earth's surface despite its buoyancy in the mantle. The granitic materials are eliminated due to crustal delamination (~1.1 km^3/year) (Clift et al. 2009), continental collision (~0.4–0.7 km^3/year) (Clift et al. 2009; Stern and Scholl 2010), and subduction at the ocean–continent subduction zone (~2.5–3 km^3/year) (Clift et al. 2009; Stern and Scholl 2010). The elimination of granitic materials is consistent with a fact that the amount of existing Archean granite is relatively small for the present Earth, although a large amount of the continental crust should have been produced during the Archean because of the high potential temperature in the mantle (Armstrong 1991; Belousova et al. 2010). On the other hand, According to geochemical studies, recycling of the subducted continental materials is suggested (e.g., Hofmann 1997). It is also suggested that there are reservoirs of continental materials in the mantle that has been effectively isolated for 2–3 Gyr (Murphy et al. 2002). Therefore, quantitative estimates on subduction of the continental materials into the deep Earth is essential to deepen the understanding of continental growth and chemical evolution of the mantle.

At the ocean–continent convergent plate boundaries, the granitic materials are conveyed through narrow subduction channels that are located between subducting slabs and mantle wedges (Cloos and Shreve 1988a, b). The thickness of the channel at the Earth's surface is estimated by geological and seismological studies as less than 2–3 km (Clift and Vannucchi 2004; Moore et al. 2007; Collot et al. 2011). The following three mechanisms are considered to form the channels (see Yamamoto et al. 2009a for review): tectonic erosion (Von Huene and Scholl 1991), sediment–trapped subduction, and direct subduction of immature oceanic arcs on oceanic micro plates (Yamamoto et al. 2009b).

On the other hand, according to a 1-D numerical simulation, the subduction channels of thickness of 2–3 km at the surface are able to reach 270 km in depth (Ichikawa et al. 2013b). This suggests that a large portion of low-density materials such as upper continental crust can sink into the deep mantle as pointed out by Dobrzhinetskaya et al. (2006), Irifune et al. (1994), and Afonso and Zlotnik (2011). In addition, experimental and theoretical studies showed that the density of continental materials such as Archean granite is higher than the mantle peridotite in between 270 and 800 km in depth (Irifune et al. 1994; Kawai et al. 2009, 2013; Kawai and Tsuchiya 2010, 2012). Therefore, the mantle transition zone is considered to be a reservoir of the continental materials (Kawai et al. 2009, 2010, 2013).

Previous studies on the subduction of continental materials have been conducted by using 1-D models (Gerya and Stöckhert 2002; Raimbourg et al. 2007; Ichikawa et al. 2013b) and 2-D or 3-D mantle convection simulations (e.g., Burov et al. 2001; Gerya et al. 2002; Sobolev and Babeyko 2005; Gerya and Stöckhert 2006; Currie et al. 2007; Babeyko and Sobolev 2008; Warren et al. 2008; Li et al. 2011; Sizova et al. 2012). The formation and exhumation of high- to ultrahigh-pressure (HP-UHP) metamorphic rocks in the continental subduction or collision zones are often the main interest in these studies, which provide evidence that these materials are able to return from the depth more than ~100 km. The exhumation from the subduction channels depends on the balance between shear traction exerted by the slab and buoyancy of the channels. When the shear traction is not strong enough, some of continental materials are expected to detach from the channel owing to their compositional buoyancy. According to 2-D simulation (Currie et al. 2007), the detachment of the channel occurs if its thickness is larger than ~350 m. Meanwhile, 1-D simulation (Ichikawa et al. 2013b) suggested that the maximum sustainable thickness of a subduction channel from surface to 270 km in depth is up to 2–3 km depending especially on the viscosity in the subduction channels. The difference between 1-D and 2- or 3-D is probably due to the difference of model geometries, such as curvature of the plate, or the difference of numerical resolution. Highly fine grids must be needed to capture the flow field in the numerical domain, where the viscosity is changed largely between mantle wedge and the subduction channels. The detachment is also controlled by the viscosity of the rock in the channels (Gerya and Stöckhert 2002; Raimbourg et al. 2007; Warren et al. 2008). Gerya and Stöckhert (2002) showed by using a 1-D analytical model that the exhumation rate is higher for a Newtonian rheology compared to a power-law rheology. Besides, sudden decrease of the viscosity by the eclogitization causes a detachment of the channels because of the decrease of maximum possible downward flow (Raimbourg et al. 2007).

Meanwhile, the effect of viscosity of the mantle wedge has not been thoroughly investigated in contrast to the effect of the viscosity of the subduction channels. Only dry rheology is used for the mantle wedge in the previous 1-D analysis (Ichikawa et al. 2013b). The viscosity of the mantle wedge can be widely affected mainly by temperature and water content. According to a 2-D simulation (Gerya et al. 2002), progressive hydration of the mantle wedge causes widening of the subduction channels. It is also considered that the sudden change of the viscosity with depth might induce the return flow of the continental materials to the surface. This is because the drag force from the mantle wedge, which is normally in the upward direction, increases as the mantle viscosity increases.

In this study, we performed a numerical simulation of a subduction channel focusing on the viscosity of the mantle wedge depending on the water content by using a 1-D model with non-Newtonian rheology, such as Gerya and Stöckhert (2002) or Ichikawa et al. (2013b). As will be shown later, flux in the channel is larger when temperature and the water content of the mantle wedge are higher.

9.2 Methods

9.2.1 Description of Model

In this study, a 1-D channel flow using a water-rich mantle rheology is calculated by using a model developed by Ichikawa et al. (2013b). In the model, a subduction channel of continental materials is located between a subducted lithospheric slab and a peridotite mantle wedge (Fig. 9.1a). On the surface, the materials are supplied by tectonic erosion or sediment subduction. The flow in the channel is restricted to the direction parallel to the slab by the relatively rigid both boundaries. In the channel, downward flow is driven by the slab subduction and impeded by their own buoyancy and forced circulation.

Figure 9.1b shows the description of the numerical model including the slab, the subduction channel, and the mantle wedge. The x- and y-directions are perpendicular and parallel to the slab, respectively. Here, only y-component of the velocity ($u(x)$) is considered in the region between $x = 0$, which is the slab-subduction channel boundary, and $x = L$. By changing the gravitational acceleration, we can also model the effect of slab dip angle (θ_d) In order to accurately reproduce the flow in the channel, L is set to 100 km.

$u(x)$ is numerically calculated at each given depth by solving the momentum equation derived by a balance between buoyancy and viscous force in the channel given by

$$\frac{\mathrm{d}\sigma}{\mathrm{d}x} + \Delta\rho g = 0, \quad (0 < x \leq D) \quad \text{and} \quad \frac{\mathrm{d}\sigma}{\mathrm{d}x} = 0, \quad (D < x \leq L), \tag{9.1}$$

where η, $\sigma = \eta\left(\frac{\mathrm{d}u}{\mathrm{d}x}\right)$, $\Delta\rho = \rho_g - \rho_m(<0)$, and $g(<0)$ are viscosity, stress, density difference between the subduction channel and the mantle (ρ_g and ρ_m are densities of granite and mantle, respectively), and gravitational acceleration, respectively. In a range, $D < x \leq L$, buoyancy force is zero so that only stress from its boundaries drives the flow in the mangle wedge. As previously described, the calculations are 1-D in the x-direction (vertical to the slab) (Fig. 9.1). Variation of a current on the y-direction is ignored because rate of change of velocity is assumed to be small and the direction of the flow is restricted from the two relatively rigid walls. We have carried out the calculation by using the finite difference method with very fine grids (10 m). $u(x)$ is set to U at $x = 0$ and U_2 at $x = L$.

We used the following three different zero-pressure density profiles for the continental crust materials in the channel: I1994, O1998, and W2009 based on Irifune et al. (1994), Ono (1998), and Wu et al. (2009), respectively (Fig. 9.2). The density of the mantle (ρ_m) is fixed to be $\rho_m = 3350\,\mathrm{kg/m^3}$ (Kawai et al. 2009). Therefore, $\Delta\rho$ is varied only by changing ρ_g. Velocity is fixed at the slab-channel boundary ($U = -8$ cm/year).

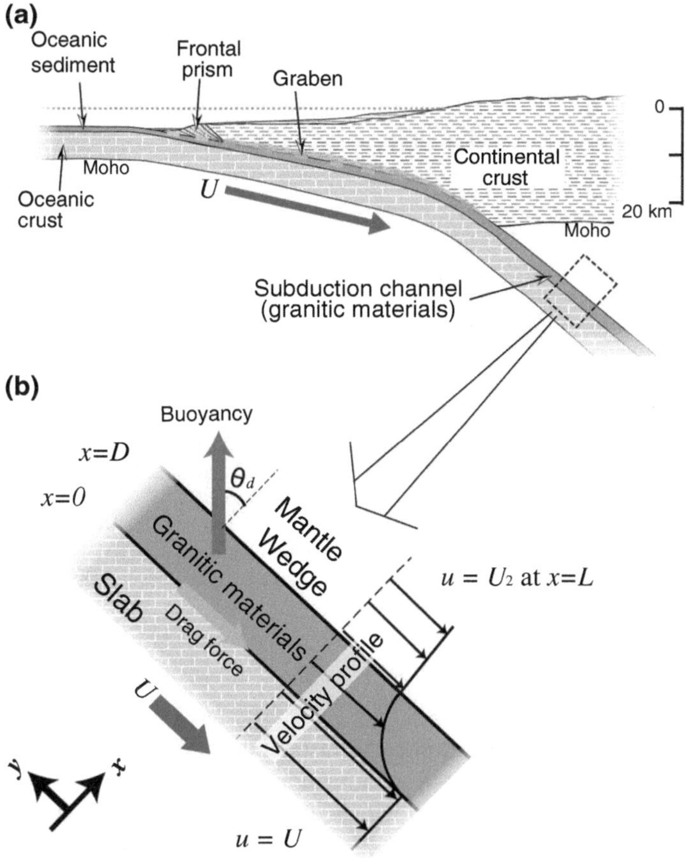

Fig. 9.1 Numerical setting modified after Ichikawa et al. (2013b). **a** Diagram of a subduction channel of continental materials with a subducting slab. Sediments and eroded continental materials form subduction channels. The flow in the subduction channel is determined by a balance between drag force from the slab and its own buoyancy. The slab dip angle is θ_d. **b** Steady one-dimensional model used in this study. The motion in the subduction channel is restricted to the y-direction (slab parallel direction). The x-direction is the slab-normal direction. The origin of the x axis is taken at the slab-subduction channel boundary. D is the thickness of the channel and L is the numerical domain size ($L = 100$ km). U and U_2 are the y-component of the velocities of the subducting slab and the mantle wedge at $x = 0$ and $x = L$, respectively

9.2.2 Rheological Parameters

Dislocation creep is assumed for the rheology of the materials because it is dominant at >80 km depth (Afonso and Zlotnik 2011). It is based on the following relations: $\dot{\varepsilon} = A\sigma^n \exp\left(\frac{-(E^*+PV^*)}{RT}\right)$, where E^*, V^*, and $\dot{\varepsilon}$, are activation energy, activation volume, and strain rate, respectively. The effective viscosity (η) is given

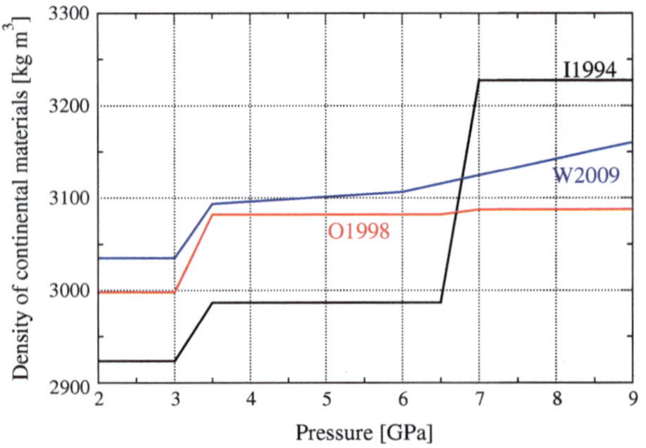

Fig. 9.2 Assumed density profiles of continental materials. Three zero-pressure density profiles are used: I1994, O1998, and W2009. The density profiles are calculated from phase proportions of experimental results: (I1994, Irifune et al. 1994; O1998, Ono 1998; W2009, Wu et al. 2009). The quartz–coesite transition causes the density jump between 3 and 3.5 GPa. A density jump at 6.5–7 GPa at I1994 is caused by the generation of wadsleyite. Continental materials are denser than the mantle density (3350 kg/m^3 in this study) at >9 GPa because of the coesite–stishovite transition

by $\eta = \frac{1}{2}(\dot{\varepsilon}_E)^{\frac{1-n}{n}}\left(A\exp\left(\frac{-(E^*+PV^*)}{RT}\right)\right)^{-\frac{1}{n}}$, where $\dot{\varepsilon}_E = \left(\frac{1}{2}\dot{\varepsilon}_{ij}\dot{\varepsilon}_{ij}\right)^{\frac{1}{2}} = \frac{1}{2}\left(\frac{du}{dx}\right)^{\frac{1}{2}}$ is the second invariant of the strain rate tensor. For subduction channels, we used quartz-controlled rheology for $P < 3.5$ GPa and polycrystalline coesite rheology for $P \geq 3.5$ GPa as in the previous study (Ichikawa et al. 2013b). Namely, $A = 6.7 \times 10^{-6}\,\mathrm{s}^{-1}\,\mathrm{Mpa}^{-n}$, $n = 2.4$, $E^* = 156\,\mathrm{kJ/mol}$, and $V^* = 0$ (Ranalli 1995, 1997) for $P < 3.5$ GPa and $A = 10^{-1.6}\,\mathrm{s}^{-1}\,\mathrm{Mpa}^{-n}$, $n = 2.9$, $E^* = 261$ kJ/mol, and $V^* = 0$ (Renner et al. 2001) for $P \geq 3.5$ GPa. For the mantle wedge, flow law parameters proposed by Kawazoe et al. (2009) are used for dry and wet mantle rheologies. $A_d = 10^{5.04}\,\mathrm{s}^{-1}\,\mathrm{MPa}^{-n}$, $n_d = 3.5$, $E_d^* = 530\,\mathrm{kJ/mol}$, and $V_d^* = 17 \times 10^{-6}\,\mathrm{m}^2/\mathrm{mol}$ for the dry mantle rheology and $A_w = 10^{2.9}\,\mathrm{s}^{-1}\,\mathrm{MPa}^{-\mathrm{n}}$, $n_w = 3.0$, $E_w^* = 470$ kJ/mol for the wet mantle rheology, where subscript d and w denote dry and wet, respectively. The strain rate is given by

$$\dot{\varepsilon} = A_d\sigma^{n_d} exp\left(\frac{-(E_d^* + PV_d^*)}{RT}\right) + A_w f_{H_2O}^r \sigma^{n_w} exp\left(\frac{-(E_w^* + PV_w^*)}{RT}\right), \quad (9.2)$$

where $f_{\mathrm{H_2O}}$ is water fugacity and r is a constant ($r = 1.2$) (Karato and Jung 2003). The water fugacity is calculated using the thermodynamics data given by Pitzer and Sterner (1995). Equations (9.1) and (9.2) are iteratively calculated to determine $u(x)$. The viscosity of the granite, dry mantle and, wet mantle as a function of temperature at 9 GPa (270 km in depth) is displayed as an example at Fig. 9.3. The figure clearly shows that the viscosity of the granite is much lower than that of the mantle wedge irrespective of its rheology. The figure also shows that the difference of the viscosity decreases as temperature increases.

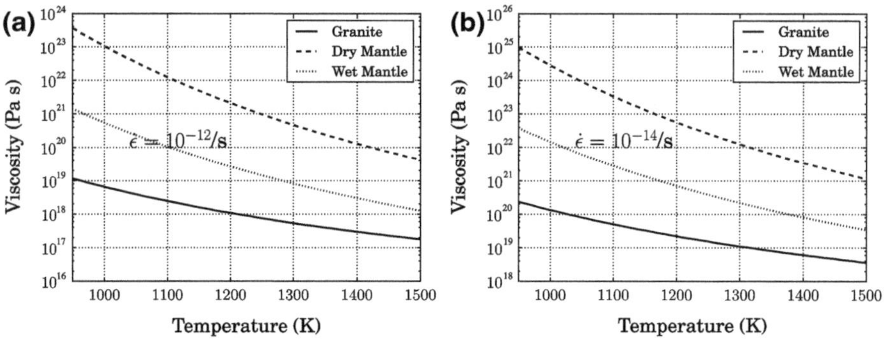

Fig. 9.3 Viscosities of dry mantle, wet mantle, and dry granite as a function of temperature at 9 GPa (270 km in depth) at **a** $\dot{\varepsilon} = 10^{-14}$/s and **b** $\dot{\varepsilon} = 10^{-12}$/s. As temperature increases, the viscosity difference between viscosity of mantle and granite becomes smaller

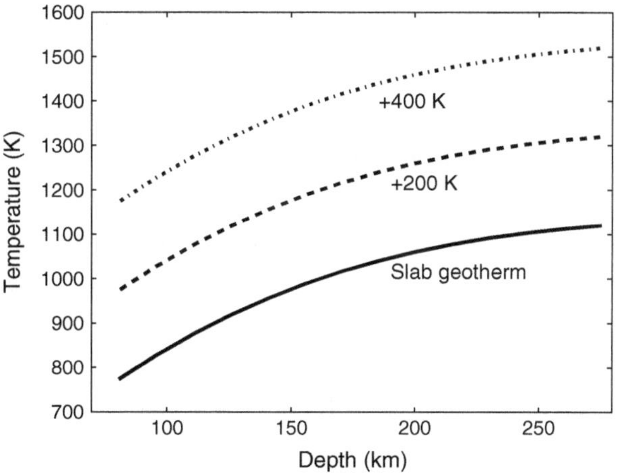

Fig. 9.4 Assumed temperature profiles. The slab geotherm is the temperature of the slab surface of a convection simulation with a fixed overriding plate (Eberle et al. 2002). Lines denoted as +200 K and +400 K are temperature profiles, which are 200 K and 400 K higher in temperature than the slab geotherm throughout the depth

9.2.3 P–T Conditions

We used same depth dependences on the temperature profiles as Ichikawa et al. (2013b) (Fig. 9.4). Slab geotherm is the slab surface temperature profiles derived from a convection simulation (Eberle et al. 2002). In order to check the dependence on temperature, two additional profiles (+200 K and +400 K) are also used. Other temperature profiles are also discussed later (Sect. 9.4.4). The calculations were carried out up to 9 GPa (~270 km depth), where the coesite–stishovite transition takes place (Zhang et al. 1996). Pressure and depth (z) are mutually exchanged by using $P = (0.0336z/\text{km}-0.236)$ GPa, which is estimated from a seismic observation (Dziewonski and Anderson 1981).

The viscosity profile of dry mantle wedge, wet mantle wedge, and dry granite (continental materials) along the slab geotherm is shown in Fig. 9.5. As seen in

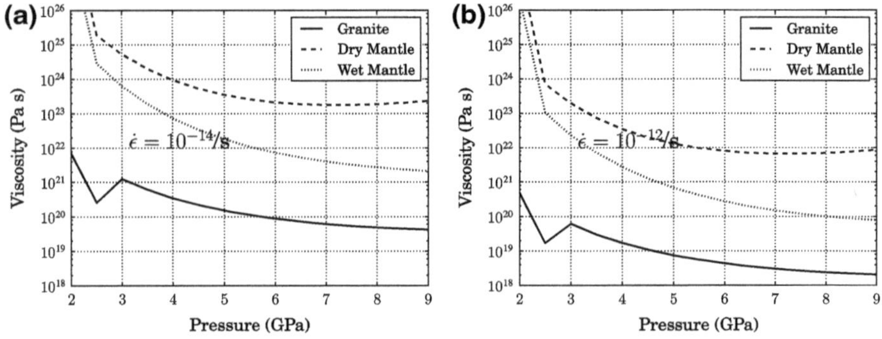

Fig. 9.5 Viscosity profiles for dry mantle, wet mantle, and dry granite along the slab geotherm at **a** $\dot{\varepsilon} = 10^{-14}$/s and **b** $\dot{\varepsilon} = 10^{-12}$/s

Fig. 9.3, viscosity of granite is smallest in the three cases. Figure 9.5 also shows that effect of the water content of the mantle wedge becomes important as the depth increases.

9.3 Analytical Treatment

In this subsection, an analytical solution at a simple boundary condition is derived. For the sake of simplicity, power-law formula can be written as, $\tau = \eta_g \left|\frac{du}{dx}\right|^{\nu_g - 1} \frac{du}{dx}$, for continental crust materials and, $\tau = \eta_m \left|\frac{du}{dx}\right|^{\nu_m - 1} \frac{du}{dx}$, for ambient mantle materials, where $\nu_g = 1/n_g$, $\nu_m = 1/n_m$, $\eta_g = \left(2A_g \exp\left(\frac{-(E_g^* + PV_g^*)}{RT}\right)\right)^{-\frac{1}{n_g}}$, and $\eta_m = \left(2A_m \exp\left(\frac{-(E_m^* + PV_m^*)}{RT}\right)\right)^{-\frac{1}{n_m}}$.

The following equations are derived by taking the integral of mechanical equation (Eq. (9.1)) with respect to x.

$$\tau = -\Delta\rho g x + A, \quad (0 < x \le D) \tag{9.3}$$

and

$$\tau = B, \quad (D < x \le L), \tag{9.4}$$

where A and B are constants of integration. The equations lead following two types of solutions depending on the boundary conditions.

1. $\frac{du}{dx} \ge 0$ in $(0 < x \le L)$
2. For D' $(0 < D' \le D)$, $\frac{du}{dx} \ge 0$ in $(0 < x \le D')$ and $\frac{du}{dx} < 0$ in $(D' < x \le L)$.

Solutions of type 2 must have a negative velocity at somewhere. We only analyze solutions of type 1, because we are interested in a solution that has the maximum downward flux of continental materials. In this case, Eqs. (9.3) and (9.4) are transformed to

$$\frac{\mathrm{d}u}{\mathrm{d}x} = \left(\frac{-\Delta\rho g x + A}{\eta_g}\right)^{1/\nu_g}, \quad (0 < x \le D), \tag{9.5}$$

and

$$\frac{\mathrm{d}u}{\mathrm{d}x} = \left(\frac{B}{\eta_m}\right)^{1/\nu_m}, \quad (D < x \le L), \tag{9.6}$$

respectively. Velocity profile is derived by integration of the above equations together with the boundary condition at $x = 0$ $(u = U)$,

$$u = \frac{\nu_g}{\nu_g + 1}\frac{\eta_g}{\Delta\rho g}\left[\left(\frac{A}{\eta_g}\right)^{\frac{\nu_g+1}{\nu_g}} - \left(\frac{-\Delta\rho g x + A}{\eta_g}\right)^{\frac{\nu_g+1}{\nu_g}}\right] + U, \quad (0 < x \le D), \tag{9.7}$$

and

$$u = \left(\frac{B}{\eta_m}\right)^{1/\nu_m} x + C, \quad (D < x \le L), \tag{9.8}$$

where C is a constant of integration. Here, we introduce downward flux of continental materials (q) as

$$q \equiv -\int_0^D u\mathrm{d}x = -\left[\frac{\nu_g^2}{(\nu_g+1)(2\nu_g+1)}\frac{\eta_g^2}{\Delta\rho^2 g^2}\left(\frac{-\Delta\rho g x + A}{\eta_g}\right)^{\frac{2\nu_g+1}{\nu_g}} + \frac{\nu_g}{\nu_g+1}\frac{\eta_g}{\Delta\rho g}\left(\frac{A}{\eta_g}\right)^{\frac{\nu_g+1}{\nu_g}} x + Ux\right]_0^D, \tag{9.9}$$

Therefore, only A is needed to evaluate the flux in the subduction channel.

On the other hand, continuity of velocity and stress at $x = D$ require the following two relations for the constants of integration.

$$B = -\Delta\rho g D + A, \tag{9.10}$$

and

$$\frac{\nu_g}{\nu_g + 1}\frac{\eta_g}{\Delta\rho g}\left[\left(\frac{A}{\eta_g}\right)^{\frac{\nu_g+1}{\nu_g}} - \left(\frac{-\Delta\rho g D + A}{\eta_g}\right)^{\frac{\nu_g+1}{\nu_g}}\right] + U = \left(\frac{B}{\eta_m}\right)^{1/\nu_m} D + C. \tag{9.11}$$

The boundary condition used in the simulation, $u = U_2$ at $x = L$, leads to complicated set of equations, which can be numerically calculated. Here, we seek a solution in the case of $\mathrm{d}u/\mathrm{d}x = 0$ at $x = L$, which is corresponding to a free-slip condition at $x = D$, because $\mathrm{d}u/\mathrm{d}x$ is constant throughout the mantel wedge

$(D < x \leq L)$ according to Eq. (9.1). In this case, the constants of integration are derived as $A = \Delta\rho g D$ and $B = 0$.

Therefore, downward flux of continental materials in the channel becomes

$$q = \frac{\nu_g}{2\nu_g + 1}\left(\frac{\Delta\rho g}{\eta_g}\right)^{\frac{1}{\nu_g}} D^{\frac{2\nu_g+1}{\nu_g}} + UD. \tag{9.12}$$

The flux is convex upward as a function of D. The maximum of q as a function of D is derived from $\mathrm{d}q/\mathrm{d}D = 0$ as follows:

$$q_{max} = \frac{\nu_g + 1}{2\nu_g + 1}\left(\frac{\Delta\rho g}{\eta_g}\right)^{-\frac{1}{\nu_g+1}} (-U)^{\frac{2\nu_g+1}{\nu_g+1}}, \tag{9.13}$$

when

$$D = \left(\frac{\Delta\rho g}{\eta_g}\right)^{-\frac{1}{\nu_g+1}} (-U)^{\frac{\nu_g}{\nu_g+1}}, \tag{9.14}$$

where $q_{\max}$ is the maximum of q. Here, we define an effective thickness of the channel (Q) as

$$Q = \frac{1}{U}\int_0^D u\mathrm{d}x, \tag{9.15}$$

Since it has a dimension of length, Q represents the effective thickness of the subduction channel. The maximum of the effective thickness ($Q_{\max}$) is written as

$$Q\max = \frac{n_g + 1}{n_g + 2}\left\{2A_g \exp\left(\frac{-(E_g^* + PV_g^*)}{RT}\right)\right\}^{-\frac{1}{n_g+1}} (\Delta\rho g)^{-\frac{n_g}{n_g+1}} (-U)^{\frac{1}{n_g+1}}. \tag{9.16}$$

This is same as the dimension analysis except the coefficient, $\frac{n_g+1}{n_g+2}$ (Ichikawa et al. 2013b). $Q_{\max}$ determines the maximum amount of the continental materials conveyed to the deep mantle. Effects of the slab dip angle can be taken into account by replacing g by $g \sin\theta_d$.

Figure 9.6 shows a simulation result and an analytical solution (Eq. 9.15) in a same parameter set, where the viscosity of the mantle is set to zero. The good agreement shows a validity of the simulation code and its numerical resolution.

In order to simplify expressions, we substitute a typical value of dislocation creep ($n_g = 3$),

$$Q_{max} \propto \left\{2A_g exp\left(\frac{-(E_g^* + PV_g^*)}{RT}\right)\right\}^{-\frac{1}{4}} (\Delta\rho g)^{-\frac{3}{4}} (-U)^{\frac{1}{4}}. \tag{9.17}$$

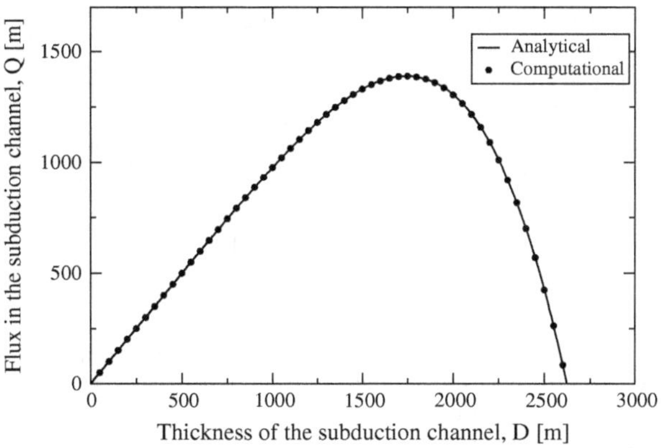

Fig. 9.6 Plots of Q. The *solid line* shows the profile given by Eq. (9.15) while *dots* show the computed one. The parameters are same as Fig. 9.7 except for the viscosity of the mantle wedge. Here, the viscosity of the mantle wedge is taken as zero

This shows that the effective thickness is not so sensitive to the slab velocity, while it is relatively sensitive to the density difference. Although the boundary condition is simplified, this equation coincides well with the results of numerical simulations (Ichikawa et al. 2013b).

9.4 Results

9.4.1 Control Experiments

We will focus our attention on the following cases: $U = -8$ cm/year, $U_2 = 0$ cm/year, $P = 6.5$ GPa (~200 km in depth), $\theta_d = 90°$, $g = -10$ m/s^2 with a density profile of I1994 (Fig. 9.2). In order to roughly incorporate the effect of higher temperature on mantle wedge, we tested an additional case using one-tenth smaller values for η_m than that of wet mantle rheology, which is attained by changing A_m. This case is named "hot wet" in figure legends. The effect of the temperature variations in x-direction is discussed later (Sect. 9.4.5). Figure 9.7a–c shows $u(x)$ for the three different values for D and the mantle rheologies. The results are very similar both for the case of dry mantle and wet mantle. For all the cases, $u(x)$ increases as x increases when $D = 500$ m (Fig. 9.7a) or $D = 1500$ m (Fig. 9.7b). $u(x)$ in the mantle wedge is almost zero at the cases of dry and wet mantle rheologies due to its relatively high viscosity. On the other hand, in the case of "hot wet," velocity in the mantle wedge is not zero and downward velocity in the subduction channel is larger. When $D = 3500$ m (Fig. 9.7c), the velocity profile has an upward component due to its buoyancy at $x \geq$ ~1000 m. This upward flow corresponds to the return flow to the surface and therefore, a channel with $D = 3500$ km cannot be sustained.

The difference between dry, wet, and "hot wet" rheologies are also clearly seen in the flux or the effective thickness of the channel [Q (Eq. 9.15)] as a function of D (Fig. 9.7d). Comparing Q of three different rheologies on the mantle wedge (Fig. 9.7d), we conclude that the viscosity of not only the channels but also the

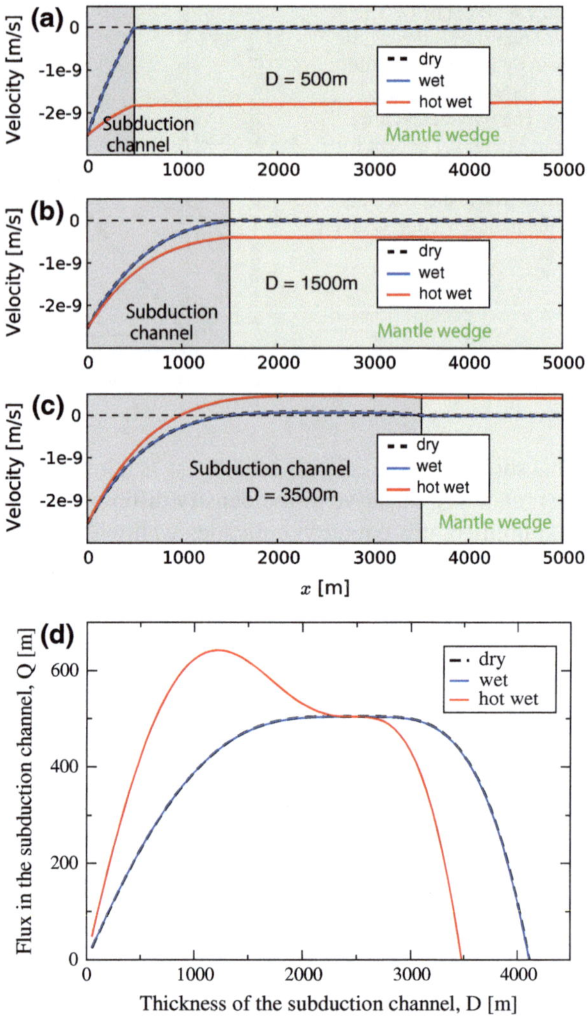

Fig. 9.7 **a–c** Velocity profiles for $U = 8$ cm/year and $U_2 = 0$ cm/year, at 6.5 GPa (200 km) for I1994 density profile for three thicknesses of the subduction channel, D, of 500 m **a**, 1500 m **b**, and 3500 m **c**. *Gray* and *green* denote subduction channel and mantle wedge, respectively. The velocity at $x = 0$ corresponds to the slab velocity of $U = 8$ cm/year ($= 2.54 \times 10^{-9}$ m/s). The *broken line* denotes a case when the mantle wedge is dry. The *blue line* denotes a case when the mantle wedge is wet. The *red line* denotes a case that the viscosity of the mantle wedge is one-tenth smaller than the wet mantle rheology by changing η_m. **d** Q (Eq. (9.15)) of the continental materials at 6.5 GPa (200 km) for I1994 density profile as a function of D for $U = 8$ cm/year and $U_2 = 0$ cm/year. The maximum of Q is about 0.5 km when the subduction channel thickness D is 2–3 km

mantle strongly affects the maximum sustainable thickness of the channels. When D is about 1000–3000 m, the slab is able to drag the continental materials of an effective thickness of about 500–600 m depending on the mantle rheology. In contrast, if viscosity of the mantle is one-tenth smaller than that of the wet mantle rheology, the maximum flux [$Q_{\max}$ (Eq. 9.16)] becomes larger than the other cases. Decrease of mantle wedge viscosity reduces the viscous drag on the subduction channel. This reduction of drag force enhances the flux in the subduction channel. Therefore, rapid increase of the viscosity of the mantle wedge with depth might induce the return flow of continental materials to the crust because of the sudden decrease of $Q_{\max}$.

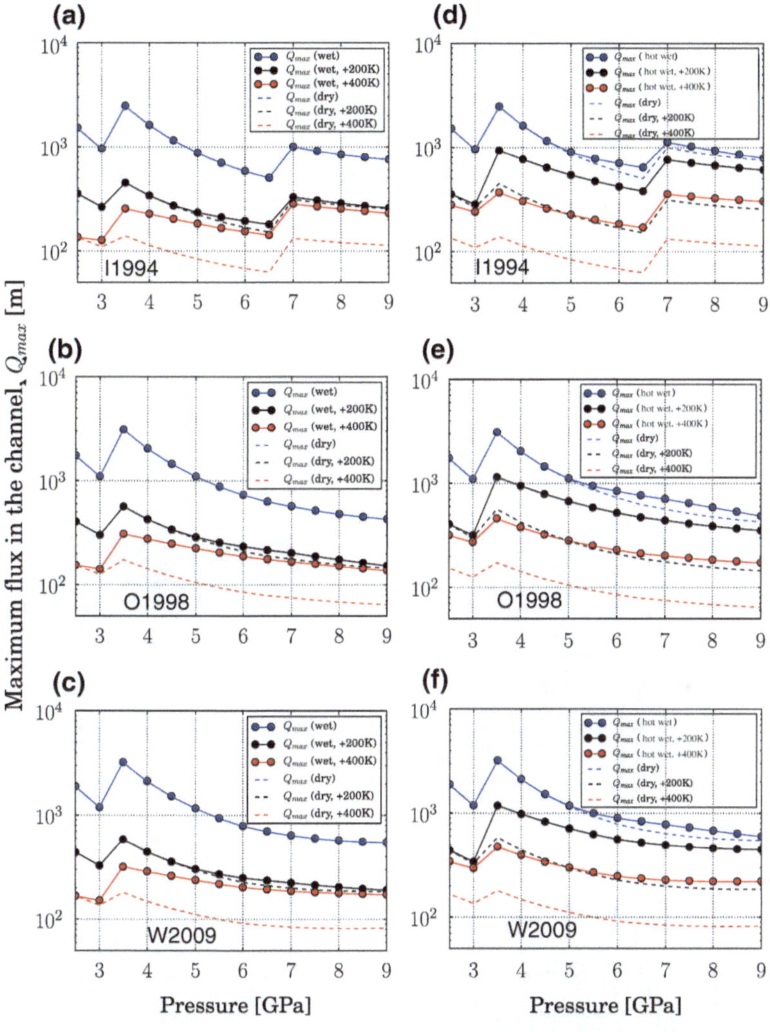

Fig. 9.8 The plots of $Q_{\max}$ (Eq. 9.16) as a function of pressure for $U = 8$ cm/year and $U_2 = 0$ cm/year with three different temperatures profiles (Fig. 9.4), three different density profiles (Fig. 9.2) of I1994 (**a**, **d**), O1998 (**b**, **e**), and W2009 (**c**, **f**), and two different mantle wedge viscosity. In **d–f**, the viscosity of the mantle wedge is one-tenth smaller than the wet mantle rheology (**a–c**) by changing η_m. The continuous lines and the broken lines denote results of dry mantle wedge and wet mantle wedge, respectively

9.4.2 Depth Dependence of Q_{max}

In Fig. 9.8a–c, we show depth dependency of $Q_{\max}$ for three different density profiles along the geotherms in Fig. 9.4. The effect of water in $Q_{\max}$ of the continental materials is important only at high temperature as shown in Fig. 9.8a–c.

The dashed and solid lines denote the results using dry and wet mantle rheology, respectively. The effect of the phase transition from quartz to coesite is clearly seen between 3 and 3.5 GPa, where Q_{max} suddenly increases. In particular, the effect of water is significant at higher pressures, where the difference in viscosity between wet mantle wedge and subduction channel is relatively small as shown in Fig. 9.5. If the viscosity of the mantle wedge is one-tenth smaller than that of the wet mantle rheology, Q_{max} becomes large at all depths (Figure 9.8d–f). However, the difference in rheology in the mantle is significant only at high temperature (geotherm +200 K and geoterm +400 K). On the other hand, the depth of the smallest Q_{max}, which determines the maximum of actual flux at 270 km depth (coesite–stishovite phase transition), is independent of the mantle rheology: 6.5 GPa (~200 km depth) for density profiles of I 1994 and 9 GPa (~270 km depth) for density profiles of O1998 and W2008.

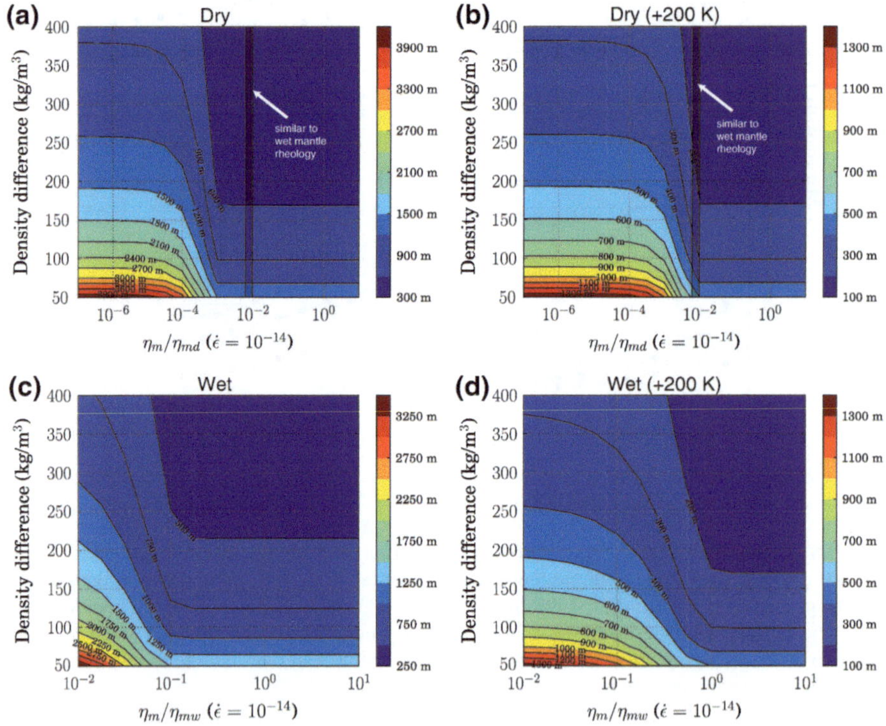

Fig. 9.9 The variations of Q_{max} (Eq. 9.16) at 9 GPa (270 km in depth) as a function of mantle wedge viscosity and density difference between the mantle wedge and the channel ($\Delta\rho$). The viscosity of the mantle wedge is varied by changing η_m from the original value. Dry and wet mantle rheology is imposed for (**a**, **b**) and (**c**, **d**), respectively. Temperature is taken as (**a**, **c**) geotherm and (**b**, **d**) geotherm +200 K (Fig. 9.4). The *shaded regions* in (**a**, **b**) roughly indicate the viscosity of the wet mantle

9.4.3 *Effects of Viscosity of Mantle Wedge and Other Parameters*

Effects of the viscosity and density difference on $Q_{\max}$ at 9 GPa (270 km in depth) using dry mantle rheology are shown in Fig. 9.9a and b. The viscosity of the mantle wedge is varied by changing η_m from the original value. The density difference is controlled from changing the density of subduction channels. The viscosity of the wet mantle rheology is roughly indicated by shaded region. Comparing Fig. 9.9a (Slab geotherm) and Fig. 9.9b (Slab geotherm +200 K), $Q_{\max}$ decreases to about one-third with a temperature increase of 200 K. This is mainly due to the weakening of the drag force resulting from decrease of the viscosity in the subduction channel, which is anticipated by Eq. (9.16). The effect of the viscosity of the mantle is rather drastic at around $\eta_m/\eta_{md} = 10^{-4}$ in Fig. 9.9a and/or $\eta_m/\eta_{md} = 10^{-2}$ in Fig. 9.9b compared to the density difference. The sudden increase of $Q_{\max}$ is caused by the reversal of ratio between viscosity of the mantle wedge and viscosity of the subduction channel. For sufficiently small η_m/η_{md}, the viscosity of the mantle wedge is smaller than that of the subduction channel, where the drag from the mantle wedge on the subduction channel is small and the approximation neglecting the viscosity of the mantle used in Sect. 9.3 is almost valid.

The sudden increase of $Q_{\max}$ observed for dry rheology can also be found for the viscosity of the wet mantle rheology. The Fig. 9.9c and d shows the variation $Q_{\max}$ for using the wet mantle rheology on the mantle wedge. Figure 9.9c, d is very similar to Fig. 9.9a, b although the parameters related to viscosity are different. The viscosity of the wet mantle rheology is very close to the viscosity in the subduction channel at high temperature (geotherm +200 K). This fact is consistent with the increase of $Q_{\max}$ at high temperature as shown in Fig. 9.8.

Figure 9.10a and b shows effects of the viscosity and mantle wedge velocity, U_2 (velocity at $x = L = 100$ km (Fig. 9.1b)) on $Q_{\max}$ using density profile of W2009 at 9 GPa (270 km in depth) for two different temperatures using dry mantle rheology. The decrease of the viscosity of the mantle wedge occasionally induces decrease of $Q_{\max}$ as shown in Fig. 9.10. The viscosity of the mantle wedge is varied by changing η_m from the original value. In this setting, nonzero U_2 mimics the corner flow in the mantle wedge. According to Fig. 9.10, $Q_{\max}$ is monotone increasing function against U_2. On the other hand, the decrease of the viscosity of the mantle wedge reduces $Q_{\max}$ when U_2 is larger than about 4 cm/year, which is about half the imposed velocity of the slab ($U = 8$ cm/year). It is because that the decrease of the viscosity of the mantle wedge cancels out the effect of U_2. That means flow in the subduction channel is not very sensitive to the velocity of the mantle wedge if the viscosity of the mantle wedge is small. As a result, $Q_{\max}$ decrease as viscosity of the mantle wedge decreases when U_2 is large. Contrary to the statement at Sect. 9.4.1, rapid decrease of the viscosity of the mantle wedge with depth might induce the return flow of continental materials to the crust because of the sudden decrease of $Q_{\max}$ when the corner flow is large enough.

The sudden changes of $Q_{\max}$ at around $\eta_m/\eta_{md} = 10^{-4}$ in Fig. 9.10a and/or at around $\eta_m/\eta_{md} = 10^{-2}$ in Fig. 9.10b are caused by the reversal of viscosity ratio

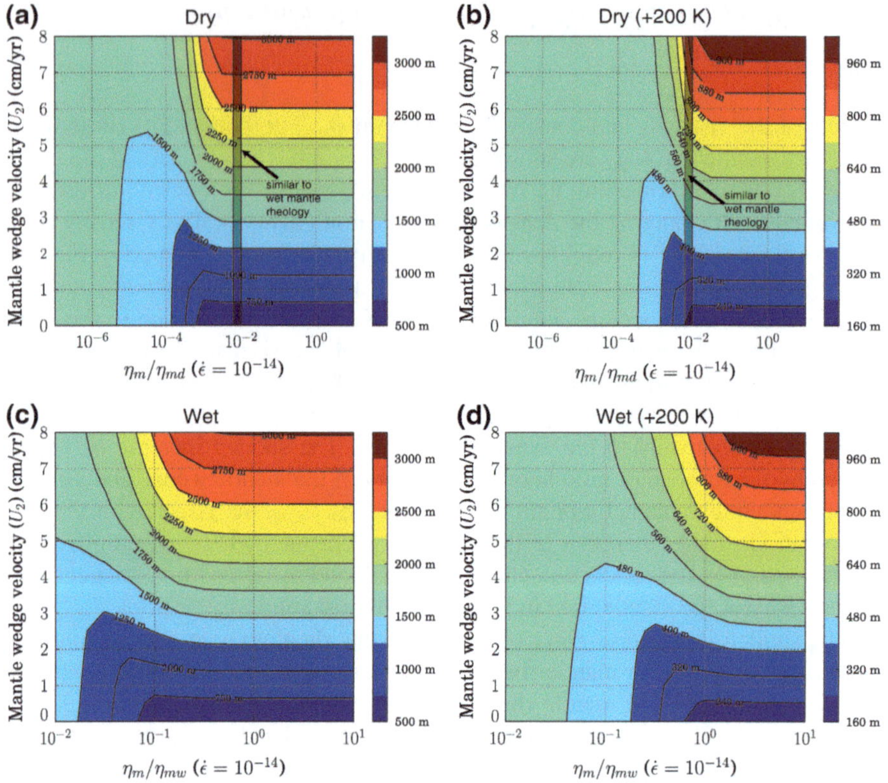

Fig. 9.10 The same as Fig. 9.9 but as a function of mantle wedge viscosity and mantle wedge velocity, U_2, using density profile of W2009 at 9 GPa (270 km in depth). The viscosity of the mantle wedge is varied by changing η_m from the original value. Dry and wet mantle rheology is imposed for (**a**, **b**) and (**c**, **d**), respectively. Temperature is taken as (**a**, **c**) geotherm and (**b**, **d**) geotherm +200 K (Fig. 9.4). The shaded regions in (**a**, **b**) roughly indicate the viscosity of the wet mantle

between the mantle wedge and the subduction channel as in the case of Fig. 9.9. Viscosity of the wet mantle rheology is roughly shown in Fig. 9.10a and b as shaded regions. The Fig. 9.10c and d shows $Q_{\max}$ for using the wet mantle rheology on the mantle wedge. Again, Fig. 9.10c and d is very similar to Fig. 9.10a and b although the parameters related to viscosity are different. In the case of high temperature (Fig. 9.10b and d), the viscosity of the wet mantle rheology is near the sudden change of $Q_{\max}$. When temperature of the mantle is large, $Q_{\max}$ is very sensitive to the viscosity of the mantle wedge.

9.4.4 Effect of Different Temperature Gradient in Depth

So far, temperature profiles in depth are assumed to be uniformly larger than that of the present slab by 200 or 400 K irrespective of depth (Fig. 9.4), although

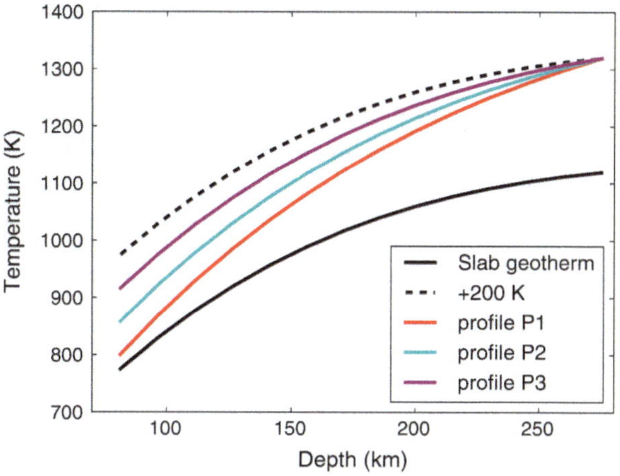

Fig. 9.11 Three additional temperature profiles (P1–P3). These are located between two temperature profiles in Fig. 9.4 and have different temperature gradients

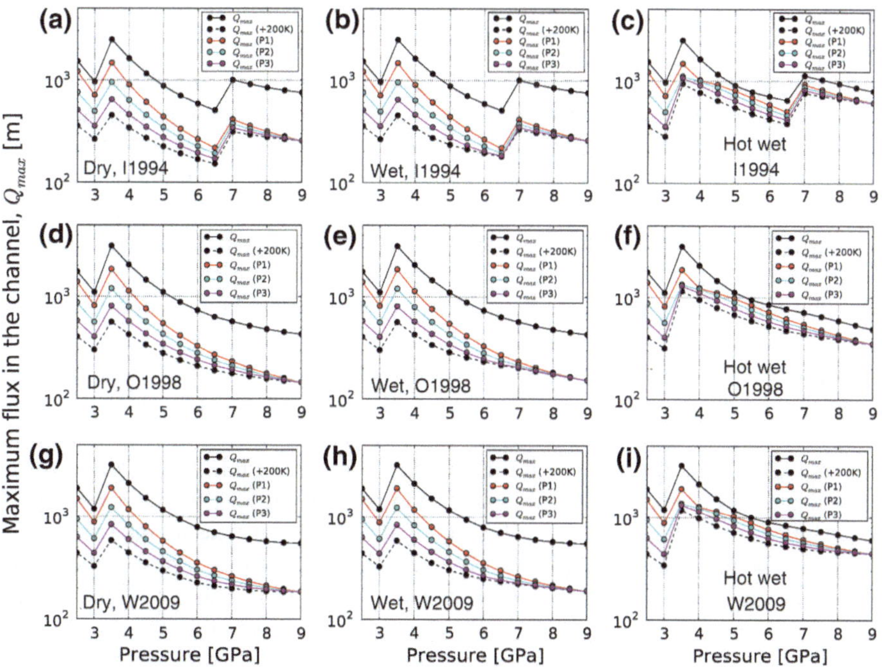

Fig. 9.12 The maximum $Q_{\max}$ (Eq. 9.16) as a function of pressure for $U = 8$ cm/year and $U_2 = 0$ cm/year with five different temperatures profiles, using density profiles of I1994 (**a–c**), O1998 (**d–f**), or W2009 (**g–i**). **a–g** $Q_{\max}$ when the rheology of mantle wedge is dry mantle rheology. **b–h** $Q_{\max}$ when the rheology of mantle wedge is wet mantle rheology. **c–i** $Q_{\max}$ when the rheology of mantle wedge is one-tenth smaller than the wet mantle rheology

temperature gradient should be larger at larger depth because of roughly constant temperature at the Earth's surface. The main result using this temperature profile is that Q_{max} in the channel is smallest at the 200 km in depth (6.5 GPa) or 270 km in depth (9 GPa), depending on the used density profiles (Fig. 9.8). In order to confirm the validity of this result, additional calculations were carried out along three additional temperature profiles (P1, P2, and P3) that have larger temperature gradient at larger depth compared to the present slab geotherm used in this study (Fig. 9.11).

The results using three additional temperature profiles (P1, P2, and P3 in Fig. 9.11) are shown in Fig. 9.12. The results show that Q_{max} in the channel are also smallest at 6.5 GPa (I1994) or 9 GPa (O1998 and W2009) as in the cases with standard temperature profiles (Fig. 9.8). Therefore, the assumption used in Sect. 9.4.3 that the actual flux at 9 GPa is Q_{max} at 6.5 or 9 GPa depending on density profiles is confirmed except for a few cases. Figure 9.12 also shows that Q_{max} along steeper temperature profiles at higher temperature condition (P1–P3) rapidly decreases compared to that along the present slab geotherm. This indicates that more HP-UHP metamorphic rocks might be expected when the mantle was hotter because of the decrease of Q_{max} with depth.

9.4.5 Effect of Slab-Normal Temperature Gradient

Slab-normal temperature gradient is considered to affect the flow in the subduction channel by reducing viscosity in the mantle wedge, although it is neglected in the simulations so far. In this subsection, the results using several slab-normal temperature profiles are shown. Temperature increase along slab-normal direction (x-direction in Fig. 9.1) is typically ~500 K at the depth of 200–300 km within $L = 0$–100 km until it connects to the ambient mantle temperature (Eberle et al. 2002). In order to evaluate the effect of the slab-normal temperature gradient, we modeled temperature profile $T(x)$ as a linear function of x, $T(x) = T(0) + \Delta Tx/L$, where ΔT is taken as, 0, 200, 400, or 600 K.

Figure 9.13 shows that Q of the continental materials at 9 GPa (270 km in depth) using W2009 density profile as a function of D for $U = 8$ cm/year and $U_2 = 0$ cm/year with four different slab-normal temperature profiles. In the case with dry mantle rheology for the mantle wedge (Fig. 9.13a), Q is almost unchanged as increasing the temperature gradient. On the contrary, the result using wet mantle rheology for the mantle wedge (Fig. 9.13b) is substantially different as increasing the temperature gradient. The increase of Q_{max} is caused by the decline of the viscosity of the mantle wedge due to increase of temperature. Comparing Q_{max} of Fig. 9.13 to Figs. 9.9a and c and 9.10a and c, the case of $\Delta T = 600$ K roughly corresponds to the reduction of a factor of ~10^{-2} in mantle wedge viscosity. This is consistent with the uniform increase of mantle wedge temperature by 300 K, where the reduction of viscosity is a factor of ~200. In summary, when slab-normal temperature gradient is incorporated, the flux of continental materials at wet mantle wedge condition may become 2–3 times larger in contrast to the dry mantle wedge condition, where the flux is almost invariant.

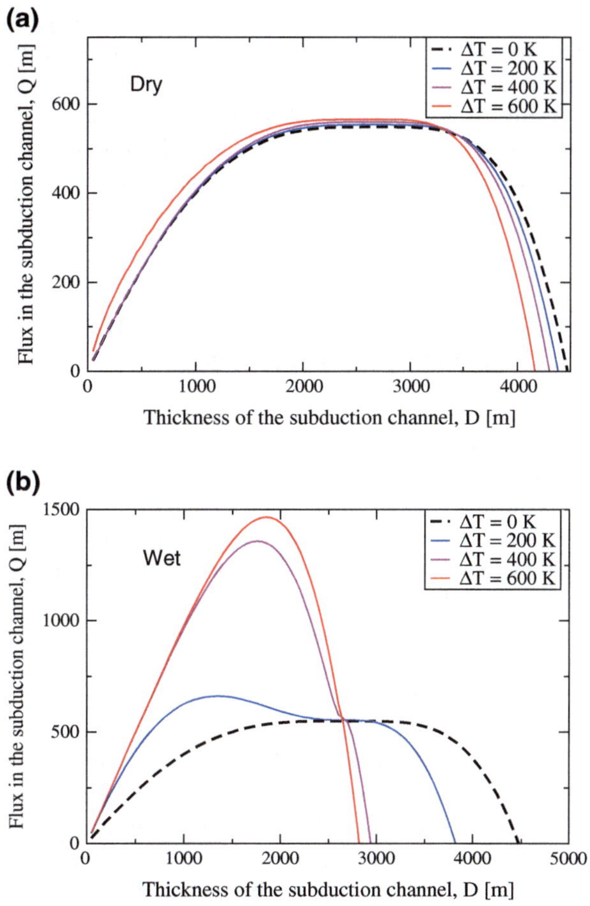

Fig. 9.13 Same as Fig. 9.7d but for using W2009 density profile with $U = 8$ cm/year, $U_2 = 0$ cm/year, and four different temperature profiles normal to the slab (*x*-direction in Fig. 9.1) at 9 GPa (270 km in depth). A linear temperature profile *T*(*x*) is imposed as $T(x) = T(0) + \Delta Tx/L$ (Ref. Figure 9.1b), where ΔT is specified in the legends. Dry (**a**) or wet (**b**) mantle rheology is imposed on the mantle wedge

9.5 Discussion and Conclusions

9.5.1 Flux in Subduction Channels

As indicated by Ichikawa et al. (2013b), when $Q_{\max}$ is smaller than the flux just above, return flow to the surface will form from the mantle wedge side in the channel (Fig. 9.7 c). If the viscosity of the mantle wedge varies with depth, the capacity of the subduction channel, $Q_{\max}$, may change with depth. Whether or not $Q_{\max}$ decreases is also dependent on the corner flow (Fig. 9.10). Even if $Q_{\max}$ is larger, a part of buoyant granitic materials detach from the slab and form diapirs in the mantle wedge when the viscosity of the mantle wedge is small. This is considered to be one of reasons for detachments of subduction channels at a depth of ~100 km indicated by 2-D simulations.

The values of $Q_{\max}$ obtained in our study suggest that the subduction channel that has $D = 2$–3 km can reach 270 km in depth (9 GPa) in terms of $Q_{\max}$ even in the case of water-rich condition on the mantle wedge. However, $Q_{\max}$ is very sensitive to the viscosity of the mantle wedge (Figs. 9.9, 9.10, 9.13). These results suggest that the viscosity of the mantle wedge is important in the ancient mantle such as Archean, where the temperature of the upper mantle has been estimated to be 170 K higher than the present upper mantle (Iwamori et al. 1995).

9.5.2 Supply Rate of Continental Materials to 270 km in Depth for Present Earth

A possible supply rate of continental materials to the deep mantle (below 270 km in depth) with wet mantle wedge rheology with constant mantle wedge temperature was estimated to be same value as that of the dry condition (Ichikawa et al. 2013b). When more realistic mantle wedge temperature distribution is taken, the supply rate will increase by a factor of ~3 (Fig. 9.13), without detachments of granitic materials induced by the weakness of the mantle wedge. The possible supply rate is estimated as a product of the following quantities (Ichikawa et al. 2013b): $Q_{\max} = \sim 500$ m at the depth of 270 km (Fig. 9.8), length of all the major subduction zones [42,250 km (Scholl and Von Huene 2007)], $U = 8$ cm/year, and the contribution of the average dip angle using Eq. (9.16), where $g \rightarrow g \sin \theta_d$, and $\theta_d = 45°$. The resulting possible supply rate to the deep mantle is 2.2 km^3/year. When the mantle wedge temperature distribution is taken into account, the supply rate becomes ~6 km^3/year.

For both the cases of dry (Ichikawa et al. 2013b) and wet mantle wedge rheologies, the possible supply rate at 270 km in depth is comparable with the reduction rate of continental crust by the subduction channel at the surface as a sum of tectonic erosion and sediment subduction (~2.5 km^3/year) (Scholl and Von Huene 2007; Stern and Scholl 2010). It is therefore highly possible that continental materials subduct to 270 km in depth with little loss through subduction channels even if viscosity of the mantle wedge changes with depth by dehydration of the slab.

On the other hand, geological studies estimated the production rate of continental materials at the surface to be 3.2–5.0 km^3/year (Clift et al. 2009) or ~3.2 km^3/year (Stern and Scholl 2010) as a sum of the contributions mainly from arc magmatism and oceanic plateaus. Therefore, the supply of the continental crust materials to the deep mantle nearly balances with the production at the surface.

9.5.3 Total Volume of Subducted Continental Materials

We will finally estimate the total volume of the subducted continental materials through the Earth's history. In the Archean, the upper mantle has been estimated to be hotter than the present upper mantle by ~200 K (Iwamori et al. 1995). Here,

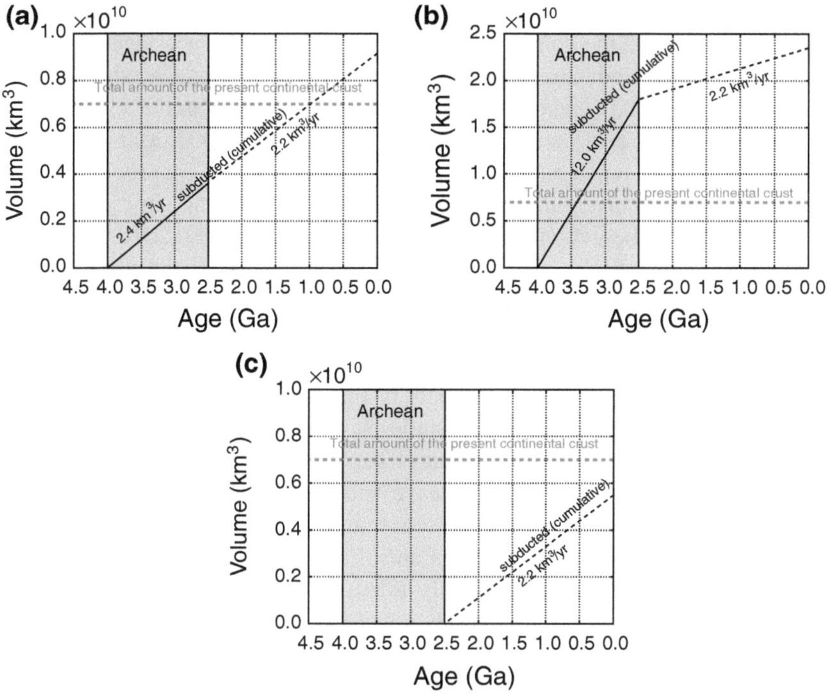

Fig. 9.14 The expected subduction rate of continental materials to 270 km in depth through subduction channels. The total amount of the present continental crust is evaluated to be 7.18×10^9 km^3 (Cogley 1984). The shaded region denotes the Archean. Following three cases are shown. **a** Plate velocity is 3 times higher and temperature is 200 K higher in Archean than these in the present mantle. **b** Same as **a** but the length of all the subduction zones in Archean is 5 times larger than that of the present mantle. **c** plate velocity is assumed to be very small at Archean

we assume two cases for the velocity of the oceanic plate at Archean: Three times higher than the present velocity (Iwamori et al. 1995; Komiya 2004; Komiya and Maruyama 2007) based on a simple estimate from the average lifetime of oceanic plates (Hargraves 1986) and velocity slower than the present one as discussed by Korenaga (2006). If we suppose the plate velocity at Archean is high, the subduction rate in the Archean is estimated to be 2.4 km^3/year (Ichikawa et al. 2013b), where temperature is assumed to be 200 K higher than that at the present, $U = 24$ cm/year, and $\theta_d = 45°$ (Fig. 9.14a). The effect of the high slab velocity is canceled out due to the effect of this high temperature. If the Archean earth is filled with oceanic micro-plates (Yamamoto et al. 2009a), the supply rate substantially increases as shown in Fig. 9.14b, where the length of the subduction zone is 5 times larger than that of the present Earth, which is roughly estimated by assuming that the aspect ratio of the convection cells of the upper mantle is unity. If plate velocity in the Archean is slower than the present velocity, on the other hand, the amount subducted during the Archean is considerably less because of its high temperature (Fig. 9.14c). In any case, the cumulative volume of the continental crust after 2.5 Ga is as much as the volume of the present continental crust. If the mantle wedge is wet and hotter than the slab, the possible subduction rate derived from the present 1-D simulation is ~3 times higher (Fig. 9.13). On the contrary, the weakness of the mantle wedge might induce detachment of the subduction channels by convection (Honda et al. 2002, 2007; Honda and Saito 2003) and/or thermal–chemical plumes (Tamura 1994; Hall and Kincaid 2001; Gerya and Yuen 2003; Obata and Takazawa 2004; Manea et al. 2005; Gerya et al. 2006; Gorczyk et al. 2007) in the mantle wedge. This problem should be clarified more quantitatively in the future by using 2-D or 3-D numerical studies.

As indicated by theoretical calculations (Kawai et al. 2009, 2010, 2013; Kawai and Tsuchiya 2010, 2012), the base of the transition zone is the place where the subducted continental materials are expected to accumulate because continental materials are denser than the ambient mantle materials in the depth range between 270 and 800 km for tonalite–trondhjemite–granodiorite (TTG) materials. The seismic velocity of the granitic materials in the depth range from 270 to 660 km is faster than those of surrounding mantle as shown by Kawai et al. (2013) and Kawai and Tsuchiya (this volume). Therefore, the subducted continental materials may cause heterogeneity on seismic wave and heat sources in the mantle. There is also a possibility that the subducted continental materials are distributed to the lower mantle at all depths via entrainment of the slab subduction. In either case, the mantle convection could be largely affected by the continental materials as indicated by a mantle convection simulation (Ichikawa et al. 2013a).

Acknowledgments The authors thank Dr. Taras Gerya and Dr. Frédéric Deschamps for their useful comments. This study was supported by Global COE program "DEEP EARTH MINERALOGY" and KAKENHI 22740297 and 24840020. Several figures have been realized using matplotlib [Hunter 2007].

References

Afonso JC, Zlotnik S (2011) The subductability of continental lithosphere: the before and after story. In: Brown D, Ryan PD (eds) Arc-continent collision, frontiers in earth sciences. Springer, Berlin

Armstrong RL (1991) The persistent myth of crustal growth. Aust J Earth Sci 38(5):613–630

Babeyko AY, Sobolev SV (2008) High-resolution numerical modeling of stress distribution in visco-elasto-plastic subducting slabs. Lithos 103(1–2):205–216. doi:10.1016/j.lithos.2007.09.015

Belousova EA, Kostitsyn YA, Griffin WL, Begg GC, O'Reilly SY, Pearson NJ (2010) The growth of the continental crust: constraints from zircon Hf-isotope data. Lithos 119(3–4):457–466

Burov E, Jolivet L, Le Pourhiet L, Poliakov A (2001) A thermomechanical model of exhumation of high pressure (HP) and ultra-high pressure (UHP) metamorphic rocks in Alpine-type collision belts. Tectonophysics 342(1–2):113–136. doi:10.1016/S0040-1951(01)00158-5

Campbell IH, Taylor SR (1983) No water, no granites—no oceans, no continents. Geophys Res Lett 10(11):1061–1064

Clift P, Vannucchi P (2004) Controls on tectonic accretion versus erosion in subduction zones: implications for the origin and recycling of the continental crust. Rev Geophys 42(2), RG2001. doi:10.1029/2003rg000127

Clift PD, Vannucchi P, Morgan JP (2009) Crustal redistribution, crust-mantle recycling and Phanerozoic evolution of the continental crust. Earth Sci Rev 97(1–4):80–104. doi:10.1016/j.earscirev.2009.10.003

Cloos M, Shreve R (1988a) Subduction-channel model of prism accretion, melange formation, sediment subduction, and subduction erosion at convergent plate margins: 1. Background and description. Pure Appl Geophys 128(3–4):455–500. doi:10.1007/BF00874548
Cloos M, Shreve R (1988b) Subduction-channel model of prism accretion, melange formation, sediment subduction, and subduction erosion at convergent plate margins: 2. Implications and discussion. Pure Appl Geophys 128(3–4):501–545. doi:10.1007/BF00874549
Cogley JG (1984) Continental margins and the extent and number of the continents. Rev Geophys Space Phys 22(2):101–122
Collot JY, Ribodetti A, Agudelo W, Sage F (2011) The South Ecuador subduction channel: evidence for a dynamic mega-shear zone from 2D fine-scale seismic reflection imaging and implications for material transfer. J Geophys Res 116(B11). doi:10.1029/2011jb008429
Currie CA, Beaumont C, Huismans RS (2007) The fate of subducted sediments: a case for back-arc intrusion and underplating. Geology 35(12):1111. doi:10.1130/g24098a.1
Dobrzhinetskaya L, Wirth R, Greenii H (2006) Nanometric inclusions of carbonates in Kokchetav diamonds from Kazakhstan: a new constraint for the depth of metamorphic diamond crystallization. Earth Planet Sci Lett 243(1–2):85–93. doi:10.1016/j.epsl.2005.11.030
Dziewonski AM, Anderson DL (1981) Preliminary reference Earth model. Phys Earth Planet Inter 25(4):297–356
Eberle MA, Grasset O, Sotin C (2002) A numerical study of the interaction between the mantle wedge, subducting slab, and overriding plate. Phys Earth Planet Inter 134(3–4):191–202
Gerya TV, Stöckhert B (2002) Exhumation rates of high pressure metamorphic rocks in subduction channels: the effect of Rheology. Geophys Res Lett 29(8):102–104. doi:10.1029/2001GL014307
Gerya T, Stöckhert B (2006) Two-dimensional numerical modeling of tectonic and metamorphic histories at active continental margins. Int J Earth Sci 95(2):250–274. doi:10.1007/s00531-005-0035-9
Gerya TV, Stöckhert B Perchuk AL (2002) Exhumation of high-pressure metamorphic rocks in a subduction channel: a numerical simulation. Tectonics 21(6):6-1–6-19. doi:10.1029/2002TC001406
Gerya TV, Yuen DA (2003) Rayleigh-Taylor instabilities from hydration and melting propel “cold plumes” at subduction zones. Earth Planet Sci Lett 212(1–2):47–62. doi:10.1016/S0012-821X(03)00265-6
Gerya TV, Connolly JAD, Yuen DA, Gorczyk W, Capel AM (2006) Seismic implications of mantle wedge plumes. Phys Earth Planet Inter 156(1–2):59–74. doi:10.1016/j.pepi.2006.02.005
Gorczyk W, Gerya TV, Connolly JAD, Yuen DA (2007) Growth and mixing dynamics of mantle wedge plumes. Geology 35(7):587–590. doi:10.1130/G23485A.1
Hall PS, Kincaid C (2001) Diapiric flow at subduction zones: a recipe for rapid transport. Science 292(5526):2472–2475. doi:10.1126/science.1060488
Hargraves RB (1986) Faster spreading or greater ridge length in the Archean? Geology 14(9):750–752
Hofmann AW (1997) Mantle geochemistry: the message from oceanic volcanism. Nature 385(6613):218–229
Honda S, Saito M (2003) Small-scale convection under the back-arc occurring in the low viscosity wedge. Earth Planet Sci Lett 216(4):703–715. doi:10.1016/S0012-821X(03)00537-5
Honda S, Saito M, Nakakuki T (2002) Possible existence of small-scale convection under the back arc. Geophys Res Lett 29(21):2043. doi:10.1029/2002GL015853
Honda S, Yoshida T, Aoike K (2007) Spatial and temporal evolution of arc volcanism in the northeast Honshu and Izu-Bonin Arcs: evidence of small-scale convection under the island arc? Island Arc 16(2):214–223. doi:10.1111/j.1440-1738.2007.00567.x
Hunter J D (2007) Matplotlib: A 2D graphics environment, Comput Sci Eng 9(3):90–95. doi:10.1109/MCSE.2007.55
Ichikawa H, Kameyama M, Kawai K (2013a) Mantle convection with continental drift and heat source around the mantle transition zone. Gondwana Res 24(3–4):1080–1090. doi:10.1016/j.gr.2013.02.001

Ichikawa H, Kawai K, Yamamoto S, Kameyama M (2013b) Supply rate of continental materials to the deep mantle through subduction channels. Tectonophysics 592:46–52. doi:10.1016/j.tecto.2013.02.001
Irifune T, Ringwood AE, Hibberson WO (1994) Subduction of continental crust and terrigenous and pelagic sediments: an experimental study. Earth Planet Sci Lett 126(4):351–368
Iwamori H, McKenzie D, Takahashi E (1995) Melt generation by isentropic mantle upwelling. Earth Planet Sci Lett 134(3–4):253–266
Karato S-I, Jung H (2003) Effects of pressure on high-temperature dislocation creep in olivine. Phil Mag 83(3):401–414. doi:10.1080/0141861021000025829
Kawai K, Tsuchiya T (2010) Ab initio investigation of high-pressure phase relation and elasticity in the $NaAlSi_2O_6$ system. Geophys Res Lett 37(17). doi:10.1029/2010gl044310
Kawai K, Tsuchiya T (2012) First principles investigations on the elasticity and phase stability of grossular garnet. J Geophys Res 117(B2):B02202. doi:10.1029/2011jb008529
Kawai K, Tsuchiya T, Tsuchiya J, Maruyama S (2009) Lost primordial continents. Gondwana Res 16(3–4):581–586. doi:10.1016/j.gr.2009.05.012
Kawai K, Tsuchiya T, Maruyama S (2010) The second continent. J Geogr 119(6):1197–1214
Kawai K, Yamamoto S, Tsuchiya T, Maruyama S (2013) The second continent: Existence of granitic continental materials around the bottom of the mantle transition zone. Geosci Front 4(1):1–6. doi:10.1016/j.gsf.2012.08.003
Kawazoe T, Karato S, Otsuka K, Jing Z, Mookherjee M (2009) Shear deformation of dry polycrystalline olivine under deep upper mantle conditions using a rotational Drickamer apparatus (RDA). Phys Earth Planet Inter 174(1–4):128–137. doi:10.1016/j.pepi.2008.06.027
Komiya T (2004) Material circulation model including chemical differentiation within the mantle and secular variation of temperature and composition of the mantle. Phys Earth Planet Inter 146(1–2):333–367. doi:10.1016/j.pepi.2003.03.001
Komiya T, Maruyama S (2007) A very hydrous mantle under the western Pacific region: Implications for formation of marginal basins and style of Archean plate tectonics. Gondwana Res 11(1–2):132–147. doi:10.1016/j.gr.2006.02.006
Korenaga J (2006) Archean geodynamics and the thermal evolution of Earth. Archean Geodynamics and Environments, vol 164. AGU, Washington, DC, pp 7–32
Li ZH, Xu ZQ, Gerya TV (2011) Flat versus steep subduction: contrasting modes for the formation and exhumation of high- to ultrahigh-pressure rocks in continental collision zones. Earth Planet Sci Lett 301(1–2):65–77. doi:10.1016/j.epsl.2010.10.014
Manea VC, Manea M, Kostoglodov V, Sewell G (2005) Thermo-mechanical model of the mantle wedge in Central Mexican subduction zone and a blob tracing approach for the magma transport. Phys Earth Planet Inter 149(1–2):165–186. doi:10.1016/j.pepi.2004.08.024
Maruyama S, Ikoma M, Genda H, Hirose K, Yokoyama T, Santosh M (2013) The naked planet Earth: most essential pre-requisite for the origin and evolution of life. Geosci Front 4(2):141–165. doi:10.1016/j.gsf.2012.11.001
Moore GF, Bangs NL, Taira A, Kuramoto S, Pangborn E, Tobin HJ (2007) Three-dimensional splay fault geometry and implications for tsunami generation. Science 318(5853):1128–1131. doi:10.1126/science.1147195
Murphy DT, Collerson KD, Kamber BS (2002) Lamproites from Gaussberg, Antarctica: possible transition zone melts of Archaean subducted sediments. J Petrol 43(6):981–1001
Obata M, Takazawa E (2004) Compositional continuity and discontinuity in the Horoman Peridotite, Japan, and its Implication for melt extraction processes in partially molten upper mantle. J Petrol 45(2):223–234. doi:10.1093/petrology/egg106
Ono S (1998) Stability limits of hydrous minerals in sediment and mid-ocean ridge basalt compositions: Implications for water transport in subduction zones. J Geophys Res 103(B8):18253–18267. doi:10.1029/98jb01351
Pitzer KS, Sterner SM (1995) Equations of state valid continuously from zero to extreme pressures with H_2O and CO_2 as examples. Int J Thermophys 16(2):511–518

Raimbourg H, Jolivet L, Leroy Y (2007) Consequences of progressive eclogitization on crustal exhumation, a mechanical study. Geophys J Int 168(1):379–401. doi:10.1111/j.1365-246X.2006.03130.x
Ranalli G (1995). Rheology of the earth, 2nd ed. Chapman and Hall, London
Ranalli G (1997) Rheology of the lithosphere in space and time. Geol Soc London Spec Publ 121:19–37. doi:10.1144/GSL.SP.1997.121.01.02
Renner J, Stöckhert B, Zerbian A, Röller K, Rummel F (2001) An experimental study into the rheology of synthetic polycrystalline coesite aggregates. J Geophys Res B Solid Earth 106(B9):19411–19429
Scholl DW, Von Huene R (2007) Crustal recycling at modern subduction zones applied to the past-issues of growth and preservation of continental basement crust, mantle geochemistry, and supercontinent reconstruction Geol Soc Amer Mem 200:9–32
Sizova E, Gerya T, Brown M (2012) Exhumation mechanisms of melt-bearing ultrahigh pressure crustal rocks during collision of spontaneously moving plates. J Metamorph Geol 30(9):927–955. doi:10.1111/j.1525-1314.2012.01004.x
Sobolev SV, Babeyko AY (2005) What drives orogeny in the Andes? Geology 33(8):617–620
Stern RJ, Scholl DW (2010) Yin and yang of continental crust creation and destruction by plate tectonic processes. Int Geol Rev 52(1):1–31
Tamura Y (1994) Genesis of island arc magmas by mantle-derived bimodal magmatism: evidence from the Shirahama Group, Japan. J Petrol 35(3):619–645. doi:10.1093/petrology/35.3.619
Von Huene R, Scholl DW (1991) Observations at convergent margins concerning sediment subduction, subduction erosion, and the growth of continental crust. Rev Geophys 29(3):279–316
Warren CJ, Beaumont C, Jamieson R (2008) Formation and exhumation of ultra-high-pressure rocks during continental collision: role of detachment in the subduction channel. Geochem Geophys Geosyst 9(4). doi:10.1029/2007GC001839
Wu Y, Fei Y, Jin Z, Liu X (2009) The fate of subducted upper continental crust: an experimental study. Earth Planet Sci Lett 282(1–4):275–284. doi:10.1016/j.epsl.2009.03.028
Yamamoto S, Senshu H, Rino S, Omori S, Maruyama S (2009a) Granite subduction: arc subduction, tectonic erosion and sediment subduction. Gondwana Res 15(3–4):443–453. doi:10.1016/j.gr.2008.12.009
Yamamoto S, Nakajima J, Hasegawa A, Maruyama S (2009b) Izu-Bonin arc subduction under the Honshu island, Japan: evidence from geological and seismological aspect. Gondwana Res 16(3–4):572–580. doi:10.1016/j.gr.2009.05.014
Zhang J, Li B, Utsumi W, Liebermann RC (1996) In situ X-ray observations of the coesite-stishovite transition: reversed phase boundary and kinetics. Phys Chem Miner 23(1):1–10

Part II
Lower Mantle

Chapter 10
The Composition of the Deep Earth

Edouard Kaminski and Marc Javoy

Abstract Earth composition models rely on three types of information: petrological sampling, geophysical sounding, and cosmochemical constraints. The relative input of a given category of information changes with the depth of the considered Earth's layer. The constraints brought by the petrological approach are dominant for the upperparts of the Mantle, whereas geophysical constraints play a crucial role for estimating deep Earth (lower mantle and core) composition. Since a direct sampling of the deep Earth is not possible, chemical constraints are mainly brought by the composition of primitive chondrites. In the more general approach, chondritic refractory lithophile elements (RLE) ratios are used to infer their content in the mantle. In addition, the family of "E-Earth models" uses not only chondritic RLE ratios, but also the bulk composition of a particular family of chondrites: enstatite (EH, EL) chondrites. These chondrites are the closest to the Earth in terms of isotopic composition as well as redox state and can be used to infer the composition of the deep Earth following a mass balance approach. In this paper, we first review the main characteristics of E-Chondrites in relation to the isotopic and redox characteristics of the Earth. We then present the main steps of the determination of the Earth composition based on a generic model of E-chondrites, and we then expand our previous results to the content of some minor and major trace elements in the deep mantle. The general characteristics of E-Earth compositions and their consequences for Earth differentiation and dynamics are discussed, and paths for further improvements of the model are presented.

Keywords Earth's composition · Enstatite chondrite · Mantle chemical heterogeneities · Early Earth

E. Kaminski (✉) · M. Javoy
Institut de Physique du Globe de Paris, Sorbonne Paris Cité, CNRS, Université Paris Diderot, 75005 Paris, France
e-mail: kaminski@ipgp.fr

© Springer International Publishing Switzerland 2015

A. Khan and F. Deschamps (eds.), *The Earth's Heterogeneous Mantle*, Springer Geophysics, DOI 10.1007/978-3-319-15627-9_10

10.1 Introduction

The knowledge of whole Earth's composition remains one of the most fundamental issues in Earth sciences. This problem bridges the planet's present-day dynamics to the planet's origins, from the characterization of its building blocks to the process and timescales at stake during its differentiation (e.g., Schubert et al. 2001). The deep Earth's composition, and particularly that of the core and lower mantle, is likely to have retained some traces of early Earth differentiation and dynamics (e.g., Tackley 2012). But a lot of controversy remains on the composition of the deepest parts of the Earth (e.g., Williams and Knittle 2005). Such controversies arise from the difficulty of current geophysical inversion methods to provide unambiguous constraints on the thermal and chemical state of the LM (e.g., Deschamps and Trampert 2004; Matas et al. 2007; Khan et al., this volume; Rost et al., this volume) and on the exact nature and concentration of light elements in the core (e.g., Poirier 1994; Alfe et al. 2003; Fiquet et al. 2008). More "direct" approaches based on primitive chondrites provide estimates of bulk Earth composition, including core and LM compositions, and can be tested against geophysical observations to define a consistent model accounting for both geochemical and geophysical constraints (Cammarano et al. 2003; Javoy et al. 2010).

A first end-member of model for the composition of the Earth's mantle is the family of models considering a homogeneous primitive mantle (e.g., McDonough and Sun 1995; Lyubetskaya and Korenaga 2007). This approach is usually considered to be the simplest, but is not sufficient to account for the chemical structure of a planet in which plate tectonics have been generating chemical anomalies, at least at the surface, for millions of years. The observation that many subducting slabs now reach the LM is often used as a key argument for a global scale mantle convection (van der Hilst et al. 1997), which should eventually lead to a global mantle chemical homogeneity. However, the detection of these same slabs, or the preservation of subducted material in super-deep diamonds (e.g., Walter et al. 2011), proves that the LM is not presently chemically homogeneous, even if some debates remain on the repartition of LM chemical heterogeneities and the ability of seismic tomography to identify them at different scales (e.g., Khan et al. 2008, 2011; Davies et al. 2012; Lekic et al. 2012; D. Rhodri Davies et al., this volume).

One of the main indications that the Earth's mantle includes different reservoirs comes from the budget of incompatible elements, which suggests that there must be at depth a reservoir enriched in these elements relative to the (depleted) upper mantle (e.g., Albarède and van der Hilst 1999). The ^{40}Ar budget has been used to argue that about 50 % of the mantle is undegassed, in line with the existence of a deep reservoir hardly affected by convective mixing (Allègre et al. 1983). However, if the significantly lower ^{40}Ar/^{39}Ar of OIB relative to MORB demonstrates that their source is less degassed, estimates of the mass of this source, which depends on the K/U ratio of the Earth, remains debated (Moreira 2013). According to Lassiter (2004), for example, the undegassed reservoir could be as small as a few percents of the mass of the mantle. Other noble gases (He, Xe, Ne) do not constrain the extent of the undegassed reservoir, but indicate also that the source of some OIB has remained undegassed since the core differentiation (e.g., Moreira et al. 1998; Mukhopadhyay 2012).

Seismological observations can be consistently interpreted in the framework of a chemically heterogeneous mantle (e.g., Kellogg et al. 1999). The relative contribution of temperature and composition to the seismic observables remains the object of debates, but a chemical contribution is required to explain the bulk sound/shear velocity anomalies decorrelation in the deepest parts of the mantle and the distribution of density inferred from normal modes and gravity anomalies (e.g., Ishii and Tromp 1999; Simmons et al. 2010; Della Mora et al. 2011). Additional constraints brought by mineral physics, either from direct experiments performed at high pressure and high temperature (e.g., Ricolleau et al. 2009) or from the inversion of seismic velocities in terms of temperature and chemical composition (e.g., Matas et al. 2007), also show that an isochemical pyrolitic mantle cannot match global 1D density and seismic velocities of PREM (preliminary reference Earth model (PREM), Dziewonski and Anderson 1981). Recent global electromagnetic studies of the mid-mantle (900–1400 km) also suggest that both temperature variations and changes in iron content are required to interpret the correlation between seismic tomography and conductivity models (Tarits and Mandea 2010). More elaborate multidisciplinary approaches, which combine seismic data with geodynamical modeling, show on the other hand that the evolution of an initially chemically heterogeneous mantle can account for various geophysical observables (e.g., Cadio et al. 2011; Deschamps et al. 2011, 2012; Tan et al. 2011; Deschamps et al., this volume).

Altogether geochemical measurements, geophysical inversions, and geodynamical modeling favor a primitive heterogeneity of the mantle. They cannot however by themselves provide a firm and precise chemical composition of the deep mantle. It is the aim of this contribution to detail how cosmochemical constraints based on the isotopic similarity between Earth and enstatite chondrites (EC) can be used to establish a reference composition for the deep Earth (LM and core) consistent with available constraints provided by geophysical and geochemical deep Earth's exploration. In the first section of the article, we summarize the isotopic similarities between the Earth and the EC and give some indications on the genetic link between the two materials, which most plausibly explains that similarity. We then show how a bulk Earth composition can be obtained from the mineralogy and composition of E-chondrites and how such a bulk composition can be used to obtain the composition of the primitive upper mantle (PUM), of the primitive lower mantle (PLoM), and of the core. In the third section, we present a model of core formation in the early Earth which can account for the chemical differences between the upper and lower primitive mantles. In the fourth section, we use that model to infer the content of some minor and trace elements in the LM. We conclude the paper by a discussion of some of the implications of the E-Earth model for the differentiation and dynamics of the Earth and on the paths that may be followed to improve our knowledge and understanding of the deep Earth composition.

10.2 The E-Earth Framework: The Link Between Enstatite Chondrites and the Bulk Composition of the Earth

The relationship between E-chondrites and the Earth is twofold: isotopic and redox. What follows about these two topics has been first put forward by (Javoy 1995), then developed at length in (Javoy et al. 2010), and discussed also in a review of isotopic compositions of chondrites by Warren (2011). We shall recall here the main points, refresh the continuing discovery of isotopic coincidences, and rather insist on the inevitable complications appearing when such complicated processes are "deeper-dug". This is for example the case of occasional isotopic discrepancies, notable although of limited extent.

The isotopic proximity between Earth and E-chondrites was recognized for oxygen close to 40 years ago after the development of triple isotope (^{16}O–^{17}O–^{18}O) analyses had allowed a higher resolution in the chondrite classification (particularly the ordinary-enstatite chondrite separation, e.g., Clayton et al. 1976). It was followed by nitrogen when Javoy et al. (1984, 1986) discovered ^{15}N-poor nitrogen in diamonds and when a systematic nitrogen-diamond survey displayed ^{15}N-poorer and poorer samples, until LM diamonds now reach the -30 to -40 ‰ range characteristic of E-chondrite nitrogen (e.g., Cartigny et al. 1997; Palot et al. 2012). That led Javoy (1995) to propose his first model of bulk Earth integral E-chondrite composition. This, little by little, led isotopic cosmochemists to pay more attention to E-chondrites, and examples of isotopic proximity, or complete identity, followed more and more rapidly: molybdenum (Dauphas et al. 2002), ruthenium (Dauphas et al. 2004), chromium (Trinquier et al. 2007), titanium (Trinquier et al. 2009), zinc (Moynier et al. 2011), nickel (Regolous et al. 2008), "solar" rare gases (Park et al. 2005; Raquin and Moreira 2009). That special proximity occurs also for more complicated tracers such as the extinct radiochronology couple, $^{146}Sm/^{142}Nd$, or even nucleosynthetic "curiosities" such as the p-process-produced ^{144}Nd nuclide for which E-chondrites are, again, the closest isotopic neighbors of the Earth's mantle (Gannoun et al. 2011).

The E-chondrite/Earth isotopic proximity is an occasion to point out the fact that, whatever the origin of isotope discrepancies, a thorough enough thermic episode is able to homogenize isotope heterogeneities to the residual level of high-temperature isotope equilibrium fractionation. Silicon, which is up to now the only element for which E-chondrites are more remote from the Earth's mantle,[1] is a good example to give an illustration of what can be that residual level.

The simplest case of isotope equilibrium, which produces the smallest isotope heterogeneity at equilibrium during condensation from a gas phase, hence adapted to the Earth/E-chondrites close proximity, is *bulk isotopic equilibrium*. In that

[1]Calcium had been signaled too (Simon and De Paolo 2010) but more recent results (Huang and Jacobsen 2012) conclude, on the contrary, that calcium, is, again, a good example of E-chondrite Earth identity.

case, isotope equilibrium between the gas (isotopic composition δ_N) and the condensate (isotopic composition δ_C) in a cooling nebula is written[2] as:

$$\delta_N - \delta_C = \Delta(T), \tag{10.1}$$

where $\Delta(T)$, the gas-condensate isotopic equilibrium fractionation, is a function of T (temperature) only. The complete treatment of that problem necessitates the knowledge of the condensate fraction as a function of temperature and is beyond the scope of this paper. We give here the main lines of the solution. Mass balance imposes that:

$$\delta_N = f_c\, \delta_C + (1 - f_c)\, [\delta_C + \Delta(T)], \tag{10.2}$$

where f_c is the condensed fraction of the element ($0 < f_c < 1$), and ($1 - f_c$) is the residual fraction of the element in the gas. For two materials condensing at equilibrium from the same nebula of isotopic composition δ_N, but at different temperatures, T_1 and T_2, Eq. (10.2) becomes

$$\delta_N = \delta_{1C} + [1 - f_c(T_1)]\, \Delta\, (T_1) = \delta_{2C} + [1 - f_c(T_2)]\, \Delta\, (T_2). \tag{10.3}$$

At high temperature, $\Delta(T) \sim A\ 10^6/T^2$, and the A coefficient, characteristic of the element, depends, among other parameters, on the relative isotopic mass variation $\Delta m/m$ and becomes very small above a certain atomic mass, like for ruthenium for example, leading to isotopic *identity*. When A is large enough however, the equilibrium fractionation between nebula and condensates may become significant even at *high temperature*. This is the case of silicon for which $A \sim -5.0$ to -5.8 ‰ (Javoy et al. 2011) which corresponds to the fractionation between the nebular mixture of SiO and SiS and the condensed silicates (pyroxenes, olivines, silica polymorphs, plagioclase) plus, for EC , trace amounts of silicon (~1–3 %) dissolved in metal. For silicon, the literature data give, for EC $\delta_{1C} = -0.62$ to -0.77 ‰, and for the Earth's mantle $\delta_{2C} = -0.29 \pm 0.08$ ‰ (Armytage et al. 2011; Fitoussi and Bourdon 2012; Savage and Moynier 2013).

The separate determination of T_1 and T_2 is impossible at that stage because the problem is under-determined. For the sake of the argument, we can chose for the Earth's material the standard $T_{50\,\%}$ condensation temperature of silicon at 10^{-4} atmosphere: $T_2 \approx 1300$ K (Wasson 1985) and $T_1 \approx 1100$ K for E-chondrites (e.g., Petaev and Khodakovsky 1986). Then, $f_2 = 0.5$ would require $f_1 = 0.75$ according to Eq. (10.3) to account for the difference in isotopic difference in Si between the two materials. A higher temperature of condensation of the Earth' building material can also explain the small amount of S in the Earth relative to E-chondrites (Javoy et al. 2010; Kaminski and Javoy 2013), by the loss at high temperature of 80–100 % of E-chondrites' sulfur by reactions such as:

[2]Isotopic compositions will be expressed in the δ notation, in ‰, unless otherwise specified as $\delta_{Sample} = 1000(R_{ech}/R_{Std} - 1)$, where the R values are the ratios, D/H, $^{15}N/^{14}N$, $^{18}O/^{16}O$, etc., and the standards, SMOW for hydrogen and oxygen, atmospheric nitrogen for nitrogen, NBS28 quartz for silicon, etc. The isotopic fractionation of 2 versus 1 is expressed as the fractionation factor $\alpha = R_2/R_1$ or the derived expressions, e.g., $\Delta = \delta_2 - \delta_1 \sim 1000\ \mathrm{Ln}\,\alpha$.

$$FeS \rightarrow Fe + 0.5\,S_2(\approx 90\,\%\ \text{of total S}), \tag{10.4}$$

$$3\,\text{Enstatite} + \text{Oldhamite} + 0.5O_2 \rightarrow \text{Forsterite} + \text{Diopside} + 0.5\,S_2(\approx 10\,\%\ \text{of total S}). \tag{10.5}$$

The redox proximity bulk Earth (i.e., core + mantle) is evident in the classical Urey-Craig diagram, where the Earth's position is very close to the FeO-free E-chondrites (Javoy 1999). The positions are even closer if one considers the possibility of silicon presence in the core, which can be related to that of FeO in the mantle by the reaction:

$$SiO_2 + 2\,Fe \rightarrow 2\,FeO + Si_{metal} \tag{10.6}$$

This reaction, first proposed by Javoy (1995), which plays a key role on the estimation of the deep mantle compositions in the E-chondrites framework (Javoy et al. 2010; Kaminski and Javoy 2013), is now acknowledged in core chemistry research (e.g., Frost et al. 2008) and is explained in detail below.

10.3 The Composition of the Primitive Upper Mantle: The Pyrolite Approach

The pioneer approach to estimate the composition of the PUM in major elements (Mg, Si, Al, Ca, Fe, Ni) was established by Ringwood (1966). It used a combination of residual peridotites (harzburgites) and basalts to define the "Pyrolite", a theoretical rock from which they both derived through partial melting. Successive versions of that approach, using for example ophiolite complexes, did not lead to a major evolution of the average pyrolitic composition, as illustrated in Table 10.1 (Green 1964; Harris et al. 1967; Carter 1970; Bickle et al. 1976; Maaloe and Aoki 1977).

Additional cosmochemical constraints on PUM composition have been introduced first by Loubet et al. (1975) and Jagoutz et al. (1979) and latter refined by Hart and Zindler (1986). The main hypothesis underlying these models is that the ratios of refractory (i.e., non volatile) elements are the same in the Earth and in the primitive chondrites (Palme and Nickel 1985). This approach allows in particular to correct terrestrial data from excess cpx, which gives non-chondritic (and higher) Ca/Al ratios in

Table 10.1 Composition of the primitive upper mantle following different approaches: using terrestrial samples (Green 1964; Ringwood 1966; Harris et al. 1967; Carter 1970; Bickle et al. 1976; Maaloe and Aoki 1977; Sun 1982) and using both terrestrial samples and meteoritics constraints

Element (wt%)	"Terrestrial" pyrolite	Hart and Zindler (1986)	Allègre et al. (1995)	McDonough and Sun (1995)	Lyubetskaya and Korenaga (2007)
Mg	22.94 ± 1.35	22.8 ± 0.6	22.8	22.8 ± 2.33	23.4 ± 0.93
Al	2.20 ± 0.45	2.15 ± 0.02	2.16	2.35 ± 0.24	1.87 ± 0.32
Si	21.47 ± 0.67	21.5 ± 0.6	21.5	21 ± 2.10	21.1 ± 0.58
Ca	2.44 ± 0.61	2.34 ± 0.02	2.31	2.53 ± 0.24	2.0 ± 0.34
Fe	6.44 ± 1.20	5.86	5.82	6.26 ± 0.63	6.22 ± 0.42
Ni	0.14 ± 0.10	0.22	0.2	0.20 ± 0.02	0.20 ± 0.02

lherzolites. Lherzolite and various meteorites (C1-enstatite-ordinary chondrites) arrays are then defined in Mg–Al–Si space, and the bulk silicate Earth (BSE) composition is taken as the intersection of the two arrays. The resulting compositions are similar to the Pyrolite for major elements, as shown in Table 10.1, and provide the composition in some refractory elements such as Nd, Sm, and Yb (see Hart and Zindler 1986). An alternative hypothesis is that the Earth rather belongs to a trend defined by carbonaceous chondrites, isotopically very remote from the Earth as they are (Allègre et al. 1995). This does not change the average PUM composition either (Table 10.1).

The approach based on the composition of chondrites has been pushed further in the family of "CI-Earth models," which assumes that the bulk composition of the Earth is similar to the composition of the CI chondrite ripped of its volatile elements and that the PUM composition is given by a volatile-free CI from which the core would have been separated. The relative amount of refractory elements in the Earth's mantle relative to CI then defines an enrichment factor (around 2.5) which can be used to estimate the composition of RLE, elements not entering the core. The model of McDonough and Sun (1995) and the related model of Lyubetskaya and Korenaga (2007) are two important examples of such "Pyrolite-CI" models which estimate both major and minor trace elements concentrations in the PUM (Table 10.1).

In the E-chondrite approach, the PUM is supposed to derive from a material with the bulk composition linked to EC rather than CI composition. Since the RLE ratios are the same in all primitive chondrites, the use of EC instead of CI would not affect the results of the pyrolitic approach for the upper mantle. However, since EC chondrites have the smallest Al/Si ratio among primitive chondrite (Al/Si $\approx$ 0.05 in EC instead of Al/Si $\approx$ 0.08 in CI), the pyrolitic composition with the smallest Al/Si ratio, i.e., that obtained by Lyubetskaya and Korenaga (2007), is used in the E-Earth models.

10.4 The Composition of the Deep Earth: Mass Balance Calculations

10.4.1 The Role of Si Dissolution in Metal

In the E-chondrite approach, the composition of the Earth is deduced from the composition of un-sulfurized ECs (see Sect. 10.2 and Javoy et al. 2010). In the first version of this model, Javoy (1995) used the average EH composition of Wasson and Kallemeyn (1988) to obtain the bulk Earth composition. In Javoy et al. (2010), we proposed a generic model to relax that hypothesis by calculating all possible E-chondrites compositions from mixtures of silicate, sulfides, and metal in the various proportions measured in EH and EL chondrites (Baedecker and Wasson 1975; Kallemeyn and Wasson 1986; Zhang et al. 1995; Kong et al. 1997; Rubin et al. 2009). Using the range of phase compositions known in "natural" EC, we developed a Monte-Carlo algorithm to randomly mix the three phase families to obtain a continuum of "synthetic" E-chondrite compositions. The result encompasses all known E-chondrite compositions, as illustrated in Fig. 10.1 and explained in more details in Javoy et al. (2010).

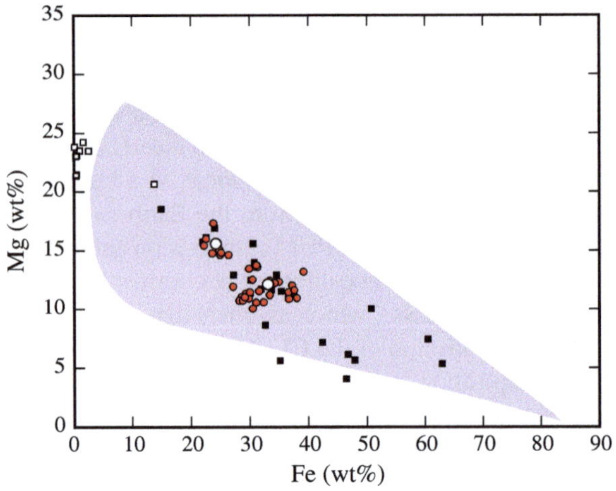

Fig. 10.1 The range of compositions (here for Fe and Mg) obtained for the "synthetic" un-sulfurized E-chondrites using the phase-mixture model of Javoy et al. (2010). The plain *red circles* correspond to the known EC in which S has been replaced by O according to Eqs. (10.4) and (10.5). *Big white circles* stand for the average composition from Wasson and Kallemeyn (1988), the *white squares* are aubrites, and the *black squares* are clasts. The *purple area* gives the range of synthetic compositions

We then separate each synthetic bulk composition between a metallic core (containing only Si and O as light elements) and a silicate mantle. The Fe content of the mantle and the Si and O contents of the core are coupled through the reactions

$$2\alpha \mathrm{Fe}_{\mathrm{metal}} + \alpha\, \mathrm{SiO}_2 = 2\alpha \mathrm{FeO} + \alpha \mathrm{Si}_{\mathrm{metal}}, \tag{10.7}$$

$$\beta \mathrm{FeO}_{\mathrm{silicae}} = \beta \mathrm{Fe}_{\mathrm{metal}} + \beta \mathrm{O}_{\mathrm{metal}}, \tag{10.8}$$

where α is the fraction of metallic Fe oxidized and β is the fraction of oxidized Fe dissolved back into the metal. Parameters α and β are randomly chosen between 0 and 1, which, for a given bulk Fe and SiO_2 content in the E-chondrite, sets the mass fraction of the planet's metallic core and its amount of light elements. Among all the resulting "planets", only the ones with terrestrial characteristics, i.e., a core forming 32.4 % of the planet and containing between 6 and 14 % of light elements (O + Si), are retained, based on estimates of (Alfè et al. 2002) and on experimental studies (e.g., Badro et al. 2007; Fischer et al. 2012). As illustrated Fig. 10.2, the resulting BSE composition is a function of the amount of FeO produced in the mantle during the formation of the core. Hence, additional "terrestrial" constraints are required to define the composition of the Earth.

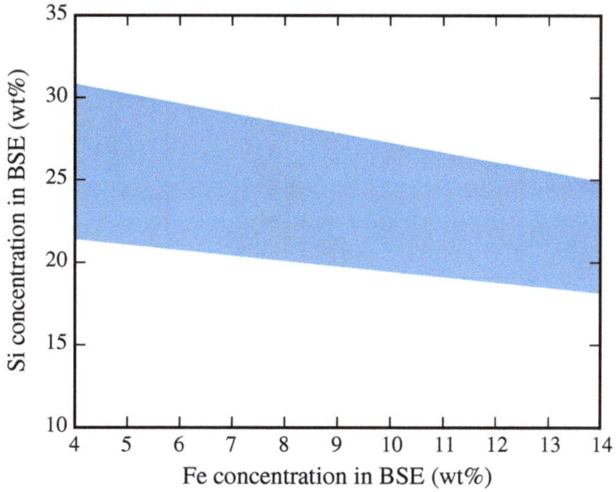

Fig. 10.2 The evolution of the composition of the BSE (for Si) as a function of the amount of FeO produced in the mantle by the dissolution of Si in the metal during the formation of the core (adapted from Javoy et al. 2010)

10.4.2 Composition of the Primitive Upper/Lower Mantles and of the Core

As noted in Sect. 10.3, we cling to the most recent pyrolitic PUM composition that of Lyubetskaya and Korenaga (2007). We then obtain the composition of the (PLoM) by a mass balance between BSE and the PUM, using a variable proportion of the PUM, between 25 % (the size of the present-day upper mantle) and 50 % (its possible maximum extension based on Ar budget). Among the resulting PLoM compositions, we retain only those compatible with two sets of geophysical and geodynamical constraints: (1) first, the PLoM FeO content must be larger than the PUM FeO content to be consistent with the framework of thermochemical convection (e.g., Davaille 1999) and (2) the PLoM MgO, FeO, and SiO_2 contents must lie within the ranges inferred from the inversion of PREM performed by Matas et al. (2007). The resulting PLoM is enriched in Si and Fe relative to the pyrolitic PUM and poorer in Mg and Ca, as shown in Fig. 10.3 and Table 10.2. The depletion in Ca (hence in Al and other RLE) in the PLoM is not related to "terrestrial" constraints, as the inversion of PREM is not very sensitive to the Ca content (Matas et al. 2007), but rather reflects the smaller concentration of RLE in E-chondrites relative to CI chondrites.

The core composition, i.e., its O and Si contents, (given in Table 10.2) is constrained by the filtering applied to the PLoM compositions. In line with the finding of Javoy (1995), Si is the dominant light element in the core (up to 7 wt%), but between 1.5 and 2.5 wt% of oxygen can be present too. These relative amounts of Si and O can be used to assess the *P*, *T* conditions which prevailed during the formation of the core and the differentiation of the upper and lower mantles.

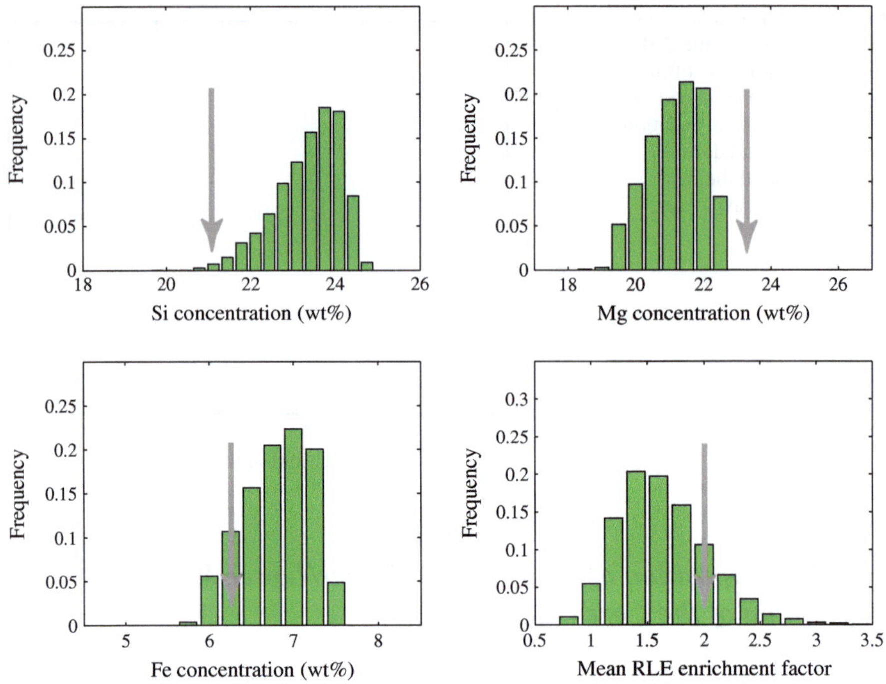

Fig. 10.3 Distribution of likely compositions for the PloM according to Javoy et al. (2010). The *gray arrow* corresponds to the average PUM composition of Lyubetskaya and Korenaga (2007). The enrichment factor is defined as the ratio of RLE (here Ca or Al) in the PloM relative to the RLE content of the CI. A smaller RLE in the PloM relative to the PUM indicates a depletion of RLE in the lower mantle

Table 10.2 Composition of the bulk earth, primitive upper mantle (PUM), the primitive lower mantle (PloM), and core, according to Javoy et al. (2010)

Element (wt%)	Bulk earth	PUM	PLoM	Core
Mg	14.7 ± 0.32	23.5 ± 0.65	21.3 ± 0.67	–
Al	0.87 ± 0.19	1.90 ± 0.32	0.98 ± 0.39	–
Si	18.2 ± 0.30	21.4 ± 0.49	24.2 ± 0.48	6.64 ± 0.51
Ca	0.92 ± 0.20	1.95 ± 0.32	1.04 ± 0.39	–
Fe	32.5 ± 0.36	6.17 ± 0.38	7.21 ± 0.39	85.5 ± 1.14
Ni	1.87 ± 0.26	0.20 ± 0.02	0.19 ± 0.13	5.35 ± 0.81
O	30.8 ± 0.24	44.4 ± 0.19	45.0 ± 0.20	1.99 ± 0.46

10.5 The Origin of the Chemical Differentiation of the Primitive Mantle and the Formation of the Core

The most common hypothesis put forward to account for potential differences between the chemical compositions of the primitive upper and lower mantles is the crystal fractionation of a deep magma ocean (e.g., Tonks and Melosh 1993). Partition coefficients between melt and perovskite (MgPv) can be used to test this hypothesis (e.g., Kato et al. 1988). These partition coefficients are much different among RLE (e.g., $0.22 \leq D_{Ca} \leq 0.33$; $0.77 \leq D_{Al} \leq 2.14$; $0.07 \leq D_{Sm} \leq 0.16$; $0.006 \leq D_{La} \leq 0.042$; Liebske et al. 2005); hence, it is not likely that chondritic ratios of RLE in the BSE would have been preserved in the PUM if it resulted from crystal fractionation (Kato et al. 1988). A maximum extent of such a deep reservoir is 13 wt% of the mantle according to Liebske et al. (2005), which will fail to account for the respective size of PUM and PLoM in the E-Earth. The alternative, we recently proposed in Kaminski and Javoy (2013), is that the chemical differentiation between the two mantles is related to changing *P*, *T* conditions of metal extraction during Earth's accretion.

10.5.1 Giant Impact, Magma Ocean and PUM Formation

The giant impact hypothesis is the favored scenario for the origin of the Moon (e.g., Cameron 2011), based in part on some isotopic identities between the Earth and the Moon (e.g., Wiechert et al. 2001; Zhang et al. 2012). One implication of the giant impact hypothesis is the formation of a magma ocean (e.g., Canup 2008) the depth of which remains the object of debates (e.g., Righter 2003). During that stage of magma ocean, metal iron (and siderophile elements dissolved within) descends through liquid silicate down to a "Metal pond" at the bottom of the molten layer and eventually down to the core as large metal blobs (e.g., Wood 2011). At first order, the average *P*, *T* conditions at the base of the molten layer control the equilibrium (i.e., respective siderophiles content) of the metal and silicate phases. Our hypothesis is that the pyrolitic PUM results from the metal-silicate separation after the giant impact.

To test the hypothesis, we consider that Earth's accretion was homogeneous and that the material falling on Earth kept the average composition obtained in Javoy et al. (2010) (Table 10.3). To separate the sinking metal from the silicate in the magma ocean, we use the "distillation" model of (Javoy 1995), in which an incoming mass $(1 + f)\mathrm{d}m$ of accreted material splits into a mass $f\mathrm{d}m$ of sinking metal and a mass $\mathrm{d}m$ of "floating" silicate, according to

$$\mathrm{d}(mC) = C_0(1 + f)\mathrm{d}m + K\,Cf\mathrm{d}m, \tag{10.9}$$

where C is the concentration of an element in the silicate, C_0 is its concentration in the accreted material, and K *is* an effective silicate/metal partition coefficient. $K = 0$

Table 10.3 Refined estimates of Earth's composition obtained when the ratios of purely lithophile elements are imposed to be the same in PUM and in PLoM

Element (wt%)	PUM	PLoM	Core
Mg	23.3 ± 0.13	20.4 ± 0.14	–
Al	1.79 ± 0.09	1.57 ± 0.09	–
Si	21.8 ± 0.15	24.2 ± 0.14	6.89 ± 0.27
Ca	1.83 ± 0.08	1.61 ± 0.08	–
Fe	6.59 ± 0.18	7.02 ± 0.24	85.5 ± 0.71
Ni	0.20 ± 0.01	0.11 ± 0.03	5.43 ± 0.67
O	44.5 ± 0.07	45.1 ± 0.09	2.19 ± 0.30

for lithophile elements (Mg, Ca, Al), whereas K depends on the P, T conditions of metal extraction for iron and elements that may be dissolved in the metal (Ni, Si, O). We furthermore consider that metal separation reached a steady state regime in the magma ocean and that the pyrolitic PUM is the time-independent solution of the distillation, i.e., the concentration of an element in the PUM (C_{PUM}) is

$$C_{\mathrm{PUM}} = C_0(1+f)/(1+Kf). \qquad (10.10)$$

As C_{PUM} and C_0 are known (from the PUM composition of Lyubetskaya and Korenaga 2007 and the bulk Earth composition of Javoy et al. 2010, Table 10.2), the model has five free parameters: f, K_{Fe}, K_{Ni}, K_{Si}, and K_{O}. To find the values of these parameters, we performed a Monte-Carlo search in which P, T are first randomly chosen in the range 20–80 GPa, 2500–4000 K and then used to estimate the partition coefficients of Ni, Si, and O from the literature data (Asahara et al. 2007; Corgne et al. 2008; Mann et al. 2009; Bouhifd and Jephcoat 2011; Siebert et al. 2012). As detailed in Kaminski and Javoy (2013), the effect of oxygen fugacity on the partition coefficients is taken into account through the use of partition coefficients normalized to Fe (Siebert et al. 2012).

The Monte-Carlo search leads to a pressure range of 45 ± 5 GPa; a temperature range of 3500 ± 500 K; corresponding partition coefficients $\log(K_{\mathrm{d}}^{\mathrm{O}}) \approx -0.04$, $\log(K_{\mathrm{d}}^{\mathrm{Si}}) \approx -2.7$, and $\log(K_{\mathrm{d}}^{\mathrm{Ni}}) \approx 0.28$ (all given within a range ± 0.15 log units); and a metal fraction of $\approx 0.38 \pm 0.02$. The resulting Si and O concentrations in the metal extracted from the magma ocean are shown in Fig. 10.4. The average content of light elements (Si + O) in the extracted metal is close to 22 wt%, with almost 10 wt% of oxygen. These values are much larger than those proposed for the present-day core and correspond to the specific P, T conditions in the magma ocean. They imply that the core did not form in one stage only. There was a proto-core already present—and with a different composition—before the giant impact.

10.5.2 Composition of the Proto-Core

The mass of the proto-core is the difference between the mass of the present-day core (32.4 % of the mass of the Earth) and the mass of metal extracted from the

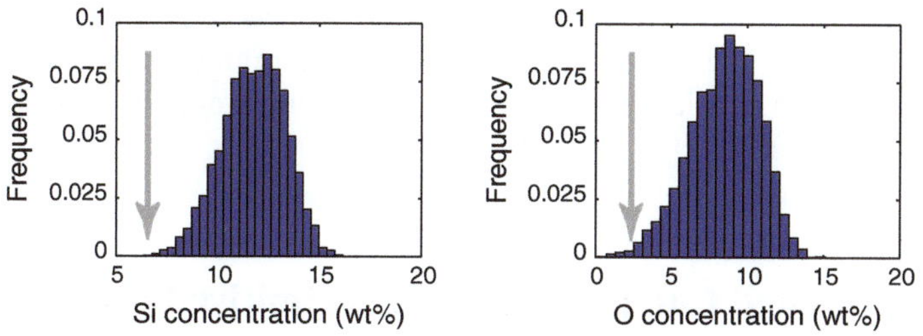

Fig. 10.4 Si and O content of the metal extracted from the magma ocean if the "residual" silicate has the pyrolitic composition of the PUM according to Kaminski and Javoy (2013). The *gray arrow* gives the average present-day core composition

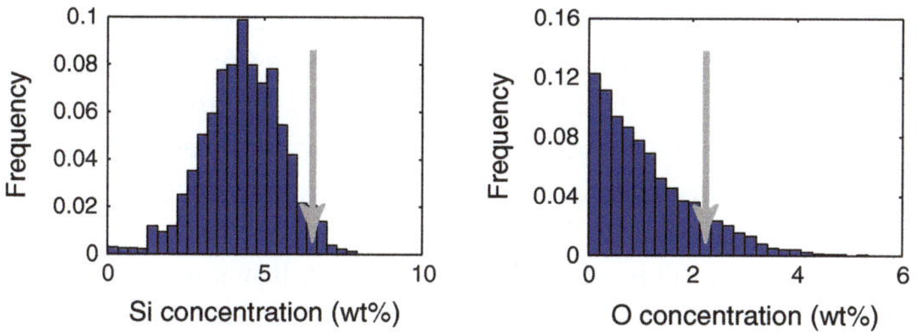

Fig. 10.5 Si and O content of the proto-core of the earth (that was formed before the giant impact) according to Kaminski and Javoy (2013). The *gray arrows* give to the average present-day core composition

magma ocean as obtained above. The depth of the magma ocean corresponding to a pressure of 45 GPa is ≈1350 km, which represents ≈44 % of the volume of the Earth or ≈30 % of its mass. The mass of metal that came from the magma ocean represented then 30 × 0.38 = 11.4 % of the mass of the Earth, and the mass of the proto-core was 21 % of the mass of the Earth or two-third of the mass of the present-day core.

The composition of the proto-core is then obtained by mass balance between the present-day core and the metal extracted from the magma ocean, using their respective masses. The result is shown in Fig. 10.5 for Si and O: The Si content of the proto-core was ~5 wt%, and it likely contained a negligible amount of oxygen. Such results favor a proto-core formation under relatively low pressure and temperature. To obtain a quantitative estimate of the *P*, *T* conditions of proto-core formation, we follow the hypothesis of homogeneous accretion and consider that the PLoM resulted from the extraction of the proto-core from an E-chondritic material with the bulk composition of Table 10.2. From the Ni, Si, and O concentrations in the PLoM and in the proto-core, we are able to calculate the partition coefficients PLoM/proto-core for these elements and get $\log(K_d^O) \approx -1.16 \pm 0.53$, $\log(K_d^{Si}) \approx -3.36 \pm 0.19$, and $\log(K_d^{Ni}) \approx 0.41 \pm 0.4$. The comparison of these values with experimental data (Mann et al. 2009; Bouhifd and Jephcoat 2011) yields a pressure of ≈20 GPa and a temperature of ≈2550 K. These conditions correspond to the expected first stage of differentiation in small planetary embryos, occurring at relatively low pressure and temperature (Ricard et al. 2009; Srámek et al. 2012).

10.6 An Updated Composition of the Primitive Lower Mantle

According to the scenario developed in the previous section, the chemical differences between the primitive upper and lower mantles result simply from the different *P*, *T* conditions that prevailed during the formation of the proto-core (relatively low *P*, *T*) and after the giant impact (high *P*, *T*). The evolution of *P*, *T* conditions between these two stages implies different efficiencies of Si and O dissolution in the metal, hence different FeO and SiO_2 contents in the upper and lower mantles. Purely lithophile elements, which do not enter the core, are not affected by the change in *P*, *T* conditions (their partition coefficient is always 0). Hence, the ratios of lithophile elements, such as Mg/Ca for example, have to be the same in the PUM and in the PLoM. This additional constraint allows us to further restrict the range of compositions proposed in Javoy et al. (2010) and leads to a refined estimate of PUM, PLoM, and core, given in Table 10.3.

Following the same line of arguments, it is possible to estimate the content of all the RLE in the PLoM, using the RLE/Ca ratio of the upper mantle and the Ca content of the PLoM. The resulting concentrations are given in Table 10.4. The same method can be used for moderately volatile lithophile elements, but the result is more speculative: Since these elements may have escaped the Earth's atmosphere more during the second stage (HP/HT) of core formation than during the differentiation of the proto-core, the method then provides only a lower bound.

The case of siderophile elements is even more complex since the change in HP/HT conditions between the two stages of core differentiation must be taken into account. For that purpose, we introduce $D^M_{\mathrm{HP/HT}}$, the metal/silicate partition coefficient of element *M* between the PUM and the metal extracted from the magma ocean, and $D^M_{\mathrm{LP/LT}}$, the metal/silicate partition coefficient of *M* between the PLoM and the proto-core. If C^M_{PUM} and C^M_{PLoM} are the concentrations of *M* in the PUM and in the PLoM, respectively, C^M_{PUM} and C^M_{PLoM} are related through

$$C^M_{\mathrm{PLoM}} = C^M_{\mathrm{PUM}}\left(1 + D^M_{\mathrm{HP/HT}}\right)/\left(1 + D^M_{\mathrm{LP/LT}}\right). \tag{10.11}$$

As stated above, D^M is replaced by the exchange coefficient $K^M_D = D^M/D^{\mathrm{Fe}}$ which is independent of oxygen fugacity (e.g., Siebert et al. 2012),

$$C^M_{\mathrm{PLoM}} = C^M_{\mathrm{PUM}}\left(1 + D^{\mathrm{Fe}}_{\mathrm{HP/HT}}\, K_D^{M_{\mathrm{HP/HT}}}\right)\Big/\left(1 + D^{\mathrm{Fe}}_{\mathrm{LP/LT}}\, K_D^{M_{\mathrm{LP/LT}}}\right), \tag{10.12}$$

Table 10.4 Composition of the primitive lower mantle including some minor and trace elements

Element	Behavior	Concentration	Element	Behavior	Concentration
Li	MVLE	1.4 ± 0.4 ppm	Ba	RLE	4.47 ± 0.9 ppm
Be	RLE	1.57 ± 0.1 ppb	La	RLE	447 ± 9 ppb
B	MVLE	0.15 ± 0.1 ppm	Ce	RLE	1179 ± 230 ppb
Na	MVLE	1950 ± 1100 ppm	Pr	RLE	178 ± 35 ppb
Mg	MVLE	20.4 ± 0.14 wt%	Nd	RLE	874 ± 170 ppb
Al	RLE	1.57 ± 0.09 wt%	Sm	RLE	285 ± 55 ppb
Si	Trans.	24.2 ± 0.14 wt%	Eu	RLE	108 ± 22 ppb
K	MVLE	167 ± 40 ppm	Gd	RLE	380 ± 74 ppb
Ca	RLE	1.61 ± 0.08 wt%	Tb	RLE	70 ± 14 ppb
Sc	RLE	11.1 ± 2.2 ppm	Dy	RLE	475 ± 92 ppb
Ti	RLE	836 ± 163 ppm	Ho	RLE	106 ± 22 ppb
Fe	Trans.	7.02 ± 0.24 wt%	Er	RLE	304 ± 60 ppb
Co	Trans.	≈48 ppm	Tm	RLE	47 ± 10 ppb
Ni	Trans.	0.11 ± 0.03 wt%	Yb	RLE	304 ± 60 ppb
Rb	MVLE	0.402 ± 0.084 ppm	Lu	RLE	47 ± 10 ppb
Sr	RLE	13.9 ± 2.7 ppm	Hf	RLE	200 ± 39 ppb
Y	RLE	2.96 ± 0.6 ppm	Ta	RLE	26 ± 6 ppb
Zr	RLE	7.41 ± 1.45 ppm	Th	RLE	55 ± 11 ppb
Nb	RLE	404 ± 171 ppb	U	RLE	15 ± 4 ppb

Refractory lithophile elements are estimated using the RLE/Ca ratio in the primitive upper mantle (Lyubetskaya and Korenaga 2007) and the Ca content of the lower mantle. The concentration of moderately refractory lithophile elements (*MVLE*) is estimated with the same method, but is more speculative and provides only a lower bound. Transitional elements (*Trans.*) are moderately volatile elements whose lithophile/siderophile behavior evolves under HP/HT conditions. Highly volatile elements are not included

where $D^{\text{Fe}}_{\text{LP/LT}} \approx 12.8$ and $D^{\text{Fe}}_{\text{HP/HT}} \approx 11.4$ are from the previous mass balance calculations.

Exchange coefficients may depend on the composition of the material used to determine them in HP/HT experiments. As no experiments have been performed yet using a composition sharing the characteristics of E-chondrites, especially its redox, the available data can be used only to get a first-order estimate of the siderophile content of the PLoM. Here, we used experimental results on Co (Siebert et al. 2012 and references within) that appear to be independent of the composition of the starting material (e.g., Siebert et al. 2011). We obtained $C^{\text{Co}}_{\text{PLoM}}/C^{\text{Co}}_{\text{PUM}} \approx 0.46$, which is close to the ratio of the Ni concentrations in the PUM and in the PLoM. The resulting smaller concentration of Co in the PLoM (Table 10.4) reflects the decrease of Co solubility in metal at higher pressure and temperature (the same explanation applies to Ni). Experimental constraints at high pressure and high temperature corresponding to the magma ocean (≈45 GPa and ≈3500 K) are however still lacking for a robust estimate of siderophile content in the PLoM.

10.7 Discussion

10.7.1 Origin and Repartition of Chemical Heterogeneities in the Deep Earth

The bulk Earth composition presented in this article falls within the ranges of E-Earth's compositions obtained in (Javoy et al. 2010). It is mainly a refinement allowed by the modeling of the core formation in two stages. Hence, it can be seen as a subclass of E-Earth compositions, whose robustness is the same as the hypotheses used to establish the model:

1. chemical heterogeneities are largely pristine in the deep mantle,
2. the formation of the Earth occurred mainly through the accretion of a homogeneous E-chondritic material latter differentiated under varying *P*, *T* conditions (Fig. 10.6).

The last hypothesis, which can be seen as the most stringent, is probably over simplified. Further modeling of core formation as a continuous process may affect the estimation of siderophile elements' concentrations, as well as the requirement of volatile elements' addition by the end of the accretion (e.g., Schonbachler et al. 2010; Marty 2012).

Other hypotheses can be put forward for the origin of chemical heterogeneities in the deep Earth, which may change the inferred composition. An opposite approach is to consider that the enrichment in Si and Fe in the LM is due to the accumulation of MORB material in the deep mantle and in particular at the CMB (e.g., Christensen and Hofmann 1994). Samuel et al. (2005) used a combined geodynamical/mineralogical and seismological approach to constrain the Si and Fe enrichment in the LM and concluded that the chemical anomaly could be created by about 1.5 Gyr-long accumulation of subducted crust in the LM. However, if only the crust is accumulated in the LM, its harzburgitic complement must be present elsewhere in the mantle, notably in the upper mantle, which would not be consistent with a pyrolitic upper mantle. Hence, a pristine origin of the Fe and Si enrichment of the LM relative to a pyrolitic upper mantle appears to be the most consistent hypothesis. This conclusion has been reached on independent grounds by Deschamps et al. (2012), which showed that MORB chemistry is not consistent with the seismic signature of the low-shear-wave velocity provinces (LLSVP) in the LM.

The size of the PLoM in the E-Earth scenario is constrained by the results of the inversion of PREM by Matas et al. (2007) and by the *P*, *T* conditions estimated for the formation of the PUM. As the uncertainties on thermo-elastic parameters used to invert PREM are large, many horizontally averaged compositional models are possible within them (e.g., Cobden et al. 2009a, b; Schaeffer and Lebedev, this volume; Rawlinson et al., this volume; Schmerr, this volume). Although the general characteristics of the PLoM compositions relative to PUM (i.e., enrichment in Si and Fe and depletion in Ca, Al, and Mg) are robust, the vertical extent and the mean composition of the PLoM will depend on the choices made on the thermo-elastic

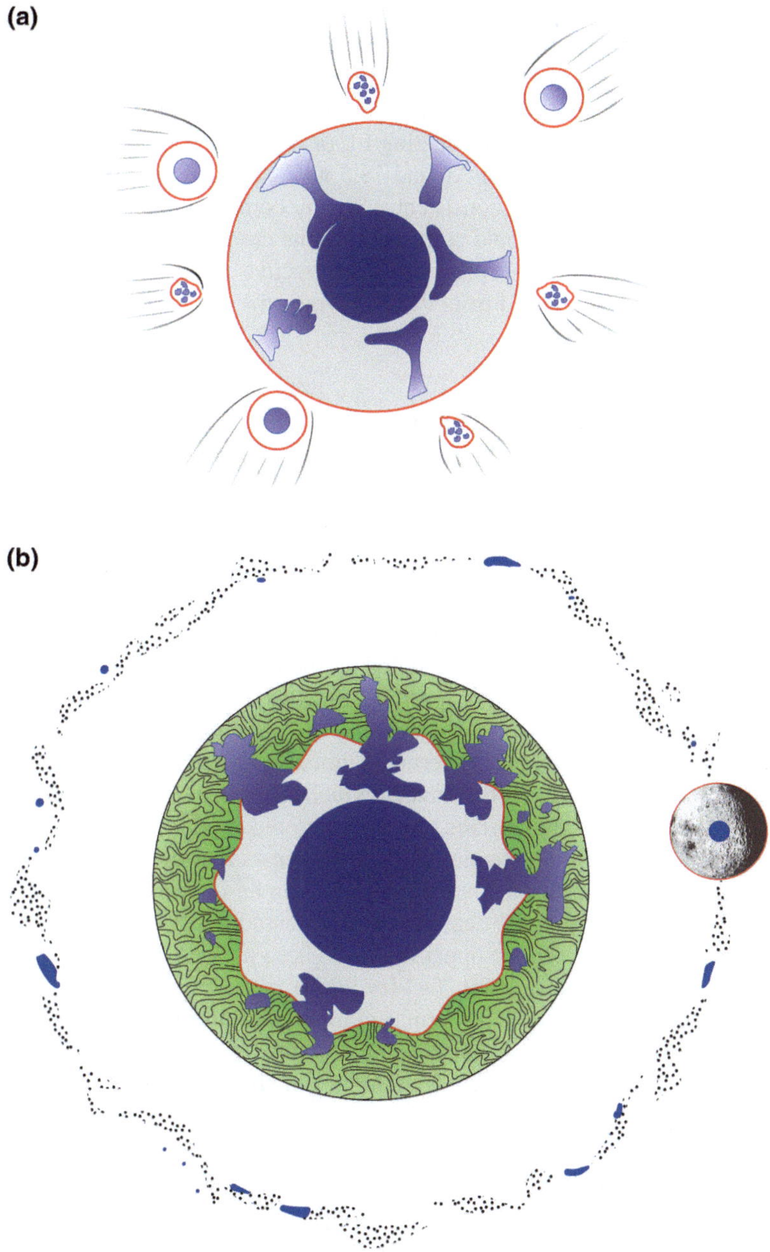

Fig. 10.6 Schematic evolution of the mantle and core in the E-Earth scenario. **a** Low P, T extraction of metal (*in blue*) from the E-chondritic accreted material producing the primitive lower mantle (*PLoM, in gray*) and the proto-core (*in blue*). **b** High P, T metal extraction associated with the giant impact and producing the (liquid) primitive upper mantle (*PUM, in green*). The composition of the core results from the merging of the metal extracted in the two stages

parameters used to invert PREM. A lower bound for the mass of the PLoM can be obtained using Ca and Al: Because the PUM is very significantly enriched in these elements relative to EH chondrites, the size of the PUM cannot be larger than 70 wt% of the primitive mantle in order to avoid negative Ca and Al concentrations in the PLoM. If the PLoM was accounting for only 30 wt% of the primitive mantle, it would indeed contain no Ca or Al, and 18.5 wt% Mg, 27.5 wt% Si, and 8.5 wt% Fe. Such a composition is not consistent with any compositional model of the LM we know of. However, this would correspond to the composition of the PLoM only, and it could have evolved, by progressive convective mixing, to the present-day composition of the LM inferred from PREM inversion.

10.7.2 Geodynamical Consequences of the E-Earth Model: Thermochemical Convection and Thermal Evolution of the Planet

In the present model, the chemical differentiation of the two mantles is related to the formation and extent of the magma ocean, and the thermal evolution of the planet is controlled by thermochemical convection. The evolution of this initial heterogeneity can furthermore be modeled as the evolution of a medium initially formed by two layers of different densities, as in Le Bars and Davaille (2004) for example.

The evolution of a two-layer thermochemical convective system depends on the mixing rate between these two layers, which itself depends on the density and viscosity contrast at their interface. As tomographic models show evidences for the presence of subducted plates in the LM, it appears that the present-day mantle convection is not a layered one, but rather occurs in an intermediate regime where regions containing "primitive" denser and undegassed material are deformed and (very) partially entrained in the upper mantle. These regions may form chemically dense domes (e.g., Davaille 1999; Tackley 2000), or rather dense and chemically more viscous pools from which thermal plumes may originate (e.g., Deschamps et al. 2012), giving rise both to the seismic anomaly (decorrelation of *V*p and *V*s) in the LM and to the geochemical signature of hot spots (e.g., rare gases implying an undegassed reservoir). Additional scales of heterogeneities are likely, such as primitive blobs (e.g., du Vignaux and Fleitout 2001) or piles of subducted dense material (e.g., Coltice and Ricard 1999). The evolution of the mantle thus results from two competing effects, the erasing of the initial chemical heterogeneity and the continuous building of heterogeneities by plate tectonics.

Tomographic models of the LM considering both thermal and chemical effects on seismic properties, either at a regional scale (e.g., Saltzer et al. 2004) or at a global scale (e.g., Simmons et al. 2010), obtain that a significant chemical contribution is required to fit the data at depths larger than ~1500 km, which remains close to the depth of the magma ocean inferred in the previous studies.

In particular, a superadiabatic temperature gradient would be required to fit the observables with a homogeneous pyrolitic mantle at these depths (e.g., Cobden et al. 2009a, b), whereas such an isochemical system heated from within displays (slightly) in fact sub-adiabatic profiles (e.g., Sotin and Labrosse 1999). It seems then that, despite the complexity of the interactions between the pristine anomalies in the deep mantle and the subducted plates that tend to deform and to erode them progressively (e.g., Kellogg et al. 1999), the convective mantle has been able to retain the gross characteristics of a "partially" layered system, where mass exchanges between the two mantles coexist with preserved deep-seated chemical reservoirs (e.g., Palot et al. 2012; Tucker et al. 2012).

Understanding the detailed evolution of a thermochemical system as complex as the Earth's mantle, and in particular the respective influence of plate subduction and remnants of a primitive layer, requires further numerical modeling in 3D spherical geometries, using two families of tracers, for the crust and for the primitive mantle, respectively (e.g., Tackley 2012; Nakagawa et al. 2012). The chemical characteristics of the E-Earth bear some specific implications that should be included in these models.

The first one is the amplitude of the density anomaly of chemical origin in the pristine reservoir: There is 1 % more Fe in the PLoM than the PREM content. This is enough to preserve the layer from massive convective entrainment, especially if its viscosity is higher due to its higher silica content, but this would not produce the unrealistically strong signature of LLSVP obtained in the model of Davies et al. (2012) by using an Fe-rich layer with more than 11 wt% FeO.

On the other hand, the RLE content of the PLoM in the E-Earth is smaller than the PUM content. This implies a smaller heat production in the PLoM, but this can still help to resolve the Urey's paradox, as the thermal boundary layer between the two mantles insulates the lower part of the mantle and delays the cooling of the Earth. McNamara and van Keken (2000) have found that a boundary layer at 1600 km depth with a uranium content of $\approx$14 ppb—which corresponds to our model results—provides a satisfactory fit to present mantle temperature and cooling rates. Before the extraction of the continental crust, however, the thermal contribution of the LM was ***smaller*** than that of the upper mantle, which may have helped to avoid a thermal catastrophe in Hadean time. Such a scenario will need to be tested using full numerical modeling. Furthermore, a thermal boundary layer between the lower and the upper mantle may not be required, because the endothermic phase transition at 660 km substantially decreases the circulation rate between the two mantles, because the plumes get thinner as they cross the 660 km boundary, and slabs are likely to be temporally stacked around that depth.

10.7.3 Future Developments

Three main categories of improvement of our knowledge of deep mantle composition and dynamics—based at least partially on E-chondrites—can be expected.

The first category is the improvement in our knowledge of the ECs' chemistry and particularly their phase composition. This step is mandatory to be able to propose a fully consistent chemical model of the LM including trace and minor elements following the approach of Javoy et al. (2010), hence to provide a robust estimate of U and Th concentrations in the primitive reservoir.

A second category is the improvement in our knowledge of partition coefficients of elements between solid silicate and melt, and between silicate and metal, at intermediate and high *P*, *T* conditions. Such a dataset is necessary to estimate the core composition in the two-stage scenario presented here. It will be also useful to better quantify the potential effect of some petrological differentiation in the mantle during the crystallization of the magma ocean. The possible change in RLE behavior at high *P*, *T*, from incompatible to moderately compatible, for example, may change the current estimates of the maximum extent of a perovskitic reservoir.

Finally, we need to develop new continuous models of Earth's differentiation, coupling the time evolution of *P* and *T* in the proto-Earth to experimentally determined partition coefficients (e.g., Rubie et al. 2011), and using E-chondrite composition as a bulk constraint on the evolution of oxygen fugacity during the coupled extraction of metal and oxidation of the mantle.

10.8 Conclusion

In year 1995, we saw three Earth models published concurrently: Mc Donough and Sun (1995), Allègre et al. (1995), and Javoy (1995). McDonough and Sun (1995) has become the reference composition for BSE. We claim here that if this composition is valid for the PUM, only the E-chondrite approach of Javoy (1995) provides a way to determine independently a LM composition, based on the solid grounds of the isotopic similarity between E-chondrites and the Earth. Javoy et al. (2010) extended that initial model to take into account the proportion and composition of phases in E-chondrites. This generic approach allows to generate a family of bulk Earth compositions which can then be refined by independent geophysical constraints. The additional modeling of the chemical differentiation between the upper and lower primitive mantles, linked to the evolution of *P*, *T* conditions during Earth's accretion, allows the determination of some minor and trace elements. The results presented above can then be seen as a subclass in the family of E-chondrites models for the bulk composition of the Earth. The main conclusions of the study shall hold, however, independently of the precise scenario used to refine the composition of the LM:

- the PLoM is enriched in Si and Fe and depleted in Mg relative to the PUM. Hence, the density of the PLoM and its viscosity were likely larger than those of the PUM, ensuring the preservation of the primitive chemical heterogeneity in the convective mantle throughout the Earth's history.

- the PLoM is depleted in Ca, Al, and other RLE relative to the PUM, due to a RLE/Si ratio smaller in E-chondrites than in CC. Hence, the production of thermal energy by internal heating is likely to have been smaller in the PLoM than in the PUM until the extraction of the continental crust. The implications of the result for the thermal evolution of the Earth and its secular cooling remain to be studied.
- Si is the main light element in the core, combined with a smaller amount of oxygen, because E-chondrites do not contain oxidized FeO: FeO in the silicate must first be produced by the dissolution of Si in the core-forming metal and only then can FeO be dissolved in the core as well. The bulk Si and O concentrations of the core resulted from the evolution of P,T conditions during the extraction of the metal from the infalling material. These conditions are likely to have evolved continuously—at least until the giant impact. Their modeling and their implication for the bulk composition of the core in terms of minor and trace elements remain to be performed and will require significant increase of the database of HP/HT partition coefficients of siderophile elements, as those presently available, for O, for Si, and for Ni, can only constrain a two-stage scenario.

Acknowledgments The author thanks an anonymous reviewer for his comments, and the Editor Frédéric Deschamps for comments and discussions and for his editorial handling. This work was supported in part by an INSU-CNRS grant from the PNP program and by ANR-11-IS04-0004.

References

Albarède F, van der Hilst RD (1999) New mantle convection model that may reconcile conflicting evidence. EOS 80:537–539

Alfe D, Gillan MJ, Price GD (2003) Thermodynamics from first principles: temperature and composition of the earth's core. Min Mag 67:113–123. doi:10.1180/0026461026610089

Alfè D, Gillan MJ, Price GD (2002) Composition and temperature of the earth's core constrained by combining ab initio calculations and seismic data. Earth Planet Sci Lett 195:91–98

Allègre CJ, Staudacher T, Sarda P, Kurz M (1983) Constraints on evolution of earth's mantle from rare gas systematics. Nature 303:762–766

Allègre CJ, Poirier JP, Humler E, Hofmann AW (1995) The chemical composition of the Earth. Earth Planet Sci Lett 134:515–526

Armytage RMG, Georg R, Savage P, Williams H, Halliday A (2011) Silicon isotopes in meteorites and planetary core formation. Geochim Cosmochim Acta 75:3662–3676. doi:10.1016/j.gca.2011.03.044

Asahara Y, Frost D, Rubie D (2007) Partitioning of FeO between magnesiowüstite and liquid iron at high pressures and temperatures: implications for the composition of the earth's outer core. Earth Planet Sci Lett 257:435–449. doi:10.1016/j.epsl.2009.09.029.D

Badro J, Fiquet G, Guyot F, Gregoryanz E, Occelli F, Antonangeli D, d'Astuto M (2007) Effect of light elements on the sound velocities in solid iron: implications for the composition of Earth's core. Earth Planet Sci Lett 254:233–238. doi:10.1016/j.epsl.2006.11.025

Baedecker PA, Wasson JT (1975) Elemental fractionations among enstatite chondrites. Geochim Cosmochim Acta 39:735–765

Bickle MJ, Hawkesworth CJ, Martin A, Nisbet EG, O'Nions RK (1976) Mantle composition derived from the chemistry of ultramafic lavas. Nature 263:577

Bouhifd MA, Jephcoat AP (2011) Convergence of Ni and Co metal-silicate partition coefficients in the deep magma-ocean and coupled silicon-oxygen solubility in iron melts at high pressures. Earth Planet Sci Lett 307:341–348. doi:10.1016/j.epsl.2011.05.006

Cadio C, Panet I, Davaille A, Diament M, Métivier L, de Viron O (2011) Pacific geoid anomalies revisited in light of thermochemical oscillating domes in the lower mantle. Earth Planet Sci Lett 306:123–135

Cameron AGW (2011) From interstellar gas to the earth-moon system. Meteor Planet Sci 36:9–22

Cammarano F, Goes S, Vacher P, Giardini D (2003) Inferring upper-mantle temperatures from seismic velocities. Phys Earth Planet Int 138:197–222

Canup RM (2008) Accretion of the earth. Phil Trans R Soc A 366:4061–4075. doi:10.1098/rsta.2008.0101

Carter JL (1970) Mineralogy and chemistry of earth's upper mantle based on partial fusion-partial crystallization model. Geol Soc Am Bull 81:2021–2031. doi:10.1130/0016-7606(1970)81[2021:MACOTE]2.0.CO;2

Cartigny P, Boyd SR, Harris J, Javoy M (1997) Nitrogen isotopes in peridotitic diamonds from china: the mantle signature. Terra Nova 94:175–179

Christensen UR, Hofmann AW (1994) Segregation of subducted oceanic crust in the convecting mantle. J Geophys Res 99:19867–19884

Clayton RN, Onuma N, Mayeda TK (1976) A classification of meteorites based on oxygen isotopes. Earth Planet Sci Lett 30:10–18

Cobden L, Goes S, Cammarano F, Connolly JAD (2009a) Thermochemical interpretation of one-dimensional seismic reference models for the upper mantle: evidence for bias due to heterogeneity. Geophy J Int 175:627–648. doi:10.1111/j.1365-246X.2008.03903.x

Cobden L, Goes S, Ravenna M, Styles E, Cammarano F, Gallagher K, Connolly JAD (2009b) Thermochemical interpretation of 1-D seismic data for the lower mantle: the significance of nonadiabatic thermal gradients and compositional heterogeneity. J Geophys Res 114:B11309. doi:10.1029/2008JB006262

Coltice N, Ricard Y (1999) Geochemical observations and one layer mantle convection. Earth Planet Sci Lett 174:125–137

Corgne A, Keshav S, Wood B, McDonough W, Fei Y (2008) Metal–silicate partitioning and constraints on core composition and oxygen fugacity during earth accretion. Geochim Cosmochim Acta 72:574–589

Dauphas N, Marty B, Reisberg L (2002) Inference on terrestrial genetics from molybdenum isotope systematics. Geophys Res Lett 296:1084–1088

Dauphas N, Davis AM, Marty B, Reisberg L (2004) The cosmic molybdenum-ruthenium isotope correlation. Earth Planet Sci Lett 226:465–475

Davaille A (1999) Simultaneous generation of hotspots and superswells by convection in a heterogeneous planetary mantle. Nature 402:756–760

Davies R, Goes S, Davies JH, Schuberth BSA, Bunge H-P, Ritsema J (2012) Reconciling dynamic and seismic models of earth's lower mantle: the dominant role of thermal heterogeneity. Earth Planet Sci Lett 353–354:253–269

Della Mora S, Boschi L, Tackley PJ, Nakagawa T, Giardini D (2011) Low seismic resolution cannot explain S/P decorrelation in the lower mantle. Geophys Res Lett 38:L12303. doi:10.1029/2011GL047559

Deschamps F, Trampert J (2004) Towards a lower mantle reference temperature and composition. Earth Planet Sci Lett 222:161–175

Deschamps F, Kaminski E, Tackley PJ (2011) A deep mantle origin for the primitive signature of ocean island basalt. Nat Geosci 4:879–882. doi:10.1038/ngeo1295

Deschamps F, Cobden L, Tackley PJ (2012) The primitive nature of large low shear-wave velocity provinces. Earth Planet Sci Lett 349–350:198–208 du Vignaux NM, Fleitout L (2001) Stretching and mixing of viscous blobs in earth's mantle. J Geophys Res 106:30893–30909

Dziewonski AM, Anderson DL (1981) Preliminary reference earth model. Phys Earth Planet Inter 25:297–356

Fiquet G, Guyot F, Badro J (2008) The earth's lower mantle and core. Elements 4:177–182. doi:10.2113/GSELEMENTS.4.3.177

Fischer RA, Campbell AJ, Caracas R, Reaman DM, Deraand P, Prakapenka VB (2012) Equation of state and phase diagram of Fe-16 Si alloy as a candidate component of earth's core. Earth Planet Sci Lett 357:268–276. doi:10.1016/j.epsl.2012.09.022

Fitoussi C, Bourdon B (2012) Silicon isotope evidence against an Enstatite Chondrite Earth. Science 335:1477–1480

Frost DJ, Mann U, Asahara Y, Rubie DC (2008) The redox state of the mantle during and just after core formation. Phil Trans R Soc A 366:4315–4337. doi:10.1098/rsta.2008.0147

Gannoun A, Boyet M, Rizao H, El Goresy A (2011) ^{146}Sm–^{142}Nd systematics measured in enstatite chondrites reveals a heterogeneous distribution of ^{142}Nd in the solar nebula. PNAS 19:7693–7697

Green DH (1964) The petrogenesis of the high-temperature peridotite intrusion in the Lizard area. Cornwall J Petrol 5:134–188

Harris PG, Reay A, White IG (1967) Chemical composition of the upper mantle. J Geophys Res 72:6359–6369. doi:10.1029/JZ072i024p06359

Hart SR, Zindler A (1986) In search of a bulk-earth composition. Chem Geol 57:247–267

Huang S, Jacobsen SB (2012) Calcium isotopic variations in chondrites: implications for planetary isotope compositions. In: 43rd lunar and planetary science conference, 1334

Ishii M, Tromp J (1999) Normal-mode and free-air gravity constraints on lateral variations in velocity and density of earth's mantle. Science 285:1231–1236

Jagoutz E, Palme H, Baddenhausen H, Blum K, Cendales M, Dreibus G, Spettel B, Lorenz V, Wänke H (1979) The abundances of major, minor and trace elements in the earth's mantle as derived from primitive ultramafic nodules. Proc Lunar Planet Sci Conf 10:2031–2050

Javoy M (1995) The integral enstatite chondrite model of the earth. Geophys Res Lett 22:2219–2222

Javoy M (1999) Chemical Earth models. CR Acad Sci 329:537–555

Javoy M, Balan E, Méheut M, Blanchard M, Lazzeri M (2011) First-principles investigation of equilibrium isotopic fractionation of O-and Si-isotopes between refractory solids and gases in the solar nebula. Earth Planet Sci Lett 319–320, 118–127, doi:10.1016/j.epsl.2011.12.029

Javoy M, Pineau F, Demaiffe G (1984) Nitrogen and carbon isotopic composition in the diamonds of Mbuji Mayi (Zaïre). Earth Planet Sci Lett 68:399–412

Javoy M, Pineau F, Delorme H (1986) Carbon and nitrogen isotopes in the mantle. Chem Geol 57:41–62

Javoy M, Kaminski E, Guyot F, Andrault D, Sanloup C, Moreira M, Labrosse S, Jambon A, Agrinier P, Davaille A, Jaupart C (2010) The chemical composition of the earth: Enstatite chondrite models. Earth Planet Sci Lett 293:259–268. doi:10.1016/j.epsl.2010.02.033

Kallemeyn GW, Wasson JT (1986) Compositions of enstatite (EH3, EH4,5 and EL6) chondrites: implications regarding their formation. Geochem Cosmochim Acta 50:2153–2164

Kaminski E, Javoy M (2013) A two-stage scenario for the formation of the earth's mantle and core. Earth Planet Sci Lett 365:97–107. doi:10.1016/j.epsl.2013.01.025

Kato T, Ringwood AE, Irifune T (1988) Constraints on element partition coefficients between $MgSiO_3$ perovskite and liquid determined by direct measurements. Earth Planet Sci Lett 90:65–68

Kellogg LH, Hager BH, van der Hilst RD (1999) Compositional stratification in the deep mantle. Science 283:1881–1884

Khan A, Connolly JAD, Taylor SR (2008) Inversion of seismic and geodetic data for the major element chemistry and temperature of the earth's mantle. J Geophys Res 113:B09308. doi:10.1029/2007JB005239

Khan A, Boschi L, Connolly JAD (2011) Mapping the earth's thermochemical and anisotropic structure using global surface wave data. J Geophys Res 116:B01301. doi:10.1029/2010JB007828

Kong P, Mori T, Ebihara M (1997) Compositional continuity of enstatite chondrites and implications for heterogeneous accretion of the enstatite chondrite parent body. Geochem Cosmochim Acta 61:4895–4914

Lassiter JC (2004) Role of recycled oceanic crust in the potassium and argon budget of the earth: toward a resolution of the "missing argon" problem. Geochem Geophy Geosy 5:Q11012. doi:10.1029/2004GC000711

Le Bars M, Davaille A (2004) Whole layer convection in a heterogeneous planetary mantle. J Geophys Res 109:B03403. doi:10.1029/2003JB002617

Lekic V, Cottaar S, Dziewonski A, Romanowicz B (2012) Cluster analysis of global lower mantle tomography: a new class of structure and implications for chemical heterogeneity. Earth Planet Sci Lett 357–358:68–77. doi:10.1016/j.epsl.2012.09.014

Liebske C, Corgne A, Frost DJ, Rubie DC, Wood BJ (2005) Compositional effects on element partitioning between Mg-silicate perovskite and silicate melts. Contrib Mineral Petrol 149:113–128. doi:10.1007/s00410-004-0641-8

Loubet M, Shimizu N, Allègre CJ (1975) Rare earth elements in alpine peridotites. Contrib Mineral Petrol 53:1–12

Lyubetskaya T, Korenaga J (2007) Chemical composition of the earth primitive mantle and its variance: 1. Method and results. J Geophys Res 112:B03211. doi:10.1029/2005JB004223

Maaloe S, Aoki KI (1977) Major element composition of upper mantle estimated from composition of lherzolites. Contrib Mineral Petrol 63:161–7173. doi:10.1007/BF00398777

Mann U, Frost DJ, Rubie DC (2009) Evidence for high-pressure core-mantle differentiation from the metal-silicate partitioning of lithophile and weakly-siderophile elements. Geochim Cosmochim Acta 73:7360–7386

Marty B (2012) The origins and concentrations of water, carbon, nitrogen and noble gases on earth. Earth Planet Sci Lett 313–314:56–66. doi:10.1016/j.epsl.2011.10.040

Matas J, Bass J, Ricard Y, Mattern E, Bukowinski MST (2007) On the bulk composition of the lower mantle: predictions and limitations from generalized inversion of radial seismic profiles. Geophys J Int 170:764–780

McDonough WF, Sun S (1995) The composition of the earth. Chem Geol 120:223–253

McNamara AK, van Keken PE (2000) Cooling of the earth: a parameterized convection study of whole versus layered models. Geochem Geophys Geosyst 1:2000GC000045

Moreira M (2013) Noble gas constraints on the origin and evolution of earth's volatiles. Geochem Perspect 2:229–413

Moreira M, Kunz J, Allègre CJ (1998) Rare gas systematics in popping rock: isotopic and elemental compositions in the upper mantle. Science 279:1178–1181

Moynier F, Paniello RC, Gounelle M, Albarède F, Beck P, Podosek F, Zanda B (2011) Nature of volatile depletion and genetic relationships in enstatite chondrites and aubrites inferred from Zn isotopes. Geochim et Cosmochim Acta 75:297–307

Mukhopadhyay S (2012) Early differentiation and volatile accretion recorded in deep-mantle neon and xenon. Nature 486:101–104

Nakagawa T, Tackley PJ, Deschamps F, Connolly JAD (2012) Radial 1-D seismic structures in the deep mantle in mantle convection simulations with self-consistently calculated mineralogy. Geochem Geophys Geosyst 13:Q11002. doi:10.1029/2012GC004325

Palme H, Nickel KG (1985) Ca/Al ratio and composition of the Earth's upper mantle. Geochem. Cosmochem. Acta 49:2123–2132

Palot M, Cartigny P, Harris JW, Kaminsky FV, Stachel T (2012) Evidence for deep mantle convection and primordial heterogeneity from nitrogen and carbon stable isotopes in diamond. Earth Planet Sci Lett 357–358:101–104. doi:10.1016/j.epsl.2012.09.015

Park J, Okazaki R, Nagao K, Baroschewitz R (2005) Noble gas study of new Enstatite SaU290 with high solar gases. Lunar Planet Sci XXXVI:1632

Petyaev M, Khodakovsky I (1986) Thermodynamic properties and conditions of formation of minerals in enstatite meteorites. Chemistry and physics of the terrestrial planets: advances in physical chemistry, vol 6. Springer, Berlin, pp 107–135

Poirier JP (1994) Light elements in the earth's outer core: a critical review. Phys Earth Planet Inter 85:319–337

Raquin A, Moreira M (2009) Atmospheric $^{38}Ar/^{36}Ar$ in the mantle: implications for the nature of the terrestrial parent bodies. Earth Planet Sci Lett 287(3–4):551–558. doi:10.1016/j.epsl.2009.09.003

Regelous M, Elliott T, Coath CD (2008) Nickel isotope heterogeneity in the early solar system. Earth Planet Sci Lett 272:330–338

Ricard Y, Sramek O, Dubuffet F (2009) A multi-phase model of runaway core-mantle segregation in planetary embryos. Earth Planet Sci Lett 284:144–150. doi:10.1016/j.epsl.2009.04.021

Ricolleau A, Fei Y, Cottrell E, Watson H, Deng L, Zhang L, Fiquet G, Auzende A, Roskosz M, Morard G, Prakapenka V (2009) Density profile of pyrolite under the lower mantle conditions. Geophys Res Lett 36:L06302. doi:10.1029/2008GL036759

Righter K (2003) Metal–silicate partitioning of siderophile elements and core formation in the early earth. Annu Rev Earth Planet Sci 31:135–174

Ringwood AE (1966) The chemical composition and origin of the earth. In: Hurley PM (ed) Advances in earth science. MIT Press, Cambridge, pp 287–356

Rubie DC, Frost DJ, Mann U, Asahara Y, Nimmo F, Tsuno K, Kegler P, Holzheid A, Palme H (2011) Heterogeneous accretion, composition and core-mantle differentiation of the earth. Earth Planet Sci Lett 301(1–2):31–42. doi:10.1016/j.epsl.2010.11.030

Rubin AE, Huber H, Wasson JT (2009) Possible impact-induced refractory-lithophile fractionations in EL chondrites. Geochem Cosmochim Acta 73:1523–1537

Saltzer RL, Stutzmann E, van der Hilst RD (2004) Poisson's ratio in the lower mantle beneath Alaska: evidence for compositional heterogeneity. J Geophys Res 109:B06301. doi:10.1029/2003JB002712

Samuel H, Farnetani CG, Andrault D (2005) Heterogeneous lowermost mantle: compositional constraints and seismological observables. In: van der Hilst RD et al (ed) Earth's deep mantle: structure, composition, and evolution, vol 160. Geophysics monograph series. AGU, Washington, DC, pp 101–116. doi:10.1029/160GM08

Savage PS, Moynier F (2013) Silicon isotopic variation in enstatite meteorites: clues to their origin and earth-forming material. Earth Planet Sci Lett 361:487–496. doi:10.1016/j.epsl.2012.11.016

Schonbachler M, Carlson RW, Horan MF, Mock TD, Hauri EH (2010) Heterogeneous accretion and the moderately volatile element budget of earth. Science 328:884–888. doi:10.1126/science.1186239

Schubert G, Turcotte DL, Olson P (2001) Mantle convection in the earth and planets. Cambridge University Press, Cambridge (940p)

Siebert J, Corgne A, Ryerson FJ (2011) Systematics of metal–silicate partitioning for many siderophile elements applied to earth's core formation. Geochim Cosmochim Acta 75:1451–1489

Siebert J, Badro J, Antonangeli D, Ryerson FJ (2012) Metal-silicate partitioning of Ni and Co in a deep magma ocean. Earth Planet Sci Lett 312–322:189–197. doi:10.1016/j.epsl.2012.01.013

Simmons NA, Forte AM, Boschi L, Grand SP (2010) GyPSuM: a joint tomographic model of mantle density and seismic wave speeds. J Geophys Res 115:B12310. doi:10.1029/2010JB007631

Simon JI, DePaolo DJ (2010) Stable calcium isotopic composition of meteorites and rocky planets. Earth Planet Sci Lett 289:457–466. doi:10.1016/j.epsl.2009.11.035

Sotin C, Labrosse S (1999) Three-dimensional thermal convection of an isoviscous, infinite-Prandtl-number fluid heated from within and from below: applications to heat transfer in planetary mantles. Phys Earth Planet Inter 112:171–190

Srámek O, Milelli L, Ricard Y, Labrosse S (2012) Thermal evolution and differentiation of planetesimals and planetary embryos. Icarus 217:339–354

Sun S-S (1982) Chemical composition and origin of the earth's primitive mantle. Geochim Cosmochim Acta 46:179–192
Tackley PJ (2000) Mantle convection and plate tectonics: toward an integrated physical and chemical theory. Science 288:2002–2007
Tackley PJ (2012) Dynamics and evolution of the deep mantle resulting from thermal, chemical, phase and melting effects. Earth-Sci Rev 110:1–25. doi:10.1016/j.earscirev.2011.10.001
Tan E, Leng W, Zhong S, Gurnis M (2011) On the location of plumes and lateral movement of thermochemical structures with high bulk modulus in the 3D compressible mantle. G-cubed 12:Q07005. doi:10.1029/2011GC003665
Tarits P, Mandéa M (2010) The heterogeneous electrical conductivity structure of the lower mantle. Phys Earth Planet Int 183:115–125
Tonks WB, Melosh HJ (1993) Magma ocean formation due to giant impacts. J Geophys Res 98:5319–5333
Trinquier A, Birck JL, Allègre CJ (2007) Widespread ^{54}Cr heterogeneities in the inner solar system. Astrophys J 655:1179–1185
Trinquier A, Elliott T, Ulfbeck D, Coath C, Krot AN, Bizzarro M (2009) Origin of nucleosynthetic isotope heterogeneity in the solar protoplanetary disk. Science 324(5925):295–424. doi:10.1126/science.1168221
Tucker JM, Mukhopadhyay S, Schilling J-G (2012) The heavy noble gas composition of the depleted MORB mantle (DMM) and its implications for the preservation of heterogeneities in the mantle. Earth Planet Sci Lett 355–356:244–254. doi:10.1016/j.epsl.2012.08.025
van der Hilst RD, Widiyantoro S, Engdahl ER (1997) Evidence for deep mantle circulation from global tomography. Nature 386:578–584
Walter MJ, Kohn SC, Araujo D, Bulanova GP, Smith CB, Gaillou E, Wang J, Steele A, Shirey SB (2011) Deep mantle cycling of oceanic crust: evidence from diamonds and their mineral inclusions. Science 334:54–57. doi:10.1126/science.1209300
Warren PH (2011) Stable-isotopic anomalies and the accretionary assemblage of the earth and Mars: a subordinate role for carbonaceous chondrites. Earth Planet Sci Lett 311:93–100. doi:10.1016/j.epsl.2011.08.047
Wasson JT (1985) Meteorites. Springer, Berlin
Wasson JT, Kallemeyn GW (1988) Compositions of chondrites. Phil Trans R Soc London A 325:535–544
Wiechert U, Halliday AN, Lee D-C, Snyder GA, Taylor LA, Rumble D (2001) Oxygen isotopes and the moon-forming giant impact. Science 294:345–348
Williams Q, Knittle E (2005) The uncertain major element bulk composition of earth's mantle. In: VanDerHilst RD, Bass JD, Matas J, Trampert J (ed) Earth's deep mantle: structure, composition and evolution, vol 160. J. Book Series: Geophysical Monograph Series, pp 187–199. doi: 10.1029/160GM12
Wood B (2011) The formation and differentiation of earth. Phys Today 64:40–45. doi:10.1063/PT.3.1362
Zhang Y, Benoit P, Sears DW (1995) The classification and complex thermal history of the enstatite chondrites. J Geophys Res 100:9417–9438
Zhang J, Dauphas N, Davis AM, Leya I, Fedkin A (2012) The proto-earth as a significant source of lunar material. Nat Geosci 5:251–255. doi:10.1038/NGEO1429

Chapter 11
Chemical Geodynamics in a Non-chondritic Earth

Guillaume Caro

Abstract Over the past 30 years, chemical geodynamic models held that terrestrial silicate reservoirs differentiated from a primitive mantle (PM) with chondritic abundances of refractory lithophile elements. This basic assumption has had major consequences for our understanding of the mantle's structure and composition. It is well established, for example, that in a chondritic Earth, the present-day continental mass does not contain a large enough budget of trace elements to balance depletion of the whole mantle. This result was taken as evidence that a hidden reservoir, either primitive or enriched in incompatible elements, segregated early in the history of the Earth and remained since then isolated from the convective system. This common view, however, is now being reconsidered, as ^{146}Sm–^{142}Nd studies of planetary and meteoritic material show that the bulk silicate Earth may in fact not have perfectly chondritic abundances of refractory lithophile elements, as has been assumed. This observation represents a challenge to compositional models of the Earth, which are systematically anchored on chondritic reference parameters. Chemical geodynamic models, based on mass balance relationships between the depleted, primitive, and enriched silicate reservoirs, would also need to be reconsidered. Here, I present a new set of reference parameters for the Sm–Nd, Lu–Hf, and Rb–Sr systems consistent with the presence of an 18 ppm ^{142}Nd excess in the bulk silicate Earth. The trace element pattern obtained for the PM using this super-chondritic Earth model (SCHEM) suggests that the proto-Earth accreted from material depleted in incompatible elements compared to chondrites, most likely as a result of preferential impact erosion of shallow crustal reservoirs during the early stage of accretion. The isotopic and trace element signature of the (non-chondritic) PM matches that observed in most high ^{3}He/^{4}He oceanic island basalts, suggesting that pristine material may have been preserved in the interior of the Earth for the past 4.45 Ga and is now sampled by modern plume magmatism. These primordial heterogeneities, however, are unlikely to represent a large reservoir, as most of the mantle appears

G. Caro (✉)
CRPG-CNRS, Université de Lorraine, 15 rue Notre Dame des Pauvres, 54501, Vandoeuvre les Nancy, France
e-mail: caro@crpg.cnrs-nancy.fr

© Springer International Publishing Switzerland 2015

A. Khan and F. Deschamps (eds.), *The Earth's Heterogeneous Mantle*, Springer Geophysics, DOI 10.1007/978-3-319-15627-9_11

to have been thoroughly degassed and depleted by crustal extraction. In a non-chondritic Earth, the compositional evolution of the "accessible" mantle appears to be essentially controlled by the continuous growth of the continental crust over the past 3.8 Ga and does not require the presence of hidden mantle reservoirs, either enriched or primitive.

Keywords Mantle differentiation · Collisional erosion · Hadean · Extinct radioactivity · Chemical geodynamics

11.1 Introduction

In the late 1970s, the pioneering applications of the ^{147}Sm–^{143}Nd system as a tracer of mantle–crust differentiation revealed that the source of mid-ocean ridge basalts (MORB) was more depleted than expected for a PM of chondritic composition (Allegre et al. 1979; DePaolo and Wasserburg 1976; Richard et al. 1976). Archean and Proterozoic mantle-derived rocks also have super-chondritic initial $^{143}Nd/^{144}Nd$ ratios, indicating the mantle evolved with a depleted composition for at least 3.8 Ga (Shirey and Hanson 1986). It was rapidly established that the only terrestrial reservoir likely to contain a sufficiently large budget of unradiogenic Nd to balance the signature observed in the MORB source was the continental crust (CC) (Taylor et al. 1983). Thus, over the past 30 years, the compositional evolution of the "accessible" mantle was interpreted as the result of progressive extraction of the continents through geological times (Armstrong 1981; DePaolo 1980; Hofmann 1988; Jacobsen 1988; Jacobsen and Wasserburg 1979).

This standard chemical geodynamic model, however, is not without caveats. First, the complementarity of the CC and depleted mantle (DM) is questionable from a mass balance point of view. It is indeed well established that in a globally chondritic silicate Earth, the continental reservoir does not contain a large enough budget of trace elements to balance the DM unless the latter represents only 25–50 % of the mass of the silicate Earth (Jacobsen and Wasserburg 1979). This observation implies either that the lower mantle (LM) (or a large reservoir within the LM) remained isolated throughout Earth's history, or an additional enriched reservoir, such as recycled oceanic crust, is needed to balance the depletion of the MORB source (Christensen and Hofmann 1994; Hofmann and White 1982; Kellogg et al. 1999). Models involving a primordial LM satisfy the constraints based on the study of noble gases, particularly Argon, which require the preservation of a deep undegassed reservoir to complement the ^{40}Ar budget of the atmosphere (Allegre 1982; Allegre et al. 1996; Arevalo et al. 2009). However, these models require the LM to have remained perfectly isolated from the upper mantle over the past 4.5 Ga, which conflicts with a wide range of geophysical, petrological, and geochemical observations (Grand 2002; Hayman et al. 2005; Montelli et al. 2004; Stachel et al. 2000; van der Hilst et al. 1997; Walter et al. 2011; Zindler and Hart 1986). Models involving a deep enriched reservoir such as recycled oceanic crust are more consistent with whole-mantle convection but do not account for the noble gas systematics of the OIB source. Thus, 35 years after the pioneering works of Richard et al. (1976) and DePaolo and Wasserburg (1976), it is still unclear which mechanism(s) caused the isotopic and chemical composition of the "accessible" mantle to deviate from that of a primitive chondritic reservoir.

Another important caveat of the standard model is that crustal growth cannot account for the isotopic evolution of the mantle unless a mass of continent roughly equivalent to that of the present day was already formed by ca. 3.8 Ga ago (Armstrong 1981). Archean cratons, however, make up less than 20 % of the present-day continents (Condie 2000), which is largely insufficient to generate significant depletion of the Earth's mantle (Caro and Bourdon 2010). This observation calls for alternative mechanisms of mantle–crust differentiation prior to 4 Ga (Caro et al. 2005; Chase and Patchett 1988; Galer and Goldstein 1991). While this crustal growth problem was often treated separately from the mass balance issue outlined above, both observations could in fact result from a single process, whereby a primordial reservoir enriched in incompatible elements was isolated from the "accessible" mantle–crust system at a relatively early stage of Earth's evolution (>4 Ga) (Tolstikhin et al. 2006).

The creation, composition, and fate of this "missing" reservoir has been the subject of intense debate fueled by the discovery of non-chondritic ^{142}Nd signatures in all terrestrial and lunar silicate reservoirs (Andreasen and Sharma 2006; Boyet and Carlson 2005, 2006; Carlson et al. 2007; Caro et al. 2008b; Nyquist et al. 1995; Touboul et al. 2009a). Since ^{142}Nd is produced by radioactive decay of the short-lived ^{146}Sm ($T_{1/2} = 68$ Ma), the radiogenic excess characterizing the silicate Earth requires that an enriched reservoir segregated early from the "accessible" mantle and was never reincorporated in the convective system. In essence, the debate centers on the following question: Is this missing reservoir hidden in the deep mantle, where it may have formed as a by-product of magma ocean crystallization (Boyet and Carlson 2005; Coltice et al. 2011; Labrosse et al. 2007; Nomura et al. 2011), or was it lost during the accretion of the proto-Earth (Bourdon and Caro 2007; Bourdon et al. 2008; Campbell and O'Neill 2012; Caro 2011; Caro et al. 2008a; Caro and Bourdon 2010; Korenaga 2009; O'Neill and Palme 2008; Warren 2008)? The impact erosion scenario is more consistent with the lack of ^{142}Nd heterogeneities in the modern mantle (Andreasen et al. 2007; Caro et al. 2006; Jackson and Carlson 2012; Murphy et al. 2010) but would imply that the BSE does not have perfectly chondritic abundances of refractory lithophile elements (Caro et al. 2008b). This would challenge a fundamental paradigm of geochemistry and, if true, would provide an entirely new perspective on geochemical signals recorded by mantle-derived rocks through geological times (Caro and Bourdon 2010; Jackson and Carlson 2011; Jackson et al. 2010).

This article reviews recent models of the evolution of the silicate Earth inferred from ^{146}Sm–^{142}Nd systematics of terrestrial, lunar, and meteoritic samples. Section 11.2 summarizes basic chronological constraints on primordial mantle differentiation processes obtained from ^{147}Sm–^{143}Nd, ^{146}Sm–^{142}Nd, and ^{176}Lu–^{176}Hf studies of the earliest rock and mineral record. In Sect. 11.3, the ^{142}Nd systematic of chondrites is reviewed and plausible causes for the apparent non-chondritic ^{142}Nd signature of the silicate Earth are discussed. Section 11.4 presents the backbone of a non-chondritic BSE compositional model anchored on ^{142,143}Nd–^{176}Hf–^{87}Sr systematics of accessible silicate reservoirs. Finally, Sect. 11.5 re-examines mass balance and isotopic constraints on mantle structure in the context of this non-chondritic Earth model. It is shown that the compositional evolution of the Earth's mantle can be accounted for by three distinct processes, namely (i) Collisional loss of crustal material at an early stage of planetary accretion, which caused significant depletion of incompatible elements in the Earth building material; (ii) early (4.45 Ga) differentiation of primordial silicate reservoirs in the aftermath of the giant impact, and their subsequent rehomogenization on a billion year timescale; and finally, (iii) continuous extraction of the CC from a well-mixed mantle.

11.2 Creation of Primordial Mantle Heterogeneities

11.2.1 ^{146,147}Sm–^{142,143}Nd Essentials

Neodymium (Nd) is a light rare earth element (REE) with 7 stable isotopes, two of which are radiogenic: ^{143}Nd is produced by α-decay of the long-lived radionuclide ^{147}Sm [$T_{1/2}$ = 106 Gyr (Audi et al. 1997)], and ^{142}Nd was produced by α-decay of now-extinct ^{146}Sm [$T_{1/2}$ = 68 Myr (Kinoshita et al. 2012)]. As all other lanthanides, Sm and Nd are refractory and lithophile elements (RLE). Consequently, their relative abundances in planetary bodies have been considered to be unfractionated compared to their chondritic abundances (McDonough and Sun 1995). Specifically, the homogeneity of ^{147}Sm/^{144}Nd and ^{143}Nd/^{144}Nd ratios among chondrite groups otherwise exhibiting variable contents of volatile elements was taken as evidence that Sm and Nd, like other RLEs, had not been subjected to significant fractionation in the solar nebula. Jacobsen and Wasserburg (1984) thus proposed a set of reference parameters for the bulk isotopic and chemical composition of the terrestrial planets based on ^{147}Sm–^{143}Nd measurements in bulk carbonaceous and ordinary chondrites (OC). For more than 30 years, these chondritic uniform reservoir (CHUR) parameters served as a baseline against which the isotopic composition of differentiated reservoirs was interpreted in terms of enrichment or depletion relative to a primitive chondritic mantle.

On a planetary scale, the fractionation of Sm from Nd is essentially restricted to mantle–crust differentiation processes (Hofmann 1988). Both elements are moderately incompatible and are enriched by approximately one order of magnitude in crustal rocks compared to their abundances in planetary mantles. Therefore, a significant budget of these elements is concentrated in volumetrically small crustal reservoirs, making the Sm–Nd system a sensitive tool for investigating the chronology and processes involved in differentiation of planetary mantles. Since Sm has a slightly smaller ionic radii compared to Nd, it behaves as a less incompatible element in magmatic systems; residual mantle from partial melting displays high

Sm/Nd ratios, while crustal reservoirs resulting from partial melting of a solid source, or from fractional crystallization of an initially molten mantle, have low Sm/Nd ratios and enriched REE abundances (Hofmann 1988). Radioactive decay in such differentiated reservoirs will generate distinct $^{143}Nd/^{144}Nd$ signatures, which, if differentiation takes place prior to ca. 4.2 Ga, will be associated with ^{142}Nd "anomalies" generated by the decay of now-extinct ^{146}Sm.

By convention, $^{142,143}Nd/^{144}Nd$ signatures in differentiated reservoirs are expressed as relative deviations compared to a reference composition, using the epsilon notation:

$$\varepsilon^{142,143}\text{Nd} = \left[\frac{\left(^{142,143}\text{Nd}/^{144}\text{Nd}\right)_i}{\left(^{142,143}\text{Nd}/^{144}\text{Nd}\right)_{\text{BSE,CHUR}}} - 1 \right] \times 10^4 \qquad (11.1)$$

The CHUR parameters serve as reference for the ^{147}Sm–^{143}Nd system, so that a chondritic reservoir would have $\varepsilon^{143}Nd$ at $=0$ all time. $^{142}Nd/^{144}Nd$ ratios are normalized to a terrestrial standard, taken to be representative of the isotopic composition of the present-day "accessible" mantle and crust. As will be discussed below in more detail, this "terrestrial" reference differs from the CHUR $^{142}Nd/^{144}Nd$ ratio by approximately 20 ppm, so that bulk chondrites have negative $\varepsilon^{142}Nd$ compared to terrestrial standards.

11.2.2 The First Terrestrial Crust

The ubiquitous presence of positive $\varepsilon^{143}Nd$ in early Archean mantle-derived rocks was established in the early 1980s, as the ^{147}Sm–^{143}Nd system became a common tool in geochronology (Shirey and Hanson 1986). Typically, initial $\varepsilon^{143}Nd$ in Eorchean mantle-derived rocks ranges from +1 to +3 (Moorbath et al. 1997), demonstrating that the Earth's mantle evolved for more than 3.8 Ga with an Sm/Nd ratio approximately 5–10 % higher than chondritic. In principle, this could indicate that the Earth's mantle was never chondritic, but this hypothesis was generally discarded as RLEs were thought to be present in chondritic abundances in the BSE. Instead, the positive $\varepsilon^{143}Nd$ values characterizing ancient mantle-derived rocks were interpreted as indicating derivation of Archean magmas from a mantle reservoir having experienced depletion at very early times (i.e., >3.8 Ga ago) (Armstrong 1981; Chase and Patchett 1988; Galer and Goldstein 1991).

It has often been suggested that early mantle depletion could have resulted from massive continental growth in the Hadean (Armstrong 1981; Jacobsen 1988). In the most extreme version of these "constant crustal mass" models (CCM), a continental mass roughly similar to the present-day forms prior to 4 Ga. Subsequent additions of juvenile magmas to the growing crust are balanced by the recycling of ancient material, so that the volume of continents remains nearly constant through time. The apparent growth peaks observed around 2.7 and 1.9 Ga (Condie 2000) would then reflect an increase of the rate of preservation of the crust, perhaps due to the formation of large cratons, rather than an actual increase in the production of juvenile crust. While the CCM is still a popular model, owing in particular to geodynamic constraints established by studies of detrital zircons (Amelin et al. 1999; Harrison 2009; Harrison et al. 2008; Hawkesworth and Kemp 2006; Hopkins et al. 2008; Mojzsis et al. 2001; Trail et al. 2007; Wilde et al. 2001), the Archean ε^{143}Nd record is consistent with many alternative scenarios. These range from crystallization of a deep magma ocean (Caro et al. 2005) to the extraction of a basaltic or komatiitic crust in the Hadean (Galer and Goldstein 1991), followed by transient (Chase and Patchett 1988) or even permanent (Tolstikhin et al. 2006) storage of such recycled crust in the deep mantle.

In the mid-1990s, Harper and Jacobsen (1992) addressed the early crustal growth problem by combining the long-lived ^{147}Sm–^{143}Nd with the short-lived ^{146}Sm–^{142}Nd chronometer (Fig. 11.1). The rationale is that if differentiation of the early Earth's mantle took place prior to 4.2 Ga, the decay of ^{146}Sm should have resulted in the creation of ^{142}Nd excesses associated with positive ε^{143}Nd values, according to the following two-stage model equations:

$$\left(\frac{^{143}\mathrm{Nd}}{^{144}\mathrm{Nd}}\right)_t^{\mathrm{DM}} = \left(\frac{^{143}\mathrm{Nd}}{^{144}\mathrm{Nd}}\right)_{t_p}^{\mathrm{PM}} + \left(\frac{^{147}\mathrm{Sm}}{^{144}\mathrm{Nd}}\right)_{t_p}^{\mathrm{PM}}\left[1-\mathrm{e}^{\lambda_{147}(t_p-t)}\right] + \left(\frac{^{147}\mathrm{Sm}}{^{144}\mathrm{Nd}}\right)_{t_p}^{\mathrm{DM}}\left[\mathrm{e}^{\lambda_{147}(t_p-t_d)} - \mathrm{e}^{\lambda_{147}(t_p-t)}\right] \tag{11.2}$$

$$\left(\frac{^{142}\mathrm{Nd}}{^{144}\mathrm{Nd}}\right)_t^{\mathrm{DM}} = \left(\frac{^{142}\mathrm{Nd}}{^{144}\mathrm{Nd}}\right)_{t_p}^{\mathrm{PM}} - \frac{\left(^{146}\mathrm{Sm}/^{144}\mathrm{Sm}\right)_{t_0}}{\left(^{147}\mathrm{Sm}/^{144}\mathrm{Sm}\right)_{t_p}}\left[\begin{array}{l}\left(\frac{^{147}\mathrm{Sm}}{^{144}\mathrm{Nd}}\right)_{t_p}^{\mathrm{PM}} \times \mathrm{e}^{-\lambda_{146}t_d} \\ +\left(\frac{^{147}\mathrm{Sm}}{^{144}\mathrm{Nd}}\right)_{t_p}^{\mathrm{DM}} \times \left[\mathrm{e}^{-\lambda_{146}t} - \mathrm{e}^{-\lambda_{146}t_d}\right]\end{array}\right] \tag{11.3}$$

where t is time running forward from the origin of the solar system ($t_0 = 0$ Ga) to present day ($t_p = 4.567$ Ga). In this simple model, the mantle experiences instantaneous depletion at time t_d from an initially PM. Stage 1 represents the closed system evolution of the PM between $t_0 = 0$ and $t = t_d$, and stage 2 is the evolution of differentiated crustal and/or DM reservoirs between t_d and any time $t > t_d$. $\lambda_{147} = 6.54 \times 10^{-3}$ Ga^{-1} and $\lambda_{146} = 10.19$ Ga^{-1} are the decay constant of ^{147}Sm (Audi et al. 1997) and ^{146}Sm (Kinoshita et al. 2012), respectively, and ^{146}Sm/^{144}Sm $= 0.0094$ is the initial abundance of ^{146}Sm at the time of formation of the solar system (Amelin and Rotenberg 2004; Boyet et al. 2010; Kinoshita et al. 2012; Prinzhofer et al. 1989, 1992).

As shown in Fig. 11.2, ^{142}Nd signatures of ancient rocks of West Greenland, South Africa, and Australia are heterogeneous, spanning a 0–20 ppm range above modern terrestrial standards (Bennett et al. 2007; Boyet et al. 2004; Caro et al. 2003, 2006). In contrast, the present-day silicate Earth was found through numerous investigations in oceanic basalts to be remarkably homogeneous (Andreasen et al. 2007; Caro et al. 2006; Jackson and Carlson 2012; Murphy et al. 2010). Anomalies measured in West Greenland rocks were not created by in situ decay of ^{146}Sm, but are inherited from a mantle source having experienced depletion at an

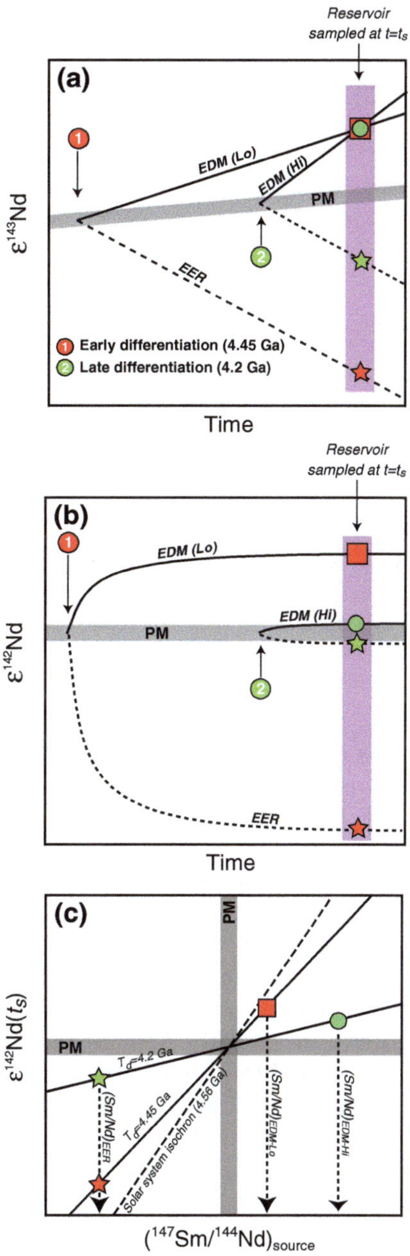

Fig. 11.1 Basic principles of $^{146,147}Sm$–$^{142,143}Nd$ systematics illustrated in the context of a two-stage model of mantle–crust evolution [modified from Caro (2011)]. The model assumes instantaneous differentiation of an early enriched reservoir (EER) and an early depleted mantle (EDM) at $t_d = 4.45$ Ga and $t_d = 4.2$ Ga. **a** The main limitation of the ^{147}Sm–^{143}Nd system lies in the fact that positive $\varepsilon^{143}Nd$ values measured in the Archean mantle can be either accounted for by very early differentiation of a moderately depleted mantle [EDM(Lo)], or by later differentiation of a highly depleted mantle [EDM(Hi)]. **b** These different models can be distinguished using the short-lived ^{146}Sm–^{142}Nd chronometer. In this simple scenario, only the earliest event generates reservoirs with large ^{142}Nd anomalies. **c** When combined, the $^{142,143}Nd/^{144}Nd$ signatures of a given reservoir (EER/EDM) at time *t* provides chronological information (t_d) as well as an independent estimate of the degree of depletion/enrichment in this reservoir (i.e., $(Sm/Nd)_{EDM/EER}$)

earlier time, when ^{146}Sm was still alive (i.e., >4.2 Ga ago). Based on the revised decay constant of Kinoshita et al. (2012), the age of mantle differentiation obtained from coupled $^{146,147}Sm$–$^{142,143}Nd$ chronometry is 4453 ± 20 Ma (Fig. 11.2), consistent with a major differentiation event near the end of Earth's accretion.

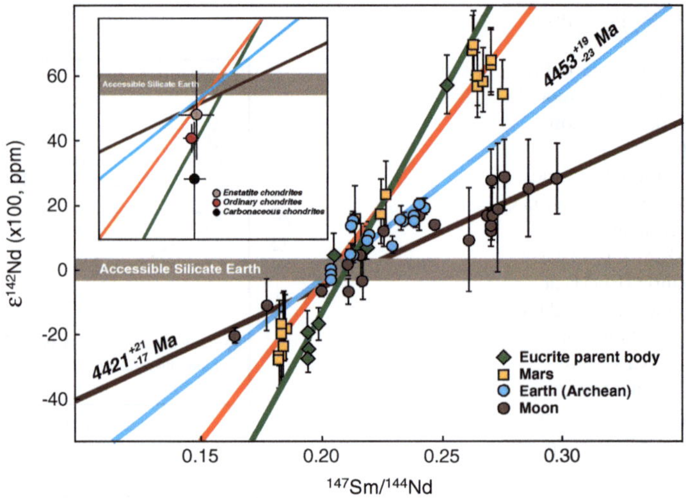

Fig. 11.2 **a** $^{146,147}Sm$–$^{142,143}Nd$ planetary isochrons for Mars (Caro et al. 2008b; Debaille et al. 2007), the Earth (Bennett et al. 2007; Caro et al. 2006), the Moon (Boyet and Carlson 2007; Brandon et al. 2009; Nyquist et al. 1995; Touboul et al. 2009a), and the Eucrite parent body (Boyet and Carlson 2005). All ages were recalculated using the ^{146}Sm decay constant and initial $^{146}Sm/^{144}Sm$ of (Kinoshita et al. 2012). **b** Range of $\varepsilon^{142}Nd$ signatures measured in bulk carbonaceous, ordinary and enstatite chondrites (Boyet and Carlson 2005; Carlson et al. 2007; Gannoun et al. 2011)

This model age of mantle depletion is similar to Lu–Hf model ages reported by Harrison et al. (2008) and Kemp et al. (2010) for the parent crustal reservoir of Jack Hills zircons (Fig. 11.3), providing further evidence that a long-lived protocrust was emplaced ca. 100 Ma after formation of the solar system (Caro 2011). The formation of this primordial crust may be related to magma ocean crystallization (Benz and Cameron 1990; Elkins-Tanton et al. 2003; Melosh 1990; Solomatov and Stevenson 1993a, b) although alternative scenarios, involving continuous crustal growth in the early Hadean, cannot be excluded on the basis of $^{142,143}Nd$ systematics alone (Bourdon and Caro 2007). The decoupling of ^{176}Hf–^{143}Nd signatures in the early Archean mantle, however, appears more consistent with an early crustal formation scenario involving segregation of Ca-perovskite in a deep magma ocean (Caro et al. 2005; Rizo et al. 2012a, b; Puchtel et al. 2013).

Further support for the magma ocean crystallization scenario has also been provided by the $^{142,143}Nd$ systematics of lunar samples. As shown in Fig. 11.2, the final stage of magma ocean solidification on the moon, obtained from coupled $^{142,143}Nd$ chronometry in mare basalts, is significantly younger (by 30–40 Ma) than the age of differentiation of the Earth's mantle (Bourdon et al. 2008; Boyet and Carlson 2007; Brandon et al. 2009; Caro et al. 2008b; Nyquist et al. 1995; Touboul et al. 2009a). While this younger age of lunar crystallization may seem surprising in the context of the giant impact model, it is in fact consistent with thermal modeling of the lunar magma ocean (LMO), showing that the emplacement of a thick anorthositic crust early in the crystallization sequence would slow

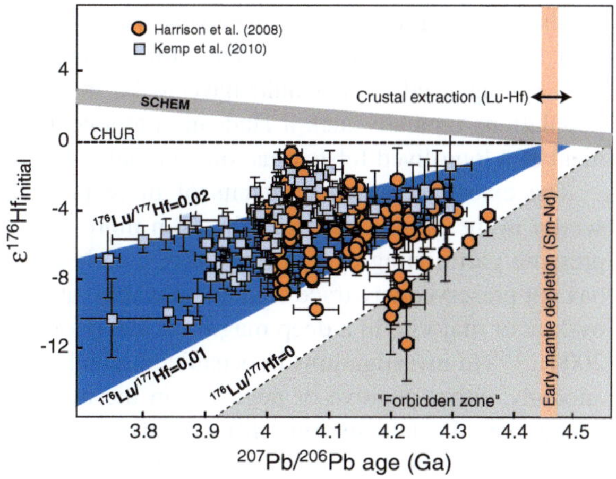

Fig. 11.3 Initial ε^{176}Hf isotopic compositions of Hadean zircons from Jack Hills (Australia) plotted against their respective U–Pb ages (Harrison et al. 2008; Kemp et al. 2010). The blue area represents the range of isotopic composition for mafic (^{176}Lu/^{177}Hf = 0.02) and felsic (^{176}Lu/^{177}Hf = 0.01) protoliths extracted from the mantle 4.45–4.5 Ga ago

down the solidification of the underlying mantle (Elkins-Tanton et al. 2011). Dates ranging from 4.3 to 4.46 Ga were reported for Ferroan anorthosites (Borg et al. 2011; Carlson and Lugmair 1988) and zircons from lunar breccias (Nemchin et al. 2009), suggesting an extended period of crustal formation on the Moon. Furthermore, a recent revision of ^{129}I–^{129}Xe systematics suggests that the Earth may have retained its atmosphere starting only 40–50 Ma after formation of the solar system, which could reflect the age of the lunar impact (Marty 2012). If correct, the first terrestrial crust would have formed between the Moon-forming event (>4.5 Ga) and the final (4.42 Ga) solidification of the lunar mantle, suggesting that silicate differentiation on both planets resulted from crystallization of global magma oceans in the aftermath of the giant impact.

11.2.3 Evolution of Primordial Mantle Heterogeneities

Perhaps the most speculative issue in the history of the silicate Earth lies in the evolution of primary mantle and crustal reservoirs created during or immediately after the final stage of accretion. Numerical simulations of mantle convection show that the characteristic mixing timescale in the hot Hadean mantle should be of the order of 100 Ma (Coltice and Schmalzl 2006), preventing long-term preservation of primordial heterogeneities in the convective system. Despite efficient convective stirring, DM domains with positive ^{142}Nd anomalies have been sampled by magmatic rocks as young as 2.7 Ga (Debaille et al. 2013) and negative ^{142}Nd anomalies were discovered in the Nuvvuagittuq supracrustal belt (Northern Quebec, O'Neil et al. 2008, 2012; Roth et al. 2013, Guitreau et al. 2013) and in rare Meso-Archean rocks from West Greenland (Rizo et al. 2012a, b), witnessing the reworking of ca. 4.4–4.5 Ga old Hadean crust within Eoarchean terranes. Overall, these observations are consistent with a lifetime of the order of 1–2 Gyr for the Hadean protocrust (Caro 2011), which could indicate formation of a "stagnant lid" following magma ocean solidification (Debaille et al. 2013). Clearly, such a long-lived crust would have to be mafic or ultramafic, as detrital zircon populations typically lack a Hadean component that would witness the emplacement of a long-lived felsic crust on a global scale prior to ca. 3.8 Ga.

The extent to which remnants of these primary silicate reservoirs were preserved in the present-day Earth is even more speculative. On the one hand, high-pressure partitioning experiments have clearly established that the modern mantle has not preserved any trace element signature attributable to the segregation of perovskite or majorite in a deep magma ocean (Corgne et al. 2005; Corgne and Wood 2004). ^{142}Nd investigations in oceanic basalts have also failed, so far, to detect any anomaly, either positive or negative, in the modern mantle, providing robust evidence that silicate reservoirs differentiated prior to 4.2 Ga were extensively rehomogenized by convective mixing (Caro et al. 2006; Jackson and Carlson 2012; Murphy et al. 2010). On the other hand, the long-term preservation of pristine reservoirs has been suggested on the basis of ^{129}Xe excesses measured in OIB lavas from Iceland and other high ^{3}He/^{4}He plumes (Allegre et al. 1987; Mukhopadhyay 2012). As Iodine is a refractory element while Xenon is volatile, ^{129}Xe heterogeneities reflect the degassing history of our planet during the first 100 Ma, when ^{129}I was still actively decaying (Allegre et al. 1987; Marty 1989; Ozima and Podosek 1999). OIBs characterized by unradiogenic ^{129}Xe/^{130}Xe ratios, however, do not carry trace element and isotopic signatures consistent with the presence of primitive chondritic material in their source (Class and Goldstein 2005). This observation has generally been taken as evidence that the behavior and/or reservoir rocks of the noble gases are decoupled from those of trace elements such as Sr, Nd, or Hf. An alternative explanation, which will be developed in the next sections, is that no evidence for primitive material is found in the OIB source because the primitive Earth's mantle may not have perfectly chondritic abundances of refractory lithophile elements. In this alternative view, primitive (but non-chondritic) mantle could possibly contribute to mantle heterogeneity but would appear as a depleted component when considered in the conventional chondritic framework.

11.3 The Missing Reservoir

11.3.1 ^{142}Nd Anomalies in Chondrites: Radiogenic or Nucleosynthetic?

While the chondritic Earth paradigm has been a cornerstone of models of mantle–crust evolution for more than 30 years, the Bulk Silicate Earth ^{147}Sm/^{144}Nd and ^{143}Nd/^{144}Nd ratios cannot be directly estimated because no terrestrial reservoir has evolved along the CHUR evolution line and because present-day accessible silicate reservoirs (i.e., the crust and OIB/MORB sources) have ^{143}Nd/^{144}Nd signatures both above and below the CHUR value. The ^{142}Nd/^{144}Nd ratio of the silicate earth, on

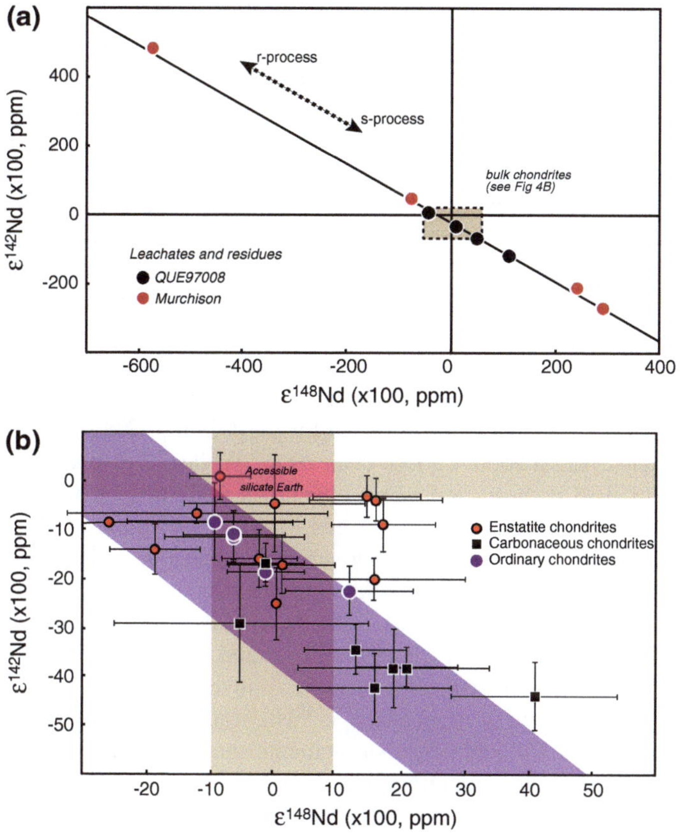

Fig. 11.4 **a** Nd isotopic composition of leachates and residues from the carbonaceous chondrite Murchison and ordinary chondrite QUE97008 (Qin et al. 2011). Leachates have large ^{148}Nd excesses and ^{142}Nd deficits consistent with the presence of a pure s-process component in the matrix of these meteorites. **b** Bulk chondrites plot on the trend defined by Murchison and QUE97008 leachates, but are shifted by approximately 20 ppm compared to the accessible silicate Earth (*red area*). With the exception of a few specimens of ECs, the variability observed in bulk chondrites can be explained by heterogeneous distribution of an s-process-rich component carried by presolar SiC. Note that while a few ECs have ^{142}Nd/^{144}Nd ratios similar to the terrestrial value, no meteorite plots in the ^{142,148}Nd field defined by terrestrial samples, suggesting that the non-chondritic terrestrial composition cannot be simply accounted for by nucleosynthetic heterogeneities in the solar nebula

the other hand, is remarkably homogeneous, allowing a straightforward comparison between CHUR and BSE. As shown in Fig. 11.4, the results of ^{142}Nd investigations in meteorites do not confirm the chondritic model, but reveal instead that the average chondritic signature is 20 ppm below the present-day ^{142}Nd/^{144}Nd ratio of the mantle–crust system (Andreasen and Sharma 2006; Boyet and Carlson 2005; Carlson et al. 2007; Gannoun et al. 2011). While this observation was first interpreted in the context of internal planetary differentiation processes (Boyet and Carlson 2005;

Labrosse et al. 2007), it has also been proposed that the bulk silicate Earth may be characterized by a super-chondritic ε^{142}Nd (Caro and Bourdon 2010; Caro et al. 2008b). This alternative interpretation suggests either that the accretion disk was isotopically heterogeneous due to incomplete mixing of nucleosynthetic products in the solar nebula, or that the Earth developed a radiogenic ^{142}Nd excess because it accreted from material with non-chondritic abundances of RLEs. Naturally, the latter scenario would require a revision of the compositional models of the BSE, which, in turn, would impact our understanding of mantle structure and evolution (Caro and Bourdon 2010; Caro et al. 2008b; Jackson et al. 2010; O'Neill and Palme 2008).

There are many difficulties in deciphering whether the non-terrestrial ^{142}Nd signature measured in meteorites is due to isotopic (i.e., nucleosynthetic) or chemical (i.e., radiogenic) heterogeneities in the accretion disk. The main challenge lies in the fact that chondrites do not have homogeneous ^{142}Nd/^{144}Nd ratios, making it more difficult to define the CHUR parameters than for the ^{147}Sm–^{143}Nd system (Fig. 11.4b) (Andreasen and Sharma 2006; Carlson et al. 2007). While OC display constant ε^{142}Nd values at -18 ± 5 ppm, carbonaceous chondrites (CC) exhibit more negative and more heterogeneous ε^{142}Nd ranging between -20 and -50 ppm. enstatite chondrites (EC) have negative ^{142}Nd anomalies, but substantial heterogeneities also exist within this class of chondrites and their ε^{142}Nd signature overlaps with the terrestrial composition (Gannoun et al. 2011). Thus, with the significant exception of OCs, ^{142}Nd heterogeneities in chondrites are larger than expected from variations of their parent/daughter ratio and, therefore, cannot be interpreted in terms of radioactive decay alone.

Following the discovery of negative ε^{142}Nd signatures in chondrites, the isotopic composition of Nd, and neighbor elements samarium and barium, was investigated at high precision in an attempt to decipher whether chondrites can provide a viable isotopic reference for the Earth building material (Andreasen and Sharma 2006; Carlson et al. 2007). ^{142}Nd is mainly produced by slow neutron capture (s-process) with a small contribution from p-process nucleosynthesis associated with photodisintegration reactions (Arlandini et al. 1999; Qin et al. 2011). ^{144}Nd and ^{146}Nd, which are used for normalizing Nd isotope ratios for instrumental mass fractionation, are produced by roughly equal contributions of s- and r-process (rapid neutron capture) nuclides, and ^{148}Nd and ^{150}Nd are pure r-process nuclides (Arlandini et al. 1999). Lastly, the parent isotope ^{146}Sm is only produced by p-process. Thus, Nd nucleosynthetic anomalies can potentially arise from heterogeneous distribution of either s-, r- ,and/or p-process nuclides in the early solar nebula.

The high-precision investigations in chondrites by Carlson et al. (2007) showed that the negative and heterogeneous ^{142}Nd signatures found in CCs were associated with excesses in the pure r-process ^{148,150}Nd nuclides (Fig. 11.4b), suggesting that ^{142}Nd variations in these meteorites are due to the presence of a component rich in s-process nuclides and with an isotopic pattern similar to that measured in Allende inclusion EK 1-4-1 (McCulloch and Wasserburg 1978). Such effects, however, were not found in OCs, leading Carlson et al. (2007) and Caro et al. (2008b) to suggest that the CHUR ^{142}Nd/^{144}Nd value was best approximated by the OC composition at ca. -18 ppm. More recently, Qin et al. (2011), using step-leaching acid procedures, revealed the presence of an s-process-rich component within the matrix of C-chondrite Murchison and the O-chondrite QUE97008 (Fig. 11.4a). The authors found that the Ba (and Sr) isotopic composition of Murchison leachates and residues was consistent with presolar SiC being the carrier of the s-process component in this meteorite.

As shown in Fig. 11.4a, leachates and residues from Murchison and QUE97008 define a negative correlation in an ε^{142}Nd versus ε^{148}Nd space, as expected from binary mixing of s- and r-process nuclides (Qin et al. 2011). This correlation passes through the analyses of bulk chondrites (Fig. 11.4b), but does not intersect the terrestrial composition (by definition at $\varepsilon^{142}\text{Nd} = 0$; $\varepsilon^{148}\text{Nd} = 0$). This observation was taken as evidence that Nd isotope variability in chondrites reflect variable addition of pure s-process nuclides to a nebular background roughly similar to the OC value, but that such nucleosynthetic effect cannot account for the high ε^{142}Nd composition of the silicate Earth (as this would translate into a detectable offset in ε^{148}Nd between OCs and terrestrial samples). Heterogeneities in p-process nuclides were reported from Murchison whole-rock and leachates, but such effects were not found in OCs (Andreasen and Sharma 2006), so that a heterogeneous distribution of ^{146}Sm can also be excluded as a plausible cause for the higher terrestrial ^{142}Nd/^{144}Nd ratio compared to this class of chondrites. This leaves radiogenic decay of ^{146}Sm as the most plausible cause for the high ^{142}Nd/^{144}Nd ratio measured in terrestrial samples. A direct implication is that the Earth, or its accessible reservoirs, must have evolved for more than 4.5 Ga with an Sm/Nd ratio higher than the canonical CHUR value.

11.3.2 Models Involving a Hidden Reservoir

The presence of ^{142}Nd anomalies in meteorites was first interpreted in the framework of the chondritic Earth model. Boyet and Carlson (2005) suggested that an early enriched reservoir (EER), with sub-chondritic ε^{142}Nd and high REE content, was needed to balance the higher ε^{142}Nd measured in all "accessible" reservoirs. The authors speculated that this dense EER could be composed of recycled KREEP-like material generated by segregation of residual melt from a magma ocean in the aftermath of the giant impact (Boyet and Carlson 2005, 2006). Previous work showed that such mechanism could explain the trace element fractionation observed in the early Archean mantle (Caro et al. 2005) and Tolstikhin et al. (2006) pointed out that the storage of such an early ultramafic crust at the CMB could reconcile geochemical requirements for a deep primordial reservoir with geophysical evidence for whole-mantle convection. From a mass balance point of view, this would require the EER to represent about 10 % of the mass of the silicate Earth (the approximate mass of D'') and concentrate 25 % of its REE budget. Tolstikhin and Hofmann (2005) further suggested that this protocrust would be recycled together with dense meteoritic material incorporated in the crust during late accretion. This meteoritic component would carry solar-implanted noble gases, thereby explaining the presence of primordial He and Ne in the recycled crustal reservoir.

An alternative scenario, first proposed by Labrosse et al. (2007), is that an EER formed directly in the deep mantle by downward migration of dense residual melt from a crystallizing magma ocean. This hypothesis is supported by tomographic observations of ultra-low velocity zones (ULVZ), indicative of partially molten regions near the CMB (Garnero and McNamara 2008). Iron-partitioning experiments also suggest that silicate melt may become denser than perovskite at 1800 km depth in the LM, thereby providing a plausible explanation for the long-term storage of residual melts in the deep mantle (Nomura et al. 2011). Note, however, that a recent study by Andrault et al. (2012) did not confirm the results of previous partitioning experiments (Nomura et al. 2011), so that the existence of a density crossover between perovskite and coexisting liquid in the LM remains uncertain.

As in the recycled crust model of Tolstikhin and Hofmann (2005), the basal magma ocean may represent a significant reservoir of primordial noble gasses (Coltice et al. 2011). While this would depend on solid/melt partition coefficients of He, Ne, and Ar at LM pressure, which are not constrained, Herzberg et al. (2013) recently presented observational evidence seemingly consistent with the presence of such volatile-rich basal magma ocean. The authors found that olivine phenocrysts in high $^{3}He/^{4}He$ lavas from several OIBs worldwide have Ni contents about 20 % higher than their MORB counterparts, which they interpreted as reflecting melting of a Ni-rich peridotitic source. Herzberg et al. (2013) proposed that such Ni enrichment may have resulted from diffusional exchange between the outer core and deep molten regions at the CMB proposed to be at the origin of these plumes. If primordial noble gasses are indeed sequestered in a molten silicate layer, as suggested by Coltice et al. (2011) and Herzberg et al. (2013), then their association with ^{129}Xe excesses (Allegre et al. 1987; Mukhopadhyay 2012) would support an origin of this deep volatile-rich reservoir as an ancient (>4.45 Ga) residue of magma ocean crystallization.

11.3.3 The Non-chondritic Earth Model

The models described above outline plausible ways of generating a hidden enriched reservoir that could possibly account for the ^{142}Nd signature of the accessible Earth and for the storage of primordial rare gasses at the CMB for the past 4.46 Ga. A major drawback, however, is that models that make an EER the repository of volatiles in the deep mantle should associate high $^{3}He/^{4}He$ values with unradiogenic ^{143}Nd, ^{176}Hf, and ^{142}Nd signatures. This requirement conflicts with the fact that the high $^{3}He/^{4}He$ mantle component is characterized by an isotopic "flavor" that has long been known to be akin to depleted rather than enriched material (Class and Goldstein 2005). This He-rich mantle end-member (sometimes labeled PHEM, PREMA, or FOZO) is difficult to define accurately, but its Sr–Nd–Hf signature is undoubtedly more depleted than CHUR and shows no obvious contribution from an ancient enriched component. Furthermore,

^{142}Nd analyses in high $^{3}He/^{4}He$ OIBs such as Hawaii and Iceland failed, so far, to detect any negative ^{142}Nd effect that would witness the presence of such ancient enriched material in their source (Caro et al. 2006; Jackson and Carlson 2012; Murphy et al. 2010).

An alternative interpretation, more consistent with the above observations, is that the Earth developed a super-chondritic Sm/Nd ratio at a very early stage of its history, as a result of preferential loss of crustal material during impacts (Bourdon et al. 2008; Caro and Bourdon 2010; Caro et al. 2008b; O'Neill and Palme 2008). This "impact erosion" scenario is supported by the models of planetary accretion, showing that "hit-and-run" collisions were the predominant collisional regime during planetary growth (Asphaug 2010). In contrast to accretionary impacts, hit-and-run collisions involve substantial mass loss from the accreting bodies. A likely outcome is that the external envelopes of the smaller bodies are stripped preferentially to their deeper layers, which would result in the accreting planets progressively evolving toward denser, volatile poor, and chemically depleted compositions (Asphaug 2010).

The consequences of collisional erosion can be observed in various forms in modern bodies of the solar system. The high iron-to-silicate ratio of Mercury may result from such collisional effect (Benz et al. 1988), although alternative models, involving fractional condensation processes, or partial mantle volatilization near the young sun, have also been proposed (see review by Benz et al. 2007). Asphaug (2010) also argued that density variations in differentiated bodies from the asteroid belt and the dwarf planets beyond Neptune may be related to variable loss of their silicate or ice mantles during successive collisions. Lastly, the asteroid 4-Vesta provides a case illustrating the preferential erosion of crustal material by small impacts. As shown in Fig. 11.2, the crust of this asteroid must have differentiated within the first few million years after formation of the solar system, and unlike Mars, the Earth, and the Moon, the $^{146,147}Sm$–$^{142,143}Nd$ isochron established from the analyses of Eucrites is consistent with Vesta being chondritic at the time of differentiation. The crust of Vesta, however, experienced substantial loss through impacts, as shown by the presence of a large crater in the southern hemisphere. According to Thomas et al. (1997), the mass of material excavated from this crater would only represent 1 % of the total mass of the planet, but this single impact would have excavated roughly a third of Vesta's crust. Since crustal reservoirs tend to have low Sm/Nd, accreting planets experiencing such collisional loss would likely evolve toward super-chondritic Sm/Nd ratios and sub-chondritic REE concentrations.

The extent to which impact erosion has affected the composition of the Earth and other terrestrial planets remains to be determined. According to O'Neill and Palme (2008), the super-chondritic Fe/Mg ratio of the bulk Earth is consistent with preferential loss of silicate material compared to metal of the order of 10 %. Note, however, that this would only affect the Sm/Nd ratio of the BSE if crustal material was also stripped preferentially to silicate mantles in similar proportions. In such a scenario, the bulk silicate Earth would be compositionally similar to a residue from crustal extraction rather than a primitive chondritic reservoir.

11.4 Redefining the Primitive Mantle: The Impact Erosion Model

While the magnitude of Sm/Nd fractionation (5–6 %) required to create an 18 ppm ^{142}Nd excess in the Earth's mantle may seem minor in comparison with volatile element depletion generated during planetary accretion, it nevertheless represents a challenge to the chondritic Earth paradigm and to the compositional models that rely on it (McDonough and Sun 1995). In this section, I propose a model for the "accessible" silicate Earth anchored on its ^{142}Nd signature. This model (labeled SCHEM for super-chondritic Earth model) was first presented in Caro and Bourdon (2010) and is now expanded and revised using a recently determined decay constant for ^{146}Sm (Kinoshita et al. 2012).

Unlike previous models, SCHEM does not assume a priori that the Earth has perfectly chondritic abundances of RLE. Instead, the model delineates the backbone of the trace element pattern of the silicate Earth using key trace element ratios (Rb/Sr, Sm/Nd, and Lu/Hf) estimated from the Sr–Nd–Hf isotope systematics of the Earth's mantle. Note that a modification of the BSE parameters for these isotopic systems changes the output of compositional models for the MORB source (Salters and Stracke 2004; Workman and Hart 2005), as these depleted mantle models (DMM) are anchored on the CHUR Sm/Nd and Lu/Hf ratios. Thus, revised estimates for the REE abundance in the MORB source consistent with the use of SCHEM are also provided. The results of this simple 3-box model suggest that the "accessible" silicate Earth is depleted in highly incompatible elements by approximately 50 % compared to their chondritic abundances. The magnitude of the depletion estimated for trace elements (e.g., REE, Hf, Sr) correlates with their degree of incompatibility in silicate melt, as predicted by the collisional erosion model of O'Neill and Palme (2008).

11.4.1 A Revision of Sm–Nd, Lu–Hf and Rb–Sr Parameters for the Silicate Earth

As discussed in Sect. 11.3, the ^{142}Nd/^{144}Nd ratios of bulk OC show little or no influence from presolar components, and this signature thus represents our best current estimate for the CHUR ^{142}Nd/^{144}Nd ratio. Thus, the SCHEM Sm–Nd parameters are anchored to the mean ε^{142}Nd value of -18 ppm measured in this class of meteorites. If one considers that the BSE evolved with a super-chondritic Sm/Nd ratio as early as 4.567 Ga, then $(^{147}\mathrm{Sm}/^{144}\mathrm{Nd})_{\mathrm{SCHEM}}$ can be estimated from Eq. (11.2) to be 0.2073 ± 0.0015 (1σ), and the corresponding ^{143}Nd/^{144}Nd ratio of the present-day BSE is then deduced from Eq. (11.1) to be 0.512962 ± 0.000045 [i.e., $\varepsilon^{143}\mathrm{Nd} = 6.3 \pm 0.9$ (1σ)].

As shown in many previous studies, the ^{87}Sr/^{86}Sr and ^{176}Hf/^{177}Hf ratios of mantle-derived rocks are closely correlated with their ^{143}Nd/^{144}Nd ratios (Fig. 11.5). Consequently, a revision of the terrestrial Sm/Nd ratio would necessarily impact the

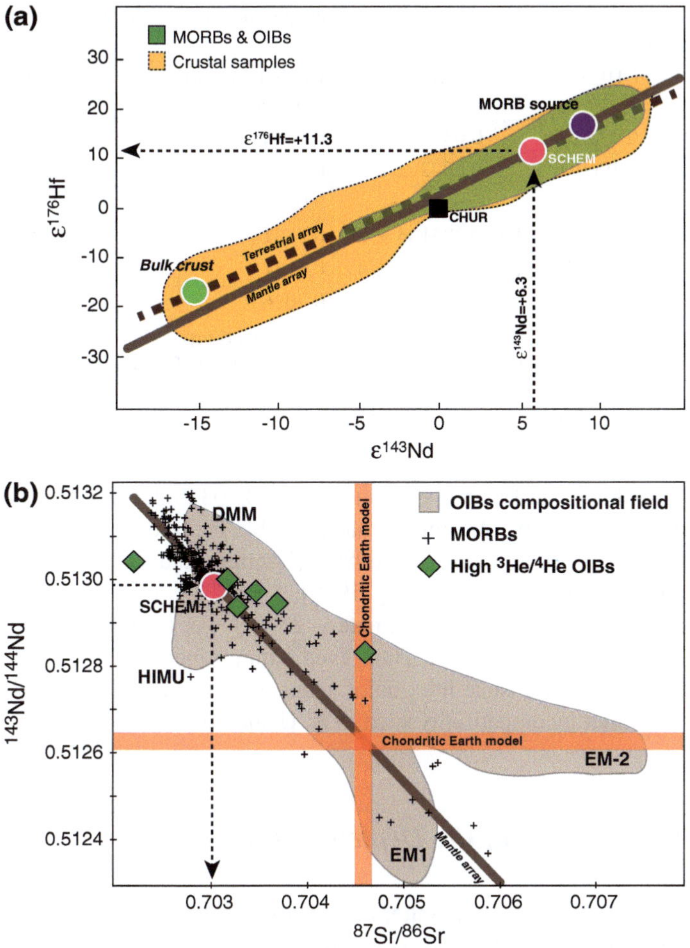

Fig. 11.5 **a** ε^{176}Hf–ε^{143}Nd relationships in the present-day mantle–crust system [modified from Chauvel et al. (2007) and Vervoort et al. (1999)]. The mantle array is obtained by regressing MORB and OIB data (Chauvel et al. 2007), and the terrestrial array includes crustal samples (Vervoort et al. 1999). The CHUR value is from Bouvier et al. (2008). **b** ^{87}Sr–^{143}Nd systematics of oceanic basalts [modified from Hofmann (2003)]. Similar to the Hf–Nd case, the ^{87}Sr/^{86}Sr composition of the bulk Earth can be obtained from the intersect of the $\varepsilon^{143}Nd_{SCHEM}$ value with the mantle array. The Sr–Nd compositions of basalts carrying the highest ^{3}He/^{4}He ratios are plotted for comparison. OIB data are from Jackson et al. (2007) and Stuart et al. (2003)

Rb/Sr and Lu/Hf ratios of the BSE. $^{176}Lu/^{177}Hf_{SCHEM}$ can be estimated by assuming that the bulk silicate Earth lies on the Hf–Nd mantle array (Fig. 11.5a), which is well constrained from numerous measurements in oceanic basalts. Considering the regression parameters of Chauvel et al. (2007) ($\varepsilon^{176}Hf = 1.59\ \varepsilon^{143}Nd + 1.28$) yields $\varepsilon^{176}Hf_{SCHEM} = 11.3 \pm 1.4$ (i.e., $^{176}Hf/^{177}Hf_{SCHEM} = 0.283106 \pm 0.000039$), from which it follows that $^{176}Lu/^{176}Hf_{SCHEM} = 0.0373 \pm 0.0004$. Similarly, the ^{87}Sr/^{86}Sr

Table 11.1 Rb–Sr, Lu–Hf and Sm–Nd reference parameters for the bulk silicate Earth and depleted mantle

	CHUR[b]	SCHEM	Depleted mantle[c] (chondritic model)	Depleted mantle (non-chondritic model)
$^{176}Lu/^{176}Hf$	0.0336 ± 1[a]	0.0373 ± 4	0.04508	0.0393 ± 7
$^{147}Sm/^{143}Nd$	0.1960 ± 4	0.2073 ± 15	0.2295	0.2181 ± 23
$^{87}Rb/^{86}Sr$	0.085 ± 5	0.064 ± 3	0.02526	–
$^{176}Hf/^{177}Hf$	0.282785 ± 11	0.283106 ± 39	0.28318	
$^{143}Nd/^{144}Nd$	0.512630 ± 11	0.512962 ± 45	0.5131	
$^{86}Sr/^{87}Sr$	0.7046 ± 3	0.70328 ± 20	0.7026	

[a]Errors, when available, are given as 1σ and refer to the last digit
[b]Sm/Nd and Lu–Hf parameters from Bouvier et al. (2008). Rb–Sr parameters from McCulloch (1994)
[c]Salters and Stracke (2004)

ratio of the bulk silicate Earth is estimated to be 0.70328 ± 0.0002 from the Sr–Nd mantle array (Fig. 11.5b). Assuming that the BSE evolved as closed system since 4.567 Ga from an initial value of 0.69897 (Lugmair and Galer 1992; McCulloch 1994; Papanastassiou and Wasserburg 1969), $^{87}Rb/^{86}Sr_{SCHEM}$ can be estimated to 0.064 ± 0.003 (Table 11.1). This estimate is 25 % lower than the currently admitted value of 0.085 (McCulloch 1994). Thus, even a minor increase in the Sm/Nd ratio of the silicate Earth will translate into major changes on the abundance of Rb and other elements of similar incompatibility such as K, U, and Th.

11.4.2 Composition of the Depleted Mantle

Following previous studies by Salters and Stracke (2004) and Workman and Hart (2005), the Lu/Hf_{DM} and Sm/Nd_{DM} ratios are estimated using a simple two-stage model of mantle–crust differentiation (Fig. 11.6). This simplification is appropriate for the Lu–Hf and Sm–Nd chronometers because both ^{147}Sm and ^{176}Lu have very long half-lives; the present-day $^{176}Hf/^{177}Hf$ and $^{143}Nd/^{144}Nd$ signatures of the DM and CC are thus independent of the growth history of the crust and can be predicted from the mean age of these reservoirs (Jacobsen and Wasserburg 1979). The Rb–Sr case is somewhat more complex due to the shorter half-life of ^{87}Rb (Workman and Hart 2005) and will not be treated here.

As shown in Fig. 11.6a, a mantle–crust isochron can be established from the $^{147}Sm–^{143}Nd$ composition of the present-day upper crust, which is well constrained from studies of worldwide loess and river sediments (Goldstein and Jacobsen 1988), and that of the bulk silicate Earth (SCHEM) estimated above. The slope of this two-point isochron yields a model age of crustal extraction of 1.97 ± 0.05 Ga, which agrees well the age distribution of modern crustal terranes (Condie 2000) as well as crustal model ages obtained from the study of worldwide loess and river sediments. If one assumes that the depletion of the MORB source

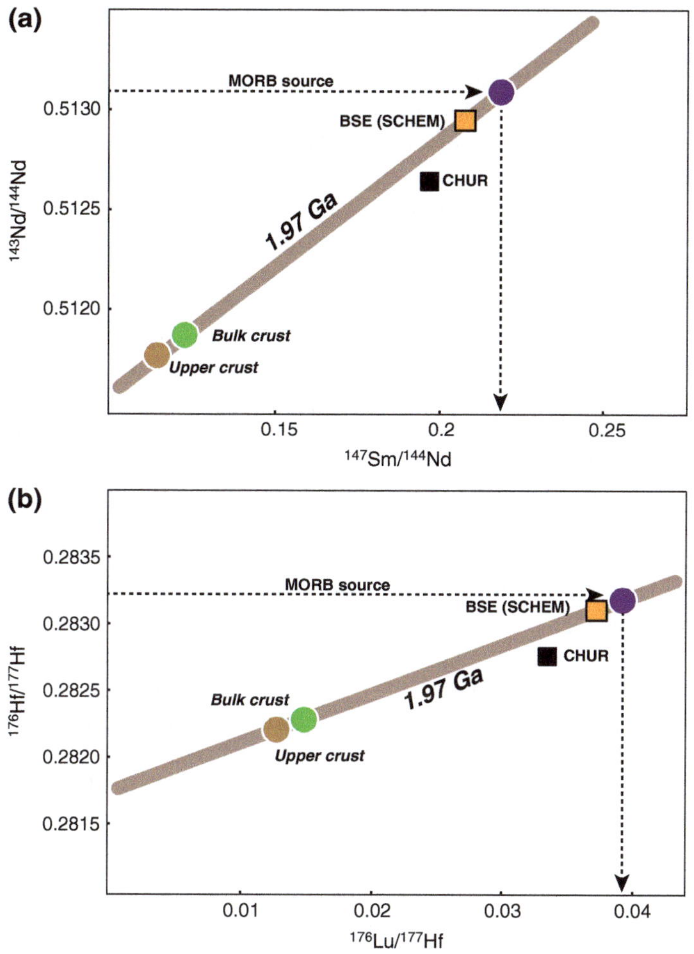

Fig. 11.6 **a** ^{147}Sm–^{143}Nd and **b** ^{176}Lu–^{176}Hf mantle–crust "isochrons" used to determine the Sm/Nd and Lu/Hf ratio of the MORB source. The upper and bulk crust compositions are from Goldstein and Jacobsen (1988) and Rudnick and Fountain (1995). Depleted mantle compositions are from Chauvel et al. (2007) and Salters and Stracke (2004)

compared to SCHEM is due solely to continental extraction, then $^{147}Sm/^{144}Nd_{DM}$ can be calculated using the intersect of this isochron with the average isotopic composition of MORBs [i.e., 0.5131, (Salters and Stracke 2004)]. This yields $^{147}Sm/^{144}Nd_{DM} = 0.2181 \pm 0.0023$, which is significantly lower than previous estimates of 0.23–0.25 based on the CHUR parameters (Salters and Stracke 2004; Workman and Hart 2005). A similar approach is then applied for calculating the ^{176}Lu/^{177}Hf ratio of the DM; the slope of the ^{176}Lu–^{176}Hf mantle–crust isochron is fixed at 1.97 ± 0.05 Ga (Fig. 11.6b), as determined from Sm–Nd systematics, and the ^{176}Lu/^{177}Hf ratio of the DM is then estimated to be 0.0393 ± 0.0007 from the average MORB ^{176}Hf/^{177}Hf signature of 0.28318 (Chauvel et al. 2007).

From the Sm/Nd and Lu/Hf ratios estimated above and the knowledge of a few key trace element ratios in MORBs, one can then derive absolute REE concentrations in the DM. To facilitate the comparison of the SCHEM-based model with previous DM estimates, I follow the approach of Salters and Stracke (2004) and anchor the DM compositional model to a CaO content of 3.5 wt%. This is justified in the context of the impact erosion model because calcium has a melt/solid partition coefficient close to unity and is therefore unlikely to be significantly depleted in the BSE (O'Neill and Palme 2008). $[Lu]_{DM}$ is estimated from the Lu versus CaO correlation in MORBs, yielding 0.063 ± 0.005 ppm, which is identical to the value obtained by Salters and Stracke (2004). Since $^{176}Lu/^{177}Hf_{DM} = 0.0393 \pm 0.0007$, this gives a Hf content of 227 ± 41 ppb for the DM.

The concentration of Sm and Nd are then calculated using the Sm/Hf ratio of the DM. While Sm/Hf is not strictly speaking a "constant" ratio, these elements show similar partitioning behavior during generation of the oceanic crust, so that MORBs display very little variation in Sm/Hf over a wide range of Hf concentration. Salters and Stracke (2004) estimated the Sm/Hf ratio of the DM to be 1.355 ± 0.12, from which it follows that $[Sm]_{DM} = 307 \pm 55$ ppb and $[Nd]_{DM} = 852 \pm 162$ ppb. Lastly, the concentrations of the heavy REE are approximated by linear interpolation between Sm and Hf, and the LREE abundances are calculated using trace element ratios of similar compatibility such as Ce/Nd and La/Ce. The resulting REE pattern is shown in Fig. 11.7 together with that estimated from conventional "chondritic" models. While relatively large uncertainties are attached to these estimates (Table 11.2), owing, in particular, to error propagation from Sm/Hf_{DM}, the results clearly show that a compositional model anchored to SCHEM would yield a MORB source significantly less depleted than CHUR-based estimates. A direct implication is that a larger depleted reservoir would be needed to balance the present-day continental crust (see Sect. 5.1).

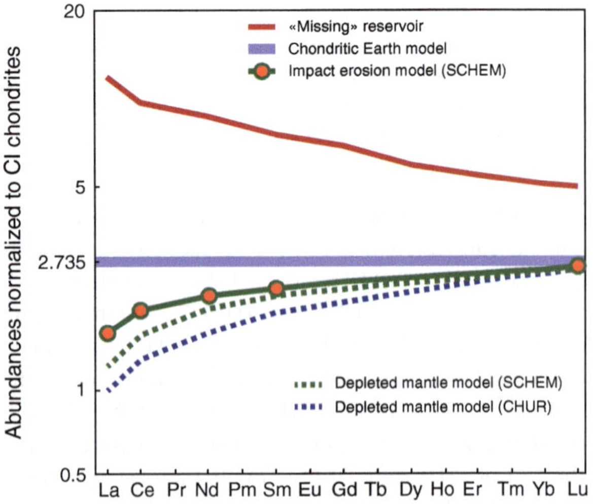

Fig. 11.7 Comparison of SCHEM- and CHUR-based estimates for the REE abundances of the MORB source. While SCHEM is associated with a depleted bulk silicate Earth, compositional models for the MORB source anchored on SCHEM parameters show a lesser degree of depletion than those obtained in chondritic Earth models (Salters and Stracke 2004; Workman and Hart 2005)

Table 11.2 Trace element abundances calculated using CHUR and SCHEM reference parameters

	Silicate Earth ("pyrolite")[b]	Silicate Earth (SCHEM)	Depleted mantle[c] (chondritic model)	Depleted mantle (non-chondritic model)
Rb	0.60 ± 0.18[a]	0.30 ± 0.09	0.088 ± 0.022	–
La	0.648 ± 0.06	0.368 ± 0.133	0.234 ± 0.033	0.92 ± 0.047
Ce	1.675 ± 0.17	1.138 ± 0.252	0.772 ± 0.115	0.2784 ± 0.304
Nd	1.25 ± 0.12	0.956 ± 0.078	0.713 ± 0.050	0.852 ± 0.162
Sr	19.9 ± 2	13.4 ± 3.5	9.8 ± 1.9	11.7 ± 4.4
Sm	0.405 ± 0.04	0.328 ± 0.029	0.27 ± 0.027	0.307 ± 0.055
Hf	0.28 ± 0.03	0.246 ± 0.024	0.199 ± 0.016	0.227 ± 0.020
Lu	0.068 ± 0.007	0.0648 ± 0.007	0.063 ± 0.005	0.063 ± 0.005

[a]Errors are given as 1σ
[b]McDonough and Sun (1995)
[c]Salters and Stracke (2004)

11.4.3 Composition of the Primitive Mantle

Hf and Nd concentrations of the Earth's PM can be obtained using a simple mass balance calculation assuming a complementary relationship between the CC and the MORB source (DM). Depending on the model, $[\mathrm{Nd}]_{\mathrm{CC}}$ and $[\mathrm{Hf}]_{\mathrm{CC}}$ are estimated to be 16–27 ppm and 3.0–4.9 ppm, respectively (Wedepohl 1995; Taylor and McLennan 1995; Rudnick and Fountain 1995). Here, I use $[\mathrm{Nd}]_{\mathrm{CC}} = 20$ ppm and $[\mathrm{Hf}]_{\mathrm{CC}} = 3.7$ ppm obtained by Rudnick and Fountain (1995), and these values should therefore be considered with a minimum uncertainty of $\pm 30\,\%$. As the CC accounts for a modest fraction ($\approx 10\,\%$) of the total Nd–Hf budget of the non-chondritic Earth, however, these uncertainties only have a marginal influence on the final SCHEM estimates.

For Nd, the mantle–crust mass balance reads as follows:

$$R_{\mathrm{DM}} + R_{\mathrm{cc}}\varphi_{\mathrm{cc}}\frac{M_{\mathrm{CC}}}{M_{\mathrm{DM}}} = R_{\mathrm{PM}}\varphi_{\mathrm{PM}}\left(1 + \frac{M_{\mathrm{CC}}}{M_{\mathrm{DM}}}\right). \tag{11.4}$$

where $\varphi_i = \frac{[\mathrm{Nd}]_i}{[\mathrm{Nd}]_{\mathrm{DM}}}$, $R_{\mathrm{CC}} = 0.511872$, $R_{\mathrm{DM}} = 0.5131$, and $R_{\mathrm{PM}} = 0.512962$. M_{DM} is the mass of the depleted mantle needed to balance the present-day continental mass (M_{CC}), and the $M_{\mathrm{CC}}/M_{\mathrm{DM}}$ ratio is given by:

$$\frac{M_{\mathrm{CC}}}{M_{\mathrm{DM}}} = \frac{1 - \varphi_{\mathrm{PM}}}{\varphi_{\mathrm{PM}} - \varphi_{\mathrm{CC}}} \tag{11.5}$$

Replacing $M_{\mathrm{CC}}/M_{\mathrm{DM}}$ by its expression in Eq. (11.4) then yields:

$$\varphi_{\mathrm{PM}} = \frac{(R_{\mathrm{DM}} - R_{\mathrm{CC}})\varphi_{\mathrm{CC}}}{(R_{\mathrm{PM}} - R_{\mathrm{CC}})\varphi_{\mathrm{CC}} - (R_{\mathrm{PM}} - R_{\mathrm{DM}})} \tag{11.6}$$

Solving Eq. (11.6) using SCHEM parameters for the PM yields $[\mathrm{Nd}]_{\mathrm{SCHEM}} = 956 \pm 78$ ppb and $[\mathrm{Hf}]_{\mathrm{SCHEM}} = 246 \pm 24$ ppb. Since $^{176}\mathrm{Lu}/^{17}$

$^7Hf_{SCHEM} = 0.0373 \pm 0.0004$ and $^{147}Sm/^{144}Nd_{SCHEM} = 0.2073 \pm 0.0015$, Sm and Lu concentrations in the BSE can then be estimated at 324 ± 29 ppb and 64.8 ± 7 ppb, respectively.

The Sr content of the silicate Earth can then be estimated from the Sr/Nd ratio of the BSE. As Sr is slightly more incompatible than Nd during crustal extraction processes (Hofmann 1988; Workman and Hart 2005), Sr/Nd_{SCHEM} must lie between the chondritic value of 15.8 (McDonough and Sun 1995) and the DM value of 13.4 (Salters and Stracke 2004). This constrains $[Sr]_{SCHEM}$ to 12.8–15.1 ppm. A similar estimate can be obtained from the weighted average of the Sr/Nd ratios of CC and DM, using an M_{CC}/M_{DM} value of 0.005 [derived from Eq. (11.5)] and an Sr/Nd_{CC} ratio of 16 (Rudnick and Fountain 1995). This gives $Sr/Nd_{SCHEM} = 14$, from which it follows that $[Sr]_{SCHEM} = 13.4 \pm 3.5$ ppm and $[Rb]_{SCHEM} = 305 \pm 91$ ppb. First-order estimates of Ce and La abundances can be obtained using a similar approach using the La/Ce and Ce/Nd ratios of the CC and DM, yielding $[Ce]_{SCHEM} = 1138 \pm 252$ ppb and $[La]_{SCHEM} = 368 \pm 133$ ppb.

The results of the above calculations are summarized in Table 11.2 and plotted in Figs. 11.7 and 11.8. They reveal that incompatible elements are variably depleted in the "accessible" silicate Earth compared to their chondritic abundances, with the most incompatible elements (Rb, La, Ce) showing a 40–50 % depletion compared to the BSE estimate of McDonough and Sun (1995). As shown in Fig. 11.9, the extent to which these trace elements are depleted in the silicate Earth appears to be essentially related to their degree of compatibility in silicate melts (expressed here as the

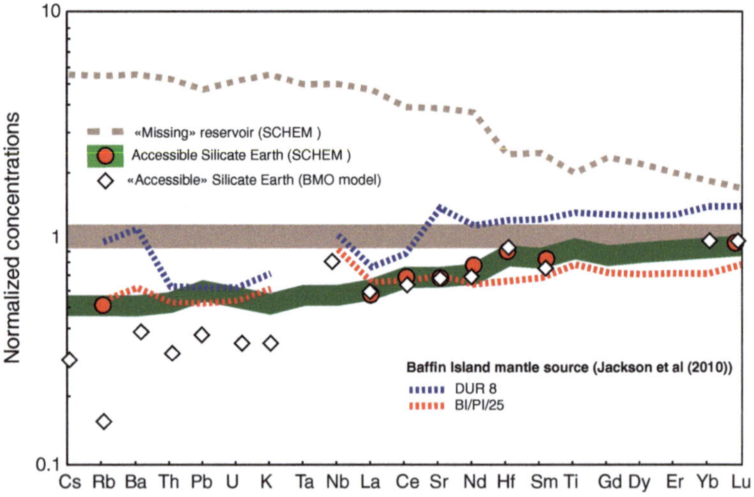

Fig. 11.8 Abundances of some trace elements in the «accessible» silicate Earth plotted as a function of their bulk partition coefficients. In panel A, normalized concentrations are plotted against D_{bulk} values of Workman and Hart (2005), which reflect element partitioning in the presence of olivine, orthopyroxene, clinopyroxene, and spinel at relatively low pressure in the upper mantle. In panel B, concentrations are plotted against partition coefficients corresponding to a high-pressure mineral assemblage composed of Ca-perovskite (5 %), Mg-perovskite (76 %), and Ferropericlase (16 %). Partition coefficients for Ca- and Mg-perovskite are from Corgne et al. (2005)

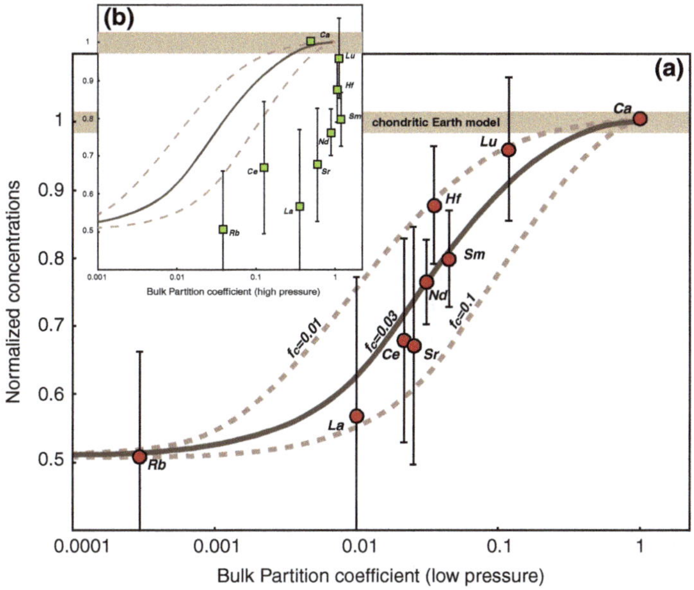

Fig. 11.9 Reconstructed trace element patterns for the mantle source of high $^{3}He/^{4}He$ lavas from the PIP (Jackson et al. 2010) compared with those calculated for the "missing" reservoir and non-chondritic BSE. The DM model is adapted from Salters and Stracke (2004)

bulk partition coefficient (D_{bulk}), see Workman and Hart (2005)). This relation demonstrates that magmatic differentiation processes are responsible for creating a non-chondritic REE pattern in the "accessible" Earth and eliminate the possibility of metal–silicate and/or volatility-controlled fractionations of the Sm/Nd ratio.

The trace element pattern estimated for the accessible Earth could reflect either crustal loss by impacts, or alternatively the long-term storage of Hadean enriched material in the deep mantle. It must be emphasized, however, that partition coefficients used in Fig. 11.9a are appropriate for partial melting in the presence of olivine, orthopyroxene, clinopyroxene, and spinel and, therefore, are only relevant to mantle–crust differentiation at relatively low pressure. When the normalized abundances of trace elements in the BSE are plotted against partition coefficients relevant to a high-pressure mineral assemblage [i.e., Ca- and Mg-perovskite (Corgne et al. 2005)], the smooth relationship with D_{bulk} is lost (Fig. 11.9b). This suggests that the non-chondritic pattern of the "accessible" Earth more likely reflects mantle–crust differentiation processes at shallow depth, rather than melt segregation in a deep magma ocean.

11.4.4 The Impact Erosion Model

A simple theoretical framework predicting trace element fractionation associated with impact erosion was presented by O'Neill and Palme (2008), and this model can now be constrained using the results obtained above. In this model, it is assumed that incompatible elements first partition between a protocrust (c) and a complementary DM (m). Assuming a batch melting process, the concentration of any element in the depleted reservoir normalized to its chondritic abundance is given by

$$C^*_{\mathrm{m}} = D_{\mathrm{E}}/(D_{\mathrm{E}} + f_{\mathrm{c}}(1 + D_{\mathrm{E}})) \tag{11.7}$$

where f_{c} is the mass fraction of crust and $f_{\mathrm{m}} = 1 - f_{\mathrm{c}}$ is the mass fraction of the complementary depleted reservoir. Collisional erosion then removes a fraction f_{E} of silicate material from the accreting planet. This eroded reservoir (E) contains both crustal (Ec) and mantle (Em) material in relative proportions f_{Ec} and f_{Em}. The composition of the eroded material is given as follows:

$$C^*_{\mathrm{E}} = f_{\mathrm{Em}} C^*_{\mathrm{m}} + f_{\mathrm{Ec}} C^*_{\mathrm{c}}. \tag{11.8}$$

and that of the silicate Earth after removal of this reservoir is given as follows:

$$C^*_{\mathrm{BSE}} = C^*_0/(1 - f_{\mathrm{e}}) - C^*_{\mathrm{E}} f_{\mathrm{e}}/(1 - f_{\mathrm{e}}) \tag{11.9}$$

For highly incompatible elements, such as Rb and La, $C^*_{\mathrm{M}} = \rightarrow 0$, so that $C^*_{\mathrm{BSE}} = f_{\mathrm{Ec}}/f_{\mathrm{c}}$. As $[\mathrm{Rb}]_{\mathrm{SCHEM}}$ and $[\mathrm{La}]_{\mathrm{SCHEM}}$ are 40–50 % lower compared to their "chondritic" abundances, the $f_{\mathrm{EC}}/f_{\mathrm{c}}$ ratio can be estimated to be 0.5–0.6. Fixing f_{E} at 0.1 to satisfy the super-chondritic Fe/Mg ratio of the Earth (O'Neill and Palme 2008) then permits solving the system for f_{c} by fitting the abundance of the moderately incompatible elements (Lu–La) (Fig. 11.9). This yields $f_{\mathrm{c}} \approx 0.03$, from which it follows that $f_{\mathrm{EC}} \approx 0.16$. Overall, this simple model shows that the sub-chondritic abundance of incompatible trace elements in the silicate Earth can be accounted for by the removal of 10 % silicate material composed of 16 % crust and 83 % DM from a planet initially composed of 3 % crust and 97 % DM. If scaled to the modern Earth, f_{E} would represent a 200-km-thick layer composed of approximately 40-km crust and 160-km DM. These estimates, however, should only be considered rough approximations, as the trace element pattern of the silicate Earth could have been produced during many successive collisions between planetesimals and planetary embryos rather than a single impact on the proto-Earth. Despite this important limitation, the model is useful in that it constrains the abundance of the RLEs in the non-chondritic PM from the knowledge of their bulk partition coefficient (Fig. 11.9). To a first order, the BSE would have abundances of moderately incompatible elements (D = 0.01–0.1) significantly fractionated from their chondritic ratios. This includes, in particular, the REE, Sr, Hf, and Ti. Highly incompatible elements (e.g., K, Th, U, Rb, Ba), whose budget is essentially dominated by crustal reservoirs, are uniformly depleted by 40–50 % compared to a chondritic Earth model. Accordingly, ratios of highly incompatible RLEs such as Th/U or Nb/Ta are expected to be chondritic, despite extensive loss caused by impact erosion. Lastly, elements more compatible than Lu should be present in both relative and absolute chondritic abundance in the bulk silicate Earth.

11.4.5 Non-chondritic Primitive Mantle Sampled by Oceanic Basalts

Since no ^{142}Nd anomaly has been detected in modern oceanic basalts, evidence that primordial reservoirs may have survived in the modern mantle essentially relies on the interpretation of ^{129}I–^{129}Xe systematics in high ^{3}He/^{4}He plumes (Mukhopadhyay 2012). The atmosphere is characterized by low ^{129}Xe/^{130}Xe—inherited from "catastrophic" degassing of the mantle prior to 4.4 Ga—while the MORB source has a radiogenic ^{129}Xe/^{130}Xe (Allegre et al. 1987; Marty 1989; Mukhopadhyay 2012). The noble gas systematics of oceanic basalts, however, cannot be simply accounted for by binary mixing between atmospheric and MORB-type Xenon (Mukhopadhyay 2012). An additional mantle reservoir is needed to account for the low ^{129}Xe/^{130}Xe ratio measured in several OIBs such as Iceland and Hawaii. This source reservoir is characterized by high ^{3}He/^{4}He and low ^{40}Ar/^{36}Ar ratios and a solar neon signature consistent with the preservation of undegassed mantle sources with low (U + Th)/He, I/Xe, and K/Ar ratios over most of the Earth's history.

The presence of ^{129}Xe excesses in reservoirs characterized by normal ^{142}Nd signatures is best explained by the preservation in the deep mantle of a pristine (undegassed and undepleted) component. While this interpretation may seem relatively straightforward, the origin of high ^{3}He/^{4}He mantle reservoirs has been the subject of debate over the past 30 years (Albarede and Rouxel 1987; Allegre et al. 1987; Class and Goldstein 2005; Coltice et al. 2011; Farley et al. 1992; Gonnermann and Mukhopadhyay 2009; Kurz et al. 1982; Moreira et al. 2001; Mukhopadhyay 2012; Stuart et al. 2003). A key observation is that oceanic island basalts carrying primordial noble gas signatures have trace element and isotopic (e.g., ^{143}Nd, ^{87}Sr, ^{176}Hf) compositions far more depleted than expected for a chondritic mantle (Class and Goldstein 2005). Models of chemical geodynamics have thus faced the challenge of reconciling seemingly contradictory constraints, whereby the noble gas systematics appears to require the presence of an ancient undegassed mantle, whereas trace elements and Sr–Nd–Hf systematics call for global depletion (and thus degassing) of the OIB source (Hilton and Porcelli 2003).

Proposed source reservoirs for the high ^{3}He/^{4}He plumes include (i) a PM isolated from mantle convection, providing primordial noble gases to a shallower depleted reservoir by diffusion or entrainment of minute amount of primitive material (Allegre et al. 1996, 1987; Moreira et al. 2001; Mukhopadhyay 2012; Stuart et al. 2003); (ii) volatile-rich but chemically depleted material preserved within the general convective circulation system (Albarede 2008; Class and Goldstein 2005; Gonnermann and Mukhopadhyay 2009); (iii) the core, leaking noble gases into the silicate Earth (Porcelli and Halliday 2001); and (iv) an early komatiitic crust, recycled together with dense meteoritic material enriched in solar-implanted noble gases (Tolstikhin and Hofmann 2005; Tolstikhin et al. 2006). Of course, these mechanisms would remain valid for explaining the noble gas systematics of the OIB source regardless of whether the Earth is chondritic or not. However, the new framework defined by SCHEM allows an alternative

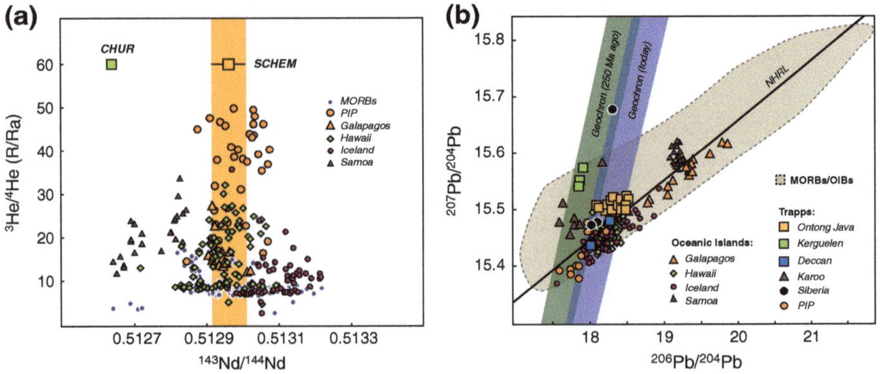

Fig. 11.10 **a** He–Nd relationships in MORBs and high $^3He/^4He$ oceanic island basalts (Abedini et al. 2006; Starkey et al. 2009). Most high $^3He/^4He$ lavas plot within error of the non-chondritic PM composition (SCHEM). **b** Pb isotopic signatures of high $^3He/^4He$ OIBs and some LIPs (Jackson and Carlson 2011; Jackson et al. 2010)

interpretation, namely that a pristine mantle reservoir with non-chondritic abundances of RLEs was preserved in the deep mantle and makes a large contribution to OIB magmatism. This hypothesis was first suggested on the basis of the similarity of $\varepsilon^{143}Nd$ compositions between high $^3He/^4He$ OIBs and the SCHEM parameters (Caro et al. 2008b) (Fig. 11.10a) and further support for the presence of pristine material in the deep Earth was provided by Pb isotope measurements in Cenozoic lavas from the proto-Iceland plume (PIP) (Baffin Island and West Greenland) and several large igneous provinces (LIPs) (Jackson and Carlson 2011; Jackson et al. 2010). Unlike many other OIBs, the PIP basalts analyzed by Jackson et al. (2010) plot near the 4.43–4.50 Ga geochron (Fig. 11.10b), suggesting that the U–Pb chronometer in their mantle source evolved as closed system since formation of the Earth (Jackson and Carlson 2011; Jackson et al. 2010). As lavas from the PIP were emplaced through old cratonic lithosphere, it is also possible that crustal assimilation shifted their Pb isotope composition from the MORB field toward the left of the geochron. However, the trace element pattern and Sr–Nd–Hf systematics of the PIP source (Jackson et al. 2010) show a good match to that obtained for the PM using the SCHEM model (Fig. 11.8), consistent with a contribution of pristine non-chondritic mantle. Furthermore, similar Nd–Pb isotope signatures are found in several other LIPs (Jackson and Carlson 2011), such as the Ontong Java Plateau (OJP) which erupted through oceanic lithosphere (Fig. 11.10b). Interestingly, olivines from PIP and OJP lavas were found to be enriched in nickel compared to olivines from MORB (Herzberg et al. 2013). As mentioned earlier, this may indicate that the He-rich component is now present as a residual melt from an early magma ocean (Coltice et al. 2011; Herzberg et al. 2013). If true, the relatively depleted trace element pattern of the PIP source (Fig. 11.8) would imply that this basal magma ocean is not enriched in incompatible elements, but, rather, retained a relatively primitive (but non-chondritic) trace element and isotopic composition.

11.5 Chemical Geodynamics in a Non-chondritic Earth

In the previous section, it was shown that impact erosion would significantly affect the abundance of trace elements in the BSE, leading to a ca. 20–50 % depletion for the moderately and highly incompatible elements. This, in turn, would modify the output of DM models anchored on CHUR parameters (Salters and Stracke 2004; Workman and Hart 2005). Chemical geodynamic constraints derived from mass balance relationships between the crust and DM would thus need to be reconsidered. This section examines more specifically the ^{40}Ar and ^{143}Nd constraints on mantle structure and evolution in the context of the super-chondritic Earth model (SCHEM) presented in Sect. 11.4. As will be shown, the ^{143}Nd evolution of the Earth's mantle can be accounted for by continuous extraction of the continental crust from a mantle now thoroughly degassed and depleted and does not require the existence of large hidden reservoirs.

11.6 Mass Balance Constraints

11.6.1 ^{147}Sm–^{143}Nd

The first major consequence of applying SCHEM parameters instead of CHUR for the BSE is a modification of the mass balance relationship between the present-day DM and the CC. In the conventional 3-box model, whereby the CC and MORB source are assumed to differentiate from a chondritic mantle (CHUR), the mass of the depleted reservoir can be simply derived from the following equation:

$$M_{\mathrm{DM}} = \frac{[\mathrm{Nd}]_{\mathrm{CC}}(R_{\mathrm{DM}} - R_{\mathrm{CC}})M_{\mathrm{CC}}}{[\mathrm{Nd}]_{\mathrm{CHUR}}(R_{\mathrm{DM}} - R_{\mathrm{CHUR}})}. \tag{11.10}$$

As $[\mathrm{Nd}]_{\mathrm{CHUR}} = 1.25$ ppm, $[\mathrm{Nd}]_{\mathrm{CC}} = 20$ ppm, $R_{\mathrm{CHUR}} = 0.512638$, and $R_{\mathrm{CC}} = 0.511883$, M_{DM} can be estimated to be 10^{24} kg, which only represents 25 % of the mass of the mantle. Since no evidence for chondritic mantle material was found in the isotopic and compositional signatures of oceanic basalts, one must either conclude that the LM is perfectly isolated from the upper mantle and has remained so for the past 4.5 Ga, or a fourth enriched reservoir lies hidden in the deep Earth and concentrates a major fraction of the Earth's trace element budget.

In a non-chondritic Earth, the MORB source would be significantly less depleted than in CHUR-based models, with an Nd content 10–20 % higher than previous estimates. In addition, the difference in ε^{143}Nd between the BSE and the DM would only be 2–3 ε-units, instead of 9 ε-units in the chondritic Earth model. These apparently minor changes result in a drastic modification of the relative proportions of depleted and enriched reservoirs in the modern mantle–crust system. Replacing R_{CHUR} and $[\mathrm{Nd}]_{\mathrm{CHUR}}$ by R_{SCHEM}, $[\mathrm{Nd}]_{\mathrm{SCHEM}}$ in Eq. (11.10) yields a value for M_{DM} of $4.2 \pm 0.9 \times 10^{24}$ kg, which is equivalent to the mass of the whole mantle. Using Taylor and McLennan's estimates for the Bulk continental crust ($[\mathrm{Nd}]_{\mathrm{CC}} = 16$ ppm,

Taylor and McLennan 1995) instead of Rudnick and Fountain's estimate ($[Nd]_{CC}$ = 20 ppm) would yield a DM mass of 3.3 * 10^{24}, or approximately 85 % of the silicate Earth. These results, therefore, are not inconsistent with the presence of deep heterogeneities containing either primitive or enriched material. However, in contrast to conventional chondritic models, mass balance constraints in the SCHEM framework suggest that the present-day MORB source may occupy most, if not the entire mantle, and could be chemically balanced solely by the continental reservoir.

11.6.1.1 ^{40}K–^{40}Ar

Similar to the Sm–Nd case, ^{40}Ar constraints depend on assumptions regarding the composition of the bulk Earth and more specifically its K content (Allegre et al. 1996; Arevalo et al. 2009). While K is not a RLE, its abundance in the silicate Earth is anchored to that of uranium, through the K/U ratio of the bulk silicate Earth (13800 ± 1300 (1σ) (Arevalo et al. 2009)). The standard chemical geodynamics model considers a "chondritic" U content of 20 ppb in the BSE (McDonough and Sun 1995), which corresponds to $[K]_{BSE} = 280 \pm 60$ (1σ) ppm. It follows that $150 \pm 32 \times 10^{18}$ g ^{40}Ar was generated over the past 4.46 Ga in the silicate Earth. The current atmosphere inventory is estimated to 66×10^{18} g (Turekian 1959) and that of the CC is unlikely to exceed 9×10^{18} g ^{40}Ar (this value assumes a K content of 17,000 ppm in the CC, no degassing over time, and a mean age of 2 Ga). In a chondritic Earth, this would leave 75–84 × 10^{18} g ^{40}Ar unaccounted for by the atmosphere and crustal reservoirs. As the MORB source is thoroughly degassed, it cannot balance the missing ^{40}Ar budget, requiring the presence of a large undegassed reservoir in the deep Earth (Allegre et al. 1996).

In a non-chondritic Earth model, the U content of the BSE would be 40–50 % lower than conventional chondritic estimates (this assumes a U partition coefficient intermediate between that of Rb and La). This corresponds to $[U]_{SCHEM}$ = 10–13 ppb, which translates into a K content of 120–145 ppm. Over the past 4.46 Ga, the Earth would have produced 65–78 × 10^{18} g ^{40}Ar, which is within errors identical to the present-day inventory of the atmosphere and the CC (66–75 × 10^{18} g ^{40}Ar). A deep reservoir containing the missing ^{40}Ar budget would thus no longer be needed, consistent with Nd isotope constraints for a mostly depleted (and thus degassed) LM.

11.6.2 Compositional Evolution of the Earth's Mantle

As pointed out in Sect. 11.4.5, the trace element pattern of the "accessible" silicate Earth (Figs. 11.8 and 11.9) is consistent with impact erosion models involving the loss of silicate material at an early stage of planetary accretion. It must be noted, however, that a trace element pattern similar to that shown in Fig. 11.8 could also be produced by segregation of an EER containing a basaltic and a depleted lithospheric component. In the basal magma ocean model, for example, the "accessible" silicate

Earth would have abundances of moderately incompatible elements similar to that estimated in the non-chondritic Earth model (Coltice et al. 2011). Some difference is seen for most incompatible elements (e.g., K, U, Rb), but this may reflect uncertainties on the parameters underlying the basal magma ocean crystallization model, rather than a diagnostic feature of the non-chondritic versus hidden reservoir models. These different hypotheses, however, could be distinguished from each other on a chronological basis. Indeed, collisional erosion could have affected the planet building material during the first few million years after formation of the solar system, as planetesimals accreted into planetary embryos (Asphaug 2010). On the other hand, segregation of an EER at the core–mantle Boundary requires core segregation to be complete, and this could not have taken place less than 30 Ma after formation of the solar system (Kleine et al. 2004; Yin et al. 2002). Using $t_d > 30$ Ma in Eqs. (11.2) and (11.3) would then yield minimum estimates of 0.2111 and +8.5 for the present-day $^{147}Sm/^{144}Nd$ and $\varepsilon^{143}Nd$ of the early DM complementary to the putative EER. As shown in Fig. 11.11b, the $\varepsilon^{143}Nd$ signature of the early depleted mantle (EDM) would have evolved toward values significantly more radiogenic than observed in most Archean mantle-derived rocks. This observation is difficult to explain in the context of the hidden reservoir model because the ^{142}Nd signature of the mantle has remained constant since 3.5 Ga, which precludes even partial remixing of the EER in the convective mantle. As the magnitude of Sm/Nd fractionation needed to create an 18 ppm excess in the EDM decreases as T_d increases, this discrepancy can only be solved by assuming that the EER formed less than 30 Ma after formation of the solar system. This event would then be contemporaneous to core formation and predate the lunar impact (Halliday 2008; Touboul et al. 2009b), an observation seemingly more consistent with the collisional erosion hypothesis.

In the context of the impact erosion scenario, the $\varepsilon^{143}Nd$ evolution of the mantle through geological times can be simply explained by continuous continental growth from a non-chondritic primitive reservoir (Caro and Bourdon 2010). The red curve in Fig. 11.11c illustrates such a scenario, with continental growth starting 3.8 Ga ago and proceeding at a constant rate until the present-day. This model shows that in the absence of segregation between the lower and the upper mantle, crustal growth would have a negligible effect on the ^{143}Nd evolution of the mantle until the Late Archean. The evolution of the Archean mantle toward increasingly radiogenic $\varepsilon^{143}Nd$ value would thus reflect the non-chondritic REE pattern of the PM, rather than early extraction of a large continental mass (Armstrong 1981; Jacobsen 1988).

The simple model presented here is unlikely to be a perfect representation of crustal growth through time, and more complex models, involving growth in the Hadean would certainly favor the development of depleted $\varepsilon^{143}Nd$ signatures at earlier times. However, as pointed out in Sect. 11.3, the long-term preservation of ^{142}Nd anomalies in the Archean mantle, combined with the lack of a widespread Hadean zircon record, appears more consistent with a mafic/ultramafic stagnant lid model rather than with early continental growth scenarios. Furthermore, as shown in Fig. 11.11c, mantle-derived rocks formed between 3.5 and 2.7 Ga appear to conform well to this simple linear growth model, where the mantle signature remains similar to the primitive SCHEM evolution line up until 2–2.5 Ga. Only the oldest rocks from the Isua greenstone belt and Itsaq gneiss complex

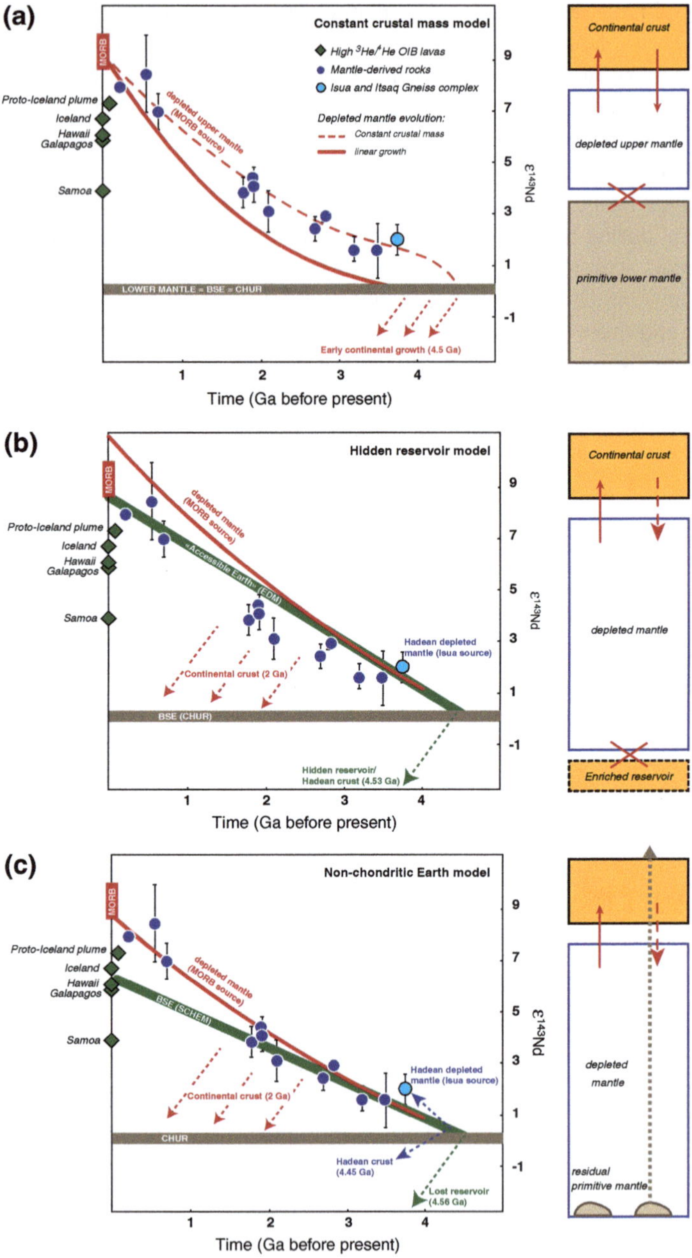
(a)
Constant crustal mass model
High ^{3}He/^{4}He OIB lavas
Mantle-derived rocks
Isua and Itsaq Gneiss complex
Depleted mantle evolution:
Constant crustal mass
linear growth
MORB
Proto-Iceland plume
Iceland
Hawaii
Galapagos
Samoa
depleted upper mantle (MORB source)
LOWER MANTLE = BSE = CHUR
Early continental growth (4.5 Ga)
Time (Ga before present)
ε^{143}Nd
Continental crust
depleted upper mantle
primitive lower mantle
(b)
Hidden reservoir model
depleted mantle (MORB source)
«Accessible Earth» (EDM)
Hadean depleted mantle (Isua source)
Continental crust (2 Ga)
BSE (CHUR)
Hidden reservoir/ Hadean crust (4.53 Ga)
depleted mantle
Enriched reservoir
(c)
Non-chondritic Earth model
BSE (SCHEM)
CHUR
Hadean crust (4.45 Ga)
Lost reservoir (4.56 Ga)
depleted mantle
residual primitive mantle

Fig. 11.11 Chemical geodynamic models viewed from the perspective of the $^{146,147}Sm$–$^{142,143}Nd$ system. **a** The standard model of mantle–crust evolution ascribes differentiation of the Earth's mantle to the extraction of the continental crust from a primitive chondritic reservoir. Mass balance considerations based on ^{40}Ar and ^{143}Nd systematics require that only the upper mantle be depleted, while the lower mantle, or a large reservoir within the lower mantle, remained pristine (i.e., undegassed and undepleted). This deep reservoir remained perfectly isolated through geological times, with only heat and Helium being transferred to the upper mantle. Note that the Archean $\varepsilon^{143}Nd$ record is inconsistent with a simple linear growth model and requires the extraction of a large continental mass prior to 4 Ga. In this context, ^{142}Nd heterogeneities measured in Archean rocks may be ascribed to the earliest stages of continental growth in the Hadean. This model, however, does not account for the non-chondritic ^{142}Nd composition of the "accessible" silicate Earth. **b** In the hidden reservoir model, the compositional evolution of the mantle is determined almost entirely by the segregation of an early enriched reservoir (*EER*) in the deep mantle. As the EER is unlikely to be older than the core, it must have formed <4.53 Ga ago. This constrains the present-day $\varepsilon^{143}Nd$ ratio of the early depleted mantle (*EDM*) to >8.5 ε-units, which is only slightly less depleted than the MORB source. This model would be consistent with whole-mantle convection but nevertheless requires the presence of an isolated reservoir in the interior of the Earth. **c** In the non-chondritic Earth model, the mantle is homogeneously depleted (and degassed) by crustal growth. The Archean mantle evolves toward positive $\varepsilon^{143}Nd$ values along the primitive evolution line, obviating the need for a large continental mass prior to 2.5 Ga. This model does not require the presence of a large hidden reservoir, either enriched or primitive, and allows for a substantial contribution of primitive mantle to the OIB source. In this model, ^{142}Nd heterogeneities in the early Archean are ascribed to the formation of long-lived differentiated mantle and crustal reservoirs from a magma ocean 4.46 Ga ago. These differentiated reservoirs, however, were probably rehomogenized prior to the extraction of the modern continental crust 2.7 Ga ago, so that no remnant of the primary differentiated silicate reservoirs survived until the present day

(3.6–3.8 Ga; Greenland) have initial $\varepsilon^{143}Nd$ that plot significantly above the SCHEM evolution line. As pointed out in Sect. 11.3, these rocks also carry radiogenic ^{142}Nd excesses, inherited from an ancient (4.45 Ga) depleted source and, therefore, are expected to have $\varepsilon^{143}Nd$ signatures more radiogenic than SCHEM. This observation is critical in the context of the non-chondritic Earth model, as it is the only remaining observation that witnesses the existence of mantle–crust differentiation processes prior to 3.8 Ga. Little is known at present on the composition and lifetime of these primordial reservoirs, as ^{142}Nd heterogeneities appear to vanish from the geological record after ca. 2.7 Ga (Debaille et al. 2013). This suggests that with the exception of primordial heterogeneities now locally sampled by plume magmatism, early differentiated mantle and crustal reservoirs were largely rehomogenized prior to the main episodes of crustal growth.

11.7 Conclusions

Over the past 30 years, the chemical and isotopic evolution of the "accessible" mantle has been largely interpreted in terms of progressive extraction of the continental crust from a primitive chondritic mantle. From the study of ^{146}Sm–^{142}Nd systematics in terrestrial and meteoritic samples, a more complex evolution scenario has emerged that can be summarized as follows:

1. A "missing reservoir," enriched in incompatible elements and segregated from the silicate Earth and/or its precursors at a very early stage of its evolution, most likely during the first 10 Ma after formation of the solar system. This reservoir must contain a trace element budget roughly similar to that of the modern CC. It has been speculated that this reservoir may have been hidden in the deep mantle or lost to space through preferential loss of crustal material during impacts on the surface of accreting protoplanets. From a geochemical viewpoint, models involving a "hidden" or a "lost" reservoir have fairly similar outcomes as they both generate elemental fractionations controlled by solid/liquid partitioning in magmatic processes. However, from a chronological viewpoint, collisional mass loss affecting the earliest planetary crusts on small differentiated bodies would be a more plausible cause for the non-chondritic ^{142,143}Nd signature of the accessible silicate Earth.
2. The Earth's mantle experienced global differentiation near the end of accretion, most likely as a result of fractional crystallization in a deep magma ocean following the giant impact. The fate of these primordial crustal and mantle reservoirs can be traced until the Neoarchean, but so far, no remnant has been reported in rocks younger than 2.7 Ga. Thus, primordial mantle heterogeneities must have been homogenized prior to the main episodes of continental growth 2.7 Ga ago. Residual primitive material may have been preserved in the deep mantle, as indicated by the noble gas systematics of oceanic island basalts. The preservation of these ancient heterogeneities may be related to the presence of dense melts near the CMB, residual from crystallization of the terrestrial magma ocean.
3. In the non-chondritic model, the evolution of the Earth's mantle only deviates from that of the PM in the Early Proterozoic, following the extraction of a large amount of CC at the Archean–Proterozoic transition. As the isotopic signature of >2.5 Ga old mantle-derived rocks matches that predicted for a primitive (non-chondritic) mantle, early crustal extraction prior to 4 Ga would no longer be required to explain the compositional evolution of the Archean mantle. Revised mass balance relationships between the crust and the modern DM suggest that the present-day continental reservoir is sufficient to balance the depletion of the MORB source even if the latter occupies the entire mantle. Likewise, the Earth's interior appears to be thoroughly degassed, and most of the terrestrial Argon budget now resides in the atmosphere. The non-chondritic Earth model would thus obviate the need for a deep enriched or primitive "hidden" reservoir and provide a straightforward explanation for the lack of chondritic chemical and isotopic signatures in modern and ancient mantle-derived rocks.

Acknowledgments This work was supported by Agence National de la Recherche (*Grant DESIR: Differentiation of the Early Silicate Earth: Isotopic constraints from the Hadean rock record*). I am grateful to the editor Frédéric Deschamps and an anonymous reviewer for a constructive review that helped improve the manuscript.

References

Abedini AA, Hurwitz S, Evans WC (2006) USGS-NoGaDat—a global dataset of noble gas concentrations and their isotopic ratios in volcanic systems. Geolog Surv Data Ser, U.S 202

Albarede F (2008) Rogue mantle helium and neon. Science 319:943–945

Albarede F, Rouxel M (1987) The Sm/Nd secular evolution of the continental crust and the depleted mantle. Earth Planet Sci Lett 82:25–35

Allegre CJ (1982) Chemical geodynamics. Tectonophysics 81:109–132

Allegre CJ, Hofmann AW, O'Nions K (1996) The Argon constraints on mantle structure. Geophys Res Lett 23:3555–3557

Allegre CJ, Othman DB, Polve M, Richard D (1979) Nd-Sr isotopic correlation in mantle materials and geodynamic consequences. Phys Earth Planet Int 19:293–306

Allegre CJ, Staudacher T, Sarda P (1987) Rare gas systematics: formation of the atmosphere, evolution and structure of the Earth's mantle. Earth Planet Sci Lett 81:127–150

Amelin Y, Lee D-C, Halliday AN, Pidgeon RT (1999) Nature of the Earth's earliest crust from Hafnium isotopes in single detrital zircons. Nature 399:252–255

Amelin Y, Rotenberg E (2004) Sm–Nd systematics of chondrites. Earth Planet Sci Lett. 223:267–282

Andrault D, Petitgirard S, Lo Nigro G, Devidal J-L, Veronesi G, Garbarino G, Mezouar M (2012) Solid–liquid iron partitioning in Earth's deep mantle. Nature 487:354–357

Andreasen R, Sharma M (2006) Solar nebula heterogeneity in p-process samarium and neodymium isotopes. Science 314:806–809

Andreasen R, Sharma M, Subbarao KV, Viladkar SG (2007) Where on Earth is the enriched Hadean reservoir? Earth Planet Sci Lett 226:14–28

Arevalo RJ, McDonough WF, Luong M (2009) The K/U ratio of the silicate Earth: insights into mantle composition, structure and thermal evolution. Earth Planet Sci Lett 278:361–369

Arlandini C, Kappeler F, Wisshak K, Gallino R, Lugaro M, Busso M, Staniero O (1999) Neutron capture in low-mass asymptotic giant branch stars: cross sections and abundance signatures. Astrophys J 525:886–900

Armstrong RL (1981) The case for crustal recycling on a near-steady-state no-continental-growth Earth. Phil Trans Roy Soc A301:443–472

Asphaug E (2010) Similar-sized collisions and the diversity of planets. Chem Erde 70:199–219

Audi G, Bersillon O, Blachot J, Wapstra AH (1997) The NUBASE evaluation of nuclear and decay properties. Nucl Phys A 624:1–124

Bennett VC, Brandon AD, Nutman AP (2007) Coupled ^{142}Nd–^{143}Nd isotopic evidence for Hadean mantle dynamics. Science 318:1907–1910

Benz W, Cameron AGW (1990) Terrestrial effects of the giant impact. In: Newsom HE, Jones JH (eds) Origin of the earth. Oxford University Press, Oxford, pp 61–68

Benz W, Slattery WL, Cameron AGW (1988) Collisional stripping of Mercury's mantle. Icarus 74:516–528

Benz W, Anic A, Horner J, Whitby JA (2007) The origin of Mercury. Space Sci Rev 132:189–202

Borg LE, Connelly JN, Boyet M, Carlson RW (2011) Chronological evidence that the Moon is either young or did not have a global magma ocean. Nature 477:70–72

Bourdon B, Caro G (2007) The early terrestrial crust. CR Geosci 339:928–936

Bourdon B, Touboul M, Caro G, Kleine T (2008) Early differentiation of the Earth and the Moon. Phil Trans Roy Soc 366:4105–4128

Bouvier A, Vervoort JD, Patchett PJ (2008) The Lu–Hf and Sm–Nd isotopic composition of CHUR: constraints from unequilibrated chondrites and implications for the bulk composition of terrestrial planets. Earth Planet Sci Lett 272:48–57

Boyet M, Blichert-Toft J, Rosing M, Storey CD, Télouk M, Albarede F (2004) ^{142}Nd evidence for early Earth differentiation. Earth Planet Sci Lett 214:427–442

Boyet M, Carlson RW (2005) ^{142}Nd evidence for early (>4.53 Ga) global differentiation of the silicate Earth. Science 214:427–442

Boyet M, Carlson RW (2006) A new geochemical model for the Earth's mantle inferred from ^{146}Sm–^{142}Nd systematics. Earth Planet Sci Lett 250:254–268

Boyet M, Carlson RW (2007) A highly depleted moon or a non-magma ocean origin for the lunar crust? Earth Planet Sci Lett 262:505–516

Boyet M, Carlson RW, Horan M (2010) Old Sm–Nd ages for cumulate eucrites and redetermination of the solar system initial $^{146}Sm/^{144}Sm$ ratio. Earth Planet Sci Lett 291:172–181

Brandon AD, Lapen TJ, Debaille V, Beard BL, Rankenburg K, Neal C (2009) Re-evaluating $^{142}Nd/^{144}Nd$ in lunar mare basalts with implications for the early evolution and bulk Sm/Nd of the Moon. Geochim Cosmochim Acta 73:6421–6445

Campbell IH, O'Neill HSC (2012) Evidence against a chodnritic Earth. Nature 483:553–558

Carlson RW, Boyet M, Horan M (2007) Chondrite barium, neodymium and samarium isotopic heterogeneity and early earth differentiation. Science 316:1175–1178

Carlson RW, Lugmair GW (1988) Yhe age of ferroan anorthosite 60025: oldest crust on a young Moon? Earth Planet Sci Lett 90:119–130

Caro G (2011) Early silicate Earth differentiation. Annu Rev Earth Planet Sci 39:31–58

Caro G, Bennett VC, Bourdon B, Harrison TM, von Quadt A, Mojzsis SJ, Harris JW (2008a) Application of precise $^{142}Nd/^{144}Nd$ analysis of small samples to inclusions in diamonds (Finsch, South Africa) and Hadean Zircons (Jack Hill, Western Australia). Chem Geol 247:253–265

Caro G, Bourdon B (2010) Non-chondritic Sm/Nd ratio in the terrestrial planets: consequences for the geochemical evolution of the mantle-crust system. Geochim Cosmochim Acta 74:3333–3349

Caro G, Bourdon B, Birck J-L, Moorbath S (2003) ^{146}Sm–^{142}Nd evidence for early differentiation of the Earth's mantle. Nature 423:428–432

Caro G, Bourdon B, Birck J-L, Moorbath S (2006) High-precision $^{142}Nd/^{144}Nd$ measurements in terrestrial rocks: constraints on the early differentiation of the Earth's mantle. Geochim Cosmochim Acta 70:164–191

Caro G, Bourdon B, Halliday AN, Quitté G (2008b) Super-chondritic Sm/Nd ratios in Mars, the Earth and the Moon. Nature 452:336–339

Caro G, Bourdon B, Wood BJ, Corgne A (2005) Trace element fractionation generated by melt segregation from a magma ocean. Nature 436:246–249

Chase CG, Patchett PJ (1988) Stored mafic/ultramafic crust and early Archean mantle depletion. Earth Planet Sci Lett 91:66–72

Chauvel C, Lewin E, Carpentier M, Arndt NT, Marini J-C (2007) Role of recycled oceanic basalt and sediment in generating the Hf–Nd mantle array. Nat Geosci 1:64–67

Christensen UR, Hofmann AW (1994) Segregation of subducted oceanic crust in the convecting mantle. J Geophys Res 99:19867–19884

Class C, Goldstein SL (2005) Evolution of helium isotopes in the Earth's mantle. Nature 436:1107–1112

Coltice N, Schmalzl J (2006) Mixing times in the mantle of the early Earth derived from 2-D and 3-D numerical simulations of convection. Geophys Res Lett 33:L23304

Coltice N, Moreira M, Hernlund J, Labrosse S (2011) Crystallization of a basal magma ocean recorded by Helium and Neon. Earth Planet Sci Lett 308:193–199

Condie KC (2000) Episodic growth models: afterthoughts and extensions. Tectonophysics 322:153–162

Corgne A, Liebske C, Wood BJ, Rubie DC, Frost DJ (2005) Silicate perovskite-melt partitioning of trace elements and geochemical signature of a deep perovskitic reservoir. Geochim Cosmochim Acta 69:485–496

Corgne A, Wood BJ (2004) Trace element partitioning between majoritic garnet and silicate melt at 25 GPa. Phys Earth Planet Int 143–144:407–419

Debaille V, Brandon AD, Yin QZ, Jacobsen B (2007) Coupled ^{142}Nd-^{143}Nd evidence for a protracted magma ocean in Mars. Nature 450:525–528

Debaille V, O'Neil C, Brandon AD, Haenecour P, Yin Q-Z, Mattielli N, Treiman AH (2013) Stagnant-lid tectonics in early Earth revealed by ^{142}Nd variations in late Archean rocks. Earth Planet Sci Lett 373:83–92

DePaolo DJ (1980) Crustal growth and mantle evolution: inferences from models of element transport and Sr and Nd isotopes. Geochim Cosmochim Acta 44:1185–1196

DePaolo DJ, Wasserburg GJ (1976) Inferences about magma sources and mantle structure from variations of $^{143}Nd/^{144}Nd$. Geophys Res Lett 3:743–746

Elkins-Tanton LT, Burgess S, Yin QZ (2011) The lunar magma ocean: Reconciling the solidification process with lunar petrology and geochronology. Earth Planet Sci Lett 304:326–336

Elkins-Tanton LT, Parmentier EM, Hess PC (2003) Magma ocean fractional crystallization and cumulate overturn in terrestrial planets: implications for Mars. Meteorit Planet Sci 38:1753–1771

Farley KA, Natland JH, Craig H (1992) Binary mixing of enriched and undegassed (primitive-questionable) mantle components (He, Sr, Nd; Pb) in Samoan lavas. Earth Planet Sci Lett 111:183–199

Galer SJG, Goldstein SL (1991) Early mantle differentiation and its thermal consequences. Geochim Cosmochim Acta 55:227–239

Gannoun A, Boyet M, Rizo H, El Goresy A (2011) ^{146}Sm–^{142}Nd systematics measured in enstatite chondrites reveals a heterogeneous distribution of ^{142}Nd in the solar nebula. PNAS 108:7693–7697

Garnero EJ, McNamara AK (2008) Structure and dynamics of Earth's lower mantle. Science 320:626–628

Goldstein SL, Jacobsen SB (1988) Nd and Sr isotopic systematics of river water suspended material: implications for crustal evolution. Earth Planet Sci Lett 87:249–265

Gonnermann HM, Mukhopadhyay S (2009) Preserving noble gases in a convecting mantle. Nature 459:560–563

Grand SP (2002) Mantle shear-wave tomography and the fate of subducted slabs. Phil Trans Roy Soc 360:2475–2491

Guitreau M, Blichert-Toft J, Mojzsis SJ, Roth ASG, Bourdon B (2013) A legacy of Hadean silicate differentiation inferred from Hf isotopes in Eoarchean rocks of the Nuvvuagittuq supracrustal belt (Québec, Canada). Earth Planet Sci Lett 362:171–181

Halliday AN (2008) A young Moon-forming giant impact at 70–110 million years accompanied by late-stage mixing, core formation and degassing of the Earth. Phil Trans Roy Soc 366:4205–4252

Harper CL, Jacobsen SB (1992) Evidence from coupled ^{147}Sm–^{143}Nd and ^{146}Sm–^{142}Nd systematics for very early (4.5 Gyr) differentiation of the Earth's mantle. Nature 360:728–732

Harrison TM (2009) The Hadean crust: Evidence from >4 Ga zircons. Annu Rev Earth Planet Sci 37:479–505

Harrison TM, Schmitt AK, McCulloch MT, Lovera OM (2008) Early (>=4.5 Ga) formation of terrestrial crust: Lu–Hf, $d^{18}O$, and Ti thermometry results for Hadean zircons. Earth Planet Sci Lett 268:476–486

Hawkesworth CJ, Kemp AIS (2006) Using Hafnium and oxygen isotopes in zircons to unravel the record of crustal evolution. Chem Geol 226:144–162

Hayman PC, Kopylova MG, Kaminsky FV (2005) Lower mantle diamonds from Rio Soriso (Juina area, Mato Grosso, Brazil). Contrib Mineral Pet 140:430–445

Herzberg CT, Asimow PD, Ionov DA, Vidito C, Jackson MG, Geist D (2013) Nickel and helium evidence for melt above the core-mantle boundary. Nature 493:393–397

Hilton DR, Porcelli D (2003) Noble gases as mantle tracers. In: Carlson RW (ed) Treatise on geochemistry, vol 2. Elsevier, Oxford, pp 277–318

Hofmann AW (1988) Chemical differentiation of the Earth: the relationship between mantle, continental crust, and oceanic crust. Earth Planet Sci Lett 90:297–314

Hofmann AW (2003) Sampling mantle heterogeneity through oceanic basalts: isotopes and trace elements. In: Holland HD, Turekian KK (eds) Treatise of geochemistry, vol 2. Elsevier, Oxford, pp 61–101

Hofmann AW, White WM (1982) Mantle plumes from ancient oceanic crust. Earth Planet Sci Lett 57:421–436

Hopkins M, Harrison TM, Manning CE (2008) Low heat flow inferred from >4 Ga zircons suggests Hadean plate boundary interactions. Nature 456:493–496

Jackson MG, Carlson RW (2011) An ancient recipe for flood basalt genesis. Nature 476:316–319
Jackson MG, Carlson RW (2012) Homogeneous superchondritic $^{142}Nd/^{144}Nd$ in the mid-ocean ridge basalt and oceanc island basalt mantle. Geochem Geophys Geosys 13
Jackson MG, Carlson RW, Kurz MD, Kempton PD, Francis D, Bulusztajn J (2010) Evidence for the survival of the oldest terrestrial mantle reservoir. Nature 466:853–856
Jackson MG, Kurz MD, Hart SR, Workman RK (2007) New Samoan lavas from Ofu Island reveal a hemispherically heterogeneous high 3He/4He mantle. Earth Planet Sci Lett 264:360–374
Jacobsen SB (1988) Isotopic constraints on crustal growth and recycling. Earth Planet Sci Lett 90:315–329
Jacobsen SB, Wasserburg GJ (1979) The mean age of mantle and crustal reservoirs. J Geophys Res 84:7411–7424
Jacobsen SB, Wasserburg GJ (1984) Sm–Nd evolution of chondrites and achondrites II. Earth Planet Sci Lett 67:137–150
Kellogg LH, Hager BH, van der Hilst RD (1999) Compositional stratification in the deep mantle. Science 5409:1881–1884
Kemp AIS, Wilde SA, Hawkesworth CJ, Coath CD, Nemchin A, Pidgeon RT, Vervoort JD (2010) Hadean crustal evolution revisited: New constraints from Pb–Hf isotope systematics of the Jack Hills zircons. Earth Planet Sci Lett 296:45–56
Kinoshita N, Paul M, Kashiv Y, Collon P, Deibel CM, DiGiovine B, Greene JP, Henderson DJ, Liang CL, Marley ST, Nakanishi T, Pardo RC, Rehm KE, Robertson D, Scott R, Schmitt C, Tang XD, Vondrasek R, Yokohama A (2012) A shorter ^{146}Sm half-life measured and implications for ^{146}Sm–^{142}Nd chronology in the solar system. Science 335:1614–1617
Kleine T, Mezger K, Munker C, Palme H, Bischoff A (2004) ^{182}Hf–^{182}W isotope systematics of chondrites, eucrites, and Martian meteorites: chronology of core formation and early differentiation in Vesta and Mars. Geochim Cosmochim Acta 68:2935–2946
Korenaga J (2009) A method to estimate the composition of the bulk silicate Earth in the presence of a hidden geochemical reservoir. Geochim Cosmochim Acta 73:6952–6964
Kurz MD, Jenkins WJ, Hart SR (1982) Helium isotopic systematics of oceanic islands and mantle heterogeneity. Nature 297:43–47
Labrosse S, Hernlund J, Coltice N (2007) A crystallizing dense magma ocean at the base of the Earth's mantle. Nature 450:866–869
Lugmair GW, Galer SJG (1992) Age and isotopic relationships among the angrites Lewis Cliff 86010 and Angra dos Reis. Geochim Cosmochim Acta 56:1673–1694
Marty B (1989) Neon and xenon isotopes in MORB: implication for the earth-atmosphere evolution. Earth Planet Sci Lett 94:45–56
Marty B (2012) The origin and concentrations of water, carbon, nitrogen and noble gases on Earth. Earth Planet Sci Lett 313–314:56–66
McCulloch MT (1994) Primitive $^{87}Sr/^{86}Sr$ from an Archean barite and conjecture on the Earth's age and origin. Earth Planet Sci Lett 126:1–13
McCulloch MT, Wasserburg GJ (1978) Barium and neodymium isotopic anomalies in the Allende meteorite. Astrophys J 220:L15–L19
McDonough WF, Sun S-S (1995) The composition of the Earth. Chemical geology 120:223–253
Melosh HJ (1990) Giant impacts and the thermal state of the early Earth. In: Newson HE, Jones JH (eds) Origin of the Earth. Oxford Univesity Press, New York, pp 69–84
Mojzsis SJ, Harrison TM, Pidgeon RT (2001) Oxygen-isotope evidence from ancient zircons for liquid water at the Earth's surface 4,300 Myr ago. Nature 409:178–181
Montelli R, Nolet G, Dahlen FA, Masters G, Engdahl ER, Hung S-H (2004) Finite-frequency tomography reveals a variety of plumes in the mantle. Science 303:338–343
Moorbath S, Whitehouse MJ, Kamber BS (1997) Extreme Nd-isotope heterogeneity in the early Archaean—fact of fiction? Case hostories from northern Canada and West Greenland. Chem Geol 135:213–231
Moreira M, Breddam K, Curtice J, Kurz MD (2001) Solar Neon in the Icelandic mantle: new evidence for an undegassed lower mantle. Earth Planet Sci Lett 185:15–23
Mukhopadhyay S (2012) Early differentiation and volatile accretion recorded in deep-mantle neon and xenon. Nature 486:101–106

Murphy DT, Brandon AD, Debaille V, Burgess R, Ballentine C (2010) In search of a hidden long-term isolated sub-chondritic $^{142}Nd/^{144}Nd$ reservoir in the deep mantle: implications for the Nd isotope systematics of the Earth. Geochim Cosmochim Acta 74:738–750

Nemchin A, Timms N, Pidgeon R, Geisler T, Reddy S, Meyer C (2009) Timing of crystallization of the lunar magma ocean constrained by the oldest zircon. Nat Geosci 2:133–136

Nomura R, Ozawa H, Tateno S, Hirose K, Hernlund J, Muto S, Ishii H, Hiraoka N (2011) Spin crossover and iron-rich silicate melt in the Earth's deep mantle. Nature 473:199–202

Nyquist LE, Wiesmann H, Bansal B, Shih C-Y, Keith JE, Harper CL (1995) ^{146}Sm–^{142}Nd formation interval for the lunar mantle. Geochim Cosmochim Acta 13:2817–2837

O'Neil J, Carlson RW, Paquette J-L, Francis D (2012) Formation age and metamorphic history of the Nuvvuagittuq Greenstone Belt. Precamb Res 220–221:23–44

O'Neil J, Carlson RW, Francis D, Stevenson RK (2008) Neodymium-142 evidence for Hadean mafic crust. Science 321:1828–1831

O'Neill HSC, Palme H (2008) Collisional erosion and the non-chondritic composition of the terrestrial planets. Phil Trans Roy Soc 366:4205–4238

Ozima M, Podosek FA (1999) Formation age of the Earth from $^{129}I/^{127}I$ and $^{244}Pu/^{238}U$ systematics and the missing Xe. J Geophys Res 104:25493–25499

Papanastassiou DA, Wasserburg GJ (1969) Initial Sr isotopic abundances and the resolution of small time differences in the formation of planetary objects. Earth Planet Sci Lett 5:361–376

Porcelli D, Halliday AN (2001) The core as a possible source of mantle helium. Earth Planet Sci Lett 192:45–56

Prinzhofer DA, Papanastassiou DA, Wasserburg GJ (1989) The presence of ^{146}Sm in the early solar system and implications for its nucleosynthesis. Astrophys J 344:L81–L84

Prinzhofer DA, Papanastassiou DA, Wasserburg GJ (1992) Samarium-Neodymium evolution of meteorites. Geochim Cosmochim Acta 56:797–815

Puchtel IS, Blichert-Toft J, Touboul M, Walker RJ, Byerly GR, Nisbet EG, Anhaeusser CR (2013) Insights into early Earth from Barberton komatiites: evidence from lithophile isotope and trace element systematics. Geochim Cosmochim Acta 208:63–90

Qin L, Carlson RW, Alexander CMOD (2011) Correlated nucleosynthetic isotopic variability in Cr, Sr, Ba, Sm, Nd and Hf in Murchison and QUE97008. Geochim Cosmochim Acta 75:7806–7828

Richard D, Shimizu N, Allegre CJ (1976) $^{143}Nd/^{146}Nd$, a natural tracer: an application to oceanic basalts. Earth Planet Sci Lett 31:269–278

Rizo H, Boyet M, Blichert-Toft J, Rosing M (2012a) Combined Nd and Hf isotope evidence for deep-seated source of Isua lavas. Earth Planet Sci Lett 312:267–279

Rizo H, Boyet M, Blichert-Toft J, O'Neil J, Rosing M, Paquette J-L (2012b) The elusive Hadean enriched reservoir revealed by ^{142}Nd deficits in Isua Archaean rocks. Nature 491:96–100

Roth ASG, Bourdon B, Mojzsis SJ, Touboul M, Sprung P, Guitreau M, Blichert-Toft J (2013) Earth Planet Sci Lett 361:50–57

Rudnick RL, Fountain DM (1995) Nature and composition of the continental crust: a lower crustal perspective. Rev Geophys 33:267–309

Salters VJM, Stracke A (2004) Composition of the depleted mantle. Geochem Geophys, Geosys 5

Shirey SB, Hanson GN (1986) Mantle heterogeneity and crustal recycling in Archean granite-greenstone belts in the Rainy Lake area, superior province, Ontario, Canada. Geochim Cosmochim Acta 50:2631–2651

Solomatov VS, Stevenson DJ (1993a) Nonfractional crystallization of a terrestrial magma ocean. J Geophys Res 98:5391–5406

Solomatov VS, Stevenson DJ (1993b) Suspension in convective layers and style of differentiation of a terrestrial magma ocean. J Geophys Res 98:5375–5390

Stachel T, Harris JW, Brey GP, Joswing W (2000) Kankan diamonds (Guinea) II: Lower mantle inclusion parageneses. Contrib Mineral Petrol 140:16–27

Starkey NA, Stuart FM, Ellam RM, Fitton RM, Basu S, Larsen LM (2009) Helium isotopes in early Iceland plume picrites: constraints on the composition of high $^{3}He/^{4}He$ mantle. Earth Planet Sci Lett 277:91–100

Stuart FM, Lass-Evans S, Fitton RM, Ellam RM (2003) High $^3He/^4He$ ratios in picritic basalts from Baffin Island and the role of a mixed reservoir in mantle plumes. Nature 424:57–63

Taylor SR, McLennan SM, McCulloch MT (1983) Geochemistry of Loess, continental crustal composition and crustal model ages. Geochim Cosmochim Acta 47:1897–1905

Taylor SR, McLennan SM (1995) The geochemical evolution of the continental crust. Rev Geophys 33:241–265

Thomas PC, Binzel RP, Gaffey MJ, Storrs AD, Wells EN, Zellner BH (1997) Impact excavation on asteroid 4 Vesta: Hubble space telescope results. Science 277:1492–1495

Tolstikhin IN, Hofmann AW (2005) Early crust on top of the Earth's core. Phys Earth Planet Int 148:109–130

Tolstikhin IN, Kramers JD, Hofmann AW (2006) A chemical Earth model with whole mantle convection: the importance of a core-mantle boundary layer (D″) and its early formation. Chem Geol 226:79–99

Touboul M, Kleine T, Bourdon B, Nyquist LE, Shih C-Y (2009a) New ^{142}Nd evidence for a non-chondritic composition of the Moon. In: 40th lunar and planetary science conference, Abstract #2269

Touboul M, Kleine T, Bourdon B, Palme H, Wieler R (2009b) Tungsten isotopes in ferroan anorthosites: implications for the age of the Moon and lifetime of its magma ocean. Icarus 199:245–249

Trail D, Mojzsis SJ, Harrison TM, Schmitt AK, Watson EB, Young ED (2007) Constraints on Hadean zircon protoliths from oxygen isotopes, Ti-thermometry, and rare earth elements. Geochem Geophys, Geosys 8

Turekian KK (1959) The terrestrial economy of helium and argon. Geochim Cosmochim Acta 17:37–43

van der Hilst RD, Widiyantoro S, Engdahl ER (1997) Evidence for deep mantle circulation from global tomography. Nature 386:578–584

Vervoort JD, Patchett PJ, Blichert-Toft J, Albarede F (1999) Relationships between Lu–Hf and Sm–Nd isotopic systems in the global sedimentary system. Earth Planet Sci Lett 168:79–99

Walter MJ, Kohn SC, Araujo D, Bulanova GP, Smith CB, Gaillou E, Wang J, Steele A, Shirey SB (2011) Deep mantle cycling of oceanic crust: evidence from diamonds and their mineral inclusions. Nature 334:54–57

Warren PH (2008) A depleted, not ideally chondritic bulk Earth: the explosive-volcanic basalt loss hypothesis. Geochim Cosmochim Acta 72:2217–2235

Wedepohl KH (1995) The composition of the continental crust. Geochim Cosmochim Acta 59:1217–1232

Wilde SA, Valley JW, Peck WH, Graham CM (2001) Evidence from detrital zircons for the existence of continental crust and oceans on the Earth 4.4 Gyr ago. Nature 409:175–178

Workman RK, Hart SR (2005) Major and trace element composition of the depleted MORB mantle (DMM). Earth Planet Sci Lett 231:53–72

Yin QZ, Jacobsen SB, Yamashita K, Blichert-Toft J, Telouk P, Albarede F (2002) A short timescale for terrestrial planet formation from Hf–W chronometry of meteorites. Nature 418:949–952

Zindler A, Hart S (1986) Chemical geodynamics. Annu Rev Earth Planet Sci 14:493–571

Chapter 12
Seismic Detections of Small-Scale Heterogeneities in the Deep Earth

Sebastian Rost, Paul S. Earle, Peter M. Shearer, Daniel A. Frost and Neil D. Selby

Abstract We report the detection of coherent scattered energy related to the phase PKPPKP ($P'P'$) in the data of medium aperture arrays. The scattered energy ($P'{\bullet}P'$) is weak and requires array processing techniques to extract the signal from the noise. The arrival time window of $P'{\bullet}P'$ is mostly free from other interfering body wave energy and can be detected over a large distance range. $P'{\bullet}P'$ has been detected in the data of large aperture arrays previously, but the detection in the data of smaller arrays shows its potential for the study of the small-scale structure of the Earth. Here, we show that $P'{\bullet}P'$ can detect scattering off small-scale heterogeneities throughout the Earth's mantle from crust to core making this one of the most versatile scattering probes available. We compare the results of $P'{\bullet}P'$ to a related scattering probe ($PK{\bullet}KP$). The detected energy is in agreement with stronger scattering, i.e., more heterogeneous structure, in the upper mantle and in an approximately 800-km-thick layer above the core–mantle boundary. Lateral variations in heterogeneity structure can also be detected through differences in scattered energy amplitude. We use an application of the F-statistic in the array processing allowing us a precise measurement of the incidence angles (slowness and backazimuth) of the scattered energy. The directivity information of the array data allows an accurate location of the scattering origin. The combination of high-resolution array processing and the scattering of $P'{\bullet}P'$

S. Rost (✉) · D.A. Frost
Institute of Geophysics and Tectonics, School of Earth and Environment,
University of Leeds, Leeds LS2 9JT, UK
e-mail: s.rost@leeds.ac.uk

P.S. Earle
United States Geological Survey, DFC, MS 966, P.O. Box 25046, Denver, CO 80225, USA

P.M. Shearer
Institute of Geophysics and Planetary Physics, Scripps Institution of Oceanography,
University of California, San Diego, USA

N.D. Selby
AWE Blacknest, Brimpton, Reading, UK

© Springer International Publishing Switzerland 2015

A. Khan and F. Deschamps (eds.), *The Earth's Heterogeneous Mantle*,
Springer Geophysics, DOI 10.1007/978-3-319-15627-9_12

as probe for small-scale heterogeneities throughout the Earth's mantle will provide constraints on mantle convection, mantle structure, and mixing related to the subduction process.

Keywords Seismic scattering · Small-scale heterogeneity · Mantle mixing · Array seismology · Core–mantle processes

12.1 Introduction

Seismology aims to resolve structures on a wide variety of scale lengths from structures spanning several thousands of kilometers studied using earthquakes and tomographic techniques to the fine-scale structures resolved in industrial-style controlled source experiments allowing the resolution of structures on scales of just a few tens of meters (Fig. 12.1). The impressive high resolution of near-surface

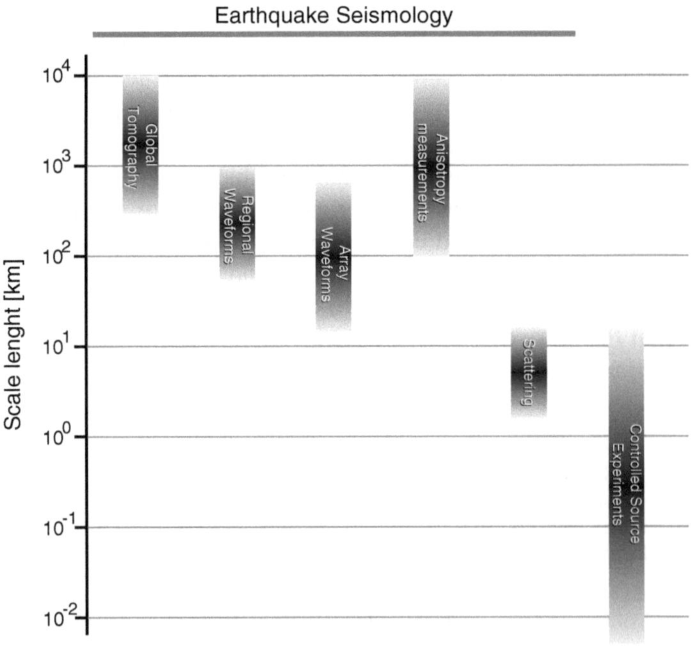

Fig. 12.1 Scale length sensitivity of different seismic probes for Earth structure. Seismic tomography resolves structures on a scale of a few hundred to a few thousand kilometers. Detailed waveform analysis using networks and arrays gives higher resolution information about Earth structure on scales as short as a few tens of kilometers. Scattering is sensitive to structure on the order of the seismic wavelength, which can reach a few kilometers in the short-period wavefield. Large-scale controlled source experiments can reach much higher resolution in the crust, but are impractical for studying the Earth's deep interior

controlled source experiments will likely always be beyond what can be achieved using global earthquake data, and only under particular circumstances is earthquake seismology able to resolve structures in the Earth's interior on scale lengths of tens of kilometers. On the other hand, geochemical analysis of mantle and crustal material typically reports heterogeneities in structure on scales as small as 10 m or less (Albarede 2005), while mineral physical studies report heterogeneities on scales comparable to the grain size and below (Stixrude and Lithgow-Bertelloni 2012). Clearly, there is a large discrepancy between the structures resolved by these disciplines studying the Earth's interior. Our understanding of the processes acting in the deep Earth and the dynamics and evolution of our planet is severely hampered by these fundamental differences in resolution.

Over the last few decades, global earthquake seismology has been very successful in mapping 3D deviations from the radially averaged 1D velocity structure that was established since the mid-twentieth century (Jeffreys and Bullen 1940; Bullen 1949; Dziewonski and Anderson 1981; Kennett and Engdahl 1991). These tomographic studies use the inversion of travel times (and to lesser extent waveforms) of seismic energy traveling along ray paths determined by the large-scale seismic velocity structure (Woodhouse and Dziewonski 1984). Nonetheless, the seismic wavefield, especially at higher frequencies, shows evidence for seismic energy not following the ray paths prescribed by the radial or large-scale seismic velocity structure. This energy, which is scattered into directions off the source–receiver plane at small-scale velocity or density variations, is most evident in the coda following the main arrivals or in dominant energy preceding certain arrivals. Codas trail the ray path traveling energy for several 10–100 s (Astiz et al. 1996; Rost et al. 2006) due to the longer path travelled by the scattered energy and seem to be incoherent between individual stations. Scattering at small-scale heterogeneities is dominant in the crust, and analysis of the coda of direct waves has widely been used to study crustal and lithospheric structure (Aki 1969; Korn 1988; Sato 1988). Nevertheless, seismic scattering has been shown to be a useful tool for studying the fine-scale structure of the Earth's deep interior (Shearer 2007).

Plate tectonics is constantly producing chemical heterogeneity through the generation of oceanic crust and the residual depleted mantle material, together forming the oceanic lithosphere, at mid-oceanic ridges and the subsequent recycling of the lithosphere into the Earth's mantle at subduction zones. The chemical heterogeneity between the crustal and mantle parts of the lithosphere is obvious from seismic studies of the crust, and seismic tomography indicates that the subducted material is traveling through the upper mantle, entering the lower mantle (Van der Hilst et al. 1991; Christensen and Hofmann 1994; Widiyantoro and Van der Hilst 1997; Li et al. 2008; McNamara and Zhong 2005; Tan and Gurnis 2005), although some slabs may remain in the upper mantle and transition zone (Fukao et al. 2009). The subducted material is likely slowly mixed into the ambient mantle (Allègre and Turcotte 1986; Olson et al. 1984) or might collect in larger volumes as indicated by tomographic studies and geodynamic modeling (Christensen and Hofmann 1994; McNamara and Zhong 2005). The mixing of the chemical heterogeneities of the subducted slabs likely leads to chemical heterogeneities throughout the mantle. We expect a range of scale lengths for the heterogeneities based on the vigor of the mantle convection and the residence time of the heterogeneity (Olson et al. 1984). Such different scale lengths have been detected in seismic studies as discussed below. Geochemical and mineral physical analysis of surface samples have provided some evidence for this process, but the imaging of their structure in situ is essential for our understanding of the evolution, composition, and dynamics of the mantle. Analysis of the scattered seismic wavefield seems a good tool to detect, characterize, and map the chemical heterogeneities in the mantle and a start in closing the gap between seismology and other deep Earth geodisciplines.

The heterogeneities in the Earth's deep interior are likely much weaker than those found in the crust, and separating the weak scattering from the Earth's interior from the effects of the seismic wavefield traveling through the strong heterogeneities of the lithosphere beneath the station is difficult. Some parts of the seismic wavefield (Fig. 12.2) are particularly suited to studying deep Earth scattering. For these specific ray geometries, the scattered energy from the deep Earth can arrive as precursors to the main arrival traveling in the source–receiver plane and is therefore separated from the lithospheric scattering that dominates the coda after the main arrival. Therefore, these seismic phases provide valuable probes for examining the small-scale structure of the Earth's interior, as discussed in the next section. Here, we focus on scattering related to the *P*-wavefield. Although scattering has also been detected in a limited number of *S*-wave studies (Shearer 2007), their ability to detect the fine-scale structure of the Earth is limited due to the typically longer periods of teleseismic *S*-waves compared to *P*-waves, making the probes sensitive to different scale lengths.

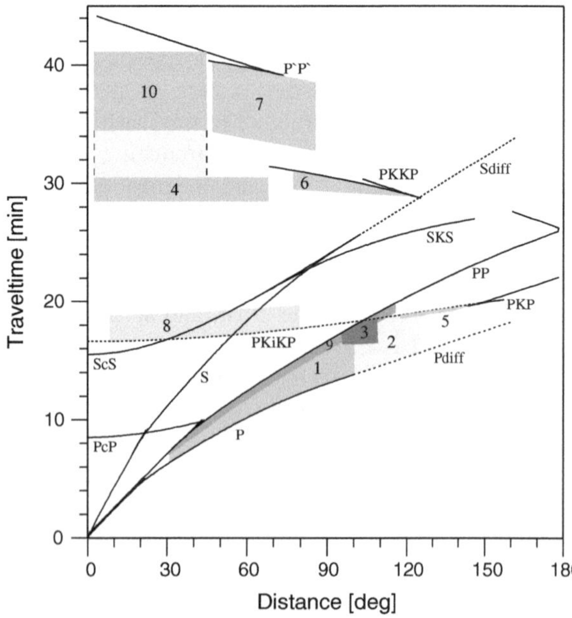

Fig. 12.2 Travel-time curves of common seismic body waves and theoretical time distance regions for several scattering probes: *1 P*-coda, *2* P_{diff} coda, *3* asymmetric *PP* precursors, *4 PK•KP* (precursors), *5 PKP* precursors, *6 PKKP* precursors, *7 P′P′* precursors, *8 PKiKP* coda, *9* symmetric *PP* precursors *10 P′•P′*. After Shearer (2007)

12.2 Scattering Probes

Due to the weak heterogeneity in the lowermost mantle compared to the lithosphere and the related change in the amplitude of the scattered energy, it is important to be able to separate the energy coming from these two scattering regions. To this end, several probes have been developed since the 1970s that allow good sampling of the lower mantle and close to the CMB.

12.2.1 PKP Precursors

PKP is a compressional wave that travels from the earthquake to the station as a *P*-wave through the mantle and core. The seismic velocity reduction at the CMB leads to a specific ray geometry where energy from the PKP_{bc} and PKP_{ab} branches, which travel through the outer core, is scattered close to the CMB and arrives as precursory energy to the *PKP* branch traversing the inner core (PKP_{df}). Ray paths for these branches are shown in Fig. 12.3a. Only scattering from heterogeneities in the deep Earth will arrive as PKP_{df} precursors. Due to the location of the *PKP* *b*-caustic, the point where the PKP_{bc} and PKP_{ab} paths merge, scattering up to approximately 1200 km above the CMB will produce precursors to PKP_{df} arriving up to 18 s before *PKP* and in an epicentral distance range from 118° to approximately 145°. These high-frequency arrivals have long been observed in the seismic wavefield (Gutenberg and Richter 1934), and their scattering origin was identified by Cleary and Haddon (1972). They have been used extensively to study scattering at the CMB to infer the structure of small-scale heterogeneities in the D″ region (Doornbos and Husebye 1972; Doornbos and Vlaar 1973; King et al. 1976; Bataille and Flatte 1988; Bataille et al. 1990; Hedlin et al. 1997; Thomas et al. 1999; Cormier 2000; Hedlin and Shearer 2000; Margerin and Nolet 2003; Cao and Romanowicz 2007; Vanacore et al. 2010). Heterogeneities higher above the CMB will not create precursors to PKP_{df} but will lead to energy with a longer travel time arriving as postcursors in the PKP_{df} coda. These are more difficult to analyze due to the simultaneously arriving energy scattered in the crust leading to a certain ambiguity in results (Hedlin and Shearer 2002) making the study of mid- and upper-mantle heterogeneity with this probe difficult.

Results have been interpreted in terms of CMB topography (Doornbos 1978) or small-scale heterogeneity close to the CMB (Haddon and Cleary 1974). Heterogeneities have been found to consist of ~1 % root-mean-square (RMS) velocity variations at scale lengths of about 8 km in a layer of several hundred kilometers thick (Bataille and Flatte 1988; Hedlin et al. 1997), although there is some trade-off between the thickness and the heterogeneity strength and recent studies report a much smaller RMS variation (Mancinelli and Shearer 2013). Alternatively, the scattering has been attributed to CMB topography with RMS

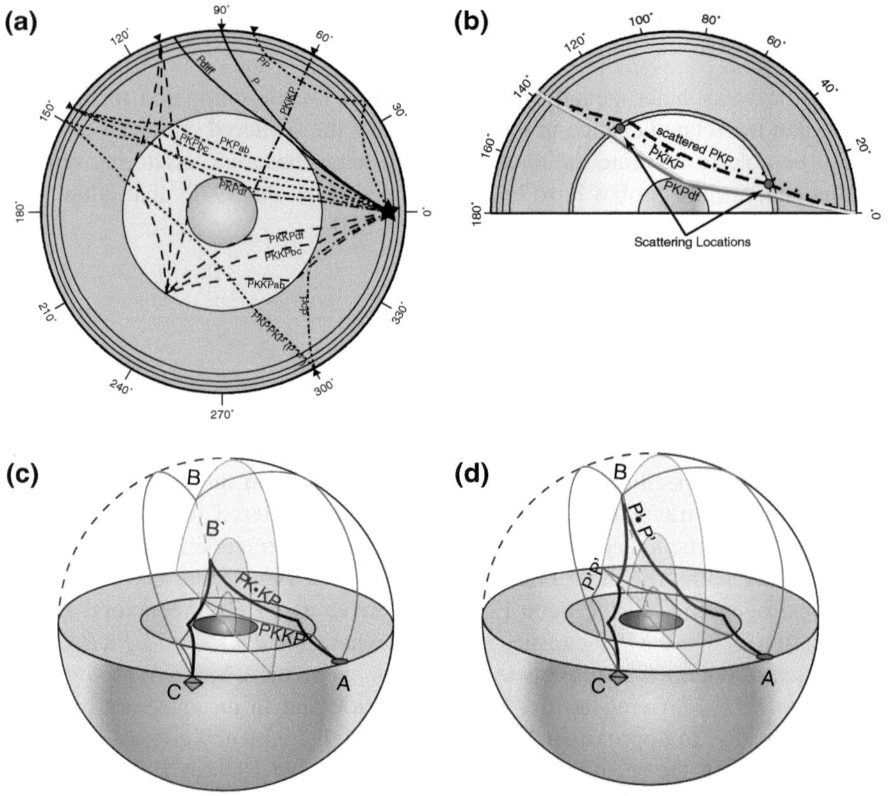

Fig. 12.3 **a** Ray paths of *PKP*, *PKiKP*, *PP*, *PKKP*, $P'P'$, P/P_{diff}, and *PcP*, all of which, have been used to study scattering in the Earth's interior. Source (star) is located at 500 km depth, and receiving stations are indicated by inverted triangles. **b** Scattered ray path related to *PKP*, producing precursors to PKP_{df}, an often-used probe for lower mantle scattering. **c** Scattered ray path (*dark solid*) related to *PKKP* (*light gray* ray path) with a scattering point (B') on the CMB at the *PKKP* CMB reflection point. Point *B* marks the surface projection of the scattering point B'. *PK•KP* scattered energy is traveling off-azimuth. The globe is cut along the *PKKP* ray path traveling on-azimuth [i.e., along the great-circle path connecting source (*A*) with receiver (*C*)]. Please note that a second mirrored scattering point can be found in the lower hemisphere. **d** Scattered ray path (*dark* ray path) related to *PKPPKP* ($P'P'$) (*light gray* ray path). Scattering is indicated near the surface at point *B*, and the scattered energy is traveling off-azimuth. The globe is cut along the *PKPPKP* ray path traveling on-azimuth between source (*A*) and receiver (*C*). Please note that a second mirrored scattering point can be found in the lower hemisphere

height of ~300 m (Bataille et al. 1990). A second group of studies located discrete scatterers at the CMB rather than treating the scattering as a statistical process (Thomas et al. 1999; Frost et al. 2013). Resolving heterogeneities high above the CMB with *PKP* is challenging, but this gap can be closed using other phases as described below.

12.2.2 *P* and P_{diff} Coda

At distances larger than about 98°, the direct *P*-wave starts to diffract around the low-velocity zone of the outer core (Fig. 12.3a). This diffracted energy is typically called P_{diff} and can be observed, in the high-frequency seismic wavefield, out to at least 115°, although it might disappear earlier (Astiz et al. 1996; Rost et al. 2006) and some observations out to 130° have been reported likely due to wave guide effects (Bataille et al. 1990; Earle and Shearer 2001). P_{diff} is often followed by long coda energy trailing the main arrival for several hundred seconds (Astiz et al. 1996; Rost et al. 2006). At large ranges, the arrivals seem emergent, reaching maximum amplitudes up to tens of seconds after the main *P* or P_{diff} arrival. The length of the coda and its emergent onset support the interpretation of the P_{diff} coda originating from multiple scattering along the diffracted path (Bataille et al. 1990). A global stack of short-period seismograms from shallow events by Earle and Shearer (2001) found evidence for 1 % RMS velocity variations on scale lengths of 2 km, smaller than generally resolved by *PKP* studies. Strong regional variations in the strength of the coda have also been observed (Rost and Thorne 2010) indicating lateral variations in scattering strength.

12.2.3 *PKKP* and *P′P′*

Core phases other than *PKP* have been used to study scattering in the deep Earth. Scattering related to *PKKP*, a core phase reflected once off the underside of the CMB (Fig. 12.3a), has been identified in several distance ranges. Early studies identified precursors to $PKKP_{df}$ between ~80° and 125° (Doornbos 1974; Earle and Shearer 1997) and have been interpreted as scattering at the CMB entry or exit points (*PKK•P* or *P•KKP*, where • denotes the location of the scattering along the ray path). The scattering process is analogous to the scattering of PKP_{bc} or PKP_{ab} leading to PKP_{df} precursors as described above with an additional reflection of the energy from the underside of the CMB. In the case of *PKKP*, most studies point toward scattering from small-scale CMB topography or roughness with amplitudes of 250–350 m on lateral scales of 7–10 km (Earle and Shearer 1997). *PKKP* scattering at distances shorter than 80° has also been detected in global stacks (Earle and Shearer 1998). The origin of this energy has later been identified as scattering of *P* to *PKP* (*P•PKP*) at the Earth's surface (Earle 2002).

An additional scattering mechanism for *PKKP* has been identified by Earle (2002) as scattering of *PKKP* at the CMB reflection point (*PK•KP*). This scattering is related to off-azimuth scattering of $PKKP_{bc}$ off volumetric heterogeneities at or above the CMB (Fig. 12.3c). This scattering geometry has been used to map lateral heterogeneities of scattering strength at the CMB and to find evidence for discrete heterogeneities likely related to subduction processes and heterogeneities at the edges of the large low shear velocity Provinces (LLSVPs) beneath the Pacific and Africa (Rost and Earle 2010).

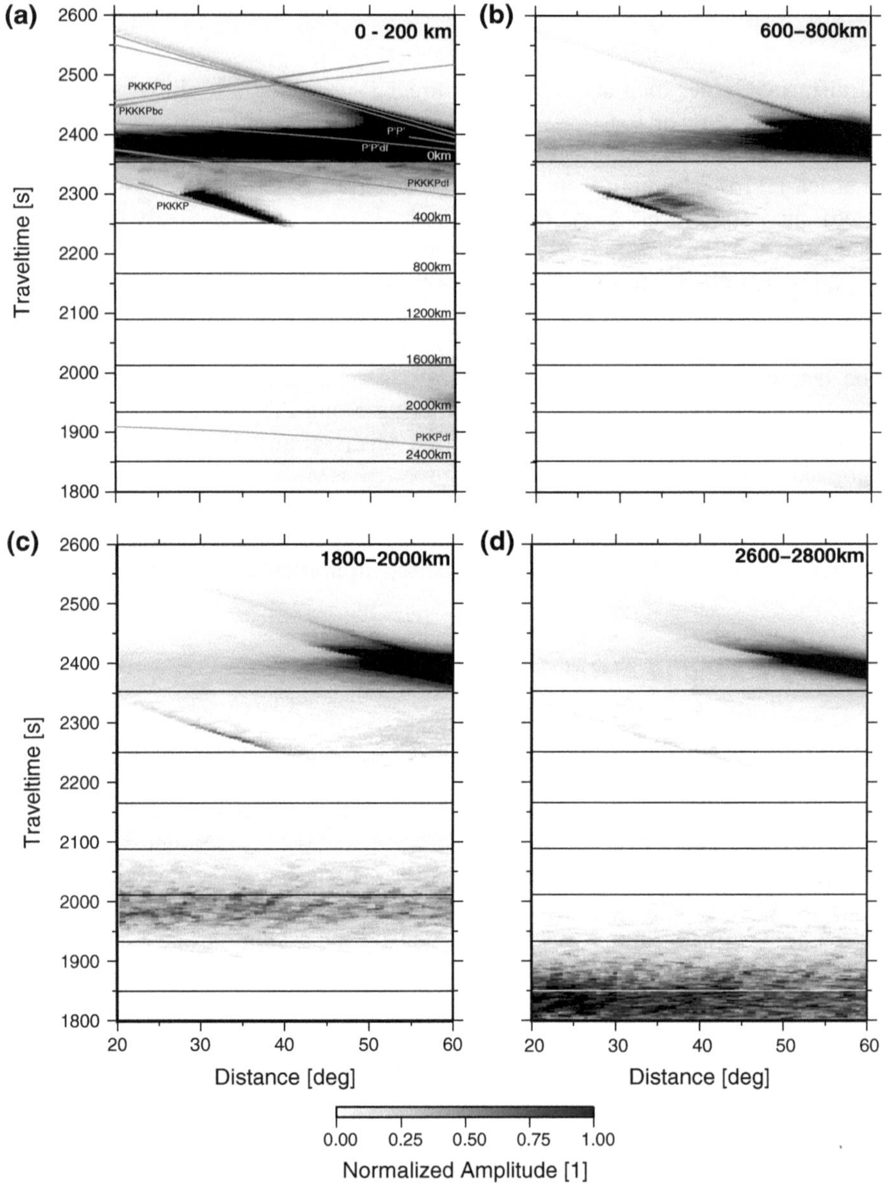

Fig. 12.4 Synthetic wavefields calculated using a Monte-Carlo phonon scattering approach (Shearer and Earle 2004, 2008). The wavefield runs from 1800 to 2600 s after the earthquake origin time and covers a distance range from 20° to 60°. Some of the main arrivals including different branches of *PKKKP* and *P′P′* (*PKPPKP*) are marked in **a**. Amplitudes are normalized to the maximum amplitude in each individual panel. Horizontal lines mark the minimum arrival times for scattering related to *P′P′* (*P′•P′*) from the surface to the D″ region. **a** Scattering in a 200 km thick layer ranging from 0 to 200 km depth with RMS velocity variations of 3 % with 4 km correlation length characterized in an exponential autocorrelation model. **b** As **a**, but for a layer ranging from 600 to 800 km depth. **c** As **a** but for a layer ranging from 1800 to 2000 km depth. **d** As **a** but for a layer ranging from 2600 to 2800 km depth. Variations in scattering amplitude are due to the normalization of each panel

PKPPKP (*P*′*P*′) is similar to *PKKP* as it is an underside reflection off the Earth's surface (Fig. 12.3a), while *PKKP* is an underside reflection off the CMB. Scattering related to *P*′*P*′ has been identified in near-podal array data (Tkalčić et al. 2006), and its timing indicates a source in the upper mantle between 150 and 220 km depth.

It has been shown that the time window from 2200 to 2500 s after the earthquake origin time for distance ranges of 30°–50° contains scattered energy related to the phase *P*′*P*′. Earle et al. (2011) identify this energy as off-great-circle path scattering of PKP_{bc} to PKP_{bc} (*P*′•*P*′). The travel time of *P*′•*P*′ is dependent on the scattering depth (Fig. 12.4) with energy scattered closer to the CMB arriving earlier in the seismogram (Earle et al. 2011). This makes *P*′•*P*′ an ideal probe to study small-scale heterogeneity from crust to core and a good supplement to scattering studies using *PKP* for which energy scattered at heterogeneities more than about 1000 km above the CMB arrives in the PKP_{df} coda, making the analysis of this energy more complex (Hedlin and Shearer 2002). With this scattering mechanism, *PK*•*KP* is an extreme example of lowermost mantle scattering of *P*′•*P*′ explaining the sensitivity of *PK*•*KP* to structure above the CMB. Here, we show evidence for mapping of whole mantle scattering using *P*′•*P*′ in the next section.

12.2.4 Other Probes (P, PcP, PP)

Other phases have been used to delineate heterogeneities at different depths throughout the Earth. Arrivals in the *P*-coda, identified as *S*-to-*P* conversions close to the source or *P*-to-*P* scattering off azimuth, have been used to map structure in the mid-mantle through analysis of array data (Castle and Creager 1999; Kaneshima and Helffrich 1999, 2003; Kaneshima 2009; Kaneshima et al. 2010; Kito et al. 2008). These detections are generally in the vicinity of current or recent subduction, outlining the penetration of slabs into the lower mantle and the recycling or deposition of crustal material in the lowermost mantle.

A similar result has been obtained by studying precursors to *PP*. The *PP* precursor wavefield shows evidence for symmetric reflections off the upper-mantle discontinuities (Shearer 1990) and for asymmetric reflections off the surface or uppermost mantle (Wright 1972; King et al. 1975; Weber and Wicks 1996). Asymmetric off-azimuth reflections off slab material arriving as *PP* precursors have been identified beneath the Mariana and Izu-Bonin subduction zones (Rost et al. 2008; Bentham and Rost 2014), outlining deep subduction beneath these subduction zones and the transport of oceanic crust far into the lower mantle. Other phases such as *PcP* have been used to image lower mantle structure through the analysis of *P*-to-*P* and *S*-to-*P* scattering (Braña and Helffrich 2004).

The outer core seems well mixed and does not show evidence for scattering away from the boundaries between mantle and core (CMB) and inner and outer core (ICB). On the other hand, there is strong evidence from the analysis of *P*-waves traversing the inner core for scattering in the inner core indicating heterogeneities of up to 1.2 % in velocity and scale lengths of 2 km in the outermost

300 km (Cormier 2007; Koper et al. 2004; Leyton and Koper 2007; Poupinet and Kennett 2004; Rost and Garnero 2004; Vidale and Earle 2000; Vidale et al. 2000). These heterogeneities are likely due to inclusion of melt in the solid core, compositional changes, or changes in the anisotropy of the core material (Vidale and Earle 2000).

12.3 *P′•P′/PK•KP* and Earth Structure

Scattering related to *P′P′* (*P′•P′*) has been identified in data of the Large Aperture Seismic array (LASA), an array of up to 625 station with an aperture of 200 km that operated in Montana from 1965 to 1978 (Earle et al. 2011). The energy, arriving at epicentral distances of 30°–50° and at times between 2300 and 2450 s after the earthquake origin, has been interpreted as off-azimuth scattering of PKP_{bc} to PKP_{bc} (Earle et al. 2011). The scattered energy recorded at the array shows a distinct off-azimuth energy pattern similar to the precursors to *PKKP* described earlier (Earle 2002; Rost and Earle 2010). Note that *P′•P′* differs from *P′P′* precursors (Fig. 12.2), which are related to underside reflections off discontinuities and through scattering at the surface reflection point and arrive primarily along the great-circle path (Cleary and Haddon 1972; Haddon et al. 1977; Chang and Cleary 1981; Cleary 1981; Tkalčić et al. 2006). The energy of *P′•P′* travels off-azimuth (i.e., the azimuth connecting source and receiver) around the inner core leading to up to two characteristic energy peaks (depending on the existence of heterogeneity in the potential scattering region) with nonzero transverse and radial slowness, indicating a CMB, lower mantle, or outer-core origin (Fig. 12.5a). The characteristic energy pattern derived using ray theory can also be identified in simulations of global seismic scattering. Figure 12.5b, c shows examples of scattered energy in a time window from 1950 to 2000 s for distance ranges of 30°–35° (Fig. 12.5b) and 40°–45° (Fig. 12.5c). Synthetics are calculated using a multiple scattering, Monte-Carlo phonon scattering approach (Shearer and Earle 2004), and the resulting energy peaks are in good agreement with the ray theoretical calculations (Earle et al. 2011).

Here, we detect the energy related to *P′•P′* in data of the medium aperture array in Yellowknife (YKA) located in northern Canada. YKA consists of 18 short-period, vertical stations deployed along two perpendicular north–south and east–west oriented legs. The maximum aperture of the array is 20 km with an interstation spacing of 2.5 km. YKA was designed to detect high-frequency *P*-waves from underground nuclear explosions (Manchee and Weichert 1968; Weichert and Whitham 1969) and is therefore well suited to study the scattered *P*-wavefield. Additionally, YKA contains 4–5 three-component broadband seismometers, but the data from these instruments are not being used in this study. Earle et al. (2011) identified *P′•P′* in data from LASA consisting of more than 600 seismometers and with an aperture of up to 200 km (Frosch and Green 1966;

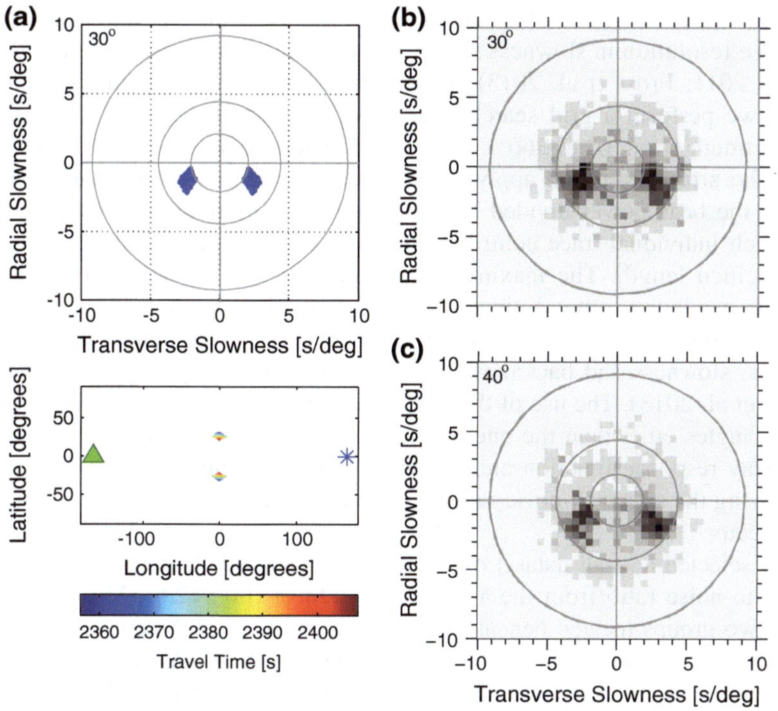

Fig. 12.5 **a** Theoretical calculation of expected slowness patterns for $P'\bullet P'$ (*top*) and travel times of scattered energy from two locations symmetrically off-azimuth (*bottom*). *Color scale* indicates travel time for scattering energy within the scattering regions. For scattering regions see also Fig. 12.3. Travel times are calculated for scattering at the surface and for a distance of 30° between source (*star*) and receiver (*triangle*). *Blue areas* in *top panel* indicate the slowness region where scattered energy for this source-receiver constellation can originate from. **b** Slowness pattern of scattered energy calculated for a distance range from 30° to 35° and in the time windows from 1950 to 2000 s after origin. Synthetics have been calculated using a Monte-Carlo phonon scattering approach (Shearer and Earle 2004). The model includes heterogeneities with RMS velocity variations of 3 % with an exponential autocorrelation length of 4 km in a 200-km-thick layer between 1800 and 2000 km depth. Two energy lobes in agreement with the theoretical calculations are visible. The difference in travel times of the scattered energy relative to **a** is due to the different scattering depth (see Fig. 12.4). **c** As in **b** but for an epicentral distance from 40° to 45°. The panels in **b** and **c** indicate the slight changes in slowness pattern of the scattered energy

Green et al. 1965). LASA therefore has much better resolution in wave number space and provides greater improvements in the signal-to-noise ratio of coherent arrivals. Here, we show the capability of smaller arrays to detect $P'\bullet P'$, opening up many more opportunities to use this energy to study Earth structure. Further testing will see whether smaller arrays than YKA (with apertures of 3–10 km, which have been built as part of the Comprehensive Test Ban Treaty) are able to detect the weak scattered energy related to $P'\bullet P'$.

To enhance detection of coherent signal power in the stacked array data and increase resolution in slowness space, we apply the *F*-statistic to the array beams (Selby 2011; Frost et al. 2013). To detect the weak scattered energy related to $P'{\bullet}P_,'$ we perform a grid search over slowness (ranging from 0 to 13 s/°) and backazimuth (from 0° to 360°). For each slowness/backazimuth combination, we form the array beam and apply the *F*-statistic, which is a dimensionless measure of the beampower divided by the variability (difference) between the beam and each individual trace contributing to the beam average over a time window of specified length. The maximum amplitude of this *F*-trace is then measured in a time window representing a specific scattering depth as discussed below. The resulting slowness/backazimuth *F*-trace beampower maps (*F*-packs) indicate the slowness and backazimuth of the incoming energy with high precision (Frost et al. 2013). The use of the *F*-statistic penalizes energy traveling with incidence angles other than the one the beam is calculated for, therefore sharpening the array response function and the wave number resolution of the array, while improving the signal-to-noise ratio of the energy traveling along the correct slowness vector.

We selected a small dataset of events with magnitude larger than 6.5 and good signal-to-noise ratio from the YKA data holdings (Fig. 12.6). Most earthquakes form two groups located beneath Kamchatka and Central America. We apply the *F*-trace stacking approach to 50-s time windows of the YKA data. Each time window represents an approximately 200-km-thick layer in the Earth. We move our analysis from the surface to the CMB (see Table 12.1 for chosen time windows). As can be seen in Fig. 12.4, each of these time slices represents scattering from a specific depth, making $P'{\bullet}P'$ an ideal probe to study the small-scale heterogeneity

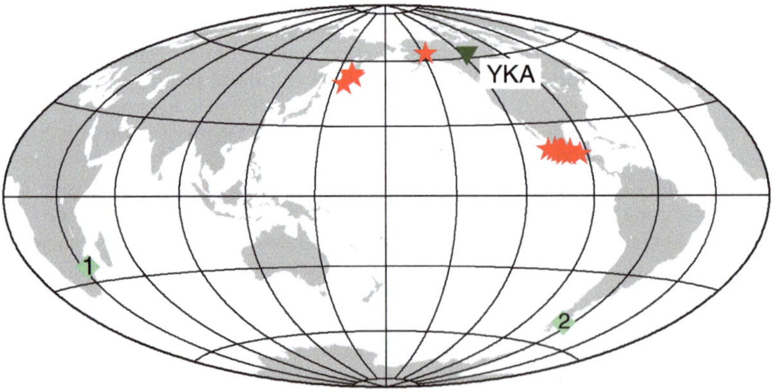

Fig. 12.6 Location of events (*stars*) analyzed in this study. The events fall into two general areas with events beneath Kamchatka and Central America. *Diamonds* show scattering regions of example event (January 28, 2002, 13:50) shown in Fig. 12.7 with the *green diamond* marked *1* showing the location in the deep mantle and the upper mantle scattering point is shown as *green circle* marked *2*

Table 12.1 Minimum travel times for $P'{\bullet}d{\bullet}P'$ scattered phases for varying scattering depths for the one-dimensional Earth model PREM (Dziewonski and Anderson 1981)

Scattering depth (km)	PREM	
	PKP caustic distance (°)	Minimum $P'dP'$ travel time (s)
0	145.05	2353.48
200	144.59	2300.36
400	144.05	2252.28
600	143.38	2208.40
800	142.55	2167.90
1200	140.49	2090.36
1600	137.87	2014.08
2000	134.44	1938.04
2400	129.46	1854.36
2600	126.14	1809.50
2800	121.40	1756.14

Time indicates absolute travel time after the origin for a surface focus. Also given is the PKP_b caustic distance for the appropriate scattering depth

from crust to core. The scattered energy from all these time windows arrives in a quiet time window with few interfering phases (*PKKKP*, with two CMB underside reflections, is the only major phase arriving in the $P'{\bullet}P'$ scattering time window, also see Figs. 12.2 and 12.4 and might be visible as on-azimuth arrival in the 200–400 km and 400–600 km depth slices in Fig. 12.8). Figures 12.7 and 12.8 show the resulting energy maps for the packed *F*-beams for the individual time windows for two example events. The energy maps show discrete and coherent $P'{\bullet}P'$ arrivals for the upper mantle (down to depths of about 400 km) and in the lowermost mantle to depths of about 900 km above the CMB (below 2000 km depth) with slowness patterns that are in excellent agreement with the ray theoretical and phonon scattering synthetics. The example event in Fig. 12.8 shows the typical two-lobed scattered energy in the uppermost mantle slice. We interpret this energy as detection of $P'{\bullet}P'$ in the YKA data. *F*-packs from the mid-mantle indicate little scattered $P'{\bullet}P'$ energy from depths of about 600–2000 km, as indicated by the absence of a focussed arrival (as, e.g., in the shallower depth slices) and a more incoherent wavefield indicating varying slowness and backazimuth in the processed time window. Tests with noise windows before the *P* arrival for these events show a random behavior of the energy maximum in these time windows, while the *F*-packs between 600 and 2000 km depth often show a predominant direction in agreement with scattering and outer core slowness, but indicating very low scattered energy. This might indicate that there is some scattered energy arriving in these time windows, but is much weaker than at the surface and in the lower mantle. Detailed modeling of this energy related to the YKA noise condition will lead to further insight into the strength of the heterogeneities at these depths but is beyond the scope of this study. The dataset studied here consists of seismicity

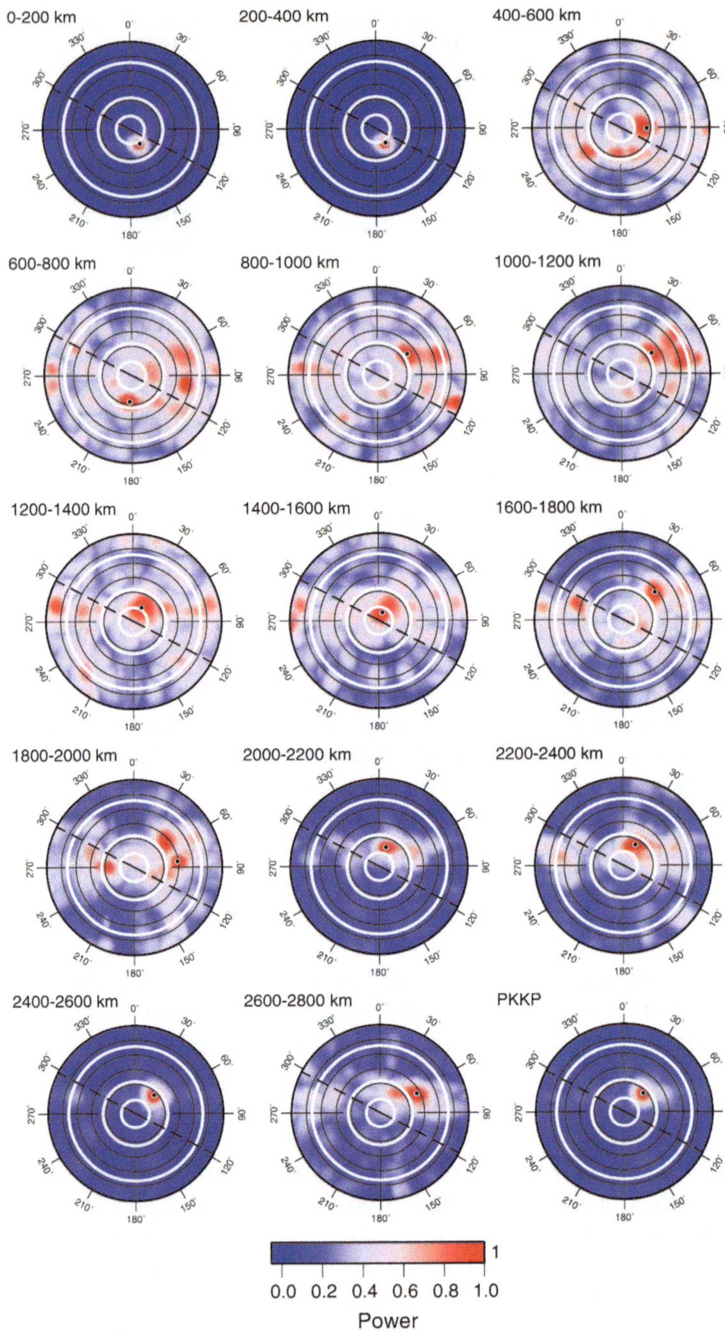
0-200 km
200-400 km
400-600 km
600-800 km
800-1000 km
1000-1200 km
1200-1400 km
1400-1600 km
1600-1800 km
1800-2000 km
2000-2200 km
2200-2400 km
2400-2600 km
2600-2800 km
PKKP
0°
30°
60°
90°
120°
150°
180°
210°
240°
270°
300°
330°
1
0.0 0.2 0.4 0.6 0.8 1.0
Power

◀ **Fig. 12.7** *F*-packs, examples for the event January 28, 2002, 13:50 in the Kamchatka region recorded at YKA. Epicentral distance for this event is 47.7°. In each *F*-pack plot, slowness is shown on the radial and backazimuth of the incident energy on the azimuthal axes. *Black circles* indicate slowness at intervals of 2 s/°. *Thick white circles* indicate slowness of 1.9, 4.4, and 9.2 s/°, indicating energy originating from the inner-core boundary, core–mantle boundary, and 660-km discontinuity, respectively. Colors indicate beampower of the *F*-beams, with each panel being normalized individually. Focussed, high beampower (as in slices for 0–200 and 200–400 km) indicates very coherent energy arriving from the related slowness and backazimuth, while distributed (non-focussed) energy from different slowness and backazimuths indicates the arrival of uncorrelated noise (as in slices from 400 to 2000 km depth). *Dashed black lines* indicate the theoretical backazimuth for this event of 297.9°. Depth slice time windows relate to Table 12.1, and the *PKKP*-labeled time window contains energy originating from the CMB (time window between 1651 and 1751 s)

beneath Kamchatka and Central America. We find that in general, the events in the Kamchatka region show evidence for scattering throughout most of the mantle (although with apparently weaker scattering in the mid-mantle), while the Central American events show only evidence for scattered energy from the upper mantle (see Fig. 12.9 for an example). This might indicate lateral variations in heterogeneity structure in the lower mantle as has been indicated before (Hedlin and Shearer 2000). Note that the lower mantle scattering (e.g., Fig. 12.7) shows different slowness and backazimuth parameters than the upper-mantle scattering in the same event. The different directivity for the scattered energy indicates different locations of the heterogeneities leading to the scattered energy. Using the slowness/backazimuth information and raytracing through a 1D Earth model allows us to infer the location of these heterogeneities. The lower mantle scattering originates from the edge of the large low shear velocity province (LLSVP) beneath Africa (scattering region 1 in Fig. 12.6 (see also Fig. 12.10)), a region of strong reductions in seismic velocities beneath Africa. Another LLSVP can be found beneath the Pacific, and combined, they form a strong degree 2 pattern of low velocities in most tomographic models (Garnero and McNamara 2008). Strong scattering from the edge of the LLSVP has been reported earlier (Wen 2000; Rost and Earle 2010; Frost et al. 2013). On the other hand, we can relocate the near-surface scattering to a region beneath South America (scattering region 2 in Fig. 12.6) where the upper mantle is strongly influenced by recent subduction. Several events from a similar source region support the detection of these scattering regions.

As shown in Figs. 12.7, 12.8 and 12.9, $P'\bullet P'$ offers a unique opportunity to probe the mantle, from crust to core, for small-scale heterogeneities using scattered waves. Due to the complex ray path of $P'\bullet P'$ and the location of most seismic arrays on the Northern Hemisphere, we expect the best sampling with this probe on the Southern Hemisphere. This will be a good complement to previous studies probing for heterogeneities close to the CMB using *PKP* (Hedlin and Shearer 2000).

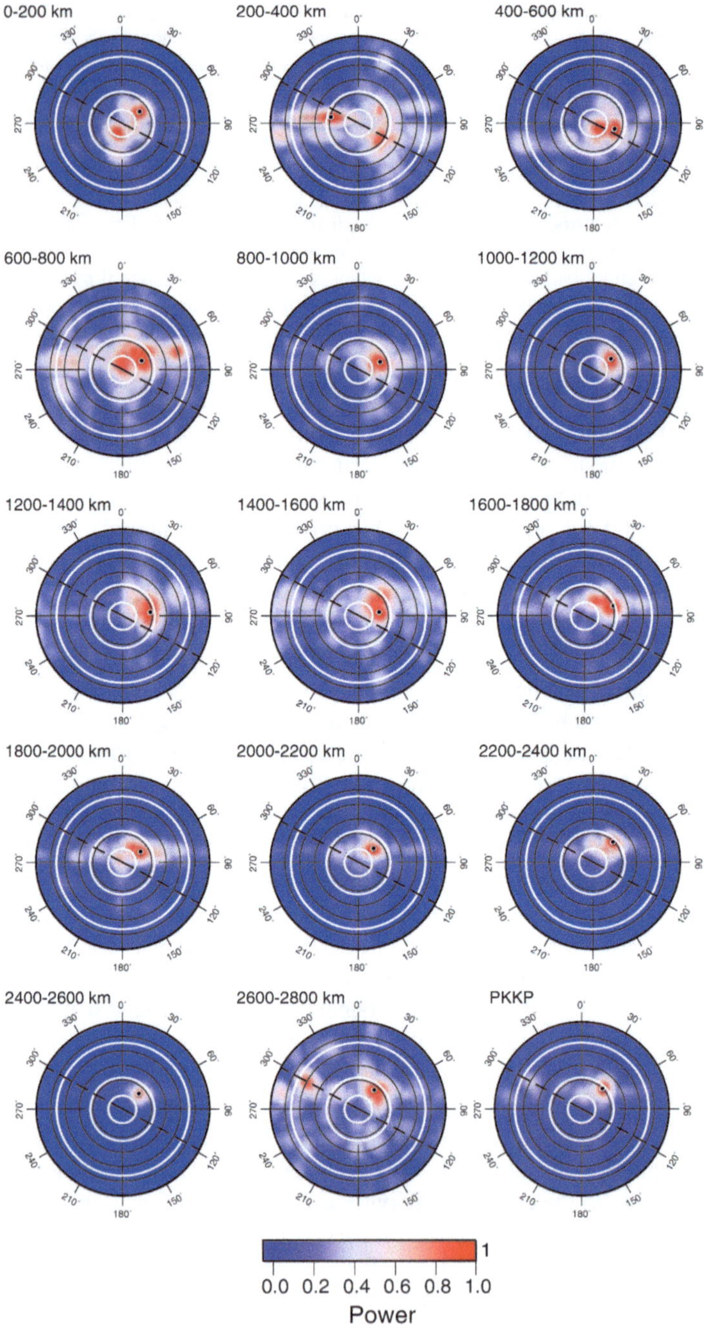
0-200 km
200-400 km
400-600 km
600-800 km
800-1000 km
1000-1200 km
1200-1400 km
1400-1600 km
1600-1800 km
1800-2000 km
2000-2200 km
2200-2400 km
2400-2600 km
2600-2800 km
PKKP
0.0 0.2 0.4 0.6 0.8 1.0
Power

Fig. 12.8 *F*-packs, examples for the event March 02, 1992, 12:29 in the Kamchatka region recorded at YKA. Epicentral distance for this event is 43.2°. In each *F*-pack, plot slowness is shown on the radial and backazimuth of the incident energy on the azimuthal axes. *Black circles* indicate slowness at intervals of 2 s/°. *Thick white circles* indicate slowness of 1.9, 4.4, and 9.2 s/°, indicating energy originating from the inner-core boundary, core–mantle boundary, and 660-km discontinuity, respectively. *Red colors* show high beampower of the *F*-beams indicating very coherent energy arriving from the related slowness and backazimuth. *Dashed black lines* indicate the theoretical backazimuth for this event of 298.3°. Depth slice time windows relate to Table 12.1, and the *PKKP*-labeled time window contains energy originating from the CMB (time window between 1651 and 1751 s)

A special case of the energy related to *P′•P′* is *PKKP* scattered energy from heterogeneities close to the CMB (*PK•KP*) as described above. This energy has been detected in both data from large aperture (Earle and Shearer 1997; Earle 2002) and medium aperture (Rost and Earle 2010) arrays. It has been used to study the lateral variations of heterogeneities along the Earth's CMB (Rost and Earle 2010). Figure 12.10 shows detected discrete scatterer locations for *PK•KP* from an earlier study (Rost and Earle 2010) and compares the locations with velocity variations derived from tomographic studies. The scatterer locations can predominantly be found at the edge of the African LLSVP, in agreement with other studies (Hedlin and Shearer 2000; Wen 2000; Frost et al. 2013) and beneath Central and South America. While the LLSVP region is seismically slow and has been interpreted as a thermochemical pile (McNamara and Zhong 2005), the other dominant scatterer region is characterized by fast seismic velocities (Fig. 12.10b). These regions are likely related to the long-lasting subduction beneath the Americas (Lithgow-Bertelloni and Richards 1998). Further scattering points, but much fewer than in these two regions, can be found in the Pacific LLSVP and in the circum-Pacific subduction ring. Together, the *P′•P′ and PK•KP* results indicate a dynamic structure of the lower mantle, where convection flows redistribute small-scale heterogeneities.

The heterogeneities from the two dominating scattering areas beneath Africa and Patagonia are likely generated by different processes. Newly introduced subducted material is likely producing the heterogeneity beneath South America. The scattering area detected beneath Patagonia agrees well with the inferred location of the Phoenix plate at the CMB (Lithgow-Bertelloni and Richards 1998). Ultra-low velocity zones (ULVZs), regions of strongly reduced seismic velocities at the CMB (Garnero 2000), have been predominantly found at the edges of LLSVPs (McNamara et al. 2010). Geodynamic modeling of the lowermost mantle indicates that dense material is likely deposited at the edges of thermochemical piles in agreement with ULVZ detection (McNamara et al. 2010). It seems possible that the heterogeneities leading to scattering at the edge of the LLSVP beneath Africa are related to dense material swept to the edge of LLSVPs by the internal convection currents of the LLSVPs. Detailed modeling of scattering from structures derived from geodynamic models might lead to better insight into these processes but is beyond the scope of this paper.

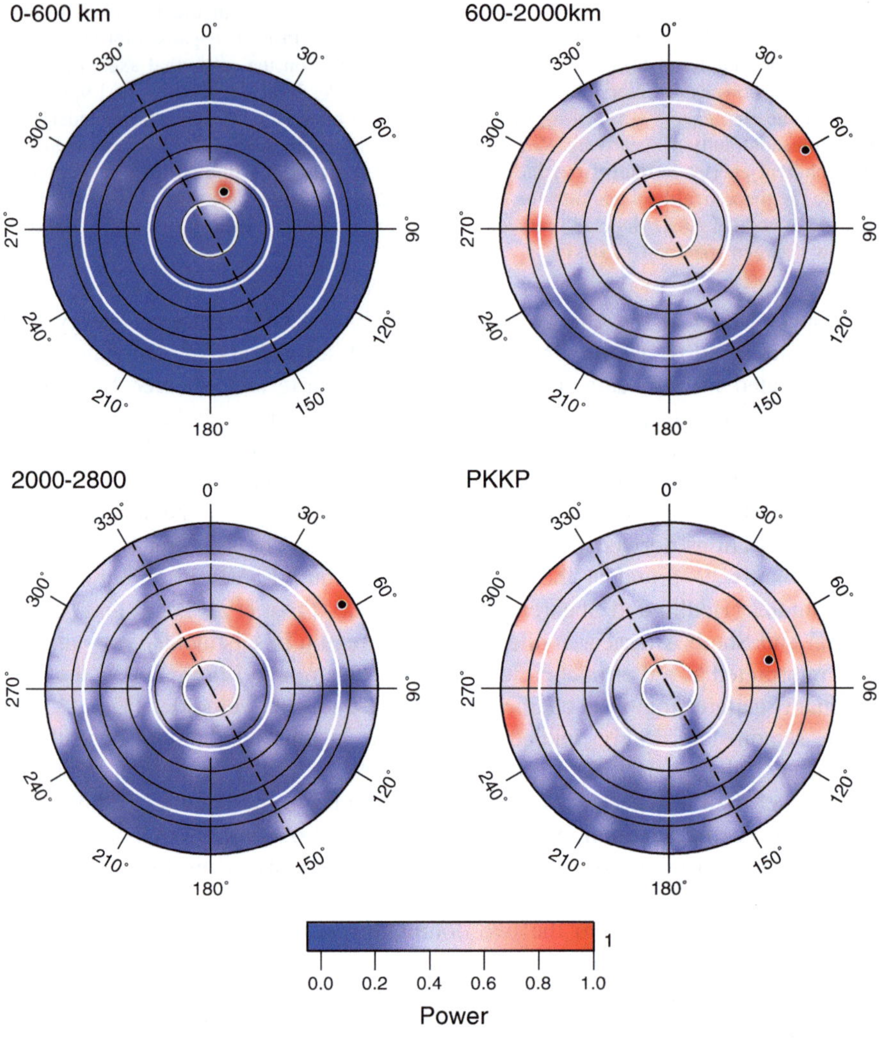

Fig. 12.9 *F*-pack results for the event October 21, 1995, 02:38 in Central America recorded at YKA with an epicentral distance of 48° and a backazimuth of 151.86° (*dashed line*). *F*-packs are shown in larger slices running from 0 to 600 km (*top left*), 600 to 2000 km (*top right*), 2000 to 2800 km (*bottom left*), and the PKKP time window (*bottom right*). *F*-packs are shown in the same way as in Fig. 12.7. Only the upper mantle time slice (*top left*) shows coherent scattering, while the other depth slices indicate mostly incoherent noise in contrast to the examples shown in Figs. 12.7 and 12.8

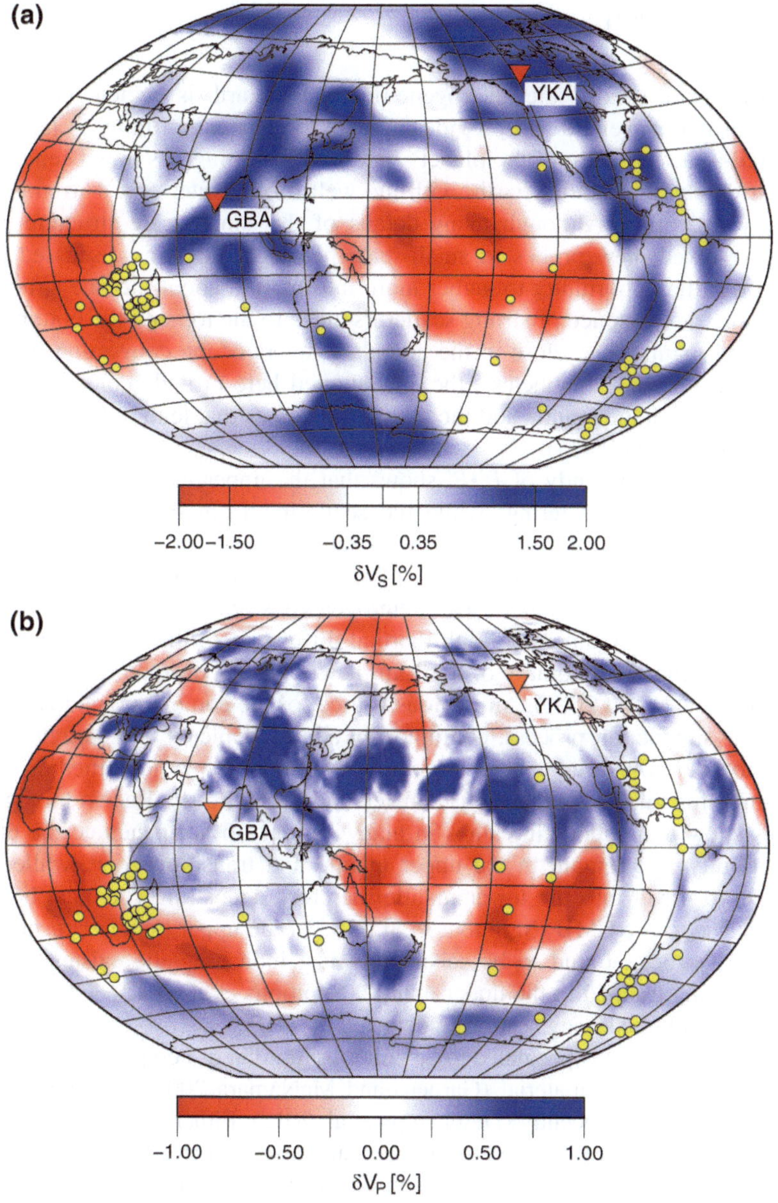

Fig. 12.10 Scatterer locations (*yellow circles*) at the CMB as found in data from 2 medium aperture arrays (*red triangles*) located in Canada (YKA) and India (GBA) (Rost and Earle 2010). **a** Background shows *S*-wave velocity at the CMB from Ritsema et al. (2011). **b** Background shows *P*-wave velocity variations from LLNL_G3Dv3 (Simmons et al. 2012)

12.4 Conclusion

Scattering from small-scale heterogeneities in the Earth is evident in the short-period seismic wavefield and provides us with the best opportunity to learn about the small-scale structure of the planet's interior. Scattering has now been detected at most depths throughout Earth, with the well-mixed outer core currently being the only exception. Although many details of this small-scale structure remain poorly resolved, new probes to small-scale heterogeneities, as discussed in this work, offer the opportunity to gather more information about the fine structure of the Earth. The existence of strong scatterers close to the top and the bottom of the convecting mantle, i.e., close to the boundary layers of the convection system, is probably expected. Nonetheless, the data studied here provide constraints on the dominant flow in the mantle and the mixing processes that tend to homogenize the mantle (Olson et al. 1984).

This preliminary study of $P'{\bullet}P'$ shows that the upper mantle and lower mantle produce strong scattering, while the scattering from the mid-mantle seems much weaker. There is also evidence for lateral variations in the heterogeneity structure as evidenced by the absence of scattering from the lower mantle in the Central American earthquakes. The heterogeneities leading to the scattering are likely compositional since the high thermal diffusivity in the solid mantle would equalize the thermal anomaly of a 10 km heterogeneity on the order of 200 kyrs (Helffrich and Wood 2001). An obvious source for the compositional heterogeneities is the subduction process, continuously adding depleted and enriched crustal and lithospheric material to the bulk mantle, which has been identified as a possible source for seismic scattering (Helffrich and Wood 2001). Recent tomographic images show that subducting slabs in many regions do not penetrate the 660-km discontinuity and remain in the upper mantle and transition zone, while others continue to travel into the lower mantle therefore depositing chemical heterogeneities close to the CMB (Simmons et al. 2012). The upper mantle beneath South America has been found to be seismically fast indicating the existence of slab material being trapped in the upper mantle (Simmons et al. 2012). This is in good agreement with the strong scattering detected by $P'{\bullet}P'$.

LLSVPs have been speculated to be either primordial or being recharged through the subducted crustal material (Garnero and McNamara 2008) although the latter has been questioned recently (Deschamps et al. 2012). Either model predicts strong structure in the LLSVPs related to the internal convection (McNamara et al. 2010). This is in good agreement with the laterally varying scattering strength close to the CMB as found in the *PK•KP* data. Although topography on the CMB has been used to explain the scattering in earlier studies (Doornbos 1978), it has been found that it cannot explain global *PKKP* scattering (Earle and Shearer 1997) and cannot explain the $P'{\bullet}P'$ scattering located away from the CMB presented here.

The new probes to small-scale heterogeneities, $P'{\bullet}P'$ and *PK•KP*, presented here and the rapid increase in dense array deployments will allow a systematic exploration of much of the Earth's mantle for small-scale heterogeneities. Using the seismic observations of scattering in the high-frequency wavefield and refined thermochemical convection models will provide further insight into the generation, distribution, and destruction of chemical heterogeneities from mixing processes in the mantle.

References

Aki K (1969) Analysis of seismic coda of local earthquakes as scattered waves. J Geophys Res 74(2):615–631

Albarede F (2005) The survival of mantle geochemical heterogeneities. Earth's Deep Mantle Struct Compos Evol 160:27–46. doi:10.1029/160GM04

Allègre CJ, Turcotte DL (1986) Implications of a two-component marble-cake mantle. Nature 323(6084):123–127. doi:10.1038/323123a0

Astiz L, Earle P, Shearer P (1996) Global stacking of broadband seismograms. Seism Res Lett 67(4):8–18. doi:10.1785/gssrl.67.4.8

Bataille K, Flatte S (1988) Inhomogeneities near the core-mantle boundary inferred from short-period scattered *PKP* waves recorded at the global digital seismograph network. J Geophys Res 93(B12):15057–15064. doi:10.1029/JB093iB12p15057

Bataille K, Wu R, Flatte S (1990) Inhomogeneities near the core-mantle boundary evidenced from scattered waves—a review. Pure Appl Geoph 132(1–2):151–173

Bentham HLM, Rost S (2014) Scattering beneath Western Pacific subduction zones: evidence for oceanic crust in the mid-mantle. Geophys J Int 197:1627–1641

Braña L, Helffrich G (2004) A scattering region near the core-mantle boundary under the North Atlantic. Geophys J Int 158(2):625–636

Bullen K (1949) An Earth model based on a compressibility-pressure hypothesis. Month Not R Astr Soc 109(6):720–720

Cao A, Romanowicz B (2007) Locating scatterers in the mantle using array analysis of PKP precursors from an earthquake doublet. Earth Planet Sci Lett 255(1–2):22–31

Castle JC, Creager KC (1999) A steeply dipping discontinuity in the lower mantle beneath Izu-Bonin. J Geophys Res 104(B4):7279–7292. doi:10.1029/1999JB900011

Chang A, Cleary J (1981) Scattered PKKP—further evidence for scattering at a rough core-mantle boundary. Phys Earth Planet Inter 24(1):15–29

Christensen U, Hofmann A (1994) Segregation of subducted oceanic-crust in the convecting mantle. J Geophys Res 99(B10):19867–19884

Cleary J (1981) Seismic-wave scattering on underside reflection at the core-mantle boundary. Phys Earth Planet Inter 26(4):266–267

Cleary J, Haddon R (1972) Seismic wave scattering near core-mantle boundary—new interpretation of precursors to PKP. Nature 240(5383):549–551

Cormier V (2000) D$''$ as a transition in the heterogeneity spectrum of the lowermost mantle. J Geophys Res 105(B7):16193–16205

Cormier V (2007) Texture of the uppermost inner core from forward- and back-scattered seismic waves. Earth Planet Sci Lett 258(3–4):442–453

Deschamps F, Cobden L, Tackley PJ (2012) The primitive nature of large low shear-wave velocity provinces. Earth Planet Sci Lett 349–350:198–208. doi:10.1016/j.epsl.2012.07.012

Doornbos D (1974) Seismic-wave scattering near caustics—observations of PKKP precursors. Nature 247(5440):352–353

Doornbos D (1978) Seismic-wave scattering by a rough core-mantle boundary. Geophys J R Astr Soc 53(3):643–662

Doornbos DJ, Husebye ES (1972) Array analysis of PKP phases and their precursors. Phys Earth Planet Inter 5:387–399. doi:10.1016/0031-9201(72)90110-0

Doornbos D, Vlaar N (1973) Regions of seismic-wave scattering in Earth's mantle and precursors to PKP. Nat Phys Sci 243(126):58–61
Dziewonski A, Anderson D (1981) Preliminary reference Earth model. Phys Earth Planet Inter 25(4):297–356
Earle P (2002) Origins of high-frequency scattered waves near PKKP from large aperture seismic array data. Bull Seism Soc Am 92(2):751–760
Earle PS, Rost S, Shearer PM, Thomas C (2011) Scattered P′P′ waves observed at short distances. Bull Seismol Soc Am 101(6):2843–2854. doi:10.1785/0120110157
Earle P, Shearer P (1997) Observations of PKKP precursors used to estimate small-scale topography on the core-mantle boundary. Science 277(5326):667–670
Earle P, Shearer P (1998) Observations of high-frequency scattered energy associated with the core phase PKKP. Geophys Res Lett 25(3):405–408
Earle P, Shearer P (2001) Distribution of fine-scale mantle heterogeneity from observations of P_{diff} coda. Bull Seism Soc Am 91(6):1875–1881
Frosch R, Green P (1966) Concept of a large aperture seismic array. Proc R Soc Lond 290(1422):368–388. doi:10.1098/rspa.1966.0056
Frost D, Rost S, Selby N, Stuart G (2013) Detection of a tall ridge at the core-mantle boundary from scattered PKP energy. Geophys J Int 195(1):558–574 (in print)
Fukao Y, Obayashi M, Nakakuki T (2009) Stagnant slab: a review. Ann Rev Earth Planet Sci 37:19–46. doi:10.1146/annurev.earth.36.031207.124224
Garnero E (2000) Heterogeneity of the lowermost mantle. Ann Rev Earth Planet Sci 28:509–537
Garnero E, McNamara A (2008) Structure and dynamics of Earth's lower mantle. Science 320(5876):626–628
Green P, Frosch R, Romney C (1965) Principles of an experimental large aperture seismic array (lasa). Proc IEEE 53(12):1821–1833
Gutenberg B, Richter C (1934) On seismic waves; I. Gerlands Beitr Geophysik 43:56–133
Haddon RAW, Cleary JR (1974) Evidence for scattering of seismic PKP waves near the mantle-core boundary. Phys Earth Planet Inter 8(3):211–234. doi:10.1016/0031-9201(74)90088-0
Haddon R, Husebye E, King D (1977) Origins of precursors to P′P′. Phys Earth Planet Inter 14(1):41–70
Hedlin MAH, Shearer PM, Earle PS (1997) Seismic evidence for small-scale heterogeneity throughout the Earth's mantle. Nature 387(6629):145–150. doi:10.1038/387145a0
Hedlin M, Shearer P (2000) An analysis of large-scale variations in small-scale mantle heterogeneity using global seismographic network recordings of precursors to PKP. J Geophys Res 105(B6):13655–13673
Hedlin M, Shearer P (2002) Probing mid-mantle heterogeneity using PKP coda waves. Phys Earth Planet Inter 130(3–4):195–208
Helffrich GR, Wood BJ (2001) The Earth's mantle. Nature 412(6846):501–507. doi:10.1038/35087500
Van der Hilst R, Engdahl E, Spakman W, Nolet G (1991) Tomographic imaging of subducted lithosphere below northwest Pacific island arcs. Nature 353(6339):37–43
Jeffreys SH, Bullen KE (1940) Seismological tables. Office of the British Association of Sciences
Kaneshima S (2009) Seismic scatterers at the shallowest lower mantle beneath subducted slabs. Earth Planet Sci Lett 286(1–2):304–315. doi:10.1016/j.epsl.2009.06.044
Kaneshima S, Helffrich G (1999) Dipping low-velocity layer in the mid-lower mantle: evidence for geochemical heterogeneity. Science 283(5409):1888–1891
Kaneshima S, Helffrich G (2003) Subparallel dipping heterogeneities in the mid-lower mantle. J Geophys Res 108(B5):2272
Kaneshima S, Helffrich G, Suetsugu D, Bina C, Inoue T, Wiens D, Jellinek M (2010) Small scale heterogeneity in the mid-lower mantle beneath the circum-Pacific area. Phys Earth Planet Inter 183(1–2):91–103. doi:10.1016/j.pepi.2010.03.011
Kennett B, Engdahl E (1991) Traveltimes for global earthquake location and phase identification. Geophys J Int 105(2):429–465
King D, Haddon R, Husebye E (1975) Precursors to PP. Phys Earth Planet Inter 10(2):103–127

King D, Husebye E, Haddon R (1976) Processing of seismic precursor data. Phys Earth Planet Inter 12(2–3):128–134

Kito T, Thomas C, Rietbrock A, Garnero E, Nippress SEJ, Heath AE (2008) Seismic evidence for a sharp lithospheric base persisting to the lowermost mantle beneath the Caribbean. Geophys J Int 174(3):1019–1028. doi:10.1111/j.1365-246X.2008.03880.x

Koper K, Franks J, Dombrovskaya M (2004) Evidence for small-scale heterogeneity in Earth's inner core from a global study of PKiKP coda waves. Earth Planet Sci Lett 228(3–4):227–241

Korn M (1988) P-wave coda analysis of short-period array data and the scattering and absorptive properties of the lithosphere. Geophys J Lond 93(3):437–449. doi:10.1111/j.1365-246X.1988.tb03871.x

Leyton F, Koper K (2007) Using PKiKP coda to determine inner core structure: 1. Synthesis of coda envelopes using single-scattering theories. J Geophys Res 112(B5):B05316

Li C, van der Hilst RD, Engdahl ER, Burdick S (2008) A new global model for P wave speed variations in Earth's mantle. Geochem Geophys Geosyst 9(5):Q05018. doi:10.1029/2007GC001806

Lithgow-Bertelloni C, Richards M (1998) The dynamics of Cenozoic and Mesozoic plate motions. Rev Geophys 36(1):27–78

Manchee E, Weichert D (1968) Epicentral uncertainties and detection probabilities from Yellowknife seismic array data. Bull Seism Soc Am 58(5):1359–1377

Mancinelli NJ, Shearer P (2013) Reconciling discrepancies among estimates of small-scale mantle heterogeneity from PKP precursors. Geophys J Int 195:1721–1729

Margerin L, Nolet G (2003) Multiple scattering of high-frequency seismic waves in the deep Earth: PKP precursor analysis and inversion for mantle granularity. J Geophys Res 108(B11):2514. doi:10.1029/2003JB002455

McNamara AK, Garnero EJ, Rost S (2010) Tracking deep mantle reservoirs with ultra-low velocity zones. Earth Planet Sci Lett 299(1–2):1–9. doi:10.1016/j.epsl.2010.07.042

McNamara A, Zhong S (2005) Thermochemical structures beneath Africa and the Pacific Ocean. Nature 437(7062):1136–1139

Olson P, Yuen D, Balsiger D (1984) Convective mixing and the fine-structure of mantle heterogeneity. Phys Earth Planet Inter 36:291–304

Poupinet G, Kennett BLN (2004) On the observation of high frequency PKiKP and its coda in Australia. Phys Earth Planet Inter 146(3–4):497–511

Ritsema J, Deuss A, van Heijst HJ, Woodhouse JH (2011) S40RTS: a degree-40 shear-velocity model for the mantle from new Rayleigh wave dispersion, teleseismic traveltime and normal-mode splitting function measurements. Geophys J Int 184(3):1223–1236. doi:10.1111/j.1365-246X.2010.04884.x

Rost S, Earle P (2010) Identifying regions of strong scattering at the core-mantle boundary from analysis of PKKP precursor energy. Earth Planet Sci Lett 297(3–4):616–626. doi:10.1016/j.epsl.2010.07.014

Rost S, Garnero E (2004) A study of the uppermost inner core from PKKP and P′P′ differential traveltimes. Geophys J Int 156(3):565–574

Rost S, Garnero E, Williams Q (2008) Seismic array detection of subducted oceanic crust in the lower mantle. J Geophys Res 113(B6):B06303. doi:10.1029/2007JB005263

Rost S, Thorne M (2010) Radial and lateral variations in mantle heterogeneity from scattered seismic waves (Invited). Abstract DI52A-08, 2010 Fall Meeting, AGU, San Francisco

Rost S, Thorne M, Garnero E (2006) Imaging global seismic phase arrivals by stacking array processed short-period data. Seism Res Lett 77(6):697–707

Sato H (1988) Temporal change in scattering and attenuation associated with the earthquake occurrence? A review of recent studies on coda waves. Pure Appl Geophys 126(2–4):465–497. doi:10.1007/BF00879007

Selby ND (2011) Improved teleseismic signal detection at small-aperture arrays. Bull Seismol Soc Amer 101(4):1563–1575. doi:10.1785/0120100253

Shearer P (1990) Seismic imaging of upper-mantle structure with new evidence for a 520-Km discontinuity. Nature 344(6262):121–126. doi:10.1038/344121a0

Shearer P, Earle P (2004) The global short-period wavefield modelled with a Monte-Carlo seismic phonon method. Geophys J Int 158(3):1103–1117
Shearer PM (2007) Deep Earth structure—seismic scattering in the deep Earth. Treatise on geophysics. Elsevier, Amsterdam, pp 695–729
Shearer PM, Earle PS (2008) Chapter 6—Observing and modeling elastic scattering in the deep Earth. In: Earth heterogeneity and scattering effects on seismic waves, vol 50. Elsevier, Amsterdam, pp 167–193
Simmons NA, Myers SC, Johannesson G, Matzel E (2012) LLNL-G3Dv3: global P wave tomography model for improved regional and teleseismic travel time prediction, J Geophys Res 117(B10). doi:10.1029/2012JB009525
Stixrude L, Lithgow-Bertelloni C (2012) Geophysics of chemical heterogeneity in the mantle. Ann Rev Earth Planet Sci 40(1):569–595. doi:10.1146/annurev.earth.36.031207.124244
Tan E, Gurnis M (2005) Metastable superplumes and mantle compressibility. Geophys Res Lett 32(20):L20307
Thomas C, Weber M, Wicks C, Scherbaum F (1999) Small scatterers in the lower mantle observed at German broadband arrays. J Geophys Res 104(B7):15073–15088
Tkalčić H, Flanagan M, Cormier V (2006) Observation of near-podal P′P′ precursors: evidence for back scattering from the 150–220 km zone in the Earth's upper mantle. Geophys Res Lett 33(3):L03305
Vanacore E, Niu F, Ma Y (2010) Large angle reflection from a dipping structure recorded as a PKIKP precursor: evidence for a low velocity zone at the core-mantle boundary beneath the Gulf of Mexico. Earth Planet Sci Lett 293(1–2):54–62. doi:10.1016/j.epsl.2010.02.018
Vidale J, Dodge D, Earle P (2000) Slow differential rotation of the Earth's inner core indicated by temporal changes in scattering. Nature 405(6785):445–448
Vidale J, Earle P (2000) Fine-scale heterogeneity in the Earth's inner core. Nature 404(6775):273–275
Weber M, Wicks C (1996) Reflections from a distant subduction zone. Geophys Res Lett 23(12):1453–1456
Weichert D, Whitham K (1969) Calibration of yellowknife seismic array with first zone explosions. Geophys J R Astr Soc 18(5):461–476
Wen L (2000) Intense seismic scattering near the Earth's core-mantle boundary beneath the Comoros hotspot. Geophys Res Lett 27(22):3627–3630
Widiyantoro S, Van der Hilst R (1997) Mantle structure beneath Indonesia inferred from high-resolution tomographic imaging. Geophys J Int 130(1):167–182. doi:10.1111/j.1365-246X.1997.tb00996.x
Woodhouse J, Dziewonski A (1984) Mapping the upper mantle—3-dimensional modeling of Earth structure by inversion of seismic waveforms. J Geophys Res 89(NB7):5953–5986
Wright C (1972) Array studies of seismic waves arriving between *P* and *PP* in distance range 90° to 115°. Bull Seismol Soc Amer 62(1):385–400

Chapter 13
Seismic Detection of Post-perovskite Inside the Earth

Laura Cobden, Christine Thomas and Jeannot Trampert

Abstract Since 2004, we have known that perovskite, the most abundant mineral in the lower mantle, has the capacity to transform to a denser structure, post-perovskite, if subjected to sufficiently high temperature and pressure. But does post-perovskite exist inside the Earth? And if it does, do we have the resources to locate it seismically? In this chapter, we present an overview of what we know about the perovskite-to-post-perovskite phase transformation from mineral physics, and how this can be translated into seismic structure. In light of these constraints, we evaluate the current lines of evidence from global and regional seismology which have been used to indicate that post-perovskite is likely present in the deep mantle.

Keywords Post-perovskite · D″ discontinuity · Seismology · Mineral physics

13.1 Introduction

The radial seismic structure of the Earth's upper mantle is characterised by regions of smoothly increasing wave speeds separated by sharp discontinuities (Fig. 13.1). These discontinuities are correlated with, and almost universally accepted as resulting from, mineralogical phase transformations in (Mg, Fe)Si_2O_4, of which about 60 % of the upper mantle is composed (concurrent transformations in less abundant phases enhance velocity gradients and influence the magnitude of the discontinuities). In contrast, in the lower mantle, wave speeds increase smoothly and monotonically with depth, with no major discontinuities, until about 300 km

L. Cobden (✉) · C. Thomas
Institut für Geophysik, Westfälische Wilhelms-Universität, Münster, Germany
e-mail: L.J.Cobden@uu.nl

L. Cobden · J. Trampert
Department of Earth Sciences, Utrecht University, Utrecht, The Netherlands

© Springer International Publishing Switzerland 2015

A. Khan and F. Deschamps (eds.), *The Earth's Heterogeneous Mantle*,
Springer Geophysics, DOI 10.1007/978-3-319-15627-9_13

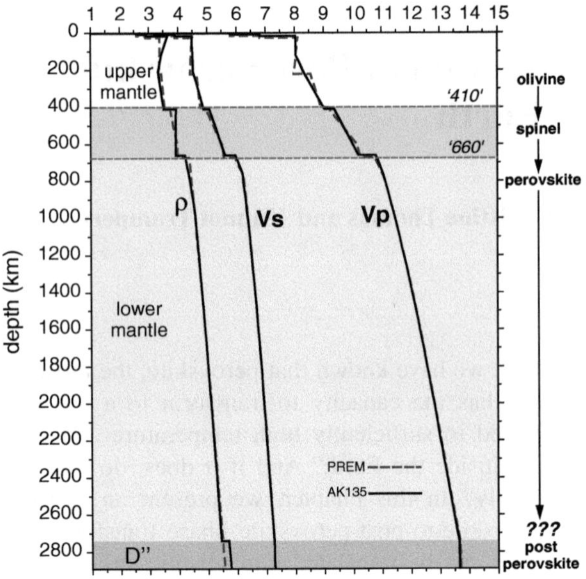

Fig. 13.1 1-D seismic velocity and density profiles for the Earth's mantle and associated major mineralogical phase transformations. The "410" discontinuity is associated with transformation of olivine to β-spinel (wadsleyite), and the "660" discontinuity is associated with transformation of γ-spinel (ringwoodite) to perovskite. These major transformations are accompanied by pyroxene to garnet, and garnet to perovskite, phase transitions which occur over broad depth intervals and which modify the depth and sharpness of the discontinuities at 410 and 660

above the core–mantle boundary (CMB), where a marked decrease in seismic velocity gradients is observed (Fig. 13.1). This region of anomalous seismic gradients was first noted by Gutenberg (1914) and referred to as the D″ layer (Bullen 1950). While globally averaged seismic profiles show only a change in the velocity gradients at the base of the mantle, regional seismic data sets have revealed that abrupt vertical changes in wave speed do occur for both *P* and *S* waves on local to regional scales at the top of the D″ region (e.g. Lay and Helmberger 1983; Baumgardt 1989; Weber 1993). These abrupt changes are usually referred to as the D″ discontinuity. By 1998, more than 40 separate studies had documented observations of a D″ discontinuity between ~100 and 450 km above the CMB (see Wysession et al. 1998 for a review) in different parts of the globe, although its apparent magnitude and depth vary significantly both within and between studies.

Given the association between upper mantle seismic discontinuities and phase transformations, it is reasonable to speculate that the D″ discontinuity may be related to another such transformation. Sidorin et al. (1999a, b) used geodynamic modelling to show that seismic observations of the D″ discontinuity could be explained by a mineral phase transition with a Clapeyron slope of about 6 MPa/K whereby lateral variations in the depth of the discontinuity would arise from lateral variations in temperature. However, at that time, no such phase transition was known to occur. The most abundant lower mantle mineral,

(Mg, Fe)SiO_3 perovskite, had demonstrated great stability at high pressures (Knittle and Jeanloz 1987; Kesson et al. 1998), and while early experiments indicated that it might transform from orthorhombic to cubic or tetragonal symmetry (Wolf and Bukowinski 1987; Wang et al. 1990) or decompose into its component oxides (Meade et al. 1995; Saxena et al. 1996, 1998), such observations were later refuted and ascribed to large temperature and pressure gradients within experimental samples. Eventually, with continuing improvements in computational and experimental techniques, several research groups independently demonstrated in 2004 that $MgSiO_3$ perovskite (Pv) will transform to a higher pressure polymorph known simply as "post-perovskite" (pPv) at pressures near to or greater than the CMB (Murakami et al. 2004; Oganov and Ono 2004; Shim et al. 2004). Subsequent investigations indicated that a similar transformation occurs in $FeSiO_3$ and Al_2O_3 perovskites (Mao et al. 2004; Oganov and Ono 2005; Caracas and Cohen 2005b). Nonetheless, large uncertainties on the precise depth of the phase transformation at high temperatures have left the question open of whether post-perovskite should exist at all within the pressure range of the Earth's mantle.

The discovery of post-perovskite has generated much excitement within geophysics. Its dense and anisotropic crystal structure (Fig. 13.2) likely gives it distinct seismic properties relative to perovskite (e.g. Wookey et al. 2005b; Wentzcovitch et al. 2006). Post-perovskite may thus provide a convenient explanation not only for the D″ discontinuity but also much of the seismic complexity observed within the D″ layer, including large- and small-scale anisotropy (Kendall and Silver 1998; Panning and Romanowicz 2006; Wookey and Kendall 2007), internal stratification of the D″ layer (Lay et al. 2006; van der Hilst et al. 2007) and anti-correlation between long-wavelength bulk and shear wave speed anomalies (Ishii and Tromp 1999; Masters et al. 2000).

The layered structure of post-perovskite (Fig. 13.2) may also render it rheologically weak (Hunt et al. 2009; Ammann et al. 2010), and this has the potential to influence the convective evolution of the mantle. Geodynamic simulations (Nakagawa and Tackley 2011; Tackley 2012) have shown that the presence or absence of post-perovskite affects how far slabs sink in the mantle: whether they

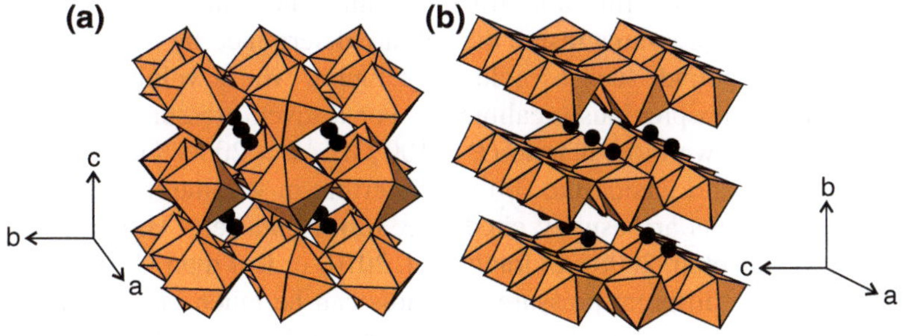

Fig. 13.2 Crystal structures of $MgSiO_3$ perovskite (**a**) and post-perovskite (**b**). *Brown octahedra* represent SiO_6 octahedra and *black spheres* represent Mg cations

pile up at the core–mantle boundary or are immediately entrained upward in the global circulation. This in turn determines the chemical composition and morphology of rising plumes, which may ultimately influence the location and intensity of intraplate volcanism. As such, ascertaining whether or not post-perovskite exists inside the mantle can have a major impact on our understanding of the Earth's dynamics.

Demonstrating the existence or absence of post-perovskite in the mantle requires a careful comparison between the experimentally and theoretically derived physical properties of post-perovskite and the observed seismic properties of the deep mantle. Accordingly, we begin with a review of what constraints can be placed on the physical properties of post-perovskite and the structure of the Pv–pPv phase boundary, as determined from mineral physics observations (Sect. 13.2). We then assess whether various seismic structures in the deep Earth are compatible with, or can be used as an indicator of, the presence of post-perovskite (Sect. 13.3).

13.2 Mineral Physics Observations of Post-perovskite

Mineral physics constraints on the seismic and thermodynamic properties of post-perovskite come from two sources: laboratory experiments, in which rock materials are compressed and heated to the pressure and temperature conditions expected in the deep mantle, and computer simulations, which predict crystal properties using quantum mechanical theory. For the lower mantle, the experimental method of choice is the diamond anvil cell, in which a small sample of a few millimetres is compressed between two opposing anvils and heated with a laser. The structural and chemical properties of the compressed material can be determined in situ via spectroscopy or, alternatively, can be analysed post-mortem using a transmission electron microscope.

Laboratory measurements suffer from uncertainties related to (1) the physical difficulty of sustaining large yet homogeneous pressures and temperatures across small crystal samples and (2) sluggish reaction kinetics, which can prevent a sample from attaining its equilibrium structure within the time frame of the experiment. Furthermore, pressures inside the cell are determined by mixing the sample with another compound, such as gold or platinum, whose volume as a function of pressure has been previously calibrated. For a given pressure standard, at the conditions of the lowermost mantle ($P > 100$ GPa, $T > 2000$ K), uncertainties in the estimated pressure inside the cell due to errors in volume measurements and temperature gradients are usually cited as ±3 to 4 GPa (e.g. Hirose et al. 2006; Hirose 2006; Ohta et al. 2008; Andrault et al. 2010) or less than ~80 km. However, the effect of using different pressure standards can be much larger. For example, pressures inferred using a platinum reference are at least 10–15 GPa higher than those inferred from gold at $P > 100$ GPa (Anderson et al. 1989; Akahama et al. 2002), while pressures estimated using MgO are about 5–6 GPa higher than for gold (Hirose et al. 2008a; Grocholski et al. 2012). Some experimentalists (Fei et al. 2004; Hirose 2006) have argued that pressure calibrations based upon MgO (Speziale et al. 2001) are the most favourable choice, because the MgO scale predicts the ringwoodite-to-perovskite and perovskite-to-post-perovskite transitions at the same depths as the seismically observed 660 km and D″ discontinuities, respectively. Added to this, uncertainties in temperature measurements are typically of the order of ±10 % at temperatures above ~2000 K (Hirose et al. 2006; Ohta et al. 2008).

Theoretical calculations (ab initio simulations) do not suffer from the physical limitations of experiments. However, they are computationally expensive, and as a result, approximations of atomic interactions must be assumed, and many simulations are "static", i.e. represent the material properties without thermal effects, at 0 K. Furthermore, it is particularly computationally difficult to obtain the physical properties of iron-bearing mineral assemblages (Cococcioni and de Gironcoli 2005). Consequently, computationally derived mineral properties are also associated with large uncertainties. For example, different molecular dynamic approximations, namely the local density approximation (LDA) and generalised gradient approximation (GGA), predict phase boundary pressures which differ by up to 15 GPa (e.g. Oganov and Ono 2004; Tsuchiya et al. 2004), even at 0 K.

13.2.1 Depth and Thickness of the Pv–pPv Phase Boundary

For a phase transition which takes place near the bottom of the mantle, where the absolute temperature is, in any case, uncertain by up to ~1500 K (on the basis of iron melting temperatures inside the core, e.g. Oganov et al. 2002; Campbell et al. 2007; Asanuma et al. 2010; Kamada et al. 2010), the combined uncertainties in experimental measurements or ab initio calculations can translate into the difference between the transition happening above the CMB (~135 GPa) or below it. We have summarised in Table 13.1 the results of 29 different studies of the Pv–pPv phase boundary for a range of chemical compositions and mineralogical assemblages. A pictorial representation of those data for which the phase boundary could be determined at 2500 K, and their relation to the D″ discontinuity, is shown in Fig. 13.3.

In pure $MgSiO_3$ (the simplest possible composition), the phase boundary should be univariant, i.e. occur at a fixed temperature for a given pressure. However, the position of the phase boundary is not tightly constrained by the available data, with a significant part of the discrepancy due to the use of different pressure standards in different studies. At 2500 K, estimates of the transition pressure range from 113 GPa (c. 2450 km) to 144 GPa (i.e. well below the CMB), although most studies place the transition within the mantle (Table 13.1; Fig. 13.3). Part of the uncertainty also comes from the discrete nature of experimental data points used to construct phase equilibria. For example, the data of Murakami et al. (2004), which are among the earliest experimental observations of

Table 13.1 Experimental and theoretical constraints on the *P*, *T* location and thickness of the Pv–pPv phase boundary, as published up to the end of 2012

References	Technique	Pressure standard	Composition	*P*, *T* constraints on Pv → pPv onset	Clapeyron slope (MPa/K)	Thickness of two-phase (Pv + pPv) region (GPa)
Murakami et al. (2004)	LHDAC	Pt-1	$MgSiO_3$	~113.5–128 GPa at 2300 K		
Shim et al. (2004)	LHDAC	Pt-2 and Ar	$MgSiO_3$	144 ± 10 GPa at T > 2500 K		
Oganov and Ono (2004)	Ab initio	–	$MgSiO_3$	~122.5 GPa at 2500 K	8.0–9.8	
Tsuchiya et al. (2004)	Ab initio	–	$MgSiO_3$	~117–128 GPa at 2500 K	7.5 ± 0.3	
Iitaka et al. (2004)	Ab initio	–	$MgSiO_3$	98 GPa at 0 K	10	
Mao et al. (2005)	Ab initio	–	$MgSiO_3$	113 GPa at 0 K		
Caracas and Cohen (2005a)	Ab initio	–	MgSiO3	107 GPa at 0 K		
Ono and Oganov (2005)	Ab initio and LHDAC	Pt-2	$MgSiO_3$	~131 GPa at 2500 K	7.0	
Hirose et al. (2006)	LHDAC	Au-1	$MgSiO_3$	113.5 GPa at 2500 K	4.7	
Hirose et al. (2006)	LHDAC	MgO	$MgSiO_3$	120 GPa at 2500 K	11.5	
Tateno et al. (2009)	LHDAC	MgO	$MgSiO_3$	121–123 GPa at 2500 K	13.3 ± 1	
Mao et al. (2004)	LHDAC	NaCl	$(Mg_{0.6}Fe_{0.4})SiO_3$	100 GPa at room T		
Mao et al. (2005)	Ab initio	–	$(Mg_{0.5}Fe_{0.5})SiO_3$	63 GPa at 0 K		
Caracas and Cohen (2005a)	Ab initio	–	$(Mg_{0.5}Fe_{0.5})SiO_3$	44 GPa at 0 K		
Mao et al. (2005)	Ab initio	–	$FeSiO_3$	pPv always stable wrt Pv at 0 K		

(continued)

Table 13.1 (continued)

References	Technique	Pressure standard	Composition	*P*, *T* constraints on Pv → pPv onset	Clapeyron slope (MPa/K)	Thickness of two-phase (Pv + pPv) region (GPa)
Caracas and Cohen (2005a)	Ab initio	–	$FeSiO_3$	pPv always stable wrt Pv at 0 K		
Shieh et al. (2006)	LHDAC	Pt-2, Au-2, NaCl	$(Mg_{0.91}Fe_{0.09})SiO_3$	109 ± 4 GPa at 2400 ± 400 K		
Tateno et al. (2007)	LHDAC	Au-1	$(Mg_{0.5}Fe_{0.5})SiO_3$	~112 GPa at 2500 K	8	>15
Tateno et al. (2007)	LHDAC	Au-1	$(Mg_{0.25}Fe_{0.75})SiO_3$	100 GPa at 2500 K		>15
Catalli et al. (2009)	LHDAC	Au-1	$(Mg_{0.91}Fe^{2+}_{0.09})SiO_3$	~111 GPa at 2500 K	6.7 ± 0.5	20 ± 5
Metsue and Tsuchiya (2012)	Ab initio	–	$(Mg_{0.9375}Fe_{0.0625})SiO_3$	111 GPa at 2500 K	10	4 (at 2500 K)
Caracas and Cohen (2005a)	Ab initio	–	Al_2O_3	119 GPa at 0 K		
Oganov and Ono (2005)	Ab initio and LHDAC	Au-3	5 mol% Al_2O_3 in $MgSiO_3$	Increases pPv-in wrt $MgSiO_3$ by 5.2 GPa		
Caracas and Cohen (2005b)	Ab initio	–	5 wt% Al_2O_3 in $MgSiO_3$	Increases pPv-in wrt $MgSiO_3$ by 1 GPa		
Akber-Knutson et al. (2005)	Ab initio	–	6.25 mol% Al_2O_3 in $MgSiO_3$	127 GPa at 2500 K		>10
				Increases pPv-in wrt $MgSiO_3$ by 7 GPa		
Tateno et al. (2005)	LHDAC	Pt-1	25 mol% Al_2O_3 in $MgSiO_3$	~142-145 GPa at 2500 K		10–30
Zhang and Oganov (2006)	Ab initio	–	3.25 mol% Al_2O_3 in $MgSiO_3$	107.5 GPa at 0 K		
				Increases pPv-in wrt $MgSiO_3$ by 5 GPa		

(continued)

Table 13.1 (continued)

References	Technique	Pressure standard	Composition	P, T constraints on Pv → pPv onset	Clapeyron slope (MPa/K)	Thickness of two-phase (Pv + pPv) region (GPa)
Tsuchiya and Tsuchiya (2008)	Ab initio	–	6 mol% Al_2O_3 in $MgSiO_3$ ("pyrolite")	Decreases pPv-in wrt $MgSiO_3$ by 0.2 GPa per mol% Al_2O_3		1 (at 2500 K)
Tsuchiya and Tsuchiya (2008)	Ab initio	–	20 mol% Al_2O_3 in $MgSiO_3$ ("MORB")			4–5 (at 2500 K)
Nishio-Hamane et al. (2007)	LHDAC	Au-1	$Mg_{0.85}Fe_{0.15}Al_{0.15}Si_{0.85}O_3$	~143–162 GPa at 2500 K		5–35
Catalli et al. (2009)	LHDAC	Au-1	$Mg_{0.9}Fe^{3+}{}_{0.1}Al_{0.1}Si_{0.9}O_3$	~112 GPa at 2500 K		30 ± 5
Andrault et al. (2010)	LHDAC	Au-3 and Re	$Mg_{0.89}Fe^{3+}{}_{0.11}Al_{0.11}Si_{0.89}O_3$	118–122 GPa at 3300 K	(Cf Catalli implies 7.5–12.5)	30–40
Andrault et al. (2010)	LHDAC	Au-3 and Re	$(FeAl)_x(MgSi)_{1-x}O_3$	142 GPa at $x = 7$ %, T = 3300 K		
				dP/dx = 0.243 at T = 3300 K		
Shieh et al. (2011)	LHDAC	Pt-3 and Au-4	$(Mg, Fe)_3Al_2Si_3O_{12}$	P ~ 124 GPa at 2500 K assuming C. slope = 7.5		≈30
				Pv-out at $P > 148$ GPa, $T > 1600$ K		
Murakami et al. (2005)	LHDAC	Au-1	KLB-1 peridotite	~113 GPa at 2500 K		
Ono and Oganov (2005)	LHDAC	Au-3	KLB-1 peridotite	~124 GPa at 2500 K	8.0	
				Decreases pPv-in wrt $MgSiO_3$ by 6 GPa		
Ohta et al. (2008)	LHDAC	Au-5	KLB-1 peridotite	~114-117 GPa at 2500 K	8 ± 4	3–10

(continued)

Table 13.1 (continued)

References	Technique	Pressure standard	Composition	P, T constraints on Pv → pPv onset	Clapeyron slope (MPa/K)	Thickness of two-phase (Pv + pPv) region (GPa)
Sinmyo et al. (2011)	LHDAC	C	$Mg_{0.84}Fe^{3+}{}_{0.10}Al_{0.10}Si_{0.96}O_3$ KLB-1 peridotite	~127 GPa at 2400 K		~3
Sinmyo et al. (2009)	LHDAC + calculations	Au-1	$(Mg_{0.91}Fe^{2+}{}_{0.09})SiO_3$ plus 30 % ferropericlase	~128 GPa at 2500 K		5
Catalli et al. (2009)	LHDAC + calculations	Au-1	$(Mg_{0.91}Fe^{2+}{}_{0.09})SiO_3$ plus 50 % ferropericlase	~131 GPa at 2500 K		2.2
Andrault et al. (2010)	LHDAC	Au-3 and Re	$Mg_{0.89}Fe^{3+}{}_{0.11}Al_{0.11}Si_{0.89}O_3$ plus ferropericlase	118 GPa at 3300 K		13
Grocholski et al. (2012)	LHDAC	Au-1	Pyrolite	140 GPa at 2500 K		30
Hirose et al. (2005)	LHDAC	Au-1	MORB	~100–114 GPa at 2500 K		
Ohta et al. (2008)	LHDAC	Au-5	MORB	~106–113 GPa at 2500 K		3–15
				pPv-in decreases wrt pyrolite by ~4 GPa		
Grocholski et al. (2012)	LHDAC	Au-1	MORB	~107 GPa at 2500 K		5–15
Grocholski et al. (2012)	LHDAC	Au-1	San Carlos olivine	~131 GPa at 2500 K		≤4

LHDAC = Laser-heated diamond anvil cell. The pressure of the phase transition is given at 2500 K where possible (on the basis of experimental data points or Clapeyron slope measurements) to facilitate comparison between different studies. Pressure scales: Ar (Ross et al. 1986); Au-1 (Tsuchiya 2003); Au-2 (Shim et al. 2002); Au-3 (Jamieson et al. 1982); Au-4 (Dewaele et al. 2004); Au-5 (Hirose et al. 2008a); C (diamond) (Akahama and Kawamura 2004); MgO (Speziale et al. 2001); NaCl (Sata et al. 2002); Pt-1 (Jamieson et al. 1982); Pt-2 (Holmes et al. 1989); Pt-3 (Fei et al. 2007); Re (Zha et al. 2004)

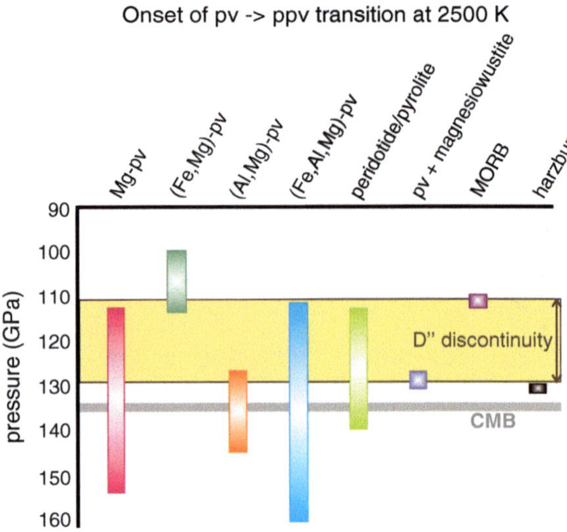

Fig. 13.3 Schematic diagram showing the experimentally observed onset of Pv to pPv transition at 2500 K for a range of chemical and mineralogical compositions, relative to the pressure range of the D″ discontinuity and the CMB. Each *vertical bar* reflects the uncertainty in pressure onset of the phase transition for a fixed composition, using the data listed in Table 13.1

pPv, constrain the boundary best at 2300 K. At this temperature, there are observations of Pv and pPv at a range of pressures, but there is a 15 GPa gap between observations of Pv and those of pPv, which puts an uncertainty of 15 GPa on the position of the phase boundary.

Adding iron and aluminium to $MgSiO_3$ changes not only the depth of the phase boundary but creates a two-phase depth interval where both perovskite and post-perovskite coexist. A thick two-phase region is seismically relevant because even if Pv and pPv have significantly differing seismic velocities (discussed below, Sect. 13.2.3), if the transformation from one to the other is gradual, it will not generate a sharp seismic discontinuity. Further, if the depth of the phase boundary becomes too shallow, it will not be consistent with the observed depths of the D″ discontinuity, and if it becomes too deep, it will place the phase transition beyond mantle pressures. Experiments and calculations on binary systems, i.e. $(Mg_{1-x}Fe_x)SiO_3$, have indicated that adding Fe^{2+} to $MgSiO_3$ will decrease the pressure of the phase boundary significantly (Mao et al. 2004, 2005; Caracas and Cohen 2005a), and measurements by different groups using different pressure standards have consistently shown that for compositions close to $(Mg_{0.9}Fe_{0.1})SiO_3$, the phase transformation to post-perovskite begins at around 111 GPa at 2500 K (Shieh et al. 2006; Catalli et al. 2009; Metsue and Tsuchiya 2012). Most studies show that adding aluminium has the opposite effect, increasing the phase boundary by ~5 GPa for ~5 mol% Al_2O_3 (Oganov and Ono 2005; e.g. Akber-Knutson et al. 2005; Zhang and Oganov 2006), but some studies predict that it will decrease the phase boundary by ~0.2 GPa per mol% Al_2O_3 (Tsuchiya and Tsuchiya 2008). Fe and Al both have the capacity to produce very thick two-phase regions: estimates of the thickness when Fe is added to the $MgSiO_3$ range from 4–20 GPa, i.e. up to 400 km, and when Al is added, between 1 and 30 GPa, i.e. up to 600 km (Table 13.1). However, in the real Earth, we expect both Fe and Al to be present simultaneously. Experimental observations of ternary systems, $(FeAl)_x(MgSi)_{1-x}O_3$, predict widely varying estimates of the onset of the phase transformation, from ~20 GPa above to ~25 GPa below, the CMB, but all studies consistently observe a broad two-phase region of about 30 GPa thickness (~600 km) before Pv completely disappears (Nishio-Hamane et al. 2007; Catalli et al. 2009; Andrault et al. 2010; Shieh et al. 2011).

Nonetheless, this does not imply that the Pv–pPv transition region is necessarily very broad inside the Earth, because the behaviour of the transformation is altered further when other minerals, especially (Mg, Fe)O magnesiowustite, are coexisting with the perovskite. Generally speaking, the presence of (Mg, Fe)O or the presence of Al-bearing minerals such as the "calcium ferrite (CF)-type phase" $(Na, Ca, Mg, Fe)_1(Al, Si, Fe, Mg)_2O_4$ will both produce a thinner two-phase region (Catalli et al. 2009; Grocholski et al. 2012) by extracting the Fe and Al from the perovskite.

The chemical composition often taken to represent the "average" mantle is pyrolite (Ringwood 1962), although a number of studies have indicated that this may not be appropriate in the lower mantle (e.g. Cammarano et al. 2005; Cobden et al. 2009; Murakami et al. 2012). Most observations of the Pv to pPv transition in pyrolitic-type compositions (often approximated with KLB-1 peridotite, Takahashi 1986) predict a relatively narrow two-phase region with a thickness of ~3–10 GPa (Table 13.1) at pressures close to D″ depths (~113–127 GPa at 2500 K) (Murakami et al. 2005; Ono and Oganov 2005; Ohta et al. 2006, 2008; Sinmyo et al. 2011). However, one recent study has observed that the phase transition in pyrolite may not occur until 140 GPa and may have a thickness of 30 GPa (Grocholski et al. 2012). Most of the studies of pyrolite-type compositions have used a gold pressure standard, so it is unlikely that pressure calibration effects would account for more than ~5 GPa of the discrepancy. Grocholski et al. (2012) noted that kinetic effects (i.e. incomplete equilibration of cation distributions) can produce differences in the position of the phase boundary inferred from forward (Pv to pPv) and reverse (pPv to Pv) experiments of up to 8 GPa in multi-phase systems, so this is likely also a contributing factor. In any case, it is clear that the behaviour of the transition in pyrolitic compositions warrants further investigation. On the other hand, experimental observations of MORB, which forms a component of subducted slabs and may be present in large quantities near the base of the mantle (e.g. Tackley 2011), are much more consistent (Table 13.1): The onset of the phase transition is predicted to occur between ~100 and 114 GPa at 2500 K, with a two-phase region between 3 and 15 GPa thick (Hirose et al. 2005; Ohta et al. 2008; Grocholski et al. 2012).

The variable behaviour of the phase boundary with respect to chemical composition and mineralogy is best understood in terms of iron and aluminium partitioning coefficients. Constraints on Fe and Al partitioning between perovskite, post-perovskite and magnesiowustite are shown in Table 13.2. Although there is a lot of discrepancy between different studies, which has been attributed to non-homogeneous cation distributions within mineral samples (Hirose et al. 2008b); complexity regarding whether Fe is present as Fe^{2+}, Fe^{3+}, high-spin state or low-spin state; and whether or not Al is present (Andrault et al. 2010), the most recent results

Table 13.2 Some constraints on iron and aluminium partitioning between perovskite (Pv), post-perovskite (pPv) and magnesiowustite (mw), as available up to end 2012

Reference	Composition	Partition coefficient			Value
Kobayashi et al. (2005)	(Mg, Fe)SiO_3 (Al-free system)	$K_{Fe^{2+}}^{(mw/Pv)}$			8.3
			$K_{Fe^{2+}}^{(mw/pPv)}$		3.3
				$K_{Fe^{2+}}^{(Pv/pPv)}$	0.4
Murakami et al. (2005)	KLB-1 peridotite (Al-bearing system)	$K_{Fe}^{(mw/Pv)}$			2.0–2.4
			$K_{Fe}^{(mw/pPv)}$		7.8 ± 2.5
				$K_{Fe}^{(Pv/pPv)}$	4
Ono and Oganov (2005)	$Al_2O_3 + MgSiO_3$ (Fe-free system)			$K_{Al}^{(Pv/pPv)}$	2.27
Zhang and Oganov (2006)	$Al_2O_3 + MgSiO_3$ (Fe-free system)			$K_{Al}^{(Pv/pPv)}$	2.67
Zhang and Oganov (2006)	$Al_2O_3 + MgSiO_3$ (Fe-bearing system)			$K_{Al}^{(Pv/pPv)}$	4.25
Hirose et al. (2008b)	$(Mg_{0.91}Fe_{0.09})SiO_3$ Al-free system			$K_{Fe}^{(Pv/pPv)}$	1.8
Andrault et al. (2010)	(Fe, Al)-bearing $MgSiO_3$ compositions			$K_{Fe}^{(Pv/pPv)}$	4.2 ± 0.5
Sinmyo et al. (2011)	$MgSiO_3$-Fe*SiO_3 (Al-bearing system)	$K_{Fe^{2+}}^{(mw/Pv)}$			2.2
			$K_{Fe^{2+}}^{(mw/pPv)}$		5.9
				$K_{Fe^{2+}}^{(Pv/pPv)}$	2.68
		$K_{Fe^*}^{(mw/Pv)}$			0.90
			$K_{Fe^*}^{(mw/pPv)}$		4.5
				$K_{Fe^*}^{(Pv/pPv)}$	5.0

$K_{Fe}^{(mw/Pv)} = \frac{(X_{Fe}/X_{Mg})_{mw}}{(X_{Fe}/X_{Mg})_{Pv}}$ where $(X_{Fe})_{mw}$ is the fraction of Fe in the (Mg, Fe)-bearing site in the mineral mw . Fe* indicates total Fe (Fe^{2+} plus Fe^{3+})

show some consistent findings: (1) Al preferentially partitions into perovskite over post-perovskite (Ono and Oganov 2005; Zhang and Oganov 2006) and (2) in the presence of Al, Fe preferentially partitions into magnesiowustite over perovskite or post-perovskite, and into perovskite over post-perovskite (Murakami et al. 2005; Hirose et al. 2008b; Andrault et al. 2010; Sinmyo et al. 2011). This means that for binary and ternary systems, both Al and Fe stabilize perovskite to higher pressures, leading to a thicker phase transition. However, in multi-phase systems, magnesiowustite will extract Fe from both the Pv and pPv and will destabilize Pv relative to pPv. This can lead to a thinner transition. Likewise, in Al-rich compositions such as MORB, where non-perovskitic Al-bearing phases (e.g. $MgAl_2O_4$) are present, the Al will preferentially partition into these phases, destabilising Pv relative to pPv and producing a thinner transition (Grocholski et al. 2012).

In summary, the position in *P-T* space and thickness of the phase boundary is a complex function of chemical composition, and there remains much uncertainty on its structural properties. On the basis of experimental constraints, it remains feasible that post-perovskite either does not occur at all within the pressure range of the mantle or is restricted to certain chemically distinct domains, i.e. it has a localised presence rather than a global one. Additionally, there is a strong possibility that the phase transformation occurs over a broad depth interval not compatible with a sharp seismic discontinuity. Even if the phase transition is broad, more work is required to determine the rate of conversion of Pv to pPv within the two-phase region, since this will affect its seismic visibility: If most of the phase transformation is completed within a narrow depth interval, it could once again produce a sharp seismic interface even if the overall two-phase region is broad. Current observations do indicate that rate of transformation is nonlinear with respect to depth and that the form of this non-linearity is dependent on composition (Catalli et al. 2009).

13.2.2 Clapeyron Slope

Many experimental observations of Pv and pPv phases are sparsely positioned in *P-T* space and leave the gradient of the phase boundary (dP/dT), i.e. the Clapeyron slope, completely unconstrained. In such cases, the phase boundary could be drawn with either a positive or negative slope. However, in a few experiments, observations of Pv and pPv are sufficiently proximal that the Clapeyron slope can be estimated (Table 13.1). The Clapeyron slope can also be calculated theoretically (Oganov and Ono 2004; Tsuchiya et al. 2004; e.g. Iitaka et al. 2004). While these observations unanimously indicate that the Clapeyron slope is positive (i.e. the phase transition occurs at higher temperatures as the pressure increases), estimates of its magnitude range from about 4 to 14 MPa/K. Within this variation, the Clapeyron slope seems to be composition independent (e.g. Hirose et al. 2006).

13.2.3 Velocity and Density Contrasts

Post-perovskite can only be detected in the Earth if its seismic properties are distinct from those of perovskite. A compilation of theoretical and experimental measurements of the changes in wave speed and density across the Pv to pPv transition is given in Table 13.3. These measurements all agree that in pure $MgSiO_3$, a density increase of 1–1.5 % is expected upon transformation to pPv at 90 GPa $< P <$ 125 GPa. Most studies also predict a significant increase in isotropic *S* wave velocity (V_S) of the order of ~1–4 % and a smaller change in *P* wave velocity (V_P) of less than ± 1 % (e.g. Iitaka et al. 2004; Oganov and Ono 2004;

Table 13.3 Some calculations and in situ experimental measurements of velocity and density changes during the Pv to pPv transition; this table is not exhaustive

References	Technique	Composition	P, T	deltaVp (%, Pv to pPv)	deltaVs (%, Pv to pPv)	delta ρ (%, Pv to pPv)
Murakami et al. (2004)	LHDAC	$MgSiO_3$				+1.0 to 1.2
Tsuchiya et al. (2004)	Ab initio	$MgSiO_3$	125 GPa, 2750 K			+1.5
Oganov and Ono (2004)	Ab initio	$MgSiO_3$	120 GPa, 0 K	+0.3	+1.9	+1.4
Iitaka et al. (2004)	Ab initio	$MgSiO_3$	98 GPa, 0 K	-0.1	+1	+1.4
Iitaka et al. (2004)	Ab initio	$MgSiO_3$ anisotropic	98 GPa, 0 K	+4 (VpH)	+ 3 to 7	
Wookey et al. (2005b)	Ab initio	$MgSiO_3$	98.7 GPa, 0 K	−0.5	+ 3 to 4	+1
Caracas and Cohen (2005a)	Ab initio	$MgSiO_3$	120 GPa, 0 K	0	+1.3	+1.5
Tsuchiya and Tsuchiya (2006)	Ab initio	$MgSiO_3$	100 GPa, 0 K	0	0	+1.5
Tsuchiya and Tsuchiya (2006)	Ab initio	$MgSiO_3$	120 GPa, 0 K	+0.7	+7.8	+1.5
Wentzcovitch et al. (2006)	Ab initio	$MgSiO_3$	125 GPa, 2500 K	+0.7	+2.8	
Wentzcovitch et al. (2006)	Ab initio	$MgSiO_3$	Along phase boundary of Tsuchiya et al. (2004)	0–0.5 with uncertainty up to ±1 %	+1.5 to +2.3 % with uncertainty up to ± 1.5	
Murakami et al. (2007)	LHDAC	$MgSiO_3$	125 GPa, 300 K		≤ 0.5	
Murakami et al. (2007)	LHDAC	$MgSiO_3$	172 GPa, 300 K		+ 2	

(continued)

Table 13.3 (continued)

References	Technique	Composition	*P*, *T*	delta*V*p (%, Pv to pPv)	delta*V*s (%, Pv to pPv)	delta ρ (%, Pv to pPv)
Ammann et al. (2010)	Ab initio	$MgSiO_3$ anisotropic	~119 GPa, ~2700 K	−0.4	+4	
Mao et al. (2005)	Ab initio	$(Mg_{(1-x)}Fe^{2+}_x)SiO_3$	130 GPa, 0 K			$\partial \ln\rho/\partial x Fe = 0.3$
Shieh et al. (2006)	LHDAC	$\left(Mg_{0.91}Fe^{2+}_{0.09}\right)SiO_3$	83–106 GPa, ?? K			−1.1 ± 2
Tsuchiya and Tsuchiya (2006)	Ab initio	$FeSiO_3$	100 GPa, 0 K	−0.8	−1.7	+1.2
Caracas and Cohen (2005a)	Ab initio	Al_2O_3	120 GPa, 0 K	−4.3	−1.4	+1.8
Tsuchiya and Tsuchiya (2006)	Ab initio	Al_2O_3	120 GPa, 0 K	−0.7	−1.4	+1.8
Ono and Oganov (2005)	LHDAC	KLB-1 peridotite				~+1
Caracas and Cohen (2005a)	Ab initio	pyrolite	120 GPa, 0 K	+0.7	+2.5	+1 to 2

Note that most measurements are made at 0 or 300 K

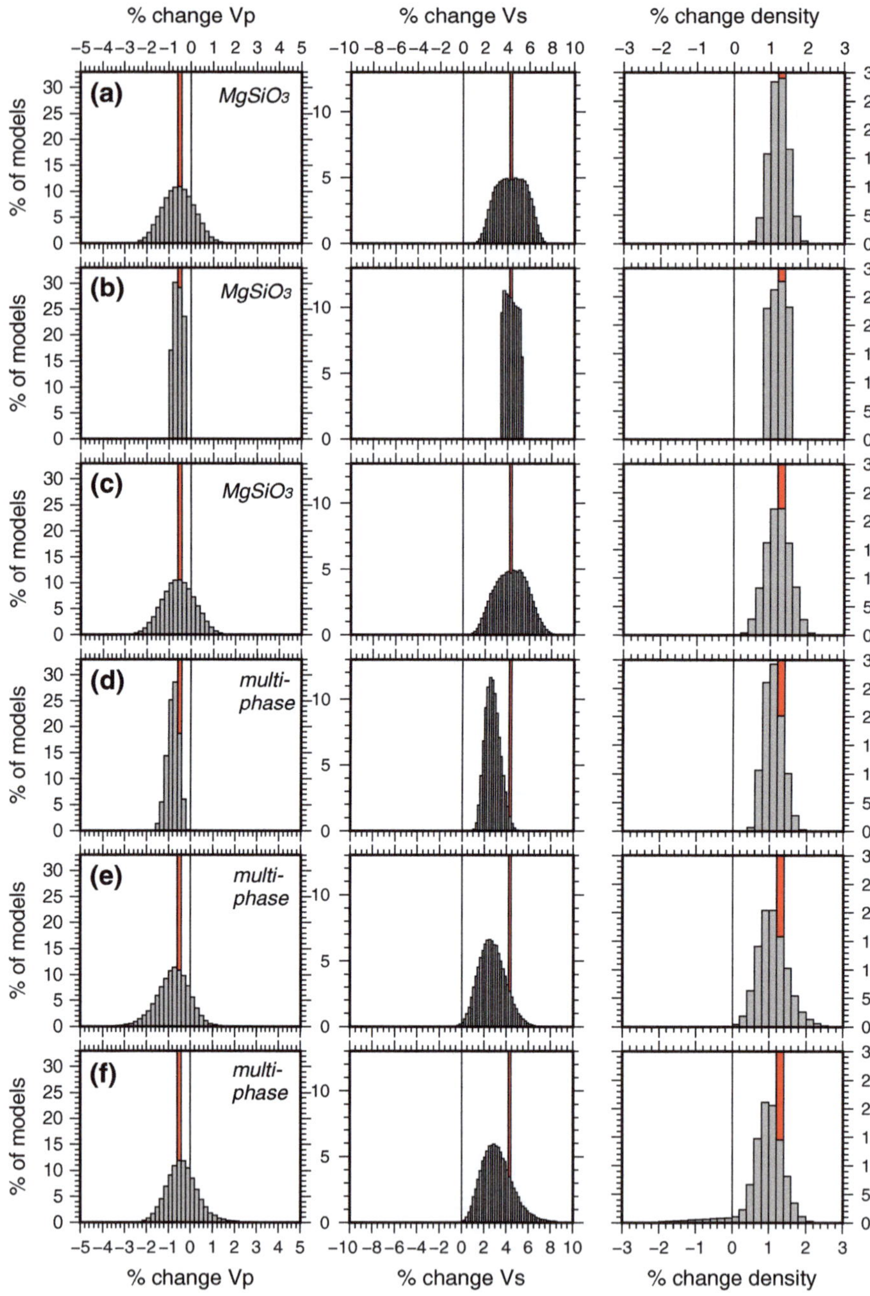
% change Vp
% change Vs
% change density
% of models
(a)
MgSiO3
(b)
MgSiO3
(c)
MgSiO3
(d)
multi-phase
(e)
multi-phase
(f)
multi-phase

Fig. 13.4 Histograms showing the sensitivity of *V*p, *V*s and density to a phase change from perovskite to post-perovskite. Calculations assume 100 % conversion of any Pv into pPv, at 2700 km depth. **a** Pure $MgSiO_3$ at a fixed temperature of 2600 K; the *red bars* indicate the value of d*V*p, d*V*s and dρ obtained using the elastic parameters published in Table A1 of Stixrude and Lithgow-Bertelloni (2011), while the *grey histograms* show the variation in dVp, dVs and dρ due to elastic parameter uncertainties, with elastic parameters varying at random within the uncertainty bounds published in Stixrude and Lithgow-Bertelloni (2011); **b** variation in dVp, dVs and dρ for pure $MgSiO_3$, with variable temperature, 1600 < T < 2600 K, and no elastic parameter uncertainties, *red bars* as in **a**; **c** pure $MgSiO_3$, 1600 < T < 2600 K, with elastic parameter uncertainties; **d** multi-mineral assemblages following the compositional ranges in Cobden et al. (2012), 1600 < T < 2600 K, no elastic parameter uncertainties; **e** multi-mineral assemblages, 1600 < T < 2600 K, with elastic parameter uncertainties; **f** as in **e** but with increased iron partitioning into magnesiowustite on conversion of Pv to pPv

Caracas and Cohen 2005a; Wookey et al. 2005b; Wentzcovitch et al. 2006). One study has however indicated that post-perovskite is not necessarily much faster in V_S than perovskite at D″ pressures (Murakami et al. 2007). Either way, direct comparison of observed seismic structures with these constraints is at best semi-quantitative: many mineral physics measurements are at 0 K or room temperature and do not necessarily represent what will happen at lower mantle temperatures, which may be between ~2000 and 4000 K near the CMB (Oganov et al. 2002; Campbell et al. 2007; Cobden et al. 2009; Kamada et al. 2010; Asanuma et al. 2010). Extrapolation of seismic properties to high temperatures requires knowledge of the temperature derivatives of the elastic properties. Furthermore, the seismic wave speeds vary with chemistry, with $FeSiO_3$ post-perovskite being ~14 % slower in V_P and 21 % slower in V_S than $MgSiO_3$, and Al_2O_3 post-perovskite being ~11 % slower in both V_P and V_S (Caracas and Cohen 2005a; Tsuchiya and Tsuchiya 2006). Nonetheless, for low concentrations (< 5–10 mol%) of Fe and Al in post-perovskite, as is expected in a pyrolitic composition, the effect of chemistry on seismic wave speeds is predicted to be small, with the effect on the phase diagram being much more significant (Caracas and Cohen 2005a).

Since direct measurements of V_P, V_S and density (as in Table 13.3) are only determined at a few temperatures and pressures, then inferring the seismic properties at other pressures or temperatures requires instead knowledge of the bulk and shear moduli (*K* and *G*, respectively) in addition to the density, at a fixed pressure (usually 0 GPa) and an equation of state which defines how to extrapolate these properties to lower mantle conditions. These parameters, derived from experimental and theoretical observations, typically have a non-negligible uncertainty which can affect seismic velocities significantly (e.g. Cobden et al. 2009). For realistic calculation of V_P, V_S and density in plausible mantle compositions, one must also take into account the presence of other mineral phases, i.e. magnesiowustite, calcium perovskite $CaSiO_3$, free SiO_2 and Al-bearing phases. It is unclear precisely how the seismic properties of individual minerals contribute to the average V_P, V_S and density of a mixed-phase assemblage, but typically a weighted averaging scheme such as the Voigt-Reuss-Hill or Hashin-Shtrikman method is assumed. In Fig. 13.4, we show the range of possible changes in V_P, V_S and density for pure $MgSiO_3$ and mixed-phase assemblages in which the perovskite converts to post-perovskite at 2700 km depth. These models were calculated using the mineral elastic parameters and uncertainty bounds on those parameters published in Stixrude and Lithgow-Bertelloni (2011), together with the equation of state of Stixrude and Lithgow-Bertelloni (2005). Uncertainties in elastic parameters alone translate into variations of ~3 %, 5 % and 1 % in V_P, V_S and density, respectively (Fig. 13.4a). Modelling the system as a mixed-phase assemblage, i.e. including the effects of iron, aluminium, calcium, and non-silicates, rather than pure $MgSiO_3$, reduces the change in V_S by about 2 % (Fig. 13.3d, e) but has a much smaller effect on V_P. Increasing the partitioning of iron into magnesiowustite by a factor of 4 or more on converting from Pv to pPv (as predicted by experiments, Table 13.2) does not appear to produce a significant change in the seismic velocities and density (Fig. 13.4f). Thus, the range of possible velocity and density changes is dominated by uncertainties in the mineral elastic parameters, and the magnitude of the change in V_S depends strongly on chemical composition.

13.2.4 Anisotropy

Both ab initio calculations (e.g. Stackhouse et al. 2005b; Wookey et al. 2005b; Wentzcovitch et al. 2006; Ammann et al. 2010; Mao et al. 2010) and experimental observations (e.g. Murakami et al. 2004; Shieh et al. 2006; Guignot et al. 2007) of post-perovskite structure under lower mantle conditions indicate that the mineral is highly anisotropic, with a maximum compressibility, and thus minimum seismic velocity, along the *b*-axis (i.e. perpendicular to the layers of SiO_6 octahedra). It is significantly more anisotropic than perovskite, and for seismic waves propagating in the crystallographic fast direction, this can enhance the velocity contrast between Pv and pPv by up to 3–7 % (Table 13.3). Calculations by Stackhouse et al. (2005a, 2006) indicate that the anisotropic structure of $MgSiO_3$ likely does not change significantly with the addition of iron, while addition of aluminium may have a small effect.

The bulk seismic anisotropy of polycrystalline post-perovskite depends not only on the inherent anisotropy of single crystals, but also on (1) its deformation history—if the individual crystals are undeformed and randomly oriented, then the bulk anisotropy will be zero—and (2) its behaviour under deformation, in particular the crystallographic plane on which slip occurs when a deviatoric stress is applied. This has been a subject of much debate, with 4 different slip planes having been proposed for $MgSiO_3$ pPv: (100), (010), (001) and (110) (see Nowacki et al. 2011 for a review). The main source of this variability is the experimental difficulty of working with $MgSiO_3$ pPv, which cannot be quenched and therefore can only be observed at very high pressures (Miyagi et al. 2011). At such high pressures, only a small sample size can be used, which limits the total amount of strain that can be imparted on the sample and makes maintaining a uniform (high) temperature across the sample challenging (see Sect. 13.2.1). Depending on the magnitude and duration of the deformation, together with the grain size in an experimental sample and the mineralogy of the starting material (i.e. if pPv is synthesised directly from Pv or from enstatite), then some experiments may observe transformation textures—texture produced during the phase transition—rather than deformation textures (Miyagi et al. 2011). To circumvent the limitations of high-pressure experiments, many studies have been done on analogue materials, primarily $CaIrO_3$ and $MgGeO_3$ post-perovskites, at lower temperatures and pressures that are easier to control and allow larger sample sizes. Experiments on $CaIrO_3$ have consistently shown that slip occurs along the (010) plane in this mineral, i.e. parallel to the layering (e.g. Yamazaki et al. 2006; Miyajima et al. 2006; Walte et al. 2007; Niwa et al. 2007; Miyagi et al. 2008; Walte et al. 2009; Niwa et al. 2012). However, differences in crystal structure between $CaIrO_3$ and $MgSiO_3$, namely bond lengths and bond angles, indicate that the former may not be a good analogue for deformation textures in the latter (Miyagi et al. 2008; Kubo et al. 2008; Miyagi et al. 2011). $MgGeO_3$ appears to be more structurally similar and a better analogue (Hirose et al. 2010; Miyagi et al. 2011). The most recent experiments on $MgSiO_3$ (at high *P* and low *T*) and $MgGeO_3$ (at moderate *P* and *T*) favour slip along the (001) plane (Miyagi et al. 2010, 2011; Okada et al. 2010; Metsue and Tsuchiya 2013), which is perpendicular to the layers of SiO_6 octahedra. Calculations suggest that 100 % alignment of pPv along the (001) plane would cause horizontally polarised shear waves to be 8–15 % faster than vertically polarised shear waves (Hirose et al. 2010), which is significantly higher than the 2 % difference estimated for slip along the (010) plane (Stackhouse and Brodholt 2007). Clearly, further work needs to be done to verify the slip plane and direction in pPv as this has a major influence on the magnitude of the bulk seismic anisotropy which can be induced in pPv.

Even knowing the slip plane(s) and slip vector(s) for pPv, it is still challenging to translate polycrystalline anisotropy measurements made in the laboratory into seismic observables, since the lowermost mantle is unlikely to be composed purely of (Mg, Fe)SiO_3, and thus, the total seismic anisotropy in a region depends also on the anisotropic properties of any other minerals present (e.g. up to ~20 % magnesiowüstite), as well as the degree of alignment (0–100 %) of the anisotropic phases. Furthermore, measurements of seismic anisotropy in the Earth cannot easily distinguish between the alignment of inherently anisotropic crystals, i.e. lattice-preferred orientation (LPO) or any extrinsic sub-(seismic-)wavelength alignment of two materials with distinct elastic properties, i.e. shape-preferred orientation (SPO), especially at long wavelengths. Sources of SPO near the CMB could be chemical layering or aligned melt inclusions (Kendall and Silver 1998). The micro- and macro-scale seismic properties of melt, as well as the amount of melt which could be present near the CMB, are currently poorly constrained (Nowacki et al. 2011).

13.3 Seismic Observations

One way to test for the presence of post-perovskite is using seismic waves that sample the lower mantle. Over the years, many different waves and wave-types have been used to investigate structures near the CMB. Here, we will give an overview of a number of these seismic probes and their ability to detect either a phase transition in or physical properties of post-perovskite. We will look at both regional and global scales.

While there are many seismic probes that are sensitive to lower mantle structure, only a few of these probes are suitable for detecting post-perovskite. These are as follows: (1) seismic discontinuities, whose depth, sharpness and impedance contrast can be compared with the Clapeyron slope, thickness and impedance change of the phase change from Pv to pPv (Sect. 13.3.1); (2) anisotropy models, whose properties can be compared with the anisotropic behaviour of pPv relative to other minerals or mineral structures (Sect. 13.3.2); and (3) tomography models of lateral variations in wave speed and density (Sect. 13.3.3). For the latter, it is essential to have models of both $\mathrm{dln}V_P$ and $\mathrm{dln}V_S$, at the same spatial resolution, in order to distinguish the trade-off between temperature and composition. Additionally, including a model of the density structure $\mathrm{dln}\rho$ at the same spatial resolution can provide even better constraints.

13.3.1 The D″ Discontinuity

Regional-scale seismic studies have repeatedly documented evidence for a seismic discontinuity near the base of the mantle (e.g. Lay and Helmberger 1983; Weber and Kornig 1990; Weber and Davis 1990; Davis and Weber 1990; Young and Lay 1990; Houard and Nataf 1992, 1993; Weber 1993; Weber and Wicks 1996; Weber et al. 1996; Kendall and Nangini 1996; Thomas and Weber 1997; Reasoner and Revenaugh 1999; Russell et al. 2001; Thomas et al. 2002, 2004a, b; Lay et al. 2004; Wallace and Thomas 2005; Lay et al. 2006; Avants et al. 2006; van der Hilst et al. 2007; Kito et al. 2007; Hutko et al. 2008, 2009; Chaloner et al. 2009). This discontinuity, usually referred to as "the D″ discontinuity", appears to have a complex structure, with both small and long wavelength lateral variations in its strength and depth. The discontinuity can be seen in both *P* and *S* waves (SH component), although not necessarily at the same time (e.g. Weber 1993), and in some regions, no discontinuity is observed at all (e.g. Neuberg and Wahr 1991; Chambers and Woodhouse 2006), implying that it is not necessarily a global structure. It manifests itself in seismic data as a triplicated arrival, due to refraction immediately above and/or below the discontinuity combined with reflection off the discontinuity (Fig. 13.5). The D″ discontinuity probably does not appear in 1-D reference models such as PREM (Dziewonski and Anderson 1981) and AK135 (Kennett et al. 1995) because of its lateral complexity and because it is detected best at a narrow range of epicentral distances, ~60°–80°, where the waves graze the discontinuity at near-critical angles and the reflection coefficient is high (Fig. 13.6). Observations of the D″ discontinuity to date are thus restricted to those regions where earthquakes and seismic stations are at the correct angular separation. Regions where a D″ discontinuity has been observed by one or more studies are mapped in Fig. 13.7. There are more observations for *S* waves than for *P* waves, and more observations have been gathered

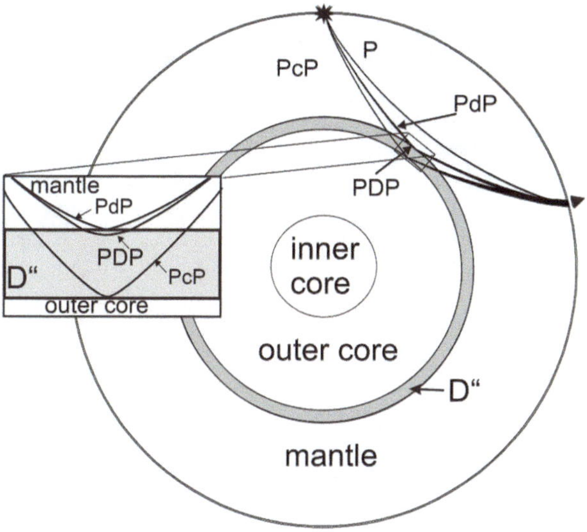

Fig. 13.5 Schematic cross section through the Earth showing ray paths for D″ phases. Earthquake indicated by *star* and seismic station indicated by *triangle*. PdP, the reflection off the top of the D″ discontinuity, is sometimes known as P_{bc}, while *PDP*, the refraction beneath the discontinuity, is sometimes referred to as P_{cd}. Refraction immediately above the D″ discontinuity may be referred to as P_{ab}. An analogous nomenclature is used for *S* waves

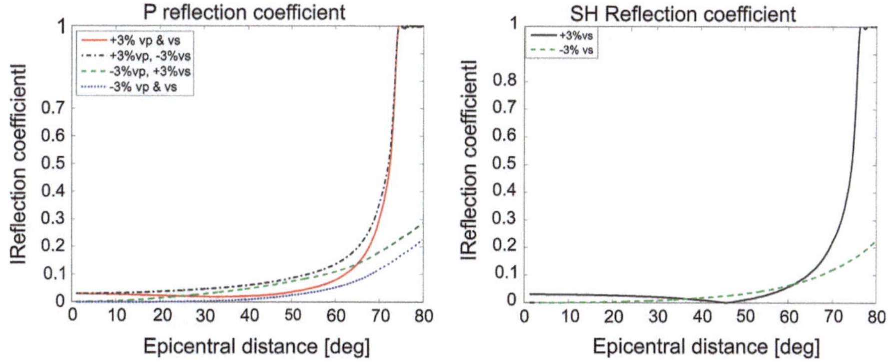

Fig. 13.6 Reflection coefficients for *P* and *S* waves at the D″ discontinuity as a function of epicentral distance

in tomographically fast regions such as the circum-Pacific than in tomographically slow regions. One possible reason for the larger amount of *S* wave observations is the generally larger amplitude of those waves compared with *P* wave triplications from the D″ discontinuity.

The D″ discontinuity triplication presents itself in seismic data as a weak phase which arrives later than the main *P* (or *S*) phase but before the

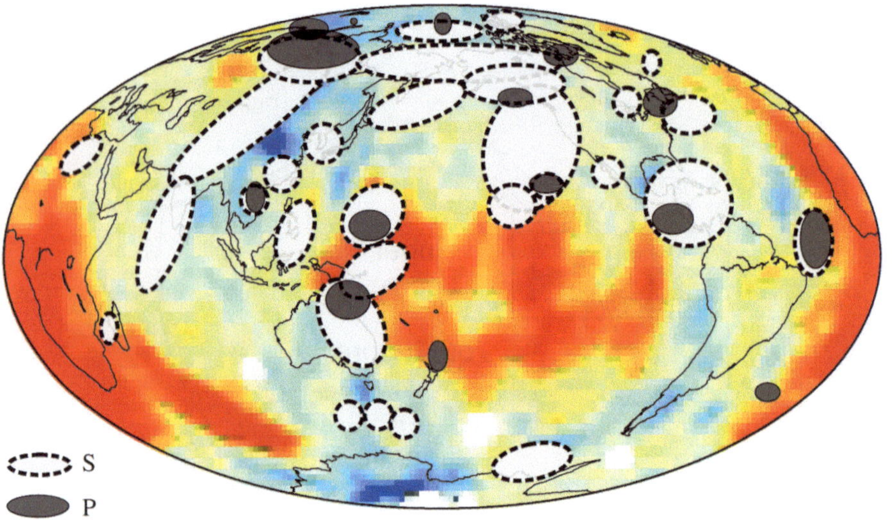

Fig. 13.7 Locations of observations of D″ discontinuity overlain on the tomography model of (Grand 2002). References for D″ observations given in Sect. 13.3.1

core-reflected phase, *PcP* (or *ScS*) (Fig. 13.5). We can be confident that the phase originates in the lowermost mantle rather than from mid-mantle or near-source scatterers (such as slabs or the Moho) if it has a slowness (move-out) intermediate between mantle refraction and core reflection (Weber 1993), and when it is seen for multiple events with different source depths (Baumgardt 1989) in the same region. This phase is given the general name *PdP* (or *SdS* for *S* waves) to indicate interaction with the D″ discontinuity, and, depending on epicentral distance, it can be a composite of a reflection off the top of the discontinuity (known as P_{bc} or sometimes *PdP*) plus a refracted wave (P_{cd} or *PDP*) turning just below the discontinuity, or a reflection alone, or a refraction alone (e.g. Lay and Helmberger 1983; Weber 1993).

Even though *PdP* and *SdS* phases can sometimes be observed in single seismic traces, they are often weak signals, and various stacking techniques have been employed to enhance their visibility. These include the following: double-array stacking (Lay et al. 2004, 2006; Kito et al. 2007; Thorne et al. 2007; Lay 2008); vespagram analysis (slant-stacking) (Thomas and Weber 1997; Thomas et al. 2002; Chaloner et al. 2009); simplified seismic migration (Thomas et al. 1999; 2004b; Kito et al. 2007; Chaloner et al. 2009); Kirchhoff migration (Hutko et al. 2006); and the generalised radon transform (GRT) (Chambers and Woodhouse 2006; van der Hilst et al. 2007). Array-based methods such as vespagrams (Davies et al. 1971; Muirhead and Datt 1976) require visual inspection of individual seismograms which is time intensive, restricting studies to small data sets; on the other hand, methods such as the GRT which allow large data sets to be processed quickly are not subjected to the same level of quality control. At least part of the apparent variability in D″ structure can be ascribed to different stacking techniques, and subjective data selection between different studies (Wysession et al. 1998; Lay and Garnero 2007).

Stacking also allows assessment of the slowness of seismic phases. In single seismogram, the move-out difference between *P*, *PcP* (or *S*, *ScS*) and *PdP* (or *SdS*) is difficult to measure. Distance-dependent displays of seismic traces, or stacking techniques such as vespagrams or frequency-wavenumber analyses (for a review see Rost and Thomas 2002 or Schweitzer et al. 2002), which allow slowness estimation, can help to identify the waves, [e.g. Lay and Helmberger 1983; Weber 1993; Thomas and Weber 1997] and distinguish them from possible Moho multiples of the *P* wave (or *S* wave) or other interfering waves.

In the following subsections, we assess the suitability of seismic observations of the D″ discontinuity for discriminating between the presence and absence of post-perovskite. This requires firstly an assessment of the compatibility between the structural properties of the discontinuity and those of the phase boundary, taking into account the likely biases and uncertainties in both seismic and mineralogical data sets. Secondly, where the data are of sufficient quality to allow such an assessment, a comparison against other thermo-mineralogical effects which may produce the D″ discontinuity is needed. For while it is tempting to interpret the D″ discontinuity in terms of a mineralogical phase change on the basis of this being the cause of the 410 and 660 discontinuities, the D″ region is the boundary layer between the solid silicate mantle and the liquid outer core, where great thermo-chemical complexity is already expected in the absence of a phase transition. Thus, the D″ discontinuity may instead represent reflections (or scattering, Scherbaum et al. 1997) off small- to regional-scale thermo-chemical heterogeneities, such as subducted slabs (Christensen 1989; Christensen and Hofmann 1994), piles or layers of dense and possibly primitive material (Davies and Gurnis 1986), and core–mantle reaction products (Knittle and Jeanloz 1989). The localised nature of these structures provides a plausible mechanism for the lateral variations in the depth, strength and visibility of the D″ discontinuity (Sidorin et al. 1998).

13.3.1.1 Discontinuity Depth

By modelling the discontinuity as a sharp interface which occurs at a single depth, it is possible to constrain its depth to within ± 20 km (e.g. Weber 1993) by measuring the timing of the *PdP*(*SdS*) arrival relative to the *P*(*S*) or *PcP*(*ScS*) and assuming a particular 1-D wave speed structure for the mantle. Slightly different depths may be obtained depending on whether the timing of the *PdP* is measured relative to the *P* (which gives the discontinuity's position as the depth below the Earth's surface) or relative to the *PcP* (which gives the position as a height above the CMB). When using the latter, care should be taken to consider the possible effects of CMB topography and the wave speed structure in between the D″ discontinuity and the CMB on the travel times.

The inferred depth depends furthermore on the frequency at which the data have been filtered, which in turn is different for *P* and *S* waves, since in reality, the discontinuity has a finite thickness to which different frequencies have variable sensitivity. *PdP* phases are usually seen at periods shorter than 10 s, while *SdS* are typically visible between ~3 and 25 s. In regions where both *PdP* and *SdS* phases are observed, the inferred depths of the discontinuity usually agree to within ~50 km. Estimates of discontinuity depth range from ~100 to 450 km above the CMB, with a global average of 250–265 km above the CMB or ~2650 km depth (Wysession et al. 1998). Overall, it is reasonable to assume that the cited depth of a D″ discontinuity in a particular study is accurate to within ~100 km. This variability, although large, is much smaller than the uncertainty on the depth of the Pv to pPv transition (Sect. 13.2; Table 13.1). As such, the depth of the D″ discontinuity is compatible with the depth of the phase boundary but cannot be used to confirm the presence of post-perovskite.

13.3.1.2 Discontinuity Thickness

Extracting the thickness of the D″ discontinuity is very difficult. In theory, it could be inferred from the frequency content of *PdP* or *SdS* phases (relative to the *PcP* or *ScS*) at near-vertical angles of incidence. In practice, such observations are extremely rare, due to the low reflection coefficient at pre-critical distances (Fig. 13.6), although one such study has indicated that *PdP* arrivals may be generated by a discontinuity as narrow as 8 km (Breger and Nataf in Wysession et al. 1998). Alternatively, the thickness can be estimated by modelling the amplitude of *PdP* or *SdS* arrivals at multiple frequencies as a function of distance: a broader transition will reduce the amplitude of pre-critical arrivals relative to post-critical ones (Lay 2008). Again this is challenging because *PdP* and *SdS* arrivals can usually only be detected in a narrow distance range just before and just after critical reflection, and data from multiple events with the same reflection point at D″ but different incidence angles are required, which are often not available. Furthermore, amplitudes are influenced by multiple factors, including 3-D discontinuity topography, and even in 1-D synthetic waveform modelling, there is a strong trade-off between the imposed width of the discontinuity and the magnitude of the velocity jump across it (see also Sect. 13.3.1.4 below). Nonetheless, these studies have indicated a maximum discontinuity thickness of ~70–90 km (e.g. Weber et al. 1996), and a recent analysis of a densely sampled region under the Cocos Plate has suggested it could be as narrow as ~30 km (Lay 2008).

The recent observation that the Pv–pPv phase transformation can, in certain multi-mineralic assemblages such as pyrolite, have a thickness of several hundred kilometres (Catalli et al. 2009; Grocholski et al. 2012) is clearly incompatible with the constraint that the D″ discontinuity is narrower than c. 90 km. This has prompted speculation that the phase change may not be responsible for the D″ discontinuity, or that if it is, it is only in regions with very specific chemistry such as basalt or harzburgite (Grocholski et al. 2012). However, we should bear in mind that even for a broad phase boundary, the rate of transformation is nonlinear (Catalli et al. 2009) and could be predominantly restricted to a narrow depth interval which is compatible with observed discontinuity thicknesses. Alternatively, in the presence of shear stresses, rapid development of aligned post-perovskite grains, i.e. a sudden change in bulk anisotropy, could be responsible for a sharp seismic discontinuity in the presence of a broad phase transition (Ammann et al. 2010). At best, the <90 km thickness of the discontinuity can be used as an indication that its origin has a significant non-thermal component, since it is difficult to sustain large-amplitude temperature anomalies over length scales of a few tens of kilometres (e.g. Kaneshima and Helffrich 1999; Ricard et al. 2005). It cannot, at present, be used to detect post-perovskite.

13.3.1.3 Discontinuity Topography

Sidorin et al. (1999a, b) used changes in the height of the D″ discontinuity in combination with long-wavelength tomographic models to deduce that if tomographic wave speed variations represented lateral variations in temperature, then the discontinuity could correspond to a phase transformation with a Clapeyron slope of +6 MPa/K. This falls within the range of Clapeyron slopes seen experimentally and theoretically for the Pv to pPv phase boundary (Table 13.1; Sect. 13.2.2), and subsequently, variations in discontinuity depth have been used to estimate temperature variations near the CMB (Lay et al. 2006; van der Hilst et al. 2007). However, given that the depth of the Pv–pPv transition can vary strongly as a function of chemistry and mineralogy, and that its visibility may also be influenced by anisotropy, it is potentially hazardous to interpret D″ topography purely in terms thermal effects on the Pv–pPv phase boundary. At the time of the modelling of Sidorin et al. (1999a, b), there were limited observations of the D″ discontinuity in the central Pacific; subsequently, studies have found that the discontinuity in this region is too shallow to be compatible with Sidorin et al.'s models, and this may be indicative of chemical effects on the phase boundary (e.g. Lay and Garnero 2007; Ohta et al. 2008). Furthermore, Hutko et al. (2006) have mapped 100-km amplitude undulations in the depth of D″ reflections over lateral distances of 200–300 km, which would require temperature variations of 500–1000 K that are geodynamically difficult to explain.

A refinement of the method of Sidorin et al. (1999a, b) was performed by Sun et al. (2006, 2007) and Sun and Helmberger (2008) using updated tomography models and allowing for simple changes in chemistry. They found that observations of D″ topography could be explained by a phase boundary with a Clapeyron slope of ~9 MPa/K, if the magnitude of the velocity jump across the discontinuity is laterally variable and if short-wavelength undulations are attributed to scattering off a buckled slab. While it is clearly possible to model D″ discontinuity topography in terms of Clapeyron slopes plus compositional heterogeneity, such modelling is not uniquely diagnostic of the presence of post-perovskite because it does not rule out the possibility that alternative structures can fit the data.

13.3.1.4 Velocity and Density Contrasts

The impedance contrast across the D″ discontinuity may be our most diagnostic tool for assessing if the discontinuity is generated by post-perovskite. In principle, the polarity of *PdP* and *SdS* waveforms indicates whether the impedance change is positive or negative, and the amplitude gives an indication of the magnitude of the change. At pre-critical distances, the waveforms are sensitive to both density and velocity, while at critical and post-critical distances where the waves graze the discontinuity obliquely (and where most observations are obtained), the waveforms are primarily sensitive to velocity changes (Lay and Garnero 2007). There are two main difficulties in measuring impedance contrasts. The first is that from a modelling perspective, there is always a trade-off between the imposed thickness of the discontinuity and magnitude of the velocity change across it. At best, it is possible to give a range of discontinuity thicknesses and velocity jumps compatible with a given seismic observation of *PdP* or *SdS*. The second—and major—issue is that 3-D topography on the discontinuity can lead to focussing and de-focussing of *PdP* and *SdS* waveforms, which may alter their amplitude significantly. To date, most estimates of velocity jumps at D″ have been based upon full-waveform reflectivity modelling in 1-D (e.g. Weber 1993; Kito et al. 2007; Chaloner et al. 2009) or ray tracing in 2-D (Thomas and Weber 1997), which cannot accurately incorporate the effect of 3-D topography. Full-waveform modelling of the D″ discontinuity in 3-D at the high frequencies at which it is observed is currently beyond available computational capacity. However, a few studies in 2.5-D have been or are being performed (e.g. Thorne et al. 2007); (Hempel et al., article in preparation). Investigations by Hempel et al. (pers. comm.) indicate that if the D″ discontinuity is dipping more steeply than ~40°, then it becomes seismically invisible along the great circle path, i.e. the direction in which data are stacked for vespagram analysis, and if it dips at shallower angles, then the amplitudes of the D″ reflection may still be strongly influenced.

Further uncertainties arise from stacking procedures, which assume that either multiple events or multiple receivers are sampling the same reflection point at D″, when in practice, they are sampling the structure, perhaps non-uniformly, within a range of several degrees, so that the image obtained is a smeared average of any 3-D variations in topography and velocity structure of the discontinuity within that region. Added to this, for a given propagating wave, the 1-D Fresnel zone at D″ depths is about 3.5° × 7° (~230 × 460 km) for *S* waves with periods of around 6 s and 2° × 4° (~130 × 260 km) for *P* waves with periods of around 1 s (Weber 1993), meaning that *P* and *S* waves for a given event are not sampling exactly the same structure and the resolution of topography within those ranges is questionable.

In general, 1-D reflectivity modelling indicates that velocity contrasts smaller than about 1 % are unlikely to be detected (e.g. Weber 1993). Thus, absence of a visible *PdP* or *SdS* arrival does not equate to the absence of a D″ discontinuity; it may well mean that the impedance contrast is too small to be seen above the noise or that the topography de-focuses the signal. Gaherty and Lay (1992) and Flores and Lay (2005) also noted that due to the distance-dependent properties of the reflection coefficient, it is easier to detect velocity increases than velocity decreases, leading to an observational bias towards detecting velocity increases in acquired data sets. We further note that the reflection coefficient is dependent on both the *P* and *S* velocity jump, and in the case of a positive *S* wave contrast and negative *P* wave contrast, the reflection coefficient of *PdP* is larger than if *P* and *S* wave speeds both increase or both decrease (Fig. 13.6). For these reasons, interpretation of discontinuity structure in terms of its apparent geographic distribution (Fig. 13.7) is not advised. For example, if the D″ discontinuity were to be observed globally, it would be likely to be caused by one single effect, with a phase change in perovskite being the most plausible option. The fact that it has only been observed in regional-scale studies leaves the possibility open that it could still be global, but seismically invisible in some regions, or that it is regional and caused by multiple different regional structures including subducted material. In addition to the bias introduced by wave-propagation complexity, the apparent observability of the D″ discontinuity also depends heavily on the limited distributions of earthquakes and seismic arrays.

Waveform polarities are a more robust seismic observable than waveform amplitudes because they do not change with epicentral distance (Fig. 13.8), and to our knowledge, they are not modified by the discontinuity thickness or velocity gradients. However, polarity measurements are most reliable when the data show a clear core-reflection phase (*PcP* or *ScS*) and when it has been verified that the polarities are not modified by the earthquake radiation pattern, i.e. that *P*, *PdP* and *PcP* all travel within the same quadrant of the focal mechanism. In such cases, reflectivity modelling (e.g. Weber 1993; Kito et al. 2007; Chaloner et al. 2009) demonstrates that the polarity of the *PdP* waveform will be the same as that of the *P* and *PcP* ("positive polarity") for a velocity increase at the D″ discontinuity, whereas the polarity of the *PdP* will be reversed relative to *P* and *PcP* for a velocity increase ("reverse polarity"; likewise for *S* wave phases)—Fig. 13.8. An example of a vespagram (slant stack) for a South American earthquake which samples D″ beneath the Caribbean is shown in Fig. 13.9. In this example, the polarity of the *PdP* indicates a negative *P* velocity jump at D″ and the polarity of the *SdS* indicates a positive *S* velocity jump. It can be shown (Cobden and Thomas 2013) that even in the absence of amplitude measurements, polarities can discriminate between different thermo-chemical origins for the D″ discontinuity, assuming that velocity contrasts greater than 1 % are needed to see a *PdP* or *SdS* phase (Fig. 13.9).

Despite the numerous pitfalls in estimating velocity changes, in some regions, consistent waveform polarities and amplitudes have been observed for multiple events. In these regions, it is perhaps reasonable (as a first-order approximation) to attempt to interpret the waveforms in terms of plausible thermo-chemical structures. Initially, such studies have been simple compatibility tests noting an agreement between the *S* and *P* velocity changes expected for conversion of $MgSiO_3$ Pv to pPv (Table 13.3) and the estimated velocity changes across the D″ discontinuity (Wookey et al. 2005b). The agreement is especially good in the Caribbean, i.e. beneath the Cocos Plate, where large increases in *S* wave speed (1–3 %) are

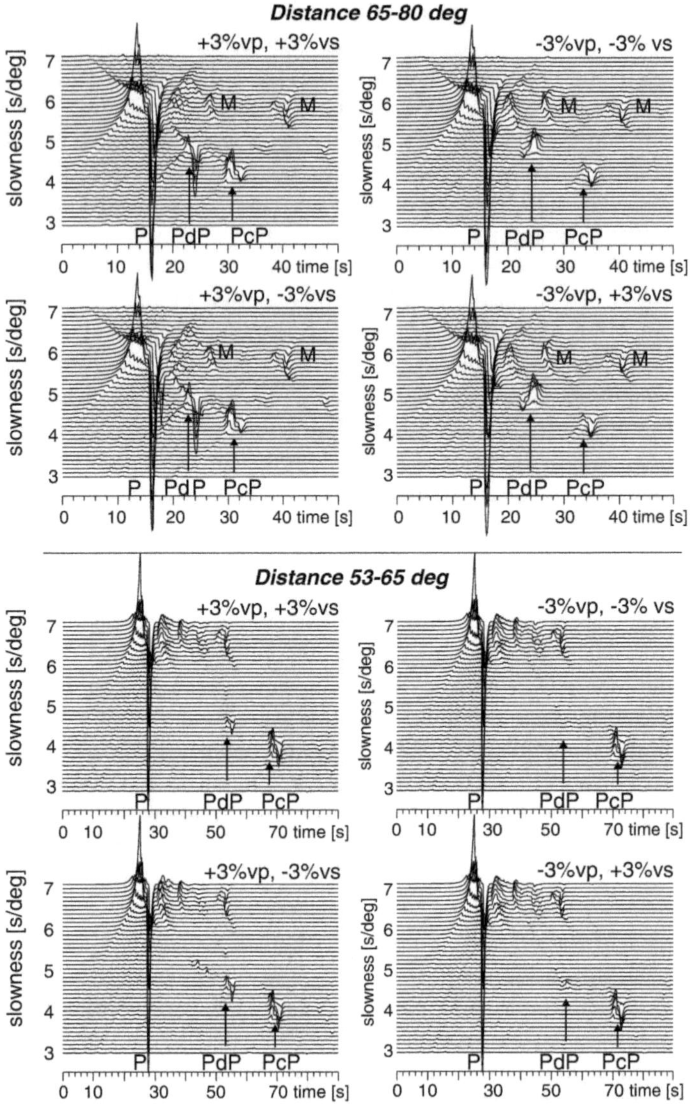

Fig. 13.8 *PdP* waveform polarity and amplitude as a function of epicentral distance, and for *P* and *S* wave speed increases and decreases. For a 1-D earth model, the polarity does not change with epicentral distance. Note the enhanced visibility of the *PdP* phase when the change in *V*p is the opposite sign to the change in *V*s, and the reversal of the *PdP* polarity for *P* wave speed decreases compared to *P* wave speed increases

seen in conjunction with negative (up to −3 %) or small positive (<1 %) changes in *P* wave speed (e.g. Lay and Helmberger 1983; Kendall and Nangini 1996; Ding and Helmberger 1997; Lay et al. 2004; Thomas et al. 2004a; Kito et al. 2007; Hutko et al. 2008). The example given in Fig. 13.9 shows synthetic vespagrams for a *P* velocity decrease of 1 % and *S* velocity increase of 2 % that are compatible with the waveform amplitudes and polarities of the real data.

It is more problematic to invoke post-perovskite for explaining observations of the D″ discontinuity underneath Siberia/Eurasia, where large-amplitude, positive polarity *PdP* phases, indicative of large (1–3 %) increases in *P* wave speed, have been documented by multiple studies (e.g. Weber and Davis 1990; Houard and Nataf 1992; Weber 1993; Scherbaum et al. 1997; Thomas and Weber 1997). Thomas et al. (2011) suggested that these large *P* wave speed increases may still be reconciled with a phase change from Pv to pPv, if the pPv grains are aligned due to flow, and the seismic waves preferentially sample the fast crystallographic direction. This hypothesis was based upon observations of opposite-polarity *PdP* waveforms for three source–receiver combinations with ray paths intersecting at 45° beneath Eurasia, compared with theoretically predicted velocity jumps due to anisotropic pPv underlying isotropic Pv. In this study, one source–receiver combination had a different geographic reflection point, so the varying polarities could also have arisen from different physical structures at the reflection points rather than anisotropy. To test this possibility, Cobden and Thomas (2013) attempted to model the polarities and amplitudes with a range of isotropic mineral structures, with and without post-perovskite. Using the mineral physics parameters compiled by Stixrude and Lithgow-Bertelloni (2011) and a Monte Carlo thermodynamic modelling procedure, they found that it is statistically difficult to generate large *P* velocity increases at the same time as large *S* velocity increases via phase change from Pv to isotropic pPv in multi-mineralic assemblages, whereas it is relatively straightforward to generate such velocity jumps using subducted oceanic crust. On the other hand, in the Caribbean, the waveform polarities and amplitudes are best fit by a phase change from Pv to isotropic pPv. For example, in the event shown in Fig. 13.9, from a sample of 1 million synthetic thermo-chemical models, the proportion of models generating a V_P decrease of 1 ± 0.5 % and V_S increase of 2 ± 1 % via conversion of Pv to pPv within a multi-mineralic assemblage is 21 %, whereas the percentage of models which generate these velocity changes via reflection off basaltic material is 0.04 %. The percentage of models which fit only the polarities (i.e. positive *SdS* and negative *PdP* regardless of the amplitude) via conversion of Pv to pPv is 91 %, compared with 4 % via reflections off basaltic material.

Overall, we can surmise that post-perovskite can fit the apparent velocity jumps at the D″ discontinuity (a) globally, if anisotropy is involved, or (b) only in specific locations, if isotropic. The latter scenario requires that the D″ discontinuity would be generated by different structures in different regions, perhaps not entirely unreasonable given the thermo-chemical complexity of the CMB region, although Occam's razor would favour the first scenario. In some regions (e.g. the Caribbean), post-perovskite appears to be the easiest way to fit the velocity

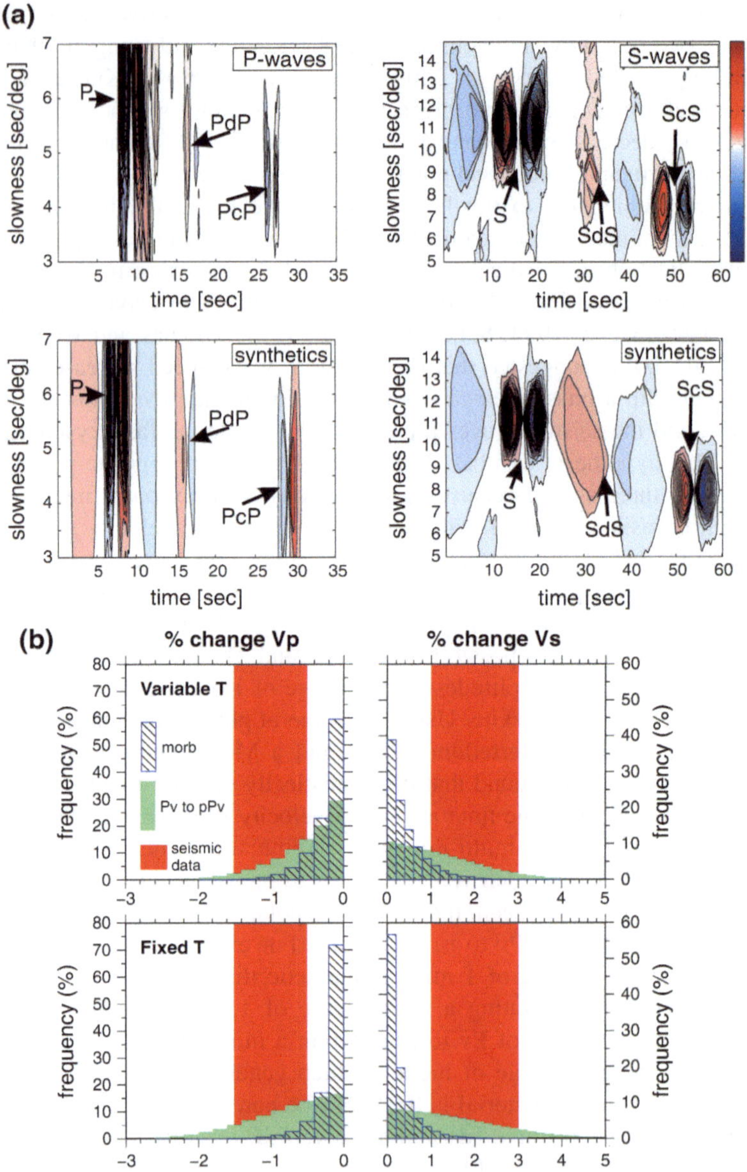
(a)
P-waves
P
PdP
PcP
slowness [sec/deg]
time [sec]
S-waves
S
SdS
ScS
positive
0
negative
synthetics
(b)
% change Vp
% change Vs
Variable T
morb
Pv to pPv
seismic data
frequency (%)
Fixed T

Fig. 13.9 **a** Waveform polarity and amplitude observations for the Caribbean, displayed as 4th-root vespagrams (Davies et al. 1971). *Top row* shows real data; the *PdP* waveform has the opposite polarity to the *P* and *PcP* waveforms, indicating a velocity decrease at D″ (see Fig. 13.8), while the *SdS* has the same polarity as the *S* and *ScS*, indicating a velocity increase at D″. The amplitude of the *PdP* relative to *P* and *PcP*, and the *SdS* relative to *S* and *ScS*, can be reproduced synthetically (*second row*) by assuming a −1 % velocity change in Vp and +2 % change in Vs at 2605 km depth. **b** deltaVp and deltaVs distributions for those thermo-chemical models from an initial set of 1 million in which both an increase in *Vs* and a decrease in *Vp* (consistent with the waveform polarities in **a**) are seen. *Green* transformation of Pv to pPv in a multi-mineralic assemblage; *black stripes* = change in MORB content across D″; *red bars* velocity changes consistent with the waveform amplitudes shown in **a**. While a large number of models in which there is a phase change from Pv to pPv can fit the waveform amplitudes, it is difficult to generate simultaneously large (>1 %) increases in *Vs* and large (>1 %) decreases in *Vp* via a change in MORB content at D″. *Upper row* results when the temperature is allowed to change across the D″ discontinuity. *Lower row* results when temperature is kept fixed across the discontinuity

changes compared to alternative thermo-chemical structures. However, given the complexity of inverting for waveform amplitudes, we should remain open to other possibilities, including the effects of 3-D scattering off folded/buckled slabs (e.g. Hutko et al. 2006) (i.e. short-wavelength topography). While the D″ discontinuity is seismically compatible with post-perovskite, it does not yet provide incontrovertible evidence for it.

13.3.1.5 Double Discontinuities

In the presence of a thermal boundary layer at the base of the mantle (Lay and Garnero 2004), in which the temperature increases rapidly, it has been suggested that a back transition of pPv to Pv may occur near the CMB (Hernlund et al. 2005), due to the positive Clapeyron slope of the phase boundary in the presence of high temperatures. If this happens, two seismic discontinuities could be generated, the upper one corresponding to the forward transformation of Pv to pPv and the lower one corresponding to the back transformation. Several studies have presented seismic evidence for such a double discontinuity beneath the Pacific and the Caribbean (Thomas et al. 2004a; Avants et al. 2006; Lay et al. 2006; Kito et al. 2007; van der Hilst et al. 2007; Hutko et al. 2008), although it is not always clear if the secondary waveform represents an out-of-plane arrival from a dipping reflector (Flores and Lay 2005; Hutko et al. 2006) or side lobes on the first arrival. Flores and Lay (2005) indicated that it is more difficult to detect a velocity decrease than an increase, and for the post-perovskite transition, a decrease in *S* wave velocity is expected for the back transformation. This means that even if double discontinuities exist, the lower reflector may not be detected. However, geodynamic modelling by Hernlund et al. (2005) predicts that the velocity contrast of the lower reflector should be much larger than that of the upper one, and theoretically, these large negative changes may be detected (e.g. Fig. 13.6). With careful analysis of waveform polarities and amplitudes, together with improved understanding of and ability to map discontinuity topography, interpretation of double discontinuities may be a fruitful area of future research. However, one should bear in mind, as for the upper discontinuity, a deeper discontinuity can have multiple causes, including ULVZs (Lay and Garnero 2007) or slabs (Thomas et al. 2004b), or thermal effects (Kawai and Tsuchiya 2009), and attributing it to post-perovskite is only appropriate with the exclusion of other explanations. If the lower discontinuity is indeed due to a back transformation of pPv to Pv, the depth between the two discontinuities should be anti-correlated (e.g. Hernlund et al. 2005), assuming they are not modified by chemical heterogeneity. Unfortunately, there are too few places yet to study this anti-correlation, but one indication of such anti-correlation can be seen in van der Hilst et al. (2007).

13.3.2 Anisotropy

The anisotropic structure of the Earth's lower mantle as derived from seismic observations has been discussed extensively in a number of excellent reviews (e.g. Kendall and Silver 1998; Wookey and Kendall 2007; Nowacki et al. 2011), and so we shall only summarise the features which may be pertinent to the detection of post-perovskite. Although post-perovskite is anisotropic for both *P* and *S* waves, the vast majority of mantle anisotropy studies are based upon *S* waves. This is because with *S* waves, there is the possibility to measure shear wave splitting, which is the difference in travel time between horizontally polarised (SH) and vertically polarised (SV) waves propagating in the same direction, and can be readily deduced from the radial and transverse components of a seismogram. Splitting occurs whenever a (polarised) transverse wave encounters an anisotropic medium; since *P* waves vibrate longitudinally, then they do not become split on encountering anisotropic media. Thus, *P* wave anisotropy can only be measured by multi-azimuthal sampling of the same target region, for which data are much sparser due to limited source–receiver configurations.

In global studies, seismic anisotropy in the Earth may be broadly divided into two categories: vertically transverse isotropy (VTI), which describes the difference in wave speed between radially (i.e. perpendicular to the CMB) and tangentially (CMB-parallel) propagating waves, and azimuthal anisotropy, which describes the difference in horizontal wave speed as a function of azimuth at a given depth. Many regional studies describe anisotropic structure in terms of tilted transverse isotropy (TTI) which, like VTI, assumes that the anisotropy is hexagonally symmetric, but while VTI assumes a vertical axis of symmetry, in TTI, the axis is (as the name suggests) inclined from the vertical. In theory, TTI could be decomposed into a VTI and azimuthal components although in practice, this is not the conventional parameterisation.

Global inversions for a 1-D anisotropy model assume that the anisotropy in the Earth is purely VTI. The degree of anisotropy at a given depth may then be expressed by the parameter $\xi = V_{\mathrm{SH}}^2/V_{\mathrm{SV}}^2$. Two studies using normal modes and/

or body waves have shown that while $\xi \approx 1$ throughout most of the lower mantle, i.e. on average the lower mantle is nearly isotropic, in the D″ region, ξ becomes greater than 1 (Montagner and Kennett 1996; Panning and Romanowicz 2006). Since post-perovskite is more anisotropic than perovskite, then qualitatively this observation is compatible with the hypothesis that the D″ discontinuity is associated with a phase transformation from Pv to pPv and that there is a significant amount of deformed pPv in the D″ region. Panning and Romanowicz (2006) suggested that such deformation could be generated by a general transition from vertical to horizontal flow associated with the mechanical boundary layer of the CMB (e.g. as sinking slabs fold at the CMB). However, a third study by Kustowski et al. (2008) found that while the average V_{SH} is greater than V_{SV} within D″, the difference is a factor of 5 smaller than that given by Panning and Romanowicz and that the inferred magnitude trades off strongly with P wave velocity in the outer core, which may not be tightly constrained. In a further study using a Monte Carlo model space search, Beghein et al. (2006) found no significant departure of ξ from PREM. However, the normal mode data did seem to require a 1-D (VTI) P wave anisotropy, although Beghein et al. (2006) noted that it trades off with density. Therefore, it is currently uncertain how robust the observation of $\xi > 1$ is.

Both global and regional seismic studies have attempted to map lateral variations in VTI at D″ depths. Although different studies do not always agree (especially in the central Pacific), and sometimes, it is not appropriate to assume the radial symmetry of VTI, a fairly robust feature is the observation that $V_{SH} > V_{SV}$ (typically by 1–3 %) in parts of the circum-Pacific (e.g. Lay and Young 1991; Matzel et al. 1996; Garnero and Lay 1997; Fouch et al. 2001; Panning and Romanowicz 2004, 2006; Usui et al. 2005; Thomas et al. 2007). These circum-Pacific regions are associated with subducted slabs. On the basis of this, Miyagi et al. (2010) applied their experimental observations of slip on the (001) plane in pPv to a geodynamic model of the lower mantle (McNamara and Zhong 2005) in order to compute the textural development in subducting slabs as they enter D″. They found that within the slabs, the (001) planes became aligned sub-parallel to the CMB, leading to the generation of $V_{SH} > V_{SV}$ anisotropy compatible with seismic observations. Similar calculations by Merkel et al. (2007) which assumed instead slip on the (010) plane could not reproduce the seismic observations. However, if it is true that slip does occur on the (001) plane, then this result is another seismic feature of the D″ region which is compatible with the presence of post-perovskite.

At present, observations of azimuthal anisotropy or TTI (e.g. Maupin et al. 2005; Wookey et al. 2005a; Rokosky et al. 2006; Thomas et al. 2007; Wookey and Kendall 2008; Long 2009; Nowacki et al. 2010) are not, per se, uniquely diagnostic of post-perovskite, because it is not clear how to distinguish LPO anisotropy in pPv from LPO anisotropy in other minerals (i.e. magnesiowüstite) or from SPO (Kendall and Silver 1998; Nowacki et al. 2011). However, we foresee that the most immediately applicable use of azimuthal anisotropy (in relation to post-perovskite) would be for studies of waveform polarities and amplitudes at the D″ discontinuity, as implemented by Thomas et al. (2011) (and as discussed in

Sect. 13.3.1.4). In order to measure anisotropy across the D″ discontinuity accurately, multi-azimuthal coverage of the same region by multiple raypaths—to determine *PdP* and *SdS* waveforms as a function of azimuth—would be required (Wookey and Kendall 2008; Nowacki et al. 2010; Thomas et al. 2011). Multi-azimuthal sampling would also help to constrain the 3-D morphology of topography on the discontinuity. Currently, such studies are unfortunately hindered by the limited distribution of seismic arrays (and earthquakes). Even if the presence and orientation of azimuthal anisotropy would be accurately constrained, in order to demonstrate that it is the result of post-perovskite, the anisotropic velocities of multi-mineralic assemblages and not only pure $MgSiO_3$ should be computed, together with calculations involving anisotropy in magnesiowustite rather than post-perovskite. Furthermore, much more work remains to be done on understanding the deformation mechanisms within D″, in addition to the seismic properties of melt inclusions which may generate SPO.

13.3.3 Long-Wavelength Seismic Structures

13.3.3.1 Presence or Absence of Post-perovskite at a Global Scale

Long-wavelength seismic tomography may also provide clues about the existence of post-perovskite in the Earth (e.g. Houser 2007; Cobden et al. 2012; Mosca et al. 2012). Since post-perovskite is most likely to occur near the CMB (Table 13.1), it is most instructive to examine tomographic models of the bottom few hundred kilometres of the mantle, and specifically, in order to distinguish thermal from chemical or mineralogical variations, then models of at least two independent seismic parameters (i.e. V_P, V_S or density) are required, at the same spatial resolution. Conventionally, tomographic data are presented as maps of lateral variations in sound speed and density with respect to a reference model (e.g. Fig. 13.10). These red-and-blue maps give a clear visual display of "fast" and "slow" regions with respect to the reference model, but details of the distributions of seismic velocities are lost due to the essentially binary parameterisation of seismic structure. Another way of analysing the models is to compile histograms of the frequency at which each value of a velocity perturbation occurs at a given depth (e.g. Deschamps and Trampert 2003; Hernlund and Houser 2008). Histograms showing the global distributions of *P* velocity perturbations ($d\ln V_P$) and *S* velocity perturbations ($d\ln V_S$) with respect to PREM (Dziewonski and Anderson 1981) at 2741 km and 2981 km depth (Mosca 2010) are shown in Fig. 13.11. The position and shape (i.e. width and skew) of these histograms are influenced by two factors: (1) the amplitude of any lateral variations in temperature and composition, and the frequency with which such variations occur and (2) errors and uncertainties in the tomographic model used to construct the histograms. Assuming that the second factor is sufficiently small relative to the first factor, we can use the histograms to place constraints on the thermo-chemical variations which are present at a given depth.

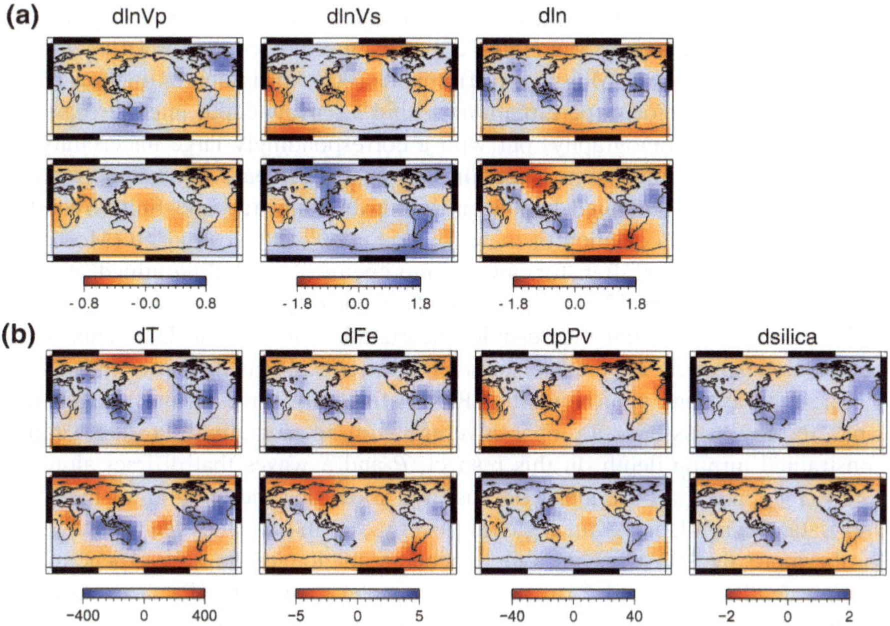

Fig. 13.10 **a** Maps of $\mathrm{dln}V_\mathrm{P}$, $\mathrm{dln}V_\mathrm{S}$ and $\mathrm{dln}\rho$ at 2600 km (*upper row*) and 2891 km (*lower row*) with respect to PREM, inferred from joint inversion of normal modes and body waves by Mosca et al. (2012). Mosca et al. used the neighbourhood algorithm to derive probability density functions (*pdf*s) of the seismic structure at 15° intervals; the maps here display the mean of each *pdf*. Only the seismic structure corresponding to even spherical harmonic degrees 0, 2, 4 and 6 was calculated. **b** Conversion of seismic structure into probabilistic maps of temperature, iron, post-perovskite and silicate variations at 2600 km (*upper row*) and 2891 km (*lower row*) by Mosca et al. (2012) using the Metropolis–Hastings algorithm (Mosegaard and Tarantola 1995). *pdf*s of the thermo-chemical structure were calculated at 15° intervals; the maps show the mean of each *pdf* within a given 15° bin. dFe refers to the volumetric change in iron-bearing minerals ($FeSiO_3$) perovskite and post-perovskite and FeO wüstite. dsilica refers to the volumetric change in Si-bearing minerals (Mg, Fe)SiO_3 perovskite and post-perovskite, $CaSiO_3$ calcium perovskite, and free SiO_2. dpPv refers to the percentage change in pPv within the (Mg, Fe)SiO_3 perovskite and post-perovskite phases. Note that at 2600 km, the seismic reference model contains ~67 ± 20 % pPv, while at 2891 km, it contains 63 ± 23 % pPv (see Fig. 13 in Mosca et al. 2012)

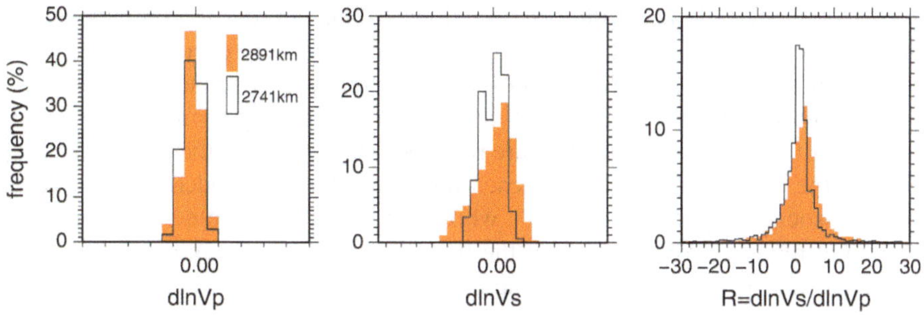

Fig. 13.11 Histograms showing the frequency distributions of lateral variations of V_P ($\mathrm{dln}V_\mathrm{P}$), V_S ($\mathrm{dln}V_\mathrm{S}$) and seismic parameter $R = \mathrm{dln}V_\mathrm{P}/\mathrm{dln}V_\mathrm{S}$ at two depths near the CMB, calculated from a global compilation of *P* and *S* wave travel times by Mosca (2010) using the path average approximation (Mosca and Trampert 2009)

The histograms in Fig. 13.11 were generated from body wave travel time observations (Ritsema and van Heijst 2002) using a procedure known as the path average approximation (Mosca and Trampert 2009). This approximation allows us to map lateral velocity perturbations within a narrow depth interval (narrower than conventional tomography) but with a correspondingly large uncertainty. By benchmarking the travel time anomalies given by the path average approximation against those generated by the more conventional ray and finite-frequency theory, the standard deviation of these uncertainties can be estimated (Mosca and Trampert 2009). Thereafter, for any thermo-chemical modelling aimed at fitting the histograms in Fig. 13.11, we can apply a correction to the synthetic models which takes into account the seismic uncertainties of the real data. This makes comparison between seismic data and thermo-chemical structures more consistent.

The path average approximation (PAVA) for body waves is most appropriate when the ray paths have a global coverage with no azimuthal bias and a tightly constrained turning depth. In this respect, P and S waves that diffract along the CMB (P_{diff} and S_{diff}) are ideal, because we know that they bottom at the CMB, and data are available with a global coverage (Ritsema and van Heijst 2002). Furthermore, the PAVA becomes more accurate the longer the waves spend at their turning depth, and P and S diffracted waves can travel long distances along the CMB before returning to shallower mantle. In Cobden et al. (2012), we attempted to construct sets of thermo-chemical variations which would reproduce the global distributions of $\mathrm{dln}V_{\mathrm{P}}$, $\mathrm{dln}V_{\mathrm{S}}$ and $R = \mathrm{dln}V_{\mathrm{P}}/\mathrm{dln}V_{\mathrm{S}}$ obtained from P_{diff} and S_{diff} arrivals (Fig. 13.11, 2891 km). In this paper, we also show results from similar calculations performed for P and S waves with bottoming depth of ~2741 km (Fig. 13.11, 2741 km). The uncertainty in the ray turning depth is greater at this depth (a slowness range of ± 0.5 s/deg is used) than at the CMB, and the ray coverage is also less dense, so the contribution of seismic uncertainties to the histograms is greater. At shallower depths than this, the ray coverage becomes too biased to represent the global structure so no thermo-chemical modelling is undertaken.

For each of the two depths considered, the seismic velocities corresponding to a given thermo-chemical structure are calculated using the mineral elastic parameters and equation of state given in (Stixrude and Lithgow-Bertelloni 2005; Xu et al. 2008; Stixrude and Lithgow-Bertelloni 2011). In order to compute seismic derivatives $\mathrm{dln}V_{\mathrm{P}}$ and $\mathrm{dln}V_{\mathrm{S}}$, we first generate a set of "reference models", whose temperature and mineralogy are chosen at random within broad but pre-defined ranges (see Cobden et al. 2012 for details), of which only those whose seismic properties fit PREM to within 1 % are retained (since this was the reference model used in the real seismic data). For each retained reference model, we then generate an associated "perturbed model", whose temperature and mineralogy are again chosen at random. We assume that the finite difference in V_{P} and V_{S} between the reference and perturbed model is approximately equal to $\mathrm{dln}V_{\mathrm{P}}$ and $\mathrm{dln}V_{\mathrm{S}}$. We compute sets of up to 800,000 initial reference and perturbed model pairs and thus 800,000 values of $\mathrm{dln}V_{\mathrm{P}}$ and $\mathrm{dln}V_{\mathrm{S}}$ from which we can construct histograms of their frequency distribution. Although the seismic properties of the synthetic models follow a much broader and more uniform distribution than the histograms in Fig. 13.11, due to the broad range of permitted thermo-chemical variations, a fitting procedure called the Metropolis–Hastings algorithm (Mosegaard and Tarantola 1995) can be used to extract a subset of these models which fit the frequency distributions of the real seismic data as closely as possible. Immediately prior to performing the Metropolis–Hastings (M-H) algorithm, seismic uncertainties following a normal distribution, with a standard deviation given by the uncertainty of the PAVA at that depth, are added to the synthetic seismic velocities.

We produced several different sets of 800,000 models. In one set, we specified that no post-perovskite should be present, and in another set, we allowed a mixture of both perovskite and post-perovskite. At the CMB, we also created a set of models in which no perovskite was present. We then compared the relative fit of the different sets to the histograms of Fig. 13.11. Although with all of the sets of models, it is possible to extract a subset which reproduces the real data histograms to more than 90 %, when we studied the seismic uncertainties of the accepted models, we found significant biases in the uncertainties of the "no post-perovskite" models. (Fig. 13.12a) Ideally, if a set of thermo-chemical models is fitting the seismic data due to its underlying thermo-chemical properties, then the mean of the uncertainties should be 0, since this was the mean of the uncertainties before data fitting. For the "no post-perovskite" models at 2891 km, the mean of the uncertainties is shifted by more than half a standard deviation towards negative values in $\mathrm{dln}V_P$ and positive values in $\mathrm{dln}V_S$ (Fig. 13.12a). At 2741 km, the bias in the uncertainties is still present, although the effect is smaller. For models containing a mixture of Pv and pPv, the uncertainties are much less biased, and at 2891 km, no significant difference can be seen for models which contain a mixture of Pv and pPv and models which contain no Pv. This result can be taken as an indication that a pPv-bearing CMB region fits the seismic data better than a pPv-free one.

For models in which both Pv and pPv were present, we allowed the percentage of pPv in the reference models to vary randomly between 0 and 100 %. The pPv contents for those (reference) models which were accepted by the M-H algorithm are shown in Fig. 13.12b. At 2891 km, the distribution has a strong negative skew with a peak at 75–80 % pPv. At 2741 km, the skew is weaker and the modal histogram bin is at 65–70 % pPv. The fact that most models favour a large percentage of pPv to be present is supportive of the hypothesis that a significant amount of pPv is present within D″. The fact that the distribution becomes more uniform at 2741 km compared to 2891 km, with a peak at a lower percentage of pPv, taken together with a smaller misfit of the *P* and *S* velocity uncertainties (Fig. 13.12a), is consistent with having a lower overall amount of pPv at a depth of 2741 km than at 2891 km. We find that it is particularly difficult to generate models which fit PREM within 1 % at 2891 km when there is zero or little post-perovskite, because the elastic parameters of the other minerals present favour very low temperatures (c. 2000 K) which are unlikely to occur near the CMB.

A related study by Mosca et al. (2012) combined body wave travel times with normal mode observations to generate a probabilistic tomography model of the lowermost mantle (Fig. 13.10). The inclusion of normal modes provides

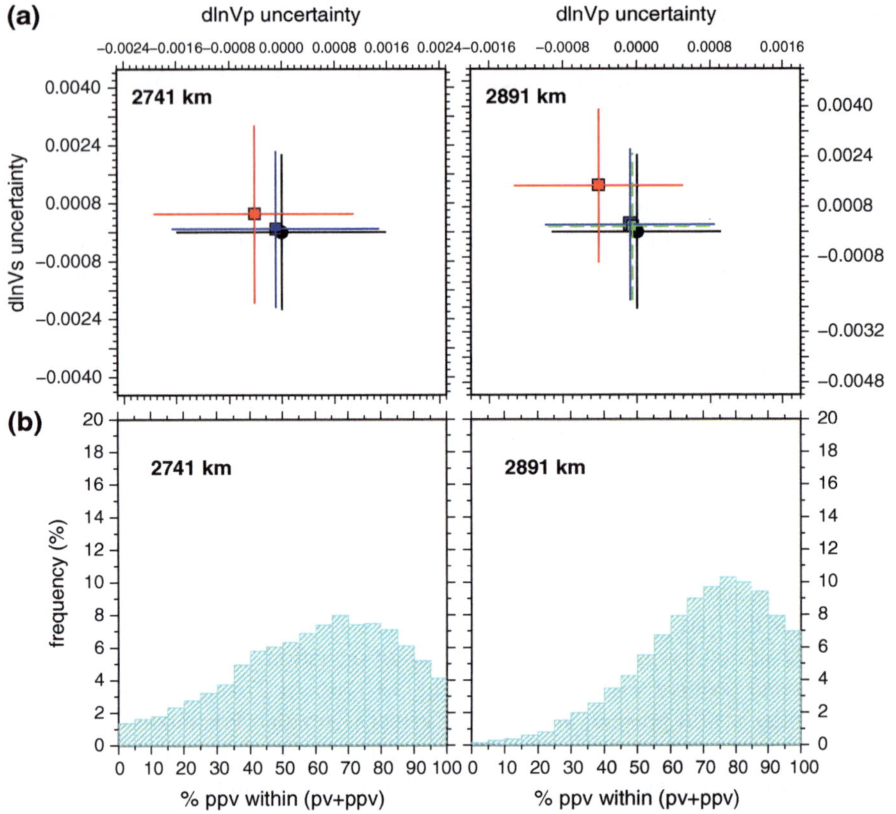

Fig. 13.12 **a** Means (*squares*) and standard deviations (*lines*) of seismic uncertainties of thermo-chemical models. *Black* uncertainty distribution before applying Metropolis–Hastings algorithm. *Colours* uncertainty distribution of accepted models after M-H. *Blue* Pv and pPv present; *red* no pPv present; *green* no Pv present. **b** Distribution of %pPv in the reference models used to generate seismic derivatives, which fit PREM within 1 % and are accepted by the M-H algorithm. %pPv refers to the volumetric percentage of pPv within the (Mg, Fe)SiO_3 perovskite and post-perovskite phases

constraints on the density structure ρ in addition to V_P and V_S. Using the M-H algorithm, Mosca et al. (2012) converted the probability density functions (*pdf*s) of V_P, V_S and ρ at each node of the tomography model into pdfs of temperature, iron, pPv and silicate variations, with the mean of the pdfs being indicative of the most likely value of dT, dFe, dpPv and dsilicate at each node (Fig. 13.10b). Mosca et al. found that the mean degree 0 structure of the models fitting the seismic data *pdf*s contained 67 % pPv at 2819 km (with a standard deviation of 20 %) and 63 % pPv (with standard deviation 23 %) at 2600 km. Given that the calculations involved a different data set (i.e. normal modes) and slightly different depth parameterisation than the PAVA calculations, these results are reasonably consistent with the PAVA calculations and again show a slight decrease in the total amount of pPv near the top of D″ compared to next to the CMB.

13.3.3.2 Lateral Variations in Post-perovskite Content

At 2891 km, we performed a simple cluster analysis (Sparks 1973) of the P_{diff} and S_{diff} travel time perturbations as a function of latitude and longitude, to divide the Earth into 4 regions possessing similar seismic characteristics (Cobden et al. 2012). One region corresponded to the position of the two large low-shear wave velocity provinces (LLSVP) under Africa and the Pacific and another correlated with fast circum-Pacific regions associated with subducted slabs. We found that in all four regions, "no post-perovskite" mantle models had a greater misfit to the seismic data than post-perovskite-bearing models, and thus, with this method, we could not detect a preference for post-perovskite to be located in a particular geographic area (such as inside or outside the LLSVPs). The tomography models of Mosca et al. (2012), however, do provide insights into the likely lateral variations of pPv content, at 2600 and 2891 km (Fig. 13.10b). These probabilistic maps show that lateral variations in pPv content correlate strongly with $\mathrm{dln}V_S$, but less so with $\mathrm{dln}V_P$ or density. At 2600 km, dpPv is most positive (and hence the total volume of pPv most likely to be highest) in the northwest and eastern margins of the Pacific, including the Caribbean, as well as throughout most of Eurasia and the Indian Ocean. dpPv is most negative (i.e. the total amount of pPv is most likely to be lowest) in the central and south-west Pacific, and beneath Africa. At 2891 km, the amplitude of negative variations decreases, but the pattern of positive and negative variations becomes more complex.

The patterns of pPv variations at 2600 km are very broadly compatible with observations of VTI anisotropy ($V_{sh} > V_{sv}$) in the circum-Pacific (Sect. 13.3.2), and the idea that post-perovskite is most likely to be found in subducted slabs (whereby the anisotropy arises from slab deformation). That the two sets of observations do not agree exactly is not surprising when differences in the spatial resolution and parameterisation between regional anisotropy studies and global tomography inversions are taken into account. In particular, the tomography study of Mosca et al. (2012) inverts only for the even spherical harmonic degrees, 0, 2, 4 and 6, while at the same time, the structure is averaged vertically over depth intervals of ~300 km. In future, inclusion of odd degrees and a finer depth resolution could allow a closer comparison between regional and global seismic studies.

Post-perovskite has been predicted to be most abundant in slabs due to their colder temperature—the positive Clapeyron slope of the phase transition could restrict pPv to only the colder parts of the D″ region (e.g. Wookey et al. 2005b). Yet, in the models of Mosca et al. (2012), the correlation between dT and dpPv is positive instead of negative. This could be indicative of a more dominant influence by chemistry than temperature on the appearance of pPv inside the Earth (although at the same time, dpPv does not correlate strongly with dFe). However, it is worth noting that one of the main restrictions of the modelling procedure of both Cobden et al. (2012) and Mosca et al. (2012) is that mineral abundances are varied randomly with no imposed phase equilibria. Thus, it is possible (and indeed likely) that we generated thermo-chemical combinations which are not thermodynamically stable. On the other hand, as we showed in Sect. 13.2.1, it is not meaningful to impose phase equilibria, when the structure of the Pv–pPv phase boundary remains so highly unconstrained, and doing so will simply force the solution towards a biased answer which does not reflect the full uncertainty of the model. Furthermore, the existence of mechanical mixtures of different compositions (Xu et al. 2008) could produce bulk mineralogies which would not actually be thermodynamically stable as a uniform assemblage. The results presented in this section rely additionally on the accuracy of the assumed mineral elastic parameters. We have used one of the most comprehensive and self-consistent data sets currently available (Stixrude and Lithgow-Bertelloni 2011), but some of the parameters are not yet tightly constrained by experiments (e.g. shear velocities of post-perovskite; see Sect. 13.2.3) and may change significantly in the future. While there is a chance that this could impact our interpretation, we do not anticipate major changes, since we have been working with velocity perturbations, which are several times less sensitive to changes in elastic parameters than absolute velocities.

13.3.4 Future Directions

The probes outlined in Sects. 13.3.1–13.3.3 have primarily been derived using classical methods (e.g. 1-D reflectivity forward modelling of waveform stacks for the seismic discontinuities and ray theory for the tomography models) which aim to fit the timing, amplitude or polarity of specific wave phases. A more accurate, but computationally demanding, approach is to reconstruct the entire seismic waveform, i.e. full-waveform inversion. This approach has the advantage that weak or overlapping phases, which are difficult to observe with classical methods, are included in the analysis, and subjective data-picking errors are avoided. For example, in traditional discontinuity studies of *PdP* and *SdS* phases, there may be a bias towards observing velocity increases over decreases (Fig. 13.6, Gaherty and Lay 1992; Flores and Lay 2005) and towards accumulating the most observations in regions with the strongest amplitude reflections (Lay and Garnero 2011). While more complete descriptions of seismic wave propagation have been developed in recent years, including the computation of finite-frequency kernels (e.g. Montelli et al. 2004; Panning and Romanowicz 2006) and full-waveform inversions (e.g. Fichtner et al. 2010), for the lowermost mantle, these methods are still in the early stages of development (e.g. Fuji et al. 2012) and have not yet been used to map 3-D seismic structure. However, they have been used to construct regional 1-D profiles of the *S* wave speed within D″ in regions of dense data coverage (e.g. Kawai et al. 2007a, b, 2010; Konishi et al. 2009; Kawai and Geller 2010; Konishi et al. 2012). These 1-D profiles are inevitably smooth and oscillatory in appearance and cannot distinguish sharp discontinuities from gradual changes in wave speed. They are qualitatively consistent with discontinuity studies (Sect. 13.3.1) in that they tend to show an increase in *S* wave speed ~200–400 km above the CMB, which is broadly compatible with a phase change from perovskite to post-perovskite. In order to be diagnostic of the presence or absence of post-perovskite, and with no information about the sharpness of the *S* wave speed changes, then in future, additional information about the *P* wave speed structure or the density structure is required, preferably in three dimensions.

13.4 Conclusions

We have seen that a wide range of seismic observations on different length scales are compatible with the presence of post-perovskite inside the Earth and with the hypothesis that the D″ discontinuity is (at least partially) associated with a phase change from perovskite to post-perovskite. Although every set of seismic observations has significant uncertainties, and although a compatibility between seismic observations and the physical properties of post-perovskite does not exclude all other physical explanations for the seismic structure, the fact that multiple data types and data inversions can be reconciled with the existence of post-perovskite, in some cases more readily than without post-perovskite, provides a strong case that the mineral is indeed present in the lowermost mantle. We anticipate that in seismology, multi-azimuthal studies of anisotropy will play an important role in further constraining deep mantle mineralogical structures, while full-waveform inversion methods may eventually have the capacity to image 3-D seismic structures in D″ at high resolution. In mineral physics, the phase equilibria between perovskite and post-perovskite are, at present, controversial. However, we expect that with the accumulation of more data, in particular, investigations of mixed-phase assemblages and calibration of the rate of conversion of Pv to pPv with increasing pressure, we will eventually converge on a consistent model of the phase equilibria, which can be applied directly to seismic observations.

Acknowledgments The authors thank Nobuaki Fuji and Frédéric Deschamps for helpful comments which improved the manuscript. This work was partially funded by the DFG (German Research Foundation) on grant number TH1530/5-1 and the NWO (Dutch Science Foundation) on grant number NWO: VIDI 864.11.008.

References

Akahama Y, Kawamura H (2004) High-pressure Raman spectroscopy of diamond anvils to 250 GPa: method for pressure determination in the multimegabar pressure range. J Appl Phys 96:3748–3751

Akahama Y, Kawamura H, Singh A (2002) Equation of state of bismuth to 222 GPa and comparison of gold and platinum pressure scales to 145 GPa. J Appl Phys 92:5892–5897

Akber-Knutson S, Steinle-Neumann G, Asimow PD (2005) Effect of Al on the sharpness of the $MgSiO_3$ perovskite to post-perovskite phase transition. Geophys Res Lett 32:L14303

Ammann MW, Brodholt JP, Wookey J, Dobson DP (2010) First-principles constraints on diffusion in lower-mantle minerals and a weak D″ layer. Nature 465:462–465

Anderson OL, Isaak DG, Yamamoto S (1989) Anharmonicity and the equation of state for gold. J Appl Phys 65:1534–1543

Andrault D, Munoz M, Bolfan-Casanova N, Guignot N, Perrillat J, Aquilanti G, Pascarelli S (2010) Experimental evidence for perovskite and post-perovskite coexistence throughout the whole D″ region. Earth Planet Sci Lett 293:90–96

Asanuma H, Ohtani E, Sakai T, Terasaki H, Kamada S, Kondo T, Kikegawa T (2010) Melting of iron-silicon alloy up to the core-mantle boundary pressure: implications to the thermal structure of the Earth's core. Phys Chem Miner 37:353–359

Avants M, Lay T, Russell S, Garnero E (2006) Shear velocity variation within the D″ region beneath the central Pacific. J Geophys Res Solid Earth 111:B05305

Baumgardt DR (1989) Evidence for a P-wave velocity anomaly in D″. Geophys Res Lett 16:657–660

Beghein C, Trampert J, van Heijst HJ (2006) Radial anisotropy in seismic reference models of the mantle. J Geophys Res Solid Earth 111:B02303

Bullen KE (1950) An earth model based on a compressibility-pressure hypothesis. Geophys J Int 6:50–59

Cammarano F, Goes S, Deuss A, Giardini D (2005) Is a pyrolitic adiabatic mantle compatible with seismic data? Earth Planet Sci Lett 232:227–243

Campbell AJ, Seagle CT, Heinz DL, Shen G, Prakapenka VB (2007) Partial melting in the iron-sulfur system at high pressure: a synchrotron X-ray diffraction study. Phys Earth Planet Inter 162:119–128

Caracas R, Cohen RE (2005a) Effect of chemistry on the stability and elasticity of the perovskite and post-perovskite phases in the $MgSiO_3$-$FeSiO_3$-Al_2O_3 system and implications for the lowermost mantle. Geophys Res Lett 32:L16310

Caracas R, Cohen R (2005b) Prediction of a new phase transition in Al_2O_3 at high pressures. Geophys Res Lett 32:L06303

Catalli K, Shim S, Prakapenka V (2009) Thickness and Clapeyron slope of the post-perovskite boundary. Nature 462:782–785

Chaloner JW, Thomas C, Rietbrock A (2009) P- and S-wave reflectors in D′ beneath southeast Asia. Geophys J Int 179:1080–1092

Chambers K, Woodhouse JH (2006) Transient D″ discontinuity revealed by seismic migration. Geophys Res Lett 33:L17312

Christensen UR (1989) Models of mantle convection—one or several layers. Philos Trans R Soc Lond Ser A Math Phys Eng Sci 328:417–424

Christensen UR, Hofmann AW (1994) Segregation of subducted oceanic-crust in the convecting mantle. J Geophys Res Solid Earth 99:19867–19884

Cobden L, Thomas C (2013) The origin of D′ reflections: a systematic study of seismic array data sets. Geophys J Int 194:1091–1118

Cobden L, Goes S, Ravenna M, Styles E, Cammarano F, Gallagher K, Connolly JAD (2009) Thermochemical interpretation of 1-D seismic data for the lower mantle: the significance of nonadiabatic thermal gradients and compositional heterogeneity. J Geophys Res Solid Earth 114:B11309

Cobden L, Mosca I, Trampert J, Ritsema J (2012) On the likelihood of post-perovskite near the core-mantle boundary: a statistical interpretation of seismic observations. Phys Earth Planet Inter 210:21–35

Cococcioni M, de Gironcoli S (2005) Linear response approach to the calculation of the effective interaction parameters in the LDA + U method. Phys Rev B 71:035105

Davies GF, Gurnis M (1986) Interaction of mantle dregs with convection: lateral heterogeneity at the core mantle boundary. Geophys Res Lett 13:1517–1520

Davies D, Kelly E, Filson J (1971) Vespa process for analysis of seismic signals. Nat Phys Sci 232:8–13

Davis JP, Weber M (1990) Lower mantle velocity inhomogeneity observed at GRF array. Geophys Res Lett 17

Deschamps F, Trampert J (2003) Mantle tomography and its relation to temperature and composition. Phys Earth Planet Inter 140:277–291

Dewaele A, Loubeyre P, Mezouar M (2004) Equations of state of six metals above 94 GPa. Phys Rev B 70:094112

Ding X, Helmberger D (1997) Modelling D″ structure beneath Central America with broadband seismic data. Phys Earth Planet Inter 101:245–270

Dziewonski AM, Anderson DL (1981) Preliminary reference earth model. Phys Earth Planet Inter 25:297–356

Fei Y, Van Orman J, Li J, van Westrenen W, Sanloup C, Minarik W, Hirose K, Komabayashi T, Walter M, Funakoshi K (2004) Experimentally determined postspinel transformation boundary in Mg_2SiO_4 using MgO as an internal pressure standard and its geophysical implications. J Geophys Res Solid Earth 109:B02305

Fei Y, Ricolleau A, Frank M, Mibe K, Shen G, Prakapenka V (2007) Toward an internally consistent pressure scale. Proc Natl Acad Sci USA 104:9182–9186

Fichtner A, Kennett BLN, Igel H, Bunge H (2010) Full waveform tomography for radially anisotropic structure: new insights into present and past states of the Australasian upper mantle. Earth Planet Sci Lett 290:270–280

Flores C, Lay T (2005) The trouble with seeing double. Geophys Res Lett 32:L24305

Fouch M, Fischer K, Wysession M (2001) Lowermost mantle anisotropy beneath the Pacific: imaging the source of the Hawaiian plume. Earth Planet Sci Lett 190:167–180

Fuji N, Chevrot S, Zhao L, Geller RJ, Kawai K (2012) Finite-frequency structural sensitivities of short-period compressional body waves. Geophys J Int 190:522–540

Gaherty JB, Lay T (1992) Investigation of laterally heterogeneous shear velocity structure in D″ beneath Eurasia. J Geophys Res Solid Earth 97:417–435

Garnero EJ, Lay T (1997) Lateral variations in lowermost mantle shear wave anisotropy beneath the north Pacific and Alaska. J Geophys Res Solid Earth 102:8121–8135

Grand S (2002) Mantle shear-wave tomography and the fate of subducted slabs. Philos Trans R Soc Lond Ser A Math Phys Eng Sci 360:2475–2491

Grocholski B, Catalli K, Shim S, Prakapenka V (2012) Mineralogical effects on the detectability of the postperovskite boundary. Proc Natl Acad Sci USA 109:2275–2279

Guignot N, Andrault D, Morard G, Bolfan-Casanova N, Mezouar M (2007) Thermoelastic properties of post-perovskite phase $MgSiO_3$ determined experimentally at core-mantle boundary P-T conditions. Earth Planet Sci Lett 256:162–168

Gutenberg B (1914) Uber Erdbebenwellen. VII A Beobachtungen an Registrierungen von Fernbeben in Gottingen und Folgerungen uber die Konstitution des Erdkorpers. Nachr d K Ges d Wiss zu Gottingen, Math Phys Klasse 125–177

Hernlund JW, Houser C (2008) The statistical distribution of seismic velocities in Earth's deep mantle. Earth Planet Sci Lett 265:423–437

Hernlund JW, Thomas C, Tackley PJ (2005) A doubling of the post-perovskite phase boundary and structure of the Earth's lowermost mantle. Nature 434:882–886

Hirose K (2006) Postperovskite phase transition and its geophysical implications. Rev Geophys 44:RG3001

Hirose K, Takafuji N, Sata N, Ohishi Y (2005) Phase transition and density of subducted MORB crust in the lower mantle. Earth Planet Sci Lett 237:239–251

Hirose K, Sinmyo R, Sata N, Ohishi Y (2006) Determination of post-perovskite phase transition boundary in $MgSiO_3$ using Au and MgO pressure standards. Geophys Res Lett 33:L01310

Hirose K, Sata N, Komabayashi T, Ohishi Y (2008a) Simultaneous volume measurements of Au and MgO to 140 GPa and thermal equation of state of Au based on the MgO pressure scale. Phys Earth Planet Inter 167:149–154

Hirose K, Takafuji N, Fujino K, Shieh SR, Duffy TS (2008b) Iron partitioning between perovskite and post-perovskite: a transmission electron microscope study. Am Mineral 93:1678–1681

Hirose K, Nagaya Y, Merkel S, Ohishi Y (2010) Deformation of $MnGeO_3$ post-perovskite at lower mantle pressure and temperature. Geophys Res Lett 37:L20302

Holmes NC, Moriarty JA, Gathers GR, Nellis WJ (1989) The equation of state of platinum to 660 gpa (6.6 mbar). J Appl Phys 66:2962–2967

Houard S, Nataf HC (1992) Further evidence for the lay discontinuity beneath northern Siberia and the North-Atlantic from short-period P-waves recorded in France. Phys, Earth Planet Inter 72

Houard S, Nataf HC (1993) Laterally varying reflector at the top of D″ beneath northern Siberia. Geophys J Int 115

Houser C (2007) Constraints on the presence or absence of post-perovskites in the lower mantle from long-period seismology. In: Hirose K, Brodholt JP, Lay T, Yuen DA (eds) Post-perovskite: the last mantle phase transition, vol 174. American Geophysical Union, Washington D.C., pp 191–216

Hunt SA, Weidner DJ, Li L, Wang L, Walte NP, Brodholt JP, Dobson DP (2009) Weakening of calcium iridate during its transformation from perovskite to post-perovskite. Nat Geosci 2:794–797

Hutko A, Lay T, Garnero E, Revenaugh J (2006) Seismic detection of folded, subducted lithosphere at the core-mantle boundary. Nature 441:333–336

Hutko AR, Lay T, Revenaugh J, Garnero EJ (2008) Anticorrelated seismic velocity anomalies from post-perovskite in the lowermost mantle. Science 320:1070–1074

Hutko AR, Lay T, Revenaugh J (2009) Localized double-array stacking analysis of PcP: D″ and ULVZ structure beneath the Cocos plate, Mexico, central Pacific, and north Pacific. Phys Earth Planet Inter 173:60–74

Iitaka T, Hirose K, Kawamura K, Murakami M (2004) The elasticity of the MgSiO(3) post-perovskite phase in the Earth's lowermost mantle. Nature 430:442–445

Ishii M, Tromp J (1999) Normal-mode and free-air gravity constraints on lateral variations in velocity and density of Earth's mantle. Science 285:1231–1236

Jamieson JC, Fritz JN, Manghnani MH (1982) Pressure measurements at high temperature in X-ray diffraction studies: gold as a primary standard. In: Akimoto S, Manghnani Murli H (eds) High-pressure research in geophysics. Centre for Academic Publications Japan, Tokyo, pp 27–48

Kamada S, Terasaki H, Ohtani E, Sakai T, Kikegawa T, Ohishi Y, Hirao N, Sata N, Kondo T (2010) Phase relationships of the Fe-FeS system in conditions up to the Earth's outer core. Earth Planet Sci Lett 294:94–100

Kaneshima S, Helffrich G (1999) Dipping low-velocity layer in the mid-lower mantle: evidence for geochemical heterogeneity. Science 283:1888–1891

Kawai K, Geller RJ (2010) Waveform inversion for localized seismic structure and an application to D″ structure beneath the Pacific. J Geophys Res Solid Earth 115:B01305

Kawai K, Tsuchiya T (2009) Temperature profile in the lowermost mantle from seismological and mineral physics joint modeling. Proc Natl Acad Sci USA 106:22119–22123

Kawai K, Geller RJ, Fuji N (2007a) D″ beneath the Arctic from inversion of shear waveforms. Geophys Res Lett 34:L21305

Kawai K, Takeuchi N, Geller RJ, Fuji N (2007b) Possible evidence for a double crossing phase transition in D″ beneath Central America from inversion of seismic waveforms. Geophys Res Lett 34:L09314

Kawai K, Geller RJ, Fuji N (2010) Waveform inversion for S-wave structure in the lowermost mantle beneath the Arctic: Implications for mineralogy and chemical composition. Geophys Res Lett 37:L16301

Kendall J, Nangini C (1996) Lateral variations in D″ below the Caribbean. Geophys Res Lett 23:399–402

Kendall JM, Silver PG (1998) Investigating causes of D″ anisotropy. In: Gurnis M, Wysession ME, Knittle E, Buffet BA (eds) The core-mantle boundary region. American Geophysical Union, Washington D.C., pp 97–118

Kennett B, Engdahl E, Buland R (1995) Constraints on seismic velocities in the Earth from travel-times. Geophys J Int 122:108–124

Kesson S, Fitz Gerald J, Shelley J (1998) Mineralogy and dynamics of a pyrolite lower mantle. Nature 393:252–255

Kito T, Rost S, Thomas C, Garnero EJ (2007) New insights into the P- and S-wave velocity structure of the D″ discontinuity beneath the Cocos plate. Geophys J Int 169:631–645

Knittle E, Jeanloz R (1987) Synthesis and equation of state of (Mg, fe)sio$_3$ perovskite to over 100 gigapascals. Science 235:668–670

Knittle E, Jeanloz R (1989) Simulating the core-mantle boundary—an experimental-study of high-pressure reactions between silicates and liquid-iron. Geophys Res Lett 16:609–612

Kobayashi Y, Kondo T, Ohtani E, Hirao N, Miyajima N, Yagi T, Nagase T, Kikegawa T (2005) Fe-Mg partitioning between (Mg, Fe)SiO_3 post-perovskite, perovskite, and magnesiowustite in the Earth's lower mantle. Geophys Res Lett 32:L19301

Konishi K, Kawai K, Geller RJ, Fuji N (2009) MORB in the lowermost mantle beneath the western Pacific: evidence from waveform inversion. Earth Planet Sci Lett 278:219–225

Konishi K, Kawai K, Geller RJ, Fuji N (2012) Waveform inversion of broad-band body wave data for the S-velocity structure in the lowermost mantle beneath the Indian subcontinent and Tibetan Plateau. Geophys J Int 191:305–316

Kubo A, Kiefer B, Shim S, Shen G, Prakapenka VB, Duffy TS (2008) Rietveld structure refinement of $MgGeO_3$ post-perovskite phase to 1 Mbar. Am Mineral 93:965–976

Kustowski B, Ekstrom G, Dziewonski AM (2008) Anisotropic shear-wave velocity structure of the Earth's mantle: a global model. J Geophys Res Sol Earth 113:B06306

Lay T (2008) Sharpness of the D″ discontinuity beneath the Cocos plate: implications for the perovskite to post-perovskite phase transition. Geophys Res Lett 35:L03304

Lay T, Garnero E (2004) Core-mantle boundary structures and processes. In: Sparkes RSJ, Hawkesworth CJ (eds) State of the planet: frontiers and challenges in geophysics, vol 150. American Geophysical Union, Washington D.C., pp 25–41

Lay T, Garnero EJ (2007) Reconciling the post-perovskite phase with seismological observations of lowermost mantle structure. In: Hirose K, Brodholt John, Lay Thorne, Yuen David (eds) Post-perovskite: the last mantle phase transition. American Geophysical Union, Washington D.C., pp 129–153

Lay T, Garnero EJ (2011) Deep mantle seismic modeling and imaging. Annu Rev Earth Planet Sci 39(39):91–123

Lay T, Helmberger D (1983) A lower mantle S-wave triplication and the shear velocity structure of D″. Geophys J Roy Astron Soc 75:799–837

Lay T, Young CJ (1991) Analysis of seismic SV waves in the cores penumbra. Geophys Res Lett 18:1373–1376

Lay T, Garnero E, Russell S (2004) Lateral variation of the D″ discontinuity beneath the Cocos plate. Geophys Res Lett 31:L15612

Lay T, Hernlund J, Garnero EJ, Thorne MS (2006) A post-perovskite lens and D″ heat flux beneath the central Pacific. Science 314:1272–1276

Long MD (2009) Complex anisotropy in D″ beneath the eastern Pacific from SKS-SKKS splitting discrepancies. Earth Planet Sci Lett 283:181–189

Mao WL, Shen G, Prakapenka V, Meng Y, Campbell A, Heinz D, Shu J, Hemley R, Mao H (2004) Ferromagnesian postperovskite silicates in the D″ layer of the Earth. Proc Natl Acad Sci USA. 101:15867–15869

Mao WL et al (2005) Iron-rich silicates in the Earth's D″ layer. Proc Natl Acad Sci USA 102

Mao WL, Meng Y, Mao H (2010) Elastic anisotropy of ferromagnesian post-perovskite in Earth's D″ layer. Phys Earth Planet Inter 180:203–208

Masters G, Laske G, Bolton H, Dziewonski AM (2000) The relative behaviour of shear velocity, bulk sound speed, and compressional velocity in the mantle: implications for chemical and thermal structure. In: Karato S, Forte A, Liebermann R, Masters G, Stixrude L (eds) Earth's deep interior: mineral physics and tomography from the atomic to the global scale, vol 117. American Geophysical Union, Washington D.C., pp 63–87

Matzel E, Sen MK, Grand SP (1996) Evidence for anisotropy in the deep mantle beneath Alaska. Geophys Res Lett 23:2417–2420

Maupin V, Garnero EJ, Lay T, Fouch MJ (2005) Azimuthal anisotropy in the D″ layer beneath the Caribbean. J Geophys Res Solid Earth 110:B08301

McNamara AK, Zhong SJ (2005) Thermochemical structures beneath Africa and the Pacific Ocean. Nature 437

Meade C, Mao HK, Hu JZ (1995) High-temperature phase-transition and dissociation of (Mg, fe)sio$_3$ perovskite at lower mantle pressures. Science 268:1743–1745
Merkel S, McNamara AK, Kubo A, Speziale S, Miyagi L, Meng Y, Duffy TS, Wenk H (2007) Deformation of (Mg, Fe)SiO(3) post-perovskite and D anisotropy. Science 316:1729–1732
Metsue A, Tsuchiya T (2012) Thermodynamic properties of (Mg, Fe$_2$+)SiO$_3$ perovskite at the lower-mantle pressures and temperatures: an internally consistent LSDA + U study. Geophys J Int 190:310–322
Metsue A, Tsuchiya T (2013) Shear response of Fe-bearing MgSiO$_3$ post-perovskite at lower mantle pressures. Proc Jpn Acad Ser B Phys Biol Sci 89:51–58
Miyagi L, Nishlyama N, Wang Y, Kubo A, West DV, Cava RJ, Duffy TS, Wenk H (2008) Deformation and texture development in CaIrO(3) post-perovskite phase up to 6 GPa and 1300 K. Earth Planet Sci Lett 268:515–525
Miyagi L, Kanitpanyacharoen W, Kaercher P, Lee KKM, Wenk H (2010) Slip systems in MgSiO$_3$ post-perovskite: implications for D$''$ anisotropy. Science 329:1639–1641
Miyagi L, Kanitpanyacharoen W, Stackhouse S, Militzer B, Wenk H (2011) The enigma of post-perovskite anisotropy: deformation versus transformation textures. Phys Chem Miner 38:665–678
Miyajima N, Ohgushi K, Ichihara M, Yagi T (2006) Crystal morphology and dislocation microstructures of CaIrO(3): a TEM study of an analogue of the MgSiO(3) post-perovskite phase. Geophys Res Lett 33:L12302
Montagner JP, Kennett BLN (1996) How to reconcile body-wave and normal-mode reference earth models. Geophys J Int 125:229–248
Montelli R, Nolet G, Dahlen FA, Masters G, Engdahl ER, Hung SH (2004) Finite-frequency tomography reveals a variety of plumes in the mantle. Science 303:338–343
Mosca I (2010) Probabilistic tomography using body wave, normal mode and surface wave data. PhD Thesis, Utrecht University, Utrecht
Mosca I, Trampert J (2009) Path-average kernels for long wavelength traveltime tomography. Geophys J Int 177:639–650
Mosca I, Cobden L, Deuss A, Ritsema J, Trampert J (2012) Seismic and mineralogical structures of the lower mantle from probabilistic tomography. J Geophys Res Solid Earth 117:B06304
Mosegaard K, Tarantola A (1995) Monte-carlo sampling of solutions to inverse problems. J Geophys Res Solid Earth 100:12431–12447
Muirhead K, Datt R (1976) N-th root process applied to seismic array data. Geophys J Roy Astron Soc 47:197–210
Murakami M, Hirose K, Kawamura K, Sata N, Ohishi Y (2004) Post-perovskite phase transition in MgSiO$_3$. Science 304:855–858
Murakami M, Hirose K, Sata N, Ohishi Y (2005) Post-perovskite phase transition and mineral chemistry in the pyrolitic lowermost mantle. Geophys Res Lett 32:L03304
Murakami M, Sinogeikin SV, Bass JD, Sata N, Ohishi Y, Hirose K (2007) Sound velocity of MgSiO$_3$ post-perovskite phase: a constraint on the D$''$discontinuity. Earth Planet Sci Lett 259:18–23
Murakami M, Ohishi Y, Hirao N, Hirose K (2012) A perovskitic lower mantle inferred from high-pressure, high-temperature sound velocity data. Nature 485:90–94
Nakagawa T, Tackley PJ (2011) Effects of low-viscosity post-perovskite on thermo-chemical mantle convection in a 3-D spherical shell. Geophys Res Lett 38:L04309
Neuberg J, Wahr J (1991) Detailed investigation of a spot on the core mantle boundary using digital PcP data. Phys Earth Planet Inter 68:132–143
Nishio-Hamane D, Fujino K, Seto Y, Nagai T (2007) Effect of the incorporation of FeAlO$_3$ into MgSiO$_3$ perovskite on the post-perovskite transition. Geophys Res Lett 34:L12307
Niwa K, Yagi T, Ohgushi K, Merkel S, Miyajima N, Kikegawa T (2007) Lattice preferred orientation in CaIrO$_3$ perovskite and post-perovskite formed by plastic deformation under pressure. Phys Chem Miner 34:679–686
Niwa K, Miyajima N, Seto Y, Ohgushi K, Gotou H, Yagi T (2012) In situ observation of shear stress-induced perovskite to post-perovskite phase transition in CaIrO$_3$ and the development of its deformation texture in a diamond-anvil cell up to 30 GPa. Phys Earth Planet Inter 194:10–17

Nowacki A, Wookey J, Kendall J- (2010) Deformation of the lowermost mantle from seismic anisotropy. Nature 467:1091–1095

Nowacki A, Wookey J, Kendall J- (2011) New advances in using seismic anisotropy, mineral physics and geodynamics to understand deformation in the lowermost mantle. J Geodyn 52:205–228

Oganov AR, Ono S (2004) Theoretical and experimental evidence for a post-perovskite phase of $MgSiO_3$ in Earth's D″ layer. Nature 430:445–448

Oganov A, Ono S (2005) The high-pressure phase of alumina and implications for Earth's D″layer. Proc Natl Acad Sci USA 102:10828–10831

Oganov AR, Brodholt JP, Price GD (2002) Ab initio theory of phase transitions and thermoelasticity of minerals. In: Gramaccioli CM (ed) Energy modelling in minerals, vol 4, pp 83–170

Ohta K, Hirose K, Sata N, Ohishi Y (2006) The sharpness and compositional effects on post-perovskite phase transition. Geochim Cosmochim, Acta 70

Ohta K, Hirose K, Lay T, Sata N, Ohishi Y (2008) Phase transitions in pyrolite and MORB at lowermost mantle conditions: implications for a MORB-rich pile above the core-mantle boundary. Earth Planet Sci Lett 267:107–117

Okada T, Yagi T, Niwa K, Kikegawa T (2010) Lattice-preferred orientations in post-perovskite-type $MgGeO_3$ formed by transformations from different pre-phases. Phys Earth Planet Inter 180:195–202

Ono S, Oganov AR (2005) In situ observations of phase transition between perovskite and $CaIrO_3$-type phase in $MgSiO_3$ and pyrolitic mantle composition. Earth Planet Sci Lett 236

Panning M, Romanowicz B (2004) Inferences on flow at the base of Earth's mantle based on seismic anisotropy. Science 303:351–353

Panning M, Romanowicz B (2006) A three-dimensional radially anisotropic model of shear velocity in the whole mantle. Geophys J Int 167:361–379

Reasoner C, Revenaugh J (1999) Short-period P wave constraints on D″ reflectivity. J Geophys Res Solid Earth 104:955–961

Ricard Y, Mattern E, Matas J (2005) Synthetic tomographic images of slabs from mineral physics. In: van der Hilst RD, Bass Jay D, Matas Jan, Trampert J (eds) Earth's deep mantle: structure, composition and evolution, vol 160. American Geophysical Union, Washington D.C., pp 285–302

Ringwood AE (1962) Model for upper mantle. J Geophys Res 67

Ritsema J, van Heijst HJ (2002) Constraints on the correlation of P- and S-wave velocity heterogeneity in the mantle from P, PP, PPP and PKPab traveltimes. Geophys J Int 149:482–489

Rokosky JM, Lay T, Garnero EJ (2006) Small-scale lateral variations in azimuthally anisotropic D″ structure beneath the Cocos plate. Earth Planet Sci Lett 248:411–425

Ross M, Mao HK, Bell PM, Xu JA (1986) The equation of state of dense argon—a comparison of shock and static studies. J Chem Phys 85:1028–1033

Rost S, Thomas C (2002) Array seismology: methods and applications. Rev Geophys 40:1008

Russell S, Reasoner C, Lay T, Revenaugh J (2001) Coexisting shear- and compressional-wave seismic velocity discontinuities beneath the central Pacific. Geophys Res Lett 28:2281–2284

Sata N, Shen G, Rivers M, Sutton S (2002) Pressure-volume equation of state of the high-pressure B2 phase of NaCl. Phys Rev B 65:104114

Saxena S, Dubrovinsky L, Lazor P, Cerenius Y, Haggkvist P, Hanfland M, Hu J (1996) Stability of perovskite ($MgSiO_3$) in the Earth's mantle. Science 274:1357–1359

Saxena S, Dubrovinsky L, Lazor P, Hu J (1998) In situ X-ray study of perovskite ($MgSiO_3$): phase transition and dissociation at mantle conditions. Eur J Mineral 10:1275–1281

Scherbaum F, Kruger F, Weber M (1997) Double beam imaging: mapping lower mantle heterogeneities using combinations of source and receiver arrays. J Geophys Res Solid Earth 102:507–522

Schweitzer J, Fyen J, Mykkeltveit S, Kvaerna T (2002) Seismic arrays (Chap. 9). In: Bormann P (ed) IASPEI new manual of seismological observatory practice. GeoForschungszentrum, Potsdam

Shieh S, Duffy T, Kubo A, Shen G, Prakapenka V, Sata N, Hirose K, Ohishi Y (2006) Equation of state of the postperovskite phase synthesized from a natural (Mg, Fe)SiO_3 orthopyroxene. Proc Natl Acad Sci USA 103:3039–3043

Shieh SR, Dorfman SM, Kubo A, Prakapenka VB, Duffy TS (2011) Synthesis and equation of state of post-perovskites in the (Mg, Fe)(3)$Al_2Si_3O_{12}$ system. Earth Planet Sci Lett 312:422–428
Shim S, Duffy T, Kenichi T (2002) Equation of state of gold and its application to the phase boundaries near 660 km depth in Earth's mantle. Earth Planet Sci Lett 203:729–739
Shim SH, Duffy TS, Jeanloz R, Shen G (2004) Stability and crystal structure of $MgSiO_3$ perovskite to the core-mantle boundary. Geophys Res Lett 31:L10603
Sidorin I, Gurnis M, Helmberger D, Ding X (1998) Interpreting D″ seismic structure using synthetic waveforms computed from dynamic models. Earth Planet Sci Lett 163:31–41
Sidorin I, Gurnis M, Helmberger D (1999a) Dynamics of a phase change at the base of the mantle consistent with seismological observations. J Geophys Res Solid Earth 104:15005–15023
Sidorin I, Gurnis M, Helmberger D (1999b) Evidence for a ubiquitous seismic discontinuity at the base of the mantle. Science 286:1326–1331
Sinmyo R, Hirose K, Muto S, Ohishi Y, Yasuhara A (2011) The valence state and partitioning of iron in the Earth's lowermost mantle. J Geophy Res Solid Earth 116:B07205
Sparks DN (1973) Euclidean cluster analysis. J R Stat Soc Ser C Appl Stat 126–130
Speziale S, Zha C, Duffy T, Hemley R, Mao H (2001) Quasi-hydrostatic compression of magnesium oxide to 52 GPa: implications for the pressure-volume-temperature equation of state. J Geophys Res Solid Earth 106:515–528
Stackhouse S, Brodholt JP (2007) High-temperature elasticity of $MgSiO_3$ post-perovskite. In: Hirose K, Brodholt JP, Lay T, Yuen D (eds) Post-perovskite: the last mantle phase transition, vol 174. American Geophysical Union, Washington D.C., pp 99–113
Stackhouse S, Brodholt JP, Price GD (2005a) High temperature elastic anisotropy of the perovskite and post-perovskite Al(2)O(3). Geophys Res Lett 32:L13305
Stackhouse S, Brodholt JP, Wookey J, Kendall JM, Price GD (2005b) The effect of temperature on the seismic anisotropy of the perovskite and post-perovskite polymorphs of $MgSiO_3$. Earth Planet Sci Lett 230:1–10
Stackhouse S, Brodholt JP, Price GD (2006) Elastic anisotropy of FeSiO(3) end-members of the perovskite and post-perovskite phases. Geophys Res Lett 33:L01304
Stixrude L, Lithgow-Bertelloni C (2005) Thermodynamics of mantle minerals—I. Physical properties. Geophys J Int 162:610–632
Stixrude L, Lithgow-Bertelloni C (2011) Thermodynamics of mantle minerals—II. Phase equilibria. Geophys J Int 184:1180–1213
Sun D, Helmberger D (2008) Lower mantle tomography and phase change mapping. J Geophy Res Solid Earth 113:B10305
Sun DY, Song TRA, Helmberger D (2006) Complexity of D″ in the presence of slab-debris and phase changes. Geophys Res Lett 33:L12S07
Sun D, Helmberger D, Song X, Grand SP (2007) Predicting a global perovskite and post-perovskite phase boundary. In: Hirose K, Brodholt J, Lay T, Yuen D (eds) Post-perovskite: the last mantle phase transition, vol 174. American Geophysical Union, Washington D.C., pp 155–170
Tackley PJ (2011) Living dead slabs in 3-D: the dynamics of compositionally-stratified slabs entering a "slab graveyarD″ above the core-mantle boundary. Phys Earth Planet Inter 188:150–162
Tackley PJ (2012) Dynamics and evolution of the deep mantle resulting from thermal, chemical, phase and melting effects. Earth-Sci Rev 110:1–25
Takahashi E (1986) Melting of a dry peridotite KLB-1 up to 14 Gpa—implications on the origin of peridotitic upper mantle. J Geophy Res Solid Earth Planets 91:9367–9382
Tateno S, Hirose K, Sata N, Ohishi Y (2005) Phase relations in $Mg_3Al_2Si_3O_{12}$ to 180 GPa: effect of Al on post-perovskite phase transition. Geophys Res Lett 32:L15306
Tateno S, Hirose K, Sata N, Ohishi Y (2007) Solubility of FeO in (Mg, Fe)SiO_3 perovskite and the post-perovskite phase transition. Phys Earth Planet Inter 160:319–325
Tateno S, Hirose K, Sata N, Ohishi Y (2009) Determination of post-perovskite phase transition boundary up to 4400 K and implications for thermal structure in D″ layer. Earth Planet, Sci Lett 277
Thomas C, Weber M (1997) P velocity heterogeneities in the lower mantle determined with the German regional seismic network. Improvement of previous models and results of 2D modelling. Phys Earth Planet Inter 101:105–117

Thomas C, Weber M, Wicks C, Scherbaum F (1999) Small scatterers in the lower mantle observed at German broadband arrays. J Geophys Res Solid Earth 104:15073–15088
Thomas C, Kendall J-, Weber M (2002) The lowermost mantle beneath northern Asia? I. Multi-azimuth studies of a D″ heterogeneity. Geophys J Int 151:279–295
Thomas C, Garnero EJ, Lay T (2004a) High-resolution imaging of lowermost mantle structure under the Cocos plate. J Geophys Res Solid Earth 109:B08307
Thomas C, Kendall JM, Lowman J (2004b) Lower-mantle seismic discontinuities and the thermal morphology of subducted slabs. Earth Planet Sci Lett 225:105–113
Thomas C, Wookey J, Simpson M (2007) D″ anisotropy beneath Southeast Asia. Geophys Res Lett 34:L04301
Thomas C, Wookey J, Brodholt J, Fieseler T (2011) Anisotropy as cause for polarity reversals of D″ reflections. Earth Planet Sci Lett 307:369–376
Thorne MS, Lay T, Garnero EJ, Jahnke G, Igel H (2007) Seismic imaging of the laterally varying D″ region beneath the Cocos plate. Geophys J Int 170:635–648
Tsuchiya T (2003) First-principles prediction of the P-V-T equation of state of gold and the 660-km discontinuity in Earth's mantle. J Geophys Res Solid Earth 108:2462
Tsuchiya T, Tsuchiya J (2006) Effect of impurity on the elasticity of perovskite and postperovskite: velocity contrast across the postperovskite transition in (Mg, Fe, Al)(Si, Al)O_3. Geophys Res Lett 33:L12S04
Tsuchiya J, Tsuchiya T (2008) Postperovskite phase equilibria in the $MgSiO_3$-Al_2O_3 system. Proc Natl Acad Sci USA 105:19160–19164
Tsuchiya T, Tsuchiya J, Umemoto K, Wentzcovitch R (2004) Phase transition in $MgSiO_3$ perovskite in the Earth's lower mantle. Earth Planet Sci Lett 224:241–248
Usui Y, Hiramatsu Y, Furumoto M, Kanao M (2005) Thick and anisotropic D″ layer beneath Antarctic Ocean. Geophys Res Lett 32:L13311
van der Hilst RD, de Hoop MV, Wang P, Shim S-, Ma P, Tenorio L (2007) Seismostratigraphy and thermal structure of Earth's core-mantle boundary region. Science 315:1813–1817
Wallace M, Thomas C (2005) Investigating D″ structure beneath the North Atlantic. Phys Earth Planet Inter 151:115–127
Walte N, Heidelbach F, Miyajima N, Frost D (2007) Texture development and TEM analysis of deformed $CaIrO_3$: implications for the D″ layer at the core-mantle boundary. Geophys Res Lett 34:L08306
Walte NP, Heidelbach F, Miyajima N, Frost DJ, Rubie DC, Dobson DP (2009) Transformation textures in post-perovskite: understanding mantle flow in the D″ layer of the Earth. Geophys Res Lett 36:L04302
Wang YB, Guyot F, Yeganehhaeri A, Liebermann RC (1990) Twinning in $Mgsio_3$ perovskite. Science 248:468–471
Weber M (1993) P-wave and S-wave reflections from anomalies in the lowermost mantle. Geophys J Int 115:183–210
Weber M, Davis JP (1990) Evidence of a laterally variable lower mantle structure from P-waves and S-waves. Geophys J Int 102
Weber M, Kornig M (1990) Lower mantle inhomogeneities inferred from PcP precursors. Geophys Res Lett 17:1993–1996
Weber M, Wicks C (1996) Reflections from a distant subduction zone. Geophys Res Lett 23:1453–1456
Weber M, Davis JP, Thomas C, Krüger F, Scherbaum F, Schlittenhardt J, Körnig M (1996) The structure of the lowermost mantle as determined from using seismic arrays. In: Boschi E, Ekström Göran, Morelli A (eds) Seismic modelling of the Earth's structure. Instituto Nazionale di Geophysica, Rome, pp 399–442
Wentzcovitch RM, Tsuchiya T, Tsuchiya J (2006) MgSiO(3) postperovskite at D″ conditions. Proc Natl Acad Sci USA 103:543–546
Wolf GH, Bukowinski ST (1987) Theoretical study of the structural properties and equations of state of $MgSiO_3$ and $CaSiO_3$ perovskites: implications for lower mantle composition. In: Manghnani Murli H, Syono Y (eds) High-pressure research in mineral physics. American Geophysical Union, Washington, D.C., pp 313–331

Wookey J, Kendall JM (2007) Seismic anisotropy of post-perovskite and the lowermost mantle. In: Hirose Kei, Brodholt John, Lay Thorne, Yuen David (eds) Post-perovskite: the last mantle phase transition. American Geophysical Union, Washington, D.C., pp 171–189

Wookey J, Kendall J- (2008) Constraints on lowermost mantle mineralogy and fabric beneath Siberia from seismic anisotropy. Earth Planet Sci Lett 275:32–42

Wookey J, Kendall JM, Rumpker G (2005a) Lowermost mantle anisotropy beneath the north Pacific from differential S-ScS splitting. Geophys J Int 161:829–838

Wookey J, Stackhouse S, Kendall JM, Brodholt J, Price GD (2005b) Efficacy of the post-perovskite phase as an explanation for lowermost-mantle seismic properties. Nature 438:1004–1007

Wysession ME, Lay T, Revenaugh J, Williams Q, Garnero EJ, Jeanloz R, Kellog LH (1998) The D″ discontinuity and its implications. In: Gurnis M, Wysession ME, Knittle E, Buffet BA (eds) The core-mantle boundary region, vol 28. American Geophysical Union, Washing D.C., pp 273–298

Xu W, Lithgow-Bertelloni C, Stixrude L, Ritsema J (2008) The effect of bulk composition and temperature on mantle seismic structure. Earth Planet Sci Lett 275:70–79

Yamazaki D, Yoshino T, Ohfuji H, Ando J, Yoneda A (2006) Origin of seismic anisotropy in the D″ layer inferred from shear deformation experiments on post-perovskite phase. Earth Planet Sci Lett 252:372–378

Young CJ, Lay T (1990) Multiple phase-analysis of the shear velocity structure in the D region beneath Alaska. J Geophys Res Solid Earth Planets 95

Zha C, Bassett W, Shim S (2004) Rhenium, an in situ pressure calibrant for internally heated diamond anvil cells. Rev Sci Instrum 75:2409–2418

Zhang F, Oganov AR (2006) Mechanisms of Al^{3+} incorporation in $MgSiO_3$ post-perovskite at high pressures. Earth Planet Sci Lett 248:69–76

Chapter 14
Thermally Dominated Deep Mantle LLSVPs: A Review

D.R. Davies, S. Goes and H.C.P. Lau

Abstract The two large low shear-wave velocity provinces (LLSVPs) that dominate lower-mantle structure may hold key information on Earth's thermal and chemical evolution. It is generally accepted that these provinces are hotter than background mantle and are likely the main source of mantle plumes. Increasingly, it is also proposed that they hold a dense (primitive and/or recycled) compositional component. The principle evidence that LLSVPs may represent thermo-chemical 'piles' comes from seismic constraints, including the following: (i) their long-wavelength nature; (ii) sharp gradients in shear-wave velocity at their margins; (iii) non-Gaussian distributions of deep mantle shear-wave velocity anomalies; (iv) anti-correlated shear-wave and bulk-sound velocity anomalies (and elevated ratios between shear- and compressional-wave velocity anomalies); (v) anti-correlated shear-wave and density anomalies; and (vi) 1-D/radial profiles of seismic velocity that deviate from those expected for an isochemical, well-mixed mantle. In addition, it has been proposed that hotspots and the reconstructed eruption sites of large igneous provinces correlate in location with LLSVP margins. In this paper, we review recent results which indicate that the majority of these constraints do not require thermo-chemical piles: they are equally well (or poorly) explained by thermal heterogeneity alone. Our analyses and conclusions are largely based on comparisons between imaged seismic structure and synthetic seismic structures from a set of thermal and thermo-chemical mantle convection models, which are constrained by ~300 Myr of plate motion histories. Modelled physical structure (temperature, pressure and composition) is converted into seismic velocities via a

D.R. Davies (✉)
Research School of Earth Sciences, The Australian National University,
Canberra, ACT, Australia
e-mail: rhodri.davies@anu.edu.au

S. Goes · H.C.P. Lau
Department of Earth Sciences and Engineering, Imperial College London, London, UK

H.C.P. Lau
Department of Earth and Planetary Sciences, Harvard University, Cambridge, MA, USA

© Springer International Publishing Switzerland 2015

A. Khan and F. Deschamps (eds.), *The Earth's Heterogeneous Mantle*,
Springer Geophysics, DOI 10.1007/978-3-319-15627-9_14

thermodynamic approach that accounts for elastic, anelastic and phase contributions and, subsequently, a tomographic resolution filter is applied to account for the damping and geographic bias inherent to seismic imaging. Our results indicate that, in terms of large-scale seismic structure and dynamics, these two provinces are predominantly thermal features and, accordingly, that chemical heterogeneity is largely a passive component of lowermost mantle dynamics.

Keywords Mantle dynamics · Earth structure · Seismic tomography · Chemical heterogeneity · LLSVP · Mantle plumes

14.1 Introduction

The nature of two large low shear-wave velocity provinces (LLSVPs) in the deep mantle beneath Africa and the South-Central Pacific (Fig. 14.1) has puzzled Earth scientists since they were first imaged in whole-mantle tomography models over 30 years ago (e.g. Dziewonski et al. 1977; Woodhouse and Dziewonski 1989). These LLSVPs cover ~20 % of the core–mantle boundary (CMB) and extend up to ~1000 km above the CMB. They are particularly prominent in *S*-velocity-sensitive data, with a weaker expression in *P* data (e.g. Woodhouse and Dziewonski 1989; Ishii and Tromp 1999; Masters et al. 2000; Hernlund and Houser 2008), hence their designation as low shear-wave velocity provinces.

There is a clear correlation between LLSVPs and the following: (i) the distribution of hotspots and the reconstructed eruptions sites of large igneous provinces (LIPs) and kimberlites (e.g. Duncan and Richards 1991; Thorne et al. 2004; Torsvik et al. 2006, 2010; Burke et al. 2008; Austermann et al. 2014); and (ii) long-wavelength geoid highs (e.g. Anderson 1982; Hager et al. 1985), indicating that these provinces are most likely hot and have a net positive buoyancy relative to the surrounding mantle (e.g. Gurnis et al. 2000; Forte and Mitrovica 2001). Furthermore, there is a strong correlation between surrounding high seismic velocity material and former subduction zones (e.g. Richards and Engebretson 1992; Bunge et al. 2002), whilst the long-wavelength geoid signature can be successfully reproduced by models where geoid highs result from a concentration of hot upwelling mantle away from ancient slabs (e.g. Ricard et al. 1993).

However, increasingly, arguments are being made that these provinces are not purely thermal features but contain a significant component of chemically dense material (see, e.g. review by Garnero and McNamara 2008; Deschamps et al. 2015). This material is proposed to be either sequestered from early in Earth's history (e.g. Allègre et al. 1987; Trieloff et al. 2000; Boyet and Carlson 2006; Labrosse et al. 2007; Jackson et al. 2010; Jackson and Carlson 2011; Deschamps et al. 2011), or predominantly composed of recycled oceanic lithosphere (e.g. Christensen and Hofmann 1994; Coltice and Ricard 1999; Tackley et al. 2005;

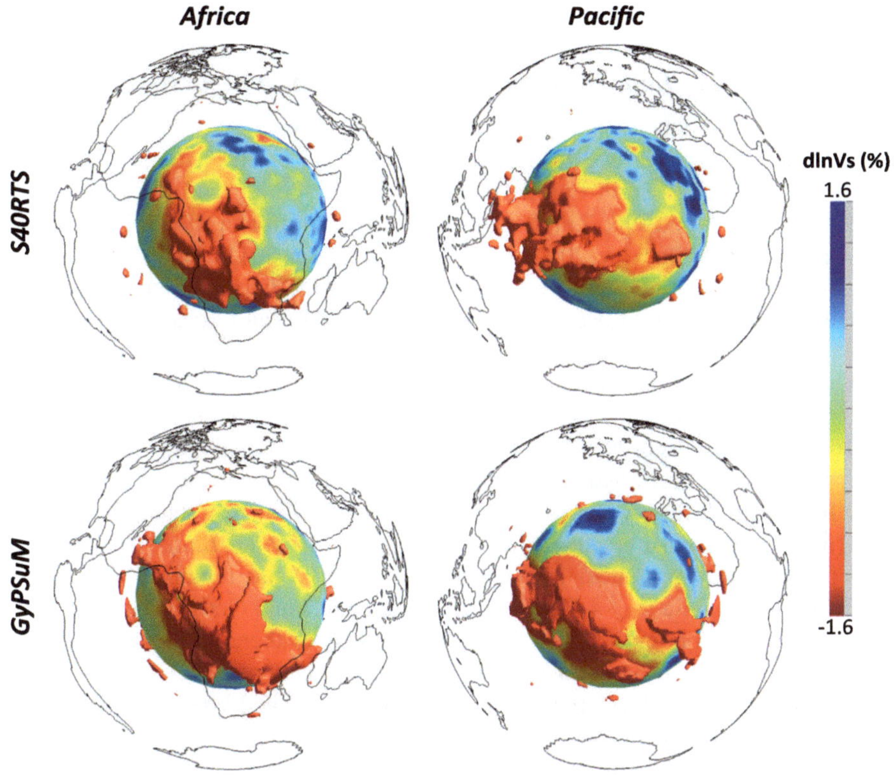

Fig. 14.1 3-D perspective of LLSVPs, outlined by shear-wave velocity perturbations beneath (*left*) Africa and (*right*) the Pacific, from tomographic models (*top*) S40RTS (Ritsema et al. 2011) and (*bottom*) GyPSum (Simmons et al. 2010). Each subfigure includes a radial surface at 2800 km depth and an isosurface at −0.9 % (S40RTS)/−1.0 % (GyPSum), clipped 1200 km above the CMB to allow for visualisation of lower-mantle features. Continental boundaries provide geographic reference

Brandenburg et al. 2008; Rapp et al. 2008). Arguments for such a compositional contribution come from geochemistry, mantle dynamics and, in particular, from seismology.

In this paper, we will review this evidence and demonstrate that the seismic observations do not require a substantial role for dense material in dictating the form of lower-mantle dynamics and its long-wavelength seismic expression. This does not imply the mantle is isochemical. Plate tectonics introduces mantle heterogeneity and a wealth of geochemical and geological data provides clear evidence for a heterogeneous mantle in major- and trace-element composition (e.g. Hofmann 1997, 2003; Davies 2011). Rather, it implies that chemical heterogeneity is largely a passive component of lower-mantle dynamics (i.e. its effect on density is outweighed by, or is secondary to, the effect of temperature) and that the seismic expression of LLSVPs is likely dominated by thermal effects.

14.2 Background

14.2.1 Key Geochemical Constraints on Mantle Structure

Geochemical observations offer important constraints on the nature of compositional heterogeneity within Earth's mantle (e.g. Wasserburg and De Paolo 1979; Allègre et al. 1980, 1996; Zindler and Hart 1986; Hofmann 1997, 2003; Boyet and Carlson 2005; Tackley 2007; Jackson and Carlson 2011), with key observations including the following: (i) Lavas sampled at the two main sites of mantle upwelling on Earth, mid-ocean-ridge basalts (MORBs) and ocean-island basalts (OIBs), are chemically distinct, with MORBs generally more depleted in incompatible elements and somewhat less heterogeneous than OIBs, and OIBs requiring several different mantle source compositions (e.g. Hofmann 1997, 2003); (ii) the isotopic ratios recorded in certain OIBs (and, occasionally, MORBs) require that some material recycled from the surface remains sequestered from the melt zone for several billion years (e.g. Zindler and Hart 1986; Hofmann 1997, 2003); and (iii) recent measurements of lead and helium isotopic ratios from a number of flood basalt provinces imply that other material has remained sequestered within Earth's mantle for ~4.5 billion years (see Hanan and Graham 1996; Jackson et al. 2010; Jackson and Carlson 2011, and references therein). This signature is observed on Baffin Island and the Ontong Java Plateau and likely exists for six of the largest recorded volcanic events over the past 250 Myr, implying that its source must be reasonably widespread.

Mass balance calculations have been used to argue that this heterogeneity requires the existence of large-scale reservoirs that differ in both depth and composition. Classical methods of estimating the bulk silicate Earth's net composition (BSE—the mantle before extraction of the continental crust) assume it has either the same composition as common chondrites or can be inferred from terrestrial samples that are supplemented with chondritic trends for refractory lithophile elements. It has been argued that relative to such BSE estimates, the MORB-source region is significantly depleted in incompatible trace elements (including the heat-producing elements U, Th and K) (e.g. McDonough and Sun 1995; Allègre et al. 2001; Javoy et al. 2010). The corollary to this is the existence of a 'hidden reservoir' (see Allègre et al. 1996; Kellogg et al. 1999; Tackley 2007, for further discussion). Such a reservoir has been proposed to comprise large parts of the lower mantle (e.g. Allègre et al. 1996; Kellogg et al. 1999) or may reside inside LLSVPs (e.g. Sramek et al. 2013).

However, the recent observation that the $^{142}Nd/^{144}Nd$ ratio is higher on Earth than in chondritic meteorites Boyet and Carlson (2005, 2006) has led to the alternative proposition that the BSE is non-chondritic (e.g. Caro and Bourdon 2010; Campbell and O'Neill 2012; Caro 2015), in which case a hidden reservoir may be unnecessary. Other studies have argued that terrestrial sample estimates of the MORB-source composition are biased towards a depleted end-member (e.g. Lyubetskaya and Korenaga 2007; Davies 2009). Their revised estimates suggest that the MORB source contains up to 100 % more incompatibles than was previously assumed, implying that, at most, a small amount of distinct material needs to remain sequestered from the convecting mantle (e.g. Lyubetskaya and Korenaga 2007; Davies 2009).

Hence, although geochemical constraints clearly require a heterogenous mantle, the temporally and spatially integrated signal provided by isotopic and trace-element trends are unable to constrain the distribution and scale of such heterogeneity.

14.2.2 Current Dynamical Views on the Nature of Compositional Heterogeneity

The geochemical data indicate that three compositions could contribute to chemically distinct LLSVPs: (i) recycled oceanic lithosphere; (ii) primitive BSE mantle, unaffected by continental crust extraction; and (iii) a reservoir comprising cumulates from an early differentiation event.

To form long-lived 'piles', material must be substantially denser than the surrounding MORB-source (pyrolitic) mantle. A number of dynamic modelling studies have investigated the conditions necessary to preserve dense material at the base of the mantle throughout Earth's history (e.g. Tackley 1998; Davaille 1999; McNamara and Zhong 2004; Deschamps and Tackley 2008, 2009; Tan et al. 2011; Bower et al. 2013). This needs to be achieved without fully layering the mantle, which is precluded by a number of geophysical observations (see Davies 1999, 2011, for detailed discussion). It was found that for densities 2–3 % higher than pyrolite, together with plausible estimates for mantle viscosity and compressibility, piles may sequester material for up to a few Ga, although they are mobile and deform through time (e.g. Zhang et al. 2010; Tan et al. 2011). Such mobile thermo-chemical piles lead to spatial and temporal variations in the CMB heat flux, which has been proposed to explain the observed variability in geomagnetic field polarity reversal intervals (Zhang and Zhong 2011; Olson et al. 2013). Material with an excess density of greater than ~3 % tends to form a dense layer that covers the entire CMB, rather than discontinuous piles. Illustrative examples of the temperature field for both layered and discontinuous pile thermo-chemical models are shown in Fig. 14.2. A similar thermo-chemical pile model is examined herein.

Mantle convection models that recycle plates form piles comprising dense basaltic material (e.g. Christensen and Hofmann 1994; Xie and Tackley 2004; Huang and Davies 2007a, b; Brandenburg et al. 2008). When the isotopic evolution of recycled material is tracked in these models, several observed geochemical trends (such as Sr, Nd, Pb and He isotope ratios) can be reproduced. Most noteworthy is that for densities within the range estimated for MORB under deep mantle conditions: (i) deep mantle piles constitute a relatively small volume fraction of the mantle (a few per cent); (ii) much of the heterogeneity that contributes to the geochemical signatures is widely distributed throughout the mantle; (iii) material

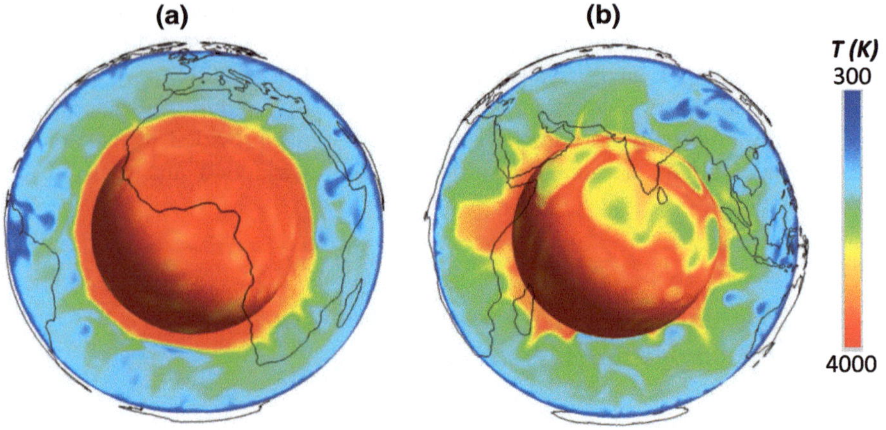

Fig. 14.2 Two conceptual models of mantle structure, from Styles et al. (2011) and Davies et al. (2012: **a** a layered case, where chemically distinct material has a density contrast of 5 % (Buoyancy Number, $B \approx 0.5$) when compared to background mantle. Such a large density contrast exceeds the upper bound expected for MORB or Fe-rich compositions relative to a pyrolite and, under such a scenario, dense material forms a hot, stable layer, which covers the entire CMB; **b** a thermo-chemical pile case, where chemically distinct material has a density contrast of 2.5 % relative to background mantle (Buoyancy Number, $B \approx 0.25$). Such a density contrast is within current estimates for the excess density of MORB and Fe-rich primitive material in comparison with pyrolite at lower-mantle depths and leads to a discontinuous chemical interface above the CMB, consisting of two piles beneath Africa and the Pacific. These piles are shaped by subducting slabs. Images include a radial surface at 2800 km depth, a cross section which provides some insight into the distribution of temperature and chemically distinct material as a function of depth, whilst continental boundaries provide geographic reference (reproduced from Styles et al. (2011: Copyright Wiley—Reprinted with permission)

with the longest residence time does not necessarily occur at the mantle's base (e.g. Brandenburg and van Keken 2007; Brandenburg et al. 2008); and (iv) piles do not govern the convective planform, but are shaped by the action of downwellings (i.e. they change shape and location over the timescale of a Wilson cycle). Furthermore, billion-year residence times, which allow isotopic ratios to evolve to the range observed in a number of OIBs (Hofmann 2003), can be achieved even if the recycled material has no excess density when compared to surrounding mantle (e.g. Huang and Davies 2007a).

Although much of the observed geochemical heterogeneity can be explained by recycled oceanic crustal material, an additional ancient mantle component is required (e.g. Jackson and Carlson 2011). This long-lived heterogeneity can remain shielded from melting either by a high intrinsic density and sequestration in the deep mantle or by a high melting temperature. Hence, both geochemical and geodynamic constraints neither require nor rule out that LLSVPs contain significant compositional heterogeneity. This, therefore, leaves the question of whether geophysical observations can distinguish between LLSVPs representing the following: (i) thermally dominated plume clusters, where chemical heterogeneity plays a negligible role in governing the underlying dynamics; (ii) chemical piles, in which substantial fractions of oceanic crust reside for hundreds of Myr; or (iii) chemical piles, which represent a reservoir of dense primitive material. Characteristics that may be geophysically observable include major-element signatures that significantly modify density and/or seismic velocity (e.g. Labrosse et al. 2007; Nakagawa et al. 2010; Mosca et al. 2012), or sequestered heat-producing elements if concentrated in large-scale piles (e.g. Sramek et al. 2013).

14.2.3 Seismic Evidence for Thermo-chemical LLSVPs

Seismology provides the most direct and detailed observations of Earth's internal structure, and hence, the debate on the nature of LLSVPs has been centred on seismological arguments, which include the correlation of seismic structure with large igneous provinces and hotspots, commonly assumed to represent the surface expression of mantle plume heads and tails, respectively (e.g. Campbell and Griffiths 1992). The principal arguments invoked in favour of compositionally distinct LLSVPs are the following:

1. LLSVPs have a longer wavelength structure than that predicted by dynamic models in which purely thermal upwelling cluster away from deeply subducted slabs (e.g. Bull et al. 2009).
2. Imaged LLSVP shear-wave velocity anomalies of -2 to -5 % (e.g. Wang and Wen 2007; Houser et al. 2008; Ritsema et al. 2011) are considered too large for purely thermal structures (e.g. Karato and Karki 2001; Brodholt et al. 2007).
3. Shear-wave velocity gradients ranging from 1.3 to 6 %/100 km have been identified along the margins (and occasionally within the interior) of both the African and Pacific LLSVPs (Ritsema et al. 1998; Ni et al. 2002; Wang and Wen 2004; To et al. 2005), indicating that these features, at least locally, have sharp sides. It is expected that purely thermal structures would be smoothed by diffusive heat loss and, hence, would be unable to explain such rapid velocity variations (e.g. Ni et al. 2002).
4. Hernlund and Houser (2008) find that the distribution of deep mantle shear-wave velocity anomalies exhibits a slow tail in several global tomographic models, whilst compressional-wave velocity anomalies form single-peaked Gaussian distributions. They attribute this tail to a combination of the post-perovskite phase transition (below ~2400 km depth) and chemical heterogeneity (at shallower depths).
5. LLSVPs have a different expression in shear- and compressional-wave-speed anomalies, reflected in high $R = \partial \ln V_S / \partial \ln V_P$ ratios in the deep mantle (e.g. Robertson and Woodhouse 1995; Ritsema and van Heijst 2002). Such elevated ratios are considered incompatible with a purely thermal origin (e.g. Karato and Karki 2001; Saltzer et al. 2001).
6. Shear-wave and bulk-sound velocity anomalies have been found to be anti-correlated within (and outside) LLSVPs and possibly over large depth ranges of the lower-mantle (e.g. Su and Dziewonski 1997; Kennett et al. 1998; Masters

et al. 2000). As shear and bulk moduli respond to temperature in the same direction, this appears inconsistent with purely thermal structures.

7. Similarly, predicted anti-correlations between shear-wave velocity and density anomalies inside (and outside) LLSVPs would preclude a thermal origin (e.g. Ishii and Tromp 1999; Trampert et al. 2004).
8. 1-D seismic reference models (e.g. Dziewonski and Anderson 1981; Kennett et al. 1995) cannot be reconciled with seismic velocities predicted for an iso-chemical and thermally well-mixed mantle and may require a superadiabatic temperature gradient or a variation in composition with depth (e.g. da Silva et al. 2000; Deschamps and Trampert 2004; Cammarano et al. 2005; Matas et al. 2007; Khan et al. 2008; Cobden et al. 2009).
9. Finally, several studies infer that hotspots and the reconstructed eruption sites of LIPs concentrate above LLSVP margins (e.g. Thorne et al. 2004; Torsvik et al. 2006, 2010; Burke et al. 2008) (Fig. 14.3). This correlation has been explained by the preferential triggering of plumes at the interface between background mantle and high density, and possibly also high bulk modulus, piles (e.g. Tan et al. 2011; Steinberger and Torsvik 2012). Such correlations have further led to the proposal that LLSVPs have remained stable in location and shape over at least 200–250 Myr and, potentially, through several Wilson cycles (Burke et al. 2008), possibly stabilised by Earth's rotation, such that they impose a bottom-up control on surface tectonics (Dziewonski et al. 2010). Such temporal and spatial LLSVP stability would require a significant contribution from excess chemical density.

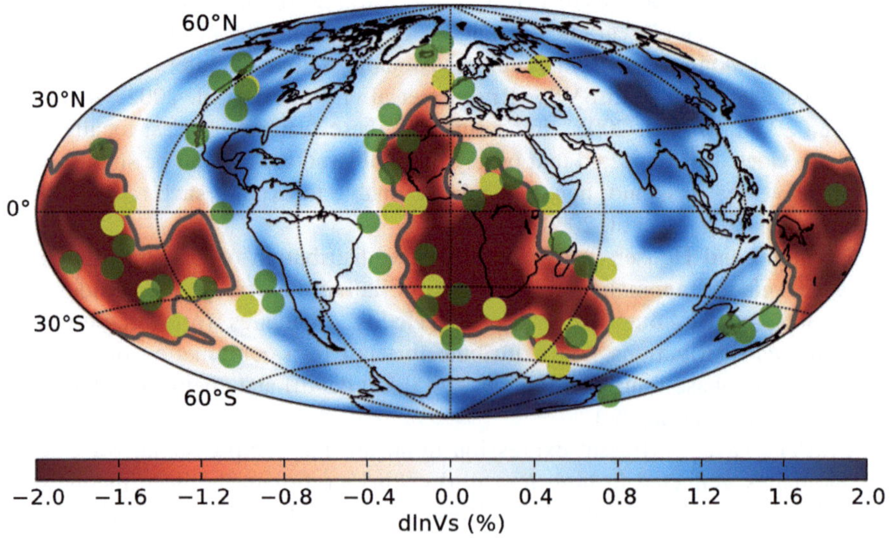

Fig. 14.3 Surface hotspot locations (*Green circles* Steinberger 2000) and the reconstructed eruption sites of large igneous provinces (*Yellow circles* Torsvik et al. 2006, 2008b) plotted above the shear-wave tomography model SMEAN (Becker and Boschi 2002), at 2800 km depth. The −1.0 % dlnV_S contour is shown in *grey*

14.3 Comparing Synthetic and Observed Seismic Structure

14.3.1 Methods

To test to what extent the seismic characteristics summarised in Sect. 2.3 require that LLSVPs constitute thermo-chemical piles and, if so, whether they can constrain pile composition, over recent years we have examined a set of global mantle convection models, in which the distribution of heterogeneity is dictated by 300 Myr of assimilated plate motion histories. We have examined a suite of simulations, with a focus on two end-member cases: (i) a purely thermal scenario, with no chemical heterogeneity; and (ii) a thermo-chemical pile scenario, where chemically dense material focuses into distinct discontinuous structures at the mantle's base, beneath Africa and the Pacific.

14.3.1.1 Dynamic Convection Models

Global temperature (*T*), pressure (*P*) and compositional (*X*) fields are generated using a modified and benchmarked version of the compressible spherical mantle convection code *TERRA* (e.g. Baumgardner 1985; Bunge et al. 1997; Davies and Davies 2009; Davies et al. 2013; Wolstencroft et al. 2009). Models incorporate compressibility, in the form of the anelastic liquid approximation, with radial reference values represented through a Murnaghan equation of state. They achieve an internally heated Rayleigh number of $Ra_{\mathrm{H}} \approx 5 \times 10^8$ and a volume averaged dissipation number of 0.5, which are similar to estimates for Earth's mantle. Surface velocities are assimilated via 300 Myr of plate motion reconstructions (Stampfli and Borel 2002; Stampfli and Hochard 2009), at discrete 1-Myr intervals, with a free-slip boundary condition specified at the CMB. Note that surface velocities are assimilated at a vigour consistent with the underlying simulation (i.e. surface RMS velocities of the free-slip stage of our simulation are consistent with those of assimilated model), to avoid excessive viscous dissipation beneath plates. Isothermal boundary conditions are prescribed at the surface (300 K) and CMB (4000 K), with the mantle also heated internally, at roughly chondritic rates (5.5×10^{-12} W kg^{-1}). Phase changes are incorporated at 410 and 660 km depth, whilst thermal conductivity increases linearly with depth, from a value of 4 W m^{-1} K^{-1} at the surface to 6 W m^{-1}K^{-1} at the CMB (e.g. de Koker 2010). Viscosity varies as a function of depth (*z*) and temperature (*T*), following the relation:

$$\mu(z,T) = \begin{cases} \mu_0 \Delta\mu_{Li} \exp\left[V_a z' - E_a T'\right] & z < 100\ \text{km} \\ \mu_0 \exp\left[V_a z' - E_a T'\right] & 100\ \text{km} \le z < 410\ \text{km} \\ \mu_0 \Delta\mu_{410} \exp\left[V_a z' - E_a T'\right] & 410\ \text{km} \le z < 660\ \text{km} \\ \mu_0 \Delta\mu_{660} \exp\left[V_a z' - E_a T'\right] & z \ge 660\ \text{km} \end{cases} \tag{14.1}$$

where T' and z' are non-dimensionalised by ΔT and mantle depth, respectively, whilst V_a and E_a are non-dimensional constants controlling the sensitivity of viscosity to depth and temperature. Our choice of V_a and E_a yields an Earth-like viscosity-depth profile, which results in model slab sinking rates that are consistent with those inferred from tomographic images.

For initial conditions, a standard convection model (with free-slip surface) is run until a thermal quasi-steady state is achieved ($\approx$1 Gyr). Early Carboniferous mantle heterogeneity is then approximated by running models with global plate configurations fixed to the reconstruction at 300 Ma, for $\approx$50 Myr, before allowing models to evolve towards the present day. The chemical field (X) is simulated via the ratio tracer particle method (Tackley and King 2003; Stegman et al. 2003), with $\approx 2.0 \times 10^9$ active tracers, of two distinct types (dense material, $X = 1$; regular material, $X = 0$). Key model parameters are provided in Table 14.1, with radial profiles of density, thermal expansivity and viscosity in Fig. 14.4.

We test two end-member scenarios: a purely thermal (isochemical) model and a thermo-chemical model with dense piles of chemical heterogeneity, which, in total, comprise 3 % of the mantle's volume. The thermo-chemical model is initiated with a 175-km-thick basal layer of excess density 2.75 % (buoyancy number, $B = \Delta\rho_X / \Delta\rho_T \approx 0.275$, where $\Delta\rho_T = \alpha_s \rho_s \Delta T_s$), which progressively deforms as the model evolves. Apart from the inclusion of dense material in the thermo-chemical model, both cases are identical. However, as we will show, their convective planforms and seismic expressions differ substantially, which allows for meaningful comparisons between each case and seismic observations of Earth's present-day mantle. It should also be noted that chemical piles influence the CMB heat flux, with the present-day CMB heat flux decreasing from ~12 TW in the purely thermal case, to around ~9 TW in the thermo-chemical case.

Our models have many Earth-like characteristics, in terms of convective vigour, thermal structure, surface heat flux and the geographic pattern of heterogeneity (controlled by the assimilated plate motion history). However, a few limitations should be noted:

(i) The parameters of our simulations are not fully self-consistent with the thermodynamic database utilised in converting from physical to seismic structure, the key difference being that we exclude dynamic effects of the post-perovskite phase transition, which has been shown to destabilise the thermal boundary layer above the CMB (e.g. Nakagawa and Tackley 2004; Tosi et al. 2010; Hunt et al. 2012).

(ii) The temperature sensitivity of viscosity in our models is lower than that on Earth (e.g. Ranalli 1995; Karato 2008), with no dependence on composition or stress. As a result, our downgoing slabs lack the strength of slabs on Earth and are likely too thick, which may alter their passage through the mantle. This manifests itself mostly within the transition zone, where, despite the inclusion of an endothermic phase transition and viscosity jump, modelled slabs stall less than is indicated by tomographic images (e.g. Li et al. 2008; Fukao and Obayashi 2013) or is predicted by recent complex rheology numerical studies (e.g. Garel et al. 2014). However, we have verified that our viscosity profile results in lower-mantle slabs being located at depths that are consistent with tomographic images.

Table 14.1 Parameters common to all models. Rayleigh numbers are calculated based upon surface reference values

Parameter	Symbol	Value	Units
Surface temperature	T_s	300	K
CMB temperature	T_{cmb}	4000	K
Internal heating rate	H	5.5×10^{-12}	W kg^{-1}
Reference viscosity	μ_0	3.0×10^{21}	Pa s
Lithospheric multiplication factor	$\Delta\mu_{Li}$	100	–
410-km Seismic discontinuity: 410-km multiplication factor	$\Delta\mu_{410}$	5	–
660-km Seismic discontinuity: 660-km multiplication factor	$\Delta\mu_{660}$	30	–
Viscosity: depth dependence	V_a	2.99	–
Viscosity: temperature dependence	E_a	4.61	–
Clapeyron slope: 410 km	Cl_{410}	1.5×10^6	M Pa K^{-1}
Clapeyron slope: 660 km	Cl_{660}	-1.0×10^6	M Pa K^{-1}
Surface density	ρ_s	3500	kg m^{-3}
CMB density	ρ_{cmb}	5568	kg m^{-3}
Surface thermal expansivity	α_s	3.8×10^{-5}	K^{-1}
CMB thermal expansivity	α_{cmb}	1.2×10^{-5}	K^{-1}
Superadiabatic temperature contrast	ΔT_s	2650	K
Total temperature contrast	ΔT	3700	K
Adiabatic footing temperature	T_{pot}	1600	K
Surface thermal conductivity	k_s	4.0	W m^{-1} K^{-1}
CMB thermal conductivity	k_{cmb}	6.0	W m^{-1} K^{-1}
Specific heat capacity	C_p	1134	J kg^{-1} K^{-1}
Surface Dissipation number	Di_s	$\approx$1.0	–
Volumetric dissipation number	Di_V	$\approx$0.5	–
Internally heated Rayleigh number	Ra_H	$\approx 5.0 \times 10^8$	–
Basally heated Rayleigh number	Ra_b	$\approx 3.0 \times 10^7$	–

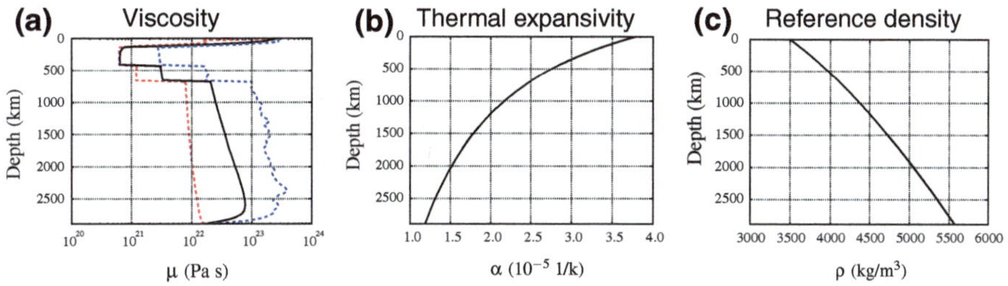

Fig. 14.4 **a** Radial viscosity profile for the isochemical model examined here in (the thermo-chemical pile model has a similar structure, with minor differences in the lowermost mantle). Note that *red*, *black* and *blue lines* denote the minimum, mean and maximum viscosities, respectively (at any given depth). The minimum viscosity is limited to 7×10^{20} Pa s, to ensure numerical stability; **b** thermal expansivity and **c** reference density profiles, for the models examined herein

(iii) In our thermo-chemical cases, piles form from an initial layer of dense material and are shaped by subduction history. Accordingly, their shape and seismic expression may differ to those formed via the recycling of oceanic plates (e.g. Brandenburg et al. 2008; Nakagawa et al. 2010), although such recycling models also create two dense piles, even without assimilating plate motion histories. We show only comparisons between a single thermo-chemical pile model and a purely thermal model, from a suite of cases that have previously been examined, where the volume fraction and excess density of chemical material are varied. Thermo-chemical models with increased volume fractions of dense material (of similar density contrast) produce piles that cover a larger portion of the CMB and extend further into the mid-mantle, resulting in stronger deep mantle seismic anomalies and a radial distribution of heterogeneity that decreases the correlation between thermo-chemical model predictions and imaged structure (Davies et al. 2012). Models with decreased volume fractions of dense material, or a lower chemical density contrast, do not generate coherent thermo-chemical piles within the 300 Myr of plate motion histories available.

(iv) Present-day synthetic LLSVP shape is sensitive to the imposed initial conditions and the plate model (and its associated reference frame) utilised as a kinematic surface velocity boundary condition (e.g. Shephard et al. 2012). Several studies have demonstrated that the most recent 120–250 Myr of plate motion histories can focus plume clusters and compositional piles into two distinct deep mantle structures beneath Africa and the Pacific (e.g. Bunge et al. 2002; McNamara and Zhong 2005; Bull et al. 2009; Schuberth et al. 2009b; Steinberger and Torsvik 2012; Bower et al. 2013). The exact shape and extent of these provinces, however, may depend on subduction histories beyond 300 Ma (e.g. Zhang et al. 2010). We note that plate motion histories prior to 150 Ma are relatively poorly constrained in the Pacific but reasonably well constrained by the continental record in the Atlantic/Eurasian domains (e.g. Torsvik et al. 2008a; Stampfli and Hochard 2009; Zhang et al. 2010).

14.3.1.2 Conversion to Seismic Structure

Predicted physical structures are converted into density and elastic parameters using a thermodynamic approach for the Na-CaFMAS database (e.g. Stixrude and Lithgow-Bertelloni 2005, 2011). We account for the effects of anelasticity following model Q4 of (Goes et al. 2004), which has a relatively low-temperature sensitivity. Uncertainties in parameters and the equation of state used in the conversion translate into uncertainties in absolute lower-mantle V_P and V_S of about $\pm$ 0.2 km/s (e.g. Cammarano et al. 2005; Cobden et al. 2009). Anelasticity enhances the temperature sensitivity of seismic velocities (e.g. Karato 1993; Goes et al. 2004), and although its importance in the lower-mantle is less than the upper mantle (e.g. Brodholt et al. 2007), it is a systematic effect that should not be neglected (Matas and Bukowinski 2007), particularly within the mantle's lower thermal boundary layer, where lateral temperature anomalies of 1000–1500 K are expected (e.g. Lay et al. 2008; Schuberth et al. 2009b; Davies et al. 2012). Uncertainties in the temperature sensitivity of seismic velocities are about +1, −2 % per 1000 K in $\partial \ln V_S/\partial T$ and +0.4, −0.5 % per 1000 K in $\partial \ln V_P/\partial T$ (Cobden et al. 2009) (Fig. 14.5).

We consider three synthetic seismic cases: a purely thermal case with a pyrolytic composition and two compositional cases, both inferred from the same dynamic thermo-chemical model, but assuming that the dense material is either recycled and has a basaltic composition, or is primitive and has an Fe-rich chondritic composition (Table 14.2). At lowermost mantle depths, the basaltic and Fe-rich compositions have similar excess densities, which justifies the use of the same dynamic model. However, they have different seismic signatures, with basalt somewhat faster in V_P and significantly faster in V_S relative to pyrolite, whilst the Fe-rich composition is slow in both V_P and V_S. Many other compositions, including peridotites, harzburgites and various chondrites have velocities that are similar to a pyrolite (Cobden et al. 2009). The only other compositional change that produces a distinct seismic expression is significant silica enrichment, as might

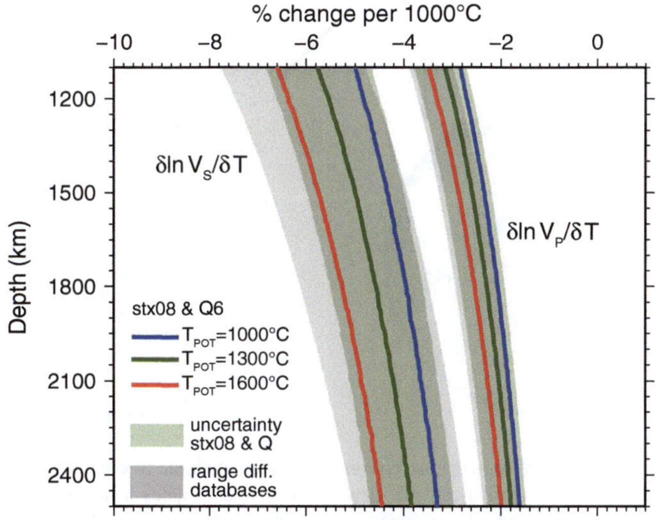

Fig. 14.5 Shear and compressional velocity sensitivity to temperature (after Cobden et al. 2009). The reference profiles use the stx08 thermodynamic database (Xu et al. 2008) and anelasticity model Q6 (Goes et al. 2004). The uncertainty in d*V*/d*T* at any given reference potential temperature (T_{pot}) is dominated by uncertainties (from Xu et al. 2008) in elastic parameters (compare uncertainties in derivatives along a 1300 °C adiabat for single database, in *light green*, and differences between databases (Cobden et al. 2009), in *grey*, with the difference between 1000 °C adiabat derivatives, where anelastic effects are minimal, and 1300 °C derivatives, where anelasticity does contribute). However, anelasticity leads to systematic shifts in d*V*/d*T* with increased sensitivity at high *T* (*red curve*) and decreased sensitivity at low *T* (*blue curve*). Neglecting anelasticity leads to systematic underestimation of the temperature sensitivity of seismic velocities, in particular V_S. Note that the seismic conversions undertaken herein utilised the stx11 database, which has a similar elastic sensitivity to stx08, and anelasticity model Q4, which has a lower temperature sensitivity than Q6

Table 14.2 Major oxide compositions (in mol %) examined herein

Component	Pyrolite	Basaltic	Fe-rich
SiO_2	38.71	51.57	40.03
MgO	49.85	14.94	43.37
FeO	6.17	7.06	11.68
CaO	2.94	13.88	3.24
Al_2O_3	2.22	10.19	1.68
Na_2O	0.11	2.18	0.0

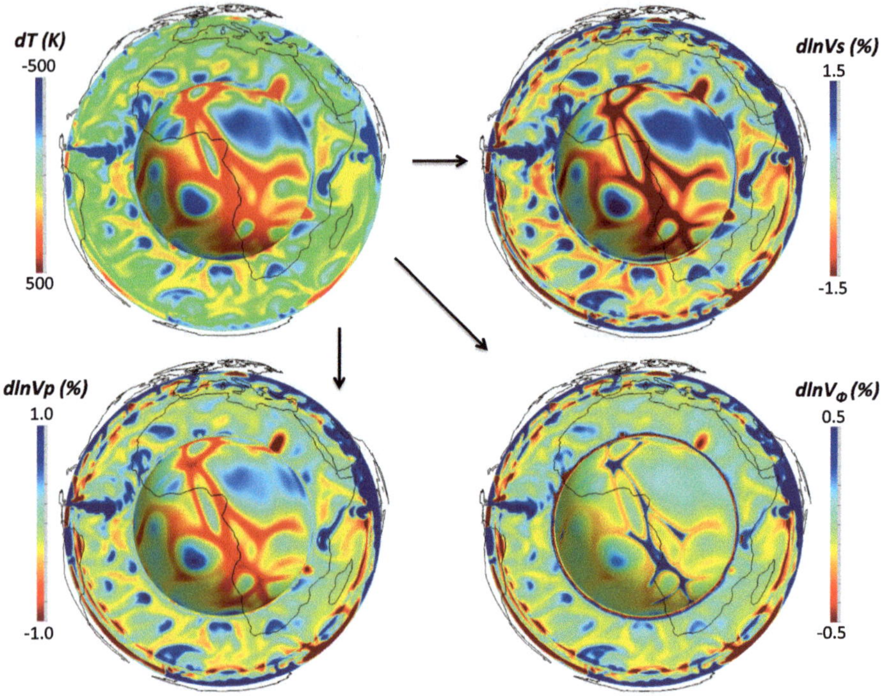

Fig. 14.6 An illustration of our seismic conversion, for an example isochemical model of pyrolitic mantle composition, in an Africa-centred view. Each image includes a radial surface at 2800 km depth and a cross section, whilst continental boundaries provide geographic reference. Modelled temperature and pressure fields are converted into elastic parameters via a thermodynamic approach (Stixrude and Lithgow-Bertelloni 2005, 2011), which accounts for the sensitivity of seismic velocities to pressure, temperature, composition and phase, inclusive of the post-perovskite phase at lowermost mantle depths. Elastic velocities are corrected for the effects of temperature- and pressure-dependent anelasticity, using model Q4 of Goes et al. (2004). Anomalies are shown relative to the model's spherical average profile. Note that no resolution filtering has been applied to the model illustrated here. Anti-correlated bulk-sound and shear-speed anomalies occur due to the presence of post-perovskite

be expected in a cumulate formed from a magma ocean (e.g. Walter et al. 2004). Although high silica content results in faster V_P and slightly faster V_S than a pyrolite, it does not increase density (e.g. Cobden et al. 2009), which is a requirement to form stable piles. An example of compressional, shear and bulk-sound velocity structures for a purely thermal model is shown in Fig. 14.6. At 2800 km depth, where the radial surface is shown, anti-correlated bulk-sound and shear-wave velocity anomalies occur due to the presence of post-perovskite (e.g. Wookey et al. 2005; Ammann et al. 2010; Stixrude and Lithgow-Bertelloni 2011).

14.3.1.3 Tomographic Resolution Filtering

Our synthetic seismic structures cannot be directly compared with tomographic models, as tomographic structures are strongly affected by resolution, due to the uneven source–receiver distribution and the type of data used, in addition to model parameterisation and regularisation. Thus, comparison with seismic constraints requires either propagation of seismic waves through the synthetic structures and direct comparison with data, or conducting a seismic resolution test with the proposed synthetic structure as input. For most of the results included here, the resolution operator of the global tomographic model S40RTS is applied to our synthetic shear-wave velocity structures (see Ritsema et al. 2007, 2011, for further details). The key limitation of this filter is that it does not account for theoretical approximations made in forward wave propagation calculations, but otherwise, it captures all imaging effects and, hence, allows for a direct comparison between tomographic and synthetic seismic structures in morphology and amplitude.

14.3.2 LLSVP Long-Wavelength Pattern

In Fig. 14.7, we show unfiltered and filtered shear-wave velocity distributions beneath the African continent for our thermal and basaltic thermo-chemical models. In both cases, the upper-mantle planform is dominated by strong downwellings in regions of present-day plate convergence. In the mid-mantle, cold downwellings are prominent beneath North America and South-East Asia, whilst remnants of older subduction are visible above the CMB. These downwellings modulate the location of hot material such that it becomes concentrated beneath Africa and the Pacific. The Pacific anomaly is reasonably circular, whilst the African anomaly, shown in Fig. 14.7, is a NW-SE trending structure, which to the north curves eastward under Europe and to the south extends into the Indian Ocean. In the purely thermal case, these structures comprise clusters of plumes and interconnected hot linear ridges, whilst they represent discontinuous chemical piles in the thermo-chemical case (Davies et al. 2012).

Our results agree with the predictions of Bull et al. (2009) that unfiltered deep mantle structure for purely thermal models (Fig. 14.7b) contains significantly more small-scale heterogeneity than thermo-chemical pile models (Fig. 14.7c). However, whether this difference in structure can be inferred from relatively smooth global tomographic models depends on the amplitude, position and spacing of heterogeneity, in addition to tomographic parameterisation and resolution. Whilst earlier tomographic models, such as the spherical harmonic degree-20

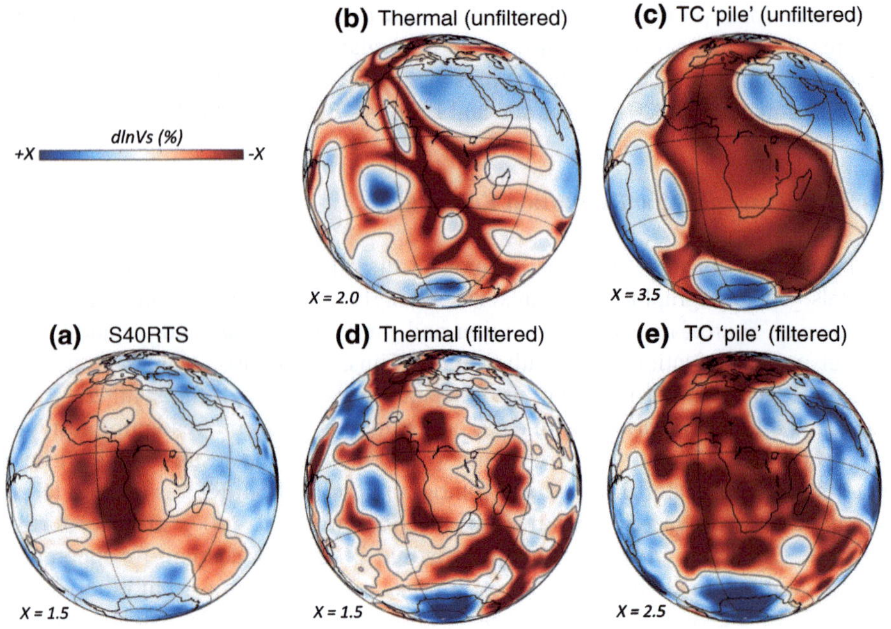

Fig. 14.7 Shear-wave velocity perturbations at 2800 km depth beneath Africa from the following: **a** tomographic model S40RTS; **b** our purely thermal model before and **d** after filtering with the resolution operator of S40RTS; and **c** our thermo-chemical model, before and **e** after filtering. In the isochemical model, (T, P, X) is converted into seismic velocity assuming a pyrolitic composition, whilst the thermo-chemical model illustrated here assumes a pyrolitic composition for background mantle and a basaltic composition for the dense chemical component. We account for the geographic bias, smearing and damping inherent to tomographic models using the resolution operator of S40RTS (Ritsema et al. 2007, 2011), thus allowing for direct comparison between our filtered models (**d**, **e**) and S40RTS (**a**). The −0.3 % grey contour approximates the margin of (**a**) tomographic and (**b**–**e**) synthetic LLSVPs. Continental boundaries provide geographic reference. Note the difference in scale between panels

S20RTS model, predicted relatively homogenous LLSVPs, higher resolution global models, such as S40RTS, S20RTS's degree-40 counterpart (Fig. 14.7a) and regional studies, recover more internal structure (e.g. Wang and Wen 2007; Houser et al. 2008; He and Wen 2009; Ritsema et al. 2011).

After application of S20RTS' resolution operator, Bull et al. (2009) concluded that both thermal plume cluster and thermo-chemical pile models shared many characteristics with imaged structure. However, they favoured the thermo-chemical pile scenario as it provided a better qualitative match to S20RTS. When we apply S40RTS' resolution operator to our thermal (Fig. 14.7d) and thermo-chemical pile (Fig. 14.7e) models, we find that with the additional structure now imaged in S40RTS, it is even more difficult to rule out thermal or thermo-chemical end-members. Focussing solely upon the length-scale and distribution of heterogeneity in Fig. 14.7d, e (i.e. neglecting the increased amplitudes of the thermo-chemical model), both cases show several morphological and geographic characteristics that closely resemble S40RTS.

14.3.3 High Amplitudes and Strong Gradients

It has been argued that a chemical contribution is required to explain the amplitude of low shear-wave velocity anomalies inside LLSVPs (e.g. Karato and Karki 2001; Brodholt et al. 2007). However, with a CMB temperature of 400 ± 200 K, as is inferred from a number of recent geodynamical, seismological and mineral physics studies (e.g. Glatzmaier and Roberts 1995; Boehler 2000; Buffett 2002; Gubbins et al. 2004; Hernlund et al. 2005; Alfè et al. 2007; van der Hilst et al. 2007; Lay et al. 2008), and plausible mantle adiabats (e.g. Brown and Shankland 1981; Cobden et al. 2009), estimates of the temperature contrast across the mantle's lower thermal boundary layer range from 1000 to 1500 K (see, e.g. Lay et al. 2008). As a result, substantial lateral temperature anomalies, on the order of 1000 K, are expected in hot regions above the CMB. Although imaged velocity anomaly amplitudes are uncertain (e.g. Becker and Boschi 2002; Ritsema et al. 2007), current estimates of shear-wave velocity sensitivity to temperature (Fig. 14.5), which consider a wide range of elastic and anelastic uncertainties (Cobden et al. 2009), imply that such temperature variations can easily account for the −2 to −5 % amplitudes of shear-wave velocity anomalies inferred for LLSVPs (e.g. Wang and Wen 2007; Houser et al. 2008; Ritsema et al. 2011). Recent studies, which apply a tomographic resolution filter to synthetic seismic structures from mantle convection models, confirm that sufficient anomaly amplitudes can be generated through thermal heterogeneity alone (e.g. Schuberth et al. 2009a; Davies et al. 2012).

In both thermal and thermo-chemical pile models, the shape of synthetic LLSVPs is modulated by subducting slabs, leading to significant local temperature gradients of 950–1200 K/100 km (see Fig. 14.7b, c), even at depths as shallow as 1500–2000 km (see Davies et al. 2012, for further details). Such thermal gradients translate into significant seismic velocity gradients of 3.5–4.5 %/100 km, which lie within the seismically inferred range of 1.3–6 %/100 km (e.g. Ni et al. 2002; Wang and Wen 2004; To et al. 2005). If LLSVPs additionally contain significant Fe-rich (and, hence, low shear-wave velocity) material, seismic velocity gradients are further enhanced, although they are lowered for a basaltic (high shear-wave velocity) composition. Note that lowermost mantle gradients can exceed 10 %/100 km, due to lateral variations in the occurrence of post-perovskite (Stixrude and Lithgow-Bertelloni 2007; Davies et al. 2012) and are further increased, in all models, if one corrects for anelastic effects using an anelasticity model with a greater temperature sensitivity. These results, therefore, suggest that chemical heterogeneity need not be invoked to explain the high seismic velocity gradients observed in the deep mantle beneath Africa and the Pacific.

14.3.4 Skewed Distributions of Deep Mantle Shear-Wave Velocity Anomalies

Hernlund and Houser (2008) analysed lower-mantle seismic heterogeneity distributions for a number of *P* and *S* wave tomographic models and demonstrated that they are well-approximated by a single Gaussian distribution at all depths, excluding V_S anomalies below ~2200 km depth, where a low-velocity tail is observed. This tail increases in both velocity amplitude and abundance towards the CMB. Figure 14.8a illustrates this tail for S40RTS, which develops a secondary 'bump' from ~2600 km downward.

Hernlund and Houser (2008) evaluated differences in resolution between the different tomographic models and, in particular, between *P* and *S* wave velocity models, concluding that the slow shear-wave velocity feature is robust, as is the absence of such a tail in compressional-wave velocity models.

We perform a similar analysis of our synthetic shear-wave velocity anomalies, filtered to the resolution of S40RTS, with results presented in Fig. 14.8b–d. The distribution of heterogeneity in our purely thermal model (Fig. 14.8b) develops

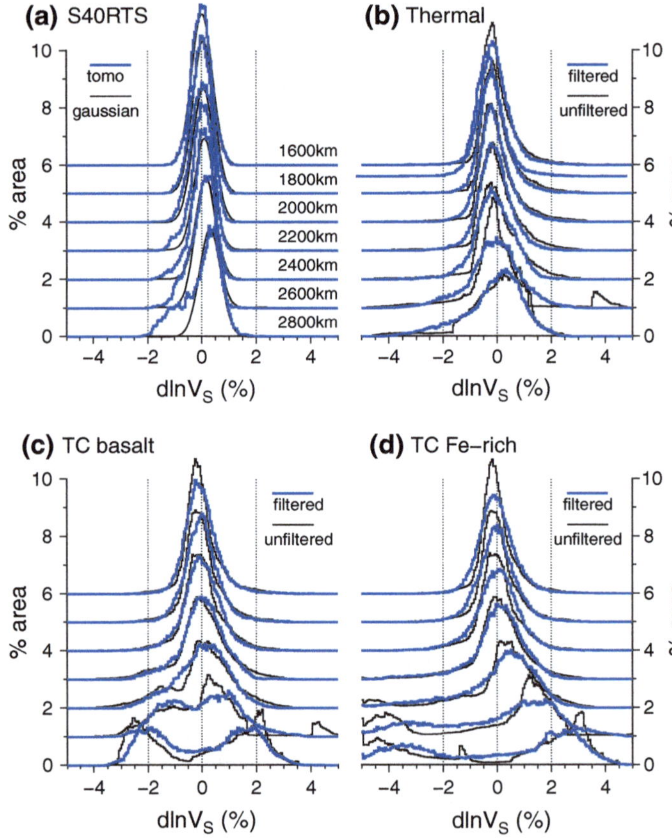

Fig. 14.8 Histograms of shear-wave velocity anomaly distributions, from 1600 to 2800 km depth, at 200-km intervals. Note that plots above 2800 km depth are each vertically offset by an extra 1 % in a real fraction to allow for comparison. Distributions are for: **a** S40RTS; **b** the purely thermal model; **c** the basaltic thermo-chemical model; and **d** the Fe-rich thermo-chemical model. *Blue* (*black*) *lines* represent S40RTS and each model, after (before) application of S40RTS' resolution operator, whilst *black lines* in panel (**a**) are visually fitted Gaussian curves, which highlight the low-velocity tail below ~2200 km depth

a low-velocity tail beneath ~2200 km depth, which increases in both velocity amplitude and abundance towards the CMB, similar to S40RTS, albeit without the slight bump found in the tomographic distributions at depths just above the CMB. Thermo-chemical models (Fig. 14.8c, d), by contrast, develop a pronounced bimodal lower-mantle heterogeneity distribution, with peaks in heterogeneity observed both at the fast and slow end of the range examined. Such a bimodal distribution is not observed in S40RTS or the other tomographic models analysed by Hernlund and Houser (2008).

Hernlund and Houser (2008) estimate that the anomalously slow shear-wave velocity material comprises a mantle volume fraction of 2 ± 0.4 %. The slow feature imaged in S40RTS is of a somewhat larger volume than the tail predicted in our purely thermal models, but is far weaker than that predicted by our thermo-chemical models. Our models therefore suggest that if the tail observed in tomographic models is due to the presence of compositional heterogeneity, it must comprise a substantially smaller volume fraction than the 3 % imposed in our thermo-chemical models. Indeed, given that at least half of the slow tail can be explained without any chemical heterogeneity, if present, dense material inside LLSVPs likely comprises no more than 1 % of the mantle's volume.

Two additional points should be noted from these distributions: (i) model anomalies are calculated relative to the model average. In thermo-chemical models, piles sequester heat in the lowermost mantle, which reduces the layer average velocity in comparison with that of the purely thermal case. Relative to this (slower) average, positive velocity anomalies are enhanced, appearing larger than they do in the purely thermal case at the same depth, despite the fact that the fast material in both cases has similar absolute velocities; (ii) Hernlund and Houser (2008) propose that at least part of the slow tail may be due to the post-perovskite phase transition. Our unfiltered distributions demonstrate that the effect of this transition would be a bump with an offset of ~2–4 % between transformed and untransformed material, which is similar to the velocity jump across the transition. This can be seen, for example, in the unfiltered distributions for our thermal and basaltic thermo-chemical models at 2600 km depth, which show a separate fast bump, relating to the small amount of cold material that has already undergone the transition to post-perovskite. Similarly at 2800 km depth, untransformed hot perovskite material results in the sharp cut-off of anomalies near -1.5 % in the thermal and Fe-rich models. Regularisation of the tomographic inversion mostly filters out these small sets of high-amplitude anomalies. The low-velocity tail in the purely thermal model thus principally results from tomographic (under-) resolution of hot, interconnected ridges (Fig. 14.7b), with strong velocity gradients.

Due to the lack of tomographic P resolution operator, we are unable to undertake a comparable analysis for V_P. However, in contrast to the significant velocity increase in V_S, the transition to post-perovskite leads to a minor decrease in V_P for pyrolitic material. Together with a smaller effect of anelasticity on V_P and overall smaller range of anomalies for the same temperature contrast (see Fig. 14.5), this results in more symmetric unfiltered thermal model distributions for P, and more compact, but nonetheless, bimodal, thermo-chemical distributions.

14.3.5 High $\partial \ln V_S / \partial \ln V_P$ and Anti-correlated $\partial \ln V_S$ and $\partial \ln V_\phi$

Elevated lower-mantle ratios of shear- and compressional-wave-speed anomalies, $R = \partial \ln V_S / \partial \ln V_P$, are regularly cited as evidence of compositional heterogeneity at depth (e.g. Masters et al. 2000; Karato and Karki 2001; Saltzer et al. 2001). Figure 14.9a shows the median value of R, as a function of depth, for our three synthetic cases, alongside seismological estimates, which infer R increases from 1.4–2.0 at 1000 km depth to 2.6–3.4 above the CMB (e.g. Robertson and Woodhouse 1995; Su and Dziewonski 1997; Kennett et al. 1998; Masters et al. 2000; Saltzer et al. 2001; Ritsema and van Heijst 2002). Note that due to the lack of complementary tomographic P and S filters, models and seismic data are not directly comparable in this figure (models have not been filtered to account for limited seismic resolution). Nonetheless, several trends are worth noting.

Above 2200 km depth, all cases are similar, showing a smooth increase in R with depth, which falls within the bounds of seismic observations. Below this depth, synthetic R varies strongly and differs between each case. Even for the purely thermal case, values within, above and below seismological estimates are predicted, with R increasing to ~5 at 2750 km depth. This is due to the post-perovskite phase transition, as relative to perovskite, post-perovskite has an increased shear-wave velocity, a decreased compressional-wave velocity and, hence, higher

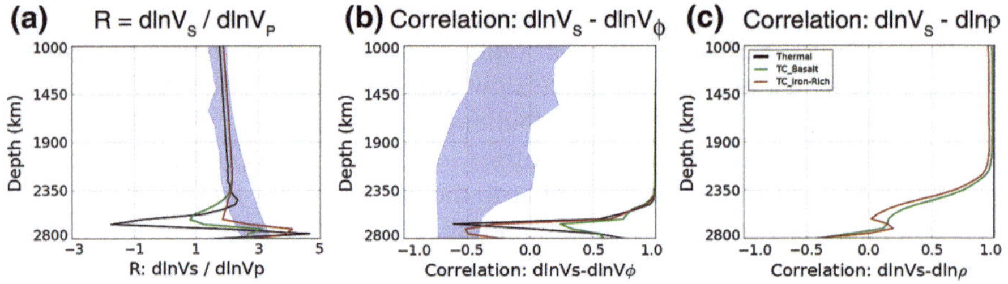

Fig. 14.9 **a** The seismic ratio, $R = \text{dln}V_S/\text{dln}V_P$, as a function of depth, for all models, compared to seismological observations (*shaded blue region* Robertson and Woodhouse 1995; Su and Dziewonski 1997; Kennett et al. 1998; Masters et al. 2000; Saltzer et al. 2001; Karato and Karki 2001; Ritsema and van Heijst 2002). R is calculated by taking the ratio of all grid nodes with nonzero values of both compressional and shear-wave velocity perturbations. A histogram of these nodal values is made and the median value chosen at each depth (following Masters et al. 2000). Unlike mean values, medians are not biased by relatively local regions with extreme values of R. **b** The correlation of $\text{dln}V_S - \text{dln}V_\phi$ as a function of depth, compared to seismological observations (e.g. Su and Dziewonski 1997; Masters et al. 2000). **c** The correlation of $\text{dln}V_S - \text{dln}V_\rho$ as a function of depth. Note that our models have not been post-processed to account for the limited resolution and non-uniqueness inherent to the seismic data and, hence, models and data are not directly comparable. Nonetheless, models predict highly variable R and anti-correlations between bulk-sound and shear-wave velocity only in the post-perovskite stability field. Furthermore, only thermo-chemical models produce an anti-correlation between $\text{dln}V_s$ and density (the purely thermal model predicts radially averaged correlations of 1, even within the post-perovskite stability field)

R (e.g. Wookey et al. 2005; Ammann et al. 2010; Stixrude and Lithgow-Bertelloni 2011; Cobden 2015). It remains to be tested how such R values, which also display strong lateral variations (Davies et al. 2012), would be resolved in long-wavelength seismic images. However, it is clear that in the presence of lower-mantle phase transformations, elevated R does not necessarily imply chemical heterogeneity.

The deep mantle anti-correlation between shear-wave velocity anomalies and bulk-sound velocity anomalies that is imaged by several studies (e.g. Wookey et al. 2005; Ammann et al. 2010; Stixrude and Lithgow-Bertelloni 2011) is closely related to high R. A smaller signature in V_P than V_S implies that the bulk modulus is less affected than, or affected in the opposite way from, the shear modulus by the physical properties of LLSVPs. Bulk-sound velocity is a factor of 2–4 less sensitive to temperature than V_S (e.g. Karato and Karki 2001; Cobden et al. 2009). However, high temperatures do lower both velocities and, hence, cannot explain the anti-correlation between bulk-sound and shear-wave velocity. As such, the imaging of anti-correlated bulk and shear-wave velocity anomalies inside LLSVPs has been taken as a strong indication for compositional heterogeneity. However, relative to perovskite, post-perovskite has a decreased bulk modulus but an increased shear modulus and, hence, this transition can also lead to anti-correlated V_S and V_ϕ (e.g. Wookey et al. 2005; Ammann et al. 2010; Stixrude and Lithgow-Bertelloni 2011) as illustrated by our models in Figs. 14.6 and 14.9b.

Significantly, there are very few compositions that have a high density (necessary to generate thermo-chemical piles) and opposite effects on bulk and shear moduli (e.g. Karato and Karki 2001; Cobden et al. 2009). The compositions examined herein do not do so at the temperature anomalies predicted. High iron content increases density but substantially lowers both moduli (e.g. Karato and Karki 2001; Trampert et al. 2001; Cobden et al. 2009; Stixrude and Lithgow-Bertelloni 2011). The presence of Ca-perovskite and stishovite in basaltic compositions increases density and shear modulus, and only has a minor effect on bulk modulus, resulting in faster shear-wave velocities than pyrolite and slightly slower bulk-sound velocities (Stixrude and Lithgow-Bertelloni 2011). At mildly increased temperatures, this could lead to fast V_S and slow V_ϕ, an anti-correlation that is opposite in sign to what is observed (Deschamps et al. 2012). However, at the temperature anomalies of 1000–1200 K that are predicted inside our thermo-chemical piles, thermal effects easily outweigh a basalt's intrinsic compositional signature, particularly when accounting for anelasticity. To reproduce imaged anti-correlations, compositions that are substantially enriched in silica, which increases bulk-sound velocities much more than it does shear-wave velocity, as well as in iron, thus producing high density, have been proposed (e.g. Trampert et al. 2004; Mosca et al. 2012; Deschamps et al. 2012). It remains to be evaluated how such a composition could evolve. For example, experimental data indicate that cumulates from a magma ocean may be silica-rich but will, most likely, have a low iron content and, hence, density (e.g. Walter et al. 2004).

Finally, it should be noted that there is substantial disagreement between different seismic models on the depth (Fig. 14.9b), geographic location, extent and even existence of the shear-bulk-sound-velocity anti-correlation (e.g. Su and

Dziewonski 1997; Kennett et al. 1998; Masters et al. 2000). For example, the study of Masters et al. (2000), which includes a wide range of data sensitive to deep mantle structure, shows a deep mantle anti-correlation that not only occurs inside LLSVPs, but also beneath the Americas and South-East Asia. Some studies question whether the seismic data require an anti-correlation (e.g. Malcolm and Trampert 2011; Schuberth et al. 2012). We note that the partial correlation between bulk-sound and shear-wave velocities inferred from imaged seismic structure (Fig. 14.9b) may be a consequence of different seismic resolution and sensitivities for *P* and *S* waves, as one would expect both structures to be differently scaled versions of the same underlying physical structure and, accordingly, either be correlated or anti-correlated.

14.3.6 Density-Shear-wave Velocity Anti-correlation

A number of studies have inferred an anti-correlation between low shear-wave velocity anomalies and high-density anomalies inside (and outside) LLSVPs, from combined inversion of seismic and geoid data (e.g. Ishii and Tromp 1999; Trampert et al. 2004; Simmons et al. 2010; Mosca et al. 2012). Indeed, such an anti-correlation can only be explained by the presence of a compositionally distinct component and not by variations in temperature or phase, as illustrated in Fig. 14.9c, where shear-wave velocity and density anomalies are fully correlated in the absence of chemical heterogeneity. However, there has been significant debate on whether lowermost mantle density heterogeneity is resolvable by the available data (e.g. Romanowicz 2001; Masters and Gubbins 2003). Furthermore, the positive geoid signature observed above LLSVPs requires that if a dense chemical component is present, the overlying mantle's net buoyancy masks this signature at the surface (e.g. Gurnis et al. 2000; Ricard et al. 2006).

14.3.7 1-D Reference Model Biases

14.3.7.1 Differences Seismic and Physical Reference Models

The seismic structure of Earth's mantle is predominantly 1-D. At depths away from phase transitions, below the lithosphere and above D'', lateral variations in seismic velocity constitute only a few per cent deviation from the mantle's spherical average. A vigorously convecting isochemical mantle, subject to internal heating, is expected to have a subadiabatic geotherm (e.g. Jeanloz and Morris 1987; Bunge 2005; Leng and Zhong 2008) and a pyrolitic (MORB-source) composition (e.g. McDonough and Sun 1995). Global seismic reference structures, such as PREM and AK135 (e.g. Dziewonski and Anderson 1981; Kennett et al. 1995), share many of the features predicted from such a reference physical profile, with jumps at depths where phase transitions are expected and a smooth lower-mantle velocity-depth profile (Fig. 14.10b).

However, there are systematic differences between the tightly constrained seismic reference profiles and those predicted from pyrolitic, subadiabatic physical structures, which exceed our current best estimates for uncertainties in mineral physics parameters and the mantle's equation of state (e.g. Cammarano et al. 2005; Cobden et al. 2009). These include differences in average upper-mantle velocities and jumps in the transition zone (Fig. 14.10b), which have been attributed to compositional gradients (1-D layering or resulting from 3-D heterogeneity) (e.g. Duffy and Anderson 1989; Cammarano and Romanowicz 2007; Cobden et al. 2008), or seismic imaging biases (e.g. Davies and Bunge 2001; Styles et al. 2011, and discussion below). Most relevant for this paper is that the seismic velocity gradients ($dV_{P,S}/dz$) below ~1500 km depth are consistently lower than those predicted from adiabatic to subadiabatic physical models (e.g. Cobden et al.

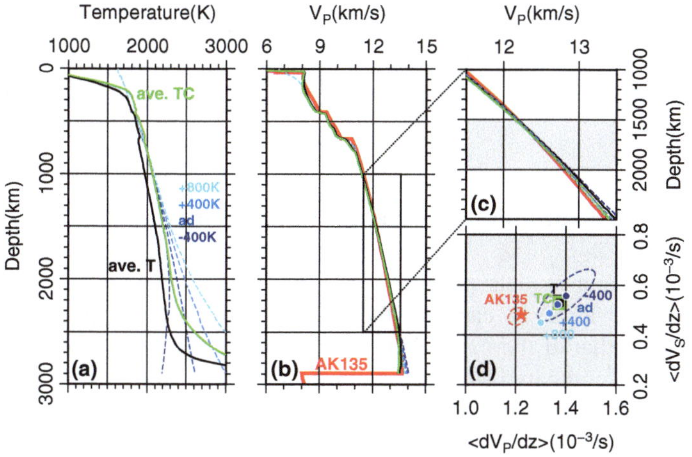

Fig. 14.10 Comparison of synthetic 1-D *P*-velocity mantle structures with seismic model AK135 (Kennett et al. 1995): **a** A range of possible mantle geotherms: average of our thermal (*black*) and basaltic thermo-chemical (*green*) models, and a set of 1-D structures with MORB-source surface temperatures of 1350 °C, ranging from subadiabatic (*dark blue* −400 K by 2890 km depth) to strongly superadiabatic (*cyan* +400 and +800 K at 2890 km depth). **b** Compressional velocity for the thermal profiles from a, compared with AK135 (*red*)—note that the uncertainty in global seismic models is no more than the width of the *red line*. **c** Zoom of lower-mantle V_P between 1000 and 2500 km depth, illustrating the systematic mismatch in gradient between seismic reference models and our mantle geotherms. **d** averaged V_P-depth gradients versus averaged V_S-depth gradients, between 1500 and 2000 km depth, for the range of models shown in (a). The *dashed red line* comprises a range of published seismic models including PREM (Dziewonski and Anderson 1981). The *dashed blue line* accounts for elastic parameter uncertainties, as computed by Cobden et al. (2009), plotted around the subadiabatic model. Although gradient mismatches between averaged 3-D models, or subadiabatic mantle and seismic models seem subtle, they fall consistently outside of the seismic and mineral physics uncertainties (Cobden et al. 2009). The differences are even more apparent in traveltime calculations (Fig. 14.11d)

2009) (Fig. 14.10c). This discrepancy has been interpreted as the result of either superadiabatic temperature gradients or variations in composition as a function of depth (e.g. da Silva et al. 2000; Deschamps and Trampert 2004; Matas et al. 2007; Khan et al. 2008; Cobden et al. 2009). For example, profiles where temperatures exceed the mantle adiabat by at least 200–400 K by 2500 km depth (such as those labelled +400 and +800 K in Fig. 14.10) can match the observed seismic gradients within the uncertainties. However, global compositional stratification is unlikely, given the absence of seismic discontinuities away from those expected for known phase transitions and evidence of slabs sinking into the lowermost mantle (e.g. Grand et al. 1997; van der Hilst et al. 1997; Tackley 2002). Hence, it has been proposed that the compositional gradients arise as a consequence of 3-D heterogeneity, in the form of discontinuous piles at the mantle's base (e.g. Cobden et al. 2009). Furthermore, without stratification, chemically stabilised piles may be required to sustain a net superadiabatic gradient (Cobden et al. 2009), because as noted previously, a vigorously convecting isochemical mantle would generate a subadiabatic mantle geotherm (e.g. Bunge 2005). As the sensitivity of seismic velocity to temperature increases at higher temperatures (due to anelasticity), the presence of hot chemical piles in the lowermost mantle may bias the average seen by seismic waves away from the actual physical average.

14.3.7.2 3-D Structure Biases to Average Mantle Structure?

Using the dynamic models presented above, we test whether or not thermal or thermo-chemical LLSVPs can bias the mantle's radially averaged seismic structure away from that of the mantle's radially averaged physical structure. In a previous paper, we compared seismic velocities for the average radial physical state with laterally averaged seismic velocities (Styles et al. 2011). We found that thermal or thermo-chemical structure that generates velocity anomalies of the same order as those inferred from seismic tomography yield an averaged lower-mantle velocity structure that matches the actual average to within 0.1 %. This difference falls within the uncertainties in seismic reference profiles and, hence, cannot account for the difference between seismically imaged structure and predicted seismic structure from a physical reference model.

However, the study by Styles et al. (2011) assumed uniform seismic sampling of mantle structure. Here, we present the results of a test of seismic sampling effects, using traveltimes for direct *P* waves. This is an extreme case, as seismic reference models are derived from a range of body-wave phases, and some models, such as PREM and AK135-f, also include normal modes (which have global sensitivity), whilst a weighting or averaging of data is often undertaken to reduce any geographic bias. We perform a relatively simple calculation, where ray paths are computed through a 1-D Earth (the model average) using TauP (Crotwell et al.

1999), whilst traveltime integration along these ray paths is done through the full 3-D structure, with CRUST2.0 (Bassin et al. 2000) superimposed above. Full 3-D ray tracing would be more accurate, but several studies have found it to have a relatively small effect, and hence, it is not commonly done in global tomographic studies (e.g. Bijwaard and Spakman 2000; Simmons et al. 2011). Inaccurate rays can lead to underestimates of the influence of small features (in particular slabs) (Zhao and Lei, 2003), but we verified that the (relatively broad) model slabs are sampled by our ray geometries. Finite-frequency effects may also influence traveltimes, but these are small for the high-frequency direct *P* wave data considered here (e.g. Montelli et al. 2004). Furthermore, we find that the predicted geographic bias is so much smaller than the discrepancy between seismic and physical reference models, that improved traveltime calculations are unlikely to change our conclusions.

Our tests demonstrate that biased geographic sampling has some systematic effects. To illustrate these, in Fig. 14.11b, c, synthetic traveltime anomalies and their means are plotted, as a function of epicentral distance, all relative to the traveltime predicted for the mean velocity structure. For comparison, Fig. 14.11a shows the scatter of ISC traveltimes around their mean, which is similar to the scatter around AK135, a model that was derived to fit these data (plus additional phases). Figure 14.11d shows the traveltime discrepancies between the seismic and physical reference models, including the thermal and basaltic thermo-chemical models examined herein and a few different adiabatic cases.

The main geographic bias is accrued in the lithosphere, above 200–300 km depth (epicentral distances <15°), where the distribution of sources and receivers leads to preferential sampling of continental keels and subducting slabs. This is a well-known bias (e.g. Davies and Bunge 2001), which means that the sampled Earth looks ~2 s faster than the actual reference structure. Below this depth, to epicentral distances of ~50°, the direct *P* waves preferentially sample slabs, leading to a further increase in the fast bias of imaged mantle structure, but only ~0.5 s. Below about 1200–1500 km depth (epicentral distances >50°), the fast bias decreases as rays sample more and more background mantle material. Note that since LLSVPs are poorly sampled by this set of rays, they have little effect on the traveltime bias, whether they are of thermal or thermo-chemical origin. Indeed, the fast bias is shown to increase further in the lowermost mantle, where rays preferentially sample faster material away from LLSVPs. As velocity anomalies are larger in the thermo-chemical model, this lowermost mantle increase in fast bias is more prominent. Interestingly, the increase and subsequent decrease of the fast slab bias through the mid-mantle have a similar trend as the difference between seismic and physical reference models. However, the resulting variations in traveltimes due to geographically biased sampling (Fig. 14.11b, c) are 3–8 times smaller (~0.5 s from $\Delta = 40$–70°) than the difference between seismic and physical models (~2 s, from $\Delta = 40$–70°, Fig. 14.11d) and, thus, cannot reconcile the two.

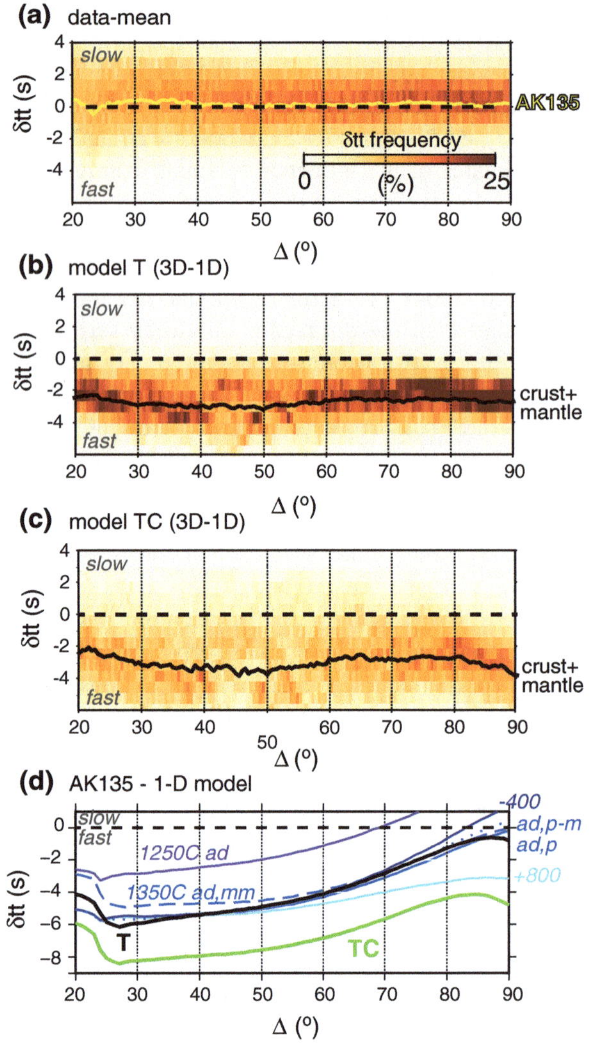

Fig. 14.11 Comparison of modelled geographic traveltime biases with scatter in ISC traveltime data and traveltime differences between physical and seismic reference model, all plotted as a function of epicentral distance Δ: **a** distribution of ISC direct *P* traveltime anomalies (*orange*) relative to catalogue mean, plus difference between catalogue mean and AK135 (*in yellow*). **b**, **c** Synthetic *P* traveltime anomalies (*orange*) relative to traveltimes through 1-D average model structure. Mean anomaly (due to 3D crust and mantle structure) in *bold black*. The offset between the 3-D mean traveltime and the time through the average 1-D structure is due to geographic sampling bias. This bias is fast throughout and mainly accumulated in the crust and shallow upper mantle ($\Delta < 20°$). **d** Traveltimes of synthetic velocity models from Fig. 14.10 relative to those for AK135 (Kennett et al. 1995), where *T* (*black*) and TC (*green*) denote the average of 3-D thermal and basaltic thermo-chemical models, respectively, 1250 °C ad (*purple*)—adiabatic profile with surface *T* of 1250 °C and pyrolitic composition, 1350 °C surface *T* profiles (*shades of blue*)—pyrolitic composition (*solid lines*): −400° subadiabatic (*dark blue*), +800° superadiabatic (*cyan*) and adiabatic (*blue*), adiabatic mechanical mixture (*blue, dashed*), adiabatic pyrolite to MORB compositional gradient (*blue, dotted*). Variations in the discrepancy between physical models and AK135 as a function of Δ clearly exceed the variation expected from geographic bias

14.3.7.3 1-D Reference Models Summary

Mismatches between seismic and physical reference models likely cannot give further insight into the thermal or thermo-chemical nature of LLSVPs. Alternative 1-D effects that have been proposed to influence deep mantle seismic structure, such as the Fe-spin transition (e.g. Badro et al. 2003), would not lead to the systematic difference between seismic and physical models that varies smoothly with depth (Cammarano et al. 2010). Rather, the mismatch may indicate that uncertainties in the mineral physic data, in particular, the pressure and temperature dependence of the shear modulus under deep mantle conditions (e.g. Jackson 1998; Deschamps and Trampert 2004; Matas et al. 2007), may exceed those previously considered (e.g. Cammarano et al. 2005; Cobden et al. 2009).

14.3.8 The Spatial Relation Between Hotspots, Reconstructed LIPs and LLSVPs

Inferred spatial correlations between the reconstructed eruption sites of LIPs and surface hotspot locations with LLSVP margins at depth are regularly invoked as firm evidence for the thermo-chemical nature of LLSVPs (e.g. Thorne et al. 2004; Torsvik et al. 2006, 2008b, 2010; Burke et al. 2008). These studies attribute the plume localisation implied by this observation to the interaction of mantle flow with stable, high-density piles, leading to preferential plume initiation near pile margins (e.g. Tan et al. 2011; Steinberger and Torsvik 2012).

However, it should be noted that if piles exist on the spatial scale implied by LLSVPs and are able to remain sequestered in the deep mantle for hundreds of millions of years, they will build up substantial excess temperatures, when compared to ambient mantle (Davies et al. 2012). Internal upwellings, in addition to those at the margins, are therefore required if piles are to release their heat (e.g. Jellinek and Manga 2002). Furthermore, even in purely thermal models, the interaction of downwellings with the mantle's lower thermal boundary layer reduces plume initiation times, facilitating plume initiation at the leading edge of deep mantle slabs (e.g. Schaeffer and Manga 2001; Tan et al. 2002; Goes et al. 2004).

Perhaps most significantly, the inferred spatial correlation between the surface expression of mantle plumes and LLSVP margins has recently been questioned. Figure 14.12 displays results from a straightforward analysis of the spatial correlation between hotspots and LLSVP margins, following the approach of Torsvik et al. (2006), but illustrating the sensitivity to the $\mathrm{dln}V_S$ contour used to delineate the LLSVP margin and the maximum plume tilt angle, angle, θ (Davies et al. 2015). This confirms that hotspots are preferentially located above the margins of the African LLSVP, although this is not the case within the Pacific domain, where they show no preference for the underlying LLSVP margin: the correlation between hotspots and LLSVP margins, therefore, is not a global feature (the same is true for reconstructed LIPs). Moreover, Monte-Carlo- based statistical analyses

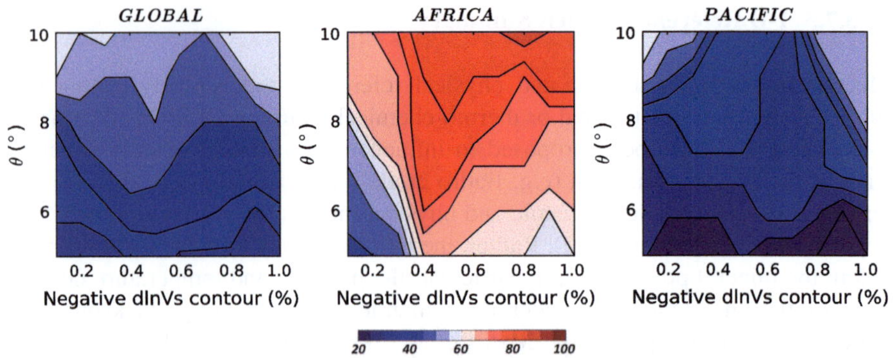

Fig. 14.12 The percentage of hotspots from the catalogue of Steinberger (2000) that are located within an angular distance (θ) of LLSVP margins, at 2800 km depth, as delineated by a specific negative dlnV_S contour in the SMEAN tomographic model (from Davies et al. 2015). Results are presented globally and regionally for hotspots that are located in the African and Pacific domains. At $\theta = 10°$ and dln$V_S = -1.0$ %, the result of Torsvik et al. (2006) is reproduced, with ~55 % of hotspots located above LLSVP margins. This spatial correlation, however, is reduced to <30 % at lower θ values and shows substantial regional variability between the African and Pacific domains, with strong correlations generally observed for the former (>80 %, at $\theta \geq 8°$) and weak correlations for the latter (<50 %, even at the highest θ values). This is consistent with the map presented in Fig. 14.3, which clearly shows that hotspots occur both above the interior (high dlnV_S amplitudes) and above the margins (low dlnV_S amplitudes) of the Pacific LLSVP

by Davies et. al. (2015), which build upon an earlier study by Austermann et al. (2014, demonstrate that the observed distribution of African and Pacific hotspots/ reconstructed LIPs are consistent with the hypothesis that they are drawn from a sample that is uniformly distributed across the entire areal extent of each LLSVP: the stronger spatial correlation with the margin of the African LLSVP is expected as a simple consequence of its elongated geometry, where greater than 75 % of the LLSVP interior lies within 10° of its margin.

Taken together, these results imply that the geographical distribution of hotspots and reconstructed LIPs does not indicate the extent to which chemical heterogeneity influences lower-mantle dynamics.

14.4 Summary

Of the nine seismic arguments commonly invoked as evidence that LLSVPs constitute thermo-chemical piles, our analyses suggest that only an anti-correlation between shear-wave velocity anomalies and density anomalies would provide unambiguous evidence for the presence of dense chemical heterogeneity. Spherically averaged lower-mantle seismic velocity-depth gradients are not sensitive to LLSVP structure, whilst all other constraints, including LLSVP morphology, shear-wave velocity amplitudes and gradients, (relative) variation of shear, compressional and bulk-sound speeds, and the relationship between LLSVPs and the distribution of intraplate volcanism, are equally well explained with thermal and thermo-chemical models. Although an anti-correlation between shear-wave velocity anomalies and density anomalies has been found in some studies (e.g. Ishii and Tromp 1999; Trampert et al. 2004), our ability to resolve lower-mantle density heterogeneity is debated (e.g. Romanowicz 2001; Masters and Gubbins 2003). Furthermore, imaged distributions of lower-mantle shear-wave velocity anomalies allow, at most, a very low volume fraction (<1 % of the mantle's volume) of dense chemical heterogeneity in deep mantle piles. Whether or not such low volume fractions are detectable remains to be seen. However, such heterogeneity, although geochemically significant, would exert control on lowermost mantle dynamics and its large-scale geophysical structure.

Acknowledgments DRD was partially funded by Fellowships from NERC (NE/H015329/1) and the ARC (FT140101262). Numerical simulations were undertaken on: (i) HECToR, the UK's national high-performance computing service, which is provided by UoE HPCx Ltd, at the University of Edinburgh, Cray Inc., and NAG Ltd, and funded by the Office of Science and Technology through EPSRC's High End Computing Program; and (ii) the NCI National Facility in Canberra, Australia, which is supported by the Australian Commonwealth Government. Authors would like to thank Lars Stixrude and Carolina Lithgow-Bertelloni for providing the lookup tables used in converting models from physical structure to seismic velocity; and Jeroen Ritsema for providing S40RTS' resolution operator. Authors benefited from discussion with Huw Davies, Jeroen Ritsema, Hans-Peter Bunge, Bernhard Schuberth, Julie Prytulak, Brian Kennett, Ian Campbell and Geoff Davies. Authors would like to thank two anonymous reviewers for constructive comments on this manuscript, as well as Frederic Deschamps for editorial input.

References

Alfè D, Gillan MJ, Price GD (2007) Temperature and composition of the Earth's core. Contemp Phys 48:63–80. doi:10.1080/00107510701529653

Allègre C, Manhes G, Lewin E (2001) Chemical composition of the Earth and the volatility control on planetary genetics. Earth Planet Sci Lett 185:49–69. doi:10.1016/S0012-821X(00)00359-9

Allègre CJ, Brevart O, Dupre B, Minster JF (1980) Isotopic and chemical effects produced in a continuously differentiating convecting Earth mantle. Philos Trans R Soc Lond Ser A 297:447–477. doi:10.1098/rsta.1980.0225

Allègre CJ, Hofmann AW, O'Nions RK (1996) The Argon constraints on mantle structure. Geophys Res Lett 23:3555–3557. doi:10.1029/96GL03373

Allègre CJ, Staudacher T, Sarda P (1987) Rare gas systematics: formation of the atmosphere, evolution and structure of the Earth's mantle. Earth Planet Sci Lett 81:127–150. doi:10.1016/0012821X(87)90151-8

Ammann MW, Brodholt JP, Wookey J, Dobson DP (2010) First-principles constraints on diffusion in lower-mantle minerals and a weak D" layer. Nature 465:251–267. doi:10.1038/nature09052

Anderson DL (1982) Hotspots, polar wander, Mesozoic convection and the geoid. Nature 297:391–393. doi:10.1038/297391a0

Austermann J, Kaye BT, Mitrovica JX, Huybers P (2014) A statistical analysis of the correlation between large igneous provinces and lower mantle seismic structure. Geophys J Int. doi:10.1093/gji/ggt500

Badro J, Fiquet G, Guyot F, Rueff J-P, Struzhkin VV, Vankó G, Monaco G (2003) Iron partitioning in Earth's mantle: toward a deep lower mantle discontinuity. Science 300:789–791. doi:10.1126/science.1081311

Bassin C, Laske G, Masters G (2000) The current limits of resolution for surface wave tomography in North America. EOS Trans AGU 81:0 F897

Baumgardner JR (1985) Three-dimensional treatment of convective flow in the Earth's mantle. J Stat Phys 39:501–511. doi:10.1007/BF01008348

Becker TW, Boschi L (2002) A comparison of tomographic and geodynamic mantle models. Geochem Geophys Geosys 3:0 2001GC000168. doi:10.129/2001GC000168

Bijwaard H, Spakman W (2000) Non-linear global P-wave tomography by iterated linearized inversion. Geophys J Int 141:71–82. doi:10.1046/j.1365-246X.2000.00053.x

Boehler R (2000) High-pressure experiments and the phase diagram of lower mantle and core materials. Rev Geophys 38:221–245. doi:10.1029/1998RG000053

Bower DJ, Gurnis M, Seton M (2013) Lower mantle structure from paleogeographically constrained dynamic Earth models. Geochem Geophys Geosys 14:44–63. doi:10.1029/2012GC004267

Boyet M, Carlson RW (2005) ^{142}Nd evidence for early (4.53 Ga) global differentiation of the silicate Earth. Science 309:576–581

Boyet M, Carlson RW (2006) A new geochemical model for the Earth's mantle inferred from ^{146}Sm/^{142}Nd systematics. Earth Planet Sci Lett 250:254–268

Brandenburg JP, Hauri EH, van Keken PE, Ballentine CJ (2008) A multiple-system study of the geochemical evolution of the mantle with force-balanced plates and thermochemical effects. Earth Planet Sci Lett 276:1–13. doi:10.1016/j.epsl.2008.08.027

Brandenburg JP, van Keken PE (2007) Deep storage of oceanic crust in a vigorously convecting mantle. J Geophys Res 112:0 B06403. doi:10.1029/2006JB004813

Brodholt JP, Hellfrich G, Trampert J (2007) Chemical versus thermal heterogeneity in the lower mantle: the most likely role of anelasticity. Earth Planet Sci Lett 262:429–437. doi:10.1016/j.epsl.2007.07.054

Brown JM, Shankland TJ (1981) Thermodynamic parameters in the Earth as determined from seismic profiles. Geophys J R Astron Soc 66:579–596

Buffett BA (2002) Estimates of heat flow in the deep mantle based on the power requirements for the geodynamo. Geophys Res Lett 29:4. doi:10.1029/2001GL014649

Bull AL, McNamara AK, Ritsema J (2009) Synthetic tomography of plume clusters and thermochemical piles. Earth Planet Sci Lett 278:152–156. doi:10.1016/j.epsl.2008.11.018

Bunge H-P (2005) Low plume excess temperature and high core heat flux inferred from non-adiabatic geotherms in internally heated mantle circulation models. Phys Earth Planet Int 153:3–10. doi:10.1016/j.pepi.2005.03.017

Bunge H-P, Richards MA, Baumgardner JR (1997) A sensitivity study of 3-D-spherical mantle convection at 10^8 Rayleigh number: effects of depth-dependent viscosity, heating mode and an endothermic phase change. J Geophys Res 102:11991–12007. doi:10.1029/96JB03806

Bunge HP, Richards MA, Baumgardner JR (2002) Mantle circulation models with sequential data-assimilation: inferring present-day mantle structure from plate motion histories. Philos Trans R Soc Lond Set A 360:2545–2567. doi:10.1098/rsta.2002.1080

Burke K, Steinberger B, Torsvik TH, Smethurst MA (2008) Plume generation zones at the margins of large low shear-wave velocity provinces on the core–mantle–boundary. Earth Planet Sci Lett 265:49–60. doi:10.1016/j.epsl.2007.09.042

Cammarano F, Goes S, Deuss A, Giardini D (2005) Is a pyrolitic adiabatic mantle compatible with seismic data? Earth Planet Sci Lett 232:227–243

Cammarano F, Marquardt H, Speziale S, Tackley PJ (2010) Role of iron-spin transition in ferropericlase on seismic interpretation: a broad thermochemical transition in the mid mantle? Geophys Res Lett 37. doi:10.1029/2009GL041583

Cammarano F, Romanowicz B (2007) Insights into the nature of the transition zone from physically constrained inversion of long period seismic data. PNAS-High Press Geosci 104:9139–9144

Campbell IH, Griffiths RW (1992) The changing nature of mantle hotspots through time: implications for the geochemical evolution of the mantle. J Geology 100:497–523
Campbell IH, O'Neill HC (2012) Evidence against a chondritic earth. Nature 483:553–558. doi:10.1038/nature10901
Caro G (2015) Chemical geodynamics in a non-chondritic Earth. In: Khan A, Deschamps F (eds) The Earth's Heterogeneous Mantle, Springer, Cham (this Volume)
Caro G, Bourdon B (2010) Non-chondritic Sm/Nd ratio in the terrestrial planets: consequences for the geochemical evolution of the mantle-crust system. Geochim Cosmochim Acta 74:3333–3349
Christensen UR, Hofmann AW (1994) Segregation of subducted oceanic crust in the mantle. J Geophys Res 99:19867–19884
Cobden L, Goes S, Cammarano F, Connolly JAD (2008) Thermochemical interpretation of one-dimensional seismic reference models for the upper mantle: evidence for bias due to heterogeneity. Geophys J Int 175:627–648. doi:10.1111/j.1365-246X.2008.03903.x
Cobden L, Thomas C, Trampert J (2015) Seismic detection of post-perovskite inside the earth. In: Khan A, Deschamps F (eds) The earth's heterogeneous mantle, Springer, Cham.(this Volume)
Cobden L, Goes S, Ravenna M, Styles E, Cammarano F, Gallagher K, Connolly JAD (2009) Thermochemical interpretation of 1-D seismic data for the lower mantle: the significance of non-adiabatic thermal gradients and compositional heterogeneity. J Geophys Res 114:B11309. doi:10.1029/2008JB006262
Coltice N, Ricard Y (1999) Geochemical observations and one layer mantle convection. Earth Planet Sci Lett 174:125–137
Crotwell HP, Owens TJ, Ritsema J (1999) The TauP toolkit: flexible seismic travel-time and ray-path utilities. Seismol Res Lett 70:154–160. doi:10.1785/gssrl.70.2.154
da Silva CRS, Wentzcovitch RM, Patel A, Price GD, Karato SI (2000) The composition and geotherm of the lower mantle: constraints from the elasticity of silicate perovskite. Earth Planet Sci Lett 118:103–109. doi:10.1016/S0031-9201(99)00133-8
Davaille A (1999) Simultaneous generation of hotspots and superswells by convection in a heterogeneous planetary mantle. Nature 402:756–760
Davies GF (1999) Dynamic Earth: plates, plumes and mantle convection. Cambridge University Press, Cambridge. ISBN 9780521599337
Davies GF (2009) Reconciling the geophysical and geochemical mantles: plume flows, heterogeneities and disequilibrium. Geochem Geophy Geosyst 10:Q10008. doi:10.1029/2009GC002634
Davies GF (2011) Dynamical geochemistry of the mantle. Solid Earth 2:159–189. doi:10.5194/se-2-159-2011
Davies DR, Davies JH (2009) Thermally–driven mantle plumes reconcile multiple hotspot observations. Earth Planet Sci Lett 278:50–54. doi:10.1016/j.epsl.2008.11.027
Davies DR, Davies JH, Bollada PC, Hassan O, Morgan K, Nithiarasu P (2013) A hierarchical mesh refinement technique for global 3D spherical mantle convection modelling. Geosci Mod Dev 6:1095–1107. doi:10.5194/gmd-6-1095-2013
Davies DR, Goes S, Davies JH, Schuberth BSA, Bunge H-P, Ritsema J (2012) Reconciling dynamic and seismic models of Earth's lower mantle: the dominant role of thermal heterogeneity. Earth Planet Sci Lett 353–354:253–269. doi:10.1016/j.epsl.2012.08.016
Davies DR, Goes S, Sambridge M (2015) On the relationship between volcanic hotspot locations, the reconstructed eruption sites of large igneous provinces and deep mantle seismic structure. Earth Planet Sci Lett 411:121–130
Davies JH, Bunge HP (2001) Seismically 'fast' geodynamic mantle models. Geophys Res Lett 28:73–76
de Koker N (2010) Thermal conductivity of MgO at high pressure: implications for the D" region. Earth Planet Sci Lett 292:392–398
Deschamps F, Cobden L, Tackley PJ (2012) The primitive nature of large low shear-wave velocity provinces. Earth Planet Sci Lett 349–350:198–208. doi:10.1016/j.epsl.2012.07.012

Deschamps F, Kaminski E, Tackley PJ (2011) A deep mantle origin for the primitive signature of ocean island basalt. Nature Geosci 4:879–882. doi:10.1038/NGEO1295

Deschamps F, Li Y, Tackley PJ (2015) Large-scale thermo-chemical structure of 1 the deep mantle: observations and models. In: Khan A, Deschamps F (eds) The Earth's Heterogeneous Mantle, Springer, Cham (this Volume)

Deschamps F, Tackley PJ (2008) Exploring the model space of thermo-chemical convection: (i) principles and influence of the rheological parameters. Phys Earth Planet Int 171:357–373

Deschamps F, Tackley PJ (2009) Searching for models of thermo-chemical convection that explain probabilistic tomography: (ii) influence of physical and compositional parameters. Phys Earth Planet Int 176:1–18

Deschamps F, Trampert J (2004) Towards a lower mantle reference temperature and composition. Earth Planet Sci Lett 222:161–175

Duffy TS, Anderson DL (1989) Seismic velocities in mantle minerals and the mineralogy of the upper mantle. J Geophys Res 940(B2):1895–1912. doi:10.1029/JB094iB02p01895

Duncan RA, Richards MA (1991) Hotspots, mantle plumes, flood basalts and true polar wander. Rev Geophys 29:31–50

Dziewonski AM, Anderson DL (1981) Preliminary reference Earth model. Phys Earth Planet Int 25:297–356

Dziewonski AM, Hager BH, O'Connell RJ (1977) Large-scale heterogeneities in the lower mantle. J Geophys Res 82:239–255

Dziewonski AM, Lekic V, Romanowicz BA (2010) Mantle anchor structure: an argument for bottom up tectonics. Earth Planet Sci Lett 299:69–79

Forte AM, Mitrovica JX (2001) Deep-mantle high-viscosity flow and thermochemical structure inferred from seismic and geodynamic data. Nature 410:1049–1056

Fukao Y, Obayashi M (2013) Subducted slabs stagnant above, penetrating through and trapped below the 660 km discontinuity. J Geophys Res 118:5920–5938. doi:10.1002/2013JB010466

Garel F, Goes S, Davies DR, Davies JH, Kramer SC, Wilson CR (2014) Interaction of subducted slabs with the mantle transition-zone: a regime diagram from 2-D thermo-mechanical models with a mobile trench and an overriding plate. Geochem Geophys Geosys 15. doi:10.1002/2014GC005257

Garnero EJ, McNamara AK (2008) Structure and dynamics of Earth's lower mantle. Science 320:626–628

Glatzmaier GA, Roberts PH (1995) A 3-D self–consistent computer simulation of a geomagnetic field reversal. Nature 377:203–209

Goes S, Cammarano F, Hansen U (2004) Synthetic seismic signature of thermal mantle plumes. Earth Planet Sci Lett 218:403–419. doi:10.1016/S0012-821X(03)00680-0

Grand S, van der Hilst RD, Widiyantoro S (1997) Global seismic tomography: a snapshot of mantle convection in the Earth. GSA Today 7:1–7

Gubbins D, Alfè D, Masters G, Price GD, Gillan M (2004) Gross thermodynamics of two–component core convection. Geophys J Int 157:1407–1414. doi:10.1111/j.1365-246X.2004.02219.x

Gurnis M, Mitrovica JX, Ritsema J, van Heijst HJ (2000) Constraining mantle density structure using geological evidence of surface uplift rates: the case of the African superplume. Geochem Geophys Geosys 1:1999GC000035

Hager BH, Clayton RW, Richards MA, Comer RP, Dziewonski AM (1985) Lower mantle heterogeneity, dynamic topography and the geoid. Nature 313:541–545. doi:10.1038/313541a0

Hanan BB, Graham DW (1996) Lead and helium isotopic evidence from oceanic basalts for a common deep source of mantle plumes. Science 272:991–995. doi:10.1126/science.272.5264.991

He Y, Wen L (2009) Structural features and shear–velocity structure of the 'Pacific anomaly'. J Geophys Res 114:B02309. doi:10.1029/2008JB005814

Hernlund JW, Houser C (2008) On the statistical distribution of seismic velocities in Earth's deep mantle. Earth Planet Sci Lett 265:423–437. doi:10.1016/j.epsl.2007.10.042

Hernlund JW, Thomas C, Tackley PJ (2005) A doubling of the post–perovskite phase boundary and structure of the Earth's lowermost mantle. Nature 434:882–886

Hofmann AW (1997) Mantle geochemistry: the message from oceanic volcanism. Nature 385:219–229. doi:10.1038/385219a0

Hofmann AW (2003) Sampling mantle heterogeneity through oceanic basalts: isotopes and trace elements. Treatise Geochem 2:61–101

Houser C, Masters G, Shearer P, Laske G (2008) Shear and compressional velocity models of the mantle from cluster analysis of long-period waveforms. Geophys J Int 174:195–212. doi:10.1111/j.1365-246X.2008.03763.x

Huang J, Davies GF (2007a) Stirring in three-dimensional mantle convection models and implications for geochemistry: heavy tracers. Geochem Geophys Geosyst 8:Q07004. doi:10.1029/2007GC001621

Huang J, Davies GF (2007b) Stirring in three-dimensional mantle convection models and implications for geochemistry: passive tracers. Geochem Geophys Geosyst 8:Q03017. doi:10.1029/2006GC001312

Hunt SA, Davies DR, Walker AM, McCormack RJ, Wills AS, Dobson DP, Li L (2012) On the increase in thermal diffusivity caused by the perovskite to post-perovskite phase transition and its implications for mantle dynamics. Earth Planet Sci Lett 319:96–103. doi:10.1016/j.epsl.2011.12.009

Ishii M, Tromp J (1999) Normal-mode and free-air gravity constraints on lateral variations in velocity and density of Earth's mantle. Science 285:1231–1236. doi:10.1126/science.285.5431.1231

Jackson I (1998) Elasticity, composition and temperature of the Earth's lower mantle: a reappraisal. Geophys J Int 134:291–311. doi:10.1046/j.1365-246x.1998.00560.x

Jackson MG, Carlson R (2011) An ancient recipe for flood basalt genesis. Nature 476:316–319. doi:10.1038/nature10326

Jackson MH, Carlson R, Kurz MD, Kempton PD, Francis D, Blusztajn J (2010) Evidence for the survival or the oldest terrestrial mantle reservoir. Nature 466:853–856

Javoy M, Kaminski E, Guyot F, Andrault D, Sanloup C, Moreira M, Labrosse S, Jambon A, Agrinier P, Davaille A, Jaupart C (2010) The chemical composition of the Earth: enstatite chondrite models. Earth Planet Sci Lett 2930(3–4):259–268. doi: 10.1016/j.epsl.2010.02.033

Jeanloz R, Morris S (1987) Is the mantle geotherm sub–adiabatic? Geophys Res Lett 143:335–338

Jellinek AM, Manga M (2002) The influence of a chemical boundary layer on the fixity, spacing and lifetime of mantle plumes. Nature 418:760–763. doi:10.1038/nature00979

Karato S-I (1993) The importance of anelasticity in the interpretation of seismic tomography. Geophys Res Lett 20:1623–1626

Karato S-I (2008) Deformation of Earth materials: an introduction to the rheology of solid Earth. Cambridge University Press, Cambridge

Karato S-I, Karki BB (2001) Origin of lateral variation of seismic wave velocities and density in the deep mantle. J Geophys Res 106:21771–21783

Kellogg LH, Hager BH, van der Hilst RD (1999) Compositional stratification in the deep mantle. Science 283:1881–1884. doi:10.1126/science.283.5409.1881

Kennett BLN, Engdahl R, Buland R (1995) Constraints on seismic velocities in the Earth from travel–times. Geophys J Int 122:108–124. doi:10.1111/j.1365-246X.1995.tb03540.x

Kennett BLN, Widiyantoro S, van der Hilst RD (1998) Joint seismic tomography for bulk sound and shear wave speed in the Earth's mantle. J Geophys Res 103:12469–12493

Khan A, Connolly JAD, Taylor SR (2008) Inversion of seismic and geodetic data for the major element chemistry and temperature of the Earth's mantle. J Geophys Res 113:B09308. doi:10.1029/2007JB005239

Labrosse S, Hernlund JW, Coltice N (2007) A crystallizing dense magma ocean at the base of Earth's mantle. Nature 450:866–869. doi:10.1038/nature06355

Lay T, Hernlund J, Buffett BA (2008) Core–mantle–boundary heat flow. Nature Geosci 1:25–32. doi:10.1038/ngeo.2007.44

Leng W, Zhong S (2008) Controls on plume heat flux and plume excess temperature. J Geophys Res 113. doi:10.1029/2007JB005155

Li C, van der Hilst RD, Engdahl ER, Burdick S (2008) A new global model for P-wave speed variations in Earth's mantle. Geochem Geophys Geosys 5. doi:10.1029/2007GC001806

Lyubetskaya T, Korenaga J (2007) Chemical composition of Earth's primitive mantle and its variance: 2. implications for global geodynamics. J Geophys Res 112:B03212. doi:10.1029/2005JB004224

Malcolm AE, Trampert J (2011) Tomographic errors from wave front healing: more than just a fast bias. Geophys J Int 185:385–402. doi:10.1111/j.1365-246X.2011.04945.x

Masters G, Gubbins D (2003) On the resolution of density within the Earth. Phys Earth Planet Int 140:159–167

Masters G, Laske G, Bolton H, Dziewonski AM (2000) The relative behavior of shear-wave velocity, bulk sound speed, and compressional velocity in the mantle: implications for chemical and thermal structure. AGU Monogr Earth's Deep Inter 171:63–87

Matas J, Bass J, Ricard Y, Mattern E, Bukowinski MST (2007) On the bulk composition of the lower mantle: predictions and limitations from generalized inversion of radial seismic profiles. Geophys J Int 170:764–780. doi:10.1111/j.1365-246X.2007.03454.x

Matas J, Bukowinski MST (2007) On the anelastic contribution to the temperature dependence of lower mantle seismic velocities. Earth Planet Sci Lett 259:51–65. doi:10.1016/j.epsl.2007.04.028

McDonough WF, Sun S-S (1995) The composition of the Earth. Chem Geol 120:223–253

McNamara AK, Zhong S (2004) Thermochemical structures within a spherical mantle. J Geophys Res 109:B07402. doi:10.1029/2003JB002847

McNamara AK, Zhong S (2005) Thermochemical structures beneath Africa and the Pacific Ocean. Nature 437:1136–1139. doi:10.1038/nature04066

Montelli R, Nolet G, Masters G, Dahlen FA, Hung S-H (2004) Global P and PP traveltime tomography: rays versus waves. Geophys J Int 158:630–654

Mosca I, Cobden L, Deuss A, Ritsema J, Trampert J (2012) Seismic and mineralogical structures of the lower mantle from probabilistic tomography. J Geophys Res 117:B06304. doi:10.1029/2011JB008851

Nakagawa T, Tackley PJ (2004) Effects of a perovskite–post perovskite phase change near core–mantle–boundary in compressible mantle convection. Geophys Res Lett 31:L16611. doi:10.1029/2004GL020648

Nakagawa T, Tackley PJ, Deschamps F, Connolly JAD (2010) The influence of MORB and Harzburgite composition on thermo-chemical mantle convection in a 3D spherical shell with self-consistently calculated mineral physics. Earth Planet Sci Lett 296:403–412. doi:10.1016/j.epsl.2010.05.026

Ni SD, Tan E, Gurnis M, Helmberger DV (2002) Sharp sides to the African superplume. Science 296:1850–1852. doi:10.1126/science.1070698

Olson P, Deguen R, Hinnov LA, Zhong SJ (2013) Controls on geomagnetic reversals and core evolution by mantle convection in the phanerozoic. Phys Earth Planet Int 214:87–103. doi:10.1016/j.pepi.2012.10.003

Ranalli S (1995) Rheology of the Earth. Chapman & Hall, London

Rapp RP, Irifune T, Shimizu N, Nishiyama N, Norman MD, Inoue T (2008) Subduction recycling of continental sediments and the origin of geochemically enriched reservoirs in the deep mantle. Earth Planet Sci Lett 271:14–23. doi:10.1016/j.epsl.2008.02.028

Ricard Y, Richards MA, Lithgow-Bertelloni C, LeStunff Y (1993) A geodynamic model of mantle mass heterogeneities. J Geophys Res 98:21895–21909

Ricard Y, Chambat F, Lithgow-Bertelloni C (2006) Gravity observations and 3-D structure of the Earth. CR Geosci 338:992–1001

Richards MA, Engebretson DC (1992) Large-scale mantle convection and the history of subduction. Nature 355:437–440. doi:10.1029/2007JB005155

Ritsema J, Ni S, Helmberger DV, Crotwell HP (1998) Evidence for strong shear-wave velocity reductions and velocity gradients in the lower mantle beneath Africa. Geophys Res Lett 25:4245–4248

Ritsema J, van Heijst HJ (2002) Constraints on the correlation of P- and S-wave velocity heterogeneity in the mantle from P, PP, PPP and PKPab traveltimes. Geophys J Int 149:482–489. doi:10.1046/j.1365-246X.2002.01631.x
Ritsema J, McNamara AK, Bull A (2007) Tomographic filtering of geodynamic models: implications for model interpretation and large-scale mantle structure. J Geophys Res 112. doi:10.1029/2006JB004566
Ritsema J, van Heijst HJ, Deuss A, Woodhouse JH (2011) S40RTS: a degree–40 shear-wave velocity model for the mantle from new Rayleigh wave dispersion, teleseismic traveltimes, and normal–mode splitting function measurements. Geophys J Int 184:1223–1236. doi:10.1111/j.1365-246X.2010.04884.x
Robertson GS, Woodhouse JH (1995) Evidence for proportionality of P and S heterogeneity in the lower mantle. Geophys J Int 123:85–116
Romanowicz B (2001) Can we resolve 3-D density heterogeneity in the lower mantle? Geophys Res Lett 28:1107–1110
Saltzer RL, van der Hilst RD, Karason H (2001) Comparing P and S wave heterogeneity in the mantle. Geophys Res Lett 28:1335–1338
Schaeffer N, Manga M (2001) Interaction of rising and sinking mantle plumes. Geophys Res Lett 28:455–458
Schuberth BSA, Bunge H-P, Ritsema J (2009a) Tomographic filtering of high-resolution mantle circulation models: can seismic heterogeneity be explained by temperature alone? Geochem Geophys Geosyst 10:Q05W03. doi:10.1029/2009GC002401
Schuberth BSA, Bunge H-P, Steinle-Neumann G, Moder C, Oeser J (2009b) Thermal versus elastic heterogeneity in high-resolution mantle circulation models with pyrolite composition: high plume excess temperatures in the lowermost mantle. Geochem Geophys Geosyst 10:Q01W01. doi:10.1029/2008GC002235
Schuberth BSA, Zaroli C, Nolet G (2012) Synthetic seismograms for a synthetic Earth: long-period P- and S-wave traveltime variations can be explained by temperature alone. Geophys J Int 200:1393–1412. doi:10.1111/j.1365-246X.2011.05333.x
Shephard GE, Bunge H-P, Schuberth BSA, Muller RD, Talsma AS, Moder C, Landgrebe TCW (2012) Testing absolute plate reference frames and the implications for the generation of geodynamic mantle heterogeneity structure. Earth Planet Sci Lett 317:204–217. doi:10.1016/j.epsl.2011.11.027
Simmons NA, Forte AM, Boschi L, Grand SP (2010) GyPSuM: a joint tomographic model of mantle density and seismic wave speeds. J Geophys Res 115. doi:10.1029/2010JB007631
Simmons NA, Myers SC, Johannesson G (2011) Global-scale p wave tomography optimized for prediction of teleseismic and regional traveltime for middle east events: 2. tomographic inversion. J Geophys Res 116:B04305. doi:10.1029/2010JB007969
Sramek O, McDonough WF, Kite ES, Lekic V, Dye ST, Zhong S (2013) Geophysical and geochemical constraints on geoneutrino fluxes from Earth's mantle. Earth Planet Sci Lett 361:356–366
Stampfli GM, Borel GD (2002) A plate tectonic model for the Paleozoic and Mesozoic constrained by dynamic plate boundaries and restored synthetic oceanic isochrons. Earth Planet Sci Lett 196:17–33
Stampfli GM, Hochard C (2009) Plate tectonics of the Alpine realm. Geol Soc London Spec Publ 327:89–111
Stegman DR, Jellinek AM, Zatman SA, Baumgardner JR, Richards MA (2003) An early lunar core dynamo driven by thermochemical mantle convection. Nature 421:143–146
Steinberger B (2000) Plumes in a convecting mantle: models and observations for individual hotspots. J Geophys Res 105:11127–11152
Steinberger B, Torsvik TH (2012) A geodynamic model of plumes from the margins of large low shear-wave velocity provinces. Geochem Geophys Geosyst 13:Q01W09. doi:10.1029/2011GC003808
Stixrude L, Lithgow-Bertelloni C (2005) Thermodynamics of mantle minerals—i. physical properties. Geophys J Int 162:610–632. doi:10.1111/j.1365-246X.2005.02642.x

Stixrude L, Lithgow-Bertelloni C (2007) Influence of phase transformations on lateral heterogeneity and dynamics in Earth's mantle. Earth Planet Sci Lett 263:45–55. doi:10.1016/j.epsl.2007.08.027

Stixrude L, Lithgow-Bertelloni C (2011) Thermodynamics of mantle minerals—ii. phase equilibria. Geophys J Int 184:1180–1213. doi:10.1111/j.1365-246X.2010.04890.x

Styles E, Davies DR, Goes S (2011) Mapping spherical seismic into physical structure: biases from 3-D phase-transition and thermal boundary-layer heterogeneity. Geophys J Int 184:1371–1378. doi:10.1111/j.1365-246X.2010.04914.x

Su WJ, Dziewonski AM (1997) Simultaneous inversion for 3-D variations in shear and bulk velocity in the mantle. Phys Earth Planet Int 100:135–156

Tackley PJ (1998) Three–dimensional simulation of mantle convection with a thermo–chemical boundary layer: D"? In: Gurnis M, Wysession ME, Knittle E, Buffet BA (eds) The core–mantle–boundary region. AGU, Washington DC, pp 231–253

Tackley PJ (2002) Strong heterogeneity caused by deep mantle layering. Geochem Geophys Geosyst 3:1024. doi:10.1029/2001GC000167

Tackley PJ (2007) Mantle geochemical geodynamics. Treatise Geophys 7: 437–505

Tackley PJ, King SD (2003) Testing the tracer ratio method for modelling active compositional fields in mantle convection simulations. Geochem Geophys Geosyst 4:8302. doi:10.1029/2001GC000214

Tackley PJ, Xie S, Nakagawa T, Hernlund JW (2005). Numerical and laboratory studies of mantle convection: philosophy, accomplishments and thermo-chemical structure and evolution. In: Earth's deep mantle: structure, composition, and evolution, vol 160. Geophysical Monograph Series, AGU, Washington DC, pp 83–99. doi:10.1029/1160GM1007

Tan E, Gurnis M, Han LJ (2002) Slabs in the lower mantle and their modulation of plume formation. Geochem Geophys Geosyst 3:1067. doi:10.1029/2001GC000238

Tan E, Leng W, Zhong S, Gurins M (2011) On the location of plumes and mobility of thermo–chemical structures with high bulk modulus in the 3-D compressible mantle. Geochem Geophys Geosyst 12:Q07005. doi:10.1029/2011GC003665

Thorne MS, Garnero EJ, Grand SP (2004) Geographic correlation between hotspots and deep mantle lateral shear–wave velocity gradients. Phys Earth Planet Int 146:47–63. doi:10.1016/j.pepi.2003.09.026

To A, Romanowicz B, Capdeville Y, Takeuchi N (2005) 3-D effects of sharp boundaries at the borders of the African and Pacific superplumes: observation and modeling. Earth Planet Sci Lett 233:137–153. doi:10.1016/j.epsl.2005.01.037

Torsvik TH, Smethurst MA, Burke K, Steinberger B (2006) Large igneous provinces generated from the margins of the large low-velocity provinces in the deep mantle. Geophys J Int 167:1447–1460. doi:10.1111/j.1365-246X.2006.03158.x

Torsvik TH, Muller RD, van der Voo R, Steinberger B, Gaina C (2008a) Global plate motion frames: towards a unified model. Rev Geophys 46:1–44. doi:10.1029/2007RG000227

Torsvik TH, Smethurst MA, Burke K, Steinberger B (2008b) Long term stability in deep mantle structure: evidence from the ~300 Ma Skagerrak-Centered Large Igneous Province (the SCLIP). Earth Planet Sci Lett 267:444–452. doi:10.1016/j.epsl.2007.12.004

Torsvik TH, Burke K, Steinberger B, Webb SJ, Ashwal LD (2010) Diamonds sampled by plumes from the core-mantle-boundary. Nature 466:352–358. doi:10.1038/nature09216

Tosi N, Yuen DA, Cadek O (2010) Dynamical consequences in the lower mantle with the post-perovskite phase change and strongly depth–dependent thermodynamic and transport properties. Earth Planet Sci Lett 298:229–243. doi:10.1016/j.epsl.2010.08.001

Trampert J, Deschamps F, Resovsky J, Yuen D (2004) Probabilistic tomography maps chemical heterogeneities throughout the lower mantle. Science 306:853–856. doi:10.1126/science.1101996

Trampert J, Vacher P, Vlaar N (2001) Sensitivities of seismic velocities to temperature, pressure and composition in the lower mantle. Phys Earth Planet Int 124:255–267. doi:10.1016/S0031-9201(01)00201-1

Trieloff M, Kunz J, Clague DA, Harrison D, Allègre CJ (2000) The nature of pristine noble gases in mantle plumes. Science 288:1036–1038. doi:10.1126/science.288.5468.1036

van der Hilst RD, de Hoop MV, Wang P, Shim SH, Ma P, Tenorio L (2007) Seismostratigraphy and thermal structure of Earth's core–mantle–boundary region. Science 315:1813–1817. doi:10.1126/science.1137867

van der Hilst RD, Widiyantoro S, Engdahl ER (1997) Evidence for deep mantle circulation from global tomography. Nature 386:578–584. doi:10.1038/386578a0

Walter MJ, Nakamura E, Tronnes RG, Frost DJ (2004) Experimental constraints on crystallisation differentiation in a deep magma ocean. Geochim Cosmo Acta 68:4267–4284. doi:10.1016/j.gca.2004.03.014

Wang Y, Wen L (2004) Mapping the geometry and geographic distribution of a very-low velocity province at the base of the Earth's mantle. J Geophys Res 109:B10305. doi:10.1029/2003JB002674

Wang Y, Wen L (2007) Geometry and P and S velocity structure of the 'African anomaly'. J Geophys Res 112. doi:10.1029/2006JB004483

Wasserburg GJ, De Paolo DJ (1979) Models of Earth structure inferred from neodymium and strontium isotopic abundances. Proc Natl Acad Sci USA 76:3594–3598

Wolstencroft M, Davies JH, Davies DR (2009) Nusselt–rayleigh number scaling for spherical shell earth mantle simulation up to a rayleigh number of 10^9. Phys Earth Planet Inter 176:132–141

Woodhouse J, Dziewonski A (1989) Seismic modeling of the Earth's large scale 3-D structure. Phil Trans Roy Soc 328:291. doi:10.1098/rsta.1989.0037

Wookey J, Stackhouse S, Kendall J-M, Brodholt J, Price GD (2005) Efficacy of the post-perovskite phase as an explanation for lowermost-mantle seismic properties. Nature 438:1004–1007. doi:10.1038/nature04345

Xie S, Tackley PJ (2004) Evolution of helium and argon isotopes in a convecting mantle. Phys Earth Planet Int 146:417–439

Xu W, Lithgow-Bertelloni C, Stixrude L, Ritsema J (2008) The effect of bulk composition and temperature on mantle seismic structure. Earth Planet Sci Lett 275:70–79

Zhang N, Zhong SJ (2011) Heat fluxes at the Earth's surface and core–mantle boundary since Pangea formation and their implications for the geomagnetic superchrons. Earth Planet Sci Lett 306:205–216. doi:10.1016/j.epsl.2011.04.001

Zhang N, Zhong SJ, Leng W, Li ZX (2010) A model for the evolution of Earth's mantle structure since the early Paleozoic. J Geophys Res 115:B06401. doi:10.1029/2009JB006896

Zhao D, Lei J (2003) Seismic ray path variations in a 3D global velocity model. Phys Earth Planet Int 141:153–166. doi:10.1016/j.pepi.2003.11.010

Zindler A, Hart S (1986) Chemical geodynamics. Ann Rev Earth Planet Sci 14:493–571. doi:10.1146/annurev.ea.14.050186.002425

Chapter 15
Large-Scale Thermo-chemical Structure of the Deep Mantle: Observations and Models

Frédéric Deschamps, Yang Li and P.J. Tackley

Abstract Seismic tomography indicates that the lowermost mantle, from 2400 km down to the core–mantle boundary (CMB) is strongly heterogeneous at large wavelengths. The most striking features are two large low-shear-wave velocity provinces (LLSVPs), where shear-wave velocity drops by a few percent compared to averaged mantle. Several seismic observations further show that lowermost mantle seismic anomalies cannot be purely thermal in origin. Compositional anomalies are required to fully explain observations like the anti-correlation between shear- and bulk-sound velocities, and the distribution of density mapped by normal modes. In the meantime, models of thermo-chemical convection indicate that reservoirs of dense, chemically differentiated material can be maintained in the lowermost mantle over long periods of time and that thermal plumes rising up to the surface are generated at the surface of these reservoirs. Model parameter searches indicate that maintaining such reservoirs requires a moderate density contrast between dense and regular material and a large thermal viscosity contrast. Current models of thermo-chemical convection also explain details revealed by travel time and seismic waveform data, in particular the LLSVP sharp edges, and the distribution of plumes at the surface of LLSVPs. A remaining question is the detailed nature of the lower mantle large-scale chemical heterogeneities. Reservoirs of dense material may result either from early partial differentiation of the mantle or recycling of oceanic crust (MORB). Seismic sensitivities inferred from a coherent mineral physics database suggest that LLSVPs are better explained by warm material enriched in iron and silicate, than by high-pressure MORB. By contrast, if colder than the surrounding mantle by ~400 K,

F. Deschamps (✉)
Institute of Earth Sciences, Academia Sinica, 128 Academia Road Sec. 2, Nangang, Taipei 11529, Taiwan
e-mail: frederic@earth.sinica.edu.tw

Y. Li · P.J. Tackley
Institute of Geophysics, Swiss Federal Institute of Technology Zurich, Sonnegstrasse 5, 8092 Zurich, Switzerland

© Springer International Publishing Switzerland 2015

A. Khan and F. Deschamps (eds.), *The Earth's Heterogeneous Mantle*, Springer Geophysics, DOI 10.1007/978-3-319-15627-9_15

high-pressure MORB explains well seismic velocity anomalies in regions where ancient slabs are expected to rest, e.g., beneath the Japan subduction zones and beneath Central and South America. The post-pervoskite phase certainly plays a significant role in explaining seismic observations, in particular the D″ discontinuity, but is unlikely to explain all seismic observations alone.

Keywords Mantle structure · Mantle composition · Mantle dynamics · LLSVPs · Thermo-chemical convection

15.1 Introduction

Our description of the Earth's deep mantle structure and dynamics has changed considerably during the past decade, aided by advances in seismology, high-pressure mineral physics, and geodynamics. The deep mantle appears much more heterogeneous than previously expected, both at small-scale (i.e., up to a few tens of kilometers; see Rost et al., this volume) and large-scale (typically 1000 km across and more). From depths of about 2400 km down to the core–mantle boundary (CMB), seismic observations such as the radial and lateral variations in seismic ratios $\mathrm{dln}V_\mathrm{P}/\mathrm{dln}V_\mathrm{S}$ and $\mathrm{dln}V_\Phi/\mathrm{dln}V_\mathrm{S}$ (van der Hilst and Kárason 1999; Deschamps and Trampert 2003), distribution histograms of the shear-wave velocity anomalies (Deschamps and Trampert 2003; Hernlund and Houser 2008; Burke et al. 2008), and the anti-correlation between shear-wave and bulk-sound phase velocity (Ishii and Tromp 1999; Masters et al. 2000; Trampert et al. 2004; Houser et al. 2008), are better explained if chemical heterogeneities are present in the deep mantle although a thermally dominated structure have also been advocated (see Davies et al., this volume). Seismic observations invalidate the hypothesis that low-shear-wave velocity provinces (LLSVPs) observed beneath Africa and beneath the Pacific result from purely thermal anomalies. Normal-mode tomography (Ishii and Tromp 1999; Trampert et al. 2004; Mosca et al. 2012) gives further constraints on the deep mantle density distribution, which further support the presence of large-scale compositional anomalies.

Due to the presence of compositional heterogeneities in the deep mantle, modeling mantle dynamics requires the use of thermo-chemical convection simulations. Compositional heterogeneities, in addition to temperature variations, cause lateral variations in density, which in turn influence the flow pattern and the heat and mass transfers. The key parameter controlling the thermo-chemical structure of the system is the density contrast between the chemical anomalies and the regular material. When an initial dense layer is present at the bottom of the system, the density contrast between the dense and regular materials controls the stability and shape of this layer (e.g., Davaille 1999; Tackley 2002; McNamara and Zhong 2004; Deschamps and Tackley 2009). Models in which reservoirs of dense material result from the segregation of recycled oceanic (e.g., Gurnis 1986; Christensen 1989; Olson and Kincaid 1991; Christensen and Hofmann 1994) also indicate that the size and stability of these reservoirs strongly vary with the density contrast between the recycled and regular materials.

Mantle petrology is based on 5 main oxides (CaO, FeO, MgO, Al_2O_3, and SiO_2) from which the most abundant mantle minerals are built. In addition, MORB requires Na_2O in weight fraction up to 2.0 %. The detailed petrology further depends on the stability of each mineral, therefore on temperature and pressure. Chemical heterogeneities in the deep mantle may thus consist of variations in the fractions of these 5 oxides compared to a reference composition, e.g., the average pyrolitic composition (Ringwood 1975). The exact mechanism responsible for generating these variations is, however, still unknown. A straightforward source of heterogeneities is the recycling of subducted slabs in the deep mantle, as suggested by tomographic images (e.g., van der Hilst et al. 1997). Compared to the hypothetical pyrolite, MORB is strongly depleted in MgO and enriched in SiO_2, CaO, and Al_2O_3, and to a lesser extent FeO (e.g., Ricard et al. 2005). Another possible source of chemical heterogeneity is the survival of reservoirs resulting from the early partial differentiation of the mantle. Possible mechanisms of differentiation include the crystallization of regular (i.e., crystallizing upwards from the CMB) or basal magma oceans (Solomatov and Stevenson 1993; Labrosse et al. 2007; Solomatov 2007), upside-down differentiation (Lee et al. 2010), and the recycling of an early crust (Boyet and Carlson 2006; Tolstikhin et al. 2006). Alternatively, if the Earth was built from enstatite chondrites, an enrichment of the lower mantle in iron and silicates may result from the a two-stage scenario of core formation, including a giant impact, and would thus not require early differentiation of the mantle (Kaminski and Javoy, this volume).

In addition to thermal and chemical anomalies, the phase change from $MgSiO_3$ perovskite to post-perovskite (Murakami et al. 2004; Oganov and Ono 2004) may be needed to fully explain available seismic observations. The transition from perovskite to post-perovskite may correspond to the D″ discontinuity, which is locally observed in the lowermost mantle (Wookey et al. 2005). It was furthermore suggested that lateral variations in the depth of the transition from perovskite to post-perovskite is responsible for the anti-correlation between shear-wave and bulk-sound velocity (Wookey et al. 2005; Hirose 2007; Hutko et al. 2008). Davies et al. (2012) proposed that seismic tomography can be fully explained by distributions of post-perovskite induced by lateral variations of temperature, but other studies (Deschamps et al. 2012; Mosca et al. 2012) arrived at the opposite conclusion.

In this chapter, we briefly review seismic observations that support the presence of large-scale chemical heterogeneities in the deep Earth mantle. We then discuss the parameters of convection models that allow reservoirs of dense material to be maintained in the deep mantle for a long period of time, and illustrate this with new models of thermo-chemical convection in spherical geometry. We finally discuss the possible nature of chemical heterogeneities and the role of the post-perovskite phase transition.

15.2 Thermo-chemical Structure: Seismological Hints

Seismic data are, so far, the most detailed information available to infer the Earth's mantle structure. Global seismic tomographic images have now reached a consensus concerning the large-scale (i.e., ~1000 km across, or more) structures (for comparison between recent models, see Schaeffer and Lebedev 2013; Ritsema et al. 2011). The strongest lateral heterogeneities are found at the top (shallower than 400 km) and at the bottom (deeper than about 2400 km) of the mantle. In between these depths, no notable large-scale structure is observed, and the RMS seismic velocity variation is small, 0.5 % or less. Below the surface, anomalies of shear-wave velocity are well correlated with surface tectonics down to depths of about 200–300 km (Schaeffer and Lebedev, this volume). The ocean–continent distribution clearly appears, with the slowest velocity being observed under mid-oceanic ridges. Within continents, tectonically active regions also appear slower than average, whereas high-velocity roots are observed beneath cratonic areas. In the lowermost mantle, the seismic tomographic images are dominated by two large low-shear-wave velocity provinces (LLSVPs) located along the equator beneath Africa and the Pacific. These regions were first observed by global tomographic models in the 1990s (Tanimoto 1990; Su et al. 1994; Li and Romanowicz 1996), and more recent models provided refined images of these structures (e.g., Antolik et al. 2003; Houser et al. 2008; Kustowski et al. 2008; Ritsema et al. 2011). Tomographic models of the mantle beneath Africa (Ritsema et al. 1999) show that the African LLSVP has globally the structure of a ridge with a NW/SE direction and is locally tilted toward the east. In contrast, the Pacific LLSVP appears roughly oval in shape (e.g., Houser et al. 2008; Ritsema et al. 2011). Importantly, LLSVPs are also observed by normal-mode tomography (Ishii and Tromp 1999; Trampert et al. 2004; Mosca et al. 2012). A cluster analysis applied to recent tomographic models indicates that LLSVPs are robust structures, not artifacts (Lekic et al. 2012). An important methodological refinement of the past decade is probabilistic tomography, first introduced by Resovsky and Trampert (2003), which allows the determination of full probability density functions (pdf) (rather than single values) of seismic velocity and density anomalies at each point of the model. This provides robust estimates of the error bars in the tomographic images. Global models of probabilistic tomography published so far (Trampert et al. 2004; Mosca et al. 2012) are based on normal-mode data. A limitation of these models is that they are restricted to spherical harmonic degrees 2, 4, and 6. Nevertheless, they indicate that LLSVPs are robust structures.

Details on the lower mantle seismic structure, in particular the shape and extent of LLSVPs, may be obtained from waveform modeling and travel times of seismic phases sampling the lowermost mantle, including S, S_{diff}, and phases reflected at the top of the outer core (ScS) or traveling through it (SKS). Keeping in mind that their interpretations involve trade-off between the amplitude, shape, and size of the velocity anomaly, these observations reveal interesting local features. The complexity of SKS waveforms and the differential travel times (relative to S waves) of ScS and SKS phases sampling the lowermost mantle beneath southern Africa are well explained by an eastward tilt of the African LLSVP (Ni et al. 2002; Ni and Helmberger 2003). Travel time delays further suggest that in this region, the African LLSVP has sharp boundaries, typically around 50 km (Ni et al. 2002). Around equatorial latitudes, S and P travel time and waveform data indicate that the African LLSVP has a bell shape with a wide (~4000 km west to east) base (Wang and Wen 2007). Additional constraints are also available for the Pacific

LLSVP. S travel times and waveforms (He and Wen 2009) and tomographic models (Ritsema et al. 2011) suggest that it may split into a western and an eastern province with a gap of several hundreds of kilometers in between. Both provinces have a trapezoidal shape, but the western region has stronger topography than the eastern region. Sharp boundaries at the southern edge of the Pacific LLSVP have been reported from the analysis of SH_{diff} waveforms (To et al. 2005), as well as at its western edge (Takeushi et al. 2008).

First hints for the presence of chemical heterogeneities in the deep mantle are provided by seismic ratios. The sharp increase of the laterally averaged ratio between compressional and shear-wave velocity anomalies ($\mathrm{dln}V_S/\mathrm{dln}V_P$) in the deep mantle (van der Hilst and Kárason 1999; Masters et al. 2000) and the strong lateral dispersion of this ratio at a given depth (Deschamps and Trampert 2003; Deschamps et al. 2007) are more easily explained if the seismic velocity anomalies are not purely thermal in origin. Furthermore, the distribution histograms of shear-wave velocities anomalies in the lowermost mantle have a double peak (Deschamps and Trampert 2003; Hernlund and Houser 2008; Burke et al. 2008), which again is better explained if chemical heterogeneities are present. A key observation from seismic tomography is the anti-correlation between shear-wave velocity (V_S) and bulk-sound velocity (V_Φ) anomalies in a layer extending from about 2400 km down to the CMB (Su and Dziewonski 1997; Masters et al. 2000; Houser et al. 2008). In particular, V_Φ is faster than average (by up to about 1 %) throughout the region spanned by LLSVPs. Because the temperature derivatives of V_S and V_Φ are both negative throughout the Earth's mantle (e.g., Trampert et al. 2001; Deschamps and Trampert 2003), purely thermal anomalies are unable to explain an anti-correlation between V_S and V_Φ anomalies, thus implying the presence of large-scale lateral variations in composition. It has recently been suggested that the anti-correlation between V_S and V_Φ anomalies may result from wave-propagation effects (Schuberth et al. 2011). However, the fact that normal-mode tomography, which is not affected by potential propagation effects of the body waves, also observes this anti-correlation (Ishii and Tromp 1999; Trampert et al. 2004; Mosca et al. 2012) strongly supports the hypothesis that it results from real structures. An additional observation pointing to the presence of compositional anomalies in the lowermost mantle is the de-correlation between seismic velocity anomalies and density anomalies mapped by normal-mode tomography (Ishii and Tromp 1999; Trampert et al. 2004; Mosca et al. 2012).

Seismic velocity anomalies result from variations in temperature and composition that one would like to map. A difficulty in solving this inverse problem is the trade-off that exists between temperature and composition. Sensitivities of shear-wave and bulk-sound velocities to one particular parameter are different in amplitude, but usually have the same sign (e.g., Trampert et al. 2001; Deschamps and Trampert 2003). For instance, both shear-wave and bulk-sound velocities decrease with increasing amount of iron. A notable exception is SiO_2, either free or in the form of perovskite, for which the sensitivity of V_S goes to zero in the deep mantle, whereas the sensitivity of V_Φ remains strongly positive (Deschamps et al. 2012). As a result, seismic velocities alone may qualitatively reveal the presence of thermal and chemical sources, but cannot quantitatively resolve the distributions in temperature and composition. A detailed mapping of thermo-chemical anomalies requires additional, independent observations, such as density distributions, which can be retrieved from normal-mode data (Ishii and Tromp 1999; Trampert et al. 2004; Mosca et al. 2012). Assuming that the chemically anomalous component consists of material enriched in iron and perovskite, and using appropriate equation of state modeling, Trampert et al. (2004) calculated distributions of temperature, iron, and perovskite in the lower mantle. Their calculations indicate that bulk-sound velocity and density anomalies are good proxies for the fraction of perovskite and the fraction of iron, respectively (Fig. 15.1). In contrast, shear-wave

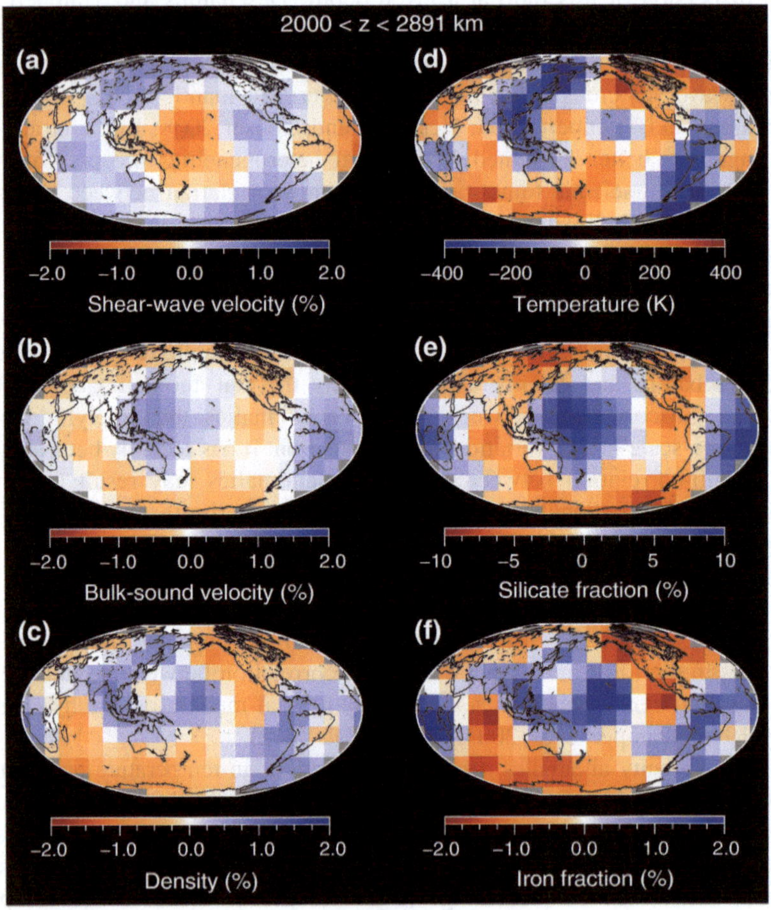

Fig. 15.1 Seismic (*left column*) and thermo-chemical (*right column*) structures from probabilistic tomography (Trampert et al. 2004) in the depth range 2000–2891 km. The seismic structure includes relative anomalies in **a** shear-wave velocity, **b** bulk-sound velocity anomalies, and **c** density anomalies. The thermo-chemical structure includes absolute anomalies in **d** temperature, **e** fraction of perovskite, and **f** fraction of iron

velocity anomalies are not a good proxy for temperature anomalies, but likely result from lateral variations in both temperature and iron. Updated models of probabilistic tomography with a higher vertical resolution and accounting for the presence of post-perovskite (Mosca et al. 2012) have confirmed these results.

Another difficulty in interpreting tomographic models is that they may be biased by uneven data coverage and a priori information (mainly damping and smoothing) imposed by inversion methods. These effects should be accounted for when comparing seismic velocity distributions deduced from models of mantle dynamics, and those from tomographic models. The simplest filter is to add random Gaussian noise based on estimated uncertainties in seismic velocity anomalies (Deschamps et al. 2012). This approach is well suited for comparison with tomographic models based on Monte Carlo methods (such as probabilistic tomography), but may also be used for comparison with other models. For a more detailed comparison with a specific tomographic model, a filter may be designed from the resolution matrix of this model (e.g., Ritsema et al. 2007; Schuberth et al. 2009). Applying such filter to the seismic velocity distribution issued from purely thermal models of convection (which are dominated by short wavelengths structures) adds substantial smearing that smooth out the input distribution (Ritsema et al. 2007; Schuberth et al. 2009). Still, this smoothed distribution hardly explains the observed tomographic model. By contrast, the synthetic seismic velocity obtained from models of thermo-chemical convection provides a better match to seismic tomography (Ritsema et al. 2007).

In addition to thermal and chemical anomalies, other effects, including the presence of fluids or partial melts and lateral variations in the depth of phase transitions, can affect seismic observations. The phase transition from perovskite to post-perovskite is of particular interest for the lowermost mantle structure (for a detailed discussion on the seismic detection and presence of post-perovskite in the deep mantle, we refer to Cobden et al., this volume). The lateral variations in the depth at which the post-perovskite phase transition occurs may play a significant role in explaining several seismic features, including the D'' discontinuity (Wookey et al. 2005) and arrival of diffracted seismic waves (Cobden et al. 2012; Cobden et al., this volume). However, both seismic sensitivities from mineral physics properties (Deschamps et al. 2012) and the most recent models of probabilistic tomography (Mosca et al. 2012) indicate that the details of seismic tomography cannot be explained only with thermal anomalies and lateral variations in the post-perovskite phase transition. We further discuss this point in Sect. 15.5.

15.3 Models of Thermo-chemical Convection

Seismic tomography indicates that large-scale compositional heterogeneities are likely present in the deep mantle, with LLSVPs being chemically distinct from the rest of the mantle. Other geophysical constraints (Torsvik et al. 2008; Dziewonski et al. 2010; Conrad et al. 2013; Sect. 15.6) further indicate that LLSVPs have remained stable during the past 300 Myr, and potentially over much longer periods of time. Models of convection describing mantle dynamics should therefore include compositional variations in addition to the thermal variations usually considered and be able to maintain large-scale thermo-chemical reservoirs during long periods of time.

A variety of experimental (e.g., Olson and Kincaid 1991; Davaille 1999; Jellinek and Manga 2002) and numerical (e.g., Hansen and Yuen 1988; Christensen and Hofmann 1994; Tackley 1998, 2002; McNamara and Zhong 2004, 2005; Tan and Gurnis 2005, 2007; Deschamps and Tackley 2008, 2009) models of thermo-chemical convection have been developed, showing strong stratification or a more complex structure, depending on the fluid properties. A major improvement was to perform calculations that account for compressible fluids, using either the anelastic (Tackley 1998), or the extended Boussinesq (Schott and Yuen 2004) approximation. In these approximations, viscous dissipation and adiabatic heating are present, and thermo-chemical properties are allowed to vary with depth through the prescription of a reference model. The decrease of thermal expansion with depth, which is well documented from mineral physics data (e.g., Anderson 1995), is particularly important because it weakens the thermal buoyancy in the deep mantle. As a result, chemical stratification or the development of stable thermo-chemical structures requires a smaller buoyancy ratio than in the classical Boussinesq approximation.

15.3.1 Stability of Thermo-chemical Reservoirs: Important Parameters

The mode of convection of a chemically heterogeneous fluid, and therefore the fluid thermo-chemical structure, is controlled by several parameters, including the buoyancy ratio (measuring the chemical density contrast relative to the top-to-bottom thermal density contrast), the fraction of dense material, and the rheology of the fluid. Among these parameters, the buoyancy ratio, B, has a particular importance because it controls the stratification of the system and the amplitude of the deformation of the dense layer (e.g., Davaille 1999; Le Bars and Davaille 2004; McNamara and Zhong 2004; Jaupart et al. 2007). Variations in B lead to different patterns, from a stable layer (large B), to a fully unstable layer (small B). The analogue experiments of Davaille (1999) observe two regimes of convection, depending on the density contrast between the dense layer and the overlying material. If the density contrast is smaller than 1 %, domes of dense material oscillate vertically, but dense and regular material remain unmixed. For larger contrasts, the system is stratified, i.e., the layer of dense material remains stable. Strong stratification is also observed by numerical models, in cases for which the buoyancy ratio is too large. A strong stratification is, however, not desirable in the case of the Earth's mantle, because it is not observed by seismic data and models. Moderate buoyancy ratios lead to large thermo-chemical piles, whose spectral content agrees well with that of thermo-chemical heterogeneities observed by probabilistic tomography

(Deschamps et al. 2007). Unfortunately, these structures are short-lived, and they do not survive convection (Deschamps and Tackley 2008). Dense and regular materials quickly mix, and the amplitude of chemical heterogeneities decreases with time. Clearly, for the Earth's mantle, the chemical density contrast between dense and regular material should not be too high to avoid strong stratification, but other parameters may control the formation and evolution of reservoirs of dense material.

A systematic search in the model space of thermo-chemical convection assuming an initial layer of dense material identified two important parameters that may enter a successful model of Earth's mantle (Deschamps and Tackley 2008, 2009): a moderate chemical density contrast (around 60–100 kg/m^3), to avoid chemical stratification, and a large thermal viscosity contrast ($\Delta\eta_T \geq 10^4$), which helps in maintaining large pools of dense material and opposes mixing. In models that combine these two properties, large reservoirs of dense material are maintained over periods of time comparable to the age of the Earth, inducing density and seismic velocity anomalies that are in good agreement with probabilistic tomography.

Another potentially important parameter is the Clapeyron slope of the phase change at 660 km depth. A negative Clapeyron slope at 660 km, as is expected for the transformation of ringwoodite to perovskite and periclase/wüstite, prevents the dense material from massively flowing into the upper mantle. An endothermic phase change thus acts as a filter, but is not required to maintain reservoirs of dense material in the lower mantle (Deschamps and Tackley 2009). One may point out that the mineralogical transition from lower to upper mantle is complex, multiple phase transitions with different Clapeyron slopes being involved (e.g., Stixrude and Lithgow-Bertelloni 2011). A positive Clapeyron slope would affect the filtering of dense material, and larger amounts of this material may penetrate in the upper mantle. If, as suggested by mineral physics experiments (Hirose 2002), hot plumes go through an exothermic phase change (from perovskite and periclase/wüstite directly to wadsleyite and/or majoritic garnet) when entering the upper mantle, the filtering of dense material at 660 km would be reduced. The presence of recycled crust in plumes may further alter the value of the Clapeyron slope of the 660 km phase transition. However, numerical experiments indicate that the amount of dense material entrained by plumes remains compatible with the entrainment expected from the lower end values of the helium ratio (^{4}He/^{3}He) in Ocean Island Basalt (Sect. 15.4.1) even for a Clapeyron slope equal to zero, provided that the buoyancy ratio of the dense material is slightly larger (Deschamps et al. 2011).

15.3.2 Stability of Primitive Reservoirs in Spherical Geometry

To check whether the conclusions obtained in 3D Cartesian geometry are also valid in spherical geometry, we conducted additional numerical experiments with StagYY (Tackley 2008). Spherical geometry is modeled by a set of Yin and Yang grids (Kageyama and Sato 2004), each grid having 64 × 192 azimuthal points. At each depth, this is equivalent to a spherical grid with 256 points in longitude and 128 points in latitude. We fixed the vertical resolution to 64 nodes, with grid refinement at the top and at the bottom of the shell, which allows resolution of the thermal boundary layers and the reservoirs of dense material. The ratio of the core-to-total radii is set to its terrestrial value, $f = 0.55$. A phase transition is imposed at $z = 660$ km with density contrast $\Delta\rho_{660} = 400$ kg/m^3. Viscosity is allowed to vary with temperature and depth, and an additional viscosity contrast $\Delta\eta_{660} = 30$ is added at the 660-km phase transition. The viscosity is fully described by

$$\eta(z,T) = \eta_0[1 + 29H(z - 660)] \exp\left[V_a\frac{z}{D} + E_a\frac{\Delta T_S}{(T + T_{\text{off}})} + K_\text{a}C\right], \tag{15.1}$$

where η_0 is a reference viscosity, H is the Heaviside step function, D the mantle thickness, ΔT_S the super-adiabatic temperature difference across the shell, and T_off the temperature offset, which is added to the temperature to reduce the viscosity jump across the top boundary layer. The reference viscosity η_0 is defined at regular composition ($C = 0$), $z = 0$, and for the surface temperature of the reference adiabat (i.e., $T_\text{as} = 0.64\Delta T_\text{S}$). The viscosity variations with depth and temperature are controlled by V_a and E_a, respectively, modeling the activation volume and energy. In all calculations discussed here, we fixed the value of V_a to 2.303, which leads to a bottom-to-top depth viscosity ratio of 300 (including the viscosity jump at 660 km, but excluding the thermally induced increase in thermal boundary layers and the increase due to adiabatic increase of temperature, which are both controlled by E_a). We define a potential thermal viscosity contrast as $\Delta\eta_\text{T} = \exp(E_\text{a})$, but due to the temperature offset, which we set to $T_\text{off} = 0.88\Delta T_\text{S}$, and to the adiabatic increase of temperature, the effective top-to-bottom thermal viscosity contrast is different from $\Delta\eta_\text{T}$. Overall, the effective contrast is smaller than $\Delta\eta_\text{T}$ by about two orders of magnitude. Finally, the viscosity variation with composition is controlled by the parameter K_a, the viscosity jump between dense and regular material being $\Delta\eta_\text{C} = \exp(K_\text{a})$.

The vigor of convection is controlled by the Rayleigh number, whose definition is non-unique due to the fact that viscosity and thermodynamic parameters vary throughout the shell. Here, we prescribed a reference Rayleigh number Ra_0, defined at the reference viscosity η_0 and surface values of the thermodynamic parameters. In all calculations, we fixed Ra_0 to 3.0×10^8. This leads to an effective Rayleigh number (defined as the Rayleigh number at the volume average viscosity) around 10^6–2.0×10^6, depending on the case.

We used a definition of the buoyancy ratio in which the reference density increases with depth following an appropriate thermodynamical model (Tackley 1998)

$$B_\text{z} = \frac{\Delta\rho_\text{c}(z)}{\alpha\rho(z)\Delta T_\text{S}}, \tag{15.2}$$

where $\Delta\rho_\text{c}$ is the chemical density contrast between dense and regular material, $\rho(z)$ the reference density at depth z, α the coefficient of thermal expansion at the surface of the Earth, and ΔT_S the super-adiabatic temperature jump. Here, we used $\Delta T_S = 2500$ K. Assuming a surface temperature of 300 K and an adiabatic increase of 1000 K, which is a typical for the Earth mantle, this leads to a temperature at the CMB of 3800 K, i.e., at the lower edge of the most recent estimates from the phase diagram of iron (Anzellini et al. 2013), but at the higher edge of the estimates from the solidus of pyrolite (Nomura et al. 2014). Calculations performed with this definition of the buoyancy ratio are equivalent to calculations made with a buoyancy ratio that refers to the surface density, B_S, but with larger value of B_S. For instance, if the density increases by 50 % from top to bottom, the structure obtained with a depth-dependent buoyancy ratio $B_z = 0.20$ is equivalent to that obtained with a surface buoyancy ratio $B_S = 0.30$. The compositional field is modeled with 25 million particle tracers, leading to an average number of tracers per cell of 15, enough to model the entrainment (Tackley and King 2003). The dense material is initially distributed in a layer at the bottom of the shell, and the initial thickness of this layer is controlled by prescribing the volume fraction of dense material, X_{DM}. Here, we fixed the volume fraction of dense material to 7 % of the volume of the shell, which is a bit large compared to the estimated volume of the LLSVPs, and did an additional experiment with $X_{DM} = 3.5$ %, corresponding to an upper estimate of the LLSVPs volume fraction (Sect. 15.3.3). The initial condition for temperature consists of an adiabatic profile with thin super-adiabatic boundary layers at the top and bottom of the shell, to which small random perturbations are added. With this setup, the experiment starts with a long transient phase (that can last up to 1.5–2.0 Gyr in dimensional time), during which the bottom dense layer heats up. After this phase, the evolution of the layer of dense material depends on the input parameters, mainly the buoyancy ratio and the thermal viscosity contrast (see below). In models that impose plate motions at the surface of the shell (e.g., McNamara and Zhong 2005), the transient phase is much shorter and thermo-chemical piles (if stable) form faster. Note that the experiments discussed here are not designed to model the detailed evolution of the Earth's mantle, for which accurate initial conditions are not yet known. Instead, they aim to identify the thermo-chemical structure obtained for a given set of parameters, and whether or not these structures are stable in time. Thus, time indication in our experiments should not be used to interpret early mantle evolution.

Figure 15.2 shows a snapshot of the case $B_z = 0.4$ and $X_{DM} = 7.0$ %. Viscosity depends on depth, but not on temperature ($E_a = 0$), and the Clapeyron slope of the 660-km depth phase transition is set to zero. For this set of parameters, the system remains stratified (Fig. 15.2, plots c and d), i.e., the layer of dense material covers the entire core. This layer has substantial topography, typically around 200 km, with highs corresponding to the foot of thermal plumes and depressions to downwellings. This layer is hot compared to the rest of the system and convection does not operate in it. Plumes are generated at its top, along a network of hot ridges (Fig. 15.2, plots a and b), but convection remains weak throughout the layer of regular material. A small fraction of dense material is, however, entrained upwards by these plumes and mixes with the regular material. This structure induces substantial chemical anomalies around the chemical discontinuity, but

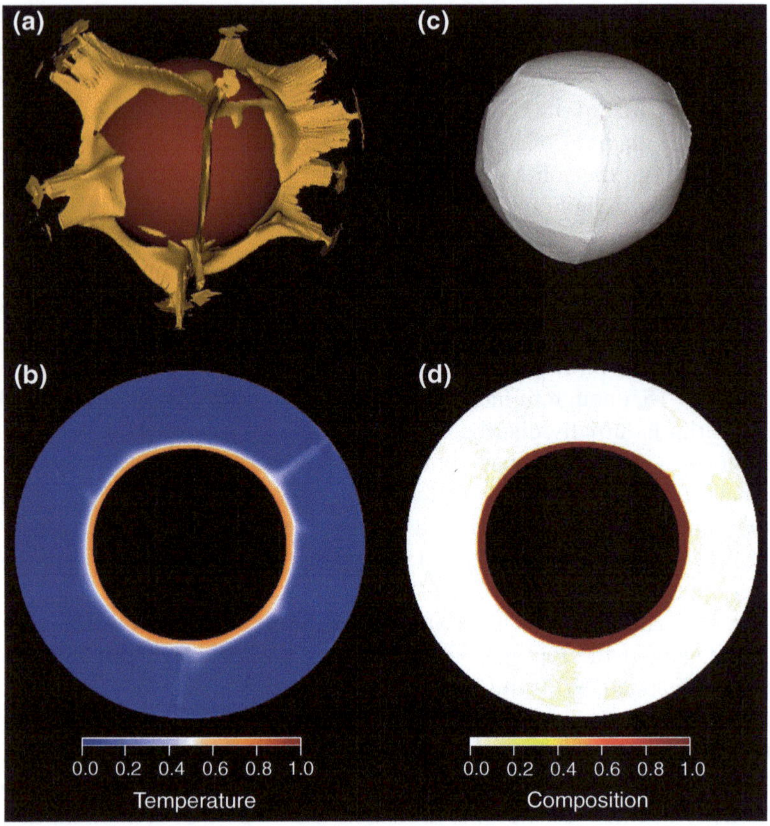

Fig. 15.2 Snapshot of a case with buoyancy ratio $B_z = 0.4$ and $X_{DM} = 7.0\ \%$. The viscosity jump at 660 km is equal to 30, and the viscosity depends neither on the temperature ($E_a = 0$, $\Delta\eta_T = 1$), nor on composition ($\Delta\eta_C = 1$). The Clapeyron slope at 660 km (Γ_{660}) is set to zero. Snapshot is taken at non-dimensional time $t = 2.12 \times 10^{-2}$ (9.0 Gyr). **a** Isosurface of the residual non-dimensional temperature with contour level $(T - \langle T \rangle) = 0.05$. **b** Polar slice of the non-dimensional temperature. **c** Isosurface of the composition with contour level $C = 0.5$. **d** Polar slice of the composition

no or very weak variations elsewhere (Fig. 15.3b). The amplitude of thermal anomalies is weak, with the largest anomalies being found at the top of the dense layer (Fig. 15.3a), indicating that the thermal boundary layer is located at the top of this layer. This reduces the temperature jump between the surface and the top of the dense layer, and thus the vigor of convection in the shell.

To avoid strong layering, we decreased the buoyancy ratio to $B_z = 0.2$ (Fig. 15.4). Other properties are similar to those of the case shown in Fig. 15.3 (in particular, viscosity does not depend on temperature). With $B_z = 0.2$, the initial layer of dense material is unstable and quickly mixes with the regular material (Fig. 15.4, plots c and d). As a result, large-scale chemical anomalies are small

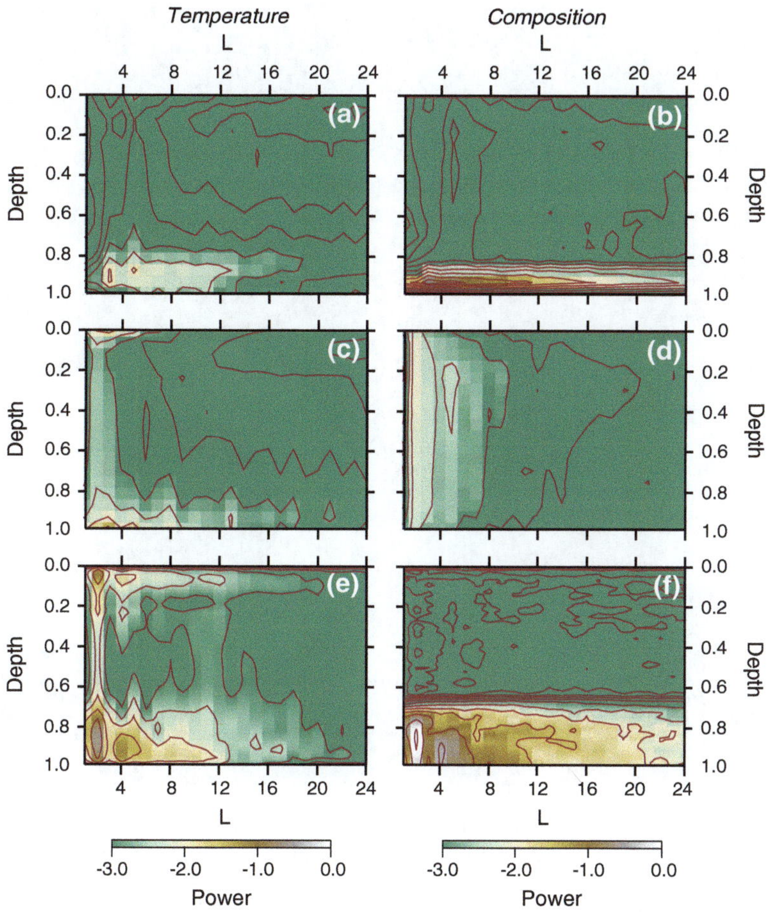

Fig. 15.3 Spectral heterogeneity map (SHM) of the anomalies in temperature (*left column*) and composition (*right column*) for three cases. *Top row* buoyancy ratio $B_z = 0.4$, $X_{DM} = 7\ \%$, $E_a = 0$ ($\Delta\eta_T = 1$), and $\Gamma_{660} = 0$, corresponding to the snapshot represented in Fig. 15.2. *Middle row* $B_z = 0.2$, $X_{DM} = 7\ \%$, $E_a = 0$ ($\Delta\eta_T = 1$), and $\Gamma_{660} = 0$, corresponding to the snapshot shown in Fig. 15.4. *Bottom row* $B_z = 0.2$, $X_{DM} = 7\ \%$, $E_a = 20.723$ ($\Delta\eta_T = 10^9$), and $\Gamma_{660} = -2.5$ MPa/K, corresponding to the snapshot represented in Fig. 15.5

throughout the mantle (Fig. 15.3d). At large scale, the mantle therefore appears homogeneous. Four large plumes are generated from a thin thermal boundary layer at the bottom of the shell (Fig. 15.4, plots a and b), inducing moderate thermal anomalies at the bottom and at the top of the system (Fig. 15.3c).

The model shown in Fig. 15.5 was obtained for $E_a = 20.723$ ($\Delta\eta_T = 10^9$, leading to an effective thermal viscosity contrast around 3.0×10^5). In addition, we imposed an endothermic phase transition at 660 km depth, with a Clapeyron slope $\Gamma = -2.5$ MPa/K. Other properties are similar to those of the cases

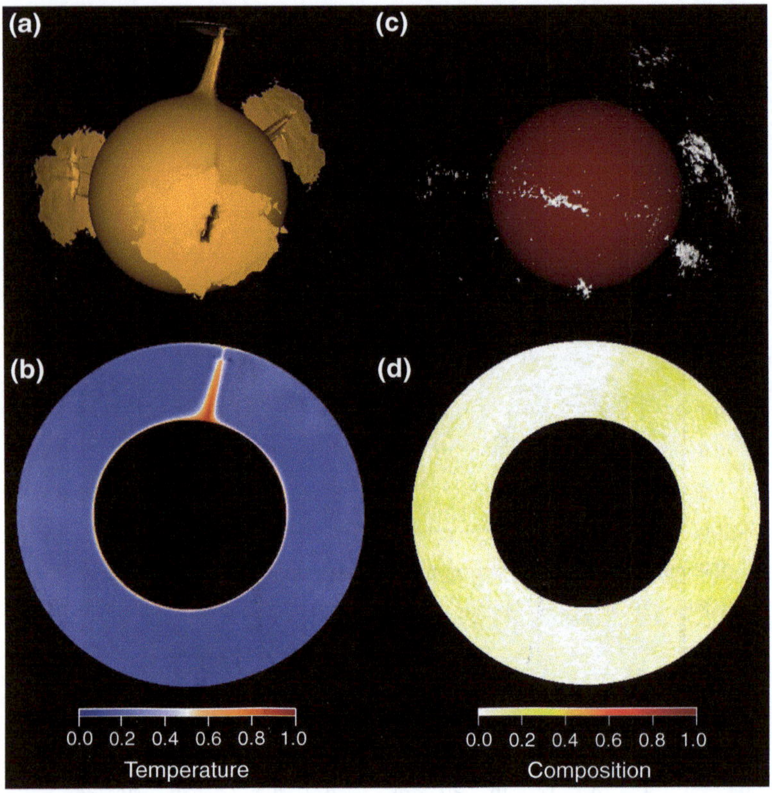

Fig. 15.4 Snapshot of the case $B_z = 0.2$, $X_{DM} = 7.0$ %, $E_a = 0$ ($\Delta\eta_T = 1$), and $\Gamma_{660} = 0$. The viscosity jump at 660 km is equal to 30, and the viscosity does not depend on composition ($\Delta\eta_C = 1$). Snapshot is taken at non-dimensional time $t = 1.06 \times 10^{-2}$ (4.5 Gyr). **a** Isosurface of the non-dimensional temperature with contour level $T = 0.55$. **b** Polar slice of the non-dimensional temperature. **c** Isosurface of the composition with contour level $C = 0.5$. **d** Polar slice of the composition

shown in Figs. 15.2 and 15.4. For this set of parameters, we observe two large reservoirs of dense material culminating at ~1000 km above the CMB (Fig. 15.5, plots c and d). These reservoirs remain stable and induce strong lateral chemical anomalies dominated by spherical harmonic degree 2 in the bottom part of the shell (Fig. 15.3f). In contrast, the rest of the shell is chemically homogeneous at large scale. Small-scale heterogeneities, due to entrainment of dense material by plumes, are, however, present locally. Plumes are generated along hot ridges at the top of the reservoirs and rise up to the surface (Fig. 15.5, plots a and b). Note that they spread beneath the 660 km and that secondary thinner plumes are generated on the top side of this phase transition. There is also evidence for plume interaction with mantle horizontal flow below the 660-km phase transition (top left and bottom right quadrants of Fig. 15.5b). The roots of these plumes are distributed

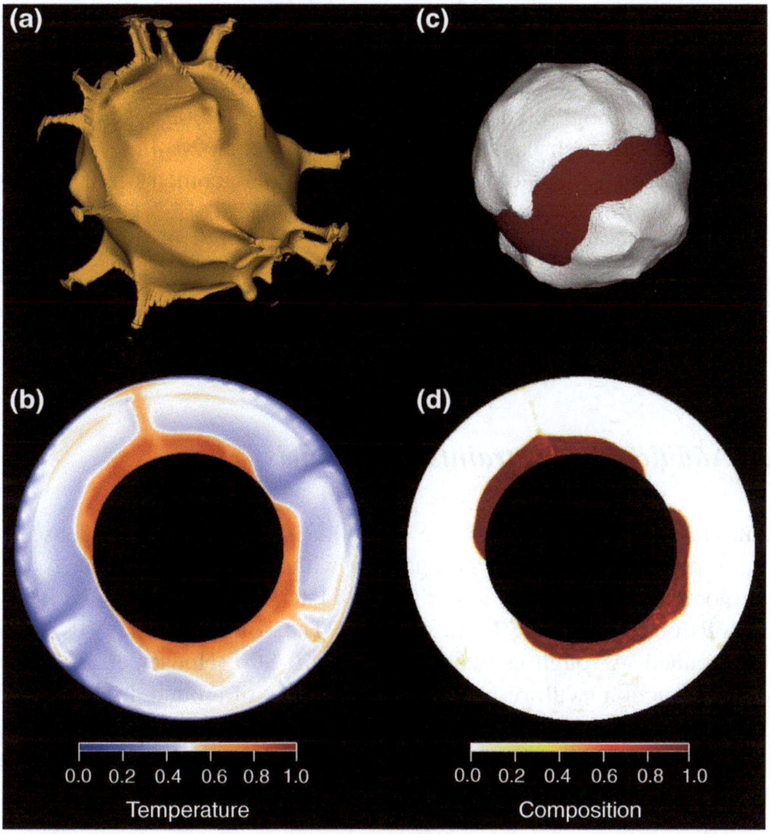

Fig. 15.5 Snapshot of the case $B_z = 0.2$, $X_{DM} = 7.0$ %, $E_a = 20.723$ ($\Delta\eta_T = 10^9$), and $\Gamma_{660} = -2.5$ MPa/K. The viscosity jump at 660 km is equal to 30, and the viscosity does not depend on composition ($\Delta\eta_C = 1$). Snapshot is taken at non-dimensional time $t = 3.18 \times 10^{-2}$ (13.5 Gyr). **a** Isosurface of the non-dimensional temperature with contour level $T = 0.55$. **b** Polar slice of the non-dimensional temperature. **c** Isosurface of the composition with contour level $C = 0.5$. **d** Polar slice of the composition

throughout the surface of the reservoirs, including at their borders. By contrast, we do not observe plumes originating from outside the reservoirs of dense material. Plumes entrain small amounts of dense material (see, e.g., the top left quadrant of Fig. 15.5d), thus eroding the dense reservoirs. Importantly, this erosion is very small, which prevent reservoirs from being completely entrained within a reasonable time (i.e., comparable to the age of the Earth). Downwellings are located between the reservoirs, suggesting that they strongly interact with the initial dense layer by deflecting it and pushing it away. This thermal structure induces strong thermal anomalies at the bottom of the shell and below its surface, where plume heads spread (Fig. 15.3e).

3D Cartesian models further indicated that the chemical viscosity contrast has an influence on the shape of reservoirs. With increasing chemical viscosity contrast, these reservoirs are thicker, cover a smaller area, and have steeper edges (McNamara and Zhong 2004; Deschamps and Tackley 2009). Additional calculations in spherical geometry confirm these trends (Li et al. 2014a).

Overall, the conclusions obtained in 3D Cartesian geometry regarding the stability and survival of reservoirs of dense material (Sect. 15.3.1) are also valid in spherical geometry. Interestingly, the chemical structure of models that include a moderate buoyancy ratio and strong thermal viscosity contrast is dominated by spherical harmonic degree 2, as in the case of the Earth's mantle, assuming that the LLSVPs observed by seismic tomography are thermo-chemical structures.

15.3.3 Additional Constraints and Models

In the model shown in Fig. 15.5, we fixed the volume fraction of dense material to 7 %, which is too large for comparison with the Earth's mantle. As a result, the reservoirs are significantly larger (both in lateral extension and thickness) than the LLSVPs observed in the Earth's mantle. The volume of LLSVPs is not very well constrained. A rough estimate from HMSL-S06 (Houser et al. 2008) leads to a volume fraction (with respect to entire mantle) of around 4.0 %. Burke et al. (2008) defined the LLSVP boundaries as the -1.0 % contour of the $d\ln V_S$ maps from the SMEAN model and found a volume fraction of 1.6 %. Estimates from the non-Gaussian anomaly in the $d\ln V_S$ distribution histograms gives values in the range 1.7–2.4 % (Hernlund and Houser 2008). Therefore, models of convection suitable for the Earth's mantle should have a fraction of dense material around 2–4 %. Interestingly, numerical experiments with a volume fraction of dense material in the range 2.5–15 % (Li et al. 2014a) show that decreasing the volume fraction of dense material does not alter the long-term stability of these reservoirs. In addition, these experiments indicate that the distribution of plumes is sensitive to the volume fraction of the dense material. For volume fractions around 5 % and smaller, plumes are mostly distributed on the edges of the reservoirs of dense material (Li et al. 2014a), whereas for larger volume fractions, they are more evenly distributed throughout the surface of these reservoirs. As an example, Fig. 15.6 shows a case with $X_{DM} = 3.5$ %, other parameters being similar to those in Fig. 15.5, except the chemical viscosity contrast, which is set to $\Delta\eta_C = 32$. The distribution of plumes along the edge of the reservoirs of dense material is globally consistent with the observation that the position of hot spots and the reconstructed positions of the large igneous provinces (LIPs) at the time they were active are preferentially located above the edges of LLSVPs (Burke and Torsvik 2004; Torsvik et al. 2008). Note that in these reconstructions, a few active hot spots and reconstructed LIPs are located within the LLSVPs. Furthermore, a recent study (Austermann et al. 2014) suggests that the correlation between LIP locations and LLSVP margins is statistically not significant and that observations

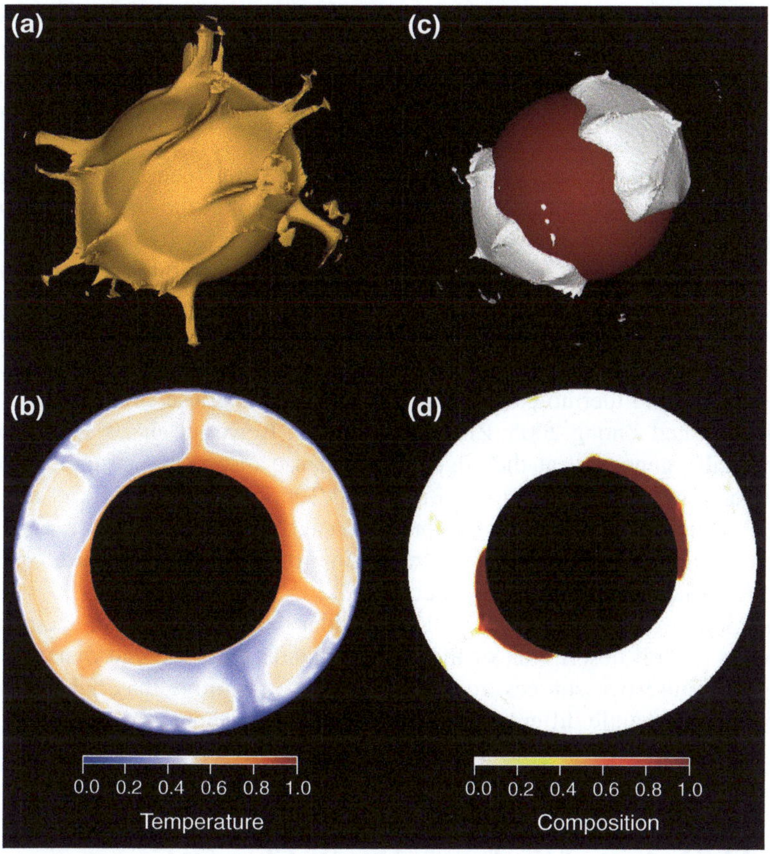

Fig. 15.6 Snapshot of the case $B_z = 0.20$, $X_{DM} = 3.5$ %, $E_a = 20.723$ ($\Delta\eta_T = 10^9$), and $\Gamma_{660} = -2.5$ MPa/K. The viscosity jump at 660 km is equal to 30, and the chemical viscosity ratio is $\Delta\eta_C = 32$. Snapshot is taken at non-dimensional time $t = 3.18 \times 10^{-2}$ (13.5 Gyr). **a** Isosurface of the non-dimensional temperature with contour level $T = 0.55$. **b** Polar slice of the non-dimensional temperature. **c** Isosurface of the composition with contour level $C = 0.5$. **d** Polar slice of the composition. Note that the view angle is different compared to that for Fig. 15.5

are also compatible with plumes originating from the LLSVP interiors. Finally, due to thermal winds and plume spreading beneath the 660-km discontinuity and beneath the surface (plots a and b in Figs. 15.5 and 15.6), the surface locations of plumes may be slightly shifted compared to the position of their root.

Because it controls the increase of density with pressure (as, for instance, in Birch–Murnaghan equation of state), the bulk modulus may influence the shape and stability of dense reservoirs (Tan et al. 2011). Density increases more rapidly with depth for material having high values of the bulk modulus. In addition, a high bulk modulus reduces the adiabatic increase of temperature, hence decreasing the thermal buoyancy of reservoirs. Models in which the bulk modulus of the dense material is larger than that of the surrounding fluid can remain stable for billions of years (Tan et al. 2011). Furthermore, these reservoirs have sharp edges, in agreement with travel and waveforms observations (Sect. 15.2), and plumes are preferentially located at their edges. Tan et al. (2011) have further investigated the interactions between downwellings and primitive reservoirs and found that downwellings tend to push plumes toward the reservoirs.

The models of convection discussed earlier in this section have a free slip mechanical surface boundary. As they reach the bottom of the shell, the downwellings generated in these models interact with the dense layer by pushing it away. In that sense, downwellings control the locations of the reservoirs of dense material. However, a detailed comparison with the present day structure of the Earth mantle requires the imposition of surface plate motions (e.g., using plate tectonic reconstruction models), which allows a better description of the interactions between downwellings and thermo-chemical piles. In models that use such reconstructions (McNamara and Zhong 2005; Zhang et al. 2010; Bower et al. 2013), plumes are preferentially generated at the edges of the dense reservoirs, in agreement with reconstructions of LIPs positions (Burke and Torsvik 2004; Torsvik et al. 2008). These models further explain the rounded shape of the Pacific LLSVP, and details of the African LLSVP observed by travel time data and seismic waveform modeling, in particular its sharp edges (Sect. 15.2).

The models discussed in this section consider only one source of chemical heterogeneity. A full description of the Earth's mantle structure and dynamics may, however, require two sources, reservoirs of dense primitive material (i.e., resulting from early mantle differentiation), and recycled oceanic crust entrained in the lower mantle by slabs. We further discuss this point in Sect. 15.4.1.

15.4 The Nature of the Lower Mantle Chemical Heterogeneities

15.4.1 Early Differentiated Material Versus Recycled MORB

The nature and origin of chemical heterogeneity detected by seismological observations is still a matter of debate. Two end-member hypotheses, the recycling of oceanic crust (MORB) and the survival of reservoir(s) of primitive or early differentiated material, are usually advocated and are both dynamically feasible.

The recycling of oceanic crust carried by slabs to the deep mantle is a straightforward source of heterogeneity. Seismic tomography clearly shows that slabs can penetrate to the deep mantle and sink down to the CMB (Fukao et al. 2001; Kárason and van der Hilst 2000; van der Hilst et al. 1997). Dynamically, numerical models of thermal convection have established that the endothermic phase transition at 660 km is not impermeable to downwelling slabs (e.g., Christensen and Yuen 1985; Machetel and Weber 1991; Tackley et al. 1993, 1994). Thus, stacking of the MORB component of slabs in the deep mantle may result in reservoirs of dense, recycled material. This hypothesis has been tested by numerical models (Christensen 1989; Christensen and Hofmann 1994) and laboratory experiments (Olson and Kincaid 1991), but recent models indicate that only small amount of oceanic crust may accumulate at the bottom of the mantle (Li and McNamara 2013). The production and recycling of MORB have also been modeled in a 2D-cylindrical shell and coupled with core evolution (Nakagawa and Tackley 2004, 2005b) and with the presence of the post-perovskite phase transition (Nakagawa and Tackley 2005a). These studies pointed out that the formation of pools of dense material and their shape depends on core properties (Nakagawa and Tackley 2005b), on the buoyancy ratio of MORB (Nakagawa and Tackley 2004), and on the Clapeyron slope of the post-perovskite phase transition (Nakagawa and Tackley 2005a). Models that include self-consistent mineralogy (Nakagawa et al. 2010) show that the shape and stability of the reservoirs of segregated MORB depend on the assumed MORB composition in main oxides. A MORB that is richer in FeO segregates more readily to a stable dense layer around the core. Finally, it has been suggested that recycled MORB may be episodically incorporated in LLSVPs (Tackley 2012). 2D Cartesian models of convection including two sources of heterogeneities have recently tested this hypothesis (Li et al. 2014b). These simulations indicate that small amounts of recycled MORB are injected in the piles of primitive material and subsequently entrained upwards by the plumes generated at the top of the piles. If the MORB is less dense than the primordial material, however, then it tends to accumulate above the dense material (Nakagawa and Tackley 2014).

The presence of an undegassed, isolated reservoir hidden in the deep mantle has long been advocated by geochemists to be one of the sources of ocean island basalt (OIB) (e.g., Farley et al. 1992; Hofmann 1997; Stuart et al. 2003). A key geochemical observation is the strong dispersion of the isotopic helium ratio ($^4He/^3He$) in OIB, ranging from 15,000 to 200,000. This suggests that the plumes at the origin of OIB sample several sources in the mantle. The low values (<30,000) of OIB helium ratio further indicate that one of the reservoirs sampled by OIB plumes is undegassed, and its lowest value (around 15,000) imposes a constraint on the maximum amount of primitive material entrained by plumes (Allègre and Moreira 2004). Other geochemical hints include the Argon budget (Allègre et al. 1996), and the high values of the $^{143}Nd/^{144}Nd$ ratio measured in the Baffin Island and western Greenland lavas (Jackson et al. 2010), which suggest that these rocks were extracted from an early-formed (~4.5 Gyr) mantle reservoir. Additional results for the Sm-Nd, Lu-Hf, and Rb-Sr systems suggest that Earth accreted from non-chondritic material depleted in incompatible elements, and that pristine material may be preserved in the deep mantle until now (see Caro, this volume). Numerical models showed that in models including a moderate buoyancy ratio and a large thermal viscosity contrast, the entrainment of dense material by plumes is less than 10 % (Deschamps et al. 2011), in agreement with geochemical estimates based on mass balance calculations (Allègre and Moreira 2004). An endothermic phase transition at 660 km further reduces the entrainment of dense material in the upper mantle, but plumes may instead go through an exothermic transition as they enter the upper mantle (Hirose 2002; Sect. 15.3.1). The large values (>100,000) of the OIB helium ratio suggest that another reservoir sampled by OIB plumes is recycled oceanic crust (e.g., Hofmann 1997). Again, this is consistent with the numerical simulations of Li et al. (2014b), which show that both primitive material and recycled MORB may be entrained by the plumes rising from the top of the reservoirs of dense material.

Several mechanisms have been proposed for the early partial differentiation of the Earth's mantle, including fractional crystallization of a regular magma ocean (Solomatov and Stevenson 1993; Solomatov 2007), crystallization of a basal magma ocean (Labrosse et al. 2007), which also explains the ultra-low seismic velocity zones (ULVZ) observed locally at the bottom of the mantle (Garnero and Helmberger 1995; Williams and Garnero 1996), 'upside-down' differentiation (Lee et al. 2010), which consists of the formation, crystallization, and sinking of dense liquids in the upper mantle during the Earth's first billion year, and the recycling of a primitive crust (Tolstikhin et al. 2006), which is supported by geochemical data (Boyet and Carlson 2006; Caro et al. 2003, 2004) and numerical models of convection (van Thienen et al. 2004, 2005). Alternatively, models of mantle composition based on E-chondrite predict that the composition in main oxides of the lower mantle is different from that of the upper mantle (Javoy et al. 2010). Kaminski and Javoy (2013) have further shown that a two-stage scenario of Earth formation, consisting of the formation of a proto-core at low pressure and temperature (around 50 GPa and 3500 K, respectively) followed by a giant impact, can explain the chemical heterogeneities observed in the lowermost mantle (see also Kaminski and Javoy, this volume).

Mineral physics experiments (Nomura et al. 2011) predict that a primitive reservoir resulting from the crystallization of a basal magma ocean would be richer in iron. It is worth noting that the Fe/Mn ratio in Hawaiian lavas indicates that the source region of the Hawaiian plume, which may be located in the lowermost mantle, is enriched in iron (Humayun et al. 2004). E-chondrite models of Earth's mantle composition also imply that reservoirs of primitive material should be enriched in iron and silicate (Javoy et al. 2010). Interestingly, the compositional anomalies inferred from probabilistic tomography (Trampert et al. 2004; Mosca et al. 2012) show that LLSVPs are enriched in iron (partitioned between Fe-perovskite and wüstite) and silicate (in perovskite). Furthermore, seismic sensitivities calculated from a coherent mineral physics database (Stixrude and Lithgow-Bertelloni 2011) indicate that LLSVPs are better explained by material enriched in iron (by about 3.0 %) and in perovskite (by about 20 %) than by recycled MORB, unless these MORB are hotter than the surrounding mantle by 1000 K or more (Deschamps et al. 2012). We further detail this point in Sect. 15.4.2.

15.4.2 Seismic Signatures of Differentiated Material and Recycled MORB

Seismic signatures provide key information about the nature of the chemical heterogeneities in the deep mantle. Linking observed seismic velocity anomalies to thermo-chemical distributions requires knowledge of seismic sensitivities to temperature and various chemical components. These sensitivities can be derived from an appropriate equation of state modeling and available mineral physics database. A Monte Carlo search allows accounting for various sources of uncertainties, including uncertainties in the mantle thermo-chemical reference model, and in the thermo-elastic and thermodynamic properties of the mantle minerals (Trampert et al. 2001; Deschamps and Trampert 2003). Alternatively, algorithms minimizing the Gibbs free energy of petrological aggregates, e.g., Perple_X (Connolly 2005), provide the mineralogical composition and thermo-elastic properties for this aggregate at given pressure, temperature, and composition in main oxides.

Seismic sensitivities to various chemical components (Trampert et al. 2001; Deschamps and Trampert 2003; Deschamps et al. 2012) show that an enrichment in iron induces a decrease in both shear-wave and bulk-sound velocities, whereas an enrichment in silicate induces an increase in bulk-sound velocity but leaves shear-wave velocity unchanged. Sensitivities to MORB further indicate that both V_S and V_Φ increase with increasing MORB fraction throughout the lower mantle. As a result, a combined enrichment in iron and silicate is a good candidate to explain LLSVPs and the anti-correlation between V_S- and V_Φ-anomalies. One may point out that because MORB is naturally richer in iron, an increase in the MORB fraction should have a seismic signature similar to that of a material enriched in iron (distributed between periclase and perovskite). An important observation, however, is that the effect of iron on V_S and V_Φ is much more pronounced in periclase (Kung et al. 2002) than in perovskite (Kiefer et al. 2002). Because periclase is not present in MORB, the seismic signature of an increase in the global volume fraction of iron on seismic velocities is significantly different from that of an increase in the volume fraction of MORB.

Figure 15.7 shows frequency histograms of shear-wave ($\mathrm{dln}V_S$) and bulk-sound ($\mathrm{dln}V_\Phi$) velocity anomalies induced either by high-pressure MORB or by material enriched in iron and perovskite in the deep (2710–2890 km) mantle. Seismic velocity anomalies for MORB and material enriched in iron and perovskite are calculated by

$$\mathrm{d}\ln V = \frac{\partial \ln V}{\partial T}\mathrm{d}T + \frac{\partial \ln V}{\partial X_{\mathrm{MORB}}}\mathrm{d}X_{\mathrm{MORB}}, \tag{15.3}$$

$$\text{and} \quad \mathrm{d}\ln V = \frac{\partial \ln V}{\partial T}\mathrm{d}T + \frac{\partial \ln V}{\partial X_{\mathrm{Pv}}}\mathrm{d}X_{\mathrm{Pv}} + \frac{\partial \ln V}{\partial X_{\mathrm{Fe}}}\mathrm{d}X_{\mathrm{Fe}}, \tag{15.4}$$

respectively, where V stands for V_S or V_Φ, and X_{MORB}, X_{Pv}, and X_{Fe} are the volume fractions of MORB, perovskite, and iron. The seismic sensitivities, $\partial \ln V/\partial X$, are from Deschamps et al. (2012). These sensitivities were obtained from a self-consistent mineral physics database (Stixrude and Lithgow-Bertelloni 2011) and an equation of state modeling that takes into account various sources of uncertainties in the thermo-elastic data and reference thermo-chemical model (Cobden et al. 2012). Seismic sensitivities are given as probability density functions (pdf) (rather than single values) with nearly Gaussian distributions. As a result, for fixed values of the temperature (dT) and compositional (dX_{MORB} or dX_{pv} and dX_{Fe}) anomalies, the seismic velocity anomalies calculated by Eqs. (15.3) and (15.4) are also pdf and can be plotted as frequency histograms. The strong dispersion in the frequency histograms calculated for MORB results from the strong dispersion in the MORB composition in main oxides. For instance, the FeO fraction in MORB can vary by as much as 0.07, depending on the sample. For a temperature anomaly of 400 K, material enriched in iron and perovskite results in $\mathrm{dln}V_S$ and $\mathrm{dln}V_\Phi$ around -2.0 and 0.5 %, respectively (Fig. 15.7g), in good agreement with the values observed in LLSVPs. By contrast, the $\mathrm{dln}V_S$ predicted by high-pressure MORB are too small in amplitude (around -0.5 %), compared to those in LLSVPs. Seismic signatures of material enriched in iron and perovskite fit well LLSVP anomalies up to temperature excess around 700 K (Deschamps et al. 2012). At higher temperatures (plots d and h in Fig. 15.7), high-pressure MORB explains the $\mathrm{dln}V_S$ observed in LLSVPs better, but induces negative $\mathrm{dln}V_\Phi$ anomalies, in contradiction with seismic tomography (Ishii and Tromp 1999; Masters et al. 2000; Trampert et al. 2004; Houser et al. 2008). High-pressure MORB colder than the surrounding mantle by 400 K induces $\mathrm{dln}V_S$ and $\mathrm{dln}V_\Phi$ around 1.0 and 0.5 %, respectively (Fig. 15.7b). These values agree well with values of $\mathrm{dln}V_S$ and $\mathrm{dln}V_\Phi$ observed by HMSL_SP06 (Houser et al. 2008) beneath the Japan subduction zone, where continuous high shear-wave velocities are observed down to the CMB (Ritsema and van Heijst 2000), and beneath Peru, Bolivia, and western Brazil. This result is consistent with the modeling of Tan et al. (2002). High-pressure MORB with larger temperature deficit (1000 K and more) induces $\mathrm{dln}V_S$ around 3.0 % (Fig. 15.7a), which is too large compared to the value observed at the expected location of stacked slabs.

Figure 15.7 further illustrates the trade-off existing between temperature and composition, since LLSVPs may be explained by either warm (hotter than the surrounding mantle by 400–700 K) material enriched in iron and perovskite, or very hot (hotter than the surrounding mantle by 1000 K) recycled MORB. In the latter case, it should, however, be noted that the predicted values of $\mathrm{dln}V_\Phi$ in LLSVPs are too small in amplitude, compared to those observed by seismic tomography. There are no direct measurements of the amplitude of the lateral variations in temperature at the bottom of the mantle. Recent estimates suggest peak-to-peak variations around 1500 K (Tackley 2012), i.e., a temperature excess (compared to average mantle) around 750 K. Another hint is given by estimates of the temperature excess in plumes, around 250–300 K (Campbell and Griffiths 1990). Such low values are consistent with the hypothesis that plumes originate from chemically distinct reservoir of dense material. For a compositional excess density between 2.0 and 5.0 %, a 250–300 K temperature excess in plumes implies temperature anomalies at the foot of the plume between 600 and 800 K (Farnetani 1997). Following this estimate, and according to Fig. 15.7, LLSVPs are better explained by warm material enriched in iron and perovskite than by hot recycled MORB. Note that other effects may contribute to reduce the temperature excess

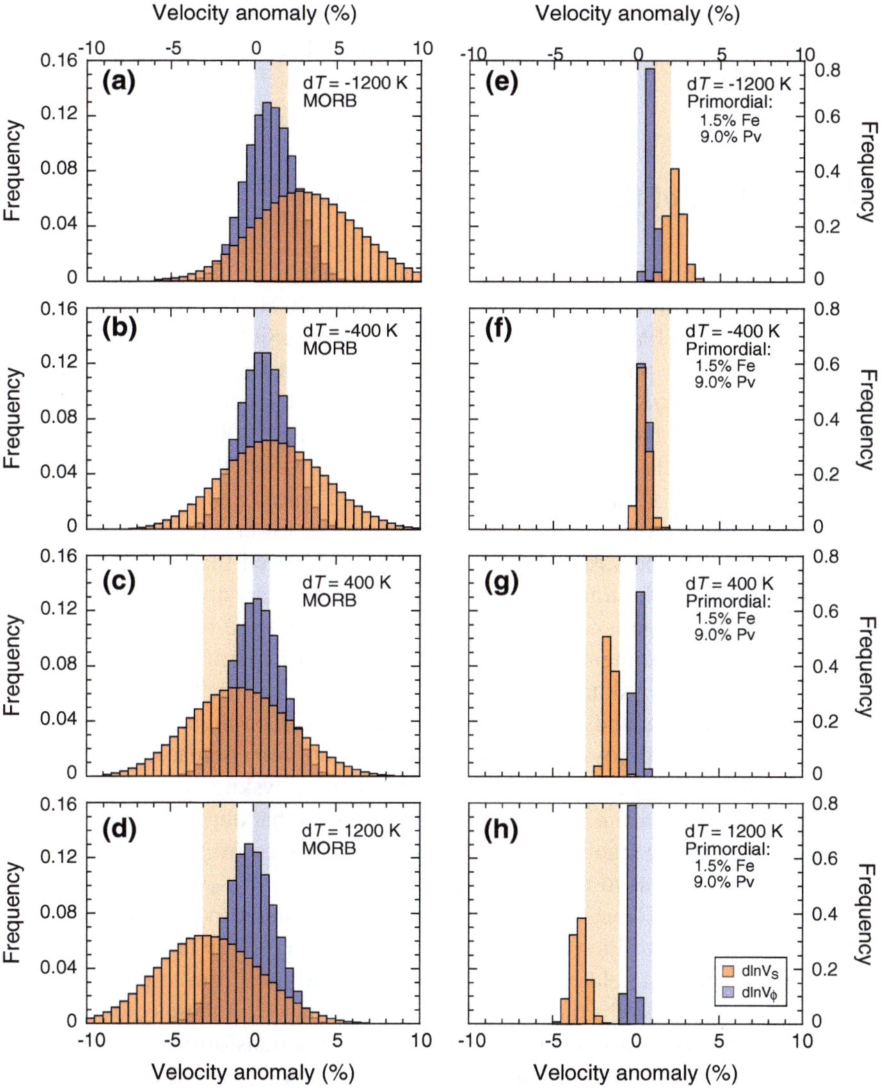

Fig. 15.7 Frequency histograms of shear-wave velocity (*orange*) and bulk-sound velocity (*blue*) anomalies for high-pressure MORB (*left column*), and for a material enriched (compared to the horizontally averaged mantle) in iron by 1.5 % and in (Mg, Fe)-perovskite by 9.0 % (*right column*). Four values of the temperature anomaly (compared to average mantle) are considered, from *top* to *bottom* $\mathrm{d}T = -1200$ K, $\mathrm{d}T = -400$ K, $\mathrm{d}T = 400$ K, and $\mathrm{d}T = 1200$. *Shaded vertical bands* in plots (**a**–**b**) and (**e**–**f**) indicate that typical values of seismic velocities observed beneath Central America and South America, where subducted slab is assumed to be present. *Shaded vertical bands* in plots (**c**–**d**) and (**g**–**h**) indicate typical values of seismic velocities in LLVSPs. Calculations are made in the layer 2710–2890 km, corresponding to the lowermost layer of tomographic model HMSL-SP06 (Houser et al. 2008)

in plumes, including internal heating (Parmentier et al. 1994), adiabatic cooling (Albers and Christensen 1996; Bunge 2005), and lateral diffusion (Mittelstaedt and Tackley 2005).

15.5 The Role of the Post-perovskite Phase

A major mineral physics discovery of the past decade is the post-perovskite phase transition (Murakami et al. 2004; Oganov and Ono 2004; Tsuchiya et al. 2004a). At pressure of 125 GPa and temperature of 2500 K, $MgSiO_3$ perovskite transforms to a post-perovskite phase. This phase transition has a large Clapeyron slope, in the range 8–10 MPa/K (Oganov and Ono 2004; Tsuchiya et al. 2004a), and even larger according to more recent estimates (Tateno et al. 2009). Furthermore, the composition is likely influencing the pressure at which the transition occurs. Experimental (Mao et al. 2004) and theoretical (Caracas and Cohen 2005) studies suggest that iron-rich perovskite transforms at lower pressures (110 GPa) than pure $MgSiO_3$ perovskite. Ohta et al. (2008) observed that compared to pure $MgSiO_3$ perovskite, the transition to post-perovskite occurs at similar pressure for a pyrolitic composition, and at slightly lower pressures for a MORB composition. Mineral physics experiments have also explored the iron partitioning between perovskite and post-perovskite. Andrault et al. (2009) found that for a temperature of 3300 K, perovskite and post-perovskite may coexist throughout the lowermost mantle, but that post-perovskite would be depleted in iron compared to perovskite, with iron partitioning between perovskite and post-perovskite around 4.2. Sinmyo et al. (2011) reached a similar conclusion and showed that during the perovskite to post-perovskite phase transition, iron changes its valence from Fe^{3+} to Fe^{2+} and preferentially partitions to ferro-periclase.

Due to the large Clapeyron slope of the phase transition to post-perovskite, the depth at which this transition occurs is expected to strongly vary laterally, depending on the temperature. In cold regions, perovskite should transform to post-perovskite at relatively shallow depths (an effect that may be enhanced if these cold regions are enriched in MORB), whereas in hot regions it may not transform at all. In cold and warm regions, double crossings are expected to happen (Hernlund et al. 2005), the thickness of the post-perovkite lens depending on the temperature at the CMB. On seismograms, the transition to post-perovskite results in precursors to core-reflected ScS waves (Hernlund et al. 2005). A double crossing induces a weaker intermediate arrival on seismograms. Waves diffracted at the CMB provide additional hints for the presence of post-perovskite. A recent study showed that post-perovskite is needed to explain shear- and compressional-wave velocity anomalies derived from P_{diff} and S_{diff} arrivals (Cobden et al. 2012). More details on the seismic detection of the post-perovskite may be found in Cobden et al. (this volume).

The thermo-elastic properties of post-perovskite are partly available (Tsuchiya et al. 2004b; Caracas and Cohen 2007; Mao et al. 2007; Oganov and Ono 2004; Stackhouse et al. 2005; Stackhouse and Brodholt 2007) and can be used to calculate seismic velocity and density sensitivities. These data indicate that shear-wave velocity in post-perovskite is larger than in perovskite by about 2 %, but that bulk-sound velocity is slightly lower, by less than 1 %. The phase transition to post-perovskite is a good candidate for the D$''$ discontinuity (Wookey et al. 2005), which has been detected in many locations (e.g., beneath North Siberia and beneath Alaska), but is not ubiquitous. Its elevation above the CMB varies between about 150 and 400 km, and the relative increase in V_S ranges from 2 to 4 % (Wysession et al. 1998). By contrast, because an excess in post-perovskite (compared to the average pyrolitic mantle) would result in positive V_S anomalies, post-perovskite is unlikely to explain LLSVPs, unless these regions are much hotter than average, in which case post-perovskite would not be stable. The distribution of post-perovskite mapped from probabilistic tomography (Mosca et al. 2012) clearly shows that post-perovskite is present only outside LLSVPs. Post-perovskite slightly modifies seismic sensitivities to temperature and other compositional parameters (Deschamps et al. 2012), which in turn influences the interpretation of seismic tomography. The most notable effect is to slightly reduce the sensitivity of V_S to temperature and to MORB.

Based on its seismic properties, it has been proposed that lateral variations in the stability field of post-perovskite triggered by variations in temperature may explain the anti-correlation between V_S and V_Φ anomalies (Hirose 2007; Hutko et al. 2008; Davies et al. 2012). Figure 15.8 displays pdf of the sensitivities of seismic velocities to the volume fraction of post-perovskite calculated following the method underlined in Sect. 15.4.2 (for details, see Deschamps et al. 2012). The mean sensitivity of V_S to post-perovskite is positive throughout the lowermost mantle (white dotted line in Fig. 15.8a), whereas the mean sensitivity of V_Φ is slightly negative (white dotted line in Fig. 15.8b), thus inducing anti-correlation between V_S and V_Φ. However, the pdf for the sensitivity of V_Φ spread around the zero value (Fig. 15.8b) and within the error bars it is not clear whether a transition to post-perovskite may trigger or not an anti-correlation between V_S and V_Φ. Furthermore, it should be remembered that post-perovskite is not stable if temperature is too high and would therefore be found preferentially in regions that are colder than average. This may compensate, at least partially, the fact that bulk-sound velocity is slightly lower in post-perovskite than in perovskite, and lead to positive anomalies in both shear-wave and bulk-sound velocities.

The stability field of post-perovskite further implies that the phase change to post-perovskite would result in an anti-correlation between V_S and V_Φ anomalies in a layer of limited thickness only, typically less than 300 km. For instance, Davies et al. (2012) do not observe any anti-correlation between their reconstructed $\mathrm{dln}V_S$ and $\mathrm{dln}V_\Phi$ at 2500 km. In tomographic models, by contrast, such anti-correlation is already present at a depth of 2500 km (Masters et al. 2000). The calculations of Davies et al. (2012) further predict sharp variations in the horizontally averaged $\mathrm{dln}V_S/\mathrm{dln}V_P$ (with values varying between -2.0 and 5.0 within a few tens of kilometers) and a strongly bimodal distribution in the values of $\mathrm{dln}V_S/\mathrm{dln}V_P$ at a given depth. Finally, the seismic velocity anomalies reconstructed from the temperature distribution predict large amplitude (up to 6 % for $\mathrm{dln}V_S$

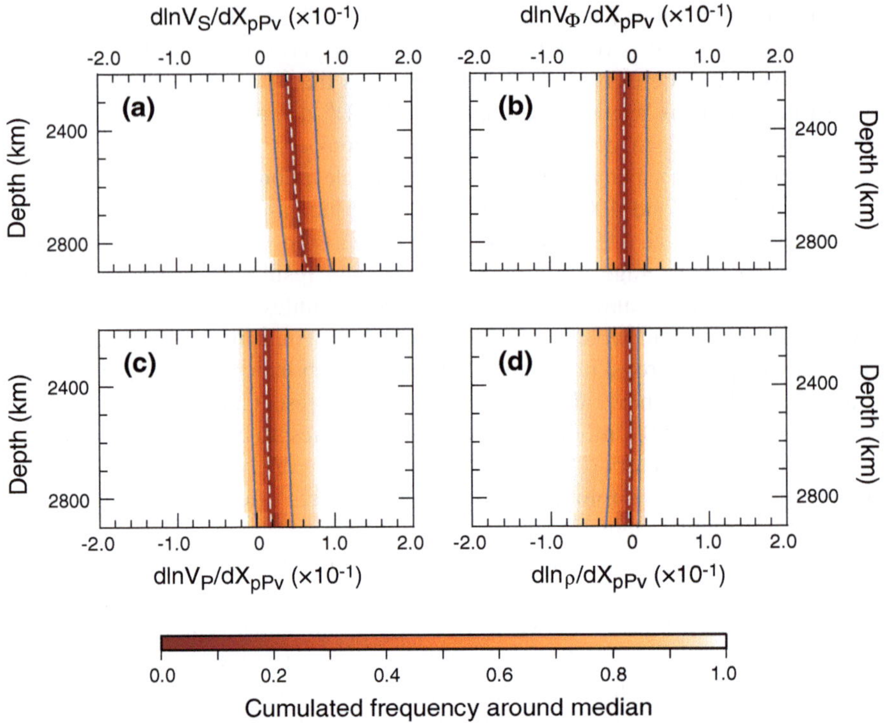

Fig. 15.8 Sensitivities of **a** shear-wave velocity, **b** bulk-sound velocity, **c** compressional-wave velocity, **d** and density to the volume fraction of post-perovskite in the depth range 2200–2900 km. The *color code* indicates the cumulated likelihood around the median value. On each plot, the *white dashed line* indicates the median value (i.e., 50 % of the explored sensitivities lie on each side of this line), and the *blue curves* represent the 0.15 and 0.85 quartiles (i.e., 70 % of the explored sensitivities lie within the area bounded by these curves)

and more than 2 % for $\mathrm{dln}V_{\Phi}$) and small-scale structures. None of these effects are observed by seismic tomography, although lateral and radial smearing may, at least partially, alter them.

Overall, lateral variations in the stability field of post-perovskite are unlikely to fully explain the available seismic observations, in particular the anti-correlation between V_S and V_{Φ} anomalies. Compositional anomalies are still needed to fully explain seismic tomography. Based on refined maps of probabilistic tomography, Mosca et al. (2012) reached a similar conclusion and pointed out that post-perovskite may be found outside LLSVPs only. The post-perovskite transition is, however, a good candidate to explain the D" discontinuity, at least in regions where the polarity of P waveforms is negative and that of S waveform is positive (Cobden and Thomas 2013). Furthermore, because it is strongly anisotropic (Wookey et al. 2005) and because it may inherit its fabrics from those of perovskite (Dobson et al. 2013), post-perovskite is a good candidate to explain anisotropy observed in the D″ layer.

Finally, even if it does not fully explain the shear-wave velocity contrast between the LLSVPs and the surrounding mantle, the presence post-perovskite outside LLSVPs certainly enhances this contrast.

15.6 A Thermo-Chemical Lowermost Mantle

Combined with mineral physics data, seismological observations reveal a lower mantle large-scale structure more complex than previously expected, involving phase transition (D″ layer), and thermo-chemical anomalies (LLSVPs). Figure 15.9 draws a simplified picture of the possible large-scale structure of the lower mantle.

Observed D″ reflections are likely due to the transition to post-perovskite in regions where P and S waveforms have different polarities, but other explanations (including chemical changes and anisotropy) are possible in regions where both P and S waveforms have a positive polarity (Cobden and Thomas 2013). According to these criteria, the nature of the D″ layer may not be unique. It may be related to the presence of post-perovskite in some regions, and to chemical and/or anisotropic effects in others. In that case, the presence of D″ reflections within the area covered by LLSVPs does not necessarily mean that post-perovskite is stable within these regions.

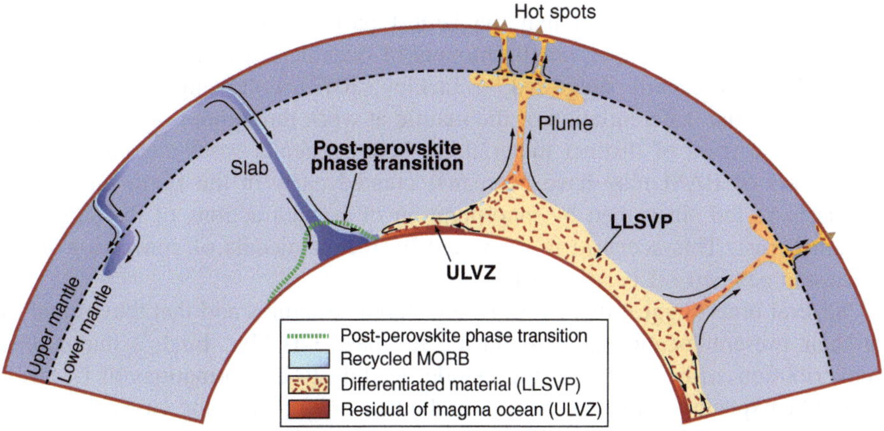

Fig. 15.9 The Earth's lower mantle? This simplified sketch shows the large-scale structures that may be present at the bottom of the Earth's mantle today. Lenses of post-perovskite are stable around slabs, with thickness that depends on the temperature. Early differentiation and possible regular re-feeding by subducted oceanic crust lead to chemically distinct structures, corresponding to the large low-shear-wave velocity provinces (LLSVPs). These structures are enriched in iron, and possibly silicates. Plumes are generated at their top and (but not uniquely) close to their edges, entraining small fraction of undegassed material and recycled oceanic crust toward the surface. Residuals of the magma ocean, strongly enriched in iron and corresponding to the ultra-low velocity zones (ULVZ), are swept out toward the edge of LLSVPs

LLSVPs clearly appear in tomographic models and cannot be explained by thermal anomalies alone. Compositional anomalies are needed to fully explain them. The exact nature of LLSVPs is still unclear, but an enrichment in iron is likely because FeO-rich minerals are seismically slower. LLSVPs are better explained by warm material enriched in iron oxide (the excess in FeO being distributed between ferro-periclase and Fe-perovskite) and in silicate (in the form of an excess in perovskite) than by recycled MORB, unless these MORBs are very hot (Deschamps et al. 2012). Enrichment in iron is further supported by mineral physics experiments showing that differentiation of mantle material lead to the formation of iron-rich reservoirs (Nomura et al. 2011). LLSVPs may thus result from partial differentiation in the Earth history, possible mechanisms of differentiation including the crystallization of a magma ocean (Solomatov and Stevenson 1993; Labrosse et al. 2007; Solomatov 2007), upside-down differentiation (Lee et al. 2010), and overturn of the early crust (Tolstikhin et al. 2006). Alternatively, an enrichment in iron is also consistent with E-chondrite models of Earth mantle (Kaminski and Javoy 2013; Kaminski and Javoy, this volume).

Available tomographic images show that the subducted slab may experience different fates (Kárason and van der Hilst 2000; Fukao et al. 2001). Some slabs are deflected and stacked around 700–1000 km depth (e.g., the Izu-Bonin slab), whereas others penetrate deeper into the lower mantle, possibly through avalanches (e.g., the Tonga and Central American slabs). Slabs entering the lower mantle may reach the CMB and stack there, in which case they would have different seismic signatures depending on their temperature (Fig. 15.7). Based on these signatures, we expect that the largest part of the subducted slab resides outside LLSVPs. However, small amounts of oceanic crust may be incorporated in LLSVPs, as recently suggested by Tackley (2012). Following this model, processes of partial differentiation of the mantle at work throughout the Earth history lead to reservoirs of distinct material, or heterogeneous basal mélange (BAM). Reservoirs of BAM may have been first created early in the Earth history and regularly re-fed since then by incorporation of small fractions of the subducted oceanic crust. This scenario is supported by recent models of convection in 2D Cartesian geometry (Li et al. 2014b).

Several hints suggest that LLSVPs are stable structures and that they remained at their present location at least during the past 320 Myr. First, a major mass redistribution at the bottom of the mantle, leading to the formations of LLSVPs, would be expected to induce changes in the true polar wandering larger than those observed during the past 200 Myr (Dziewonski et al. 2010). Second, the reconstructions of the geographic positions of most LIPs during the past 320 Myr are located above LLSVPs, mostly along their edges (Burke and Torsvik 2004; Torsvik et al. 2008). Therefore, if LLSVPs are the source regions of the LIP plumes, they must have remained at their present location since at least the past 320 Myr. Finally, mantle flow patterns inferred from tectonic plate motions reconstruction indicate that upwellings are present beneath Africa and the Pacific (i.e., above the LLSVPs) and have been stable around these locations during at least the past 250 Myr (Conrad et al. 2013). Models of thermo-chemical convection with an initial layer of dense material, including those discussed in this chapter (Figs. 15.5 and 15.6), show that reservoirs of dense material can remain stable for a period of time comparable to the age of the Earth without mixing substantially with the surrounding mantle. The plumes generated at the top of the reservoirs entrain small amounts of dense material, but the erosion rate is small enough to prevent mixing during the past 4.5 Gyr.

Assuming that they formed early in the Earth history, the reservoirs of dense material should be mostly undegassed. Following the BAM hypothesis (Tackley 2012) and 2D Cartesian numerical models (Li et al. 2014b), regular re-feeding by oceanic crust introduces some small patches of degassed material in these reservoirs. Thermal plumes are generated at the top of the primordial reservoirs and are entraining small amounts of dense material and, if present, recycled oceanic crust toward the surface, thus providing a probe to the chemistry of the deep reservoirs. This is consistent with OIB geochemical signatures, which indicates that OIB plumes sample both undegassed and recycled materials (e.g., Hofmann 1997). Mass balance calculations further show that the fraction of undegassed material entrained by OIB plumes is limited, 10 % or less (Allègre and Moreira 2004), a value that is well explained by plume entrainment derived from models of convection (Deschamps et al. 2011).

The slow fractional crystallization of a basal magma ocean (Labrosse et al. 2007), from which LLSVPs may result, indicates that some pockets of partial melt strongly enriched in iron may exist today at the bottom of the mantle. Good candidates for such pockets are the ultra-low velocity zones (ULVZ). ULVZ, first observed by Garnero and Helmberger (1995), are thin (between 5 and 40 km thick) structures lying above the CMB, in which compressional-wave (V_P) and shear-wave (V_S) velocities drop by about 10 %. They are usually interpreted as regions with a strong increase in density (due to chemical differentiation, e.g., an enrichment in iron) and/or partial melting (Williams and Garnero 1996; Rost et al. 2005). McNamara et al. (2010) noted that they are found predominantly in or at the edge of LLSVPs and conducted high-resolution experiments of thermo-chemical convection with three chemical components (for regular material and two types of dense material) indicating that the densest material (modeling the ULVZ) is swept out toward the edges of the reservoirs formed by the intermediate material (modeling the LLSVPs) and concentrates there.

15.7 Conclusion and Perspectives

Results obtained in seismology, mineral physics, and geodynamics during the past decade indicate that the large-scale structure of the lowermost mantle, between 2400 km depth and the CMB, likely results from several effects. Thermal variations, compositional anomalies of different sources (early mantle partial differentiation and regular recycling of oceanic crust), and the post-perovskite phase transition may all play a role and combine together to induce the observed seismological features, including the D″ discontinuity and the lower mantle tomography.

Refined models of lower mantle structure, composition, and dynamics require additional observables and modeling. The electrical conductivity of mantle rocks, like their thermo-elastic properties, is sensitive to temperature and composition. Interestingly, experimental data indicate that the electrical conductivity of Earth's lower mantle minerals increases with temperature (e.g., Xu et al. 2000; Vacher and Verhoeven 2007), whereas seismic velocities and density all decrease with increasing temperature. This should help breaking the trade-offs existing between temperature and composition. Mantle electrical conductivity tomography may be obtained from time variations of the magnetic field, but to date, available data do not allow sampling the mantle at depths greater than 1600 km (e.g., Semenov and Kuvshinov 2012). Lateral variations in the CMB heat flux influence the magnetic field pattern at the CMB, for which observations exist. They may thus be used as an additional constrain on the lowermost mantle thermo-chemical structure. Preliminary calculations indicate that if LLSVPs are hotter than average and enriched in iron, a belt of high electrical conductivity should be present in the lowermost mantle (Deschamps, 2015). Current estimates of heat flux lateral variations used in models of geodynamo are based on temperature distributions derived from shear-wave velocity tomography (e.g., Aubert et al. 2007; Amit et al. 2010), thus neglecting the effects of chemical anomalies. Different models of dynamics and thermo-chemical structure further result in different dynamic topographies at the CMB, which may be used as an additional constraint. Inferring the CMB topography from seismic data is, however, a difficult task, and so far no robust map of CMB topography is available. Finally, models of thermo-chemical convection that include two sources of chemical heterogeneities, i.e., recycled oceanic crust and primordial undegassed material may provide a better explanation of OIB geochemistry (Li et al. 2014b) and seismic tomography.

Acknowledgments We are grateful to Dan Bower and two other anonymous colleagues for their detailed and constructive reviews that helped improving the first version of this chapter. This work was funded by Academia Sinica (Taiwan) grant AS-102-CDA-M02, National Science Council of Taiwan (NSC) grant 101-2116-M-001-001-MY3, and Swiss National Science Fundation (SNF) grants 200021_129510 and 200021_149625. Models of thermo-chemical convection shown here were calculated on the ETH Linux cluster *brutus*.

References

Albers M, Christensen UR (1996) The excess temperature of plumes rising from the core-mantle boundary. Geophys Res Lett 23:3567–3570

Allègre CJ, Moreira M (2004) Rare gas systematic and the origin of oceanic islands: the key role of entrainment at the 670 km boundary layer. Earth Planet Sci Lett 228:85–92

Allègre CJ, Hofmann AO, Nions K (1996) The Argon constraints on mantle structure. Geophys Res Lett 23:3555–3557

Amit H, Aubert J, Hulot G (2010) Stationary, oscillating or drifting mantle-driven geomagnetic flux patches? J Geophys Res 115. doi:10.1029/2009JB006542

Anderson OL (1995) Equations of state of solids for geophysics and ceramic sciences. Oxford University Press, Oxford 405 pp

Andrault D, Muñoz M, Bolfan-Casanova N, Guignot N, Perrillat J-P, Anquilanti G, Pascarelli S (2009) Experimental evidence for perovskite and post-perovskite coexistence throughout the whole D″ region. Earth Planet Sci Lett 293:90–96

Antolik M, Gu YJ, Ekström G, Dziewonski AM (2003) J362D28: a new joint model of compressional and shear velocity in the Earth's mantle. Geophys J Int 153:443–466

Anzellini S, Dewaele A, Mezouar M, Loubeyre P, Morard G (2013) Melting of iron at Earth's inner core boundary based on fast X-ray diffraction. Science 340:464–466

Aubert J, Amit H, Hulot G (2007) Detecting thermal boundary control in surface flows from numerical dynamos. Phys Earth Planet Inter 160:143–156

Austermann J, Kaye BT, Mitrovica JX, Huybers P (2014) A statistical analysis of the correlation between large igneous provinces and lower mantle seismic structure. Geophys J Int 197:1–9

Bower DJ, Gurnis M, Seton M (2013) Lower mantle structure from paleogeographically constrained dynamic Earth models. Geochem Geophys Geosys 14. doi:10.1029/2012GC004267

Boyet M, Carlson RW (2006) ^{142}Nd evidence for early (>4.53 Ga) global differentiation of the silicate Earth. Science 309:576–581

Bunge H-P (2005) Low plume excess temperature and high core heat flux inferred from non-adiabatic geotherms in internally heated mantle circulation models. Phys Earth Planet Inter 153:3–10

Burke K, Torsvik TH (2004) Derivation of large Igneous Provinces of the past 200 million years from long-term heterogeneities in the deep mantle. Earth Planet Sci Lett 227:531–538

Burke K, Steinberger B, Torsvik TH, Smethurst MA (2008) Plume generation zones at the margins of large low shear velocity provinces on the core–mantle boundary. Earth Planet Sci Lett 265:49–60

Campbell IH, Griffiths RW (1990) Implications of mantle plume structure for the evolution of flood basalts. Earth Planet Sci Lett 99:79–93

Caracas R, Cohen RE (2005) Effects of chemistry on the stability and elasticity of the perovskite and post-perovskite phases in the $MgSiO_3$–$FeSiO_3$–Al_2O_3 system and implications for the lowermost mantle. Geophys Res Lett 32. doi:10.1029/2005GL023164

Caracas R, Cohen RE (2007) Effect of chemistry on the physical properties of perovskite and post-perovskite. In: Hirose K et al (eds) Post-perovskite the last mantle phase transition, vol 174. Geophysical Monograph. American Geophysical Union, pp 115–128

Caro G, Bourdon B, Birck JL, Moorbath S (2003) ^{146}Sm-^{142}Nd evidence from Isua metamorphosed sediments for early differentiation of the Earth's mantle. Nature 423:428–432

Caro G, Bourdon B, Wood BJ, Corgne A (2004) Trace-element fractionation in Hadean mantle generated by melt segregation from a magma ocean. Nature 436:246–249

Christensen UR (1989) Models of mantle convection: one or several layers. Phil Trans R Soc London A 328:417–424

Christensen UR, Hofmann AW (1994) Segregation of subducted oceanic crust in the convecting mantle. J Geophys Res 99:19867–19884

Christensen UR, Yuen DA (1985) Layered convection induced by phase transitions. J Geophys Res 90:10291–10300

Cobden L, Thomas C (2013) The origin of D″ reflections a systematic study of seismic array data sets. Geophys J Int 194:1091–1118

Cobden L, Mosca I, Trampert J, Ritsema J (2012) On the likelihood of post-perovskite near the core-mantle boundary: a statistical interpretation of seismic observations. Phys Earth Planet Inter 210–211:21–35

Connolly JAD (2005) Computation of phase equilibria by linear programming: a tool for geodynamic modeling and its application to subduction zone decarbonisation. Earth Planet Sci Lett 236:524–541

Conrad CP, Steinberger B, Torsvik TH (2013) Stability of active mantle upwelling revealed by net characteristics of plate tectonics. Nature 498:479–482

Davaille A (1999) Simultaneous generation of hotspots and superswells by convection in a heterogeneous planetary mantle. Nature 402:756–760

Davies DR, Goes S, Davies JH, Schuberth BSA, Bunge H-P, Ritsema J (2012) Reconciling dynamic and seismic models of Earth's lower mantle: the dominant role of thermal heterogeneity. Earth Planet Sci Lett 353–354:253–269

Deschamps, (2015) Lower mantle electrical conductivity inferred from probabilistic tomography. Terr Atmos Ocean Sci 26:27–40. doi:10.3319/TAO.2014.08.19.03(GRT)

Deschamps F, Tackley PJ (2008) Exploring the model space of thermo-chemical convection I—principles and influence of the rheological parameters. Phys Earth Planet Inter 171:357–373

Deschamps F, Tackley PJ (2009) Searching for models of thermo-chemical convection that explain probabilistic tomography. II. Influence of physical and compositional parameters. Phys Earth Planet Inter 176:1–18

Deschamps F, Trampert J (2003) Mantle tomography and its relation to temperature and composition. Phys Earth Planet Inter 140:277–291

Deschamps F, Trampert J, Tackley PJ (2007) Thermo-chemical structure of the lower mantle: seismological evidence and consequences for geodynamics. In Yuen DA et al (eds) Superplume: beyond plate tectonics. Springer, Berlin, pp 293–320

Deschamps F, Kaminski E, Tackley PJ (2011) A deep mantle origin for the primitive signature of Ocean Island Basalt. Nat Geosci 4:879–882

Deschamps F, Cobden L, Tackley PJ (2012) The primitive nature of large low shear-wave velocity provinces. Earth Planet Sci Lett 349–350:198–208

Dobson DP, Miyajima N, Nestola F, Alvaro M, Casati N, Liebske C, Wood IG, Walker AM (2013) Strong inheritance of texture between perovskite and post-perovskite in the D″ layer. Nat Geosci 6:575–578

Dziewonski AM, Lekic V, Romanowicz B (2010) Mantle anchor structure: an argument for bottom up tectonics. Earth Planet Sci Lett 299:69–79

Farley KA, Natland JH, Craig H (1992) Binary mixing of enriched and undegassed (primitive?) mantle components (He, Sr, Nd, Pb) in Samoan lavas. Earth Planet Sci Lett 111:183–199

Farnetani CG (1997) Excess temperature of mantle plumes: the role of chemical stratification across D″. Geophys Res Lett 24:1583–1586

Fukao Y, Widiyantoro S, Obayashi M (2001) Stagnant slabs in the upper and lower transition regions. Rev Geophys 39:291–323

Garnero EJ, Helmberger DV (1995) A very slow basal layer underlying large-scale low-velocity anomalies in the lower mantle beneath the Pacific: evidence from core phases. Earth Planet Sci Lett 91:161–176

Gurnis M (1986) The effect of chemical density differences on convective mixing in the Earth's mantle. J Geophys Res 91:11407–11419

Hansen U, Yuen DA (1988) Numerical simulations of thermo-chemical instabilities at the core-mantle boundary. Nature 334:237–240

He Y, Wen L (2009) Structural features and shear-velocity structure of the "Pacific Anomaly". J Geophys Res 114. doi:10.1029/2008JB005814

Hernlund J, Houser C (2008) On the statistical distribution of seismic velocities in Earth's deep mantle. Earth Planet Sci Lett 265:423–437

Hernlund J, Thomas C, Tackley PJ (2005) A doubling of the post-perovskite phase boundary and structure of the Earth's lowermost mantle. Nature 434:882–886

Hirose K (2002) Phase transitions in pyrolitic mantle around 670-km depth: implications for upwelling of plumes from the lower mantle. J Geophys Res 107. doi:10.1029/2001JB000597

Hirose K (2007) Discovery of post-perovskite phase transition and the nature of the D″ layer. In: Hirose K et al (eds) Post-perovskite the last mantle phase transition, vol 174. Geophysical Monograph. American Geophysical Union, pp 19–35

Hofmann AW (1997) Mantle geochemistry: the message from oceanic volcanism. Nature 385:219–229

Houser C, Masters G, Shearer P, Laske G (2008) Shear and compressional velocity models of the mantle from cluster analysis of long-period waveforms. Geophys J Int 174:195–212

Humayun M, Qin L, Norman ND (2004) Geochemical evidence for excess iron in the Mantle beneath Hawaii. Science 306:91–94

Hutko AR, Lay T, Revenaugh J, Garnero EJ (2008) Anticorrelated seismic velocity anomalies from post-perovskite in the lowermost mantle. Science 320:1070–1074

Ishii M, Tromp J (1999) Normal-mode and free-air gravity constraints on lateral variations in velocity and density of Earth's mantle. Science 285:1231–1236

Jackson MG, Carlson RW, Kurz MD, Kempton PD, Francis D, Blusztajn J (2010) Evidence for the survival of the oldest terrestrial mantle reservoir. Nature 466:853–856

Jaupart C, Molnar P, Cottrell E (2007) Instability of a chemically dense layer heated from below and overlain by a deep less viscous fluid. J Fluid Mech 572:433–469

Javoy M, Kaminski E, Guyot F, Andrault D, Sanloup C, Moreira M, Labrosse S, Jambon A, Agrinier P, Davaille A, Jaupart C (2010) The chemical composition of the Earth: enstatite chondrite model. Earth Planet Sci Lett 293:259–268

Jellinek AM, Manga M (2002) The influence of a chemical boundary layer on the fixity, spacing and lifetime of mantle plumes. Nature 418:760–763

Kageyama A, Sato T (2004) "Yin-Yang grid": an overset grid in spherical geometry. Geochem Geophys Geosyst 5. doi:10.1029/2004GC000734

Kaminski E, Javoy M (2013) A two-stage scenario for the formation of the Earth's mantle and core. Earth Planet Sci Lett 365:97–107

Kaminski E, Javoy M (2014) The composition of the deep Earth. In: Khan A et al (eds) The heterogeneous Earth mantle (in press)

Kárason H, van der Hilst RD (2000) Constraints on mantle convection from mantle tomography. In: Richards MA et al (eds) The history and dynamics of global plate motions, vol 121. Geophysical Monograph. American Geophysical Union, pp 277–288

Kiefer B, Stixrude L, Wentzcovitch RM (2002) Elasticity of (Mg, Fe)SiO_3 perovskite at high pressures. Geophys Res Lett 29. doi:10.1029/2002GL014683

Kung J, Li B, Weidner DJ, Zhang J, Liebermann RC (2002) Elasticity of (Mg 0.83, Fe 0.17) O ferropericlase at high pressure: ultrasonic measurements in conjunction with X-radiation techniques. Earth Planet Sci Lett 203:557–566

Kustowski B, Ekström G, Dziewonski AM (2008) Anisotropic shear-wave velocity structure of the Earth's mantle: a global model. J Geophys Res 113:B06306. doi:10.1029/2007JB005169

Labrosse S, Hernlund JW, Coltice N (2007) A crystallizing dense magma ocean at the base of the Earth's mantle. Nature 450:866–869

Le Bars M, Davaille A (2004) Whole layer convection in a homogeneous planetary mantle. J Geophys Res 109. doi:10.1029/2003JB002617

Lee C-T, Luffi P, Hoink T, Li J, Dasgupta R, Hernlund J (2010) Upside-down differentiation and generation of a 'primordial' lower mantle. Nature 463:930–933

Lekic V, Cottaar S, Dziewonski AM, Romanowicz B (2012) Cluster analysis of global lower mantle tomography: a new class of structure and implications for chemical heterogeneity. Earth Planet Sci Lett 357–358:68–77

Li M, McNamara AK (2013) The difficulty for subducted oceanic crust to accumulate at the Earth's core-mantle boundary. J Geophys Res 118:1807–1816. doi:10.1002/jgrb.50156

Li X-D, Romanowicz B (1996) Global mantle shear velocity model developed using nonlinear asymptotic coupling theory. J Geophys Res 101:22245–22272

Li Y, Deschamps F, Tackley PJ (2014a) The stability and structure of primordial reservoirs in the lower mantle: insights from models of thermo-chemical convection in 3-D spherical geometry. Geophys J Int (submitted)

Li M, McNamara AK, Garnero EJ (2014b) Chemical complexity of hotspots caused by cycling oceanic crust through mantle reservoirs. Nat Geosci 7:366–370

Machetel P, Weber P (1991) Intermittent layered convection in a model mantle with an endothermic phase change at 670 km. Nature 350:55–57

Mao W, Shen G, Prakapenka VB, Meng Y, Campbell AJ, Heinz D, Shu J, Hemley RJ, Mao HK (2004) Ferromagnesian perovskite silicates in the D″ layer of the Earth. Proc Natl Acad Sci USA 101:15867–15869

Mao W, Campbell AJ, Prakapenka VB, Hemley RJ, Mao H-K (2007) Effect of iron on the properties of post-perovskite silicate. In: Hirose K et al (eds) Post-perovskite the last mantle phase transition, vol 174. Geophysical Monograph. American Geophysical Union, pp 37–46

Masters G, Laske G, Bolton H, Dziewonski AM (2000) The relative behavior of shear velocity, bulk sound speed, and compressional velocity in the mantle: implication for thermal and chemical structure. In Karato S-I et al (eds) Earth's deep interior: mineral physics and tomography from the atomic to the global scale, vol 117. Geophysical Monograph Series. American Geophysical Union, Washington, DC, pp 63–87

McNamara AK, Zhong S (2004) Thermochemical structures within a spherical mantle. J Geophys Res 109. doi:10.1029/2003JB002847

McNamara AK, Zhong S (2005) Thermochemical structure beneath Africa and the Pacific ocean. Nature 437:1136–1139

McNamara AK, Garnero EJ, Rost S (2010) Tracking deep mantle reservoirs with ultra-low velocity zones. Earth Planet Sci Lett 299:1–9

Mittelstaedt E, Tackley PJ (2005) Plume heat flow is much less than CMB heat flow. Earth Planet Sci Lett 241:202–210

Mosca I, Cobden L, Deuss A, Ritsema J, Trampert J (2012) Seismic and mineralogical structures of the lower mantle from probabilistic tomography. J Geophys Res 117:B06304. doi:10.1029/2011JB008851

Murakami M, Hirose K, Kawamura K, Sata N, Ohishi Y (2004) Post-perovskite phase transition in $MgSiO_3$. Science 304:855–858

Nakagawa T, Tackley PJ (2004) Effects of thermo-chemical convection on the thermal evolution of the Earth's core. Earth Planet Sci Lett 220:207–219

Nakagawa T, Tackley PJ (2005a) The interaction between the post-perovskite phase change and a thermo-chemical boundary layer near the core-mantle boundary. Earth Planet Sci Lett 238:204–216

Nakagawa T, Tackley PJ (2005b) Deep mantle heat flow and thermal evolution of Earth's core in thermo-chemical multiphase models of mantle convection. Geochem Geophys Geosyst 6. doi:10.1029/2005GC000967

Nakagawa T, Tackley PJ (2014) Influence of combined primordial layering and recycled MORB on the coupled thermal evolution of Earth's mantle and core. Geochem Geophys Geosyst 15:619–633. doi:10.1002/2013GC005128

Nakagawa T, Tackley PJ, Deschamps F, Connolly JAD (2010) The influence of MORB and harzburgite composition on thermo-chemical mantle convection in a 3D Spherical shell with self-consistently calculated mineral physics. Earth Planet Sci Lett 296:403–412

Ni S, Helmberger D (2003) Ridge-like lower mantle structure beneath South Africa. J Geophys Res 108. doi:10.1029/2001JB001545

Ni S, Tan E, Gurnis M, Helmberger D (2002) Sharp sides to the African superplumes. Science 296:1850–1852

Nomura R, Ozawa H, Tateno S, Hirose K, Hernlund J, Muto S, Ishii H, Hiraoka N (2011) Spin crossover and iron-rich silicate melt in the Earth's deep mantle. Nature 473:199–203

Nomura R, Hirose K, Uesugi K, Ohishi Y, Tsuchiyama A, Miyake A, Ueno Y (2014) Low core-mantle boundary temperature inferred from the solidus of pyrolite. Science 343:522–525

Oganov AR, Ono S (2004) Theoretical and experimental evidence for a post-perovskite phase of $MgSiO_3$ in Earth's D″ layer. Nature 430:445–448

Ohta K, Hirose K, Lay T, Stat N, Ohishi Y (2008) Phase transitions in pyrolite and MORB at lowermost mantle conditions: implication for a MORB-rich pile above the core-mantle boundary. Earth Planet Sci Lett 267:107–117

Olson P, Kincaid C (1991) Experiment on the interaction of thermal convection and compositional layering at the base of the mantle. J Geophys Res 96:4347–4354

Parmentier EM, Sotin C, Travis BJ (1994) Turbulent 3-D thermal convection in an infinite Prandtl number, volumetrically heated fluid—implications for mantle dynamics. Geophys J Int 116:241–251

Resovsky JS, Trampert J (2003) Using probabilistic seismic tomography to test mantle velocity-density relationships. Earth Planet Sci Lett 215:121–134

Ricard Y, Mattern E, Matas J (2005) Synthetic tomographic images of slabs from mineral physics. In: van der Hilst RD et al (eds) Earth's deep mantle: Structure, composition and evolution, vol 160. American Geophysical Union, Geophysical Monograph, pp 285–302

Ringwood AE (1975) Composition and petrology of the Earth mantle. McGraw-Hill, New York, 618 pp

Ritsema J, van Heijst H (2000) Seismic imaging of structural heterogeneity in Earth's mantle: evidence for large-scale mantle flow. Sci Progress 83:243–259

Ritsema J, van Heijst H, Woodhouse JH (1999) Complex shear-wave velocity structure imaged beneath Africa and Iceland. Science 286:1925–1928

Ritsema J, McNamara AK, Bull AL (2007) Tomographic filtering of geodynamic models: implications for model interpretation and large-scale mantle structure. J Geophys Res 112. doi:10.1029/2006JB004566

Ritsema J, Deuss A, vanHeijst H-J, Woodhouse JH (2011) S40RTS: a degree-40 shear-velocity model for the mantle from new Rayleigh wave dispersion, teleseismic traveltime and normal-mode splitting function measurements. Geophys J Int 184:1223–1236

Rost S, Garnero EJ, Williams Q, Manga M (2005) Seismological constraints on a possible plume root at the core–mantle boundary. Nature 435:666–669

Schott B, Yuen DA (2004) Influences of dissipation and rheology on mantle plumes coming from the D″-layer. Phys Earth Planet Inter 146:139–145

Schuberth BSA, Bunge H-P, Ritsema J (2009) Tomographic filtering of high resolution mantle circulation models: can seismic heterogeneity be explained by temperature alone? Geochem Geophys Geosyst 10. doi:10.1029/2009GC002401

Schuberth BSA, Zaroli C, Nolet G (2011) Synthetic seismograms for a synthetic Earth: long-period *P*- and *S*-wave traveltime variations can be explained by temperature alone. Geophys J Int 188:1393–1412

Semenov A, Kuvshinov A (2012) Global 3-D imaging of mantle conductivity based on inversion of observatory *C*-responses—II. Data analysis and results. Geophys J Int 191:965–992

Sinmyo R, Hirose K, Muto S, Ohishi Y, Yasuhara A (2011) The valence state and partitioning of iron in the Earth's lowermost mantle. J Geophys Res 116. doi:10.1029/2010JB008179

Solomatov VS (2007) Magma oceans and primordial mantle differentiation. In: Stevenson DJ (ed) Treatise on geophysics, vol 9., Earth formation and evolutionElsevier B.V., Amsterdam, pp 91–119

Solomatov VS, Stevenson DJ (1993) Suspension in convective layers and style of differentiation of a terrestrial magma ocean. J Geophys Res 98:5375–5390

Stackhouse S, Brodholt JP (2007) The high-temperature elasticity of $MgSiO_3$ post-perovskite silicate. In: Hirose K et al (eds) Post-perovskite the last mantle phase transition,vol 174. Geophysical Monograph. American Geophysical Union, pp 99–114

Stackhouse S, Brodholt JP, Wookey J, Kendal J-M, Price GD (2005) The effect of temperature on the seismic anisotropy of the perovskite and post-perovskite polymorphs of $MgSiO_3$. Earth Planet Sci Lett 230:1–10

Stixrude L, Lithgow-Bertelloni C (2011) Thermodynamics of mantle minerals—II. Phase equilibria. Geophys J Int 184:1180–1213

Stuart FM, Lass-Evans S, Fitton JG, Ellam RM (2003) High $^3He/^4He$ ratios in picritic basalts from Baffin Island and the role of a mixed reservoir in mantle plumes. Nature 424:57–59

Su W-J, Dziewonski AM (1997) Simultaneous inversion for 3-D variations in shear and bulk velocity in the mantle. Phys Earth Planet Inter 100:135–156

Su W-J, Woodward RL, Dziewonski AM (1994) Degree 12 model of shear velocity heterogeneity in the mantle. J Geophys Res 99:6945–6980

Tackley PJ (1998) Three-dimensional simulations of mantle convection with a thermo-chemical CMB boundary layer: D″? In: Gurnis M et al (eds) The core-mantle boundary region, vol 28. Geodynamical series, pp 231–253

Tackley PJ (2002) Strong heterogeneity caused by deep mantle layering. Geochem Geophys Geosyst 3. doi:10.1029/2001GC000167

Tackley PJ (2008) Modelling compressible mantle convection with large viscosity contrasts in a three-dimensional spherical shell using the yin-yang grid. Phys Earth Planet Inter 171:7–18

Tackley PJ (2012) Dynamics and evolution of the deep mantle resulting from thermal, chemical, phase and melting effects. Earth Sci Rev 110:1–25

Tackley PJ, King SD (2003) Testing the tracer ratio method for modelling active compositional fields in mantle convection simulations. Geochem Geophys Geosyst 4:Q08302

Tackley PJ, Stevenson DJ, Glatzmaier GA, Schubert G (1993) Effects of an endothermic phase transition at 670 km depth in a spherical model of convection in the Earth's mantle. Nature 361:699–704

Tackley PJ, Stevenson DJ, Glatzmaier GA, Schubert G (1994) Effects of multiple phase transition in a 3-D spherical model of convection in Earth's mantle. J Geophys Res 99:15877–15901

Takeushi N, Morita Y, Xuyen ND, Zung NQ (2008) Extent of the low-velocity region in the lowermost mantle beneath the western Pacific detected by the Vietnamese Broadband Seismograph Array. Geophys Res Lett 35. doi:10.1029/2008GL033197

Tan E, Gurnis M (2005) Metastable superplumes and mantle compressibility. Geophys Res Lett 32. doi:10.1029/2005GL024190

Tan E, Gurnis M (2007) Compressible thermo-chemical convection and application to the lower mantle. J Geophys Res 112. doi:10.1029/2006JB004505

Tan E, Gurnis M, Han L (2002) Slabs in the lower mantle and their modulation of plume formation. Geochem Geophys Geosyst 3. doi:10.1029/2001GC000238

Tan E, Leng W, Zhong S, Gurnis M (2011) On the location of plumes and lateral movement of thermochemical structures with high bulk modulus in the 3-D compressible mantle. Geochem Geophys Geosyst 7. doi:10.1029/2011GC003665

Tanimoto T (1990) Long-wavelength S-wave velocity structure throughout the mantle. Geophys J Int 100:327–336

Tateno S, Hirose K, Sata N, Ohishi Y (2009) Determination of post-perovskite phase transition boundary up to 4400 K and implications for thermal structure in D" layer. Earth Planet Sci Lett 277:130–136. doi:10.1016/j.epsl.2008.10.004

To A, Romanowicz B, Capdeville Y, Takeuchi N (2005) 3D effects of sharp boundaries at the border of the African and Pacific superplumes: observations and modeling. Earth Planet Sci Lett 233:137–153

Tolstikhin IN, Kramers JD, Hofmann AW (2006) A chemical Earth model with whole mantle convection: the importance of a core–mantle boundary layer (D″) and its early formation. Chem Geol 226:79–99

Torsvik TH, Steinberger B, Cocks LRM, Burke K (2008) Longitude: linking Earth's ancient surface to its deep interior. Earth Planet Sci Lett 276:273–282

Trampert J, Vacher P, Vlaar N (2001) Sensitivities of seismic velocities to temperature, pressure and composition in the lower mantle. Phys Earth Planet Inter 124:255–267

Trampert J, Deschamps F, Resovsky JS, Yuen DA (2004) Probabilistic tomography maps significant chemical heterogeneities in the lower mantle. Science 306:853–856

Tsuchiya T, Tsuchiya J, Umemoto K, Wentzcovitch RM (2004a) Phase transition in $MgSiO_3$ perovskite in the Earth's lower mantle. Earth Planet Sci Lett 224:241–248

Tsuchiya T, Tsuchiya J, Umemoto K, Wentzcovitch RM (2004b) Elasticity of post-perovskite $MgSiO_3$. Geophys Res Lett 31. doi:10.1029/2004GL020278

Vacher P, Verhoeven O (2007) Modelling the electrical conductivity of iron-rich minerals for planetary applications. Planet Space Sci 55:455–456

van der Hilst RD, Kárason H (1999) Compositional heterogeneity in the bottom 1000 kilometers of Earth's mantle: towards a hybrid convection model. Science 283:1885–1888 van der Hilst RD, Widiyantoro S, Engdahl ER (1997) Evidence for deep mantle circulation from seismic tomography. Nature 386:578–584

van Thienen P, van den Berg AP, Vlaar NJ (2004) Production and recycling of oceanic crust in the early Earth. Tectonophysics 386:41–65

van Thienen P, van Summeren J, van der Hilst RD, van den Berg AP, Vlaar NJ (2005) Numerical study of the origin and stability of chemically distinct reservoirs deep in Earth's mantle. In: van der Hilst RD et al (eds) Earth's deep mantle: structure, evolution and composition, vol 160. American Geophysical Union, Geophysical Monograph, pp 117–136

Wang Y, Wen L (2007) Geometry and P and S velocity structure of the "African anomaly". J Geophys Res 112. doi:10.1029/2006JB004483

Williams Q, Garnero EJ (1996) Seismic evidence for partial melt at the base of the Earth mantle. Science 273:1528–1530

Wookey J, Stackhouse S, Kendall JM, Brodholt J, Price GD (2005) Efficacy of the post-perovskite phase as an explanation for lowermost-mantle seismic properties. Nature 438:1004–1007

Wysession ME, Lay T, Revenaugh J, Williams Q, Garnero EJ, Jeanloz R, Kellogg LH (1998) The D″ discontinuity and its implications. In: Gurnis M et al (eds) The core-mantle boundary region, vol 28. Geodynamical series, pp 231–253

Xu Y, Shankland TJ, Brent TP (2000) Laboratory based electrical conductivity in the Earth's mantle. J Geophys Res 105:27865–27875

Zhang N, Zhong S, Leng W, Li Z-X (2010) A model for the evolution of the Earth's mantle structure since the early paleozoic. J Geophys Res 115. doi:10.1029/2009JB006896

Index

© Springer International Publishing Switzerland 2015

A. Khan and F. Deschamps (eds.), *The Earth's Heterogeneous Mantle*,
Springer Geophysics, DOI 10.1007/978-3-319-15627-9

T

MIX
Papier aus verantwortungsvollen Quellen
Paper from responsible sources
FSC® C105338
FSC
www.fsc.org

If you have any concerns about our products,
you can contact us on
ProductSafety@springernature.com

In case Publisher is established outside the EU,
the EU authorized representative is:
Springer Nature Customer Service Center GmbH
Europaplatz 3, 69115 Heidelberg, Germany

Printed by Libri Plureos GmbH
in Hamburg, Germany